W9-BEU-183

Sixth Edition

PHYSICS

Its Methods and Meanings

Sixth Edition

PHYSICS
Its Methods and Meanings

Alexander Taffel, Ph.D.

Formerly Teacher of Physics and Principal,
Bronx High School of Science,
New York, New York

PRENTICE HALL   Needham, Massachusetts ■ Englewood Cliffs, New Jersey

Alexander Taffel, Ph.D., spent many years as a physics teacher and department head in various New York City high schools before he became Principal of the Bronx High School of Science from which he has since retired. During these years he was active in numerous professional organizations related to science education and received, among other recognitions, the NBC Award for Public Service. He is the author of three textbooks in physics and of numerous professional articles.

Dedication and Thanks

The author dedicates this book to his wife, Mildred, who has been a full partner in creating it. He also expresses deep appreciation to the many teachers and supervisors who have kindly offered their evaluations of the book and suggestions for its improvement. These have been most useful in preparing the sixth edition.

Staff Credits

Editorial Director	**Robert J. Hope**
Senior Editor	**Lois Arnold**
Marketing Director	**Arthur C. Germano**
Associate Product Manager	**Michael D. Buckley**
Art Director and Cover Art Direction	**L. Christopher Valente**
Production Editor	**Shyamol Bhattacherya**
Book Design/Production Coordinator	**Jonathan Pollard**
Photo Research Manager	**Russell Lappa**
Production/Manufacturing Coordinator	**Bill Wood**

Outside Credits

Editorial Services	**Sylvia Gelb**
Design Services	**Judith Pinkham-Cataldo**
Sixth Edition Illustrations	**S. T. Associates, Inc.**
Sixth Edition Photo Research	**Suzi Howard**

Cover Design | **Martucci Studio**

Snow Goose courtesy Peabody Museum, Salem, Massachusetts

Lear Jet model 55C courtesy Lear Jet Corporation

PRENTICE HALL
A Division of Simon & Schuster
Englewood Cliffs, New Jersey 07632

ISBN: 0-13-666868-2

Printed in the United States of America

8 9 00

To the Student

Why Study Physics?

The study of physics can significantly enrich our lives. It helps us to understand the laws and principles that influence every aspect of the physical world. It also introduces us to the attitudes and methods of science as approaches to problem solving in all areas of human experience.

Physics: Its Methods and Meanings

The text has been designed to enable you to get the most out of your study of physics. It has as a unifying theme the historical unfolding of the growth and fusion of the fundamental concepts of matter and energy. To assist you in understanding the ideas presented, each topic has been developed in a careful progression from simpler to more complex ideas and is presented in a clear, direct manner.

A variety of study aids is provided to help reinforce and deepen your understanding. A list of "Aims" appears at the beginning of each chapter. Problem-solving skills are strengthened through the step-by-step solution of "Sample Problems." Periodic "Test Yourself" problems offer an opportunity for you to check on your progress. The "Chapter Review" at the end of each chapter contains a detailed chapter summary as well as two groups of "Questions" and "Problems"—a basic "Group 1" and a more challenging "Group 2." There is also an "Applying Physics" section that contains suggested activities that can be done at home in order to gain direct, hands-on experience with physics. (Laboratory experiments are provided in the *Laboratory Manual.*)

The important role played by physics in our world is addressed throughout the text and in special features. A wide range of physics-based applications and careers is presented in special sections at the end of each of the seven units. The societal impact of advances

in science and technology is the subject of the essays that open the units, and of several of the "Physics Plus" essays that appear at the end of selected chapters.

Newly Revised Sixth Edition

The experience and advice of the many teachers and students who have used the earlier editions have gone into this expanded and updated edition of *Physics: Its Methods and Meanings*. In areas such as mechanics, new material has been added in order to bring greater depth to the presentation. The scope of the text has been broadened through the addition of important topics such as thermodynamics. In fast-changing fields such as particle physics, up-to-the-minute developments have been incorporated. Other enhancements include the use of the International System (SI) for units of measurement, and changes in the "Physics Plus" features. Topics such as black holes and the Hubble space telescope have been brought up to date; new essays on such crucial societal issues as global warming have been added.

All of us who have contributed to this latest edition of *Physics: Its Methods and Meanings* hope you find it an enjoyable, thought-provoking, and illuminating introduction to the study of physics.

CONTENTS

UNIT ONE

Methods of Science and Measurement   1

UNIT TWO

Force, Motion, and Energy   47

UNIT THREE

Heat and the Structure of Matter   203

UNIT FOUR

Wave Motion, Sound, and Light   293

UNIT FIVE

Electricity   417

UNIT SIX

Electromagnetism   507

UNIT SEVEN

Quantum Theory and Nuclear Physics   581

UNIT 1

Methods of Science and Measurement

The circle of stone dolmens at Stonehenge, England, is the remains of an ancient measuring instrument that was probably used to keep track of the movements of the sun, moon, and stars.

Everything in the physical world can be described in terms of two related concepts—matter and energy. Physics is the study of these two concepts, and every unit in this book will make some contribution to our understanding of matter and energy. In this unit we review the processes that have been useful in advancing scientific knowledge. Basic to these processes is the acquisition of accurate information about the things we observe. To obtain such information, we have developed the art of measurement to a high degree.

Fundamental also to the search for new knowledge and understanding is the freedom of the scientist to seek the truth objectively and to follow the paths into which it leads without fear of punishment or persecution. Scientists have had to struggle for that freedom in the past and have made great gains since the days of Copernicus and Galileo.

Today, however, when science plays such a vital role in national defense, new controls and restraints are being imposed by governments on their scientists. It is becoming clear that only in a world of peace and cooperation among nations can scientists have the full freedom to learn all that research can reveal.

Advancing our knowledge of energy and matter has had more than scientific interest. The applications of that knowledge to the problems of everyday life have vastly changed the character and quality of life on our planet. On the one hand, science has freed us from the backbreaking toil of an earlier day and has greatly increased the production of the basic necessities of life. On the other, science has made our lives more complicated. Advanced industrialization has brought with it the pollution of our environment. Advanced communications have made all peoples of the earth our neighbors, our competitors, and sometimes our adversaries.

The lesson for scientists is that they cannot be content solely with the discovery of new knowledge. They must also join their fellow citizens in establishing controls and policies that will assure that scientific advances shall be used exclusively for the benefit of humankind.

1 Physics and the Methods of Science

Aims

1. To introduce the concepts of matter and energy.
2. To understand some of the methods and activities by which scientists study matter and energy.

Forms of Matter and Energy

1-1 What Is Matter?

All matter has mass and occupies space.

All the substances of which the objects about us are made, such as wood, stone, iron, water, and air, are examples of *matter.* Note that matter occurs in three forms: *solid, liquid,* and *gas.* At ordinary room temperatures, iron, stone, and wood are examples of solids; oil, alcohol, and water are examples of liquids; and air, natural gas, and hydrogen are examples of gases.

All bodies of matter, whether solid, liquid, or gas, have two properties in common. They have *mass* and they occupy *space.* Some substances, such as gold and platinum, are relatively heavy. Other substances, such as air and helium, are relatively light. The lightest substance known is the gas hydrogen, which is therefore useful as a filler for balloons.

The quantity of space which a body occupies is called its *volume.* *Solids* are distinguished from other forms of matter in that they have both a definite shape and a definite volume. *Liquids* have a definite volume but no definite shape. They assume the shape of their containers. *Gases* have neither definite shape nor definite volume. They can be kept intact only in a closed container, and they spread out to all parts of any container into which they are put.

1-2 What Is Energy?

Energy is the ability to do work.

Today at the flick of a switch or the movement of a control lever, agents such as electricity, heat, water power, chemicals, fuels, and nuclear power are immediately at our disposal. They provide transportation in ships, trains, airplanes, elevators, motorcars, and trucks. They provide rapid means of communication by radio, television, telephone, and telegraph. They work in factories, mines, and farms to produce abundantly the products that fulfill our daily needs. They serve in the home to cook, to clean, to wash clothes, to refrigerate foods, to sew, to provide light and heat, and even to supply entertainment.

The capacity that enables these agents to do work is called *energy*. As a preliminary definition, we may say that *energy is the ability to do work.*

1-3 How Energy Does Work

In sawing a piece of wood by hand, we move the saw back and forth over the wood. We say we are exerting a force on the saw. A *force* means simply *a push or a pull.*

When we use an electrically driven saw, we no longer have to supply the force that moves the saw. Instead, the saw's electric motor pushes the saw for us. Electrical energy, acting through an electric motor, now provides the force that formerly we had to supply by hand.

This illustrates an important characteristic of electrical energy as well as of all other forms of energy. When used in suitable devices, all forms of energy can supply forces that can move objects. When forces move the objects on which they act, they are said to do *work.*

In the above example, the electric motor was the means whereby electrical energy furnished the force for sawing the wood. As a second example, let us note that the engine of the family car performs a similar service for heat energy. When we burn gasoline in the car's engine, the heat produced causes the engine to run and to supply the force that ultimately pushes the car forward.

Fig. 1-1. The energy of the wind supplies the force that pushes the sailboat forward.

1-4 Forms of Energy

A list of some of the different kinds of energy we use includes magnetism, electrical energy, heat, light, sound, and nuclear energy. While all of these forms of energy can supply forces and do useful tasks, they are not all equally effective in doing jobs requiring large forces, such as driving trains or lifting elevators.

Some forms of energy usually supply only small forces. But even these can be very useful. This is the case with sound energy. We are not ordinarily aware of the fact that sounds cause things to move. However, when there is a very loud sound like that of an explosion, we notice that many objects around us are set into vibration. Sometimes windows are actually shattered by such vibrations. Here it is clear that a sound is providing the forces that cause these objects to move. While the forces accompanying most ordinary sounds are much smaller than these, they are most important in our daily lives. It is these tiny forces that make our eardrums vibrate and cause us to hear the sounds from which they came.

1-5 Sources of Energy

The *sun* is the ultimate source of most of the energy we use today. This energy comes to us in the form of heat, light, and other rays, all of which are given the common name *radiation,* or *radiant*

Fig. 1-2. The sun is our major supplier of energy. Here, the supply is temporarily cut off during an eclipse.

energy. The sun's radiation supplies the energy that makes plants grow and thus insures our food supply. It warms the earth and is responsible for clouds, winds, rain, and all other weather phenomena (fee·NAH·min·ah). It is the agent that replenishes the water supply in the streams and rivers from which we obtain water power—one of our major large-scale sources of energy.

The sun is also responsible for the earth's supply of coal, oil, and natural gas, upon which we depend so heavily. These fuels are called *fossil fuels* because they are the remains of plants that lived on the earth a long time ago. Thousands of years ago, great changes must have taken place in the earth's crust as a result of which many plants were buried under vast masses of rock, water, and earth. Here, under great heat and pressure, they were gradually transformed into the coal, oil, and natural gas now stored in the earth for our use.

1-6 Energy Outlook for the Future

The ever-increasing use of fossil fuels is hastening the day when the world's reserves of these fuels will no longer be able to fill our needs. We shall then have to turn to other sources of energy. The most obvious one is the sun. It is estimated that if all the heat and light that the sun casts on the roof of a typical one-family house could be collected and stored for use, it would provide enough energy to fill all that home's normal electrical and heating needs. However, we have only begun to develop cheap and efficient ways of collecting and using the sun's vast energy resources. Scientists and engineers all over the world are working on this problem. Already, solar collectors capable of furnishing a major part of home needs for heat and hot water have proven practical and are being widely used. Another promising development is the solar cell that converts sunlight directly into electricity.

Other possible future energy sources are water power harnessed from the daily tides, the huge reservoirs (REZ-ur-vwawrz) of heat in the oceans and in the interior of the earth, the energy of winds, and nuclear energy. (See page 12.) While all of these are being explored with intense interest, nuclear energy has received by far the most development. It is today a significant and growing source of energy used in the development of electric power. However, its progress has been impeded by severe problems of safety, pollution, and the disposal of nuclear wastes. The major lesson we can draw from the current world energy picture is to conserve and use wisely the remaining energy sources at our disposal.

1-7 Energy or Matter

Nuclear energy is obtained from a process in which small amounts of matter are changed into very large quantities of energy. We shall study the process later in more detail. Here it is enough to note that our brief introduction to energy has led us right back to matter as a source of energy. This is but the first of many occasions when we shall find that the properties of matter and those of energy are inextricably (in·EKS·trik·ub·lee) interwoven. Let us now look at the methods by which the physicist explores and studies those properties.

Methods of Science

1-8 The Goal Is Understanding

The goal of physicists is to understand the world about us. What does this mean? It means two things: first, knowing exactly what is happening, and second, being able to explain why it is happening. Let us illustrate what it means to understand a set of natural events like the falling of objects when they are dropped from a height. Here, knowing exactly what is happening involves getting all the facts we can about the way in which objects fall. These would include such information as the way in which the speed of a falling object changes as it falls, and the effects of its size and weight on the rate of fall. Such facts are obtained by careful observation and experimentation.

To explain why objects fall as they do, we must not only identify the force that controls their motion but we must also know how this force produces the specific motion we observe. The force that causes objects to fall has been identified as the force of gravity that pulls all objects toward the center of the earth. Since we also know exactly how this force controls the motion of the objects on which it acts, we have attained the goal of understanding why objects fall as they do.

1-9 Methods

The combination of activities in which scientists engage to achieve the understandings they seek is sometimes called the *scientific method*. There is however no single method of science, but rather a variety of activities in which scientists engage in different combinations, in solving different problems. Scientific activities include recognizing and defining problems, observing, measuring, experimenting, making hypotheses and theories, and communicating with other scientists.

The scientific method is not one scientific activity but a combination of scientific activities.

1-10 Defining Problems

Scientific activity usually begins when one observes something that presents a problem needing solution. The first step in the solution is to define the problem as specifically as possible. This is often done by asking questions about it in such a way as to break up the problem into simpler parts and also to suggest ways of obtaining the answers. The answers then lead to the solution of the original problem.

Here is an illustration. Suppose your desk lamp goes out while you are studying. The problem is: *Why did this happen?* To solve it we ask specific questions, such as: *Did the bulb burn out? Did the fuse blow out? Was the power turned off?* Each question suggests an activity that will answer it. We may answer the first question by testing the bulb in another fixture known to be in good order. If the bulb lights, we know it is all right, and go on to the second question. We may answer this by replacing the fuse with a new one. If this does not solve the problem, we go on to the next question, and so forth.

Defining problems by asking the right kinds of questions is one of the most effective skills of scientists. It requires imagination as well as creative thinking.

1-11 Observation

The facts we need to solve a problem are generally obtained by observation. In the simplest cases, observation involves merely using our senses—seeing, hearing, tasting, smelling, and touching. However, the range of observations we can make directly with our senses is limited. Our unaided eyes cannot see into the world of objects of microscopic size or into the far reaches of space. Our unaided ears cannot hear sounds that are too weak or beyond the range of human hearing. To expand our ability to observe, we have developed microscopes, telescopes, sound detectors and amplifiers, and many other kinds of instruments. These instruments open new worlds for exploration. The scanning electron microscope, for example, enables us to observe the tiny world of atoms and molecules.

1-12 Measurement

The most useful observations are those that can be expressed precisely in terms of numbers. Such observations are obtained by measurement. We can easily observe that as a body falls from a height to the ground, its speed increases. This becomes more meaningful when we can also state exactly what the rate of fall is from instant to instant. To obtain this kind of information, we need instruments to measure the distance the body falls as well as the time required.

Thus measurement is an important partner of observation, and devising schemes and instruments for making more accurate measurements is a major activity of the scientist. Among the great physicists whose fame rests upon their brilliance in devising ways of making accurate measurements are the following. Albert A. Michelson (MY·kel·son) developed precise methods for measuring the

a

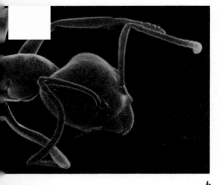

b

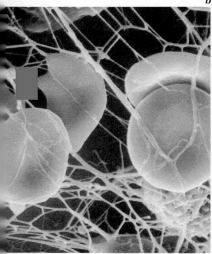

Fig. 1-3. Scanning electron micrographs of (**a**) a velvety tree ant (20X); (**b**) a clump of human red blood corpuscles (8000X).

speed of light. Robert A. Millikan (MIL·i·kun) found an ingenious way of measuring the charge on a single electron. Ernest Rutherford (RUTH·ur·furd) devised a technique for measuring the size of atomic nuclei. You will learn of many more as you continue your study.

1-13 Experimentation

One way of obtaining the facts about occurrences and also the relationships between them is the *controlled experiment.* In the simplest experiments of this kind, scientists control the factors that influence the particular situation under investigation so that all but one are kept constant while the one is permitted to vary. They can then see more clearly what specific effect that factor has on the situation under study.

For example, suppose a scientist wishes to know what effect the shape of a body has on its rate of fall through a vacuum. He or she must design an experiment in which all factors except the shapes of the falling bodies remain the same. One way of doing this is to make a series of objects out of the same material, each having the same weight but a different shape. The objects are then dropped one by one from the same height in an evacuated glass tube, and the time each takes to reach the bottom is measured. Since the only way in which the objects differ is their shape, this experiment will reveal the effect of shape on rate of fall.

1-14 Making Hypotheses

In every step of the scientific process, whether it is defining the problem, devising experiments to solve it, or coming to tentative solutions, the chief input of the scientist is thinking. Often this thinking consists of making an informed guess at the solution of the problem under consideration. Then comes a test to find out if the guess is correct. Such a guess is called a *hypothesis* (hy·PAH·thuh·sis). It is rarely a wild guess, but rather one that is based on the scientist's previous experiences and insight into the problem.

For example, let us consider the problem of why certain solid substances float in water while others sink. We know from experience that it is generally lighter solids, such as wax and wood, that float in water, while heavier ones, such as lead and iron, sink. It seems reasonable to guess that the *density* or *weight per unit of volume,* of a substance compared to that of water may be the key to the problem. We hypothesize that all solids that are less dense (lighter) than water float in it, while all solids that are more dense (heavier) than water sink when immersed in it.

We must now test the hypothesis with as many samples of different solid substances as possible. If we are right in every case, the

Fig. 1-4. The experimental work of Dr. Chien Shiung Wu at Columbia University has contributed to the theory of elementary particles and the structure of the nuclei of atoms.

A hypothesis is an educated guess at the solution of a problem.

hypothesis is confirmed. If we are wrong in even a single case we must find a way to explain the exception or modify the hypothesis to include it.

1-15 Developing Theories

A successful theory can explain known facts and predict new ones.

The climax of the scientist's efforts to understand the world about us is the discovery of logical, comprehensive schemes that explain the physical world and account for innumerable related observations and facts. Such schemes are called *theories* (THEE·uh·reez). A good theory not only accounts for and organizes the set of known facts it purports to explain, but also predicts new facts not yet observed. Thus it points the way to new knowledge and deeper understanding.

As methods of observation and experimentation improve, new facts are discovered. The existing theory must be able to explain these as well as the older facts. If it cannot, it must be modified, or a new theory must be developed that can explain all the facts, both new and old.

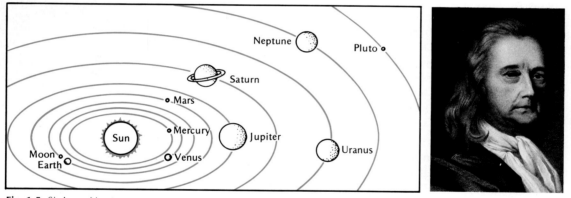

Fig. 1-5. Sir Isaac Newton discovered the law of universal gravitation which governs the movements of the planets and other bodies in the solar system.

1-16 Example of a Great Theory

One of the most brilliant theories of physics is Isaac Newton's *theory of gravitation.* According to this theory, all bodies in the universe exert a force of attraction on each other. It is this force of gravitation that causes objects dropped from a height to be pulled toward the earth. This force also keeps the moon revolving about the earth, and keeps the earth and other planets revolving about the sun.

Newton described the exact nature of the force of gravitation. This knowledge has enabled us not only to compute the precise manner in which a body falls, but also to compute correctly the orbits of the moon, the earth, the planets, and other heavenly bodies. The theory has also enabled us to explain the tides of the oceans and to land spacecraft on the moon. Thus the theory has explained successfully both earthly and astronomical effects of the force of gravitation.

When it was found in the late 18th century that the orbit of the planet Uranus (YUR·an·us) deviates slightly from the path predicted for it by Newton's theory, an explanation was sought in terms of the theory. It was then predicted that the deviation is caused by the gravitational pull of a nearby and hitherto undiscovered planet. Astronomers were advised where to look for the new planet and soon discovered it. It was named Neptune (NEP·choon). Thus, the theory fulfilled a major requirement of a good theory by predicting a hitherto unknown fact.

The theory of gravitation was used to predict the existence of previously unobserved planets.

Soon after Neptune's discovery, deviations were observed in its predicted orbit similar to those that had been observed in the orbit of Uranus. Once again, the theory predicted the existence of a hitherto unknown planet, which was soon discovered and named Pluto (PLOO·toh).

1-17 Communicating

The rapid pace at which science has advanced is made possible by the commitment of scientists to share what they have learned with all others who are interested. When scientists make discoveries or do significant experiments, they promptly report the results and methods to others by having them published in scientific journals. Thus the work is made available to all other scientists. This makes it possible for them to check the work and verify its accuracy. It also enables other scientists to build on work that has already been accepted without having to repeat it. Thus each scientist can take up the advance of science where others have left off.

In recent years the freedom of scientists to communicate with others has been seriously restricted for reasons of national security. This has held up the advance of scientific knowledge in many areas. It is hoped that the speedy establishment of the conditions for a peaceful world will soon enable scientists to resume fully the free exchange of ideas that is so vital to the advance of knowledge.

1-18 Science, Good or Evil?

The uses to which the discoveries of scientists have been put have not been uniformly beneficial. Increased understanding of energy and matter has made possible the development of machines, engines, and manufacturing processes that have lightened our work and made many of the physical conditions of our lives easier and more pleasant. At the same time, these developments have had unfortunate effects. The waste products from automobiles and factories have polluted the atmosphere and the waters in lakes, rivers, and oceans. The development of atomic energy has brought into the world weapons capable of destroying the human race. As the

Fig. 1-6. An important part of scientists' work is publishing the results of research in scientific journals so that others may learn of their methods and findings. These journals are bound and retained in libraries throughout the world.

shadow over our future deepens, people are asking, *Is science good or evil?*

What is the answer? Science of itself is neither good nor evil. Only the uses to which it is put can be good or evil, and those uses are determined by people and their governments. It is therefore up to the people and their governments to assure that scientific advances are used solely to benefit humanity. Every citizen has a responsibility for realizing this goal. Today, individual scientists as well as organizations of scientists are recognizing this responsibility and taking a leading part in helping their governments to apply the advances of science to the needs of us all.

CHAPTER REVIEW

Summary

Physics is the study of matter and energy. **Matter** is characterized by the fact that it *has mass and occupies space* or has *volume*. It occurs in three forms: *solid, liquid,* and *gas.* **Energy** is the ability to *do work.* It occurs in several forms, such as electrical energy, heat, light, and sound. Each form of energy can be used in a suitable device to *exert forces* upon bodies and to control their motion. This book will explore our present ideas of matter and energy.

The goal of physicists is to understand the world about us. They try to achieve this understanding by engaging in various **scientific activities.** These include *recognizing and defining problems, observing, measuring, devising* **controlled experiments,** *making* **hypotheses** *and* **theories,** *and communicating* with other scientists.

Questions

Group 1

1. (a) What are the two distinguishing characteristics of ordinary matter? (b) How can it be shown that air is matter?

2. What are two examples of material substances that are ordinarily in (a) solid form; (b) liquid form; (c) gas form?

3. (a) What is the main characteristic of energy? (b) How do we know that light and heat are forms of energy? (c) Name two household appliances in which electrical energy is used to move a specific object or the parts of machinery.

4. (a) What are the chief energy sources of the world today? (b) Why are the common fuels referred to as *fossil fuels?*

5. (a) Why is it necessary to look for new sources of energy? (b) What known reserves of energy are still to be developed? (c) Why have they not been developed thus far?

6. An observer saw a moving object flash for a moment in the night sky and then die out. The time was noted and also the part of the sky where the flash occurred. These observations and a description of the flash were reported to a nearby astronomical observatory. The observer guessed the object was a meteor. List the scientific activities involved.

7. (a) Name two devices used by scientists to extend the range of their observations. (b) Explain how each enables one to observe events not observable without them.

8. (a) State two advantages that a successful theory gives the scientist. (b) What should be done with a theory that can explain all but two or three of the known facts?

9. How does the free communication of scientists with each other advance the progress of science?

10. What is your reaction to the charge made by some people that scientific advances do more harm than good?

Group 2

11. A motorist finds that an hour after driving the car over some broken glass one of the tires went flat. (a) State a likely hypothesis that will explain this occurrence. (b) State another hypothesis that is possible but less likely.

12. A youngster notices that a rubber balloon filled with hot air will rise when released, while the same balloon filled to the same volume with cold air will sink to the ground. The child wishes to know why this happens and states the problem in two different ways:

    (a) Why does the balloon rise when filled with hot air and sink when filled with cold air?

    (b) What differences in properties between hot and cold air account for the fact that the balloon rises when filled with hot air and sinks when filled with cold air?

    Which is a better statement of the problem? Why?

13. Two pupils are asked to compare the lengths of two steel bars, A and B. The first looks at the bars and reports that A is a little longer than B. The second measures them and reports that A is 27.5 centimeters long and B is 25.0 centimeters long. In what two ways is the second report superior to the first?

14. In daily life it is said, *Honesty is the best policy*. In communications among scientists, honesty is an absolute necessity. Explain why.

Applying Physics

1. Measure the time it takes you to drill a 5-mm hole in a block of wood about 5 cm thick. Next, measure the time it takes to drill a similar hole in the same block of wood with an electric drill. Compare the rate at which the energy to do this job is provided by you with the rate at which electricity provides the necessary energy.

CAUTION: If you are not familiar with the safe use of drills, check with someone who is before you do this activity.

2. How good are your eyes as instruments for estimating lengths? Test them with these figures. (a) Judge whether the length of AB is less than, equal to, or greater than that of BC. (b) Judge whether the length of EF is less than, equal to, or greater than that of GH. Now measure these four lengths with a ruler and see if your judgments made with the unaided eye are correct. What does this experiment tell you about depending upon your eyes alone to obtain precise information about the lengths of objects?

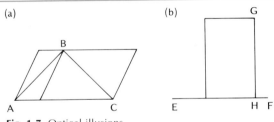

Fig. 1-7. Optical illusions.

Developing Energy Sources for the Future

The search is on to find substitutes for the dwindling supplies of coal, oil, and natural gas that have powered our factories and vehicles but are now choking the atmosphere with pollutants. A two-fold approach to the problem is necessary: until usable replacements can be found, we must conserve these fossil fuels and do what we can to combat the air pollution, acid rain, and global warming to which they contribute.

Renewable, nonfossil energy sources could increase the world's energy supply with minimal environmental damage. However, none may yet be ready to step into the leading role now played by fossil fuels. Of the many possible alternative energy sources, only nuclear, biological (or "biomass"), and solar sources are expected to become

Geothermal plants use the earth's internal heat to produce steam for heating buildings.

competitive with fossil fuels before the year 2000. And there are still many hurdles to be cleared before this can happen.

The expansion of nuclear power has been slowed down by public concern over reactor safety, disposal of radioactive waste, and escalating construction and operating costs. The potential of biomass is limited, since most of the world's farm land must be used to grow food crops rather than the special "fuel crops" that can be fermented to produce clean-burning alcohols. The most promising application of solar power is photovoltaic systems, which convert light into electricity. Although still relatively expensive (with a unit cost about four times that of today's U.S. average), as such systems become more efficient, they will be quite attractive due to their exceptional flexibility. Photovoltaic systems, unlike most other energy sources, can be set up on a large scale in a centralized facility or installed on an individualized basis.

Other naturally occurring energy sources include the wind, the heat produced deep within the earth ("geothermal"), and water ("hydropower"). Although none is yet being used on a worldwide basis, each has found application in limited regions. For example, in Iceland most homes are heated by steam from the country's many hot springs, a form of geothermal energy. In addition, much of Iceland's electricity is produced by hydropower from falling water. In fact, Iceland's hydropower resources are so great that the export of electricity to nearby Scotland is a possibility.

Wind power is being used quite extensively in both California and Hawaii, where it is viewed as an alternative to expensive foreign oil. Wind power has its disadvantages, too, since it is intermittent and the rows of

Solar panels on the roof supply energy by converting sunlight to electricity.

Modern windmills are used to produce electricity.

modern windmills both mar the scenery and may interfere with radio and TV transmission. To overcome such problems, Britain is planning to build the first offshore wind power operation. Of course the difficulty of construction at sea combined with the need for miles of undersea cable adds considerably to the cost of such an undertaking.

Beyond the approaches already in use, scientists throughout the world are investigating innovative methods of energy production. These include the tapping of geothermal energy by pumping water into cracks leading into the hot, molten rock called magma. The resultant steam can be used for heating buildings. Another way of using water to generate power is ocean thermal energy conversion. Such a plant produces electricity by taking advantage of the difference in temperature between the warmer surface water and the colder water below the surface. The warmer water is used to evaporate a liquid with a relatively low boiling point, and the cooler water is used to condense the resultant gas. The gas drives a turbine that generates electricity.

With the importance of energy to the economic, environmental, social, and political future of the world, it is certain that scientists will continue to search for better means of producing it.

2 Measuring Length and Time

Measurement Standards and Systems

2-1 Measurement in Your Life

Do you realize how often we measure things in everyday life? When we buy foods like butter, potatoes, and flour, we are measuring *weight*. When we buy some cloth or determine how far we have traveled in the family car, we are measuring *length*. Again, when we record the number of hours we have worked, keep an appointment, or tell our ages, we are measuring *time*.

Turning to the professions and occupations, we find the doctor measuring the temperatures, pulse rates, and blood pressures of patients, the farmer measuring the quantity of milk obtained from a herd of cows, the cook measuring out the ingredients needed to bake a cake, and the scientist measuring such extremes as the size of the universe and the diameter of a single proton.

2-2 Measurement Standards

Measurement depends on the establishment of standards.

We measure a dimension or a property of an object by comparing it with something we have agreed to accept as a *standard unit*. For example, when we measure the length of a swimming pool in meters, we are comparing the pool's length with that of a meterstick. Here we use the length of the meterstick as the standard unit. If we find that the pool's length is equal to the distance spanned by 30 metersticks placed end to end, we describe the length of the pool as 30 meters, meaning it is 30 times as long as the standard meter.

Similarly, in timing a runner in a race with a stopwatch, we are comparing the time required to run the race with the second, which is the standard unit of time we have agreed upon. If the stopwatch ticks off 10 seconds from the beginning to the end of the race, we know that the runner has taken 10 times as long to run the race as the standard second. Thus before we can make any measurements, we must agree upon standards like the meter and the second with which we can compare them.

For this reason, our government and many other governments throughout the world maintain special agencies. The National Bureau of Standards in Washington, D.C. is one such agency. It is the job of these agencies to prepare and maintain standards for all scientific and commercial measurements.

2-3 The Meter, a Typical Standard

To illustrate how a standard unit of measurement is defined and maintained, let us briefly review the story of the *meter*. Originally, the French commission that created the metric system decided that the basic unit of length should be one ten-millionth (10^{-7}) of the distance from the pole of the earth to the equator along the meridian passing through Paris. This length was carefully determined and marked off by means of two parallel scratches on a bar of platinum-iridium alloy. The distance between these two scratches when the bar is kept at a temperature of 0° Celsius (SEL·see·us) was then officially declared to be the standard meter. It was necessary to specify that the temperature be 0° Celsius because a metal bar expands when the temperature rises and contracts when the temperature falls. The distance between the scratches on the platinum-iridium bar is therefore exactly one standard meter long only when the temperature is 0° Celsius.

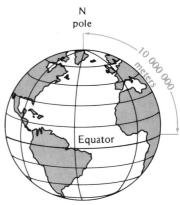

Fig. 2-1. The meter was originally defined as one ten-millionth of the distance from the equator to the pole.

To preserve this standard meter, the original platinum-iridium bar has been kept in the International Bureau of Weights and Measures at Sèvres (SEHVR), France. Copies of it have been made for the United States and other countries and have been kept in their bureaus of standards for the use of their scientists and industries.

Because an error crept into the original determination, the standard meter turned out not to be exactly one ten-millionth of the earth's equator-to-pole distance. However, this in no way affected its precision as a measuring unit. As was noted earlier, the particular length selected as a standard of measurement is entirely a matter of choice. All that was needed to make the particular length marked off on the platinum-iridium bar at Sevres the standard meter was for the world to agree to accept this particular length as the standard unit.

As science and technology advanced, it became increasingly urgent to redefine the standard meter so as to make more precise measurements possible. For this reason, the standard meter was redefined by the International Conference on Weights and Measures in 1960 in terms of the wavelength of orange light from the gas krypton-86 (KRIP·ton). This made it possible to measure distance to a precision of about 4 parts per billion. However, this standard proved inadequate to meet the demand for even greater precision. Therefore, the standard meter has recently been redefined as the distance traveled by light in $\dfrac{1}{299\ 792\ 458}$ second.

Fig. 2-2. The above apparatus was used by researchers at the U.S. National Bureau of Standards to make new measurements of certain light frequencies. In conjunction with the new definition of the meter, the measurements make it possible to determine the standard length far more precisely than previously.

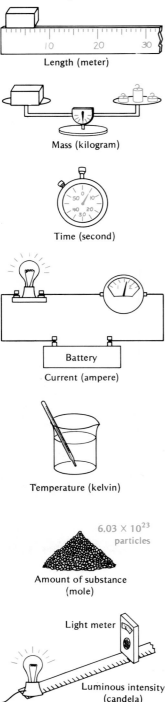

Fig. 2-3. The seven basic units of the *SI* metric system.

Length (meter)

Mass (kilogram)

Time (second)

Battery

Current (ampere)

Temperature (kelvin)

6.03×10^{23} particles

Amount of substance (mole)

Light meter

Luminous intensity (candela)

2-4 Need for a Single System of Measurement

To set up a system of measurement, all that is needed is a set of standard units, like the meter, in terms of which all measurements will be made. It does not matter what we select as standard units as long as we agree upon them. Each standard unit should have two important characteristics (KA·ruk·tuh·RIS·tiks). (1) It should remain constant so that it will give very nearly the same result whenever we use it to repeat the same measurement. (2) It should be readily duplicated so that it can be available to all who need to use it.

Throughout history, many different systems of measurement have developed in different countries. This led to confusion not only in international trade but also in the exchange of scientific knowledge among countries with different measuring systems. There was a clear need for a single measuring system that would be used the world over.

A great step forward in this direction was taken in 1791 by a commission of French scientists who created the *metric system* of measurement. Designed so that it is simple, precise, and practical, the metric system was long ago adopted by the scientific world. In addition, it has been adopted for general use by most of the nations of the world.

There is another system of measurement long used in the United States and the British Commonwealth of Nations called the English system. Since the metric system is the system of measurement that is generally used in science, we shall use only the metric system in this book. Here, we shall only mention that the English system uses the *foot* as the unit of length, the *pound* as the unit of weight, and the *second* as the unit of time.

2-5 The Metric System

This system was originally designed to deal principally with the measurement of motion and the forces that control it. For this it needed only three fundamental units, the *meter* which is the unit of length, the *kilogram* which is the unit of mass, and the *second* which is the unit of time. These units gave the system the descriptive name of the meter-kilogram-second or *MKS* system. As physics grew into the areas of heat, sound, light, magnetism, electricity, and atomic and nuclear energy, additional units were introduced as needed for measurement in these fields.

A major advantage of the metric system is that it is a *decimal* system. Any unit is related to a larger or smaller unit by some power of ten. The meter, for example, is divided into 100 smaller units called *centimeters*. Each centimeter, in turn, is divided into 10 *millimeters*. Thus, it is easy to change meters into centimeters or millimeters, and vice versa, by simply moving the decimal point. For example, to change a measurement of 2.567 meters into centimeters, we multiply by 100 by moving the decimal point two places to the right, giving 256.7 centimeters. To change this reading into millimeters, we note that a centimeter is equal to 10 millimeters and multiply by 10, giving 2567 millimeters.

2-6 Prefixes in the Metric System

To show the relationship between units in the metric system, the prefixes listed in Table 2.1 are commonly used. For example, a billionth (10^{-9}) of a second is called a nanosecond and abbreviated ns. A million (10^6) meters is a megameter, abbreviated Mm.

2-7 The International System (*SI*)

As greater skill in measurement has been developed and more precise measuring instruments devised, the metric system has been refined and improved. For this purpose an international body of scientists, called the General Conference on Weights and Measures (*CGPM*), has been meeting from time to time. In 1960, the Eleventh General Conference of the *CGPM* was held. It reorganized the metric system by establishing the *International System of Units.* This system, also known by its French name, *Le Système Internationale d'Unités,* is abbreviated *SI*. The *International System* adds four basic units to the three earlier ones of length, mass, and time in the *MKS* system. Combinations of these seven basic units are sufficient to make all physical measurements whether in mechanics, heat, sound, light, magnetism, electricity, or atomic phenomena. The *SI* basic units are listed in Table 2.2.

All other units used in measurement are combinations of two or more of these basic units and are called *derived units.* Speed, for example, is a derived unit. To measure the average speed of a car, we measure the distance it travels and the time it takes. We then divide the distance by the time. The measurement of speed is thus derived from a measurement of length and a measurement of time. If the distance is measured in meters (m) and the time is measured in seconds (s), the speed is obtained in meters per second (m/s).

Table 2.1 Metric Prefixes

PREFIX	MEANING	SYMBOL
pico (PEE·koh)	10^{-12}	p
nano (NAN·oh)	10^{-9}	n
micro (MY·kroh)	10^{-6}	μ
milli (MIL·ee)	10^{-3}	m
centi (SEN·tee)	10^{-2}	c
kilo (KIL·oh)	10^3	k
mega (MEG·uh)	10^6	M
giga (JIG·uh)	10^9	G
tera (TEHR·uh)	10^{12}	T

Table 2.2 *SI* Metric Basic Units

QUANTITY	UNIT	SYMBOL
length	meter	m
mass	kilogram	kg
time	second	s
electric current	ampere	A
temperature	kelvin	K
amount of a substance	mole	mol
luminous intensity	candela (kan·DEL·uh)	cd

Measurement of Length, Area, and Volume

Fig. 2-4. A metric micrometer caliper used to measure small lengths accurately.

2-8 How We Measure Length

The direct way to measure length is by means of the straight edge of a ruler or meterstick. The ruler is placed alongside the object to be measured, and the number of unit intervals of the ruler equal to the length of the object is then noted.

We can use a ruler to make direct measurements of length only if the dimensions of the object to be measured are neither much smaller than the smallest unit interval on the ruler nor much larger than the ruler itself. Thus, we cannot measure the thickness of a piece of paper or the distance from the earth to the moon directly with a ruler. For such measurements we must use instruments or methods that combine the use of a ruler with a knowledge of geometry, of physical laws and facts, and any other pertinent information.

For example, to measure the thickness of a piece of paper, we use a *micrometer caliper* (my·KROM·i·tur KAL·i·pur). This instrument is, in effect, a combination of a screw and a small ruler. It utilizes the geometry of the screw to enable it to measure paper-thin thicknesses.

A measurement of such a length as the distance from the earth to the moon or to a planet requires considerable ingenuity. One method for doing this, called *radar*, sends a radio signal from the earth to the moon where the signal is reflected back to the earth. The travel time taken by the signal to make the round trip to the moon and back is measured. The total distance traveled by the signal is then obtained by multiplying the known speed of the radio signal, which is 300 megameters per second, by the travel time. Half of this round-trip distance is the distance from the earth to the moon.

Note that in making this measurement we have enlisted our knowledge of radio transmission and reception. We have also had

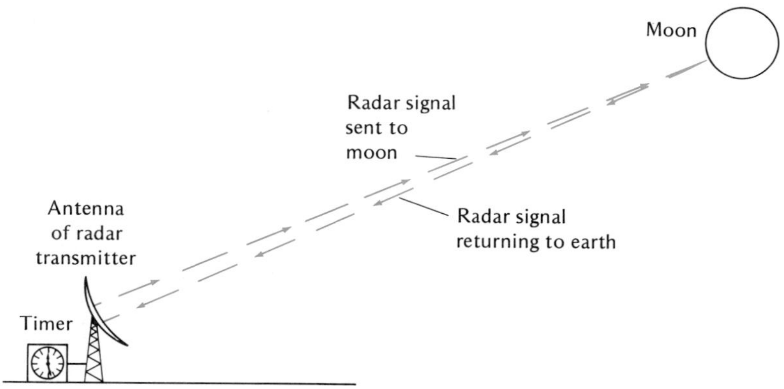

Moon

Radar signal
sent to
moon

Antenna
of radar
transmitter

Radar signal
returning to earth

Timer

Fig. 2-5. Measuring the distance to the moon by radar signals.

to assume that the radio signal traveled to the moon and back in straight lines and that its speed remained constant throughout the entire trip. Our knowledge of radio makes these assumptions reasonable.

Thus we see that the measurement of distance is a science requiring both imagination in inventing measuring methods and instruments and skill in using them. We shall see that this is true of all measurement.

Centimeters

Fig. 2-6. This 10-cm part of a meterstick shows subdivisions into millimeters.

2-9 Units of Length

The meter (m) which is the standard unit of length in the *SI* metric system has already been defined. It is divided into 100 centimeters (cm) and 1000 millimeters (mm). For long distances we generally use 1000 meters, or one kilometer (km), as the unit of measure. An idea of the size of these units can be obtained from the scale shown in Fig. 2-6. It is one tenth of a meter long and is divided into centimeters and millimeters.

The techniques for measuring length enable us to measure area and volume, both of which are derived from length measurements.

The meter (m) is the SI unit of length.

2-10 Area

The unit of area is a square having a standard unit of length as a side. In the *SI* metric system it is a square 1 meter on a side. It is called a *square meter* and is written 1 m². To measure smaller areas, we may use a square 1 centimeter on a side. This is called a *square centimeter* and is written 1 cm². One square meter contains 10 000 square centimeters. The number of square meters or square centimeters that fit into any surface is a measure of the area of that surface.

To measure the areas of the surfaces of some objects, we frequently make use of geometry. For example, geometry tells us that the area of a rectangle is equal to its *base times its height*. So we can measure the area of a rectangle by measuring the length of its base and its height and multiplying these quantities together. The area of a rectangle having a base of 3 centimeters and a height of 2 centimeters is 3 cm × 2 cm = 6 cm².

To measure the area of a circle, we note from geometry that the area of a circle is equal to πr^2, where π (Greek *pi*) equals 3.14 and r is the radius of the circle. Thus the area of a circle whose radius is 10.0 meters is $\pi \times (10.0 \text{ m})^2 = 314 \text{ m}^2$.

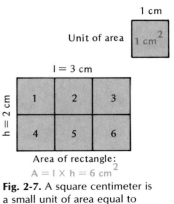

Fig. 2-7. A square centimeter is a small unit of area equal to 1/10 000 of a square meter.

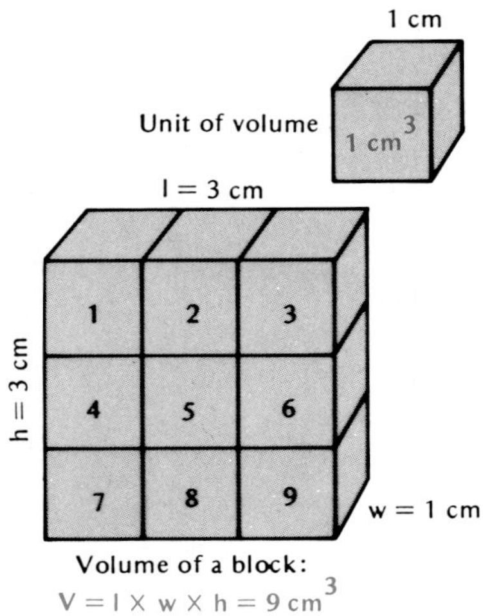

1 cm

Unit of volume

$1\ cm^3$

l = 3 cm

h = 3 cm

w = 1 cm

Volume of a block:
$V = l \times w \times h = 9\ cm^3$

Fig. 2-8. A cubic centimeter is a small unit of volume equal to one millionth of a cubic meter. It also equals 1 milliliter.

2-11 Volume

The volume of a body means the space it occupies. The unit of volume is the space occupied by a cube having the standard unit of length for its edge. In the *SI* metric system it is the space occupied by a cube one meter on an edge and called a *cubic meter*. In symbols it is written $1\ m^3$. To measure smaller volumes, we use as the unit a cube one centimeter on an edge. It is called one *cubic centimeter* and is written $1\ cm^3$. There are one million cubic centimeters in a cubic meter. Another common unit of volume is the *liter* (LEE·tur) which is defined as exactly 1000 cubic centimeters. The liter is abbreviated L.

2-12 Using Geometry to Obtain Volumes

The volumes of regular objects, such as a rectangular block, a cylinder, or a sphere, can be obtained with the aid of a ruler and geometry.

From geometry we know that the volume of a rectangular block is:

$$V = l \times w \times h$$

where *l* is the length, *w* is the width, and *h* is the height of the block.

The volume of a right cylinder is given by:

$$V = \pi r^2 h$$

where *r* is its radius and *h* is its height.

The volume of a sphere is given by:

$$V = \frac{4}{3}\pi r^3$$

where *r* is its radius.

Sample Problem

What is the volume of a block 3 cm long, 1 cm wide, and 3 cm high?

Solution:

$l = 3\ cm \qquad w = 1\ cm \qquad h = 3\ cm$
$V = l \times w \times h$
$V = 3\ cm \times 1\ cm \times 3\ cm = 9\ cm^3$

Test Yourself Problems

1. Find the volume of a rectangular stone block 2.0 m long, 1.5 m wide, and 0.20 m thick.
2. Find the volume of a sphere of radius 6.00 cm.

2-13 Volume of Liquids and Irregular Bodies

The volume of a liquid can be measured by means of a *graduated cylinder* (GRAD·yoo·way·ted SIL·in·dur). This is a tall, cylindrical, glass container with numbered graduations etched upon it. The liquid to be measured is poured into the graduated cylinder and its volume is then read by noting the number opposite the graduation that coincides with the surface of the liquid. Graduated cylinders used in science laboratories are generally marked off in milliliters. A *milliliter* (mL) is one thousandth of a liter and equal to one cubic centimeter.

A graduated cylinder may also be used to measure the volume of a solid of irregular shape, such as a piece of coal. The method is as follows.

Fill the graduated cylinder about half full of water and note the reading. Let us say it is 50 milliliters. Now lower the piece of coal into the cylinder until it is completely under water. The water level will rise because the space occupied by both water and coal is greater than that occupied by the water alone. Note the new water level. Suppose it is 82 milliliters. The volume of the coal is the difference between the combined volume of the water and the coal (82 mL) and the volume of the water by itself (50 mL). Hence, subtracting, we have 82−50 = 32 milliliters, or 32 cubic centimeters, for the volume of the coal.

This method of measuring volume by displacement is limited to solids that are not porous and that do not dissolve in water.

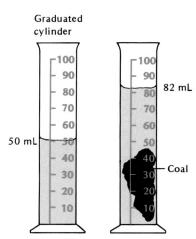

Fig. 2-9. Measuring the volume of an irregular solid by measuring the volume of water it displaces.

Questions

Group 1

1. Cloth tape measures are used commercially to make coarse measurements of length. Why are such tape measures unsuitable for precise measurement?

2. In using the platinum bar at Sèvres, France, as the standard meter, why is it necessary to specify that the temperature must be 0°C?

3. (a) How can the volume of a rectangular room be measured? (b) How can the volume of a door key be measured?

4. (a) Explain how a ruler is used to measure the inside volume of an open cylindrical can. (b) How can the same volume be measured with a graduated cylinder?

5. How can the volume of a basketball be determined?

Group 2

6. (a) How will the change-over by the United States from the English to the metric system of measurement be an advantage in international trade? (b) Describe two difficulties that might arise in an industry, such as car manufacturing, in changing from English to metric units.

7. The odometer (oh·DAHM·i·tur) measures the distance traveled by a car by counting the number of complete turns made by one of the wheels and multiplying it by the circumference of the wheel. If the odometer reads the distance traveled accurately when the car's tires are new, why will the reading not be accurate when the tires become worn?

8. A small porous solid is in the shape of a cube. Its volume is determined in two ways. First, it is computed from a measurement of the cube's edge. Then it is obtained by indirect measurement using a graduated cylinder partly filled with water. (a) Why might there be a considerable difference between the two values of the volume thus obtained? (b) What does this difference represent?

9. The sides of two squares are in the ratio of 2 to 1. (a) What is the ratio of their areas? (b) If the corresponding dimensions of any two similar plane figures are in the ratio of 2 to 1, explain why their areas will be in the same ratio as that of the two squares.

10. (a) If the length of each edge of a cube is doubled, what happens to the volume? (b) An enlarged model of a small solid statue is made by doubling each dimension of the original. How does the volume of the enlarged model compare with that of the original?

Problems

Group 1

Note. In several of the problems below you are asked to round answers to the nearest cm², cm³, m², m³, etc. The reason for this rounding will be explained in the next section. Use 3.14 for π.

1. Convert to m: (a) 425 cm; (b) 4.25 cm; (c) 4.25 mm; (d) 4.250 km; (e) 0.425 km.

2. Convert to mm: (a) 52.3 cm; (b) 5.23 cm; (c) 0.423 m; (d) 0.042 cm; (e) 10.008 m.

3. Convert to km: (a) 725.9 m; (b) 725 m; (c) 5.258 m; (d) 5 000 000 cm; (e) 5275 cm.

4. A box is 3.2 cm wide, 4.0 cm long and 2.1 cm high. Find (a) the area of its base to the nearest cm²; (b) its volume to the nearest cm³.

5. Find to the nearest m² the area of a circle whose radius is found to be 15.0 m.

6. When ten similar coins are dropped into a graduated cylinder, the level of the water rises from 51 mL to 75 mL. Find to the nearest tenth of a mL the average volume of each coin.

7. Find to the nearest m³ the volume of a cylinder with a radius of 5.00 m and a height of 4.00 m.

8. Find to the nearest cm³ the volume of a sphere whose radius is 3.0 cm.

Group 2

9. How many m² are there in (a) 1 km²; (b) 20 000 cm²?

10. How many m³ are there in (a) 2 km³; (b) 50 million cm³?

11. A circle whose diameter measures 10.0 cm is cut out from a square sheet of metal that measures 10.0 cm on a side. Find to the nearest tenth of a cm² the area of the part of the metal sheet that is left over.

12. A cubical box 10.0 cm on an edge is filled with water to a height of 5.0 cm. When a piece of iron is completely immersed in the water, the level rises to 5.9 cm. Find the volume of the iron to the nearest cm³.

13. A coin has a radius of 1.0 cm and a thickness of 1.5 mm. Find its volume to the nearest hundredth of a cm³.

14. How does the quantity of steel needed to make a ball bearing having a radius of 8 mm compare with that needed to make a ball bearing having a 2-mm radius?

Arithmetic of Measurement

2-14 Measurements Are Approximate

Every measurement is an approximation.

Before we continue with the study of measurement, it is important to note that every measurement is an approximation. It represents the best value that can be obtained for a given measurement with the instruments used.

Suppose we measure the thickness of a plate of glass, first by means of a meterstick and then by means of a micrometer caliper. The smallest unit on the meterstick is a millimeter. Using the meterstick, we find the thickness of the glass to be considerably more than three millimeters but less than four millimeters. We therefore estimate its thickness at 3.7 millimeters.

Turning to the micrometer caliper, we find that it can measure a length to the nearest one-hundredth of a millimeter. With it we find the thickness of the glass to be 3.65 millimeters.

Here, we have two measurements of the same quantity. Both are approximations of the true value of the thickness of the glass, but the second is a better approximation than the first. We describe the precision of our measurement in each case by noting how many digits were obtained by actual measurement. Each of these digits is called a *significant figure*. The number of significant figures in a measurement indicates how precise it is. The meterstick measurement of 3.7 millimeters gives two significant figures. We can be sure that the 3 in this measurement is correct. The 7, however, is doubtful because we cannot read tenths of a millimeter on the meterstick directly. We have to estimate them.

The more precise measurement of 3.65 millimeters obtained with the micrometer caliper has three significant figures. We can be sure that the 3 and 6 of this measurement are correct. The 5, however, is doubtful. It could be a little less or a little more than 5, which is simply the closest approximation that can be made with this instrument. Note that, in general, the last significant figure of a measurement is always doubtful.

> The more significant figures in a measurement, the more precise it is.

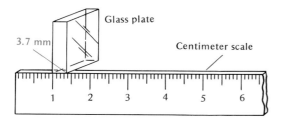

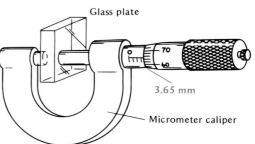

Fig. 2-10. Measuring the thickness of a glass plate with a centimeter ruler and a micrometer caliper.

While the above example concerns the measurement of length, all measurements of physical quantities give only approximate values of the quantity measured. The closeness of the approximation is determined by the number of significant figures in the measurement. The greater the number of significant figures, the more precise is the measurement.

2-15 Significant Figures

The number of significant figures in a measured number is found by counting the doubtful digit and all the digits to the left of it up to and including the last digit that is not zero. For example, suppose a certain length is measured as 0.000 310 2 meter. The doubtful digit is the final 2. Counting to the left from the 2, we include the 0, the 1, and the 3 which is the last non-zero digit to the left of 2. This makes four significant figures.

In determining the number of significant figures, zeros at the right end of a measured number are counted only if they were obtained by actual measurement. If you measure the length of a sheet of paper and find that it is 23.30 centimeters long, the 0 is significant because you got it by actual measurement. Thus, 23.30

centimeters has 4 significant figures. In this book, when zeros appear at the right end of a measured number, you are to assume that they were obtained by measurement and that they are significant.

2-16 Arithmetic with Measured Quantities

When we add, subtract, multiply, or divide measured quantities, the result cannot be more precise than the least precise of these quantities. For example, suppose we are finding the sum of two measured lengths of wire of 4.58 and 85.9 meters respectively. The first wire is measured to the nearest hundredth of a meter, but the second wire is measured only to the nearest tenth of a meter. The sum of the two lengths of wire can therefore be precise only to the nearest tenth of a meter. Before finding their sum we therefore round the 4.58 meters to the nearest tenth of a meter, giving 4.6 meters. Now adding 4.6 meters and 85.9 meters gives 90.5 meters as the sum.

In making calculations involving measured quantities, we retain in the answer only those figures that are significant. This can be done by using the following rules.

1. In adding or subtracting measured quantities, first round each of them to the number of decimal places in the quantity having the least number of decimal places. Then add or subtract the rounded quantities.

In rounding a number, the last digit retained is raised by 1 if the digit that follows it is more than 5 or is 5 followed by numbers other than zero. It is not changed if the digit that follows it is 5 or less than 5. Thus, 3.853 rounded to three significant figures becomes 3.85. Rounded to two significant figures, it becomes 3.9.

2. In multiplying or dividing two measured numbers, retain in the product or quotient the same number of significant figures as the less precise of the two numbers. For example, let us find the area of a rectangle whose sides measure 1.21 meters by 2.4 meters. Note that the 1.21-meter side is measured to three significant figures while the 2.4-meter side is measured to only two significant figures. Only two significant figures are therefore retained in the area obtained by multiplying 1.21 meters by 2.4 meters. The calculation gives 2.904 square meters which is rounded to two significant figures, giving 2.9 square meters as the area of the rectangle.

2-17 Range of Lengths We Measure

The lengths we measure range from very great to very small distances. Among the very great is the distance to the outermost galaxies—about 10 000 000 000 000 000 000 000 000 meters. Among the very small is the diameter of a proton—about

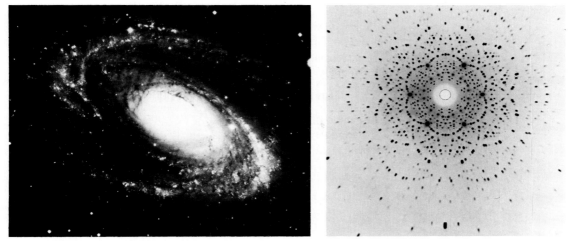

0.000 000 000 000 001 meter. In between, we measure such distances as: the distance from the earth to the sun, 149 000 000 000 meters; the equatorial radius of the earth, 6378 kilometers; the altitude of the tallest mountains, 8 kilometers; the thickness of a sheet of paper, 0.0001 meter; and the radius of the orbit of the electron in a hydrogen atom, 0.000 000 000 0529 meter.

Fig. 2-11. (a) Spiral galaxy M-81 in Ursa Major whose distance is measured in millions of light years. (b) X-ray diffraction photograph showing the arrangement of atoms in a beryl crystal whose sizes are measured in billionths of a meter or nanometers.

2-18 Scientific Notation

To simplify working with numbers as large as the distance to a galaxy or as small as the diameter of a proton, we usually write them in an abbreviated form as powers of ten. This convenient method of writing very large and very small numbers is called exponential (eks·poh·NEN·shul) or *scientific notation*. Before attempting to write numbers according to this system, let us review the relationships in Table 2.3.

Table 2.3 Powers of Ten

$$1\ 000\ 000 = 10 \times 10 \times 10 \times 10 \times 10 \times 10 = 10^6$$
$$100\ 000 = 10 \times 10 \times 10 \times 10 \times 10 = 10^5$$
$$10\ 000 = 10 \times 10 \times 10 \times 10 = 10^4$$
$$1000 = 10 \times 10 \times 10 = 10^3$$
$$100 = 10 \times 10 = 10^2$$
$$10 = 10 = 10^1$$
$$1 = 1 = 10^0$$
$$0.1 = 1/10 = 10^{-1}$$
$$0.01 = 1/100 = 1/10^2 = 10^{-2}$$
$$0.001 = 1/1\ 000 = 1/10^3 = 10^{-3}$$
$$0.000\ 1 = 1/10\ 000 = 1/10^4 = 10^{-4}$$
$$0.000\ 01 = 1/100\ 000 = 1/10^5 = 10^{-5}$$
$$0.000\ 001 = 1/1\ 000\ 000 = 1/10^6 = 10^{-6}$$

Table 2.3 can be continued to 10^7, 10^8, 10^9, etc., in the direction of larger numbers, and to 10^{-7}, 10^{-8}, 10^{-9}, etc., in the direction of the smaller numbers.

Now let us note that

$$6\ 000\ 000 = 6 \times 1\ 000\ 000 = 6 \times 10^6$$
$$6\ 700\ 000 = 6.7 \times 1\ 000\ 000 = 6.7 \times 10^6$$

Scientific notation is a convenient way of writing very large or very small numbers.

Another possible way of writing 6 700 000 is $67 \times 100\ 000 = 67 \times 10^5$. However, it is customary in this notation to have only one integer to the left of the decimal point in the number in front of the power of ten.

The number 6×10^6 is read: "Six times ten to the sixth power"; 6.7×10^6 is read: "Six point seven times ten to the sixth power."

Numbers smaller than one can be written as follows:

$$0.000\ 3 = 3 \times 0.000\ 1 = 3 \times 10^{-4}$$
$$0.000\ 35 = 3.5 \times 0.000\ 1 = 3.5 \times 10^{-4}$$

These numbers are read: "Three times ten to the minus fourth power," and "Three point five times ten to the minus fourth power" respectively.

2-19 General Procedure

We can now state a general procedure for writing any number or measured quantity in scientific notation.

1. Move the decimal point of the given number to the left or right until there is only one digit to the left of the decimal point.

2. Count the number of places you moved the decimal point. Call the number positive if you moved the decimal point to the left and negative if you moved it to the right. Raise ten to the power of this number.

3. Multiply the number obtained in step 1 by this power of ten.

Applying this procedure to the radius of the orbit of an electron of hydrogen mentioned above, we have 0.000 000 000 052 9 meter $= 5.29 \times 10^{-11}$ meter. Here, we moved the decimal point eleven places to the right. For the equatorial radius of the earth, we have 6378 kilometers $= 6.378 \times 10^3$ kilometers. Here we moved the decimal point three places to the left.

Note that when a measurement is expressed properly in scientific notation, all the digits that appear before the power of ten are significant. Thus, 5.29×10^{-11} meter has three significant figures while 6.378×10^3 kilometers has four significant figures.

2-20 Addition and Subtraction of Numbers Expressed in Scientific Notation

Numbers can be added or subtracted provided the powers of ten are equal. If they are not equal, they must be made equal by a suitable movement of the decimal point. The procedure is as follows: (1) make sure that the powers of ten of the two numbers are equal; (2) add or subtract the numbers before the powers of ten; (3) retain the same power of ten in the answer.

Sample Problems

Find the value of (a) $5 \times 10^3 + 2 \times 10^3$; (b) $4.0 \times 10^4 + 5 \times 10^3$; (c) $2.0 \times 10^5 - 4 \times 10^4$

Solution:

(a) $5 \times 10^3 + 2 \times 10^3 = 7 \times 10^3$

(b) $4.0 \times 10^4 + 5 \times 10^3 = 4.0 \times 10^4 + 0.5 \times 10^4 = 4.5 \times 10^4$

(c) $2.0 \times 10^5 - 4 \times 10^4 = 2.0 \times 10^5 - 0.4 \times 10^5 = 1.6 \times 10^5$

Test Yourself Problems

Find the value of (a) $6 \times 10^5 + 3 \times 10^5$; (b) $8.0 \times 10^5 + 5.0 \times 10^6$; (c) $3.0 \times 10^8 - 9.0 \times 10^7$; (d) $4 \times 10^{-3} + 3 \times 10^{-3}$; (e) $4.0 \times 10^{-7} - 2.0 \times 10^{-8}$

2-21 Multiplication of Numbers Expressed in Scientific Notation

To multiply one number by another, (1) multiply the two numbers preceding the powers of ten; (2) add the exponents to obtain the exponent of the power of ten in the answer.

Sample Problems

Find the value of (a) $2 \times (3 \times 10^4)$; (b) $(2 \times 10^5) \times (3 \times 10^2)$; (c) $(2 \times 10^5) \times (4 \times 10^{-3})$

Solution:

(a) $2 \times (3 \times 10^4) = 6 \times 10^4$

(b) $(2 \times 10^5) \times (3 \times 10^2) = 2 \times 3 \times 10^5 \times 10^2 = 6 \times 10^{5+2} = 6 \times 10^7$

(c) $(2 \times 10^5) \times (4 \times 10^{-3}) = 2 \times 4 \times 10^5 \times 10^{-3} = 8 \times 10^{5-3} = 8 \times 10^2$

Test Yourself Problems

Find the value of (a) $(6.0 \times 10^8) \times (1.2 \times 10^6)$; (b) $(2.4 \times 10^{-3}) \times (4.0 \times 10^5)$; (c) $(8.0 \times 10^{-4}) \times (2.0 \times 10^{-5})$; (d) $3.0 \times (2.5 \times 10^9)$

2-22 Division of Numbers Expressed in Scientific Notation

To divide one number by another, (1) divide the number preceding the power of ten in the numerator, or dividend, by the number preceding the power of ten in the denominator, or divisor; (2) subtract the exponent of the power of ten in the denominator from the exponent of the power of ten in the numerator; this gives the exponent of the power of ten in the answer.

Sample Problems

Find the value of:

(a) $\dfrac{7.5 \times 10^5}{3.0 \times 10^2}$ (b) $\dfrac{6 \times 10^4}{2 \times 10^{-2}}$ (c) $\dfrac{6}{2 \times 10^5}$

Solution:

(a) $\dfrac{7.5 \times 10^5}{3.0 \times 10^2} = \dfrac{7.5}{3.0} \times \dfrac{10^5}{10^2} = 2.5 \times 10^{5-2}$

   $= 2.5 \times 10^3$

(b) $\dfrac{6 \times 10^4}{2 \times 10^{-2}} = \dfrac{6}{2} \times \dfrac{10^4}{10^{-2}} = 3 \times 10^{4-(-2)} = 3 \times 10^6$

(c) $\dfrac{6}{2 \times 10^5} = \dfrac{6 \times 10^0}{2 \times 10^5} = 3 \times 10^{0-5} = 3 \times 10^{-5}$

Test Yourself Problems

Find the value of

(a) $\dfrac{8.4 \times 10^8}{2.1 \times 10^5}$;   (b) $\dfrac{5.0 \times 10^{-3}}{2.5 \times 10^4}$;   (c) $\dfrac{6.5 \times 10^4}{1.3 \times 10^{-5}}$;   (d) $\dfrac{1.8 \times 10^{-3}}{9.0 \times 10^{-6}}$

2-23 Orders of Magnitude

Order of magnitude is the power of ten nearest to a given measurement.

The power of ten that is the nearest approximation of a measurement is called the *order of magnitude* of that measurement. For example, the radius of the earth is about 6378 kilometers. This equals 6.378×10^3 kilometers or 6.378×10^6 meters. This number is between 10^6 and 10^7 meters. However, since it is nearer to 10^7 meters, the order of magnitude of the earth's radius is 10^7 meters.

The distance from the earth to the sun is about 1.5×10^8 kilometers or 1.5×10^{11} meters. Since this is nearer to 10^{11} meters than it is to 10^{12} meters, the order of magnitude of this distance is 10^{11} meters.

The length of one wave of yellow light from a sodium lamp is 5.9×10^{-7} meter. This length is between 1×10^{-7} and 10×10^{-7}, or 10^{-6} meter. Since it is nearer to 10^{-6} meter than it is to 10^{-7} meter, its order of magnitude is 10^{-6} meter.

Orders of magnitude are particularly useful when it is desired to make a quick estimate of the sizes or of the relative sizes of different measured quantities. From their orders of magnitude, we can see at a glance that the ratio of the distance of the earth from the sun to the radius of the earth is roughly 10^{11} m divided by 10^7 m or 10^4. That is, the first distance is about 10 000 times as large as the second. Similarly, the fact that the order of magnitude of the wavelength of yellow sodium light is 10^{-6} meter tells us that waves of this light are about one millionth of a meter long.

The concept of orders of magnitude is used not only with measurements of length but with all measured quantities. Generally, to obtain the order of magnitude of a quantity, we express it in scientific notation and then decide which is the power of ten nearest to it.

Sample Problems

1. What is the order of magnitude of the height of a person who is 1.8 m tall?

   Solution:
   1.8 m = 1.8×10^0 m
   The nearest power of 10 is 10^0. The order of magnitude is therefore 10^0 m.

2. What is the order of magnitude of a radio wave 824 m long?

   Solution:
   824 m = 8.24×10^2 m
   The nearest power of 10 is 10^3. The order of magnitude is therefore 10^3 m.

Test Yourself Problems

What is the order of magnitude of (a) a building 78 m tall; (b) a particle of diameter 0.0006 m; (c) the distance between Jupiter and the sun, 7.78×10^8 km?

Measuring Time

2-24 Standard of Time

The *second* is the official standard unit of time. Until 1956, it was defined by using the earth as a clock to measure the mean solar day. This is the yearly average of the time it takes the earth to make one complete turn on its axis. The mean solar day was divided into 24 hours. Each hour was divided into 60 minutes and each minute into 60 seconds. Thus, there were 86 400 seconds per day and the second was defined as 1/86 400 of the mean solar day.

This definition has proven unsatisfactory because the rate of rotation of the earth is gradually becoming slower. As a result, the mean solar day and the second that is based upon it increased

The second(s) is the *SI* unit of time.

Fig. 2-12. This cesium atomic clock measures time to an accuracy equivalent to a loss of less than one second in 1000 years.

slightly from year to year. To establish a more nearly constant unit, scientists redefined the second in 1956 as a fraction of a year rather than of a mean solar day. They chose the year 1900 and defined the second as 1/31 556 925.974 7 of that year.

Since that time, the atomic clock, a highly precise instrument, has been developed. The atomic clock uses the fact that atoms or parts of atoms vibrate at extremely rapid but regular rates to make precise measurements of time. In 1964, the International Committee on Weights and Measures selected an atomic clock operated by the vibrations of cesium-133 atoms as the standard means of measuring time. The second was then officially defined as the time taken by a cesium-133 atom to make 9 192 631 770 vibrations. This is now the accepted value of the standard second.

2-25 Instruments for Measuring Time

Any device that undergoes changes at a regular rate can serve as a clock.

Any device that revolves, rotates, vibrates, moves, or changes at a regular rate can be used to measure time. Thus, a candle burning down at a regular rate can show the passage of time against a scale and thus serve as a clock. So can grains of sand trickling through the narrow neck of an hour glass or egg timer. Some mechanical clocks use the vibrations of a pendulum to measure time, while household electric clocks depend upon constant-speed motors to move their clock hands around the dial at the correct rate. A mechanical watch generally measures time by means of the regular vibrations of a balance wheel that is attached to a fine hair spring and controls the motion of the watch hands through a set of gears.

The time intervals in which we are interested extend from such long intervals as 10^{17} seconds—the order of magnitude of the age of the earth's oldest rocks—down to intervals as short as 10^{-20} second—the order of magnitude of the time it takes certain electrons to revolve about the nuclei of their atoms. For many of these measurements we need ingenious time-measuring methods. For example, for long periods of time like the age of the earth or the time elapsed since humans first appeared on earth, we use the *radioactive clock*. This device measures time by the rate at which the atoms of certain radioactive substances, such as uranium, disintegrate.

For measuring periods of time as small as a billionth of a second, we may use an electronic clock in which time is measured by the rate at which a beam of electrons speeds across a screen. To measure still smaller time intervals we turn to *atomic clocks*. In these, the extremely rapid atomic vibrations provide time units small enough to measure such tiny intervals. The cesium-133 clock, for example, can measure time to a precision of one part in ten billion.

2-26 Multiple-Flash Photography

Frequently, we wish to measure the very small time interval during which a bullet or some other rapidly moving object crosses a small distance. One way of doing this is by multiple-flash photography. Here, a camera is mounted in a darkened room with its shutter open ready to take a picture. Next to the camera is a brilliant light source that can give momentary flashes at a known rapid but regular rate. This is called a stroboscopic (stroh·boh·SKAH·pic) light source, or *strobe light*.

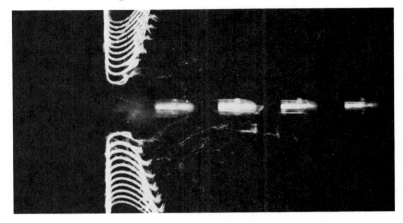

Fig. 2-13. Multiple-flash photograph of a bullet breaking a string. If the time between flashes is 1/10 000 of a second, about how fast is the bullet moving? Assume it is 0.025 m long.

To use this setup to observe a bullet, the bullet is fired in front of the camera and the strobe light is turned on. Each time the light flashes, it illuminates the bullet, and the camera takes a picture. This results in a series of pictures showing the bullet in successive positions. Since the time between light flashes is known, the time taken by the bullet to move from any one position to any other is easily determined. For example, suppose the time between flashes is 1/1000 of a second. The time taken by the bullet to move the distance from its position in one picture to its position in the next picture will then also be 1/1000 of a second. With this information, we can readily determine from the successive pictures how long it took the bullet to move over each part of its path.

The rate of flashing can be made slower or faster depending upon the speed of the object to be photographed. For a high-speed object like a bullet, the rate of flashing may be as high as 12 000 flashes per second. For a body moving more slowly, such as a falling body, the rate may be as slow as 50 flashes a second.

CHAPTER REVIEW

Summary

Measurement consists of comparing the item being measured with a **standard unit** of measurement. The **International System of Units (SI)** defines seven standard units of measurement: the **meter (m)** is the unit of *length*, the **kilogram (kg)** is the unit of *mass*, the **second (s)** is the unit of *time*, the **kelvin (K)** is the unit of *temperature*, the **ampere (A)** is the unit of *electric current*, the **candela (cd)** is the unit of *light intensity*, and the **mole (mol)** is the unit of *quantity*. These seven units are called *basic units* because all other physical quantities can be measured in combinations of these units.

A major advantage of this improved metric system is that it is a **decimal system** and permits easy conversion from smaller to larger units and vice versa.

All measurements are approximations· whose precision is indicated by the number of **significant figures** they contain. The last of these is always doubtful. It is customary to express measurements involving very large or very small numbers in exponential or **scientific notation.** The **order of magnitude** of a measurement is the power of ten that is closest to it.

Questions

Group 1

1. Using a ruler on which the smallest unit was 1 mm, two students measured the thickness of a wooden block. The first recorded a value of 4.73 cm while the second recorded a value of 4.75 cm. Why may these two values be considered to be equally precise?
2. Which of the following list of lengths are of the same order of magnitude: 45 m, 65 m, 105 m, 300 m?
3. Why do the vibrations of the cesium-133 atom provide a better way of defining the second than the rotation of the earth?
4. (a) What property of the pendulum makes it suitable for measuring time? (b) Why isn't your pulse a precise means of measuring time?

Group 2

5. Even though the lengths of two objects are of the same order of magnitude, one of the objects may be four times as long as the other. Explain this statement by illustrating it with an example.
6. In adding a measured length of 3.162 m to one of 18 m, a student wrote the sum as 21.162 m. How should the sum have been written? Explain.
7. The length and width of a rectangular box are each measured to four significant figures, while the height of the box is measured to three significant figures. The area of the base and the volume are now computed. How many significant figures will there be in (a) the computed value of the area of the base; and (b) the computed value of the volume?
8. Three scientists measured the same time interval. The first recorded the result as 4.1 s; the second, as 4.10 s; and the third, as 4.100 s. Is there any difference between these measurements? Explain.

Problems

Group 1

1. Express each of the following as powers of ten: (a) 8 000 000 000 000; (b) 43 456; (c) 7; (d) 25; (e) 7654.321
2. Express each of the following as powers of ten: (a) 0.35; (b) 0.0035; (c) 0.000 305; (d) 0.300 05; (e) 12.636
3. Add: (a) $4 \times 10^2 + 3 \times 10^2$; (b) $4.1 \times 10^3 + 3 \times 10^2$; (c) $4 \times 10^{-2} + 3 \times 10^{-2}$; (d) $4 \times 10^{-3} + 3.1 \times 10^{-2}$.
4. Subtract: (a) $8 \times 10^3 - 3 \times 10^3$; (b) $2.2 \times 10^4 - 3 \times 10^3$; (c) $8.5 \times 10^{-4} - 4.9 \times 10^{-4}$; (d) $8.53 \times 10^{-3} - 3.2 \times 10^{-4}$.
5. Multiply: (a) $(2 \times 10^7) \times (3 \times 10^4)$; (b) $(4 \times 10^2)(2 \times 10^3)$; (c) $(5 \times 10^5)(7 \times 10^3)$; (d) $(4 \times 10^{-2})(2 \times 10^{-3})$; (e) $(3 \times 10^5)(2 \times 10^{-3})$.
6. Divide: (a) $(8 \times 10^6) \div (2 \times 10^4)$; (b) $(6 \times 10^3) \div (3 \times 10^5)$; (c) $(9 \times 10^{-2}) \div (3 \times 10^4)$; (d) $(2.4 \times 10^{-6}) \div (3 \times 10^{-2})$; (e) $(2.4 \times 10^{-6}) \div (3 \times 10^{-12})$.
7. Arrange the following numbers in order of size beginning with the smallest number: 2.1×10^6; 3.9×10^{-7}; 1.2×10^8; 4.3×10^{-9}; 3.0×10^{-10}.
8. What is the order of magnitude of each of the following numbers: (a) 3.21×10^{10}; (b) 7.5×10^4; (c) 4.99×10^6; (d) 87; (e) 11 000?

9. What is the order of magnitude of each of the following numbers: (a) 1.25×10^{-4}; (b) 9.2×10^{-4}; (c) 8.0×10^{-9}; (d) 0.015; (e) 0.0069?

10. What is the number of significant figures in each of the following measurements: (a) 1502 m; (b) 15.02 m; (c) 0.0023 m; (d) 2.59×10^{-4} m; (e) 4.0×10^{3} m?

Group 2

11. Measurement of the three sides of a triangular field gave the following values: 25.27 m, 18.432 m and 20.9 m. What is the perimeter of the field to the best accuracy obtainable with these measurements?

12. What is the area of a rectangle whose sides measure 10.1 cm by 6.21 cm?

13. Assuming that an atom of hydrogen occupies a cubical space that is 5.0×10^{-11} m on an edge, what is the order of magnitude of the volume of the atom?

14. How long does it take light traveling at 3×10^{8} m/s to cross a room 6 m long?

15. A series of pictures of a bullet in motion is made by a flash-photography setup in which the light flashes occur at the rate of 1000/s. The pictures show that the bullet moved 0.5 m between successive flashes. What was the bullet's speed?

Applying Physics

1. Count out 100 sheets of paper and pile them in a rectangular block. Determine the average thickness of one sheet of paper by dividing the thickness of the block of paper by 100. Now measure the length and width of one sheet of paper and determine the volume of a sheet of paper.

2. Make a pendulum by hanging a weight from a string the upper end of which is fastened so that the weight can swing back and forth freely. Adjust the string so that the pendulum is about 45 cm long. Start it swinging through a small arc and, using a clock or watch with a second hand, see how many vibrations it makes in 30 seconds. A vibration is one complete swing forward and back to the starting point. Keeping the arc of swing small, repeat the experiment several times. Determine from your observations how long it takes the pendulum to make one vibration during each of the trials. Is your pendulum dependable as a means of measuring time?

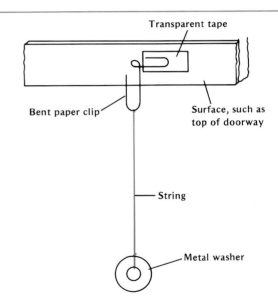

Fig. 2-14. A homemade pendulum.

3

Measuring Mass and Weight

Aims

1. To identify mass and weight as properties of all matter and to learn how to measure them.
2. To learn the difference between mass and weight.

A body's mass depends on the quantity of matter it contains. Its weight is the force with which the earth attracts it.

An astronaut can jump much higher on the moon than on the earth because the astronaut's weight on the moon (that is, the moon's gravitational pull on the astronaut) is only one-sixth that on the earth. The astronaut's mass, however, is the same on both the earth and the moon.

Fig. 3-1. Prototype kilogram No. 20 is the national standard of mass. The platinum-iridium cylinder is about 39 mm high and 39 mm in diameter.

3-1 Mass Versus Weight

Mass and weight are quite different quantities. *The mass of a body depends on the quantity of matter it contains.* It is a property that determines how hard or how easy it is to change the motion of that body. The larger the mass of a body, the harder it is to set that body moving when it is at rest or to slow it down or speed it up once it is moving. A train going at 50 kilometers per hour is harder to stop than a car moving at the same speed because the train has a larger mass than the car.

When a body is at rest or is moving at speeds much smaller than the speed of light, its mass is practically constant. However, as Albert Einstein (1879-1955) has shown, the mass of a body does not actually remain constant but increases as the speed of the body increases. We do not usually observe this because the increase in mass is negligibly (NEG · lij · ib · lee) small for all speeds except those very near the speed of light. At those very high speeds, it becomes important and must be taken into account.

The weight of a body is simply the force with which the earth pulls that body toward its center. This force arises from the fact that every body in the universe exerts an attraction on every other body in the universe, called the *force of gravitation* (grav · i · - TAY · shun). It is the earth's gravitational pull on all the bodies on or near it that causes them to have weight.

The weight of a body depends upon two factors: *its mass and its distance from the center of the earth.* The greater the mass of a body, the harder the earth pulls it downward and the greater the body's weight becomes. Also, the further from the center of the earth a body goes, the smaller the earth's gravitational pull becomes and the less the body weighs. Thus a body high up on a mountain top or in an airplane has a smaller weight than it does at sea level. Its mass, however, remains the same regardless of its location.

3-2 Units of Mass

The standard of mass of the *MKS* or International System (*SI*) is the mass of a platinum-iridium cylinder kept in the Bureau of Weights and Measures at Sèvres, France, and called the *kilogram*.

34

Fig. 3-2. The cruise ship and tugboats may be moving at the same speed, but it would take a much greater force to stop the ship due to its much greater mass.

The kilogram (symbol, kg) is divided into 1000 units each of which is called a *gram* (g). Each gram is divided into 1000 units called *milligrams* (mg). Copies of the standard kilogram have been reproduced and are kept for convenient use in bureaus of weights and measures in many parts of the world.

The kilogram (kg) is the *SI* unit of mass.

An idea of the size of these mass units may be obtained from the following facts.

1. An American nickel has a mass of about 5 grams.
2. A liter of water has a mass of about 1 kilogram.
3. A mid-sized car has a mass of about 1500 kilograms.

3-3 Units of Force and Weight

The standard unit of force in the International System of measurement is the *newton*. Its symbol is N. The newton fulfills the essential requirements of a proper unit of measurement in that it is constant and remains constant independent of the location in which it is measured.

The newton (N) is the *SI* unit of force and weight.

All forces are measured in newtons. Since the weights of bodies are forces, they are also measured in newtons. However, weights differ from other forces in that they act only downward toward the center of the earth. All other forces can act in any direction.

The newton will be precisely defined in Chapter 4. Here we shall only give a rough idea of its size by noting that the downward force exerted by the earth's gravity on a mass of 1 kilogram at sea level is 9.8 newtons, or roughly 10 newtons. Put another way, a newton is about one-tenth the weight of a kilogram. However, while the weight of a kilogram may vary with location, the newton remains a constant.

Knowing the weight of a 1-kg mass at sea level, you can find the weights at sea level of other masses. For example, the weight of a mid-sized car is about 14 700 N, and the weight of an American nickel is about 0.049 N.

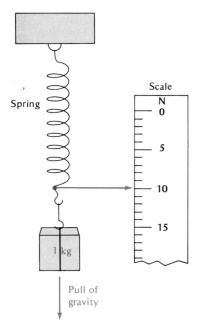

Fig. 3-3. A spring scale. A mass of one kilogram weighs about ten newtons.

3-4 Measuring Weight or Force

Weight is often measured by means of a *spring scale,* or *spring balance.* A spring scale consists of a spring, a fixed scale, and a pointer fastened to the bottom of the spring. The object to be weighed is suspended from the spring. The earth's downward pull on the object causes the spring to stretch until the resisting upward force of the spring just balances the downward pull of gravity. When the two forces are balanced, the position of the pointer against the scale shows the weight of the object. The greater the weight of the object, the more it stretches the spring and the greater the reading on the scale.

In the preceding example, the spring scale is measuring weight, a downward force exerted by gravity. However, it can measure a force exerted in any direction. Thus, to measure the force exerted by a person pulling a cart by a rope, the hook end of the spring is attached to the rope while the person pulls on the other end of the spring. The pointer then shows how much force is exerted.

3-5 Hooke's Law

The accuracy of spring scales depends on a law discovered by the Englishman, Robert Hooke (1635-1703). He noted that when an object such as a spring is stretched by a force, the amount of elongation of the spring is proportional to the force that produces it. That is, if a 1-newton force makes the spring stretch 1 centimeter, a 2-newton force will stretch the spring 2 centimeters, and so forth. Ordinarily when the force is removed from the spring, the spring regains its original length.

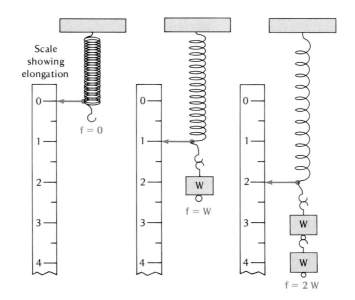

Fig. 3-4. Hooke's law: the elongation (strain) is proportional to the stretching force (stress).

The ability of a spring to regain its original form after it has been stretched and released is called *elasticity* (ih·las·TIS·i·tee). There is, however, a limit beyond which the spring must not be stretched if it is to regain its original form when released. This is called the *elastic limit* of the spring. When a weight suspended from a spring is great enough to stretch the spring beyond its elastic limit, the spring will not regain its original length after the weight is removed. Instead, it will be permanently stretched or deformed.

In elastic bodies, strain is proportional to stress.

Hooke's law applies not only to springs but to all elastic bodies when they are acted upon by forces. Hooke called the force acting upon a body, the *stress*, and the elongation or deformation it produces in the body, the *strain*. He then stated his law as follows: *As long as the elastic limit of a body is not exceeded, the strain produced is proportional to the stress causing it.*

3-6 Expressing a Proportion

The fact that two quantities are proportional to each other is expressed by setting their ratio equal to a constant. Thus, if y is proportional to x, it follows that $y/x = k$ where k is called the *constant of proportionality* (pruh·POR·shun·AL·i·tee). This relationship may be written $y = kx$, which is the general equation expressing that y is proportional to x. The numerical value of k may be found by substituting any value of x and its corresponding y in the relationship $y/x = k$.

To illustrate, suppose we have several bags containing different numbers of nickels. Neglecting the masses of the bags, we find that the mass of each bag of nickels is proportional to the number of nickels in it. We may therefore write $M = kn$, where M is the mass of the nickels and n is the number of nickels in any bag. It is evident that k is simply the mass of one nickel.

To find the value of k, note that $k = M/n$ and divide the mass of any one of the bags by the number of nickels in it. Thus, suppose a bag containing 100 nickels has a mass of 500 grams. The mass of one nickel is $k = M/n = 500$ grams/$100 = 5.00$ grams. The relationship $M = kn$, now becomes $M = (5.00 \text{ grams})n$, from which we can find the mass of any of the bags of nickels from the number of nickels in it. For example, a bag containing 150 nickels has a mass of 5.00 grams per nickel $\times$ 150 nickels = 750 grams.

3-7 Graph of a Direct Proportion

The proportion expressed by Hooke's law is usually written $f = kx$ where f is the force stretching the spring and x is the elongation produced by it. Since $k = f/x$, it follows that k is the amount of force it takes to produce one unit of elongation. Table 3.1 lists the results of an idealized experiment in which various weights are hung from a spring and the elongation produced by each is meas-

The graph of a direct proportion is a straight line.

Table 3.1

f = WEIGHT (N)	x = ELONGATION (cm)
0.0	0.0
1.0	0.5
2.0	1.0
3.0	1.5
4.0	2.0

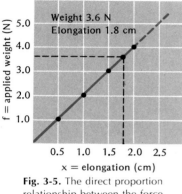

Fig. 3-5. The direct proportion relationship between the force and elongation is shown graphically by a straight line.

ured. Note that the weights and the elongations are proportional to each other.

The graph in Fig. 3-5 plots the applied weights f along the vertical axis and the elongation x along the horizontal axis. The vertical distance of each plotted point from the horizontal axis is called its *ordinate* (OR·din·it). The horizontal distance of each plotted point from the vertical axis is called its *abscissa* (ab·SIS·uh). Thus, for the last plotted point D, the ordinate f is 4.0 newtons and the abscissa x is 2.0 centimeters. From these corresponding values of f and x, we have $k = f/x = 4.0$ newtons divided by 2.0 centimeters, or 2.0 newtons per centimeter. We can therefore write the proportional relationship between f and x for this spring as $f = (2.0$ newtons per centimeter$)x$. Note that the graph is a straight line. This is typical for a direct proportional relationship.

Once we have the graph, we can predict the elongation that will be produced by any applied weight, provided the spring is not stretched beyond its elastic limit. In Fig. 3-5, we predict the elongation that will be produced by a weight of 3.6 newtons by drawing a horizontal line at the 3.6-newton level until it intersects the graph in point C. The ordinate of C is the predicted elongation of 1.8 centimeters.

3-8 How Mass Is Measured

The mass of a body is measured by comparing its mass with a set of known standard masses. This is done with a *beam balance,* which consists essentially of a uniform beam having a pan suspended from each of its ends. The beam is delicately balanced at its center on a knife-edge. A long pointer attached to the center of the beam extends downward over a scale and makes equal swings about the center of the scale when the beam is perfectly balanced.

To find the mass of an object, it is put into the left pan and different combinations of standard masses are then put into the right pan until the pointer shows that the beam is balanced. The mass of the object in the left pan is equal to the sum of the standard masses in the right pan.

3-9 The Beam Balance and Constancy of Mass

The operation of the beam balance depends upon the fact that, in any one place, the earth's downward pull on all bodies is proportional to their masses. If the mass in the left pan is larger than the total of standard masses in the right pan, the earth pulls the left pan down harder than it does the right pan, and the beam goes down on the left. When the masses in the left and right pans are equal, the earth pulls on them equally and the beam is balanced.

If the mass of an object is measured in a beam balance and then the balance and the equal masses in its pans are taken to the top of a mountain, the earth's pull on all the masses will become smaller. The object being measured in the left-hand pan will weigh less than before, and so will the standard masses in the right pan. However, the weight lost by the object will be exactly the same as the total weight lost by the standard masses. As a result, the beam will remain balanced, showing that the mass of the object continues to be equal to the same combination of standard masses as before. Thus the balance shows that, under ordinary conditions, *the mass of an object remains constant regardless of its position on the earth.*

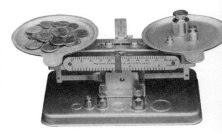

Fig. 3-6. A two-pan, equal-arm beam balance.

3-10 Range of Masses We Measure

We are interested in masses ranging from that of the tiny electron to those of the most massive stars. The mass of the electron is 9.11×10^{-31} kilogram; that of the earth is 5.983×10^{24} kilograms; while that of the sun, which is a middle-sized star, is 1.98×10^{30} kilograms. Between these extremes are the masses of the common objects with which we deal in everyday life, such as the mass of a nickel which is about 5 grams and the mass of a liter of milk which is about 1 kilogram.

The very tiniest and the very largest masses usually require special methods for their measurement. To measure the mass of the electron, for example, we need to make use of electromagnetic theory as well as the laws of motion. To measure the mass of the earth or of the sun, we need the help of the theory of gravitation. Thus, once again, we note that measurement is often a very complicated business.

3-11 Density

From everyday experience we know that substances like iron and lead are "heavier" than substances like aluminum and wood. In physics we state this fact more precisely by saying that iron is denser than aluminum. Exactly what does this mean? It means that a given volume of iron has more mass than the same volume of aluminum. Since the weight of all objects in any one place is proportional to their masses, it also means that the weight of a given volume of iron is greater than the weight of the same volume of aluminum.

The density of a substance is its mass per unit of volume. The density of water is 1 gram per cubic centimeter. This means that 1 cubic centimeter of water has a mass of 1 gram. It is no accident that the density of water has this simple unit value. The French commission that established the metric units of measurement planned originally to take the mass of 1 cubic centimeter of water as its standard and to call it 1 gram. Since the maintenance of this water standard presented too many difficulties, the standard of mass finally selected was the platinum-iridium kilogram now kept at Sèvres, France, with the standard meter.

Fig. 3-7. With the above single-pan balance, direct readings to tenths of a milligram can be rapidly obtained.

The density of aluminum is 2.7 grams per cubic centimeter. Thus, a cubic centimeter of aluminum has 2.7 times as much mass as a cubic centimeter of water. In any one place, therefore, a sample of aluminum weighs 2.7 times as much as an equal volume of water.

The density of water may also be expressed as 1 kilogram per liter and as 1000 kilograms per cubic meter. Since aluminum is 2.7 times as dense as water, its density is 2.7×10^3 kilograms per cubic meter.

3-12 Measuring the Density of a Substance

To measure the density of a substance, we obtain a sample of that substance and measure its mass and its volume. The density of the substance is then found from the relationship which follows from the definition:

$$\text{Density} = \frac{\text{Mass}}{\text{Volume}} \quad \text{or} \quad D = \frac{M}{V}$$

Thus, if a given sample of metal is found to have a mass of 210 grams and a volume of 20.0 cubic centimeters, the density of that metal is:

$$D = \frac{M}{V} = \frac{210 \text{ g}}{20.0 \text{ cm}^3} = 10.5 \text{ g/cm}^3$$

A measurement that is particularly useful when we wish to compare the densities of different substances is that of specific gravity (speh·SIF·ik GRAV·i·tee). *The specific gravity of a substance is the number of times it is as dense as water.* The specific gravity of water is, of course, 1. The specific gravity of sulfur is 2. Thus, sulfur is twice as dense as water. In any particular location on the earth it means also that a volume of sulfur weighs twice as much as an equal volume of water. The specific gravity of alcohol is 0.8. Alcohol, therefore, is only 0.8 times as dense as water.

3-13 Physical Units Combine Algebraically

In the problems that follow and those in the rest of this book, you will see that in performing algebraic operations, the units also obey the rules of algebra. Thus, we know from algebra that two quantities, *a* and *b*, may make these typical combinations:

$$a \times a = a^2$$
$$a \times a \times a = a^3$$
$$a \times b = ab$$
$$\frac{a^2 \times b}{a} = ab, \text{ and so forth}$$

Units of measure combine in the same way. Thus, in computing the volume of a box whose dimensions are 3 meters by 2 meters by 1 meter,

since

$$V = l \times w \times h$$

we have:

$$V = 3 \text{ m} \times 2 \text{ m} \times 1 \text{ m}$$

Now,

$$\text{m} \times \text{m} \times \text{m} = \text{m}^3$$

Therefore,

$$V = 6 \text{ m}^3$$

Again, in computing the mass of a body whose volume is 2 cm³ and whose density is 8 g/cm³, we have

$$D = M/V \quad \text{or} \quad M = D \times V$$
$$M = 8 \text{ g/cm}^3 \times 2 \text{ cm}^3 = 16 \text{ g}$$

Notice that cm³ cancels to leave only grams, the proper unit for the mass.

Units combine or cancel algebraically to give the correct units for answers.

CHAPTER REVIEW

Summary

The **mass** of a body depends on the quantity of matter it contains. It is a property that determines how hard it is to change the motion of that body. It is constant under ordinary conditions. The **weight** of a body is the force with which the gravitational pull of the earth attracts it. *The weight of a body is directly proportional to its mass.* However, it is not constant, but decreases as the body moves away from the center of the earth.

Weights or forces are commonly measured by means of the **spring balance**. The spring of the balance obeys **Hooke's law** according to which *the strain produced in the spring is proportional to the stress* causing it. The law holds as long as the elastic limit of the spring is not exceeded. The basic unit of weight or force is the **newton (N).**

Masses are commonly measured with a **beam balance.** The units of mass are the **kilogram (kg)** and the **gram (g).**

The **density** of a substance is its *mass per unit volume.* It can be found from the relationship $D = M/V$. The **specific gravity** of a substance is the number of times it is as dense as water.

Questions

Group 1
1. (a) Distinguish between the mass of a body and its weight. (b) In what units is each measured?
2. (a) What is the relationship between the mass of a body and the force needed to change its motion? (b) Under what conditions does the mass of a body remain constant?
3. (a) When an object is taken up in a balloon,

what changes, if any, take place in its mass as the balloon rises? (b) What changes, if any, take place in its weight?

4. Why is the weight of a body not a good measure of its mass?

5. (a) How is a spring balance used to measure weight? (b) How does Hooke's law determine how much elongation the spring will suffer when different weights are hung from it? (c) What property of the spring assures that it will return to its original length after the weight that stretched it is removed?

6. (a) How is a beam balance used to measure the mass of an object? (b) Explain why a beam balance gives the same value for the mass of a body whether the measurement is done at sea level or on a mountain top.

7. At sea level, a spring scale shows that one body weighs twice as much as a second body. What does this tell us about the masses of the two bodies?

8. Suppose that the two bodies in question 7 are taken up to a high altitude and weighed again with the same spring scale. (a) Compare their weights now with their weights at sea level. (b) Compare the ratio of the weights of the bodies with their ratio of 2 : 1 at sea level. (c) What information does the answer to (b) give about the ratio of the masses of the two bodies?

9. (a) What is meant by the density of a substance?

(b) How may the density of a substance be determined? (c) What is meant by the specific gravity of a substance?

10. (a) What is wrong with saying: Gold is heavier than aluminum? (b) Correct the statement by changing only one word in it.

Group 2

11. (a) Write an equation representing the fact that the weight of a copper wire of constant cross section is directly proportional to the length of the wire. (b) What does the proportionality factor in this equation mean? (c) What does the graph of this relationship look like?

12. (a) Write an equation stating the fact that the mass of any sample of a substance is proportional to the volume of that sample. (b) What is the proportionality factor in this equation?

13. It takes 1 kg of metal to make a certain solid cube. What mass of metal is needed to make a cube of the same metal whose edge is (a) twice as big as the first; (b) three times as big as the first; (c) n times as big as the first?

14. Two cubes are made of silver and have equal edges. One of the cubes is hollow. Explain how to determine how much of the volume of this cube is empty.

15. What effect will an increase in temperature have on (a) the mass of a block of iron; (b) the density of this same block of iron. Explain.

Problems

Group 1

1. Convert to kg; (a) 540 g; (b) 5400 g.
2. Convert to g: (a) 2.402 kg; (b) 0.0465 kg; (c) 23 mg.
3. A body is weighed at a certain location where a mass of 1 kilogram weighs 9.80 N. If the weight of the body is 29.4 N, what is its mass?
4. A spring has a proportionality constant of 0.20 N per cm. How much will a 1.50-N force make the spring stretch?
5. The measured elongations of a spring produced by each of a series of forces is recorded in Table 3.2.
   Make a graph of the force versus the elongation. From the graph, predict the elongation produced by a force of (a) 4.70 N; (b) 11.0 N, (c) What is the value of the proportionality factor k?

Table 3.2	
FORCE (N)	ELONGATION (cm)
0.00	0.00
2.00	1.00
4.00	2.00
6.00	3.00
8.00	4.00
10.00	5.00

6. What is the mass of water that fills a rectangular tank 5.0 m long, 4.0 m wide and 2.0 m high?
7. A graduated cylinder is filled with 50.0 cm³ of water. When a piece of copper is dropped into the cylinder, the water level rises to 63.5 cm³. If the density of copper is 8.90 g/cm³, what is the mass of the piece of copper?

8. Find the density of a sulfuric acid solution of which 50 cm³ has a mass of 65 g.
9. The density of aluminum is 2.7×10^3 kg/m³. What is the mass of a slab of aluminum 1.00 m long, 0.50 m wide, and 0.10 m thick?
10. The density of mercury is 13.6 g/cm³. What is the volume of 680 g of mercury?
11. (a) What is the mass of 12 cm³ of water? (b) What is the mass of 12 cm³ of alcohol having a specific gravity of 0.80?
12. The specific gravity of iron is 7.9. Find its density in (a) g/cm³; (b) kg/m³.
13. The density of silver is 10.5 g/cm³. What is the volume of a solid silver statue having a mass of 210 g?
14. The specific gravity of a certain rock is 2.5. What is the volume of 1.0 kg of this rock?
15. A 1.00-kg body weighs 9.80 N at sea level. What is the weight of a 12.0-kg body?

Group 2

16. When a force of 20 N is applied to an elastic spring, its length increases by 0.24 m. (a) Assuming that the spring is not stretched beyond its elastic limit, what is its proportionality constant? (b) If a force of 6.0 N is applied to the spring, what will be its elongation?
17. Table 3.3 shows the elongation of a spring when a series of increasing forces is applied to it. (a) Make a graph of the force versus the elongations produced by them. (b) What evidence is there that the elastic limit of the spring has been exceeded? (c) What was the value of the proportionality factor k before the elastic limit was exceeded?

Table 3.3

FORCE (N)	ELONGATION (m)
0.0	0.00
1.0	0.03
2:0	0.06
3.0	0.09
4.0	0.13
5.0	0.18
6.0	0.24

18. (a) Assuming that a hydrogen atom occupies a cubical space 1.0×10^{-10} m on an edge, what is its volume? (b) Given that the hydrogen atom has a mass of 1.7×10^{-27} kg, find the average density of the atom.
19. Ernest Rutherford showed that most of the space inside an atom is empty and that practically all the atom's mass is located in a central part called its nucleus. Assuming that the nucleus of a hydrogen atom occupies a cube 10^{-14} m on a side, and that practically the entire mass of the atom, 1.7×10^{-27} kg, is located in this cube, find the density of the matter making up the nucleus.
20. The volume of the earth as calculated from its radius is about 1.1×10^{21} m³. Assuming that the average density of the matter composing the earth is 5.5×10^3 kg per m³, estimate as closely as you can the mass of the earth.
21. When a constant mass M of a gas is heated at constant pressure, both its volume V and its density D change with its temperature T. If the relationship between V and T is given by $V = kT$ where k is a constant, what is the relationship between D and T?

Applying Physics

1. Find the mass of a single sheet of paper by finding the mass of 100 sheets and then dividing that value by 100. What assumption is made in finding the mass of the single sheet in this way? Repeat the procedure with a heavier grade of paper. How great a difference between the papers do you find?
2. Hang a series of weights of increasing size from a rubber band and measure the elongation produced by each weight. Plot the elongations against the weights producing them. Does the rubber band obey Hooke's law?
3. Estimate the density of sugar using a 1-pound carton of granulated sugar as follows. First measure the length, width, and height of the box in centimeters and multiply to obtain the approximate volume of the sugar. Then, knowing that the mass of 1 pound of sugar is 454 grams, find the density from the relationship, density = mass/volume.

The Tools of the Surveyor

Surveyors measure distances, directions, and angles between points and lines on the surface of the land. What instruments do they make these measurements with?

The device for measuring short distances is the steel tape. This is a steel wire or flat steel ribbon, graduated in length units. Tapes are usually 100 feet long in the United States, but they are available in greater lengths.

A method of measuring distance you may have seen is the stadia technique. This requires what is called a transit or theodolite mounted on a tripod, and a graduated stadia rod. The rod is held upright at a point and it is sighted through a transit telescope from the other point. The length of stadia rod visible between two horizontal hairs in the telescope viewer allows calculation of the distance between the two points.

Distance can be measured electronically too. Microwave or infrared signals are sent from one point to another, and reflected back.

The distance between the two points is then calculated from the time required for the signal to make the round trip.

The transit is the instrument most commonly used for measuring angles. A theodolite is a transit equipped to provide more precise angular measurements. These instruments consist basically of a telescope that can be moved a measured amount vertically or horizontally. Calibrated scales built into the instrument give the vertical and horizontal angles of rotation.

Directions are usually measured by magnetic compasses or celestial observations. Many transits are equipped with compasses and one can usually be mounted on a theodolite. Direction is read from the compass card in relation to magnetic north. Measuring direction by celestial observation involves observations of the sun or stars using a transit or theodolite. In many parts of the northern hemisphere an accurate determination of true

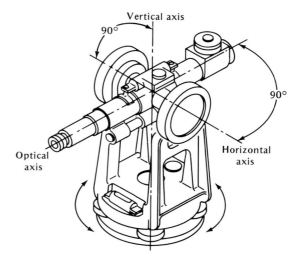

north can be made by sighting on Polaris, the North Star. Directions are then given in relation to true north.

Surveyors and Related Fields

Land surveys are usually carried out by a team of workers. The results of survey measurements are recorded and verified. They are then used to prepare sketches, maps, reports, and legal descriptions for deeds, leases, and other documents.

Some subspecialties in the field of surveying are as follows: land surveyors locate and map the boundaries of tracts of land; topographic surveyors determine the contours of the land and pinpoint features such as buildings, roads, forests, and rivers; geodetic surveyors measure large areas of the Earth's surface, often using satellite observations; marine surveyors determine shorelines, features of the bottom, and the depths of bodies of water. Some other occupations that use accurate measurements of land areas, coastlines, and natural and constructed features are cartographic drafters, cartographers, topographical drafters, and geodesists. For additional information on these and other related occupations, consult the career information files in your school's guidance department.

Much of the work of a surveyor is done outdoors, in all types of weather. Team members must often be on their feet for many hours, walk long distances, and carry heavy packs of instruments.

Opportunities for surveyors and related occupations are expected to grow about as fast as the average for all occupations through the year 2000. Jobs will be available in housing and other building construction, street layout, the development of recreation areas, and highway development and construction.

To become a licensed surveyor, you will need some postsecondary education and extensive on-the-job training. Most people enroll in a junior college or technical institute program before seeking employment. The quickest path to a surveyor's license is usually a four-year college degree and two to four years of work experience. By the year 2000, most or all states may require a Bachelor's degree.

UNIT 2
Force, Motion, and Energy

(Left) Launch of the Space Shuttle Atlantis in May 1989. During this mission, the Magellan space probe (above) was launched to Venus.

In this unit we learn that energy often takes the form of matter in motion, and that forces control the motion of bodies. Our study of matter and energy thus leads directly to the study of force and motion.

Three great figures stand out as the major contributors to our understanding of force and motion: Galileo Galilei, Isaac Newton, and Albert Einstein. Galileo (1564-1642) explored the nature of motion experimentally as well as theoretically and showed how to use mathematical methods in describing and analyzing motion. Newton (1642-1727) investigated how forces control the motion of bodies and formulated the relationships between force and motion in the three laws that bear his name. Newton's laws of motion were fantastically successful in explaining the motion of earthly as well as heavenly bodies and remained unchallenged until the close of the nineteenth century. At that time, new experimental knowledge revealed that Newton's laws failed when applied to bodies moving at velocities close to that of light. That failure brought about a re-examination of the nature of force and motion by Einstein (1879-1955). His theory of relativity introduced utterly new ideas about the nature of mass and its relation to motion and established the fact that mass is actually a form of energy.

The development of ideas about the nature of force and motion was accompanied by a parallel development of ideas about the nature of energy. This came to a climax in the mid-nineteenth century with the establishment of the law of conservation of energy, a major triumph in our understanding of the physical world. The law states that energy is indestructible; that it is neither created nor destroyed in any process, but merely changes from one form to another. With the identification of mass as a form of energy, the law applies to mass as well as to other forms of energy. It is one of the most powerful investigative tools in physics.

A major outcome of our understanding of force, motion, and energy has been a sharp reduction in the size of our world. By providing us with swift means of transportation, it has made all peoples of the earth our close neighbors. Our future and theirs depend on how well we live and work together.

4

Vectors, Force, and Motion

Aims

1. To note that energy is often associated with matter in motion and that motion is controlled by forces.
2. To learn how to represent forces and motion by vectors.
3. To learn how to obtain the combined effect of two or more vectors acting upon the same point of a body.
4. To understand how a vector can act in directions other than its own.

Vectors and Scalars

4-1 Change, Motion, and Force

Practically all of the changes we see in the world about us are the result of *motion*. Day and night are caused by the rotation of the earth on its axis. The winds and their effects are caused by the motion of air. The conversion of raw materials into the products we need in everyday living is brought about by various motions. These include transporting the materials to factories, combining them, shaping them into finished products, and transporting the products to market. Change and motion go hand in hand.

Motion is controlled by forces.

When a car stalls, it must be pushed to get it going. When the engine is running, it is the engine which pushes the car forward. In either case, a *force* is being used to change the state of the car from rest to motion. In order to make the car slow down or to stop altogether, the brakes are applied. Again, a force is being applied to change the motion of the car. These examples illustrate that *motion is controlled or changed by means of force.*

4-2 Displacement Is a Vector

Motion generally involves a change of position of the object being moved. A change of position is called a *displacement*. Suppose a body is moved from a point *A* to a second point *B* 10 meters to the northeast of *A*. How shall we describe this displacement? It is not enough to say that the body has moved 10 meters from *A*. Its final position could then be any point on the circumference of a circle centered on *A* and having a 10-meter radius. To state exactly how the position of the body changed, we must also state in what direction it was moved. Thus, the displacement from *A* to *B* is described as one of 10 meters northeast.

Quantities such as displacements are called *vectors* (VEK·torz). A vector is characterized by the fact that it *has both a magnitude or size and a direction. Quantities having only magnitude,* such as the mass or length of an object, are called *scalars* (SKAY·lurz).

A vector has magnitude and direction.

4-3 Velocity and Force Are Vectors

Two other important vectors related to the study of motion are *velocity* (vel·AH·si·tee) and *force.* To say that an airplane is going 500 kilometers (KIL·oh·mee·turz) per hour does not describe its motion completely. It tells us the speed of the airplane but not the direction in which it is going. To tell exactly how the airplane is moving at a given moment, we give its velocity. By velocity we mean not only the speed of a body but also the direction in which it is moving. Thus, *velocity is a vector whose magnitude is the speed of the body and whose direction is the direction of motion of the body.* If the airplane is traveling westward, its velocity is stated as 500 kilometers per hour westward.

It is evident that a force is a vector, since the effect a force has on a body depends not only on the size of the force but also on the direction in which it acts. A force of 10 newtons pushing north and a force of 10 newtons pushing south will obviously produce opposite effects when applied to a body. Therefore, in describing any force, we must tell not only its magnitude but also its direction.

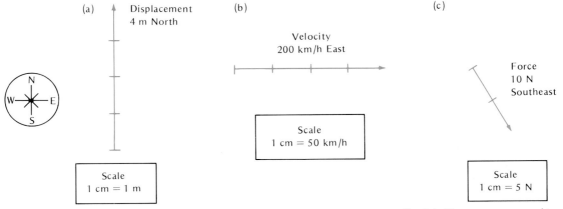

Fig. 4-1. How arrows are used to represent vectors.

4-4 Representing a Vector

A vector is represented by an arrow drawn to some selected scale. The length of the arrow shows the magnitude of the vector. The direction of the arrowhead shows the direction of the vector. To represent a displacement of 4 meters to the north we first select a scale which, in this case, we take as 1 centimeter = 1 meter. Now we draw a north-south line 4 centimeters long to represent 4 meters. Finally, we put an arrowhead on top of this line to show the direction of the displacement. This vector represents a displacement or change of position of 4 meters to the north. See Fig. 4-1(*a*).

Fig. 4-1 (*b*) shows the vector way of representing a velocity of 200 kilometers per hour eastward. Here the scale used is 1 centimeter = 50 kilometers per hour. The vector is therefore 4 centimeters long.

In Fig. 4-1 (*c*) a force of 10 newtons southeast is represented by a vector to a scale of 1 centimeter = 5 newtons.

How Vectors Combine

4-5 Resultant of Two Vectors

The resultant is the vector sum of two vectors acting at the same point.

Often we are dealing with two or more vectors that act upon the same body. In a soccer game, two opposing players may kick the ball at exactly the same instant, one to the east and one to the north. The ball therefore is acted upon by two forces, one pushing it eastward and one pushing it northward. What will be the combined effect of these two forces on the ball? To find out, we must know how to add the effects of the two vectors representing the forces in order to obtain their combined effect or *resultant* (reh·ZUL·tunt). *The resultant or vector sum of two vectors acting at the same point is that single vector whose effect on a body is equal to the combined effects of the two given vectors on that body.*

4-6 Vectors Acting in the Same Direction

When an airplane is traveling in a wind, its displacement or change of position is the result of two displacements. One of these is caused by the wind and the other by the pull of the airplane's engine.

Suppose the engine of an airplane pulls it northward a distance of 600 meters while at the same time the wind carries the airplane northward another 200 meters. The net result, or resultant, of these

Fig. 4-2. When an airplane fires a rocket, the resultant velocity of the rocket is equal to the sum of its own velocity and that of the airplane.

two northward displacements is a total displacement of 600 + 200 = 800 meters to the north. This example illustrates that the resultant of two vectors acting in the same direction is a vector whose magnitude is equal to the sum of their magnitudes and acts in the same direction as they do.

To represent the combination of the two displacements we select 0.5 centimeter = 100 meters as a scale. The 600-meter displacement is represented as a northward vector $6 \times 0.5 = 3$ centimeters long. The 200-meter displacement is represented by a second northward vector $2 \times 0.5 = 1$ centimeter long. To find the resultant of these two displacements, we attach the tail of one displacement to the head of the other as shown in Fig. 4-3. Note that there are two ways of doing this, but both give the same answer. The resultant is the vector formed by joining the two displacements. Here it is a northward vector $3 + 1 = 4$ centimeters long, representing a displacement of 800 meters north.

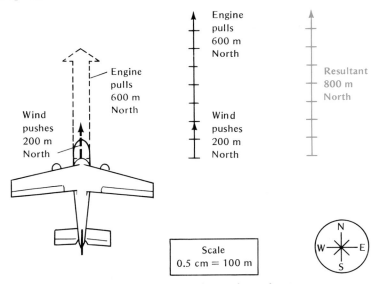

Scale
0.5 cm = 100 m

Fig. 4-3. The resultant of two vectors acting in the same direction equals their sum.

4-7 Vectors Acting in Opposite Directions

Suppose the wind reverses its direction so that it blows the airplane 200 meters southward while the engine of the plane pulls it 600 meters northward. The resultant of these two displacements is 600 − 200 = 400 meters northward. That is, the combination of the two opposing displacements will put the airplane 400 meters north of its starting position.

The two displacements are shown as vectors in Fig. 4-4 to a scale of 0.5 centimeter = 100 meters. As before, the vectors are combined by attaching the tail of one vector to the head of the other. The resultant is the vector joining the tail of the first vector to the head of the second. Here it is a northward vector $4 \times 0.5 = 2$ centimeters long representing a displacement of 400 meters north.

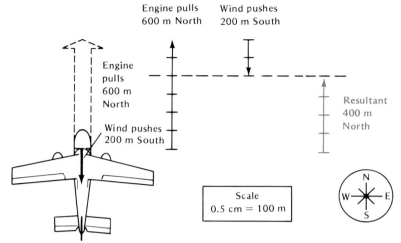

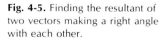

Fig. 4-4. The resultant of two vectors acting in opposite directions equals their difference.

This example illustrates that the resultant of two vectors acting in opposite directions is a vector whose magnitude is the difference of their magnitudes and which acts in the direction of the greater vector. A force of 50 newtons acting upward, for example, combines with a force of 60 newtons acting downward to produce a resultant of $60 - 50 = 10$ newtons acting downward.

4-8 Vectors Acting in Any Direction

Usually the wind's direction and the airplane's direction are in neither the same nor the opposite direction, but make an acute or obtuse angle with each other. Consider the case in which the engine pulls the airplane 600 meters north while the wind blows it 200

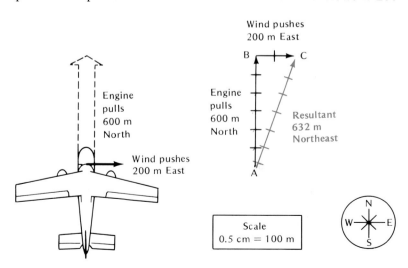

Fig. 4-5. Finding the resultant of two vectors making a right angle with each other.

meters east. The two displacements which the airplane undergoes
are drawn to the same scale as before in Fig. 4-5. To find their
resultant, consider the effect of each one separately. Starting at
point A, the engine pull moves the airplane 600 meters north to
point B, while the wind moves the airplane 200 meters east of B
to point C. The net effect of the two displacements is the same as
if the plane moved directly from A to C. Thus the displacement AC
is the resultant of the displacements AB and AC. Measuring the
length of AC, we find it is 3.16 centimeters long which, on the
scale of 0.5 centimeter = 100 meters, represents a displacement of
632 meters. The combined effect of having the airplane moved 600
meters northward and 200 meters eastward is a total displacement
of 632 meters in the northeasterly direction shown in Fig. 4-5.

4-9 General Method of Finding a Resultant

The above example suggests the general procedure for finding the
sum or resultant of two vectors. First, draw one of the vectors to
scale. Then put the tail of the second vector at the head of the first
vector and draw it to scale. The resultant is now found by joining
the tail of the second vector to the head of the first. The length of
this resultant is measured and the scale is used to determine its
magnitude. The direction of the resultant is from the tail of the
second vector toward the head of the first.

In finding the resultant of two vectors, it makes no difference
whether we attach the first to the second vector or the second to
the first. The resultant will be the same. Thus, in the case of the
airplane in our example, it makes no difference whether the airplane is moved 600 meters northward by its engine and then 200
meters eastward by the wind or whether it is first moved 200 meters
eastward by the wind and then moved 600 meters northward by
its engine. The final result is the same as shown in Fig. 4-6.

There is a simple way of identifying vectors. The northward
displacement of 600 meters is represented by the symbol $\vec{a}$. The
arrow above the letter indicates that it refers to a vector. Thus $\vec{b}$
is used to represent the eastward 200-meter displacement while $\vec{c}$
represents the resultant. In triangle I, $\vec{c}$ is found by attaching $\vec{b}$ to
the head of $\vec{a}$; that is, $\vec{c} = \vec{a} + \vec{b}$. In triangle II, $\vec{c}$ is found by
attaching $\vec{a}$ to the head of $\vec{b}$; that is, $\vec{c} = \vec{b} + \vec{a}$. Notice that the
resultant $\vec{c}$ is exactly the same in both cases. Observe that the
accuracy of an answer obtained by this graphical method depends
upon how precisely we draw the diagrams and measure the vectors.
In Section 4-17 we shall see how to solve vector problems more
accurately by trigonometry.

We shall now apply the graphical method to find the resultant
of two velocities.

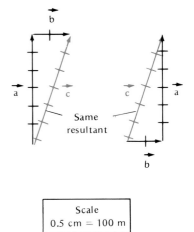

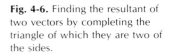

Scale
0.5 cm = 100 m

Fig. 4-6. Finding the resultant of
two vectors by completing the
triangle of which they are two of
the sides.

Sample Problem

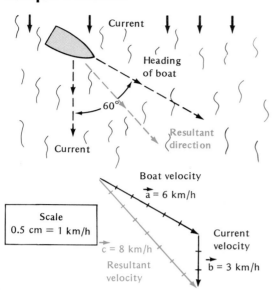

Scale
0.5 cm = 1 km/h

Fig. 4-7. Finding the resultant of two velocities.

A motorboat is heading across a river at 6 km/h in a direction making an angle of 60° to the current. At the same time, the current is moving the boat downstream at 3 km/h. What is the resultant velocity of the boat?

Solution:
The boat has two velocities, $\vec{a}$ supplied by its motor, and $\vec{b}$ supplied by the current. Fig. 4-7 shows how to find their resultant. Using a scale of 0.5 cm = 1 km/h, draw $\vec{a}$ in the direction of the boat's heading. Make it 6 × 0.5 cm long to represent the 6 km/h velocity given to the boat by its motor. To the head of $\vec{a}$, attach $\vec{b}$ drawing it in the direction of the current. Make $\vec{b}$ 3 × 0.5 = 1.5 cm long to represent the 3 km/h velocity given to the boat by the current.

Now draw $\vec{c}$ from the tail of $\vec{a}$ to the head of $\vec{b}$. The vector $\vec{c}$ is the resultant of $\vec{a}$ and $\vec{b}$. Measure its length. It is 4 cm. On the scale of 0.5 cm = 1 km/h, this represents a speed of 8 km/h. Thus the 6-km/h velocity given the boat by its motor and the 3-km/h velocity given by the current combine to give the boat a velocity of 8 km/h in the direction shown.

Test Yourself Problems

1. An airplane is heading eastward at 300 km/h while the wind is blowing it northward at 80.0 km/h. Find the resultant velocity.

2. A force of 150 N and a force of 200 N act at the same point of an object and make an angle of 45° with each other. Find their resultant.

4-10 How Do We Know?

We have just found the resultant of two velocities by *vector addition*. How do we know that velocities really add up in this way? The question, *How do we know?* is one we must ask ourselves frequently to be sure we are not assuming relationships that are not always true.

This question is generally answered by appeal to experience or experiment. We test the relationship in many actual situations. The more tests and the more varied tests we make, the better. If the relationship actually works in all of them, we can have confidence that it will also work in many other similar tests even though we have not actually made them.

However, we cannot be absolutely sure it works in any situation in which we have not actually tested it. If the relationship works in most of the tests but not in certain ones, we must restate the relationship so that its use is limited to those situations where it applies.

A common method of testing a relationship is to use it to make a *prediction*, and then to see if actual observation agrees with the prediction. For example, in the case of the addition of the two velocities acting on the boat, we can predict from our construction of the vector triangle in Fig. 4-7 that the boat should have a resultant of 8 kilometers per hour in the direction shown. We can test this prediction by having a boat cross the river under the given conditions and measuring its actual resultant velocity. If we find that it is 8 kilometers per hour in the predicted direction, we have another demonstration that velocities add like vectors.

However, we must test the vector addition of velocities in many, many other cases before we can be confident that it works generally. When we do this, we find that velocities do combine like vectors in the vast majority of cases. But there is an important group of exceptions.

When we are dealing with velocities very near to the velocity of light, we find that they no longer obey the rules of vector addition. According to the *theory of relativity* proposed by Albert Einstein, *no body can have a velocity greater than that of light*. It follows that no combination of two velocities, no matter how large, can ever give a body a resultant velocity greater than the velocity of light.

It happens that most of the bodies with which we will deal in the next chapters move at much slower rates than the velocity of light. Therefore, we shall not be concerned at present with this exception to the vector method for adding velocities.

4-11 Effect of the Angle Between Two Vectors on the Resultant

As the angle between two vectors increases, their resultant decreases. This is shown in Fig. 4-8 where the resultant made by a 4.0-newton force and a 5.0-newton force is constructed for the cases where the angle between them is 0°, 45°, 90°, 135°, and 180°. Note that for an angle of 0°, the forces act in the same direction and have a resultant of 4.0 + 5.0 = 9.0 newtons in the same direction as both forces. This is the largest resultant a 4.0-newton and a 5.0-newton force can have.

For an angle of 180°, the forces act in opposite directions and their resultant is 5.0 − 4.0 = 1.0 newton in the direction of the 5.0-newton force. This is the smallest resultant a 4.0-newton and a 5.0-newton force can have. As the angle between the forces increases from 0° to 180°, the resultant decreases from its maximum value at 0° to its minimum value at 180°.

4-12 Resultant of Three or More Vectors

To find the resultant of three vectors, we extend the method used for two vectors as follows. Draw each of the vectors to scale, then attach to the head of one of the vectors the tail of a second. To the head of the second vector, attach the tail of the third. The resultant

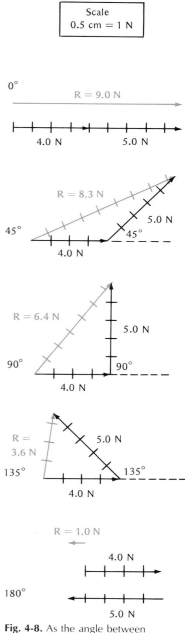

Fig. 4-8. As the angle between two vectors increases, their resultant decreases. At what angle does it reach minimum value?

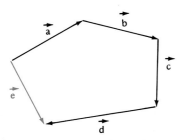

Fig. 4-9. Vector $\vec{e}$ is the resultant of vectors $\vec{a}, \vec{b}, \vec{c}, \vec{d}$.

vector is drawn by joining the tail of the first vector to the head of the third vector. The magnitude of this vector is determined from the scale.

To add four or more vectors, we continue the process of arranging the vectors in a head-to-tail sequence as above. The resultant is again the vector joining the tail of the first vector to the head of the last vector in the sequence.

4-13 Demonstrating that Forces Combine Vectorially

To see whether forces actually combine like vectors to form a resultant, we use the testing method we have just discussed for velocities. First, we use the vector method to predict the resultant of two forces. We then find the actual resultant of the two forces by experiment and see if our prediction agrees with actual observation.

Here is a way of doing this. From two nails set about a meter apart in a wall, hang two spring balances. Tie the ends of a string to the hooks of the balances and tie a weight of 20.0 newtons to the string as shown in Fig. 4-10. The spring balances now show that a force $\vec{a}$ of 10.0 newtons is pulling along the direction OA and a force $\vec{b}$ of 17.3 newtons is pulling along OB. Since these two forces combine their pulls to support the 20.0-newton weight, their actual resultant must be an upward force of 20.0 newtons.

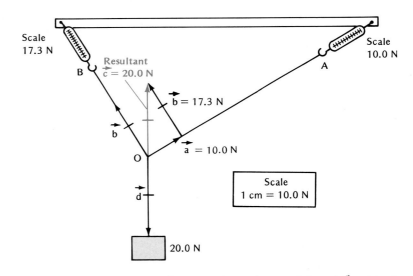

Fig. 4-10. Demonstrating that forces combine vectorially.

Now let us compare this actual resultant of $\vec{a}$ and $\vec{b}$ with the resultant predicted by the vector method. To determine this resultant, put a piece of paper under O and on it draw the directions of the forces $\vec{a}, \vec{b}$, and $\vec{d}$. Using as a scale 1 centimeter = 10 newtons, draw $\vec{a}, \vec{b}$, and $\vec{d}$ to scale. Then find the resultant of the forces $\vec{a}$ and $\vec{b}$ by attaching $\vec{b}$ to $\vec{a}$ as shown, and drawing $\vec{c}$. Measure the length of resultant $\vec{c}$. You will find it to be 2.00 centimeters long which represents a force of 20.0 newtons. Note also that $\vec{c}$ acts vertically upward directly opposite to $\vec{d}$.

Thus, the predicted resultant $\vec{c}$ is equal to the actual resultant holding up the 20.0-newton weight. Of course, this demonstrates that the vector method of predicting the resultant of two forces is correct in this one case only. However, many experiments done with all kinds of forces at all possible angles confirm that this method works generally.

Fig. 4-11. The essential condition for the success of any delicate balancing act is that all forces be in equilibrium.

4-14 Forces in Equilibrium

In dealing with forces, we often want the forces acting on a body to neutralize each other. Thus, for a picture to remain hanging on a wall, it is necessary to support its weight. This is done by having the nail from which it is hung supply a net upward force just equal to the downward force due to the weight of the picture.

The simplest way to neutralize a force acting on any point of a body is to apply an equal and opposite force to that point. Thus, if two teams of equal strength are playing tug-of-war, their equal and opposite forces will neutralize each other and neither team will be able to make the other move. Forces that cancel out each other's effect are said to be in *equilibrium* (ee·kwil·LIB·ree·um). Since forces in equilibrium have no net effect on the body on which they act, their resultant is zero.

Forces in equilibrium neutralize each other.

4-15 The Equilibrant

The single force that can neutralize the effects of two other forces acting on a given point of a body is called their equilibrant (ee·-KWIL·i·brunt). To find the equilibrant of two such forces, first find their resultant by vector addition in the usual way. Then, from the tail of the resultant construct a force equal to it in magnitude but opposite in direction. This force is the desired equilibrant. It is in equilibrium with the two given forces because it neutralizes their resultant. Note that in Fig. 4-12, the force $\vec{d}$ is the equilibrant of $\vec{a}$ and $\vec{b}$

Sample Problem

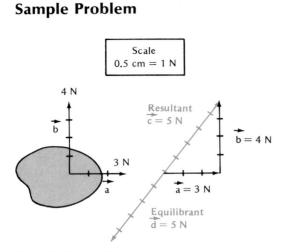

Fig. 4-12. The equilibrant of two forces is equal and opposite to their resultant.

A force $\vec{a}$ of 3 N and a force $\vec{b}$ of 4 N act at an angle of 90° with each other at a point O of an object. See Fig. 4-12. Find their equilibrant.

Solution:
Using 0.5 cm = 1 N as a scale, first find $\vec{c}$, the resultant of $\vec{a}$ and $\vec{b}$. Measure its length. It is 2.5 cm long, representing a magnitude of 5 N. Now draw the vector $\vec{d}$ equal and opposite to $\vec{c}$. The vector $\vec{d}$ is the equilibrant of $\vec{a}$ and $\vec{b}$.

Note: Since $\vec{a}$, $\vec{b}$, and $\vec{d}$ are in equilibrium, any one of these forces is holding the other two in equilibrium. Thus $\vec{a}$ is the equilibrant of $\vec{b}$ and $\vec{d}$ while $\vec{b}$ is the equilibrant of $\vec{a}$ and $\vec{d}$.

Test Yourself Problem

Find the equilibrant of a 10-N force acting eastward and a 12-N force acting 60° north of east.

4-16 Short Method of Finding an Equilibrant

Since the forces $\vec{a}$, $\vec{b}$, and $\vec{d}$ in the above sample problem are in equilibrium, their resultant is zero. This means that when these three forces are combined to find their net resultant, they must form a closed triangle, as shown in Fig. 4-13 (a), so that there will be no gap between the tail of the first vector in the sequence and the head of the last vector in the sequence.

This suggests a general way of finding the equilibrant of any number of forces acting on the same body. Draw each force to scale, and attach its tail to the head of a previous force as is done to obtain their resultant. Then, from the head of the last force to the tail of the first force, draw a vector closing the polygon formed by the force vectors. This vector represents the equilibrant of the forces. Notice that it is equal and opposite to the resultant of the forces.

The procedure is illustrated in Fig. 4-13 (b) where $\vec{e}$ is shown to be the equilibrant of $\vec{a}$, $\vec{b}$, $\vec{c}$, and $\vec{d}$.

4-17 Relationships in Triangles

The precision of *graphical* solutions to vector problems depends upon how precisely we construct and measure the vector triangles. We can obtain more accurate solutions by *computation* (KAHM·pew·TAY·shun). For this we need the following relationships between the parts of a triangle.

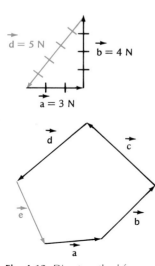

Fig. 4-13. Direct method for finding the equilibrant of two or more forces.

(1) **Right Triangle.** The sine and the cosine of any acute angle may be defined in terms of the sides of a right triangle in which the angle is contained. The sine of the angle is given by the ratio of the side opposite the angle to the hypotenuse. The cosine of the angle is given by the ratio of the adjacent side to the hypotenuse. In the right triangle in Fig. 4-14:

$$\sin B = \frac{b}{c}$$

$$\cos B = \frac{a}{c}$$

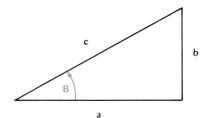

Fig. 4-14. In a right triangle, the sine of either acute angle equals the opposite side over the hypotenuse and the cosine equals the adjacent side over the hypotenuse.

By constructing right triangles in which angle B takes on all possible values, a, b, and c can be measured and the value of all sines and cosines of angles from $0°$ to $90°$ can be obtained from the above definitions. Values of sines, cosines, and tangents are given in the table on page 703.

Frequently, we are interested in finding the sides of a right triangle when we know one of its acute angles and the hypotenuse. From the above relationships, we have:

$$b = c \sin B$$
$$a = c \cos B$$

Another useful relationship connecting the sides of a right triangle is the *Pythagorean theorem:*

$$c^2 = a^2 + b^2$$

(2) **Any Triangle.** In any triangle, the sides have the following relationship:

$$c^2 = a^2 + b^2 - 2\ ab \cos C$$

In Fig. 4-15, the sides and the angle C are shown first for the case where C is less than $90°$ and then for the case where C is more than $90°$. When C is more than $90°$, its cosine is negative and is numerically equal to the cosine of its supplement, that is, $180° - C$. Thus, $\cos 120° = -\cos (180° - 120°) = -\cos 60°$. The table indicates that $\cos 60° = 0.500$. Therefore $\cos 120° = -0.500$.

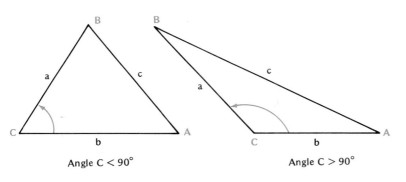

Angle C < 90°   Angle C > 90°

Fig. 4-15. General acute and obtuse triangles.

Another set of relationships that applies in all triangles is

$$\frac{a}{\sin A} = \frac{b}{\sin B} = \frac{c}{\sin C}$$

The above relationships are applied in the sample problems which follow and in those on pages 63-64.

Sample Problems

1. *Combining vectors acting at right angles to each other.*

What is the resultant of $\vec{a}$, a 25-N force pushing a body east and $\vec{b}$, a 20-N force pushing the same body north?

Solution:

In the vector triangle shown in Fig. 4-16, $\vec{c}$ is the resultant of $\vec{a}$ and $\vec{b}$.

$$c^2 = a^2 + b^2$$
$$c^2 = 25^2 + 20^2$$
$$c^2 = 625 + 400 = 1025$$
$$\vec{c} = \sqrt{1025} = 32 \text{ N in the direction shown}$$

2. *Combining vectors forming any angle.*

What is the resultant of a 12.0-N force $\vec{a}$ and a 9.00-N force $\vec{b}$ pulling an object and forming an angle of 60° with each other?

Solution:

Refer to Fig. 4-17 where $\vec{c}$ is the required resultant.

$$c^2 = a^2 + b^2 - 2\,ab\,\cos C$$

Note that $\cos C = \cos 120° = -0.50$. Then

$$c^2 = (12.0)^2 + (9.00)^2 - 2 \times 12.0 \times 9.00 \times (-0.50)$$
$$c^2 = 333$$
$$\vec{c} = \sqrt{333} = 18.2 \text{ N in the direction shown}$$

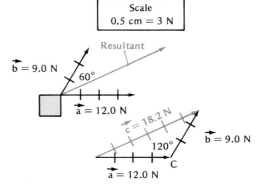

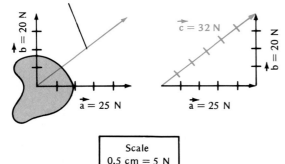

Fig. 4-16. Sample Problem 1.

Fig. 4-17. Sample Problem 2.

Test Yourself Problem

Compute the resultants in the two Test Yourself Problems in Section 4-9. Compare the computed answers with those obtained graphically.

Questions

Group 1

1. (a) Describe two instances in which it is evident that a change in the world about us is the result of motion. (b) Describe two instances in which

forces are used to start, stop, or change the motion of a body.

2. (a) What is the difference between a vector and a scalar? Which of the following are vector quantities: (b) the mass of an object; (c) the weight of an object; (d) the length of an object; (e) the velocity of an object; (f) the volume of an object; (g) the displacement of an object?

3. (a) In the graphical representation of a vector, how are its magnitude and direction shown? (b) What is the resultant of two vectors?

4. (a) When are forces said to be in equilibrium? (b) What is the equilibrant of two forces acting at the same point of a body?

5. A force having a magnitude of 10 N acts at the same point as a second force having a magnitude of 8 N. The directions of these forces can be changed at will. (a) When will these forces produce their largest resultant? (b) What will it be? (c) When will these forces produce their smallest resultant? (d) What will it be?

6. (a) As the angle between the two forces of question 5 is steadily made smaller, what happens to the magnitude of their resultant? (b) Is there an angle for which these two forces keep the body on which they act in equilibrium? Explain.

7. What limitation does the behavior of velocities near the velocity of light put on our use of the vector triangle in finding the resultant of two velocities?

Group 2

8. An airplane travels north for 1 h, east for 2 h, and northeast for 3 h. Which of the following are vectors: (a) the travel time for each leg of the flight; (b) the distance traveled on each leg; (c) the displacement during each leg?

9. Three forces acting at the same point of a body are in equilibrium. When two additional forces are now applied to the body, it remains in equilibrium. What must be true about the two additional forces?

10. Two or more of the following sets of displacements, when carried out in the order shown, have the same resultant. Which are they?
(a) 2 m east; 3 m north; 4 m west
(b) 2 m north; 3 m west; 4 m east
(c) 2 m east; 4 m west; 3 m north
(d) 3 m north; 2 m east; 4 m west

11. (a) What is the effect on their resultant of changing the order in which two or more displacements are combined? (b) Does this conclusion also apply to other vector quantities such as velocities and forces?

12. Explain why the resultant of two forces of unequal magnitude can never be zero.

Problems

Solve each of the following problems by making vector diagrams drawn accurately to a suitable scale. Use a protractor to construct the required angles. In the case of the starred (*) problems, check answers by computation.

Group 1

*1. A 60.0 km/h wind is blowing due north. What is the magnitude and direction of the velocity of an airplane traveling at an air speed of 150 km/h when it is heading (a) north; (b) south; (c) east?

*2. An automobile travels 40 km due east and then 60 km due north. (a) What is the magnitude of the resultant displacement of the automobile? (b) Determine its direction with a protractor.

*3. A force of 30 N and a force of 40 N are applied to the same point of a box. What is the resultant force on the box when the angle between the two forces is (a) 0°; (b) 30°; (c) 60°; (d) 90°? (e) What is the direction and magnitude of the equilibrant in (a), (b), (c), and (d)?

*4. A boat whose speed with respect to the water is 5.0 km/h travels downstream at a 45° angle with the current whose speed is 3.0 km/h. What is the resultant speed and direction of the boat?

*5. A bullet is fired due east from a gun mounted in an airplane traveling due northeast. If the speed of the bullet is 600 m/s and the speed of the airplane is 150 m/s, what is the resultant velocity of the bullet?

*6. Find the equilibrant of two 10.0-N forces acting upon a body when the angle between the forces is (a) 90°; (b) 120°; (c) 150°; (d) 180°.

*7. An airplane heading 45° south of east at an air speed of 120 m/s is being blown eastward by a 30.0 m/s wind. What is the plane's resultant ground speed?

8. (a) A force of 5 N, a force of 7 N, and a force of 12 N acting together on a body keep it in equilibrium. Show the arrangement of the forces by a diagram. (b) A force of 5 N, a force of 7 N, and a force of 2 N acting on a body are in equi-

librium. Show the arrangement of the forces by means of a diagram.

9. Determine from problem 8 whether each of the following combinations of forces can be in equilibrium if their directions are properly selected: (a) 5 N, 7 N, 13 N; (b) 5 N, 7 N, 9 N; (c) 5 N, 7 N, 1 N.

Group 2

10. A 3-N force is pushing an object due east. Also pushing the object at the same point are a 4-N force and a 6-N force whose directions can be changed as desired. Determine by constructing a vector triangle the directions in which the 4-N and the 6-N forces must act to keep the object in equilibrium.

*11. A passenger standing on the deck of a ship sailing eastward at 15 km/h feels a wind. (a) If there is no true wind, what is the speed and direction of this apparent wind? (b) If there is a true wind having a velocity of 20 km/h southward, what is the speed and direction of the apparent wind felt by the passenger?

12. An airplane flies 2000 m north, then 1500 m due northeast, and finally 3000 m due south. Determine graphically the distance between the

airplane's starting and ending positions. Then determine the resultant direction in which the airplane was displaced.

13. Four forces of 10 N each act at the same point of a body. One force acts due north, the second acts due northeast, the third acts due east, and the fourth acts due southeast. (a) Determine graphically the direction and magnitude of the resultant of the forces. (b) Determine the equilibrant of the forces.

*14. An electron has two forces acting on it, one of 1.5×10^{-16} N and a second of 2.5×10^{-16} N. What is the resultant force acting on the electron when the two forces act (a) in the same direction; (b) in opposite directions?

*15. (a) What is the resultant force acting on the electron in problem 14 when the two forces make a 30° angle with each other? (b) What is the equilibrant of these two forces?

16. A boat is heading directly across a river at 10.0 km/h. The current is 6.0 km/h and the river is 1.0 km across. (a) How long will it take the boat to cross the river? (b) How much further downstream will the boat be on arriving at the opposite bank? (c) What is the resultant velocity of the boat?

Resolving Vectors Into Components

4-18 Single Vectors May Be Broken into Components

We have been adding or combining two or more vectors to find their resultant. The vectors combined in this way are called *components* (kom·POH·nunts). Frequently, we are interested in the opposite process, namely, that of starting with a single vector which we may think of as a resultant, and finding its components in certain directions. In sailing it is quite easy to make the boat go with the wind. However, when we wish to sail in some other direction, we must know how to break up the force of the wind into components in such a way that one of them will push the boat in the desired direction. *We call the process of dividing or resolving a vector into components, resolution* (reh·zoh·LOO·shun).

Dividing a vector into components is called resolution.

4-19 Resolving a Vector into Components

Usually, we are interested in finding the effects of a vector in two specific directions. We then resolve the given vector into two components in these directions. The method is opposite to that of com-

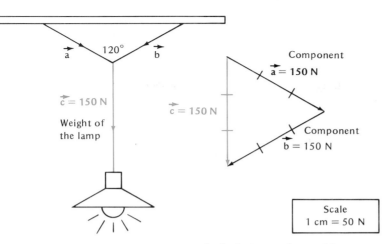

Fig. 4-18. Resolving the force $\vec{c}$ into two components $\vec{a}$ and $\vec{b}$.

bining two component vectors to find their resultant. Now, we start with the resultant and then find its components. The procedure is as follows.

Construct a given force to scale and from one end of it draw a line in one of the directions in which a component is desired. From the other end, draw a second line in the second direction in which a component is desired. A triangle is thus formed in which one side is the given vector and the other two sides are the required components. Measure the components to determine their magnitudes and put arrowheads on them to show their direction.

Let us apply this method to the street lamp hanging from two cords making an angle of 120° to each other in Fig. 4-18. As the weight of the lamp pulls downward, its force breaks up into two components pulling downward along each cord. If the lamp weighs 150 newtons, how hard does each of the cords pull on the point to which it is fastened?

Using as a scale 1 centimeter = 50 newtons, draw the downward force $\vec{c}$ exerted by the 150-newton lamp to scale. From the tail of $\vec{c}$, draw a line parallel to the left cord. From the head of $\vec{c}$, draw a line parallel to the right cord thus forming a vector triangle. The vectors $\vec{a}$ and $\vec{b}$ are the required components of $\vec{c}$. Measuring $\vec{a}$ and $\vec{b}$, we find each to be 3 centimeters long and therefore representing a force of 150 newtons each. Thus, the downward 150-newton pull of the lamp is resolved into two 150-newton forces, each pulling along one of the cords in the direction shown.

4-20 Resolving Vectors into Perpendicular Components

Most of the time, we resolve a vector into two components that are perpendicular to each other. There is an important advantage in doing this because *perpendicular components* are completely independent of each other. Each therefore gives the total effect of the given vector in its direction.

Consider the child pulling the sled in Fig. 4-19. The applied force acts along the inclined rope but the sled does not move in the direction of the rope. Instead, the force makes it move horizontally forward. In addition, the force pulls the sled vertically upward. We do not notice this upward pull because it is usually neutralized by part of the weight of the loaded sled and therefore does not move the sled upward. Thus, the applied force is resolved into two components, one pulling the sled horizontally forward and the second pulling it vertically upward.

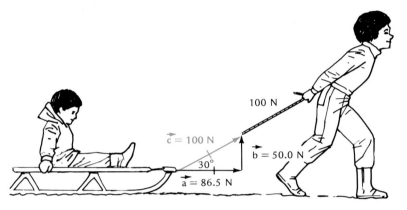

Fig. 4-19. The horizontal and vertical components acting on a sled.

Assuming that the child's pull is 100 newtons and that the rope makes an angle of 30° with the horizontal, we can now find its components in the horizontal and vertical directions. At an angle of 30° with the horizontal we draw the vector $\vec{c}$ to the scale of 1 centimeter = 50 newtons to represent the applied force. From the tail of $\vec{c}$ we draw a horizontal line, and from the head of $\vec{c}$ we draw a vertical line, thus forming a vector triangle. The vector $\vec{a}$ is the forward component of $\vec{c}$ and the vector $\vec{b}$ is the upward component. Measuring these components, we find that $\vec{a}$ is 1.73 centimeters long, representing a force of 86.5 newtons, and $\vec{b}$ is 1.00 centimeter long, representing a force of 50.0 newtons. The child's 100-newton pull along the rope is resolved into an 86.5-newton force pulling the sled horizontally and a 50.0-newton force pulling the sled upward.

4-21 Effect of the Angle

The child in the above example is usually interested in getting as large a forward component for the 100-newton force as is convenient because that component is the one that pulls the sled in the desired direction. To increase this component, should she increase or decrease the angle the rope makes with the horizontal? Fig. 4-20 shows that the forward components of the child's force for rope angles of 0°, 30°, 60°, and 90° vary from 100 newtons to 0 newtons. It is evident that the smaller the angle between the given force and

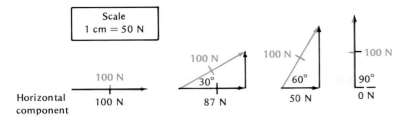

Fig. 4-20. The horizontal and vertical components of a force change as its direction changes.

the direction in which we desire the object we are pulling to move, the greater will be the component of the force in this direction.

4-22 Resolution of a Velocity

As a final example, consider the airplane pilot who is attempting to fly northward in an 80 kilometer-per-hour wind blowing directly to the northeast. The pilot wants to know what part of the wind's velocity will blow the plane north and what part will blow it east. Fig. 4-21 solves the problem by resolving the velocity of the wind $\vec{c}$ into two perpendicular components, an eastward velocity $\vec{a}$ and a northward velocity $\vec{b}$. Note that both components of the 80-kilometer-per-hour wind have magnitudes of 57 kilometers per hour. Thus, the given wind has the same effect as two winds each of 57 kilometers per hour, one blowing to the north and the other blowing to the east.

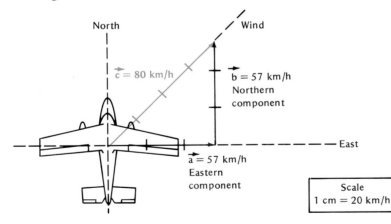

Fig. 4-21. Resolving the velocity of the wind into two perpendicular components acting on an airplane.

4-23 Computing Components of a Vector

The components into which a vector is resolved may also be computed using the relationships in triangles summarized in Section 4-17 as illustrated in the following examples.

Sample Problems

1. *Resolving a vector into perpendicular components.*

A 60 km/h wind is blowing in a direction 30° north of east. What is the component of the wind velocity to the east? To the north?

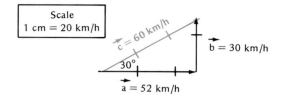

Solution:

In Fig. 4-22, $\vec{c}$ is the given 60 km/h wind velocity, $\vec{a}$ is its eastern component and $\vec{b}$ is its northern component.

$$a = c \cos 30°; \quad \cos 30° = 0.87$$
$$a = 60 \times 0.87; \quad \vec{a} = 52 \text{ km/h east}$$

$$b = c \sin 30° \quad \sin 30° = 0.50$$
$$b = 60 \times 0.50; \quad \vec{b} = 30 \text{ km/h north}$$

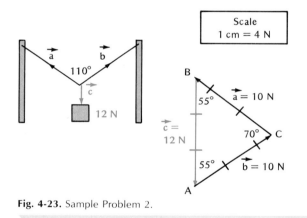

Fig. 4-23. Sample Problem 2.

2. *Resolving a vector into nonperpendicular components.*

A horizontal wire is stretched between two poles. When a 12-N weight is hung at its center, the wire sags until the angle between its halves is 110°. What forces are exerted by each half of the wire in supporting the weight?

Solution:

The weight $\vec{c}$ and the two forces $\vec{a}$ and $\vec{b}$ in the wire are in equilibrium. These vectors must therefore form a closed triangle as shown in Fig. 4-23. Here $\vec{c}$ is drawn to scale as a downward 12-N force. The triangle is then completed by drawing $\vec{a}$ and $\vec{b}$ parallel to the forces they represent. In this triangle, angle $A = 55°$, angle $B = 55°$, angle $C = 70°$, and $c = 12$ N.

$$\frac{a}{\sin A} = \frac{c}{\sin C}$$

$$\frac{a}{\sin 55°} = \frac{12 \text{ N}}{\sin 70°}; \quad \frac{a}{0.82} = \frac{12 \text{ N}}{0.94}$$

$$\vec{a} = \frac{12 \text{ N} (0.82)}{0.94} = 10 \text{ N}$$

in the direction shown in the diagram. Since triangle ABC is isosceles, $\vec{b}$ is also a 10-N force in the direction shown.

Test Yourself Problems

Solve these problems by making vector diagrams and also by computation.

1. A child pulls a sled by means of a cord making an angle of 20° to the horizontal. If the child's force is 50 N, what are its components in (a) the horizontal direction; (b) the vertical direction?

2. A picture weighing 10.0 N is suspended from a nail in a wall by two wires each of which makes an angle of 45° with the vertical. What is the component of the picture's weight along each of the wires?

CHAPTER REVIEW

Summary

A **vector** is a quantity having both magnitude and direction. Quantities having only magnitude are called **scalars.** *Displacements, velocities,* and *forces* are typical vector quantities. Two or more vectors acting at the same point may combine to form a **resultant** whose effect is equal to the combined effects of the two vectors. The resultant of two vectors may be found graphically by drawing a vector triangle, or it may be computed from a knowledge of the vectors and the angle between them. As the angle between vectors of fixed magnitude is increased, their resultant decreases. Two vectors have the largest resultant when they act in the same direction and the smallest resultant when they act in opposite directions.

The **equilibrant** of two forces acting at the same point of a body is the single force that neutralizes the effects of those forces. The equilibrant of the two forces is equal in magnitude to their resultant but opposite to it in direction. The net resultant of the two forces and their equilibrant is zero. When two or more forces have a zero resultant, they are said to be in **equilibrium.**

A vector may act in two or more directions other than its own. Its effect in each of these directions is called a **component** and the process of determining the components of a vector is called **resolution.** Usually, a vector is resolved into only two components. These may be found graphically by constructing a vector triangle or the may be computed.

Questions

Group 1

1. Show that when a vector is resolved into perpendicular components, the magnitude of each component must be smaller than that of the given vector.

2. A gardener pushes a lawn mower whose handle makes a 45° angle with the horizontal. The force along the handle may be resolved into two components, one in the horizontal direction in which the lawn mower is moving, and the second vertically downward. (a) What effect will decreasing the angle between the lawn mower handle and the ground have on the magnitude of each of these components? (b) Will this make it easier or harder to move the lawn mower forward?

3. A student pulls a window downward by means of a window pole making a 30° angle with the top of the window. (a) Show by means of a diagram how the force applied along the pole is resolved into a component pulling the window downward and a component pulling the window horizontally inward. (b) If the window pole is held at 0° to the window instead of 30°, what will be the downward component of the applied force? (c) If the window pole is held at 90° to the window, what will be the downward component of the force?

4. Which one or more of the following pairs can be the components of a velocity of 30 m/s east?

(a) 20 m/s east;   10 m/s east
(b) 35 m/s east;   5 m/s west
(c) 20 m/s east;   10 m/s west
(d) 15 m/s east;   15 m/s west

5. One or more of the following pairs can be the components of a 10-N force if the components are given the proper direction. Which are they?

(a) 2 N; 9 N   (b) 3 N; 6 N
(c) 5 N; 5 N   (d) 10 N; 10 N

Explain your answer with the aid of diagrams.

Group 2

6. A mover is pulling a heavy box across a floor with a rope fastened to a point near the bottom of the front side of the box. It is easier to move the box when the rope is longer than when it is shorter. Explain.

7. An airplane is flying north when a wind begins to blow it toward the east. (a) If the airplane maintains a constant speed, how should it change its heading in order to continue its motion northward? (b) Can the airplane continue its northward motion by changing its speed but not its heading? Explain.

8. (a) Look at Fig. 4-18. Show by a diagram that when the angle between the cords holding up the lamp is increased, the component of the force exerted by the weight of the lamp along each rope increases. (b) Show that for all angles above 120°, each of these components is greater than the weight of the lamp.

9. A metal wire was able to support a weight of 100 N hung vertically from it. The ends of the wire were then fastened to two vertical posts and the wire was stretched tightly until it was nearly horizontal. A weight of only 10 N hung at its center then caused the wire to snap. Explain.

10. Make a diagram showing a pendulum at one end of its swing. Draw a vector triangle showing how the weight of the bob is resolved into a component along the string and a second component perpendicular to the string. (a) Which component produces the tension in the string? (b) Which moves the bob? (c) What happens to the magnitude of each of these components as the bob moves toward its lowest position?

Problems

Group 1

Solve each of the problems in Groups 1 and 2 by making vector diagrams drawn accurately to a suitable scale. The starred (*) problems should also be done by computation.

*1. A student pulls a box across a floor by means of a rope making an angle of 60° with the horizontal. If the force applied is 60 N, what are the components of force in (a) the horizontal direction; (b) the vertical direction? (c) If the angle of the rope is reduced to 30° and the same force applied, what will now be the magnitudes of the horizontal and vertical components?

*2. A 50 km/h wind is blowing due northwest. What are the components of the wind in (a) the northward direction; (b) the westward direction?

*3. A gardener pushes a lawn mower by applying a force of 150 N to the handle. Find the horizontal component of this force when the lawn mower handle makes an angle of (a) 30°; (b) 45°; (c) 60° with the horizontal.

*4. An electric sign weighing 200 N is supported by two slanting cables. Each cable is attached to the same point of the sign and makes a 45° angle with the horizontal. What is the component of the force exerted by the sign along each of its cables?

*5. The pilot of an airplane wishes to reach a point 300 km east of the starting point. To avoid a storm area, the pilot decides to proceed to the destination in two hops. On the first hop, the pilot flies due northeast and on the second hop due south. (a) How many kilometers should be flown to the northeast? (b) How many kilometers southward?

*6. As a result of a combination of its own motion and that of a river current, a boat actually moves directly across the river at 8 km/h. If the boat is heading in a direction making an angle of 135° to the current, what is the magnitude of the current?

*7. The resultant motion of a boat on a river is 20 km/h in a direction making an angle of 60° with the current. If the river is 1.5 km wide, how long does it take the boat to cross the river?

*8. Point B is 5.0 km east of point A. A cyclist leaves A and goes 3.0 km in a direction 30° to the north of east. (a) In what direction must the rider go from this point to reach B? (b) What distance must be covered?

Group 2

*9. A car weighing 15 000 N rests on a 15° hill. Find the components of the force exerted by the car's weight (a) at right angles with the hill; (b) parallel to the hill. (c) If the car's brakes were released, which of these components would move the car downhill?

10. A weight of 50.0 N is hung from the center of a stretched horizontal wire 10.0 m long. It causes the wire to sag 1.00 m at its center. What is the force exerted by each half of the wire in supporting the weight? (Hint: Make a scale diagram of the wire with the weight hung from it.)

*11. A pendulum consists of a 10-N bob hung from a string 20 cm long. (a) Find the tension produced in the string by the bob at the moment the pendulum reaches its highest point and makes an angle of 30° with the vertical. (b) What is the compnent of force acting at right angles to the string?

12. A rifle bullet fired at a 60° angle with the horizontal leaves the gun with a speed of 800 m/s. Resolve the velocity of the bullet into two components, one upward and the other horizontal. Assuming that the horizontal component remains constant while the bullet is in flight, how far will the bullet have moved forward exactly 2 s after it leaves the gun?

13. An airplane's speed in still air is 240 km/h. (a) If the pilot wishes the resultant motion of the plane to be due north when a 60 km/h wind is blowing the plane due east, in what direction should the plane be headed? (b) What will then be the resultant speed of the airplane over the ground?

14. An electric sign is supported by a wire making a 45° angle with a boom as shown in Fig. 4-24. (a) If the sign weighs 1000 N, what is the outward thrust exerted by the boom? (b) What force is being exerted by the wire in helping to support the sign?

15. When a boat heads across a river at an angle of 120° with the current at a speed of 10 km/h, its resultant motion is at right angles to the current. What is the magnitude of the current?

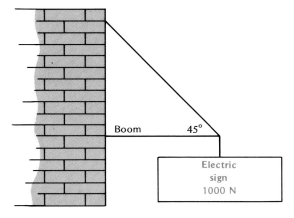

Applying Physics

1. Cut a heavy rubber band to make a strip of rubber about 15 to 20 cm long. To the middle of the rubber strip fasten an object like a box of granulated sugar and support the box by holding the two free ends of the rubber strip vertically. The box selected should be heavy enough to stretch each half of the rubber strip about 1 cm. Each half of the rubber strip now supports about half the weight of the box. Have a friend measure the length of each half of the rubber strip in this position.

   Now move the hands apart so as to increase the angle between the two halves of the rubber strip, first to about 45°, then to 90°, and finally to 135°. In each case, have a friend measure and record the increase in the length of each half of the rubber strip. These increases in length are approximate measures of the increases in the forces that are exerted by your hands to hold up the box. Compare the increases in the forces you must exert as the angle between the two halves of the rubber strip changes from 45° to 90° to 135°.

2. Measure the breaking strength of a cotton thread as follows. Tie a known weight such as that of a 1-kg mass to one end of the cotton thread. Suspend the weight from a nail so that it hangs like a pendulum. Mount a protractor behind the nail so that the angle which the thread makes to the horizontal can be read. Finally, tie a piece of string around the weight leaving a free end of string that can be pulled.

   Pull the weight away from its vertical position by slowly pulling the free end of the string horizontally. As the weight moves sideward, read on the protractor the angle through which the cotton thread moves. Continue to pull the string horizontally until the cotton breaks. Note on the protractor the angle at which this happens.

   At the moment of breaking, the downward force exerted by the 1-kg mass, the horizontal force exerted by you on the string, and the force exerted upward along the cotton thread are in equilibrium. Knowing the downward force to be the weight of 1 kg, or 9.8 N, and also knowing the directions of the other two forces, draw the vector triangle representing these forces to scale. From this triangle, determine the magnitude of the force exerted along the thread. This is the approximate breaking strength of the thread.

   Repeat the experiment and take the average of several values of the breaking strength as the best value.

Motion in a Straight Line

1. To understand that the motion of a body is relative and can only be described with respect to some other body.
2. To learn how to describe and analyze the motion of a body moving in a straight line.

Speed, Velocity, and Acceleration

5-1 Motion Is Relative

Would it surprise you to learn that at this very moment you are moving at a speed of more than 100 000 kilometers per hour? The explanation is simple. Since you are on the earth, it carries you along as it speeds around the sun in its orbit. Therefore you share the earth's orbital speed which is more than 100 000 kilometers per hour.

You do not usually think of yourself as having this motion. That is because, in everyday life, when you say that a body is moving, you mean that it is moving with respect to the surface of the earth. You can notice this motion because the moving body is increasing or decreasing its distance from objects that are fixed on the earth, such as buildings or trees.

You are right, therefore, in thinking yourself at rest with respect to the earth when you simply sit in your chair. The fact that, at the same time, you share the earth's motion through space, illustrates that *motion is relative.* This means that an object can be moving with respect to one body and at the same time be at rest or be moving at a different speed with respect to a second body.

5-2 Frames of Reference

When you are sitting in a bus traveling at 40 kilometers per hour, you are moving with respect to the road, but not with respect to the seats, floor, or walls of the bus. Your speed with respect to the road is 40 kilometers per hour. Your speed with respect to the floor of the bus is zero. If another bus traveling 40 kilometers per hour should be coming toward you, your speed with respect to that bus would be 80 kilometers per hour. Here you have three different relative motions depending upon whether you consider yourself to be moving with respect to your own bus, the road, or the oncoming bus.

This illustrates that, in dealing with the motion of a body, it is

Fig. 5-1. The relative speed of the planes with respect to each other is zero.

important to state with respect to what other body or *frame of reference* its motion is being described. For many of the motions we shall discuss, our frame of reference will be attached to some point on the earth. In the simple case in which a body moves in a straight line, we may choose any point on the line as a *reference point*. Then the movement of the body can be described by telling how far it is to one side or the other of the reference point and how fast it is moving toward or away from the reference point. In the next chapter we shall discuss briefly how frames of reference used to describe nonlinear motions are selected.

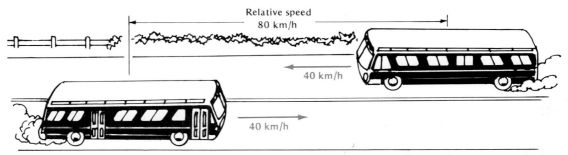

Fig. 5-2. The relative speed of each bus with respect to the other is 80 km/h.

5-3 Speed and Velocity

The *speed* of a body tells us how far it travels during every unit of time that it maintains that speed. A typical automobile goes 72 kilometers per hour or 20 meters per second. This means that the automobile travels a distance of 20 meters during every second that it maintains that speed. Speeds are commonly measured in kilometers per hour, meters per second, and centimeters per second. The fastest speed possible is the speed of light. It is 3×10^8 meters per second, or 300 megameters per second (300 Mm/s).

In physics we give a more complete description of the motion of a body by telling its velocity instead of its speed. As we noted in Chapter 4, *velocity* is a vector. It tells us two things about a moving body: its speed and its direction of motion. Thus, the velocity of an airplane is stated as 300 kilometers per hour, westward.

Velocity is a vector describing direction as well as speed.

5-4 Uniform and Accelerated Motion

All the different kinds of motion we notice in the world about us can be classified as one of two main types, uniform (YOO·ni·form) motion and accelerated (ak·SEL·ur·ay·ted) motion.

In *uniform motion,* both the speed and direction of the moving body remain the same. It is therefore motion at constant velocity. When you are riding in your car at a steady speed of 50 kilometers per hour on a straight road, you are in uniform motion. (See Fig. 5-3.)

You cannot usually travel any great distance in a car at constant velocity. Changing road and traffic conditions make it necessary

Uniform motion has a constant velocity; accelerated motion has changing velocity.

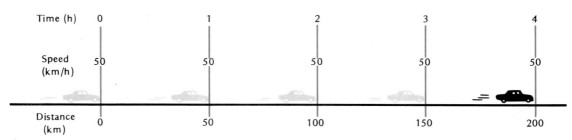

Fig. 5-3. Motion at a constant velocity of 50 km/h.

for you to keep changing your direction, your speed, or both your direction and speed. Your velocity therefore keeps changing. Motion with changing velocity is called *accelerated motion*. In everyday usage, acceleration generally means speeding up. The accelerator pedal of your car, for example, is the one you press to increase its speed. In physics however, we must remember that acceleration refers to any change of velocity. It may therefore mean the changing of direction, the changing of speed (either increasing or decreasing), or the changing of both speed and direction.

5-5 Units of Acceleration

In the remainder of this chapter, we shall limit our discussion to those accelerations that involve increasing or decreasing the speed of a body without changing its direction. The *acceleration* (ak·sel·ur·AY·shun) of a body is then simply the *rate at which its speed is changing.* If the speed of a body increases or decreases at a regular rate, its acceleration can be found by dividing the change that takes place in its speed by the time during which the change occurred.

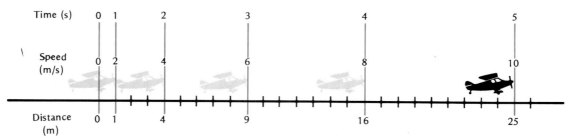

Fig. 5-4. Motion at a constant acceleration of 2 m/s².

Suppose that the speed of an airplane increases steadily during the first 5 seconds of take-off from zero to 10 meters per second. The change in its speed is then $10 - 0 = 10$ meters per second. Its acceleration is therefore $10 \div 5 = 2$ meters per second per second. This acceleration means that during each second of take-off the airplane increased its speed by 2 meters per second. At the end of the first second, its speed increased from 0 to 2 meters per second. During the second second, the speed increased another 2 meters per second, making its new speed $2 + 2 = 4$ meters per second. By the end of the third second, the airplane's speed increased another 2 meters per second, giving it a speed of $4 + 2 = 6$ meters per second, and so on. (See Fig. 5-4.)

If the speed of the airplane were steadily decreasing as it does on landing, its acceleration would be found in the same way but it would be given a negative sign.

The metric unit of acceleration is *meters per second per second*. In symbols this is written m/s². Acceleration may also be expressed in centimeters per second per second, or cm/s².

The *SI* unit of acceleration is meters per second per second (m/s²).

5-6 Straight-Line Motion

The study of the motion of bodies that travel in a straight line, is important for two reasons. First, many objects, such as freely falling bodies, actually move in straight-line paths or in paths that are very nearly straight lines. Second, many complicated motions of bodies can be considered combinations of two or more straight-line motions and therefore can be analyzed in terms of straight-line motions.

When a body moves in a straight line, its velocity is still a vector, but it can have only two directions on that line. We can simplify our study of this motion by considering only the speed of the body. We shall denote this by the letter v as distinguished from the velocity vector $\vec{v}$. We shall take account of the direction of motion by calling v positive when the body is moving in one direction and negative when it is moving in the opposite direction.

Similarly, while the acceleration of a body moving in a straight line is also a vector, it can only have two directions. In one of these, it acts in the same direction as the velocity of the body and makes the body move faster. In the other, it acts in the direction opposite to the velocity of the body and makes the body move more slowly, or even reverses its motion.

We shall simplify our study of acceleration in straight-line motion by considering only the magnitude of the acceleration or the rate at which the speed of the body changes. This rate will be denoted by the letter a as distinguished from the acceleration vector $\vec{a}$. Generally, we shall take account of the direction of motion by calling the acceleration positive when it is increasing the speed of the body and negative when it is decreasing the speed of the body.

5-7 Average Speed and Instantaneous Speed

When you take a trip in your car and go 100 kilometers in 2 hours, you averaged $100 \div 2 = 50$ kilometers per hour. The *average speed* is found by dividing the distance traveled by the travel time.

The average speed is a very useful idea. Frequently we do not know the speeds a body such as an airplane actually has from moment to moment during a trip because the airplane will be helped or hindered by weather and wind conditions. However,

Average speed is determined over an interval of time.

Fig. 5-5. Stroboscopic photo of a golfer driving a ball, taken at 100 flashes per second. At what part of the swing was the head of the golf club moving fastest?

from experience we know what average speed it can maintain. It is then a simple matter to estimate how far the plane will go in a given time.

If a small plane can average 400 kilometers per hour, we can predict that in 2 hours it will fly $2 \times 400 = 800$ kilometers; in 3 hours it will fly about 1200 kilometers, and so on. We are using the relationship,

$$\text{distance} = \text{average speed} \times \text{time}$$

$$d = \bar{v}t$$

where d is the distance traveled, $\bar{v}$ (read VEE·bahr) is the average speed, and t is the time of travel.

Instantaneous speed refers to any particular instant.

The average speed of a car tells us nothing about the speeds it may have from moment to moment during a trip. When you watch the speedometer of your car, you notice that as traffic conditions change, the speed of the car keeps increasing or decreasing. The speedometer needle shows the *instantaneous* (in·stun·TAY·nee·us) *speed* of the car at any given moment. A speedometer reading of 60 kilometers per hour or 1 kilometer per minute means that the car is at that moment moving at such a speed that it will travel 1 kilometer during every minute that it maintains that speed.

In the case of a body that moves at constant speed in a straight line, the instantaneous speed and the average speed are equal at all times. For such a body, the distance it travels is simply its constant speed v multiplied by the time of travel t. That is,

$$d = vt$$

5-8 Uniformly Accelerated Motion

The simplest type of accelerated motion is that of a body moving in a straight line with constant acceleration. In this case, the body will speed up or slow down at a constant rate. If the body speeds up, we say that the acceleration is positive. If it slows down, we say the acceleration is negative.

A very important example of motion at constant acceleration is that of a *body falling freely* in a vacuum. Galileo (gal·il·AY·oh) showed that a body speeds up as it falls with an acceleration of about 9.80 meters per second per second. This means that during every second, its speed increases by 9.80 meters per second. After falling from rest for one second, the speed of the body will be 9.80 meters per second. After 2 seconds, its speed will have increased another 9.80 meters per second and will be 9.80 + 9.80 = 19.60 meters per second. After 3 seconds, its speed will have increased another 9.80 meters per second, and so on.

Because it is the gravitational pull of the earth that causes bodies to fall, their downward acceleration is called the *acceleration of gravity.* It is designated by the letter *g*, and has the same value for all bodies in the same location. The value of *g* varies a little from place to place on earth, depending upon the latitude and the altitude of the place, and on the composition and structure of the local rocks. In this book we shall use as an approximate value, $g = 9.80$ meters per second per second, or 980 centimeters per second per second.

When a body is thrown upward, it is also acted on by the acceleration of gravity. In this case, the acceleration of gravity acting downward steadily reduces the speed of the body. For example, a ball thrown upward with a speed of 29.40 meters per second loses speed at the rate of 9.80 meters per second every second until it comes to rest. It therefore continues to rise for 29.40 ÷ 9.80, or 3.00 seconds. After that the pull of gravity makes the ball reverse direction. It then takes another 3.00 seconds for the body to fall to its starting point and regain its original speed of 29.40 meters per second.

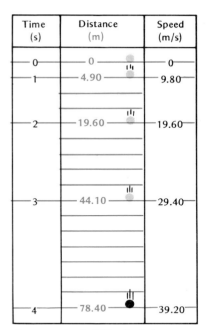

Time (s)	Distance (m)	Speed (m/s)
0	0	0
1	4.90	9.80
2	19.60	19.60
3	44.10	29.40
4	78.40	39.20

Fig. 5-6. A freely falling body is accelerated at 9.80 m/s².

Fig. 5-7. A sky diver in free fall is accelerated rapidly until reaching terminal velocity.

Mathematical Analysis of Motion

5-9 Relating Acceleration, Speed, and Time

We have defined acceleration as the rate at which the velocity of a body changes. For a body moving in a straight line with constant acceleration, we can find the acceleration by dividing the change in the speed of the body that took place during a given time by the time. If v_o is the speed of a body at the start and v is the final speed acquired by the body after being uniformly accelerated for a time t, the constant acceleration a, is:

$$\mathbf{a} = \frac{\mathbf{v} - \mathbf{v}_o}{\mathbf{t}} \qquad (\mathbf{A})$$

Solving for v, we have $v - v_o = at$; whence,

$$\mathbf{v} = \mathbf{v}_o + \mathbf{at} \qquad (\mathbf{A})$$

Notice that the speed of a uniformly accelerated body at a given time t after the acceleration began is made up of two parts, its initial speed v_o and the increase or decrease in speed produced by the acceleration. This increase or decrease is given by the product at.

Sample Problems

1. What is the acceleration of a car whose speed increases steadily from 15 m/s to 25 m/s in 5 s?

   Solution:
   $v = 25$ m/s $v_o = 15$ m/s $t = 5$ s

   The change in speed of the car is $v - v_o = 25$ m/s $- 15$ m/s $= 10$ m/s. Since this change took place in 5 s, the acceleration is:

   $$a = \frac{v - v_o}{t}$$

   $$a = \frac{10 \text{ m/s}}{5 \text{ s}} = 2 \text{ m/s}^2$$

2. An airplane flying 60 m/s is accelerated uniformly at the rate of 0.50 m/s². What is its speed at the end of 10 s?

   Solution:
   $v_o = 60$ m/s $a = 0.50$ m/s² $t = 10$ s
   $$v = v_o + at$$
   $$v = 60 \text{ m/s} + 0.50 \text{ m/s}^2 \times 10 \text{ s}$$
   $$v = 60 \text{ m/s} + 5.0 \text{ m/s} = 65 \text{ m/s}$$

   Note that the final speed of the airplane consists of two parts, its initial speed of 60 m/s and the added speed of 5.0 m/s that it acquired as the result of the acceleration.

Test Yourself Problems

1. What is the acceleration of an airplane whose speed decreases steadily from 250 m/s to 200 m/s in 10.0 s?

2. A car traveling at 18.0 m/s accelerates at the rate of 0.25 m/s². What is its speed at the end of 20 s?

5-10 Average Speed and Distance Traveled During Constant Acceleration

Suppose you accelerate your car at a constant rate for 20 seconds so that its speed rises steadily from 30 kilometers per hour to 50 kilometers per hour. The average speed of your car during those 20 seconds was 40 kilometers per hour, or midway between the initial and final speeds. In general, the average speed $\bar{v}$ of a body undergoing constant acceleration for a given time t is midway between its initial speed v_o and its final speed v, and is given by:

$$\bar{\mathbf{v}} = \frac{\mathbf{v}_o + \mathbf{v}}{2}$$

Note from the relationship (**A**) that $v = v_o + at$. Substituting in the above we have:

$$\bar{v} = \frac{v_o + (v_o + at)}{2} = v_o + \tfrac{1}{2}\,at$$

From this relationship, we can calculate the distance traveled by the body in time t. Noting that the distance traveled by a body is equal to its average speed times the travel time, we have:

$$d = \bar{v}t = (v_o + \tfrac{1}{2}\,at)\,t$$

whence,

$$d = \mathbf{v}_o t + \tfrac{1}{2}\,\mathbf{a}t^2 \qquad (\mathbf{B})$$

Note that d consists of two parts: $v_o t$, which is the distance the body would have traveled at its initial speed v_o if it had not been accelerated; and $\tfrac{1}{2}\,at^2$, which is the increase or decrease in the distance traveled because of the acceleration. When the moving body is speeding up, $\tfrac{1}{2}\,at^2$ is positive and represents an increase; when the body is slowing down, $\tfrac{1}{2}\,at^2$ is negative and represents a decrease in the distance traveled.

Sample Problem

A plane flying at 80.0 m/s is uniformly accelerated at the rate of 2.00 m/s². What distance will it travel during a 10.0-s interval after the acceleration begins?

Solution:

$v_o = 80.0$ m/s $a = 2.00$ m/s² $t = 10.0$ s

$d = v_o t + \tfrac{1}{2}\,at^2$

$d = (80.0 \text{ m/s})(10.0 \text{ s}) + \tfrac{1}{2}\,(2.00 \text{ m/s}^2)(10.0 \text{ s})^2$

$d = 800 \text{ m} + 100 \text{ m} = 900 \text{ m}$

Thus, the distance traveled by the plane consists of two parts: 800 meters, equal to $v_o t$, is the distance it would have traveled if it had not been accelerated; and 100 meters, equal to $\tfrac{1}{2}\,at^2$, is the extra distance traveled by the plane because it was accelerated.

Test Yourself Problem

A car traveling at 6.0 m/s is uniformly accelerated at 0.50 m/s². What distance will it travel during an 8.0-s interval after the acceleration begins?

5-11 Summary of Relationships in Uniformly Accelerated Motion

All problems involving bodies moving with constant acceleration in a straight line can be solved by using either or both of the following relationships:

$$v = v_o + at \quad \text{or} \quad a = \frac{v - v_o}{t} \tag{A}$$

$$d = v_o t + \tfrac{1}{2} at^2 \tag{B}$$

Solving for t in (A) and substituting in (B) gives another useful relationship:

$$v^2 - v_o^2 = 2\,ad \tag{C}$$

In using these relationships, the acceleration is taken positive when the body is speeding up and negative when it is slowing down. For bodies that are accelerated by gravity the relationships take the special form in which $a = g = 9.80$ meters per second per second, or 980 centimeters per second per second.

When the body starts from rest, $v_o = 0$, and the above relationships are simplified as follows:

$$v = at \quad \text{or} \quad a = \frac{v}{t} \tag{A'}$$

$$d = \tfrac{1}{2} at^2 \tag{B'}$$

$$v^2 = 2\,ad \tag{C'}$$

Sample Problems

1. A stone dropped from a cliff hits the ground 3.00 s later. Find (*a*) the speed with which the stone hits the ground, and (*b*) the distance it fell.

Solutions:

(*a*) $a = g = 9.80$ m/s² $t = 3.00$ s
       $v = at = gt$
       $v = 9.80$ m/s² $\times 3.00$ s $= 29.4$ m/s

(*b*) $d = \tfrac{1}{2} at^2 = \tfrac{1}{2} gt^2$
       $d = \tfrac{1}{2} (9.80$ m/s²$)(3.00$ s$)^2 = 44.1$ m

2. A bullet leaves the barrel of a gun 0.50 m long with a muzzle velocity of 500 m/s. Assuming that the bullet was uniformly accelerated, find (*a*) its acceleration and (*b*) the length of time it was in the barrel of the gun.

Solutions:

(*a*) $d = 0.50$ m $v_o = 0$ $v = 500$ m/s

$$v^2 = 2\,ad \quad \text{or} \quad a = \frac{v^2}{2d}$$

$$a = \frac{(500 \text{ m/s})^2}{2 \times 0.50 \text{ m}} = \frac{250\,000 \text{ m}^2/\text{s}^2}{1.0 \text{ m}}$$

$$= 2.5 \times 10^5 \text{ m/s}^2$$

(*b*) $d = \tfrac{1}{2} at^2$ or $t^2 = \dfrac{2d}{a}$

$$t^2 = \frac{2(0.50 \text{ m})}{2.5 \times 10^5 \text{ m/s}^2} = \frac{1}{0.25 \times 10^6} \text{ s}^2$$

$$t = \sqrt{4.0 \times 10^{-6} \text{ s}^2} = 2.0 \times 10^{-3} \text{ s}$$

$$= 0.0020 \text{ s}$$

3. A car moving at 16 m/s is brought to a stop in 8.0 s by applying the brake. Assuming uniform negative acceleration, find (*a*) the acceleration and (*b*) the distance it traveled after the brake was applied.

Solutions:

(*a*) $v_o = 16$ m/s $v = 0$ m/s $t = 8.0$ s

$$a = \frac{v - v_o}{t} = \frac{0 - 16 \text{ m/s}}{8.0 \text{ s}} = -2.0 \text{ m/s}^2$$

The minus sign shows that the car is being slowed down.

(b) $d = v_o t + \frac{1}{2} at^2$

$d = (16 \text{ m/s})(8.0 \text{ s}) + \frac{1}{2}(-2.0 \text{ m/s}^2)(8.0 \text{ s})^2$

$= 128 \text{ m} - 64 \text{ m} = 64 \text{ m}$

An alternative solution is to find the average speed of the car during the time it was being braked.

$$\bar{v} = \frac{v_o + v}{2} = \frac{16 \text{ m/s} + 0 \text{ m/s}}{2} = 8.0 \text{ m/s}$$

Now, $d = \bar{v} \times t = 8.0 \text{ m/s} \times 8.0 \text{ s} = 64 \text{ m}$

Test Yourself Problems

1. An object dropped from an airplane hits the ground 8.00 s later. How high is the airplane?

2. A car traveling at 21.0 m/s is uniformly slowed down at the rate of 0.400 m/s². What distance will it travel during a 5.00-s interval?

Questions

Group 1

1. Two cars are moving in the same direction at the same speed. What is the speed of either car with respect to the other?
2. Two buses are moving north, the first at 30 km/h and the second at 20 km/h. (a) What is the relative speed and direction of motion of the first bus with respect to a person seated in the second bus? (b) What is the relative speed and direction of motion of the second bus with respect to a person seated in the first bus?
3. (a) Distinguish between a car's average speed and its instantaneous speed. (b) Which speed does the speedometer show?
4. Describe the motion of the speedometer needle of a car when the car is in (a) uniform straight-line motion; (b) uniformly accelerated straight-line motion.
5. A car is traveling at a constant speed of 20 m/s. What distance will it have covered at the end of each of the first 3 s?
6. A ball rolling down a hill is uniformly accelerated at 3 m/s². Assuming it starts from rest, what is its speed at the end of each of the first 3 s?
7. Which of the following are examples of accelerated motion? (a) A train moving on a straight-line track at constant speed. (b) A train moving on a curved track at constant speed. (c) A train moving on a straight track at variable speed.

8. Describe the motion of a car whose acceleration is negative.
9. Describe the motion of a car whose acceleration is zero but whose speed is not zero.

Group 2

10. With respect to the motion of a car, describe a situation in which the car has zero speed but has a definite acceleration.
11. Describe the motion of a car which is being accelerated although its speed remains constant.
12. Prove that for a uniformly accelerated body starting from rest, the acceleration is numerically equal to twice the distance traveled by the body in the first second.
13. Show that the average velocity of a body undergoing constant acceleration and starting from rest is one-half of its final velocity.
14. A youngster throws a ball vertically upward. As the ball rises to its highest point and then falls, what changes, if any, take place in (a) its velocity; (b) its acceleration?
15. A bullet shot vertically upward with muzzle velocity of 600 m/s rises and then falls back to its starting point. Explain why, neglecting air friction, its final velocity will be equal to its initial velocity.

Problems

Group 1

1. A young couple is walking toward the front of a train at a speed of 2 m/s. (a) What is their speed with respect to the ground when the train speed is 30 m/s? (b) What is their speed with respect to someone in a train on a parallel track

moving at 30 m/s in the opposite direction?

2. An automobile moving at 25 m/s passes a highway patrol car moving at 23 m/s in the same direction. (a) What is the relative speed of the automobile as measured in the highway patrol car? (b) How far ahead of the patrol car will the automobile be 10 s after passing it?

3. A plane flies a distance of 1200 km in 2.50 h. What is its average speed for the flight?

4. In a 4-hour period, a hiker walked 5.0 km during the first hour and 4.4 km during the second hour. After resting for an hour, the hiker walked 4.6 km during the fourth hour. What was the average speed during (a) the first two hours; (b) the first three hours; (c) the entire 4-hour period?

5. A ball rolls down a hill with a constant acceleration of 3.0 m/s². (a) If it starts from rest, what is its speed at the end of 4.0 s? (b) How far did it move?

6. (a) How long will it take a freely falling body to acquire a speed of 58.8 m/s? (b) How far will it have fallen during that time?

7. A car moving on a straight road increases its speed at a uniform rate from 10 m/s to 20 m/s in 5.0 s. (a) What is its acceleration? (b) How far did it go during the 5.0 s?

8. A car starting from rest is accelerated at the constant rate of 3.0 m/s². (a) What is its speed at the end of 8.0 s? (b) How far did it travel during the first 8.0 s?

9. A car moving at 10 m/s is uniformly accelerated at the rate of 0.50 m/s². (a) What is the speed of the car 6.0 s after the acceleration begins? (b) How far does the car move during the 6.0 s?

10. A coin dropped from the top of a precipice takes 5.00 s to hit the ground. How high is the precipice?

11. A ball thrown vertically upward returns to the hand of the thrower 6.00 s later. (a) For how many seconds did the ball fall after reaching its high point? (b) How high did the ball go?

12. A ball rolls down an incline 12 m long in 3.0 s. (a) Assuming that the ball is uniformly accelerated, what is its average speed? (b) What is its final speed?

13. An object falls to the floor from a shelf 2.5 m high. With what speed does it strike the floor?

Group 2

14. Two airplanes are 10.0 km apart and are moving at constant speeds in the same direction. The second overtakes the first in 2.00 h. (a) What is the relative speed of the first plane with respect to the second? (b) If the ground speed of the second plane is 400 km/h, what is the ground speed of the first plane?

15. Two cars are moving in the same direction at constant speeds. The first is traveling 30 km/h and the second is traveling 40 km/h. If the second car is 4.0 km behind the first car, how long will it take it to catch up with the first car?

16. (a) With what speed does a freely falling body dropped from a height of 88.2 m hit the ground? (b) How long does the body take to fall this distance?

17. A car traveling at 25.0 m/s is brought to rest at a constant rate in 20.0 s by applying the brake. (a) What is its acceleration? (b) How far did it go after the brake was applied?

18. A bullet shot vertically upward has an initial speed of 588 m/s. (a) How long does it take before the bullet stops rising? (b) How high does the bullet go during this time?

19. A bullet leaves the muzzle of a gun at a speed of 400 m/s. The length of the gun barrel is 0.50 m. Assuming the bullet was uniformly accelerated, (a) what was its average speed inside the barrel? (b) How long was the bullet in the gun after it was fired?

20. An object dropped from a balloon descending at 4.2 m/s lands on the ground 10 s later. What was the altitude of the balloon at the time the object was dropped?

21. A bullet shot vertically upward leaves the gun at a speed of 655 m/s. How far above the muzzle will the bullet be 1.00 s after it is fired?

22. At the moment car A is starting from rest and accelerating at 4.0 m/s², car B passes it, moving at a constant speed of 28 m/s. How long will it take car A to catch up with car B?

23. Pressing the brake of a car caused it to slow down from 30.0 m/s to 20.0 m/s in 8.00 s. How far did the car travel during these 8.00 s?

24. An electron is accelerated uniformly from rest to a speed of 2.0×10^7 m/s. (a) If the electron traveled 0.10 m while it was being accelerated, what was its acceleration? (b) How long did it take to attain its final speed?

25. During a 30.0-s interval, the speed of a rocket rose steadily from 100 m/s to 500 m/s. How far did the rocket go during this time?

Graphical Analysis of Motion

5-12 Distance-Time Graph of Motion at Constant Speed

The use of a graph is helpful in analyzing the motion of a body. To illustrate, let us apply it to an airplane moving at constant speed in a straight course at 100 meters per second. Table 5.1 lists the distance traveled by the airplane at the end of each of five 1-second intervals.

For each of the positions of the airplane, we plot a point on the graph in Fig. 5-9. The distance traveled by the plane is shown on the vertical axis and the travel time is shown on the horizontal axis. Point A represents the starting time of 0 seconds. Point B represents the 100-meter distance traveled by the plane at the end of 1.00 second. Its ordinate, or vertical distance from the horizontal axis, is 100 meters on the distance scale. Its abscissa, or horizontal distance from the vertical axis, is 1.00 second on the time scale. Point C represents the 200-meter distance traveled by the airplane at the end of 2.00 seconds. Its ordinate is 200 meters and its abscissa is 2.00 seconds. Points D, E, and F are obtained in a similar manner.

Now, connecting the six points, we notice that they all fall on a straight line. As we have seen earlier, a straight-line graph between two quantities shows that the quantities are proportional to each other. In this case, the distance-time graph shows that the distance traveled by the body is directly proportional to the travel time. That is, as the travel time is doubled, the distance is doubled; as the travel time is tripled, the distance is tripled, and so on.

Fig. 5-8. Contrails can be used to determine the path of this airplane in space and time.

Table 5.1	
TRAVEL TIME (s)	**DISTANCE TRAVELED (m)**
0.00	000
1.00	100
2.00	200
3.00	300
4.00	400
5.00	500

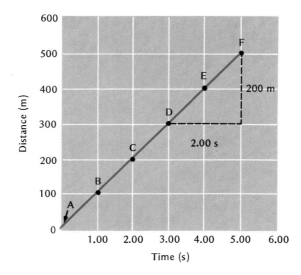

Fig. 5-9. A distance-time graph of motion at constant speed is a straight line.

5-13 Slope Represents the Speed

The slope or slant of the distance-time graph represents the speed, in this case, 100 meters per second. Let us show this.

In geometry, the slope of a line tells how much it is inclined to the horizontal axis. The slope is found by taking any two points on the line and dividing the difference between their ordinates by the difference between their abscissas.

For example, let us find the slope of the distance-time line from the points F and D. On the vertical scale, the ordinate of F represents a 500-meter distance and that of D represents a 300-meter distance. This gives a 200-meter difference in distance traveled. On the time scale, the abscissa of F represents a time 5.00 seconds after starting and the abscissa of D represents a time 3.00 seconds after starting. This gives a difference of 2.00 seconds in travel time. Dividing the 200-meter difference in the ordinates of D and F by the 2.00-second difference of their abscissas, we have 200 meters ÷ 2.00 seconds = 100 meters per second. This checks with the given speed of the airplane.

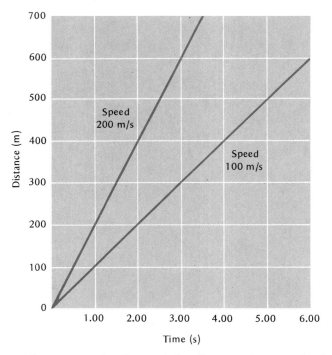

Fig. 5-10. The slope of the distance-time graph represents the speed.

The greater the slope of the distance-time graph, the greater is the speed it represents. In Fig. 5-10 are shown the distance-time graphs for two planes, one traveling at 100 meters per second and the other traveling at 200 meters per second. Note that the slope of the line representing the faster plane is steeper than that representing the slower plane.

5-14 Speed-Time Graph of Motion at Constant Speed

In Fig. 5-11, the constant speed of the plane flying at 100 meters per second is plotted against the time of travel. The speed is shown on the vertical axis and the time is shown on the horizontal axis. Since the speed is constant, all the points on the speed-time graph are the same distance above the horizontal axis and the line drawn through them is parallel to the horizontal axis.

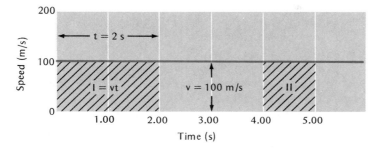

Fig. 5-11. A speed-time graph of motion at constant speed is a line parallel to the horizontal axis.

A particularly useful feature of this graph is the fact that *the area between the speed-time line and the horizontal axis represents the distance traveled by the body up to that time.* This is evident from rectangle I. Its vertical side is the speed v, and its horizontal side is the time of travel t. Its area is therefore $v \times t$ or vt, which is equal to the distance traveled by a body moving at constant speed v for a time t. The area of rectangle I gives us the distance traveled by the plane in 2.00 seconds. Since v is 100 meters per second and t is 2.00 seconds, the area is $100 \times 2.00 = 200$. This represents a distance of 200 meters. The area of rectangle II shows the distance covered in the time between the fourth and fifth seconds. This is $100 \times 1.00 = 100$ meters.

5-15 Distance-Time Graph of Uniformly Accelerated Motion

As an example of graphic analysis of uniformly accelerated motion, consider the motion of a body starting from rest and moving with an acceleration of 2.00 m/s². Table 5.2 lists the distances traveled by the body at the end of each of the first five seconds.

In Fig. 5-12, the distance is plotted as an ordinate and the time of fall as an abscissa for each second. Unlike the case of the body moving at constant speed, the distance-time graph is not a straight line but a curve. This curve, known as a *parabola* (puh·RAB·uh·luh) shows graphically the fact that the distance is directly proportional to the square of the time. It expresses the relationship $d = \frac{1}{2} at^2$.

5-16 Finding Speed from the Slope

The slope of the tangent to the distance-time graph at any point P is the instantaneous speed at that point. This can be seen from Fig.

Table 5.2	
TIME (s)	DISTANCE (m)
0.00	0.00
1.00	1.00
2.00	4.00
3.00	9.00
4.00	16.00
5.00	25.00

5-12 where P_1 represents a position of the body a short time t_1 before it reaches P, and P_2 represents the position of the body a short time t_2 after it leaves P. The slope of the line P_1P_2 is the distance traveled by the body in going from P_1 to P_2 divided by the time, $t_2 - t_1$. It therefore represents the average speed of the body between P_1 and P_2. If P_1 and P_2 are taken closer and closer to P, the line P_1P_2 coincides with the tangent to the curve at P. Its slope then is equal to the instantaneous speed of the body at P.

·Notice that the slope of the tangent at Q is steeper than that at P, which, in turn, is steeper than the slope of the tangent at N. This shows how the speed increases steadily as we go from N to P to Q.

Thus, an increasing slope in the distance-time graph indicates that the body is being positively accelerated. The numerical value of the slope of the tangent to the curve at any point may be determined exactly as was done in Section 5-13 by taking any two points on the tangent line and dividing the difference of their ordinates by the difference of their abscissas. Figure 5-12 shows how this is done for point P, at which the slope of the tangent turns out to be 12.00 meters divided by 2.00 seconds, or 6.00 meters per second. Using this procedure, the speeds at the end of each of the first 5 seconds have been determined and are shown in Table 5.3.

Table 5.3

TIME (s)	SPEED (m/s)
0.00	0.00
1.00	2.00
2.00	4.00
3.00	6.00
4.00	8.00
5.00	10.00

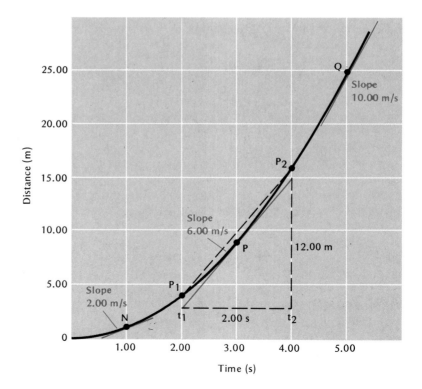

Fig. 5-12. Distance-time graph of uniformly accelerated motion.

5-17 Speed-Time Graph of Uniformly Accelerated Motion

In Fig. 5-13, the speeds listed in Table 5.3 are plotted as ordinates against the time as abscissas. The speed-time graph is a straight line showing that the speed is proportional to the travel time. As in the case of motion at constant speed, the area under the speed-time line gives the distance covered by the body. This can be seen from the shaded triangle. Its horizontal leg is the time t. Its vertical leg is the speed at time t, which is equal to $v = at$. Its area is one-half the product of its legs, or $\frac{1}{2} at \times t = \frac{1}{2} at^2$. This is also the expression for the distance d which the body moves in time t.

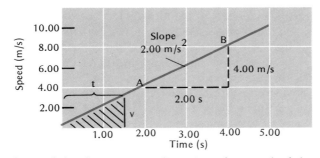

Fig. 5-13. A speed-time graph of uniformly accelerated motion is a straight line.

Just as the slope of the distance-time line gives the speed of the body at each point, so *the slope of the speed-time line gives the acceleration of the body at each point.* This can be seen by finding the slope from the points A and B in Fig. 5-13. The difference in the ordinates of the points A and B is $8.00 - 4.00 = 4.00$ meters per second. The difference of the abscissas of A and B is $4.00 - 2.00 = 2.00$ seconds. Hence, the slope is equal to 4.00 meters per second $\div$ 2.00 seconds = 2.00 meters per second per second. This is the acceleration of this body.

5-18 Distance-Time Graph of Any Motion

Using the graphic method, we can now analyze any motion, whether uniformly or not uniformly accelerated. We usually have a record of the distance traveled by the moving body during a given time. We can plot this record as a distance-time graph. The slope of the tangent to this graph at any point will then tell us the instantaneous speed of the body at the time represented by that point.

To illustrate, the distance-time graph of a moving body is shown in Fig. 5-14. The straight-line section AB shows that the body moved at constant speed from $t = 0$ to $t = 2$ seconds. This speed is given by the slope of AB, which can be found from points A and B as follows.

The ordinate of B is 4 meters and that of A is 0 meters. The difference in ordinates is therefore $4 - 0 = 4$ meters. The abscissa of B is 2 seconds and that of A is 0 seconds. The difference of abscissas is $2 - 0 = 2$ seconds. The slope of AB is the difference

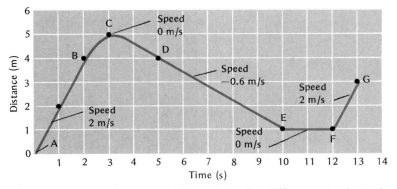

Fig. 5-14. Analysis of motion by means of a distance-time graph.

of the ordinates of A and B divided by the difference in their abscissas, or 4 meters ÷ 2 seconds = 2 meters per second. The speed of the body from A to B was 2 meters per second.

Between B and D the distance traveled increased to 5 meters at C, then decreased to 4 meters at D. This means that the body slowed up, stopped, and then reversed its direction of motion. The speed of the body at C is given by the slope of the tangent at C. Since this tangent is parallel to the time axis, it has zero slope. This shows that the instantaneous speed of the body at C is zero, or that the body was momentarily at rest at C.

From D to E the body again moved at constant speed, as shown by the slope of the straight-line segment DE. We obtain this slope from the points D and E as follows. The difference of the ordinates of E and D is 1 meter − 4 meters = −3 meters. The difference in the abscissas of E and D is 10 seconds − 5 seconds = 5 seconds. The slope of DE is therefore −3 meters divided by 5 seconds or −0.6 meter per second. Since the slope is negative, the speed it represents is opposite in direction to the body's original direction of motion. Thus, between D and E the body moved back toward its starting point at the speed of 0.6 meter per second.

At E, the body stopped moving and remained at rest between E and F. Finally, from F to G the body reversed direction again and moved at a constant speed of 2 meters per second.

The graph also gives information concerning the acceleration of the body. On each of the straight-line sections AB, DE, EF, and FG, the speed was constant or zero. The acceleration of the body while on each of these straight sections was therefore zero. Between B and D, the changing slope of the curve shows that the body underwent a negative acceleration that first decreased its speed to zero and then sped it up in the opposite direction.

5-19 Speed-Time Graph of Any Motion

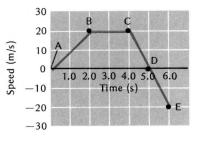

In Fig. 5-15, the speed-time graph of a typical body is shown. It is seen that the slope of AB is 10 meters per second per second, the slope of BC is zero, and the slope of CD is −20 meters per second per second. This indicates that the body was uniformly accelerated at 10 meters per second per second during the first 2.0 seconds. It had no acceleration during the next 2.0 seconds and it was slowed up or negatively accelerated at −20 meters per second per second during the final 2.0 seconds. At *D,* it reversed its direction.

The distance traveled during the first 2.0 seconds is given by the area under *AB.* This can be obtained from the figure by counting the number of small rectangles contained between *AB* and the time axis. Since the area of each of these rectangles is equal to its vertical side, representing 10 meters per second, times its horizontal side, representing 1.0 second, each small rectangle represents a distance of 10 meters. The area under *AB* contains a total of two small rectangles and it therefore represents a distance of 20 meters.

In the same way, the distance traveled during the next 2.0 seconds is given by the area under *BC.* This area contains four small rectangles and represents a distance of 4 × 10, or 40 meters. The distance traveled during the next 2.0 seconds is the area under *CD,* which contains a total of one small rectangle and represents a distance of 10 meters. Finally, the distance traveled between the fifth and sixth second is the area between *DE* and the time axis which is −10 meters. The minus sign indicates that this distance is opposite to the previous direction of motion.

Motion of Falling Bodies

5-20 Thought Experiment

The efforts to describe and understand the motion of falling bodies occupied men's minds for many centuries before the time of Galileo. These efforts were important for two reasons. First, they led to an understanding of the general laws of motion. Secondly, in the hands of Galileo, Newton, and their successors, they resulted in the development of methods of experimenting and thinking that became the foundations of modern science. One of these methods is the *thought experiment,* in which an experiment is imagined under simplified conditions not attainable in the laboratory. Such experiments frequently uncover a principle or law obscured by factors that cannot be practically removed. Because Galileo's work on falling bodies illustrates this approach, we shall take a little time to discuss it.

5-21 Galileo's Analysis

Galileo noted, as we all do, that, in reality, objects do not fall in air with the same acceleration. When a sheet of paper and a brick are dropped together from the same height, the brick falls much faster. Many experiences of this sort convinced scholars before the

Fig. 5-16. Galileo's observations and experiments with falling bodies established a new approach to the study of moving bodies.

days of Galileo that the law obeyed by falling bodies is: the heavier an object, the faster it falls.

However, an ingenious thought experiment led Galileo to suspect that this supposed law was not true. He guessed that the true law was obscured by the fact that the friction of the air changes the rate of fall of different bodies in different ways. Since there were no vacuum pumps in his day, Galileo had no way of removing the air from a glass tube and actually observing how bodies of different weights fall when dropped through the vacuum inside such a tube. He therefore resorted to a thought experiment in which he imagined what would happen if bodies were allowed to fall in a vacuum.

Suppose, Galileo reasoned, three exactly similar blocks of iron, A, B, and C, are dropped from the same height at the same time. Since they are similar in every respect, they will fall in exactly the same way, remaining together throughout their fall and striking the ground at exactly the same time. Now, bring the blocks A and B side by side and attach them to each other with a weightless chain. Repeat the experiment of dropping all the blocks at the same time. The blocks A, B, and C will again fall at the same rate and in the same way as before but A and B, fastened together by the chain, now make a body twice as heavy as C. Thus, a quantity of iron twice the weight of C falls at the same rate as C.

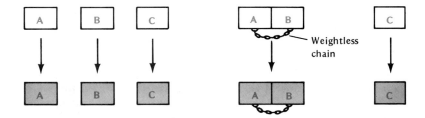

Fig. 5-17. Galileo's thought experiment: bodies A, B, and C fall in the same way whether separated or joined together.

The argument is readily extended to show that a body consisting of three equal iron blocks or four equal iron blocks or any number of equal iron blocks chained together will fall no faster than one iron block. Thus, it may be inferred that in a vacuum, any two bodies made of iron will fall at the same rate regardless of their weight.

Now, repeat this experiment in imagination with blocks of wood, stone, brass and all other materials. We reach this general conclusion: two bodies made of the same material fall at the same rate in a vacuum regardless of their weight.

5-22 Actual Experiment

Galileo's thought experiment suggested that, in a vacuum, any two bodies of the same material would fall at the same rate. However, his thought experiment did not tell whether bodies of two different materials, such as a piece of iron and a piece of lead, would also

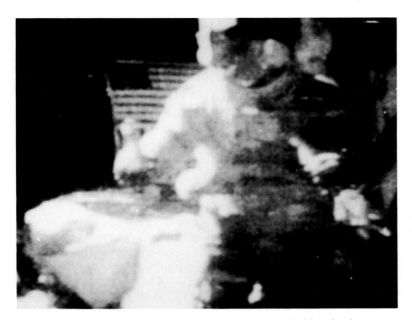

Fig. 5-18. Proof that objects of different mass fall at the same rate in a fixed gravitational field without air. This blurred photograph of a TV monitor screen shows Astronaut David Scott on the moon dropping a hammer from his right hand and a feather from his left hand simultaneously from waist height. The two objects struck the moon's surface at the same time.

fall at the same rate. To answer this question Galileo had to turn to *actual experiment.* Noting that air friction does not affect the fall of massive bodies as much as that of light ones, he dropped two massive pieces of different substances together and compared their rates of fall. He noted that there was no measurable difference. Combining this observation with the result of his thought experiment he inferred that, in a vacuum, any two bodies, regardless of their composition, size or mass, would also fall at the same rate. In this way, by a combination of thought experiments, actual experiments, and ingenious reasoning, Galileo was led to conclude that, in a vacuum, all bodies near the surface of the earth fall at the same rate.

5-23 Test of Galileo's Conclusion

The actual experimental test of Galileo's conclusion could only be made after the vacuum pump was invented years later by Otto Von Guericke (fun·GAIR·ik·uh). Today it is often demonstrated by removing the air from a closed 1.5-meter glass cylinder containing a coin and a feather. The cylinder is held vertically and then rapidly turned upside down so that the coin and feather start falling together. It is noticed that both objects fall at the same rate and hit the bottom of the tube at the same time.

5-24 Motion of Falling Bodies in Air

As we have seen, an object falling through air is not a freely falling body, because the air resists its motion and tends to slow it down. When a falling object is light in weight and has a large surface

Fig. 5-19. Sport parachutists make use of air resistance to slow their parachutes down to a safe landing speed.

exposed to the air, like a feather or a sheet of paper, the effect of the air in slowing down its rate of fall is immediately seen. Parachutists take advantage of this slowing action of the air on the large surface of their parachutes to reduce their rate of fall to a safe speed.

When a falling object is fairly dense and has a comparatively small surface, like a stone or a solid block of iron, the slowing action of the air is small and is not readily noticed. Such an object falling through a small height will be accelerated by gravity very much like a freely falling body. However, when such a body falls from a great height, the air resistance upon it increases steadily, as its speed of fall increases.

After a time, the air resistance may become so large that it prevents any further increase in the speed of the body. The body then continues the remainder of its fall at constant speed. Thus, a body dropped from an airplane from a very great height will keep speeding up until it reaches a certain maximum velocity, called its *terminal velocity*. After that it will continue to fall at this maximum velocity without further change.

CHAPTER REVIEW

Summary

To describe the motion of a body, a **reference point** or **frame of reference** must first be chosen. The position of the body at any instant is then given by the body's displacement from the reference point. Its motion is given by its velocity with respect to the reference point.

A body may be in either uniform or accelerated motion. It is in **uniform motion** when its velocity is constant in both magnitude and direction. It is in **accelerated motion** when its velocity is changing in either direction or magnitude or both. The **acceleration** of a body at any instant is defined as the rate of change of its velocity. It is a vector quantity.

Motion in a straight line is particularly important because all more complicated motions can be considered to be composed of combinations of straight-line motions. A body in uniform motion moves in a straight line at constant speed. Such a body travels equal distances during equal intervals of time.

A body in **uniformly accelerated motion** in a straight line is one that has constant acceleration. Such a body either increases or decreases its speed at a constant rate.

For a body having an initial speed v_o and a constant acceleration a, the speed v at the end of a time interval t is given by:

$$\mathbf{v = v_o + at}$$

The distance d traveled by such a body is given by:

$$\mathbf{d = v_o t + \tfrac{1}{2} a t^2}$$

These two relationships combine to give:

$$\mathbf{v^2 - v_o^2 = 2\, ad}$$

For a body starting from rest, these relationships become:

$$v = at$$
$$d = \frac{1}{2} at^2$$
$$v^2 = 2\,ad$$

A **body falling freely** under the force of gravity undergoes uniformly accelerated motion. The constant **acceleration of gravity** (g) is equal to 9.80 meters per second per second, or 980 centimeters per second per second.

Bodies do not fall freely through air. They encounter frictional resistance that slows down their rate of fall and prevents their speeding up beyond a certain **terminal velocity.**

Questions

Group 1

1. In the distance-time graph of a body's motion, what does the slope at any point tell about the body's motion?
2. In the speed-time graph of a body's motion, what do each of the following tell about the body's motion: (a) the area under the speed-time line; (b) the slope of the speed-time line at any point?
3. The distance-time graph of a body's motion is a straight line making an acute angle with the time axis. What does this tell you about (a) the body's speed; (b) its acceleration?
4. The speed-time graph of a body is a straight line making an acute angle with the time axis. What does this tell you about (a) the body's speed; (b) its acceleration?
5. A sheet of paper and a pencil are pushed off a table at the same time and fall to the floor. (a) Why are the rates of fall different? (b) What is meant by the terminal velocity of a falling body?

(c) Under what conditions would the downward acceleration of these two bodies be the same?
6. Explain Galileo's thought experiment showing that freely falling bodies of the same material should fall in exactly the same way.

Group 2

7. (a) For a certain time interval, the distance-time graph of a body's motion is parallel to the time axis. What does this tell about the body's motion during that time interval? (b) Why is it not possible for any part of the distance-time graph of a body to be a straight-line segment parallel to the distance axis? (*Hint:* What would such a straight-line segment indicate about the time taken to travel the distance it represents?)
8. (a) For a certain time interval, the speed-time graph of a body's motion is parallel to the time axis. What does this tell about the body's motion during that time interval? (b) Is it possible for any part of the speed-time graph to be parallel to the speed axis? Explain.

Problems

Group 1

1. The distance traveled by a car over a 10-s time interval increased as shown in Table 5.4.
   (a) Plot the distance-time graph of the motion. (b) What does it reveal about the motion of the car? (c) Determine the distance traveled by the car at the end of 5.5 s. (d) Determine the speed of the car.
2. A car is moving at a constant speed of 20 m/s. (a) Plot its speed-time graph for a 10.0-s interval.

Table 5.4	
TIME (s)	DISTANCE (m)
0.0	0
2.0	35
4.0	70
6.0	105
8.0	140
10.0	175

(b) Make a table of the distances traveled by the car at the end of 0.0, 2.0, 4.0, 6.0, 8.0, and 10.0 s, and plot the distance-time graph of the car.

3. Each of two cars is moving at constant speed, the first at 15 m/s, the second at 20 m/s. (a) on the same piece of graph paper, plot the speed-time graph of each of these cars over a 10.0-s interval. (b) On a second piece of graph paper, plot the distance-time graph of each car assuming that they started from the same point. (c) What is the distance between the cars at the end of 6.0 s?

4. A car moving at a constant speed of 15 m/s is 20 m ahead of a second car moving at a constant speed of 20 m/s. (a) Plot the distance-time graph of each of the cars on the same piece of graph paper. Determine from the graphs (b) how long it takes the second car to catch up with the first; and (c) how far the second car travels in this time.

5. The speed of an airplane increased during a 4.0-s interval according to the data in Table 5.5. (a) Make a speed-time graph of the motion. (b) Find the distance traveled by the airplane during the first 3.0 s. (c) Find the acceleration of the plane at the end of 2.0 s. (d) How does the acceleration obtained in (c) compare with the acceleration at the end of 3.0 s?

6. A ball rolling down an incline is accelerated uniformly at the rate of 4.0 m/s². (a) Assuming the ball starts from rest, make a table of its speeds at the end of each of the first 5 s of its

motion. (b) Plot the speed-time graph of its motion. (c) Determine from the graph how far the ball travels in 3.5 s.

7. The distance a piece of ice slid down a 7.5-m slope at the end of each second after its release is given by Table 5.6.
(a) Make a distance-time graph of this motion. (b) How far will the ice have traveled at the end of 3.5 s? (c) Determine the speed of the ice at that time from the graph-

8. An airplane moving at 200 m/s is accelerated uniformly at 5.00 m/s² for 4.00 s. It then continues its flight at the speed it has attained at that time. (a) Plot the speed-time graph for the first 8.00 s of the motion. Determine from the graph how far the airplane traveled during (b) the first 4.00 s; (c) the first 8.00 s.

9. A train moving at a constant speed of 15 m/s travels for 2.0 s and is then uniformly accelerated at the rate of 2.0 m/s² for 3.0 s. From the speed-time graph of the train's motion determine (a) its speed at the end of 4.0 s; (b) the distance it has traveled in that time.

Group 2

10. A car traveled for 0.10 h at 30 km/h and for 0.20 h at 40 km/h. It then stopped for 0.10 h and finally completed its trip in an additional 0.30 h at 20 km/h. From a speed-time graph of the car's motion, determine how far the car had traveled at the end of (a) 0.30 h; (b) 0.70 h; (c) What was the car's average speed during (a) and (b)?

11. Refer to the distance-time graph in Fig. 5-20 and determine (a) the distance traveled by the body between the first and third hours; (b) the distance traveled by the body during the first 3.0 h; (c) the average speed of the body during the first 3.0 h; (d) its average speed in the first 4.0 h.

Table 5.5	
TIME (s)	SPEED (m/s)
0.0	30
1.0	40
2.0	50
3.0	60
4.0	70

Table 5.6	
TIME (s)	DISTANCE (m)
0.0	0.0
1.0	0.3
2.0	1.2
3.0	2.7
4.0	4.8
5.0	7.5

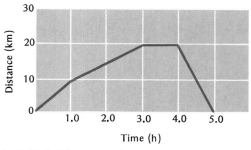

Fig. 5-20. Problems 11 and 12.

12. From Fig. 5-20 find the speed of the body (a) during the first hour; (b) between the first and third hours; (c) between the third and fourth hours; (d) between the fourth and fifth hours.

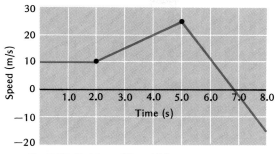

Fig. 5-21. Problems 13 and 14.

13. Use the speed-time graph of Fig. 5-21 to find how far the body traveled (a) between $t = 0.0$ s and $t = 2.0$ s; (b) between $t = 2.0$ s and $t = 5.0$ s. (c) What is the average speed of the body for the first 5.0 s?

14. Use the speed-time graph of Fig. 5-21 to find the acceleration at (a) $t = 1.0$ s; (b) $t = 4.0$ s; (c) $t = 6.0$ s. (d) Make a graph plotting the acceleration of this body against the time.

15. In Fig. 5-22, (a) at what instants after $t = 2.0$ s is the body at rest? (b) At what instants does it reverse its direction? (c) Between what values of t is it moving at constant speed? (d) Between what values of t is it being positively accelerated?

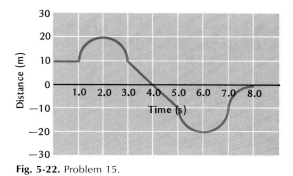

Fig. 5-22. Problem 15.

Applying Physics

1. On an automobile trip, make a record of the distance reading of the odometer at two-minute intervals. Use these data to make a distance-time graph of the motion of the car for the first twenty minutes of the trip. From the graph determine the average speed of the car (a) during the first 5 minutes; (b) during the entire 20 minutes.

2. To estimate the height to which you can throw a ball, throw it vertically as hard as you can. Measure the time that elapses between the time the ball leaves the hand and the time it returns. Half this time interval was taken by the ball in falling from the highest point back to the ground. Determine the distance fallen from $d = \frac{1}{2} gt^2$.

3. What effect does air resistance have on the rate of fall of a light object having a large surface? Fill a toy balloon with air until it is about 20 cm in diameter and tie off the mouth of the balloon. Let the balloon be dropped from a measured height of 2.5 m. Time its fall and determine its average speed. Compute the average speed of a freely falling body when it falls 2.5 m in a vacuum from the fact that $\bar{v} = v/2$. What loss in the average speed for the falling balloon can be attributed to air resistance?

Repeat the experiment with the balloon blown up until it is only 10 cm in diameter. What difference do you observe?

Two-Dimensional Motion

Aims

1. To learn how to describe and analyze the motion of bodies that are confined to a plane.
2. To apply that knowledge to projectiles, planets, and pendulums.

Motion of Projectiles

6-1 Two-Dimensional Motion

Many bodies remain in the same plane while they are moving. Their motion can therefore be described in two dimensions. Particularly important among such bodies are the planets in their orbits around the sun and the moon in its orbit around the earth. On the earth itself, for short distances, the surface of the earth may be considered to be nearly a plane and the motions of bodies on it are therefore approximately two-dimensional. The motion of a projectile (pruh·JEK·tul) fired from a gun is also approximately two-dimensional. Neglecting the effect of air resistance, a projectile moving over short distances remains in the same vertical plane. Another example of two-dimensional motion in a vertical plane is that of the simple pendulum.

6-2 Frames of Reference

We often use the graph technique in describing the motion of a body on a plane. A point O is selected on the plane in which the motion takes place and perpendicular axes, OX and OY, drawn through this point, serve as a *frame of reference*. A point is then located by giving its distances from each of these axes. The distance from the OY axis is called x and that from the OX axis is called y. For point A in Fig. 6-2, $x = 3$ and $y = 2$. These values of x and y for A are called the *coordinates* (koh·AWR·din·uts) of A.

For points above the OX axis, y is taken positive, and for points below the OX axis, y is taken negative. For points to the right of the OY axis, x is taken positive, and for points to the left of the OY axis, x is taken negative. Thus, point B has the coordinates $x = -5$, $y = +2$. Point C has the coordinates $x = +2$, $y = -4$.

6-3 Vectors Represent Velocity Changes

In studying the motion of a body on a surface, we must take into account the fact that the motion changes not only in speed but also in direction. In Fig. 6-3 (a), a car is shown in two positions as it

Fig. 6-1. Two-dimensional motion in a vertical plane is illustrated by the bouncing ball.

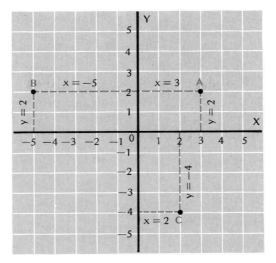

Fig. 6-2. A point in a plane is located by two coordinates.

goes around a turn. The car is being accelerated because both its speed and its direction are changing. Assuming that the speedometer reading of the car in the first position is 40 kilometers per hour and in the second position is 50 kilometers per hour, the vectors $\vec{v}$ and $\vec{v}_1$ drawn to scale in Fig. 6-3 (b) represent the two velocities of the car.

Let us now introduce the symbol Δ, which is the Greek letter *delta*, and is used in mathematics to mean the *change of* or *difference of*. Then $\Delta\vec{v}$, which is read "delta $\vec{v}$," is the vector change that has taken place in the velocity $\vec{v}$ to give the car the new velocity $\vec{v}_1$.

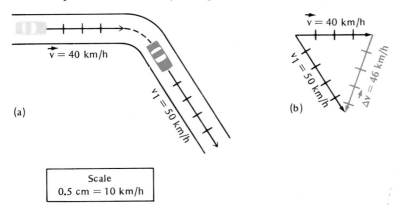

(a)

(b)

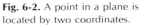

Scale
0.5 cm = 10 km/h

Fig. 6-3. The vector change in velocity when $\vec{v}$ changes to $\vec{v}_1$ is $\Delta\vec{v}$.

To find $\Delta\vec{v}$, note that $\vec{v}_1$ is the resultant of $\vec{v}$ and $\Delta\vec{v}$. We therefore resolve $\vec{v}_1$ into components as shown in Fig. 6-3 (b). From the same point we draw $\vec{v}$ and $\vec{v}_1$ to a scale of 0.5 centimeter = 10 kilometers per hour. The vector triangle is then completed by drawing $\Delta\vec{v}$ from the head of $\vec{v}$ to the head of $\vec{v}_1$. Next we measure $\Delta\vec{v}$. It is 2.3 centimeters long and represents a velocity of 46 kilometers per hour in the direction shown. Thus, when the car's velocity changed from 40 kilometers per hour eastward to 50 kilometers per hour southeastward, the vector change in its velocity was 46 kilometers per hour in the direction of $\Delta\vec{v}$.

6-4 Average Acceleration

Average acceleration is a vector change in velocity over a period of time.

As in the case of straight-line motion, the acceleration is the rate of change of the velocity. But now we must allow for the fact that the direction of the velocity also changes. The *average acceleration* is therefore a vector given by:

$$\text{average } \vec{a} = \frac{\Delta\vec{v}}{\Delta t}$$

where $\Delta\vec{v}$ is the vector change in the velocity and Δt is the change in time or the time interval during which this velocity change took place.

In the above example, we found that the change $\Delta\vec{v}$ in the velocity of the car was 46 kilometers per hour in the direction shown. Assuming that this change took place in 10 seconds, the average acceleration during the 10 seconds when the car was moving from the first to the second position was:

$$\text{average } \vec{a} = \frac{\Delta\vec{v}}{\Delta t} = \frac{46 \text{ km/h}}{10 \text{ s}}$$

$$\text{average } \vec{a} = 4.6 \text{ kilometers per hour}$$

$$\text{per second in the direction of } \Delta\vec{v}$$

6-5 Instantaneous and Average Acceleration

Instantaneous acceleration is the rate of change in velocity at any instant.

Just as the instantaneous (in·stan·TAY·nee·us) velocity of a car may change from moment to moment, so its *instantaneous acceleration* may change in direction, magnitude, or both from moment to moment. The instantaneous acceleration at any time t is approximately equal to the average acceleration of the car over a very small time interval Δt that begins a tiny fraction of a second before t and ends a tiny fraction of a second after t. This is so because, when Δt is very small, the instantaneous acceleration of the car during this tiny time interval will not have time to change very much. Its value during Δt will therefore be nearly constant and nearly equal to the average acceleration during this time interval.

As the time interval around t is made vanishingly small, the actual difference between the average acceleration over this interval and the instantaneous acceleration at t will become negligible. We may therefore regard the instantaneous acceleration at time t as the limiting value which the average acceleration during a small time interval Δt (including t) approaches as Δt is made to approach zero.

6-6 Independence of Motions

In our study of the resolution of a vector into components, we noted that when the components are taken perpendicular to each other, they are independent. We can show that this is true of the components of a velocity.

When a bullet is fired horizontally, the gun gives it a forward velocity while the force of gravity pulls it and accelerates it downward. You might think that because a bullet is traveling forward at high speed, it falls less rapidly than it would fall if it simply dropped out of the muzzle of the gun. But this is not true.

When an object like the bullet has two perpendicular motions, each motion is independent of the other. The forward motion of the bullet is not affected by its downward motion and the downward motion of the bullet is not affected by its forward motion. Though speeding forward, the bullet will be accelerated downward at the rate of 9.80 meters per second per second like any other falling body. Thus, if one bullet is fired horizontally from a given height above the ground and a second bullet is dropped from the same height at the same time, you will find that both bullets will fall at the same rate and hit the ground at the same time.

You can demonstrate this roughly by bending a 3-centimeter width of sheet metal around one end of a hacksaw blade to form two small platforms as shown in Fig. 6-4. Hold the other end of the hacksaw blade against the corner edge of a table. Put a coin on each platform of sheet metal. Bend the free end of the hacksaw blade back and then release it suddenly. One coin will be shot forward; the other will begin to fall at the same time. You will find that both coins hit the ground at the same time.

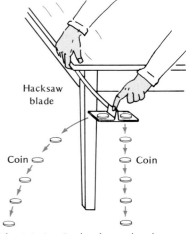

Fig. 6-4. A coin shot forward and a coin dropped vertically undergo the same downward acceleration.

6-7 Projectile Fired Horizontally

Any velocity can be broken down into perpendicular components independent of each other. We can therefore analyze the motion of a body by analyzing these independent component motions. This simplifies matters because the component motions are in a straight line and involve no change in direction.

Let us apply this procedure to the motion of a bullet fired horizontally from a gun from a height of 78.4 meters above the ground. We have chosen this height for simplicity in our calculations. As we shall see, the bullet shot from this height will fall to earth under the influence of gravity in exactly 4.00 seconds. We note that the motion of the bullet is a combination of two independent motions. One of these is the horizontal motion given to the bullet by the gun. Let us assume that the bullet leaves the muzzle of the gun at 200 meters per second. Then it will continue to move horizontally at a velocity of 200 meters per second.

The second motion is the vertically downward movement given to the bullet by the pull of gravity which accelerates it at 9.80

Fig. 6-5. A ski jumper follows a parabolic path similar to that of a projectile fired horizontally.

meters per second per second. Let us consider each of these motions separately. Although the frictional resistance of the air has a significant influence on these motions, we shall for simplicity neglect its effect.

The horizontal motion of a projectile is at constant velocity.

1. Horizontal Motion. This is motion at the constant velocity of 200 meters per second. Therefore, the horizontal distance x traveled by the bullet during any time t is $x = vt$, where v is the constant horizontal speed. The second column of Table 6.1 shows the distances traveled by the bullet during the first 4 seconds.

Table 6.1

TIME	$x = vt$	$d = \frac{1}{2} gt^2$	$y = 78.4 - d$
(seconds)	(meters)	(meters)	(meters)
0.00	0.00	0.00	78.4
1.00	200	4.90	73.5
2.00	400	19.6	58.8
3.00	600	44.1	34.3
4.00	800	78.4	0.00

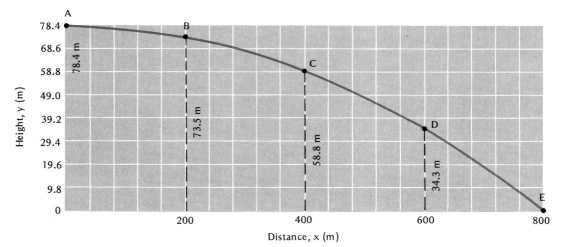

Fig. 6-6. The path of a projectile fired horizontally. (Vertical scale exaggerated.)

The vertical motion of a projectile is uniformly accelerated downward at 9.80 m/s².

2. Vertical Motion. In this motion the bullet is falling from a height of 78.4 meters at the same rate as it would fall if it had no forward motion. The distance d fallen by a body starting from rest is given by $d = \frac{1}{2} gt^2$, where g is the acceleration of gravity. Substituting $g = 9.80$ meters per second per second and the different values of t gives the third column in the table which shows the distances fallen by the bullet during the first 4 seconds. The fourth column lists the height above the ground y after each second of fall. This height is obtained by subtracting d from 78.4 meters.

In Fig. 6-6 we plot the altitude y of the bullet at each instant of time against its horizontal distance from the starting point x. At the beginning, the bullet is at a height of 78.4 meters, shown by point *A*. At the end of the first second, it has fallen 4.90 meters to an altitude of 73.5 meters and has moved 200 meters forward giving point *B*. At the end of 2.00 seconds, it has dropped to an altitude of 58.8 meters and moved 400 meters forward, giving point *C*, and so on for points *D* and *E*. The path of the bullet is the curve drawn through these five points. The curve is a *parabola* and is the characteristic path of a projectile fired over a small distance.

6-8 Projectile Fired at an Angle

Consider the case of a projectile that leaves the muzzle of a gun at an angle of 30° with the horizontal and a velocity of 39.2 meters per second. To study its motion, we first resolve its initial velocity into two components, one moving the projectile forward and the other moving it upward. In Fig. 6-7 we have done this to scale and found that the upward component $\vec{v}_y$ is 19.6 meters per second, and the forward component $\vec{v}_x$ is 33.9 meters per second. Now consider each of these motions separately.

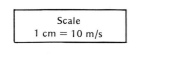

Scale
1 cm = 10 m/s

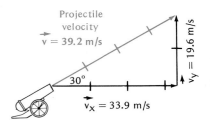

Fig. 6-7. Horizontal and vertical components of the initial velocity of the projectile.

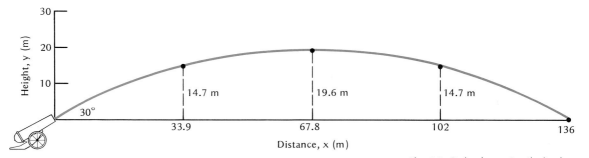

Fig. 6-8. Path of a projectile fired at an angle of 30° above the horizontal.

    1. Horizontal Motion. The projectile is moving forward at a constant speed of $v_x = 33.9$ meters per second. The horizontal distance traveled at the end of *t* seconds is $x = v_x t = 33.9\,t$. The second column in Table 6.2 lists the distances traveled at the end of each of the first four seconds.

    2. Vertical Motion. In the third column, *y* represents the altitude of the projectile above the ground. To obtain it, note that the projectile begins with an upward velocity v_y of 19.6 meters per second, but gravity accelerates it downward at the rate of $g = 9.80$ meters per second per second. Now, in a time *t*, the body's upward velocity v_y will carry it upward a distance, $v_y t = 19.6\,t$. However, at the same time, gravity will pull the body down a distance $\frac{1}{2}\,gt^2 = \frac{1}{2}(9.80)t^2$. The net distance traveled upward by the body in time *t* is therefore $y = v_y t - \frac{1}{2}\,gt^2$, or:

$$y = 19.6\,t - \frac{1}{2}(9.80)t^2$$

Table 6.2		
t (s)	*x* (m)	*y* (m)
0.00	0.00	0.00
1.00	33.9	14.7
2.00	67.8	19.6
3.00	102.	14.7
4.00	136.	0.00

Using this relationship, we have computed the value of y at the end of each of the first four seconds of flight. These are listed in the third column of Table 6.2. Notice in the table that, as the projectile rises, gravity slows its upward motion. It rises for 2.00 seconds, reaching a height of 19.6 meters; then it falls down again for 2.00 seconds.

In Fig. 6-8, we obtain the path of the projectile by plotting its altitude y against the horizontal distance x covered at the end of each second. The curve drawn through these points is the path of the projectile, and once more it is a parabola.

6-9 Horizontal Distance Traveled by a Projectile

A projectile goes farthest when fired at a 45° angle to the horizontal.

For short ranges, the distance traveled by a projectile depends only on the forward component of its velocity and the time the projectile remains in the air. In the above case, the projectile went upward during the first 2 seconds and downward during the next 2 seconds, remaining in the air a total of 4.00 seconds. The forward distance it traveled is therefore its forward velocity of 33.9 meters per second × 4.00 seconds = 136 meters. Galileo was the first to show mathematically that, for a given muzzle velocity, a projectile would go furthest when fired at a 45° angle with the horizontal.

Sample Problem

The vertical component of the velocity of a projectile fired from a cannon is 14.7 m/s. The horizontal component is 300 m/s (a) How long does it take the projectile to reach its highest point? (b) How high does it rise? (c) How far forward does it go?

Solution:

(a) $v_x = 300$ m/s $v_y = 14.7$ m/s

As the projectile rises, gravity reduces its vertical speed by 9.80 m/s every second. It will therefore stop rising and reach its highest point in:

$$\frac{14.7 \text{ m/s}}{9.80 \text{ m/s}^2} = 1.50 \text{ s}$$

(b) The projectile takes 1.50 s to reach its highest point and another 1.50 s to fall back to earth. The distance it falls from its highest point is:

$$d_y = \tfrac{1}{2}gt^2 = \tfrac{1}{2}(9.80 \text{ m/s}^2)(1.50 \text{ s})^2 = 11.0 \text{ m}$$

(c) The projectile remains in the air for a total of 3.00 s. The distance traveled forward in that time is:

$$d_x = v_x t = 300 \text{ m/s} \times 3.00 \text{ s} = 900 \text{ m}$$

Test Yourself Problems

1. A projectile is fired horizontally from a height of 44.1 m at a speed of 500 m/s. (a) How long after it was fired did it hit the ground? (b) How far did it travel?

2. The upward component of the velocity of a projectile fired from a gun is 98.0 m/s. The horizontal component is 400 m/s. (a) How long does it take the projectile to reach its highest point? (b) How long does it remain in the air? (c) How far does it travel?

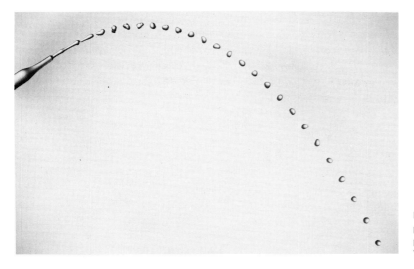

Fig. 6-9. Parabolic path of a projected water drop photographed by Harold M. Waage, Princeton University.

6-10 Actual Motion of a Projectile

In the discussion of projectile motion, we have made several simplifying assumptions. Among these are the following: (1) The effect of air resistance can be neglected. (2) The earth may be considered a flat rather than a curved surface when the distance traveled by the projectile is small. (3) For these small distances, the direction in which the earth's gravity accelerates the projectile remains virtually the same. (4) The projectile remains in the same vertical plane throughout its flight.

In reality, none of these assumptions is exactly true. In predicting the actual motion of a projectile, allowances have to be made for deviations from the ideal conditions represented by these assumptions. The slowing affect of air resistance, for example, causes the path of the projectile to turn downward from its ideal path and thus to shorten its range.

Circular Motion

6-11 Circular Motion at Constant Speed

A very important type of motion is that of a body moving in a circle at constant speed. The earth and some of the planets approximate this kind of motion in their orbits around the sun. Every object standing on the surface of the earth has this kind of motion because the earth rotates at a nearly constant rate on its axis.

The time taken by a body moving in a circle to go around the circle once and return to its starting point is called the *period*. The earth, which moves in an almost circular orbit around the sun, has a period of one year.

The period of a body moving at constant speed v in a circle depends upon the speed of the body and the radius r of the circle. The distance traveled by the body during one period T is its speed times the period, or $v \times T$. But, during a period, the body goes around the circle once and therefore travels a distance equal to the circumference $2\pi r$. It follows that $v \times T = 2\pi r$, whence,

$$T = \frac{2\pi r}{v} \quad \text{or} \quad v = \frac{2\pi r}{T}$$

Sample Problem

Assuming that the earth moves around the sun at constant speed in a circle whose radius is 1.5×10^8 km and that its period is 1.0 y, what is its orbital velocity?

Solution:

$$r = 1.5 \times 10^8 \text{ km}$$

$$T = 1.0 \text{ y} = 8.8 \times 10^3 \text{ h}$$

$$v = \frac{2\pi r}{T}$$

$$v = \frac{2\pi \times 1.5 \times 10^8 \text{ km}}{8.8 \times 10^3 \text{ h}}$$

$$v = 1.1 \times 10^5 \text{ km/h}$$

Test Yourself Problem

An object moving at constant speed in a circle of radius 1.0 m has a period of 1.0 s. What is its orbital velocity?

6-12 Centripetal Acceleration

Although moving at constant speed, a body moving in a circle is changing direction and is therefore undergoing an acceleration. In the next section, we shall prove that the magnitude of the acceleration a of a body moving in a circular path of radius r at constant speed v is:

$$a = \frac{v^2}{r}$$

As the body continues to move in a circle, this acceleration keeps changing its direction so that it is always directed toward the center of the circle. An acceleration directed radially toward a center in this fashion is called a *centripetal* (sen·TRI·pi·tul) *acceleration.*

Sample Problem

What is the centripetal acceleration of a body moving in a circle of radius 10 m at a speed of 5 m/s?

Solution:

$$v = 5 \text{ m/s} \quad r = 10 \text{ m}$$

$$a = \frac{v^2}{r}$$

$$a = \frac{(5 \text{ m/s})^2}{10 \text{ m}} = 2.5 \text{ m/s}^2$$

Test Yourself Problem

What is the centripetal acceleration of a car moving in a circle of radius 20.0 m at a speed of 10.0 m/s?

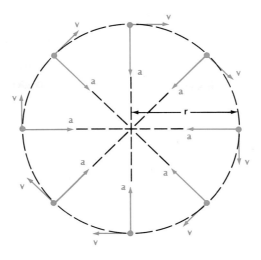

Fig. 6-10. A body in circular motion at constant speed v undergoes a centripetal acceleration of constant magnitude, $a = v^2/r$.

6-13 Proof of Centripetal Acceleration Formula

Optional proof for honor students

Let A and B in Fig. 6-11 be two positions of a body moving in a circle of radius r at constant speed v. The vector $\vec{v}_1$ is tangent to the circle at A and represents the velocity of the body at A. The vector $\vec{v}_2$ similarly represents the velocity of the body at B. In magnitude, $\vec{v}_1$ and $\vec{v}_2$ are both equal to v. Let $\Delta\vec{v}$ be the vector change that took place in the velocity as the body went from A to B. To find $\Delta\vec{v}$, construct $\vec{v}_1$ and $\vec{v}_2$ to scale from the common point D as shown in Fig. 6-11 (b) and complete the vector triangle DEF. Note that, in this triangle, $\vec{v}_2$ is the resultant of adding $\Delta\vec{v}$ to $\vec{v}_1$. If Δt is the time taken by the body to go from A to B, the average acceleration $\vec{a}$ of the body during this time is:

$$\text{average } \vec{a} = \frac{\Delta\vec{v}}{\Delta t}$$

(b)

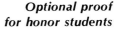

Now triangle I and triangle II are isosceles because $OA = OB = r$ and $DF = DE = v$. They are also similar triangles because the corresponding sides of angle O are perpendicular to the corresponding sides of angle D, making these angles equal. In these triangles therefore,

$$\frac{AB}{r} = \frac{\Delta v}{v}$$

where Δv is the magnitude of $\Delta\vec{v}$.

Now we can choose A and B as close to each other as we please. If we take them close enough, the arc AB is very nearly equal to the chord AB. Hence, we may write:

$$\frac{\text{arc } AB}{r} = \frac{\Delta v}{v} \text{ (approximately)}$$

But arc AB is the distance traveled by the body at speed v during

(a)

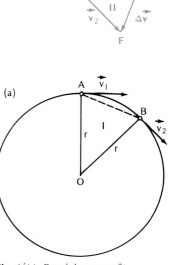

Fig. 6-11. Proof that $a = v^2/r$.

the time Δt and is equal to $v \times \Delta t$. Substituting this value for the arc AB in the above relationship, we have:

$$\frac{v\Delta t}{r} = \frac{\Delta v}{v} \text{ (approximately)}$$

whence,

$$\frac{\Delta v}{\Delta t} = \frac{v^2}{r} \text{ (approximately)}$$

As Δt is made smaller and smaller A and B come closer and closer together until eventually they coincide. The arc AB is then the same as the chord AB and the average acceleration over AB will then become equal to the instantaneous acceleration at the coinciding points A and B. The magnitude of the average and instantaneous accelerations at that time is exactly $\Delta v/\Delta t$ which is then exactly equal to v^2/r. Therefore,

$$\mathbf{a} = \frac{\mathbf{v}^2}{\mathbf{r}}$$

This gives only the magnitude of the acceleration. We must now find its direction. The direction of $\vec{a}$ is the same as the direction of $\Delta \vec{v}$. In the limit, $\Delta \vec{v}$ is perpendicular to $\vec{v}_1$ (or v_2 which coincides with it). Since $\vec{v}_1$ is perpendicular to r, $\Delta \vec{v}$ is in the same direction as r and points toward the center of the circle, thus making it a centripetal acceleration.

6-14 Centripetal Acceleration and Period

The centripetal acceleration of a body in circular motion at constant speed v can be expressed in terms of its period T and the radius of its path r. We have seen in Section 6-11 that $v = 2\pi r/T$. Substituting this value in:

$$a = \frac{v^2}{r}$$

we have:

$$a = \left(\frac{2\pi r}{T}\right)^2 \frac{1}{r}$$

$$a = \frac{4\pi^2 r}{T^2}$$

6-15 Centripetal Acceleration of the Moon

To illustrate the use of the above relationship, let us estimate the centripetal acceleration that keeps the moon moving in a circular path around the earth. The distance from the moon to the earth is the radius of the path and is 3.8×10^8 meters. The moon's period is 2.3×10^6 seconds (27.3 days).

$$a = \frac{4\pi^2 r}{T^2} = \frac{4\pi^2 \times 3.8 \times 10^8 \text{m}}{(2.3 \times 10^6 \text{ s})^2}$$

whence;

$$a = 2.8 \times 10^{-3} \text{ m/s}^2$$

This acceleration is directed toward the earth and is provided by the gravitational pull of the earth on the moon. Note how very small it is compared to the acceleration $g = 9.80$ meters per second per second which the earth gives to all freely falling bodies near its surface. Thus, it appears that as a body moves away from the earth, the acceleration with which the earth makes it fall back decreases. This fact was an important clue for Sir Isaac Newton in his discovery of the law of univeral gravitation to be studied in Chapter 8.

Newton used this calculation of the moon's acceleration to test his theory of gravitation.

6-16 Two Views of Planetary Motions

For centuries people have watched the motions of the heavenly bodies. They have noticed that the stars seem to be fixed with respect to each other. The relative arrangement of groups of stars such as the Big Dipper and Orion (oh·RY·on) remains unchanged and can be recognized century after century. They have also noted that during the course of each night, the entire sky with all its stars rotates. In the northern hemisphere each star moves in a circular path having its center very near the North Star. (See Fig. 6-12).

Now there are different ways of describing these motions, depending on the body we select as a frame of reference. If we take the earth as the body to which we attach our frame of reference, we consider ourselves fixed in space. We then describe the motions of the stars by saying that they move as though they were attached to a very large rotating sphere having the earth at its center. This is the *earth-centered* or *geocentric* (JEE·oh·SEN·trik) *theory*, which was accepted by the ancients. Its development is associated with the name of the astronomer Claudius Ptolemy (TOH·luh·mee), who lived about 150 A.D.

Another point of view is to consider the stars to be on a fixed sphere in space with the sun at its center serving as the fixed body to which we attach our frame of reference. The observed changes in the positions of the sun and the stars can then be regarded as the result of the rotation of the earth on its axis and its annual revolution about the sun. This *sun-centered* view of the universe is called the *heliocentric* (HEE·lee·oh·SEN·trik) *theory*. It is associated with the Polish astronomer Nicolaus Copernicus (koh·PER·ni·kus) (1473–1543).

Fig. 6-12. One-hour time exposure of stars in the northern sky showing their apparent paths around the North Star.

6-17 Which View Is Correct?

The answer is, *both*. The geocentric theory describes the motions of the heavenly bodies from the point of view of an observer on the earth. The heliocentric theory describes these same motions from the point of view of an observer stationed on the sun. The differences in their descriptions therefore represent only the difference in the body used as a frame of reference. How then can we decide which of these two points of view to adopt?

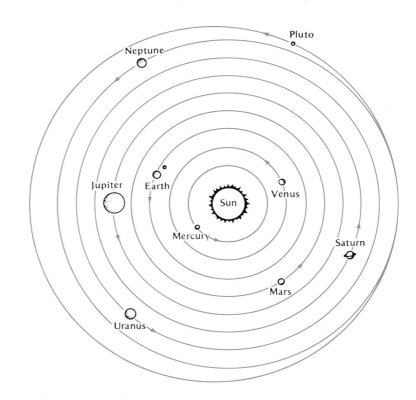

Fig. 6-13. Orbits of the planets around the sun.

The answer is that we prefer that point of view that gives us the simplest picture of the universe. We choose the theory that is easier to work with in solving such problems as the prediction of the paths and future positions of the planets, the prediction of eclipses, phases of the moon, and so forth. On this basis, the Copernican or heliocentric theory has displaced the geocentric theory. A major test for each theory was the description of the motion of the planets. While the geocentric theory required complicated geometrical curves to describe and deal with the motions of the planets, the heliocentric theory was able to show that the planetary motions could be accurately described by simple curves such as the circle and the ellipse.

6-18 Kepler's Laws

The heliocentric theory received further support from the work of Johannes Kepler (1571–1630), who made a detailed study of the motions of the planets. Kepler had at his disposal the careful and accurate astronomical observations made by his teacher, Tycho Brahe (TEE·koh BRAH·hee). In analyzing these observations Kepler discovered the following three laws which apply to the motions of the planets:

1. *Every planet moves about the sun in an elliptical orbit having the sun at its focus. For most of the planets, the elliptical orbits are almost circular.* (See Fig. 6-13).

2. *If a line is drawn from the sun to a planet, it will pass over equal areas in equal intervals of time.* (See Fig. 6-14).

3. *The cube of the average orbital radius R of each planet, divided by the square of its period T, is a constant,* That is,

$$\frac{R^3}{T^2} = K$$

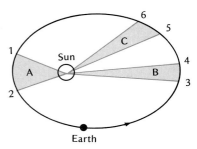

Fig. 6-14. Because the orbit of the earth is slightly elliptical, the time taken by the earth to go from point 1 to 2 is the same as the time taken to go from 3 to 4 and from 5 to 6. According to Kepler's second law, areas A, B, and C are equal.

The constant K turns out to be 3.35×10^{18} m³/s².

Although Kepler discovered these remarkable regularities in the motions of the planets, he made no effort to explain them. That task remained for Isaac Newton, whose work will be discussed in the next chapters.

As will be shown by the following problem, Kepler's third law may be applied to *artificial planets* launched around the sun as well as to the natural planets.

Sample Problem

What will be the period of a satellite moving around the sun between the orbits of the Earth and Mars at a mean distance of 2.00×10^{11} m from the sun?

Solution:
$R = 2.00 \times 10^{11}$ m $K = 3.35 \times 10^{18}$ m³/s²

From Kepler's law, $\frac{R^3}{T^2} = K$. Whence,

$$T^2 = \frac{R^3}{K}$$

$$T^2 = \frac{(2.00 \times 10^{11} \text{ m})^3}{3.35 \times 10^{18} \text{ m}^3/\text{s}^2}$$

$$T^2 = \frac{8.00 \times 10^{33}}{3.35 \times 10^{18}} \text{ s}^2$$

$$T = 4.89 \times 10^7 \text{ s}$$

Test Yourself Problem

What will be the period of a satellite moving around the sun at a mean distance of 8.00×10^{11} m?

Simple Harmonic Motion

6-19 Vibrations and Oscillations

Many motions take the form of *vibrations* or *oscillations* (ah·sil·AY·shuns). Although these motions take place in a straight line, they are closely related to uniform circular motion, and we shall therefore study them here. In these motions, a body or the parts of a body move back and forth over and over again. A common example is the motion of the bob of a pendulum. The parts of a violin string also move back and forth regularly when the string is plucked or bowed. Similarly, the parts of a suspension bridge undergo small vibrations when cars and trucks pass over it and sometimes even when the wind blows upon it.

In describing the vibrational motion of a body, we generally refer to its amplitude, its period, and its frequency. The *amplitude* is the greatest distance that a vibrating body goes from its middle or equilibrium position. For example, the amplitude of a pendulum is the distance between the center or lowest position of its bob and the highest position at either side of center. The *period* is the time taken by the vibrating body to make one complete vibration. In the case of the pendulum, the period is the time taken by the bob to go from one end of its swing to the opposite end and then to return to its starting point.

The *frequency* is the number of vibrations made by a body per unit of time. The frequency is equal to 1 divided by the period. A pendulum that has a period of ½ second vibrates 1 divided by ½ or 2 times a second. Its frequency is therefore 2 vibrations per second or in symbols, 2 s^{-1}. The unit of frequency in the International Metric System is the *hertz* (HURTS), symbol Hz. *A hertz is one vibration or one cycle per second.* A body that vibrates 100 times a second has a frequency of 100 hertz (100 Hz).

6-20 Circular Motion and Simple Harmonic Motion

In simple harmonic motion, the acceleration is proportional to the displacement of a body from the center of vibration and directed toward that center.

A particularly important kind of vibrational motion is that in which the acceleration is proportional to the displacement of the body from its equilibrium position and always directed toward that position. This is called *simple harmonic motion* and is related to circular motion at constant speed in a very simple way.

In Fig. 6-15, a body A is moving around the circle at constant velocity v. For this body, the centripetal acceleration, $a = v^2/r$, is constant in magnitude but always changing direction so that it remains directed toward the center of the circle. Now let a second body B move on the diameter PQ in such a way that B is always at the foot of the perpendicular dropped from A. The figure shows the successive positions of B at equal time intervals as A moves around the upper half of the circle from P to Q and then around the lower half of the circle from Q to P. Note that each time A goes around the circle, B moves forward and back on PQ, and so makes one complete vibration.

The motion of B is the horizontal component of the motion of A, and the acceleration of B at any moment is therefore the horizontal component of the acceleration of A at that moment. In Fig. 6-15, the acceleration vector of A is drawn to scale along the radius, and its horizontal component $\vec{a}_x$ and its vertical component $\vec{a}_y$ are obtained by means of the small vector right triangle I. This triangle is similar to right triangle OAB, and $a/OA = a_x/OB$, where a and a_x are the magnitudes of $\vec{a}$ and $\vec{a}_x$.

But OA is the radius of the circle r and OB is the displacement x of B from the middle point of the oscillation. Hence, $a_x/x = a/r$, or $a_x = (a/r)x$. Since a and r are constants, the acceleration a_x is at each moment proportional to the displacement x and directed toward point O. By definition, therefore, B is in simple harmonic motion.

Note that B's acceleration is zero when x is zero and B is at point O. Also, B's acceleration is a maximum when B is at P or Q. For these positions $x = r$, and $a_x = (a/r)r$, or $a_x = a$. Thus, the accelerations of B and A are equal when B is at either end of its vibration.

6-21 Period of Simple Harmonic Motion

From the above description of the motions of A and B, it is clear that the period of a body B, in simple harmonic motion, is the same as the period of the body A, whose circular motion corresponds to it. In Section 6-14 we have seen that for uniform circular motion $a = 4\pi^2 r/T^2$. Hence, the period of both A and B is given by the expression $T^2 = 4\pi^2 r/a$.

We shall now apply the above relationship to two examples of simple harmonic motion: the period of a mass hung from a spring, and the simple harmonic motion of a pendulum. In considering these, we should note that r is equal to the amplitude of B, and a is equal to the maximum value of B's acceleration.

6-22 Period of a Mass Hung from a Spring

Suppose a body hung from a vertical spring is pulled downward a small distance and then released. The action of the spring causes it to vibrate up and down around its equilibrium position. If the mass of the spring is negligibly small compared to the mass of the body, this vibration is a simple harmonic motion.

The magnitude of the acceleration of such a body at any given displacement x is known to be equal to $(k/m)x$ where k is a constant equal to the force needed to produce unit elongation of the spring and m is the mass of the body. Let r be the amplitude of the

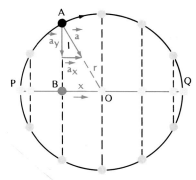

Fig. 6-15. As A moves around the circle at constant speed, B moves from P to Q and back again in simple harmonic motion.

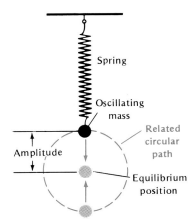

Fig. 6-16. Simple harmonic motion of a mass hung from a spring.

vibration and therefore the maximum value of x. Then, the maximum value of $a_x = a = (k/m)r$. Substituting this value of a in $T^2 = 4\pi^2 r/a$, we have:

$$T^2 = \frac{4\pi^2 r}{\frac{k}{m}r} = 4\pi^2 \frac{m}{k}$$

Whence,

$$T = 2\pi\sqrt{\frac{m}{k}}$$

Note that the period of this motion does not depend on the amplitude of the motion. It makes no difference whether the body vibrates over a larger distance or a smaller one. The period remains the same.

Sample Problem

A mass of 1.00 kg is hung from a spring whose constant k is 100 N/m. (This means it takes a force of 100 N to make the spring stretch 1 m, or a force of 1 N to make the spring stretch 1 cm.) What is the period of oscillation of this mass when it is allowed to vibrate?

Solution:

$$m = 1.00 \text{ kg} \qquad k = 100 \text{ N/m}$$

$$T = 2\pi\sqrt{\frac{m}{k}}$$

$$T = 2(3.14)\sqrt{\frac{1.00 \text{ kg}}{100 \text{ N/m}}}$$

$$T = 6.28\left(\frac{1}{10}\right) \text{ s} = 0.628 \text{ s}$$

Note that $\sqrt{\dfrac{\text{kg}}{\text{N/m}}} = \sqrt{\dfrac{\text{kg} \cdot \text{m}}{\text{N}}} = \text{s}$

To show this, refer to Section 7-7 where the newton is defined as $N = \dfrac{\text{kg} \cdot \text{m}}{\text{s}^2}$. Substitute this value of N in the equation above:

$$\sqrt{\frac{\text{kg} \cdot \text{m}}{\text{N}}} = \sqrt{\frac{\text{kg} \cdot \text{m}}{\frac{\text{kg} \cdot \text{m}}{\text{s}^2}}} = \sqrt{\text{s}^2} = \text{s}$$

Test Yourself Problem

A mass of 2.00 kg is hung from a spring having a constant k of 50 N/m. What is the period of vibration when the spring is stretched and then released?

6-23 Simple Harmonic Motion of a Pendulum

When a pendulum is swinging over a small arc, its bob is in simple harmonic motion. To show this, let us prove that the acceleration of the bob is proportional to its displacement from its center position and always acts toward that center.

Consider position A of the pendulum in Fig. 6-17. If the bob were free to fall, gravity would pull it downward with acceleration $\vec{g}$. Since the string compels it to move along arc AB, the acceleration moving it toward B is the component of $\vec{g}$ acting along the tangent to the arc AB at A. Resolving the vector $\vec{g}$ into two perpendicular components, one parallel to the string and one tangent to arc AB

at A, we get components $\vec{g}_l$ and $\vec{g}_t$ in right triangle I. This right triangle is similar to right triangle II since $\vec{g}_l$ is parallel to l and $\vec{g}$ is parallel to h, making the angles included between these sides equal. Hence, $g_t/x = g/l$ or $g_t = (g/l)x$.

If the arc ABC is small enough, it practically coincides with chord AC, and g_t may then be considered to be acting along x on chord AC. Thus, the acceleration of the bob along the straight segment AC is very closely equal to $g_t = (g/l)x$. Since g and l are constant, this shows that the acceleration of the bob is proportional to its displacement x from its center position and always acts toward that center. The bob is therefore in simple harmonic motion.

6-24 Period of a Pendulum

We can now determine the period of a pendulum from the relationship $T^2 = 4\pi^2 r/a$, where r is the amplitude of the motion and a is its maximum acceleration. Let x be the maximum displacement or amplitude of the pendulum. Then the maximum acceleration acting on it is $g_t = (g/l)x$. Substituting in,

$$T^2 = \frac{4\pi^2 r}{a}$$

we have,

$$T^2 = \frac{4\pi^2 x}{\left(\dfrac{g}{l}\right)x} = 4\pi^2 \frac{l}{g}$$

Whence,

$$T = 2\pi\sqrt{\frac{l}{g}}$$

Thus, the period of a pendulum depends only on its length. It does not depend on the mass of the bob or upon the amplitude of the swing, provided that the swing is small.

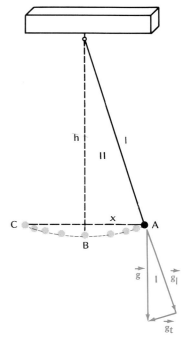

Fig. 6-17. For small swings, the bob of a pendulum moves with simple harmonic motion.

Sample Problem

What is the period of a pendulum 0.392 meter long?

Solution:

$l = 0.392$ m $g = 9.80$ m/s² $\pi = 3.14$

$$T = 2\pi\sqrt{\frac{l}{g}}$$

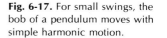

$$T = 6.28\sqrt{\frac{0.392 \text{ m}}{9.80 \text{ m/s}^2}} = 1.26 \text{ s}$$

Test Yourself Problems

1. What is the period of a pendulum that is 0.490 m long?

2. What is the length of a pendulum that has a period of 0.500 s?

CHAPTER REVIEW

Summary

When a body moves in a plane, its velocity and acceleration generally change in both magnitude and direction. The motion of such a body may be analyzed by resolving it into two component straight-line motions at right angles to each other. Each of these motions then may be analyzed separately by the methods presented in Chapter 5. **Projectile** and **circular motions** are readily analyzed in this way.

To move in a circle at constant speed, a body must have a **centripetal acceleration**. Such an acceleration is constant in magnitude, but is steadily changing in direction in such a way that it is always directed toward the center of the circle. The centripetal acceleration a of a body moving at constant speed v in a circle of radius r is given by:

$$a = \frac{v^2}{r}$$

As the planets move around the sun, they undergo centripetal accelerations, but these are not constant in magnitude. The planets therefore do not move around the sun in circular orbits. Instead, as **Kepler** discovered, the following three **laws** apply to the orbits of the planets. 1. *Every planetary orbit is an ellipse having the sun at its focus. 2. A line drawn from the sun to a planet passes over equal areas in equal intervals of time. 3. The cube of the average orbital radius R of each planet divided by the square of its period T is a constant K. That is: $R^3/T^2 = K$.*

Simple harmonic motion is related to circular motion at constant speed. In this motion, a body moves back and forth on a straight line in such a way that its acceleration is proportional to its displacement from its equilibrium position and is directed toward that position. Examples of simple harmonic motion are the vibrations of a body attached to a spring and the vibrations of the bob of a pendulum.

The period of a body of mass m attached to a spring of constant k is:

$$T = 2\pi \sqrt{\frac{m}{k}}$$

The period of a pendulum of length l is:

$$T = 2\pi \sqrt{\frac{l}{g}}$$

where g is the acceleration of gravity.

Questions

Group 1

1. A car's speedometer indicates a speed of 30 km/h. If the car is undergoing acceleration, what two changes may be occurring in its motion?
2. Describe an experiment that shows that the rate at which an object falls is not affected by any horizontal motion it may have.
3. When a projectile is fired horizontally, which of the components of its motion is (a) motion at constant speed; (b) uniformly accelerated motion? (c) If the initial velocity of the projectile is increased, what effect will this have on the time it takes the projectile to fall to the ground?
4. (a) Cite an example of a body that is undergoing centripetal acceleration. (b) Upon what two factors does the centripetal acceleration depend?

(c) In what direction does the centripetal acceleration act?

5. A train moves on a circular track at constant speed. (a) What is the direction of its acceleration? (b) As the train changes its position on the circle, what happens to the magnitude of the acceleration? (c) If the train doubles its speed, what happens to the magnitude of its acceleration?

6. (a) Compare the geocentric and the heliocentric views of the planetary motions. (b) Why has the heliocentric view been preferred?

7. (a) What is simple harmonic motion? (b) Describe two cases of simple harmonic motion. (c) Show by a diagram how simple harmonic motion is related to circular motion at constant speed.

8. (a) In the motion of a pendulum having a small amplitude, toward what point is the pendulum bob accelerated at all times? (b) At what points of its swing does the bob have its greatest acceleration? (c) At what point of its swing does the bob have its greatest velocity?

Group 2

9. A projectile, fired at an angle of 45° to the ground, takes 3 s to reach the top of its path. (a) How long does it travel before it hits the ground? (b) What acceleration does the projectile have? (c) What effect will increasing the angle at which the projectile is fired have on the horizontal component of the velocity of the projectile? (d) What effect will increasing the angle at which the projectile is fired have on its acceleration?

10. Two bodies are moving at the same speed in circular paths having different radii. Compare their accelerations when the radius of one of the paths is twice as great as the radius of the other.

11. Show that when the velocity and radius are expressed in SI units, the centripetal acceleration $a = v^2/r$ comes out in units m/s².

12. A satellite is sent into a circular orbit around the sun and its period is measured. How can the radius of the orbit now be determined?

13. (a) According to the relationship giving the period of a pendulum, what effect does increasing the length of a pendulum from 0.1 m to 0.4 m have on its period? (b) What effect does using a bob of greater mass for a given length of pendulum have on its period? (c) What effect does increasing the amplitude of vibration have on its period? Assume that the new amplitude is still very small compared to the length of the pendulum.

14. (a) Referring to the simple harmonic motion of a body suspended from a spring, state what effect increasing the mass of the suspended body has on the period of the motion. (b) What effect does using a tighter spring (one having a larger constant k) have on the period of the motion?

15. In pulling out of a dive, the pilot of an airplane may momentarily experience an acceleration of 2 or 3 g. (a) What is responsible for this acceleration? (b) What determines its magnitude?

Problems

Group 1

1. An auto traveling east at a constant speed of 20 m/s makes a 90° turn to the north and continues at the same speed. From a scale drawing, obtain the magnitude and direction of the change in the car's velocity.

2. An airplane moving north at 150 m/s turns 30° to the east of north and increases its speed to 200 m/s. (a) From a scale drawing, obtain the magnitude and direction of the change in velocity of the airplane. (b) If the change took place in 20.0 s, what was the average acceleration of the airplane during that time?

3. The change in the velocity of an automobile during a 5.0-s period is 20 m/s in the due easterly direction. What is the magnitude and direction of its average acceleration?

4. A bullet fired horizontally from the top of a building has a muzzle velocity of 500 m/s. A similar bullet dropped from the top of the same building takes 4.00 s to reach the ground. How far forward does the first bullet go before it hits the ground?

5. A ball is thrown horizontally from the roof of a building at a speed of 20 m/s and hits the ground 3.0 s later. (a) Make a table showing how far the ball falls during each second of its fall. (b) In a parallel column, show how far the ball moved forward during each second of its motion. (c) Plot the path of the ball.

6. A package in an airplane moving horizontally at 150 m/s is dropped when the altitude is 490 m. (a) How long does it take the package to fall to the ground? (b) How far forward from the spot over which it was dropped does the package land? (c) What kind of path does the package follow?

7. A boy standing on top of a hill throws a stone horizontally. The stone hits the ground at the foot of the hill 2.5 s later. How high is the hill?

8. The muzzle velocity of a projectile fired from a gun has an upward component of 49.0 m/s and a horizontal component of 60.0 m/s. (a) Make a table showing the vertical and horizontal displacements of the projectile at the end of each second of its flight and plot its path. (b) How long does it take for gravity to reduce the upward component of the projectile's velocity to zero? (c) How far upward does the projectile go? (d) How far forward does it go?

9. A projectile is shot upward at a 60° angle with the ground and a speed of 200 m/s. (a) Obtain the vertical and horizontal components of its velocity. (b) How far has the projectile gone horizontally at the end of the first 4.00 s?

10. A car moves around a circular section of a road having a radius of 50 m at a constant speed of 20 m/s. What is the centripetal acceleration of the car?

11. A stone attached to a string 2.0 m long is whirled in a horizontal circle. At what speed must the stone move for its centripetal acceleration to be equal to the acceleration of gravity g?

12. A toy electric train moving at constant speed on a circular track that has a radius of 1.0 m goes around the track every 10 s. What is the centripetal acceleration of the train?

13. An airplane flying at constant speed in a circular course of radius 5000 m is observed to complete each round trip in 400 s. (a) What is the speed of the airplane? (b) What is its centripetal acceleration?

14. What is the period of a pendulum that is 0.098 m long?

15. How long should a pendulum be to have a period of exactly one second?

Group 2

16. A bullet is fired horizontally from a height of 78.4 m and hits the ground 2400 m away. With what velocity does the bullet leave the gun?

17. A projectile remains in the air for 6.00 s after it is fired. The horizontal component of its velocity is 100 m/s. (a) How far did it move forward before it hit the ground? (b) How long after firing did it reach the highest point of its path? (c) What was the altitude of the highest point in its path?

18. A projectile fired from a ship comes down 10.0 s after it was fired and hits the water at a distance of 6000 m from the ship. (a) What was the horizontal component of its initial velocity? (b) At what time did it reach the highest point of its path? (c) What was its altitude at that time? (d) What was the vertical component of its initial velocity?

19. A bullet is fired from a gun at a 30.0° angle with the horizontal and a muzzle velocity of 600 m/s. What is the range of the gun for this angle of elevation?

20. An electron moves in a circular path of radius 0.10 m at a constant speed of 2.0×10^6 m/s. (a) What is the period of its motion? (b) What is its centripetal acceleration?

21. The blade of a fan is 0.20 m long and makes 20 revolutions per second. (a) What is the period of the motion of the blade? (b) What is the centripetal acceleration of a particle located at the end of the blade?

22. A satellite will remain in a circular orbit near the surface of the earth when it travels at such a speed that its centripetal acceleration is equal to the acceleration of gravity. Assuming that the radius of the satellite's orbit is equal to 6.37×10^6 m, the radius of the earth, compute the speed at which the satellite must travel to remain in orbit.

23. A satellite is launched into an orbit of radius 3.00×10^{12} m around the sun. From Kepler's third law, predict the period of the satellite's motion.

24. A 4.00-kg mass hung from a spring whose constant k is 100 N/m is pulled downward a short distance and then released. (a) What is the period of its vibration? (b) If the 4.00-kg mass were replaced by a mass of 1.00 kg, what change would take place in the period of vibration of the spring and the attached mass?

25. If we measure l and T for a pendulum in a given place, we can compute the value of g at that place. What is the value of g in a place where a pendulum 0.990 m long has a period of 2.00 s?

Applying Physics

1. Find the speed with which water comes out of a jet, such as the nozzle of a garden hose. With the nozzle held horizontally, measure its height above the ground y and the horizontal distance x the water travels before hitting the ground. (See Fig. 6-18.) The time t that the water remains in the air is the same as the time it takes the water to fall the distance y. It therefore can be found from $y = \frac{1}{2}gt^2$. Now, the distance that the water moves forward in this time is $x = v_x t$, where v_x is the speed of the jet. From the known values of x and t, compute v_x.

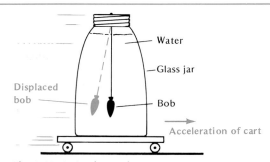

Fig. 6-19. A simple accelerometer.

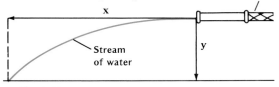

Fig. 6-18.

2. Determine the value of g in your neighborhood by means of a pendulum. Make a pendulum about a meter long and set it oscillating over a small arc. Measure its length l and its period T. Noting that $T = 2\pi\sqrt{l/g}$ can be put into the form $g = 4\pi^2 l/T^2$, substitute the values of l and

T and compute g. Repeat this experiment for pendulums of different lengths and compare the values of g obtained.

3. A simple device that can detect the acceleration of a moving body and indicate its magnitude may be made by hanging a heavy pendulum bob from the lid of a large water-filled glass jar as shown in Fig. 6-19. When this device is on an accelerating body such as a moving car, the pendulum will be displaced from the vertical in a direction opposite to that of the acceleration. The greater the acceleration, the greater will be the displacement of the bob.

Black Holes

Scientists have detected places in the universe where stars have shrunk to such small dimensions that they have extraordinarily great densities and consequently overwhelming gravitational force. In such places gravity is so strong that it allows nothing to escape—neither matter nor light or any other form of energy. Since we observe objects by the light they emit or reflect, these shrunken stars have become truly invisible. Appropriately, they are called *black holes.*

A black hole is formed when a star that has a mass more than three times that of our sun exhausts the fuel it uses to generate light and heat. The star no longer has sufficient internal pressure to support its surface layers. It collapses until even the very atoms of its gases have broken down into their constituent subatomic particles. Under intense gravitational force, these particles become tightly packed. In less than a hundredth of a second, the star fades to invisibility to become a black hole.

So how have scientists been able to "see" black holes? They search for a region in space where it appears matter is being pulled forcefully from the surface of a visible star by the gravitational force of an unseen object—a black hole. Matter spiralling into a black hole is accelerated, compressed, and heated to such high temperatures that high-energy X rays are emitted. By studying the strength of these X rays and the motion of the visible star, scientists can estimate the mass and size of the black hole.

One of the first X-ray sources to be discovered, Cygnus X-1, appears to be a binary star system in which a giant visible star is being tugged on by an unseen object, a black hole. In 1989 astrophysicists discovered what may be another black hole in the Swan constellation, V404 Cygni.

Astrophysicists are also searching for black holes at the centers of galaxies. Scientists hypothesize that at some time between 10 000 and 100 000 years ago, an exploding star, called a *supernova*, may have disturbed a cloud of interstellar gas that was orbiting near the center of our galaxy, the Milky Way. The resulting black hole is believed to have a mass several million times that of our sun. There is evidence that a thin stream of gas is spiralling into this black hole.

In this drawing an artist has portrayed a stream of gas from a giant blue star being pulled into the black hole in Cygnus X-1.

Laws of Motion

7

Aims

1. To learn Newton's three laws of motion, which describe how forces control motion.
2. To understand the limitations of Newton's laws and the role of Einstein's theory of relativity in dealing with them.
3. To understand the nature of friction and its role in opposing the motion of bodies.
4. To understand and apply the concept of momentum and the law of conservation of momentum.

7-1 Work of Galileo and Newton

The science of describing motion is called *kinematics* (kin· ih·MAT·iks). It was largely developed by Galileo (1564–1642). Galileo did not investigate the causes of motion. That task was left for his great successor, Sir Isaac Newton (1642–1727).

Newton asked himself what makes bodies move, stop moving, or change their motion. He concluded that forces are the agents that control the motion of bodies. The study of the manner in which forces do this is called *dynamics* (dy·NAM·iks). Newton's system of dynamics is embodied in his three laws which have an amazing range of usefulness and success. They apply to the motions of astronomical bodies such as planets, as well as to the motions of earthly objects such as cars and airplanes.

Newton was born in the year in which Galileo died.

First Law of Motion

7-2 Law of Inertia

Although this law is generally credited to Newton, it was first recognized by Galileo. It states:

> When no net or resultant force acts upon it, a body at rest remains at rest and a body in motion continues to move in the same direction in a straight line with constant speed.

According to this law, a body naturally tends to continue in whatever state of motion or rest it is at any instant and tends to resist any changes. *The extent to which a body resists changes in its state of rest or motion is characteristic of the body and is called its inertia.* We frequently experience the effects of inertia in trips in the family car. When the car starts suddenly, we are thrown

Inertia causes buildings to collapse during an earthquake because the buildings tend to stay at rest while the earth under them moves.

117

backward against the seats. Since we were at rest, we tend to remain at rest even when the car surges forward. When the car is moving and slows down or stops suddenly, we find ourselves thrown forward. Our inertia tends to keep us moving forward at the same speed, even though the car is slowing down. Again, when a car makes a sharp right turn, we are thrown to the left.

7-3 Thought Experiment Leading to the First Law

The condition in which no net force acts upon a body can often be approximated but never attained. Every body we observe has a net force acting on it. Even a body at rest on the earth's surface is acted upon by the unbalanced gravitational force of the sun, which keeps it and the earth orbiting around the sun.

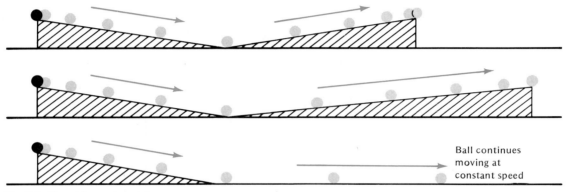

Fig. 7-1. Thought experiment that suggested the law of inertia.

How then can we find out how a body would move when there is no net force acting on it? Here, physicists imagine a situation in which the unattainable conditions exist and experiment with it in thought. Galileo used this approach in arriving at the principle of inertia.

First Galileo observed that a ball allowed to roll down one hill and up a second hill reaches practically the same horizontal level from which it started. (See Fig. 7-1). He attributed the small difference in the starting and ending levels of the ball to frictional resistance. He assumed that, in the absence of friction, the ball would roll up the second hill to exactly the same horizontal level from which it started.

Next Galileo imagined what would happen when this experiment is repeated while the second hill is made less and less steep. After rolling down the first hill, the ball would lose speed more slowly as it moved up the less steep second hill. Furthermore, it would travel a longer distance uphill before it reached the level from which it started. It follows that, when the second hill finally coincides with the horizontal, the ball should continue to move over it indefinitely without loss of speed.

Thus, Galileo inferred that an object started moving on a frictionless horizontal surface would continue to move in a straight line without loss of speed.

7-4 Friction and Inertia

When a smooth block is set sliding on a smooth horizontal floor, we note that it moves in a straight line but gradually slows down and comes to a stop. Why does it not obey the law of inertia and continue at constant speed? The answer is that there is an unbalanced force acting upon it. This is the force of *friction* exerted upon it by the floor. *Friction may be defined as the force that resists a body when it moves over or through another body.* Thus a projectile, once set moving, is slowed down by friction as it moves through the air. The block in our example is slowed down by friction as it moves over the floor.

Friction explains why a car, once set moving, will not continue to move at constant speed on a horizontal road without further help from the engine. The frictional resistance of the road and the air to its forward motion and of its axles to the motion of its wheels over them gradually brings it to a halt once the engine is turned off. It follows that, to keep the car moving at constant speed, the engine must supply just enough force to overcome the retarding force of friction. With the force of friction neutralized in this manner, there are no unbalanced forces acting on the car and it obeys the law of inertia.

Second Law of Motion

7-5 Law of Force and Acceleration

The fact that, in the absence of a net force acting upon it, a body simply continues its state of rest or uniform motion suggests that the presence of a force acting upon a body will change its state of rest or motion. Forces are thus agents that change the motion of a body. They can do this by causing a body at rest to move, by speeding up or slowing down a body already in motion, or by changing the direction of a moving body. In short, *a force accelerates* the body on which it acts. Newton examined the relationship between a force and the acceleration it produces in a body and stated it as follows:

> When an unbalanced force acts upon a body, it accelerates that body in the direction of the force. The acceleration produced is directly proportional to the force.

It follows that, if a force $\vec{f}_1$ produces an acceleration $\vec{a}_1$ when applied to a body, and a force $\vec{f}_2$ applied to the same body produces an acceleration of $\vec{a}_2$,

$$\frac{f_1}{a_1} = \frac{f_2}{a_2} = \text{constant}$$

In this relationship, f_1, f_2, a_1, and a_2 are the magnitudes of the vectors $\vec{f}_1$, $\vec{f}_2$, $\vec{a}_1$, and $\vec{a}_2$ respectively.

Fig. 7-2. This multiple-flash photograph of a bowler delivering a ball was taken at intervals of 1/50 second. The distances between successive pictures of the ball show that it was accelerated only while it was in the bowler's hand.

Thus, if a 1-newton force gives a body an acceleration of 5 meters per second per second, a 2-newton force will give the same body twice as much acceleration, or $2 \times 5 = 10$ meters per second per second; a 3-newton force will give the same body three times as much acceleration, or $3 \times 5 = 15$ meters per second per second, and so on. This means that for each body the ratio between the force and the acceleration it produces is a constant. Note in the above example that the ratio of force to acceleration is $1/5$ when the 1-newton force is acting, $2/10 = 1/5$ when the 2-newton force is acting, and $3/15 = 1/15$ when the 3-newton force is acting. No matter what force acts on this particular body, the ratio of the force to the acceleration it produces is always the constant $1/5$.

7-6 Effect of Mass

When we apply forces to other bodies, we again find that, for each body, the ratio of the force f to the acceleration a it produces is a constant that is usually different for different bodies. When we list these f/a constants for many bodies, and compare them with the masses of those bodies as measured with an equal-arm balance, we find that they are proportional to the masses. They may be set equal to the masses by selecting appropriate units for f and a.

Thus, the law of acceleration may be restated as follows in the form in which it commonly appears.

> The ratio of the force acting upon a body to the acceleration it produces is equal to the mass of the body. In symbols:

$$\frac{f}{a} = m \quad \text{or} \quad f = ma$$

where f = force, m = mass, and a = acceleration.

It follows that the mass of a body is a measure of its inertia or its resistance to changes in its motion. The larger the mass of a body, the smaller is the acceleration or change in its motion that a given force will be able to produce. If a force gives a mass of 1 kilogram an acceleration of 10 meters per second per second, that same force applied to a 2-kilogram mass will give it an acceleration only half as great, or 5 meters per second per second.

The second law of motion explains why a much larger force is needed to accelerate a train than is needed to give a car with a much smaller mass the same acceleration. Similarly, to decelerate a moving train and bring it to a stop requires a much larger force than is needed to decelerate a car moving at the same rate.

The larger the mass of a body, the smaller is the acceleration imparted to it by a given force.

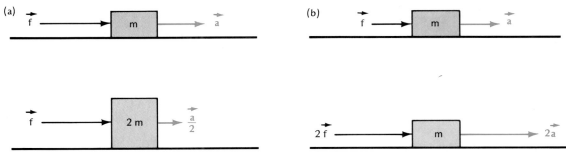

Fig. 7-3. (a) The larger the mass of a body, the smaller the acceleration imparted to it by a given force. (b) The acceleration imparted to a given mass increases directly with the force producing it.

7-7 Definition of the Newton

Using the relationship, $f = ma$, we can now define the newton precisely. Imagine a 1-kilogram mass resting on a perfectly frictionless horizontal surface. The force that, when applied to this 1-kilogram mass, gives it an acceleration of 1 meter per second per second is a newton. Substituting in $f = ma$, it follows that *1 newton equals 1 kilogram times 1 meter per second per second.* Expressed in symbols:

$$1 \text{ N} = 1 \text{ kg} \cdot \text{m/s}^2$$

Note that the newton is a combination of mass units (kg), length units (m), and time units (s).

The relationship, $f = ma$, enables us to predict the motion of a body once we know its mass and the force acting on it. Before we illustrate this by some examples, let us stress the importance of using the proper combination of units. In using $f = ma$, forces are expressed in newtons, masses in kilograms, and accelerations in meters per second per second.

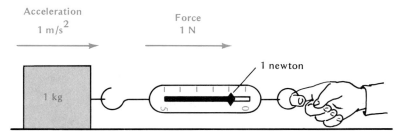

Fig. 7-4. Calibrating a spring scale in newtons.

Sample Problems

1. What acceleration does the application of an 8-N force give to a 2-kg mass?

Solution:

$$f = 8 \text{ N} \qquad m = 2 \text{ kg}$$

$f = ma$ whence $a = \dfrac{f}{m}$

$$a = \frac{8 \text{ N}}{2 \text{ kg}} = 4 \frac{\text{N}}{\text{kg}} = 4 \text{ m/s}^2$$

Note that N/kg = m/s² since, by definition, 1 N = 1 kg·m/s².

2. What force is needed to accelerate an electron (mass = 9.1×10^{-31} kg) from rest to a speed of 3.0×10^8 m/s in 10 s?

Solution:

We know the mass of the electron but not its acceleration. However, since it starts from rest and is uniformly accelerated, its acceleration is equal to its final speed v divided by the time of travel t.

$$v = 3.0 \times 10^8 \text{ m/s} \qquad t = 10 \text{ s}$$

$$a = \frac{v}{t}$$

$$a = \frac{3.0 \times 10^8 \text{ m/s}}{10 \text{ s}} = 3.0 \times 10^7 \text{ m/s}^2$$

We substitute this value of a and $m = 9.1 \times 10^{-31}$ kg in the formula $f = ma$

$$f = (9.1 \times 10^{-31} \text{ kg})(3.0 \times 10^7 \text{m/s}^2)$$

$$f = 2.7 \times 10^{-23} \text{ N}$$

Test Yourself Problems

1. What force must act on a 50.0-kg mass to give it an acceleration of 0.30 m/s²?

2. A 1500-kg car starting from rest attains a speed of 25.0 m/s in 50.0 s. (*a*) What is its acceleration? (*b*) What force is acting upon it?

7-8 Acceleration and Freely Falling Bodies

Near the surface of the earth, the pull of gravity on a body is practically constant and every falling body acquires a constant acceleration equal to g. The force that accelerates the body downward is its weight w. It follows that if m is the mass of the body, $f = ma$ becomes:

$$w = mg$$

This relationship enables us to compute the weight of a body from its mass or the mass of a body from its weight.

For example, assuming $g = 9.8$ meters per second per second in a given place, a 1.0-kilogram mass weighs $w = mg = 1.0 \times 9.8$

= 9.8 newtons, a 2.0-kilogram mass weighs 2.0 × 9.8 = 19.6 newtons, and so forth. It follows that to support a 1.0-kilogram mass, we must exert an upward force on it equal to its weight, or 9.8 newtons. To accelerate it upward, as we do with rockets, we must exert an upward force greater than 9.8 newtons. The difference between the upward force we exert and the weight of the body is the net force that produces the upward acceleration. Let us show how this applies to a rocket.

Sample Problem

A 5.0-kg rocket is acted on by an upward force of 59 N supplied by its engine. (*a*) What is the net upward force acting on it? (*b*) What is its acceleration?

Solution:

(*a*) $m = 5.0$ kg $g = 9.8$ m/s^2
$w = mg = 5.0$ kg × 9.8 m/s$^2 = 49$ N

The net upward force acting on the rocket is 59 N − 49 N = 10 N.

(*b*) $f = 10$ N $m = 5.0$ kg

$$a = \frac{f}{m} = \frac{10 \text{ N}}{5.0 \text{ kg}} = 2.0 \text{ m/s}^2$$

Test Yourself Problem

A rocket having a mass of 100 kg is pushed upward by its engine with a force of 1470 N. (*a*) What is the net upward force acting upon it? (*b*) What acceleration does this force produce?

7-9 Two Kinds of Mass

We have shown that the mass of a body can be obtained by two quite different methods. In Chapter 3, we obtained the mass of a body by using the equal-arm balance. There, we used the fact that the weight of a body is proportional to its mass, and we said that two bodies were of equal mass when the weight of one balanced the weight of the other.

In this chapter, we measured the mass of a body by using the relationship $f/a = m$. We took the mass to be measured, applied a known force to it, and measured the acceleration it produced. We then divided the known force by the acceleration to get the mass.

In the first case, we measured the mass of a body by measuring the pull that gravity has on it. To identify the mass obtained in this way, let us call it the *gravitational mass* of the body. In the second case, we measured the mass of a body by measuring the acceleration that a known force gives to it. Since here the mass tells how much inertia the body has, or how much it tends to resist change in its motion, it is called the *inertial mass* of the body.

We have already noted that if appropriate units are selected, the gravitational and inertial masses of a body always come out equal to each other. Nevertheless, because the two kinds of masses are measured by such different methods, Newton believed that they are different entities. Einstein disagreed. He asserted that the equality of gravitational and inertial mass is not a coincidence but an

Inertial mass equals gravitational mass.

important clue to the nature of matter itself. This clue was one of the important starting points for the development of Einstein's theory of motion, known as the *general theory of relativity.*

We shall not distinguish further between the two kinds of mass because their values are always the same. However, we should note the important part they played in Einstein's revision of our ideas of matter and motion.

Relativity

7-10 Failure of the Second Law at High Velocities

For more than 200 years, the applications of the law of acceleration met with almost unbroken successes in explaining the motions of heavenly bodies as well as of everyday objects and vehicles. However, at the beginning of the twentieth century it became evident that the law fails when applied to bodies moving at velocities close to that of light. To explain this, Albert Einstein developed the *special theory of relativity,* a theory of motion that applies to all bodies regardless of their velocities.

Einstein's theory is based on the fact that all observers, whether they are moving or at rest with respect to a source of light, obtain the same value of 3×10^8 meters per second when they measure the velocity of light from that source. This is illustrated in Fig. 7-5, where light from a star is traveling toward two observers A and B.

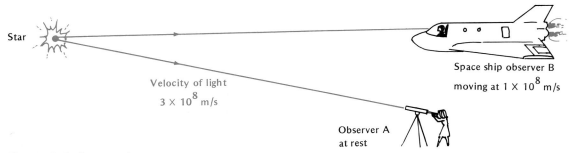

Star

Velocity of light
3×10^8 m/s

Space ship observer B
moving at 1×10^8 m/s

Observer A
at rest

Fig. 7-5. Both observers obtain the same value for the velocity of light.

Observer A is at rest while observer B is in a space ship speeding toward the star at the enormous velocity of 1×10^8 meters per second. Both observers measure the velocity of the light coming from the star. A finds it to be 3×10^8 meters per second. B expects to find it to be the relative velocity obtained by adding his own velocity to that of the light, or $1 \times 10^8 + 3 \times 10^8 = 4 \times 10^8$ meters per second. However, B finds that despite his own motion, he obtains the same value for the velocity of the light coming from the star as does A.

This illustrates that the combination of the velocity of the observer and the velocity of the light he is measuring always comes out equal to the constant value of the velocity of light. It suggests that no observer or other body can travel faster than the velocity of light, and thus contradicts the second law, $f = ma$. According to this law, we can give a body any velocity we wish if we apply a large enough force and use it to accelerate the body for a long enough time. Now it appears that no force, no matter how great and how long applied, can increase the velocity of a body beyond the velocity of light.

7-11 Dependence of Mass on Velocity

Einstein showed that one of the reasons for the failure of the law of acceleration for bodies moving at high velocities is that the mass of a body does not remain constant as Newton believed it did. Instead the mass m of a body increases with increasing velocity in accordance with the relationship:

A mass increases with velocity, becoming infinite at the velocity of light.

$$m = m_o \left(\frac{1}{\sqrt{1 - \dfrac{v^2}{c^2}}} \right)$$

where m_o is the mass of the body at rest, called its rest mass, and c is the velocity of light. Note that when the body is at rest, $v = 0$ and $m = m_o(1) =$ rest mass m_o.

In general, the increase in mass is negligibly small when the velocity of a body is small compared to that of light. Even a rocket which attains a velocity as high as 10 kilometers per second undergoes an increase in mass of only about one part in a billion. That is why changes in mass are not noticed in bodies traveling at ordinary velocities. It also explains why Newton's laws work so well at these low velocities.

However, in the atomic and subatomic world, electrons, protons, and other particles often acquire velocities comparable to that of

Table 7.1		
v as a percent of c	$\left(\sqrt{1 - \dfrac{v^2}{c^2}}\right)$	$\left(\dfrac{1}{\sqrt{1 - \dfrac{v^2}{c^2}}}\right)$
10%	0.995	1.005
50%	0.87	1.15
90%	0.43	2.3
99%	0.14	7.1
99.9%	0.045	22.3

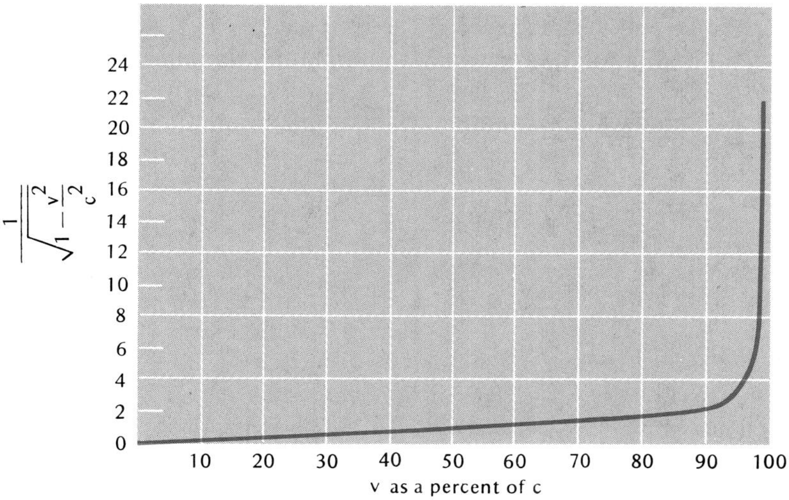

Fig. 7-6. Graph showing how the mass of a body varies with its velocity. The increase in mass becomes noticeable only as the body's velocity approaches the velocity of light.

light. Table 7.1 and Fig. 7-6 indicate what happens to the masses of such bodies as their velocities approach that of light. The table gives the value of $1/\sqrt{1 - v^2/c^2}$ for a series of increasing values of v expressed as a percentage of the velocity of light. Note that the factor $1/\sqrt{1 - v^2/c^2}$ indicates how many times m is larger than m_o. It is seen that as v increases from 10% to 99.9% of the speed of light, m increases from 1.005 m_o to 22.3 m_o. As v continues to increase, m increases more and more rapidly until, when v comes infinitesimally close to the velocity of light, m becomes infinitely large.

The velocity of a body cannot exceed the velocity of light.

This fact explains why a body cannot be accelerated to a velocity beyond that of light. According to the relationship, $f = ma$, as m becomes infinitely large at velocities very close to the velocity of light, an infinitely large force is needed to accelerate m further. Since there are no infinite forces, a body cannot be accelerated to a velocity greater than that of light.

7-12 Length and Time Also Not Constant

Two other sources of error in Newton's formulation of the law of acceleration arose from his assumption that a given length and a given time interval are constant quantities. Einstein showed that just as mass measurements are influenced by the velocity of the object being observed or by the velocity of the observer, so, too, are measurements of length and time. Here too, the effect of the relative velocity of the observer with respect to the body or the events he is observing becomes significant only when it is near that of light.

7-13 Dependence of Length on Velocity

Imagine that you are observing a spaceship moving away from the earth at a velocity close to that of light. In the ship is a meterstick placed on a table in a direction parallel to that in which the spaceship is moving. According to Einstein, you will notice that the stick is shorter than it was when at rest at the beginning of the trip. Moreover, the greater the relative velocity of the spaceship with respect to you, the shorter the meterstick appears to be.

What is true of the meterstick is also true of the spaceship and of all other bodies in it. They all appear to you to become shorter in the direction of their relative motion with respect to you in accordance with the relationship:

$$l = l_o \sqrt{1 - \frac{v^2}{c^2}}$$

Fig. 7-7. In accelerators such as this one, subatomic particles are accelerated to velocities approaching that of light.

Here l is the length of a body moving at velocity v with respect to the observer, l_o is the length of the body when at rest, and c is the velocity of light. As expected, when the body is at rest, $v = 0$ and $l = l_o$.

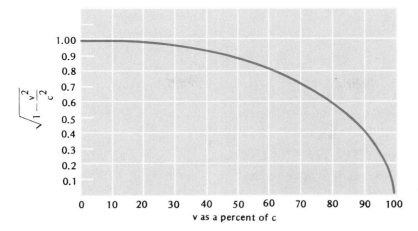

Fig. 7-8. Graph showing how time and the length of a body vary with velocity, as given in column 2 of Table 7-1.

As long as v remains much smaller than c, $\sqrt{1 - v^2/c^2}$ is practically equal to 1, and l is practically equal to l_o. However, as v increases steadily and approaches c, the value of $\sqrt{1 - v^2/c^2}$ decreases and approaches zero, and so does l. (See Table 7.1, second column, and Fig. 7-8.) At this point, the length of the body would appear to the observer to be shrinking to nothing in the direction of its motion.

The change in length of a moving body is seen only by an observer with respect to whom it is moving. The occupant of the spaceship notices no change in the length of bodies in the ship because they are at rest with respect to that observer. Furthermore, the change in length of a moving body takes place only in the direction of its relative motion with respect to the observer. Thus, if the occupant of the spaceship picked up the meterstick and held it perpendicular to the direction of the motion, the earthly observer would see the stick regain the same length it had on earth. At the same time, a proportionate decrease in the thickness of the stick would be seen because it is now parallel to the direction of the motion.

A body's length in the direction of its motion decreases as its velocity increases.

7-14 Dependence of Time on Velocity

Let us now see what happens to the measurement of time as we observe events in a rapidly moving system such as the spaceship just discussed. Suppose that before the spaceship departs, we synchronize two clocks, put one on the spaceship, and keep the other on earth. As the spaceship zooms into space at a speed approaching that of light, we observe its clock. We find that it has slowed down and is no longer synchronized with our clock on earth. When our clock shows that an hour has passed by, the clock in the spaceship shows that less than one hour has passed by there. The faster the spaceship travels, the more its clock slows down.

The manner in which time is thus observed to slow down in a

Time in a moving system slows up as the velocity increases.

rapidly moving system by an outside observer is expressed by the relationship:

$$t = t_0 \sqrt{1 - \frac{v^2}{c^2}}$$

A person in a system moving at the velocity of light would appear not to age to an observer on the earth.

Here t_0 is the interval of time measured by the stationary clock of the outside observer, t is the corresponding interval of time observed from outside to have elapsed on a clock in a system moving with velocity v relative to the outside observer, and c is the velocity of light.

Note that the relationship between the time intervals t and t_0 is similar to that between the lengths, l and l_0. As before, as long as v is much smaller than c, $\sqrt{1 - v^2/c^2}$ is practically equal to one, and t is practically equal to t_0. However, as v increases and approaches the speed of light, t decreases more and more rapidly and approaches zero. Thus, the outside observer notices that time in the moving system is passing more and more slowly.

All events in the moving system, whether clock movements, heartbeats, or the growth of plants, appear to the outside observer to slow down like a slow motion picture film. As v comes infinitesimally close to c the observer sees time in the moving system coming to a complete halt. All objects in the system will then be coming to rest and no further change will be able to take place in the system.

7-15 Einstein's Theory Versus Newton's

In the formulation of his special theory of relativity, Einstein took the dependence of mass, length, and time measurements on velocity fully into account. His theory is more general than that of Newton because it applies to all velocities. This does not mean that we should discard Newton's work. It happens that for ordinary velocities, Einstein's theory reduces to Newton's law of acceleration so that we may continue to use this powerful law without error in the ordinary range of velocities where it has been so successful.

Third Law of Motion

7-16 Law of Action and Reaction

Since forces are the agents whereby we can control the motion of bodies, it is important to know the conditions under which we may apply a force to a body. The third law of motion tells us this. It is called the *law of action and reaction*. It states:

> Every action or force is accompanied by an equal and opposite reaction or force.

Forces can only act on bodies that resist them.

According to this law, whenever one body exerts a force on

another body, the second body exerts an equal and opposite force on the first. Thus, when you sit in a chair, your weight exerts a downward force or action on the chair. In order to support you, the chair must exert an equal and opposite force on you.

When you walk forward in the classroom, the floor is actually pushing you in that direction. With each step that you take, one foot pushes backward against the floor. The floor reacts by exerting an equal and opposite forward force on your foot. It is the floor's reaction that actually pushes you forward. This becomes evident when the floor is made slippery by means of a coat of wax or grease. Now the condition of the floor prevents you from pushing it backward with your foot. The floor, in turn, cannot react and supply the force needed to push you forward.

7-17 Action and Reaction on Different Bodies

The third law tells us that forces always occur in pairs. However, *both members of the pair do not act upon the same body.* If they did, they would cancel each other and have no resultant effect. To clarify this important point, let us consider a few examples.

Attach the hooks of two spring scales together and have two students pull them apart until one of the scales reads 20 newtons. (See Fig. 7-10.) Note that the other scale also reads 20 newtons. The first student is pulling the second with a force of 20 newtons. Call this the action. The second student reacts by pulling the first with a force of 20 newtons. Thus, one of the 20-newton forces acts on one student and the other acts on the second. Let one of the students increase the pull to 40 newtons. Note that the second one automatically increases the pull to 40 newtons, as shown by the scale.

Fig. 7-9. Reaction to the gases expelled rapidly from the rocket engines drives the Saturn rocket upward.

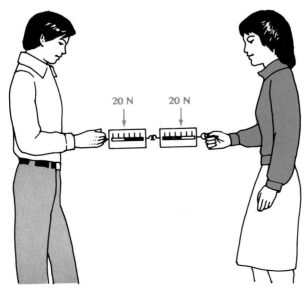

20 N   20 N

Fig. 7-10. Illustrating Newton's third law: action and reaction are equal.

When a bullet is fired from a gun, the bullet is shot forward. The gun recoils in the opposite direction. The gun acts upon the bullet and pushes it forward. The bullet reacts upon the gun with equal force and pushes the gun backward.

In jet and rocket engines, a fuel is burned inside a combustion chamber, forming a large quantity of hot gases. A high pressure thus builds up in the combustion chamber and drives the hot gases rearward out of a jet. The engine therefore exerts a rearward force on the gases that expels them from the jet. The hot gases react by exerting an equal and opposite forward force on the jet engine.

Friction and Motion

7-18 Static Friction

Newton's laws tell us how forces affect motion. A major force that opposes motion is friction. We shall now examine the nature of this force and how it may be controlled.

When you try to slide a heavy piece of furniture across the floor of a room, the force you exert is opposed by the force of friction between the legs of the furniture and the floor. Unless the force you exert exceeds a certain minimum value, the furniture will not move. The effect of friction up to this point is to keep the furniture stationary. This type of friction is therefore called *static friction.*

Static friction tends to keep an object stationary until the force exerted on the object exceeds a certain minimum value.

Static friction may be measured by a spring scale, as shown in Fig. 7-11, where a block resting on a table is being pulled by the force f. There are four forces acting on the block:

(1) the downward force w exerted by the weight of the block;

(2) the upward force f_n, which is called the normal force and is equal in magnitude to the weight of the block but opposite to it in direction. The normal force is the table's reaction to the weight of the block and is the force that presses the table and block together;

(3) the force f that pulls the block to the right; and

(4) the force of friction f_f that opposes f by pulling the block to the left.

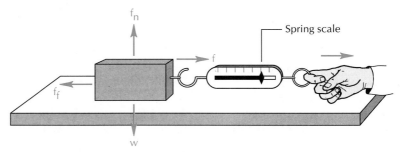

Fig. 7-11. Setup used to measure the force of friction.

Imagine that the person is not pulling the spring scale, so that it reads "zero," indicating that $f = 0$. Then, the force of friction, f_f, must also be zero, since the block remains at rest. Now suppose the person pulls to the right with a steadily increasing force. The spring scale shows that f is getting larger and larger. At the same time, the force of friction must become larger and larger. In fact, as long as the block does not move, the forces f and f_f must be equal and opposite to each other. Thus, static friction enables the block to remain at rest because it neutralizes the forces that tend to move the block.

Fig. 7-12. Static friction enables a car to be parked on a hill without rolling down.

7-19 Starting Friction and Sliding Friction

Returning to Fig. 7-11, suppose the person continues to pull harder. The force of friction will continue to increase but will still be equal and opposite to the force exerted by the person, which can be read from the spring scale. Eventually, however, the force of friction will reach a maximum value and will stop increasing. As the force exerted by the person exceeds this value, the block will begin to move. The reading on the scale when the block just begins to move is the magnitude of the maximum frictional force, which is called the *starting friction,* $f_{f(\text{start})}$.

Once the block has begun to slide, the frictional force drops sharply. As a result, the force needed to keep the block sliding at a constant rate is less than the force needed to start it sliding. The reading on the scale now shows the magnitude of the frictional force that acts as the block slides at a constant rate. This force, which is equal and opposite to the force being exerted by the person, is called the *sliding friction,* $f_{f(\text{slide})}$. Thus, the setup shown in Fig. 7-11 enables us to measure both the starting friction and the sliding friction.

Sliding friction works in your favor when you wish to stop or slow down a moving object. It is the sliding friction in the brake system of a car that enables the driver to stop the car or slow it down. When sliding friction is low, as it is on an icy road, people slip and fall and cars skid because they slide too easily. Sprinkling sand on such a road increases the friction and makes it safer for walking or riding.

7-20 Factors That Affect Friction

Friction is a complicated subject about which our knowledge is incomplete. What we know about it comes mainly from our experiments with friction in the laboratory and our experiences in everyday life. As a result, the quantitative relationships used to describe friction, though useful, are generally approximate rather than precise. You should keep this in mind as you now begin to study these relationships.

Let us repeat the experiment shown in Fig. 7-11 but this time we will turn the block on its end as shown in Fig. 7-13. How will the starting and sliding friction for the block in this position compare with their values when the block is on its side in Fig. 7-11? Although the side of the block has a larger area than its end, experiment shows that the starting and sliding friction in both cases are the same. That starting and sliding friction do not depend on the areas of the surfaces in contact is understandable if it is noted that the surfaces in contact actually touch only at a limited number of points that project from each surface. Increasing the area of contact apparently makes little difference in the number of projecting points that actually touch. The nature of the contact between the surfaces remains essentially the same.

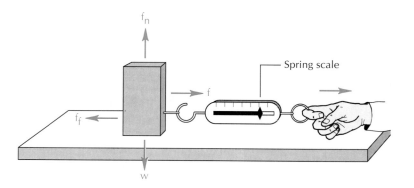

Fig. 7-13. Starting and sliding friction do not depend on the areas in contact.

Starting and sliding friction do not depend on the areas of the surfaces in contact, but both these frictional forces are proportional to the weight of the object being pushed or pulled.

If the area of contact has no effect on the frictional forces, what about the weight of the block and its reaction force, the normal force that presses the block and table together? To find out, suppose we put a small weight on the block in Fig. 7-13 and determine the forces of starting and sliding friction. We find that both are greater for the weighted block than for the block alone. If we put a heavier weight on the block and repeat the experiment, we find that both frictional forces have increased still further. It turns out that both frictional forces are proportional to the total weight pressing the block against the table.

Friction also depends on the roughness of the surfaces. You must pull harder to slide a block across a rough surface such as sandpaper than across a smooth surface such as glass. However, when the surfaces in contact are polished to an extreme degree of smoothness, the frictional forces become greater again. To explain this increase, it is theorized that when the distance between two surfaces is small enough, the attractive forces between molecules come into play. These attractive forces cause the frictional forces between extremely smooth surfaces to increase.

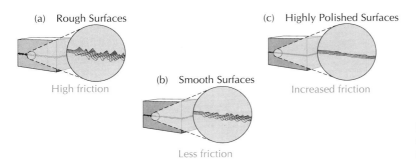

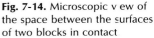

Fig. 7-14. Microscopic view of the space between the surfaces of two blocks in contact

7-21 Coefficient of Friction

We have seen that the forces of both starting and sliding friction are proportional to the weight pressing the block against the table. This weight is equal and opposite to the normal force f_n. We may therefore say that the forces of friction are proportional to the normal force, or that $f_f = \mu f_n$, where μ (the Greek letter "mu"), the proportionality constant, is the *coefficient of friction* between the two surfaces. It follows that:

$$\mu = \frac{f_f}{f_n}$$

This relationship applies to both starting and sliding friction:

$$\mu_{start} = \frac{f_{f(start)}}{f_n} \qquad \mu_{slide} = \frac{f_{f(slide)}}{f_n}$$

Since $f_{f(start)}$ is usually greater than $f_{f(slide)}$ for a particular pair of surfaces, μ_{start} is generally larger than μ_{slide} for those surfaces. For example, for a steel block on a steel surface, the coefficient of starting friction is 0.74, while the coefficient of sliding friction is 0.57. It follows that it takes more force to start a steel block moving across a steel floor than to keep it sliding at a constant rate. The coefficients of friction for different substances are shown in Table 7.2.

Table 7.2
COEFFICIENTS OF FRICTION

	μ_{start}	μ_{slide}
steel on steel	0.74	0.57
copper on steel	0.61	0.47
glass on glass	0.94	0.40
wood on wood	0.3–0.5	0.20
teflon on teflon	0.04	0.04
metal on metal (lubricated)	0.15	0.06

Sample Problem

It takes a force of 35 N to start a wooden block weighing 50 N sliding across a table top, while it takes only 20 N to keep it sliding. Find the coefficient of (a) starting friction and (b) sliding friction.

Solution:

(a) Keeping in the mind that the magnitude of the normal force f_n is equal to the weight of the block, $w = f_n = 50$ N, and that $f_{f(\text{start})} = 35$ N, we have:

$$\mu_{\text{start}} = \frac{f_{f(\text{start})}}{f_n}$$

$$\mu_{\text{start}} = \frac{35 \text{ N}}{50 \text{ N}} = 0.70$$

(b) Noting that $f_n = 50$ N and $f_{f(\text{slide})} = 20$ N, we have

$$\mu_{\text{slide}} = \frac{f_{f(\text{slide})}}{f_n}$$

$$\mu_{\text{slide}} = \frac{20 \text{ N}}{50 \text{ N}} = 0.40$$

Test Yourself Problems

1. The coefficient of sliding friction for a particular sample of steel on steel is 0.65. What force is needed to keep a steel box weighing 2500 N sliding across a steel floor?

2. It takes 200 N to start a 350-N crate sliding across a floor. Find the coefficient of starting friction.

7-22 Rolling Friction

When an object rolls over another, it encounters rolling friction. Rolling friction is usually much smaller than sliding friction. This means that it is much easier to move a heavy crate by rolling it on wheels than it is to slide the crate along the floor. Ball bearings provide rolling friction. The rolling friction between the ball bearings and each surface is much smaller than the sliding friction would be between the same two surfaces.

It is not always desirable to reduce rolling friction. In order for a car to move forward, there must be enough rolling friction between the tires and the road to give the tires traction. If the rolling friction on a road is too low (as on an icy road), there is too little traction. Then the tires will slip too readily.

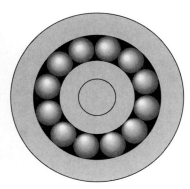

Fig. 7-15. Ball bearings have rolling friction, which is usually much smaller than sliding friction.

7-23 Fluid Friction

Objects moving through fluids (liquids and gases) encounter retarding forces called *fluid friction*. Open the freezer door of a refrigerator and watch the cold, moist air *pour* out and down to the floor. Fluid friction is exerted on moving cars and airplanes by the surrounding air. Fluid friction is exerted on boats as they move through water. As the speed of a body moving through a fluid increases, fluid friction increases. An object dropped in the air encounters greater and greater fluid friction as its speed increases. If an object is dropped from a high enough altitude, the force of fluid friction will eventually become equal to the weight of the object. From that point on the object falls with a constant speed called its *terminal velocity.*

Fluid friction is greater for objects with greater surface area. When an object attached to an opened parachute falls through the air, the large surface area of the parachute provides a greater force of fluid friction at any particular speed. This greater force enables the object to reach terminal velocity at a slower speed.

Shape is another factor that affects fluid friction. To reduce fluid friction, cars, planes, and boats are built to have streamlined shapes.

The frictional resistance offered by liquids and gases also depends on a property called *viscosity.* Viscosity is related to the rate at which a liquid or gas can flow. Highly viscous fluids like molasses flow much more slowly than less viscous fluids like water and air. A fluid with high viscosity offers much more frictional resistance to objects moving in it than does a fluid with low viscosity.

Momentum and Its Conservation

7-24 Momentum

You have probably noticed that a train is harder to stop than a car moving at the same speed. We state this fact by saying that the train has more momentum than the car. *By momentum we mean the product of the mass of an object and its velocity.* That is,

momentum = mass × velocity

Since velocity is a vector quantity, *momentum is also a vector quantity and has the same direction as the velocity.* From the definition, we can see that a moving body can have a large momentum if its mass is large, if its velocity is large, or if both its mass and velocity are large. The train has much more momentum than the car moving at the same speed because its mass is much greater than that of the car. On the other hand, an object with a small mass, such as a bullet, may also be given a large momentum if its speed is made very great. The high momentum of a rifle bullet makes it hard to stop and enables it to penetrate iron, wood, and other solid objects.

Momentum is a measure of how hard it is to stop a moving body.

7-25 Momentum Form of the Second Law

In Section 5-9 we noted that when a body is moving with constant acceleration in a straight line, $a = (v - v_o)/t$, where a is the acceleration, v_o is the initial velocity, and v is the velocity at time t. Substituting this value for a in $f = ma$, we have:

$$f = \frac{m(v - v_o)}{t}$$

or,

$$ft = mv - mv_o$$

Since mv is the momentum of the body at time t and mv_o is the initial momentum of the body, the quantity $mv - mv_o$ is therefore the change in the momentum. Thus, the second law of motion can be restated as follows:

> The change in momentum produced by a force acting upon a body is in the same direction as the force and is proportional to the force and to the time that it acts.

The product, $f \times t$, is generally referred to as the *impulse*.

7-26 Application of the Law of Momentum

We apply the law of momentum in the "follow through" that is used in many sports, such as baseball, tennis, and football. In baseball, for example, the batter follows through when hitting the ball. In this way the force of the bat is applied to the ball for a longer time, thus giving the ball a greater impulse. The result is a greater final momentum and a higher speed for the struck ball. The tennis player uses the same principle in getting over a fast serve, while the kicker in football applies it to get off a high punt.

The long-range gun is generally equipped with a long barrel to take full advantage of the time factor in the law of momentum. When the gunpowder is burned in the gun, it produces the hot expanding gases of combustion that exert a force on the projectile and begin to push it out of the gun. The longer the barrel of the gun, the longer the expanding gases of combustion will push the projectile before it leaves the gun and the greater the momentum given the projectile.

Sample Problems

1. How long does it take a net upward force of 100 N acting on a 50-kg rocket to increase its speed from 100 m/s to 150 m/s?

Solution:

$f = 100$ N $m = 50$ kg
$v = 150$ m/s $v_o = 100$ m/s

$$ft = mv - mv_o = m(v - v_o)$$
$$100 \text{ N} \times t = 50 \text{ kg}(150 \text{ m/s} - 100 \text{ m/s})$$
$$t = \frac{50 \text{ kg}(150 \text{ m/s} - 100 \text{ m/s})}{100 \text{ N}}$$

Since 1 N = 1 kg · m/s², we have

$$t = \frac{2500 \text{ kg} \cdot \text{m/s}}{100 \text{ kg} \cdot \text{m/s}^2} = 25 \text{ s}$$

2. What braking force is needed to bring a car having a mass of 1000 kg and moving at 30.0 m/s to a stop 60.0 s after the brake is applied?

   Solution:

   $t = 60.0$ s $m = 1000$ kg
   $v = 0.00$ m/s $v_o = 30.0$ m/s

$$ft = m(v - v_o)$$

$$f \times 60.0 \text{ s} = 1000 \text{ kg}(0.00 - 30.0) \text{ m/s}$$

$$f = \frac{1000 \text{ kg} \times (-30.0 \text{ m/s})}{60.0 \text{ s}} = -500 \text{ N}$$

The negative sign shows that the force is retarding the car instead of speeding it up.

Test Yourself Problems

1. How long does it take a 250-N force to increase the speed of a 100-kg rocket fired horizontally from 10.0 m/s to 200 m/s?

2. The speed of a 1200-kg car increases steadily from 5.00 m/s to 29.0 m/s in 12.0 s. What force is accelerating the car?

7-27 Law of Conservation of Momentum

An important result of the second and third laws of motion is the *law of conservation of momentum.*

Suppose that two bodies that are moving toward each other collide. During the collision each body exerts a force upon the other and, by the law of action and reaction, these forces are equal and opposite to each other. According to the law of momentum, each of these forces changes the momentum of the body on which it acts by an amount proportional to the force and to the time that it acts. Since the two forces are equal and opposite and act for the same time, they produce equal but opposite changes in the momentum of the two bodies. This means that whatever momentum one body loses during the collision is gained by the second body. We have here an illustration of the law of conservation of momentum which states:

> In a system consisting of bodies on which no outside forces are acting, the total momentum of the system remains the same.

Any momentum that is lost by one body in the system is gained by one or more of the other bodies in the system.

As another illustration of this law, consider a loaded gun. Before it is fired, the gun and the bullet in it have zero velocity and therefore zero momentum. Since no forces outside of the gun are acting to accelerate either the gun or the bullet, it follows that, after firing, the total momentum of the gun and the bullet must remain zero. This means that the forward momentum given to the bullet must be equal and opposite to the backward momentum given to the gun.

Fig. 7-16. The decrease in momentum of the heavy iron ball swung from the crane is equal to the increase in momentum of all the pieces of masonry dislodged by the ball.

Sample Problem

A bullet of mass 0.050 kg leaves the muzzle of a gun of mass 4.0 kg with a velocity of 400 m/s. What is the velocity of recoil of the gun?

Solution:
The momentum of the bullet is 0.050 kg × 400

m/s = 20 kg·m/s. The momentum of the gun is 4.0 kg × v, where v is the velocity of recoil of the gun. As explained above, these two momenta must be equal and opposite in direction. Hence,

$$4.0 \text{ kg} \times v = 20 \text{ kg} \cdot \text{m/s}$$
$$v = 5.0 \text{ m/s}$$

Test Yourself Problem

A toy railroad engine having a mass of 1.0 kg and a speed of 2.0 m/s collides with a similar engine at rest. On colliding the two engines lock and remain together. (a) What is the momentum of each engine before the collision? (b) What is the magnitude and direction of the momentum of both engines after the collision? (c) What is the velocity of the pair of engines after the collision?

7-28 General Form of the Momentum Conservation Law

In its most general form, this law takes into account the fact that momentum is a vector quantity having direction as well as magnitude. It tells us that *the vector sum of the momenta of all the bodies in a system remains unchanged in both magnitude and direction* (no matter what collisions or other changes take place in the system) *as long as no forces outside the system are permitted to act upon it.*

To illustrate, imagine a red ball of mass 1.0 kilogram moving on a very smooth floor with a velocity of 4.0 meters per second eastward. Let this ball strike a stationary black target ball of unknown mass and after the collision let the directions of the red ball and the black target ball be those shown in Fig. 7-17. Our task is to find the momentum of each ball after the collision.

First we obtain the total momentum of the red ball-black ball system before the collision. Since the target ball is at rest, the total momentum of the system before the collision is simply the momentum of the red ball. This is $m\bar{v}$ = 1.0 kilogram × 4.0 meters per second = 4.0 kilogram-meters per second east.

By the law of conservation of momentum, the total momentum of this system must remain the same. Therefore, after the collision the vector combination of the new momentum of the red ball and the momentum of the black target ball must give a resultant equal to 4.0 kilogram-meters per second east.

Making a vector diagram, we resolve the known total momentum into two components, one along each of the directions taken by the red ball and the target black ball after the collision. This is done in Fig. 7-17 to a scale of 1.0 centimeter = 1.0 kilogram-meter per second. The vector $\bar{a}$ represents the total momentum of the system, or 4.0 kilogram-meters per second. This has been resolved into two components, $\vec{b}$ in the direction of the target black ball and $\vec{c}$ in the direction taken by the red ball. By measurement, we find the magnitude of $\vec{b}$ is 3 centimeters, representing a momentum of 3.0 kilogram-meters per second in the direction of $\vec{b}$. The magnitude of $\vec{c}$ is 2 centimeters, representing a momentum of 2.0 kilogram-meters per second in the direction of $\vec{c}$. Thus, we have determined the momentum of each ball after the collision.

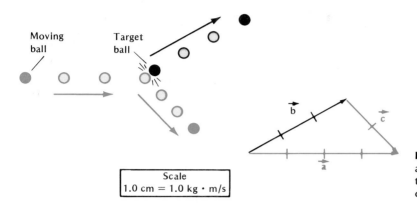

Moving ball

Target ball

Scale
1.0 cm = 1.0 kg · m/s

Fig. 7-17. The total momentum after the collision is equal to the total momentum before the collision.

7-29 Determining Mass or Velocity

Since we know the mass of the red ball is 1.0 kg and we have just determined that its momentum after the collision is 2.0 kilogram-meters per second, we can find its velocity after the collision by dividing its momentum $m\vec{v}$ by its mass m. This gives $\vec{v} = 2.0/1.0 = 2.0$ meters per second in the direction of $\vec{c}$.

In the case of the target black ball, we can also find its velocity if we are given its mass. On the other hand, if we are not given the mass but can measure its velocity, we can find its mass by dividing its momentum by its velocity. Thus, suppose we find by measurement that the target black ball moves at 1.5 meters per second after the collision. Dividing its 3.0-kilogram-meters per second momentum by its 1.5-meters per second velocity, we get $m\vec{v}/\vec{v} = m = 3.0/1.5 = 2.0$ kilograms.

7-30 Conservation of Momentum in Nuclear Physics

The colliding balls in the preceding example illustrate a method used to study the collisions of the particles of nuclear physics. By means of Wilson cloud chambers, bubble chambers, or photographic plates, pictures are made of collisions between tiny particles such as electrons, protons, neutrons, and other atomic and subatomic particles. The law of conservation of momentum then enables us to interpret these collisions and obtain much information from them.

Fig. 7-18. Paths of alpha particles in a cloud chamber containing hydrogen gas. The forked track at upper left was made when an alpha particle collided with a hydrogen atom of the gas.

In Fig. 7-18, the straight lines are the paths in a cloud chamber made by alpha particles that have been fired from left to right. An alpha particle is the nucleus of an atom of helium. Alpha and other charged particles make such paths on passing through a cloud chamber containing moist air in the same way as an airplane flying at high altitude makes a vapor trail. Note that one alpha particle has struck an unknown body and has rebounded upward while the unknown body has moved off in a downward direction.

In this case we know the mass of the alpha particle and its velocity, hence we can find its momentum. This is the total momentum before the collision, which must be equal to the vector sum of the momentum of the alpha particle and the momentum of the unknown body after the collision. By making a vector diagram exactly as in the case of the colliding balls, we determine the momentum of the unknown body as well as the momentum of the alpha particle after the collision. Since the velocity of the unknown body can be determined by other means, we can compute its mass and thus identify it. It turns out to be an atom of hydrogen.

7-31 Importance of the Law of Conservation of Momentum

This law is one of the most powerful in all physics. Its importance stems from the fact that it applies regardless of what kinds of collisions or other interactions take place among the bodies of a system. Thus, it makes no difference whether the bodies collide and then rebound or whether they collide and stick together. The total momentum of the system will remain the same.

Newton discovered the relationships between forces and motion and expressed them in the three laws of motion. The first law, called the **law of inertia,** states that, when no resultant force acts upon a body, a body at rest remains at rest and a body in motion continues to move in the same direction in a straight line with constant speed.

The second law, called the **law of acceleration,** states that, when an unbalanced force acts upon a body, it accelerates that body in the direction of the force, and the acceleration produced is directly proportional to the force. This relationship is expressed as $f = ma$ where f is the force, m is the mass of the body being acted upon, and a is the acceleration produced.

The third law, called the **law of action and reaction,** states that every action or force is accompanied by an equal and opposite reaction or force. The action is the force exerted by the first body on the second. The reaction is the force exerted by the second body on the first.

Friction is the force that resists the motion of one body over or through another. Friction retards motion and can be reduced by making the rubbing surfaces smoother, by lubricating them, by substituting rolling friction for sliding friction, and by streamlining bodies that move through fluids. The coefficient of friction, μ, between two rubbing surfaces is the ratio of the force of friction f_f to the normal force f_n pressing the two surfaces together. That is,

$$\mu = \frac{f_f}{f_n}$$

The **momentum** of a body is the product of its mass and its velocity. It is a vector quantity having the same direction as the velocity. The second law of motion may be restated as the **law of momentum.** In this form it states: the change in momentum produced by a force acting upon a body is in the same direction as the force and is proportional to the force and to the time that it acts. This relationship is expressed as $ft = mv - mv_o$, where f is the force, t is the time, mv_o is the initial momentum of the body, and mv is the final momentum of the body.

Momentum obeys the very important **law of conservation of momentum.** It states that in any system consisting of bodies on which no outside forces are acting, the total momentum of the system remains the same.

Newton's laws of motion apply only to bodies whose velocities are small compared to the velocity of light. For such bodies, measurements of mass, length, and time are not significantly affected by their motion, and remain practically constant. As the velocity of a body approaches that of light, this is no longer true. Measurements of length, mass, and time become increasingly dependent on the velocity of the body, and change markedly as it changes. Newton's laws no longer apply, and Einstein's *special theory of relativity* must be applied instead.

CHAPTER REVIEW

Summary

Questions

Group 1

1. How does the law of inertia explain each of the following: (a) Passengers standing in a train are thrown backward when the train suddenly moves forward. (b) They are thrown forward when the train suddenly stops. (c) They are thrown to one side when the train rounds a curve.

2. A highly polished block of wood is given a push that sets it sliding across a long polished floor. (a) What path does it take? (b) Why does it gradually slow down? (c) What change would take place in the motion of the block of wood if the friction between it and the floor were increased?

3. (a) Why is it not possible to test the law of inertia by an actual experiment? (b) How did Galileo come upon the law by a thought experiment?

4. A certain force applied to a body gives it an acceleration of 10 m/s² in the easterly direction. (a) What is the direction of the force? (b) If the force is doubled without changing its direction, what is the new acceleration of the body? (c) If the direction of the force in (b) is now changed so that it acts northward, what is the magnitude and direction of the acceleration it produces?

5. Using the law of acceleration, define the newton.

6. A truck has ten times the mass of a car. Both start from rest and are uniformly accelerated for the same interval of time. At the end of the time both truck and car are moving at 40 km/h. (a) Compare the acceleration of the truck with that of the car. (b) Compare the force that accelerates the truck with the force that accelerates the car.

7. A force applied to a rocket gives it an acceleration equal to 2 g. (Note that g stands for the acceleration due to gravity.) The same force applied to a second rocket gives it an acceleration equal to 6 g. Compare the masses of the two rockets.

8. When a certain braking force is applied to a car moving at a given speed, the car is brought to rest in 10 s. (a) If the braking force is doubled, what happens to the negative acceleration given to the car? (b) How long does it now take the car to come to a halt?

9. Describe how to measure (a) the gravitational mass of a body; (b) the inertial mass of a body.

(c) What is the relationship between these two masses?

10. (a) Under what conditions should the special theory of relativity be used instead of the law of acceleration? (b) Describe the changes that occur in the mass of a body as its velocity nears that of light.

11. How does the law of action and reaction explain each of the following: (a) the recoil of a gun when it fires a bullet; (b) the inability of a car to start moving on a flat, ice-covered road?

12. Two teams playing tug-of-war are momentarily in equilibrium. The first team pulls the rope with 250 N of force. (a) What is the force exerted by the second team? (b) If it takes 260 N of force to break the rope, will it break under these conditions? Explain.

13. To slide a heavy bureau across a room, a person first removes all the drawers. Why is it easier to move the bureau once the drawers have been removed?

14. Describe a situation in which high static friction is an advantage.

15. (a) List three ways in which the sliding friction between two surfaces can be reduced. (b) List two ways in which sliding friction can be increased.

16. As the normal force pressing two surfaces together increases, what happens to the coefficient of friction between those surfaces?

17. (a) How is the momentum of a body obtained? (b) How is the momentum of a body affected by an increase in its speed? (c) How is it affected by a change in the direction of motion of the body?

18. A body undergoes a certain change of momentum when a force acts on it for 1 s. (a) If the force is doubled, how long will it have to act upon the body to produce an equal change in momentum? (b) If half the original force is applied for 4 s, how will the change of momentum produced compare with that produced in (a)?

19. (a) State the law of conservation of momentum. (b) While a small laboratory cart is moving over a table at constant velocity, a 10-kg mass is gently deposited on it. Assuming that the mass had zero momentum to begin with, what hap-

pens to the velocity of the cart? (c) How does the initial momentum of the cart compare with the final momentum of the cart and the mass together?

20. A shell fired upward explodes and bursts into two unequal pieces at the moment it comes to rest at the highest point of its path. (a) Prove that the two pieces of the shell must go off in exactly opposite directions. (b) Prove that if one of the pieces has twice the mass of the other, it will have only half the velocity of the other.

Group 2

21. An electron is being accelerated by a force acting on it. If the acceleration is to remain constant, what change, if any, must be made in the applied force when the velocity of the electron is (a) small compared to the velocity of light; (b) nearing the velocity of light? (c) Why can't the velocity of the electron ever become greater than the velocity of light?
22. What characteristic of the force of gravity explains why all bodies, regardless of mass, fall with the same acceleration in the same locality?
23. A 500-N girl standing on a bathroom scale jumps upward. Why does the scale read more than 500 N at the moment she jumps?

24. A 750-N fellow stands on a bathroom scale in an elevator. What will the scale read while the elevator is (a) descending at constant speed; (b) descending with half the acceleration of gravity?
25. In a collision, a car is brought to a stop in 2 s. In a second collision, a similar car moving at the same speed is brought to a stop in 4 s. Compare the forces which act upon these two cars during the collisions.
26. Explain why it is dangerous for a person to jump from a 5-m height to a concrete sidewalk while it is quite safe to jump from the same height into a pile of hay.
27. Two skaters are facing each other a little distance apart on an ice pond. The first skater throws a heavy medicine ball that is caught by the second one. Assuming the ice to be frictionless, describe the motion of each a moment after the ball is caught.
28. Two toy trains are moving toward each other at the same speed on the same track. Upon colliding, the trains lock together. If one train has twice the mass of the other, what will be the ratio of the speed of the trains after the collision to their speed before the collision?

Problems

Group 1

1. What force is needed to give a mass of 25 kg an acceleration of 20 m/s²?
2. A force of 6 N acts upon a 2-kg mass. What is the acceleration of the mass?
3. It takes a force of 50 N to give a body an acceleration of 10 m/s². What is the mass of the body?
4. (a) What is the weight of a 10-kg piece of iron? (b) What is the mass of an object that weighs 490 N?
5. A car has a mass of 1600 kg. If a braking force of 500 N is applied to the car while it is in motion, what is the direction and magnitude of the acceleration it produces?
6. An 8.0-kg rocket fired horizontally encounters a force of air resistance of 4.9 N. The force supplied by the rocket's engine is 60.9 N. (a) What is the net force accelerating the rocket? (b) What is the acceleration imparted to the rocket?

7. An electron is accelerated uniformly from rest to a speed of 4.0×10^6 m/s in 2.0 s. (a) What is the acceleration of the electron? (b) If the mass of the electron is 9.1×10^{-31} kg, what is the force producing this acceleration?
8. Assuming that at sea level the value of g is 9.8 m/s², find the gravitational force exerted on a proton of mass 1.7×10^{-27} kg.
9. (a) An object having a mass of 8.0 kg slides down a smooth inclined board with an acceleration of 2.0 m/s². What force is acting upon it? (b) How does this force compare with the weight of the body?
10. A spring balance shows that it takes 5.0 N to accelerate a body across a smooth (nearly frictionless) flat surface at the rate of 0.40 m/s². (a) What is the inertial mass of this body? (b) If the gravitational mass of this body is now measured using an equal arm balance, what value may it be expected to have?

11. It takes a force of 300 N to keep a 500-N box sliding across a floor. Find the coefficient of sliding friction.

12. The coefficient of starting friction for steel on steel is 0.75. What force is needed to start a steel box weighing 3000 N moving across a steel floor?

13. A force acting on a 5.0-kg body increases its speed uniformly from 2.0 m/s to 8.0 m/s in 3.0 s. (a) What is the initial momentum of the body? (b) What is the final momentum of the body? (c) What is the force acting on it?

14. The driver of a 1.5×10^3-kg car traveling on a straight road at 10 m/s increases its speed to 30 m/s in 15 s. Find (a) the initial momentum; (b) the final momentum; (c) the force acting on the car.

15. A braking force is applied to a 300-kg motorcycle to reduce its speed from 20.0 m/s to 16.0 m/s in 10.0 s. Find (a) the change that has taken place in the momentum of the motorcycle; (b) the braking force.

16. How long must a 50-N force act on a 400-kg cart to raise its speed from 10 m/s to 12 m/s?

17. A shell having a mass of 25.0 kg is fired horizontally eastward from a cannon with a velocity of 500 m/s. If the mass of the cannon is 1000 kg, what is the magnitude and direction of the velocity of recoil?

18. A bomb having a mass of 8 kg explodes into two pieces that fly out horizontally in opposite directions. One piece is found to have a mass of 6 kg and the other a mass of 2 kg. What was the ratio of the speeds with which the two pieces moved apart immediately after the explosion occurred?

19. A toy railroad engine having a mass of 1.5 kg and moving on a straight track at 0.20 m/s collides with a similar engine ahead of it moving in the same direction at 0.12 m/s. On colliding, the engines lock and remain together. (a) What is the momentum of each engine before the collision? (b) What is the magnitude and direction of the momentum of both engines after the collision? (c) What is the velocity of the pair of engines after the collision?

20. When a car of mass 2.0×10^3 kg moving at 9.0 m/s collides head on with a second car having a mass of 1.5×10^3 kg, the cars lock and come to rest at the point of collision. (a) What was the

momentum of the second car before the collision? (b) What was the velocity of the second car before the collision?

Group 2

21. (a) What is the mass of a rocket weighing 196 N? (b) What force in excess of the weight of the rocket is needed to accelerate it upward at 4.0 m/s²? (c) What is the total upward force that must be exerted on the rocket to give it this acceleration?

22. (a) What force in excess of the weight of a 500-kg rocket is needed to accelerate it upward at a rate equal to that of gravity, or 9.80 m/s²? (b) What is the total force acting on the rocket in (a)?

23. A meterstick, a 1-kg block, and a clock are placed aboard a spaceship which blasts into space at 99% of the speed of light. (a) Find the length of the meterstick and mass of the block as seen from the earth. (b) When an hour has passed on earth, how much time would be seen to have passed on the spaceship clock? (See Table 7.1.)

24. A 1500-kg car starting from rest is accelerated by a force of 2000 N during the first 15 s of its motion. What speed will the car have at the end of that time?

25. What force must be exerted on a proton (mass $= 1.7 \times 10^{-27}$ kg) that is moving at 5.0×10^7 m/s to bring it to rest in a distance of 0.25 m?

26. A measurement of the momentum of a proton yields a value of 6.8×10^{-21} kg·m/s. From the fact that the mass of a proton is 1.7×10^{-27} kg, compute the speed of this proton.

27. A ball of mass 0.20 kg is pitched to the batter at 30 m/s and hit back by the batter at 20 m/s. If the ball and bat were in contact for 0.025 s, what force was exerted by the bat on the ball?

28. A spaceship is equipped with a rocket motor that can increase or decrease its speed by exerting a force on it of 1.5×10^4 N for 100 s. The spaceship has a mass of 3.0×10^3 kg and is moving at 2.0×10^3 m/s. If the rocket motor is used to decrease the velocity of the spaceship without changing its direction, what is (a) the momentum and (b) the velocity of the spaceship after the rocket motor has fired?

29. A gun fires a 0.50-kg projectile which acquires a muzzle velocity of 500 m/s. If the projectile

takes 5.0×10^{-3}s to travel the length of the gun barrel, what is the average force exerted by the gun on the projectile? (Assume the projectile undergoes uniform acceleration while in the gun.)

30. A 1-kg body moving forward at 5 m/s hits a second body at rest. After the collision, the 1-kg body reverses its direction and moves at 1 m/s while the second body moves forward at 2 m/s. (a) What is the direction and magnitude of the momentum of each of the bodies after the collision? (b) What is the mass of the second body?

31. A proton of mass 1.7×10^{-27} kg, moving forward at a speed of 4.0×10^6 m/s, hits a stationary unknown particle. After the collision, the proton reverses its direction and moves at a speed of 2.0×10^6 m/s while the unknown particle moves forward at 1.0×10^6 m/s. What is the mass of the unknown particle?

32. Body A of mass 4.0 kg and speed 1.0 m/s collides with a similar stationary body B. After the collision, A moves in a direction 30° to the left of its original direction while B moves in a direction 60° to the right of A's original direction. (a) From a vector diagram drawn to scale, determine the momentum of each of the bodies after the collision. (b) What are the speeds of A and B after the collision?

Applying Physics

1. This experiment gives an insight into the law of action and reaction. Hang a book from a rubber band and measure the length of the stretched rubber band. This length is a measure of the tension in the rubber band. Repeat this procedure with the same book and a second, similar rubber band. Now, tie the second rubber band to the bottom of the first to form a line of the two rubber bands. Fasten the first rubber band to a support and hang the book from the bottom of the second rubber band. With both rubber bands supporting the weight of the book, measure the length of each rubber band. How does each length compare with the length of that rubber band when it alone was supporting the book? How does the law of action and reaction explain your observations?

2. To observe what changes take place in your weight when you are accelerated, stand on a bathroom scale and note its reading. Now, let your body be accelerated downward by doing a single rapid knee bend. What happens to the reading of the scale while you are descending? Now, from the squat position, come up quickly to the erect position. What happens to the scale reading at the moment you accelerate your body upward? Explain the changes in the scale reading in terms of Newton's second law of acceleration.

3. Follow the procedure illustrated in Fig. 7-17 to make a vector triangle of the momenta of the particles in the collision shown in Fig. 7-18. Note that the alpha particle has the thicker track and that it has four times the mass of the hydrogen atom. From this, determine the relative speeds of the alpha particle before and after the collision and the relative speed of the hydrogen atom.

Law of Gravitation and Planetary Motion

1. To see how the application of Newton's laws of motion led to the law of gravitation.
2. To understand that the law of gravitation applies to all matter in the universe.
3. To apply the law of gravitation in explaining the motion of planets, satellites, and falling bodies.

8-1 Newton's Law Applied to Astronomy

One of the great triumphs of the law of acceleration was its success in explaining precisely the motion of the planets in their orbits around the sun. This required knowing what forces were acting on the planets and thus controlling their motion. Newton provided that information when he discovered the law of gravitation. Using the law of gravitation and the law of acceleration, it became possible to compute correctly the motions not only of the planets, but also of the moon, comets, and other heavenly bodies. Today, Newton's discoveries and methods continue to be applied successfully in launching satellites and other space vehicles.

8-2 Centripetal Force

Centripetal force keeps a body moving in a circular path by pulling it toward the center of the circle.

Tie a small stone to one end of a string and, holding the other end, swing the stone in a horizontal circle. You will find that you must keep pulling the stone toward the center of the circle. If you let the string go, the stone flies off on a tangent to the circle at the point where it was released. You are experiencing the fact that *to keep a body moving in a circle you must continually exert a force pulling it toward the center of the circle.* This is called *centripetal force* (sen·TRI·pi·tul FORS).

We have seen in Section 6-12 that a body moving in a circle of radius r at constant speed v is continually being accelerated toward the center of the circle with an acceleration $a = v^2/r$. This acceleration is produced by the centripetal force. Hence, if m is the mass of the body, the centripetal force needed to keep it in a circle of radius r is:

$$f_c = ma = \frac{mv^2}{r}$$

We express m in kilograms, v in meters per second and r in meters. The force f_c is then in newtons.

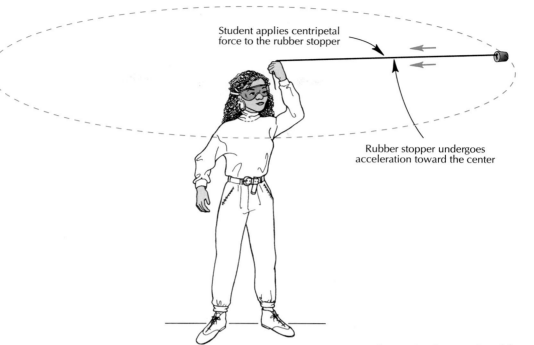

Student applies centripetal force to the rubber stopper

Rubber stopper undergoes acceleration toward the center

Fig. 8-1. It takes centripetal force to keep the rubber stopper moving in a circle.

Since, as we have seen in Section 6-14, centripetal acceleration for a body in uniform circular motion is $a = 4\pi^2 r/T^2$, where r is the radius of the path and T the period, the centripetal force on such a body is:

$$\mathbf{f}_c = \mathbf{ma} = \frac{4\pi^2\mathbf{mr}}{\mathbf{T}^2}$$

Sample Problems

1. A student attaches a mass of 0.5 kg to one end of a rope. The student then swings the mass in a horizontal circle having a radius of 1 m so that its tangential speed is 4 m/s. What centripetal force must be exerted on the mass to keep it moving in a circle?

Solution:

$m = 0.5$ kg $v = 4$ m/s $r = 1$ m

$$f_c = \frac{mv^2}{r}$$

$$f_c = \frac{0.5 \text{ kg} \times (4 \text{ m/s})^2}{1 \text{ m}} = 8 \text{ N}$$

2. An artificial satellite has a period of 5.6×10^3 s and an orbital radius of 6.8×10^6 m. If its mass is 2.0×10^3 kg, what is the centripetal force keeping it in orbit?

Solution:

$T = 5.6 \times 10^3$ s $r = 6.8 \times 10^6$ m
$m = 2.0 \times 10^3$ kg

$$f_c = \frac{4\pi^2 mr}{T^2}$$

$$f_c = \frac{4(3.14)^2 \times (2.0 \times 10^3 \text{ kg}) \times (6.8 \times 10^6 \text{ m})}{(5.6 \times 10^3 \text{ s})^2}$$

$$f_c = 1.7 \times 10^4 \text{ N}$$

Test Yourself Problems

1. A student swings a mass of 0.120 kg tied to one end of a cord in a horizontal circle of radius 1.00 m so that its tangential speed is 2.00 m/s. What is the centripetal force acting on the mass?

2. An iron flywheel 1.50 m in radius is rotating at 5.00 turns per second. (*a*) What is the period of the motion? (*b*) What centripetal force is acting on each kg of iron on the outer rim of the flywheel?

8-3 "Centrifugal Force"

Centrifugal force does not exist.

If the student in Fig. 8-1 lets go of the string, there will no longer be a centripetal force compelling the stone to move in a circle. In accordance with the law of inertia, the stone will then fly off at constant velocity in a straight line tangent to its original circular path. The fact that, at the same time, the stone moves away from the center of the circle is often erroneously attributed to a so-called "centrifugal force" (sen·TRIF·yoo-gul FORS) that accelerates the stone away from the center of the circle. Actually, there is no such thing as a centifugal force acting on the stone. Before the stone is released, the only force acting upon it (other than gravity) is the centripetal force supplied by the student to keep it moving in a circle. After the stone is released, no force is acting upon it and inertia then causes it to fly off at a tangent to its circular path. This behavior is typical of any body moving in a circle when the centripetal force acting on it is suddenly removed. It explains many practical devices and experiences.

8-4 Applications of Centripetal Force

As a speeding car makes a sharp left turn, a passenger is forced over toward the right side of the car until pressed up against it. The motion and the force acting on the passenger are explained in this way. When the car begins to make the left turn, the passenger at first continues to move in the original direction of the car in accordance with the law of inertia. However, as the car continues to move toward the left, the straight-line motion of the passenger soon brings him up against the right side of the car, which blocks the continuation of the straight-line motion. At the same time, it exerts a force pushing the passenger toward the center of the car's circular path. This is the centripetal force that now compels the passenger to travel around the turn.

On a rainy day, the tires of a bicycle pick up water and carry it around with them as they turn. However, the adhesive force with which the water clings to the tires is not great enough to provide the centripetal force needed to keep the water moving in a circle. Inertia therefore causes the water to break away and fly off the tires at a tangent. To protect the riders, bicycles are therefore equipped with mudguards.

Centripetal force supplied by the gravity of the sun keeps the planets in their orbits.

In constructing high speed roads and turnpikes, engineers bank the road on the turns so that the outside of the road is higher than the inside. This slope of the road resists the inertial tendency of a car to leave the road at a turn and go off on a tangent. In doing so, the road supplies the centripetal force needed to keep the car going around the turn.

8-5 Problem of Gravitation

We can now return to the problem Newton set for himself of seeking a law of gravitation. Newton once stated that he got the basic idea for the law when he saw an apple fall from a tree. He had been thinking about the problem for a long time and this observation seems to have been the stimulus that crystallized his thoughts.

Specifically, the problems that Newton was trying to solve were these: (1) What is the nature of the force that accelerates the apple or any other body toward the earth? (2) What is the nature of the force exerted by the sun that gives the planets the centripetal acceleration needed to keep them in their orbits?

8-6 Mutual Attraction Between Bodies

What clues were available to Newton as to the nature of this force? A major one was the fact that all freely falling bodies are accelerated toward the earth with constant acceleration g. According to the second law of motion, this acceleration must be produced by a force. Since all bodies fall toward the earth, the earth must be exerting an attractive force on them. Furthermore, since every action has a reaction, a body attracted downward by the earth reacts by attracting the earth upward toward itself. Each attracts the other.

Now the fact that all bodies are attracted by the earth and attract it in turn suggests that this property of mutual attraction is a characteristic of all matter. That is, every body attracts and is attracted by every other body. This mutual attraction Newton called the *force of gravity*. It simplified the two problems with which he began by making them one. The force of attraction between the earth and the apple and the force that the sun exerts on a planet are the same kind of force. Both are simply gravitational forces.

8-7 Factors Affecting Gravitational Force

It now remained to find out what factors determine how great the attractive force is between any two bodies. It seemed evident that, since the gravitational attraction of the earth for a body depends upon its mass, two of the factors must be the *masses* of each of the two attracting bodies. It also seemed reasonable to expect that the attractive force between two bodies depends upon the *distance between them;* that is, the greater the distance, the less the attraction.

But exactly what is the relationship between the force of attraction, the distances between the two bodies, and their masses? At this point, the logical step would have been to set up a series of

Fig. 8-2. What holds the passengers tightly in their seats as this roller coaster goes through a full loop?

experiments and actually measure the attractive forces between bodies of different masses separated by different distances. These experiments would have revealed the desired relationship. However, it was not until more than one hundred years after Newton's time that experimental methods precise enough to measure the tiny gravitational attraction between two ordinary bodies were developed. Newton therefore had to turn to ingenious thinking and the evidence furnished by astronomy to guide him to the law he was seeking.

8-8 Law of Gravitation

Since the gravitational attraction between the earth and any body is proportional to the mass of the body, Newton inferred that the gravitational force between any two bodies is proportional to their masses. This would explain why the earth with its large mass exerts an easily noticed attractive force on a falling apple while the attraction between an apple and another apple is so small that it escapes detection.

Newton found in Kepler's third law a clue to the law of gravitation.

For a clue as to how the distance between two bodies affects the force between them, Newton turned to the planets and worked backward from Kepler's third law. In Section 8-2, we found that the centripetal force acting on a planet of mass m, period T, and orbital radius r is $f_c = 4 \pi^2 mr/T^2$. From Section 6-18, we find that Kepler's third law may be restated as $T^2 = r^3/k$: Substituting this value for T^2 in f_c, gives:

$$f_c = \frac{4 \pi^2 mrk}{r^3} = 4 \pi^2 mk \left(\frac{1}{r^2} \right)$$

This is the centripetal force supplied by the gravitational attraction of the sun that is responsible for keeping the planet moving in its orbit around the sun. It is evident that the attractive force of the sun on a planet depends on $1/r^2$; that is, it is inversely proportional to the square of the distance of the planet from the sun.

This clue led Newton to infer that the force of gravitation between any two bodies is also inversely proportional to the square of the distance between them. He was now ready to state the law of gravitation.

Every body attracts every other body in the universe with a force that is directly proportional to the masses of the two bodies and inversely proportional to the square of the distance between them. In symbols:

$$f \propto \frac{m_1 m_2}{d^2}$$

where f is the gravitational force between the masses m_1 and m_2

whose distance apart is *d*. Changing the proportion to an equality, we have,

$$f = G \frac{m_1 m_2}{d^2}$$

where *G* is a constant that is the same for all bodies in the universe.

8-9 Value of *G*

The universal constant *G* must not be confused with *g*, the acceleration of gravity. It was not until 1797, seventy years after Newton's death, that the first experimental determination of *G* was made by Henry Cavendish (KAV·en·dish). His equipment is shown schematically in Fig. 8-3. It consisted of a pair of small, similar lead balls attached to the end of a bar suspended at its center by a delicate wire. Two large lead balls were brought up into place so that each came close to one of the small balls.

Cavendish demonstrated that even small bodies exert measurable gravitational attractions on each other.

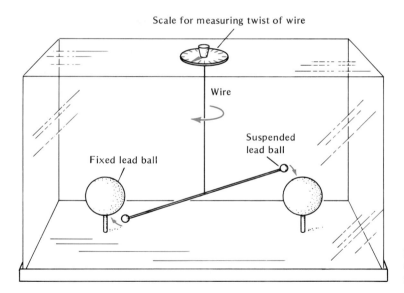

Scale for measuring twist of wire

Wire

Suspended lead ball

Fixed lead ball

Fig. 8-3. Apparatus used by Cavendish to measure the universal gravitational constant *G*.

The gravitational attraction between the large and small balls caused the bar to turn and twist the supporting wire until the resistance of the wire brought the bar to rest. The angle through which the wire was twisted was measured. Since this angle is proportional to the force that twisted the wire, it served as a measure of the force between the large and small balls that produced the twist. Thus, Cavendish measured the force between each large and small ball. He also measured their masses and the distance between them. Substituting these values in the law of gravitation, he solved for *G*. It turned out to be 6.67×10^{-11} m³/kg·s².

Sample Problem

What is the force of gravitational attraction between two 1.00-kg masses that are 1.00 m apart?

Solution:

$m_1 = m_2 = 1.00$ kg $d = 1.00$ m
$G = 6.67 \times 10^{-11}$ m³/kg·s²

$$f = G \frac{m_1 m_2}{d^2}$$

$$f = 6.67 \times 10^{-11} \text{ m}^3/\text{kg} \cdot \text{s}^2 \times \frac{1.00 \text{ kg} \times 1.00 \text{ kg}}{(1.00 \text{ m})^2}$$

$$f = 6.67 \times 10^{-11} \text{ kg} \cdot \text{m/s}^2 = 6.67 \times 10^{-11} \text{ N}$$

This force is far too small to be detected in everyday experience. That is why we do not notice it.

Test Yourself Problem

The mass of the earth is 6.0×10^{24} kg. The mass of the moon is 7.4×10^{23} kg. The distance between them is 3.8×10^8 m. What is the gravitational attraction between them?

8-10 Computing the Earth's Mass

The law of gravitation can be used to determine the mass of the earth. Consider a mass of 1.00 kilogram on the surface of the earth. Its distance from the center of the earth is the earth's radius, which is 6.37×10^6 meters. Since its mass is 1.00 kilogram, it is attracted to the earth by a force equal to its weight which is $mg = 1.00$ kilogram $\times$ 9.80 meters per second per second, or 9.80 newtons. We now have

$$m_1 = \text{mass of earth}$$
$$m_2 = 1.00 \text{ kg}$$
$$d = 6.37 \times 10^6 \text{ m}$$
$$G = 6.67 \times 10^{-11} \text{ m}^3/\text{kg} \cdot \text{s}^2$$
$$f = 9.80 \text{ N}$$

Substitute these values in the formula for gravitational force:

$$f = G \frac{m_1 m_2}{d^2}$$

$$9.80 \text{ N} = 6.67 \times 10^{-11} \text{ m}^3/\text{kg} \cdot \text{s}^2 \times \frac{m_1 \times 1.00 \text{ kg}}{(6.37 \times 10^6 \text{ m})^2}$$

Solving for m_1, the earth's mass, we find:

$$m_1 = \frac{9.80 \times (6.37 \times 10^6)^2}{6.67 \times 10^{-11}} \text{ kg} = 5.97 \times 10^{24} \text{ kg}$$

8-11 Law of Gravitation and Weight

According to the law of gravitation, the earth's attraction for a body is inversely proportional to the square of the distance between the body and the center of the earth. This means that the weight of a body, which is the pull of the earth on it, decreases as the body moves away from the earth.

Fig. 8-4. The gravitational pull of Saturn keeps each of the tiny particles that make up its rings orbiting around it.

For example, at the earth's surface, a rocket is about 6000 kilometers from the earth's center. When a rocket reaches a point 12 000 kilometers from the earth's center, its distance from the center of the earth has doubled. The earth's attraction upon it, or the rocket's weight, therefore decreases to $\frac{1}{2}^2$ or $\frac{1}{4}$ of what it was at the surface. When the rocket's distance is 18 000 kilometers from the center of the earth, or three times its initial distance, the weight of the rocket will decrease to $\frac{1}{3}^2$ or $\frac{1}{9}$ of what it was at the surface, and so forth.

Even over the surface of the earth, there are small but significant differences in the weight of a body, depending on its distance from the center of the earth. For this reason, bodies weigh less on mountain tops than they do at sea level. Also, since the rotation of the earth causes it to bulge a little at the equator, a body is a little further from the center of the earth when it is at the equator than when it is at either of the poles. It therefore weighs a little less at the equator than it does at either pole.

8-12 Confirming the Law of Gravitation

Once Newton had formulated the law of gravitation, it could be tested in many ways. The procedure was to use the law to predict some aspect of the motion of a planet or other heavenly body, and then to see if the prediction agreed with the observed behavior of that body.

One immediate success involved the moon, which revolves about the earth in the same way that a planet revolves around the sun. Newton knew that the moon is about sixty times as far from the center of the earth as is a body resting on the surface of the earth. The earth's pull should therefore accelerate the moon toward the earth at a rate $1/60^2 = 1/3600$ as great as the acceleration g it gives to bodies on its surface. This is 9.80 meters per second per second $\div\ 3600 = 2.72 \times 10^{-3}$ m/s². Since this acceleration is the centripetal acceleration that keeps the moon in its orbit, it can be determined independently from the moon's period and the known value of its orbital radius. We have done this in Section 6-15. Comparing these two values of the moon's acceleration toward the earth, we find that they are about equal, thus confirming the law of gravitation.

8-13 Law of Gravitation and Acceleration of Gravity

The earth's pull on a body decreases as the body gets farther from the earth. The acceleration g, which the earth's pull gives a body free to fall, therefore also decreases as the body moves away from the earth. At points near the earth, g is practically constant and is equal to 9.80 meters per second per second. The value of g at points far above the surface of the earth may be found from the law of gravitation.

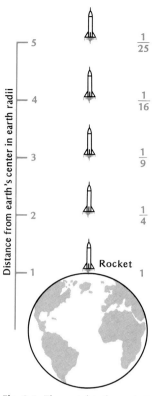

Fig. 8-5. The weight of a rocket varies inversely as the square of its distance from the center of the earth.

The earth's pull on a body of mass m at a distance R from its center is:

$$f = \frac{GMm}{R^2}$$

where M is the mass of the earth. According to the law of acceleration, this force applied to m will give it a downward acceleration g such that:

$$f = mg$$

Setting these two values of f equal to each other we find:

$$mg = \frac{GMm}{R^2}$$

or,

$$g = \frac{GM}{R^2}$$

The acceleration due to gravity decreases with distance from the earth's center as $1/R^2$.

Now substituting,

$$G = 6.67 \times 10^{-11} \text{ m}^3/\text{kg} \cdot \text{s}^2$$
$$M = 5.97 \times 10^{24} \text{ kg}$$

we have,

$$g = \frac{3.98 \times 10^{14}}{R^2} \text{ m/s}^2$$

Thus, the gravitational acceleration given by the earth to a body may be computed for any value of R. At the surface of the earth, R is the radius of the earth: 6.37×10^6 meters. Substituting this value for R, we get:

$$g = \frac{3.98 \times 10^{14}}{(6.37 \times 10^6)^2} \text{ m/s}^2$$

$$g = 9.80 \text{ m/s}^2$$

Thus our computed value of g checks with the known measured value of g at the surface of the earth.

Sample Problem

Compute the value of g for an object, such as a satellite, when it is at an altitude of 200 000 m.

Solution:
The body's distance from the earth's center will be the earth's radius plus the altitude: 6.37×10^6 m + 200 000 m = 6.37×10^6 m + $.20 \times 10^6$ m = 6.57×10^6 m.

Substitute this value of R in the expression:

$$g = \frac{3.98 \times 10^{14}}{R^2} \text{ m/s}^2$$

$$g = \frac{3.98 \times 10^{14}}{(6.57 \times 10^6)^2} \text{ m/s}^2$$

whence,

$$g = 9.21 \text{ m/s}^2$$

Thus, at 200 000 m above the earth, the acceleration of gravity has dropped from the 9.8 m/s² value it has at the surface to about 9.2 m/s².

Test Yourself Problem

Compute the value for g at a point (*a*) 8.00×10^6 m from the center of the earth, and (*b*) at a point twice as far.

8-14 Gravitational Fields

A body exerts its gravitational force on a second body through the space between them. There is no contact between the two bodies, yet each exerts a force upon the other. The *gravitational field* idea attempts to explain how this force is transmitted through space.

We can think of the space around a body as being modified by the presence of that body in such a way that masses placed at all points in that space are acted upon by forces pulling them toward the body. *The space around the body in which it sets up these forces is called its gravitational field of force. The strength of the gravitational field at any point is equal to the force it exerts per unit of mass at that point.*

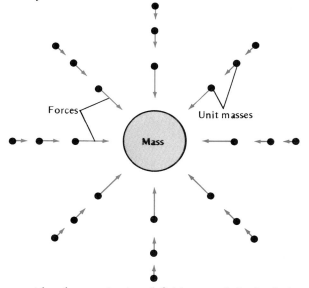

Forces   Unit masses

Mass

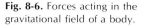

Fig. 8-6. Forces acting in the gravitational field of a body.

To illustrate, consider the gravitational field around the body in Fig. 8-6. Placed in the field are several bodies each of unit mass. We know from experiment that each of these unit masses is acted upon by a force attracting it toward the body. The forces are greater for the unit masses near the body and smaller for those farther away, as shown by the vectors attached to them. The totality of these forces, acting at every point in the space around the body, make up its gravitational field.

According to the field idea, a body does not exert its gravitational attraction on a second body directly. It merely sets up a gravitational field around itself and the field then acts upon the second body. In the same manner, the second body sets up its own gravitational field which acts with equal and opposite force on the first body. Each body is therefore acted upon by the gravitational field of the other.

8-15 Field Strength of the Earth

An object at any point in the space around the earth is acted upon by the gravitational field of the earth with a force f that is equal to its weight w at that point. Since the weight of a body of mass m at a place where the acceleration of gravity is g is equal to mg, it follows that $f = mg$, or:

$$g = \frac{f}{m}$$

By definition, f/m, the force exerted per unit mass at any point in a gravitational field, is the strength of the gravitational field at that point. Hence, *the strength of the earth's gravitational field at any point in space is the value of g at that point and is a vector in the same direction as g.*

Sample Problem

In the sample problem in Section 8-13 we found that g changes from 9.8 m/s² at the surface of the earth to 9.2 m/s² at an altitude of 2×10^5 m. Compare the forces exerted by the earth's gravitational field on a mass of 10 kg placed at each of these points.

Solution:
g_1 at surface = 9.8 m/s² g_2 = 9.2 m/s²
m = 10 kg

The force exerted by the earth's gravitational field is:

$$f = mg$$

At the surface this is:

$$f = 10 \text{ kg} \times 9.8 \text{ m/s}^2 = 98 \text{ N}$$

At the altitude of 2×10^5 m this is:

$$f = 10 \text{ kg} \times 9.2 \text{ m/s}^2 = 92 \text{ N}$$

Test Yourself Problems

1. What force is exerted on a mass of 20.0 kg at a point in space where $g = 8.50$ m/s²?
2. The gravitational force exerted by the earth on a 50.0 kg mass at a certain point in space is 450 N. What is the strength of the earth's gravitational field at that point?

8-16 Artificial Satellites

Newton predicted that an artificial satellite could be put into orbit around the earth if it were fired from the top of a mountain and given a horizontal velocity of about 8 kilometers per second. The launching of such a satellite would have been a direct demonstration that the law of gravitation and the laws of motion apply to astronomical bodies as well as to earthly ones. However, rocketry was not sufficiently developed in Newton's day to make such a test, and it was not until recent times that Newton's prediction could be tested.

Newton's calculation is made as follows. The centripetal force needed to keep a satellite m in a circular orbit of radius R is mv^2/R, where v is the orbital speed of the satellite. This force is supplied by the earth's pull on the body and is simply the weight of the body. Since the weight of the body is equal to mg,

$$mg = \frac{mv^2}{R}$$

$$\mathbf{g} = \frac{\mathbf{v}^2}{\mathbf{R}}$$

whence,

$$\mathbf{v} = \sqrt{\mathbf{Rg}}$$

Now, at the surface of the earth $R = 6.37 \times 10^6$ meters and $g = 9.8$ meters per second per second. Substituting, we find that:

$$v = \sqrt{6.37 \times 10^6 \times 9.8 \, \frac{m^2}{s^2}}$$

$$v = 7.9 \times 10^3 \text{ m/s}$$

Thus, the launching velocity the satellite must be given to stay in orbit is 7900 meters per second, or nearly 8 kilometers per second.

To compute the velocity needed to launch a satellite at some distance above the earth, we must know what the acceleration of gravity g is at that altitude. This we can find from the relationship $g = GM/R^2$ in Section 8-13, where we found that g for a distance of 200 000 meters above the earth is 9.2 meters per second per second. The magnitude of the velocity with which a satellite must be launched into orbit at this level is:

$$v = \sqrt{Rg}$$

where,

$$R = 6.57 \times 10^6 \text{ m} \qquad \text{and} \qquad g = 9.2 \text{ m/s}^2$$

$$v = \sqrt{(6.57 \times 10^6 \text{ m}) \times (9.2 \text{ m/s}^2)}$$
$$v = 7.8 \times 10^3 \text{ m/s}$$

Note that as the satellite is launched at a higher altitude, a smaller launching velocity is needed to put it into orbit.

8-17 Launching a Satellite

To launch a satellite, it is first projected to an altitude of 150 or more kilometers. It is then fired as nearly horizontally as possible at the velocity required to keep it in the orbit that has previously been selected for it.

Satellites are generally launched with multistage rockets. The simplest form of this device consists of two stages—a large rocket to which is attached a smaller rocket. The satellite to be put into orbit is usually in the nose of the smaller rocket. When the large

This is how Newton calculated the velocity needed to put a satellite into a circular orbit.

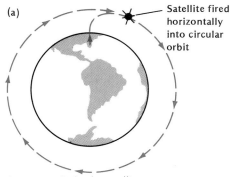

(a)

Satellite fired
horizontally
into circular
orbit

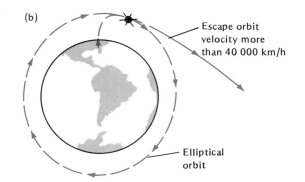

(b)

Escape orbit
velocity more
than 40 000 km/h

Elliptical
orbit

Fig. 8-7. (**a**) Path of a satellite launched into a circular orbit around the earth. (**b**) Paths of satellites launched into an elliptical orbit and into an escape orbit.

rocket is fired, it carries the smaller rocket and the satellite to a planned altitude. The large rocket then detaches itself from the second stage, eventually falling back to earth. The smaller rocket continues upward until it reaches the altitude selected for the orbit. At this level it fires horizontally, giving the satellite in its nose the necessary velocity to put it into orbit. An especially important orbit is the *geosynchronous orbit,* in which a satellite circles the earth from west to east within a period of one day. Since the earth rotates from west to east within the same period, the satellite stays in the same position in relation to the earth. Such "stationary" satellites, spaced around the earth and equipped to receive and transmit television and radio signals, form a communication network for sending such signals over long distances.

8-18 Escape Velocity

If a satellite is given a higher initial horizontal velocity than the one that is needed to put it into a circular orbit, it goes into an elliptical orbit. The higher the launching velocity, the longer is the ellipse and the farther from the earth the satellite goes before turning back. When the launching velocity exceeds a value known as the *escape velocity,* the satellite goes out so far that the earth's gravity can no longer pull it back. It then "escapes," that is, travels off into space. The earth's escape velocity is about 11.2 kilometers per second. It has been readily achieved by spacecraft sent to the moon and the planets.

8-19 Effect of Air Resistance

Actually, a satellite launched at an altitude of a 150 kilometers will not remain in orbit indefinitely because of the resistance of the air. Although only a tiny trace of air is left at this high altitude, it provides enough frictional resistance to keep slowing the satellite down. As the satellite loses speed, gravity pulls it closer to the earth. In doing so, gravity speeds the satellite up until its speed is high enough to remain in an orbit smaller than its original orbit.

In the smaller orbit, the satellite encounters more air resistance and is once more pulled closer to the earth and accelerated to a still higher speed than it had originally. This process continues with the result that, over a period of time, the satellite spirals toward the earth with ever increasing speed. On its way, it enters the denser levels of the atmosphere where increased air friction soon causes it to become white hot. Eventually it burns up completely, leaving only its ashes to drift back to the earth.

8-20 Weightlessness

Passengers in a satellite orbiting around the earth and all the objects inside the satellite appear to be weightless. In this condition, the slightest push sends one floating through the air. Objects released from one's hand do not fall toward the floor of the satellite, and water in a glass does not pour out when the glass is inverted.

To understand this condition of apparent weightlessness, imagine that you are standing on a weighing scale in an elevator. When the elevator is at rest, the earth pulls you downward with a force equal to your weight. The scale reacts by pushing you upward with an equal force and therefore reads your weight. It is the upward supporting force of the scale on you that gives you the sensation of having weight.

Now, suppose the elevator is allowed to fall freely. The scale on which you are standing is no longer supporting you because it is falling away from you as fast as you are falling toward it. The scale will therefore read zero. If you release a ball from your hand, the ball will not fall to the floor of the elevator but will remain the same distance above it. The floor is falling away from the ball as fast as the ball is falling toward the floor. You and every object in the elevator are still being pulled toward the earth by the force of gravity. However, this force is not noticed by you because you and all the objects around you are unsupported and are falling freely.

Fig. 8-8. Astronauts Kathryn D. Sullivan, left, and David C. Leestma performing an extravehicular activity (EVA) in October 1984. This was the first EVA in which an American woman participated.

All objects inside an orbiting satellite are also freely falling bodies and therefore appear to be weightless. To understand this, note that a satellite has two independent motions. The first is its orbital speed, which keeps moving it away from the earth on a straight line tangent to the orbit. The second is its vertical motion, which causes it to fall freely toward the earth with acceleration g. This falling motion is caused by the gravitational pull of the earth.

When the satellite is in a circular orbit, its orbital speed takes it away from the earth during each second just as far as the earth's gravitational pull causes it to fall toward the earth. Thus, although the satellite remains the same distance from the earth, it is constantly falling freely. All objects in it are therefore also falling freely and, because of this, appear to be weightless.

The sensation of weightlessness has an effect on the mental and physical functioning of those who go up in satellites. It is therefore the subject of much study and research in space medicine.

CHAPTER REVIEW

Summary

The force needed to keep a body moving in a circular path at constant speed is called a **centripetal force.** It is directed toward the center of the circle and is equal to mv^2/r, where m is the mass of the body, v is its speed, and r is the radius of its circular path.

Newton discovered the **law of gravitation** which states: Every body attracts every other body in the universe with a force directly proportional to the masses of the two bodies and inversely proportional to the square of the distance between them. For two masses m_1 and m_2 separated by distance d the force of attraction is given by:

$$f = G \frac{m_1 m_2}{d^2}$$

where G is a universal constant equal to 6.67×10^{-11} m³/kg·s².

The weight of a body is the **gravitational force** with which the earth attracts it. It therefore decreases as the body moves further away from the surface of the earth and increases as the body comes closer to the surface of the earth.

The earth's gravitational pull upon a body determines the acceleration g with which the body will fall. The value of g is largest near the surface of the earth and decreases steadily at points further and further away from the earth.

According to the field idea of gravitation, every body modifies the space around it in such a way that the space exerts gravitational forces on all bodies located within it. The space around a body is therefore called its **gravitational field.** A body does not exert a gravitational attraction on a second body directly but does so through its gravitational field. Similarly, the second body does not exert its attraction on the first body directly but through its gravitational field.

The intensity of a gravitational field at any point is the force exerted on a unit mass placed at that point. The intensity of the gravitational field at a given point in the space around the earth is equal to the value of g at that point. The force on a body of mass m placed at that point is $f = mg$ and is equal to the weight of the body at that point.

Questions

Group 1

1. (a) Why is a centripetal force needed to keep a body moving in a circular path? (b) In what direction does a centripetal force accelerate the body on which it acts? (c) Upon what three factors does the centripetal force keeping a body moving in a circular path depend?

2. Suppose you swing a stone at the end of a string in a horizontal circle over your head. (a) What happens to the magnitude of the centripetal force you apply as you allow the radius of the circle to increase but keep the speed constant? (b) How does the centripetal force change when you increase the speed of the stone while the radius of the circle is unchanged? (c) How will the stone move if the string breaks?

3. (a) Why do water drops fly tangentially off the wheels of a rapidly moving car on a rainy day? (b) How does banking a road on a turn help a speeding car make the turn safely?

4. (a) Why do we not notice the force of gravitational attraction between two ordinary objects such as a pair of books? (b) What effect does doubling the distance between two bodies have on the force of gravitational attraction between them?

5. How does the law of gravitation explain why the weight of a rocket changes as it moves further away from the earth?

6. (a) What is the relationship between the force exerted by the gravitational field of the earth and the acceleration of gravity at that point? (b) If the mass put at a given point in the earth's gravitational field is doubled, what will happen to the force exerted by the field upon it? (c) In what direction do the forces at all points in the gravitational field of the earth act?

7. (a) What provides the centripetal force that keeps a satellite in an orbit? (b) How is an artificial satellite launched? (c) On what factors does the velocity with which a satellite must be launched in an orbit depend?

8. Explain why a satellite launched two hundred kilometers above the earth will gradually spiral in toward the earth.

9. Explain why objects in a satellite orbiting around the earth appear to be weightless.

10. How is the value of the gravitational constant G determined?

Group 2

11. (a) A car moving at constant speed travels around two successive turns of a winding road. The first turn has twice the radius of the second turn. How does the centripetal force acting on the car on the first turn compare with that acting on the car on the second turn? (b) If the car doubles its speed, what happens to the centripetal force acting upon it as it makes each turn?

12. What effect does the rotation of the earth have upon the weight of someone standing (a) on the equator; (b) on the north pole?

13. Suppose you are standing on a weighing scale in an elevator. When the elevator is at rest, the scale gives your weight in kilograms. (a) If the elevator now falls freely with acceleration g, what happens to the reading of the scale? (b) If the elevator descends at a constant acceleration less than g, what happens to the reading of the scale? (c) If the elevator ascends with a constant acceleration less than g, what happens to the reading of the scale?

14. (a) Explain the statement that the field intensity at any point in the earth's gravitational field is equal to g. (b) Why does the value of g change as a body moves away from the earth?

15. Two bodies, A and B, are small fixed spheres having masses in the ratio of 4 to 1. At what point on the line joining the centers of A and B should a third small sphere C be placed so that the gravitational forces of A and B on C are in equilibrium?

Problems

Note. Unless otherwise specified in these problems, assume g at the surface of the earth to be 9.8 m/s²; G to be 6.7×10^{-11} m³/kg·s²; and π^2 to be 9.9.

Group 1

1. What centripetal force is needed to keep a 2-kg mass moving at a constant speed of 4 m/s in a circle having a radius of (a) 4 m; (b) 8 m?

2. An auto having a mass of 1500 kg makes a turn on a banked circular track of radius 200 m at a speed of 20 m/s. What centripetal force does the track exert on the auto?

3. A 1-kg body is kept revolving in a circle of radius 1 m by a centripetal force of 9 N. (a) At what speed is the body moving? (b) How does the centripetal force on this body compare with its weight?

4. What is the centripetal force acting on each kilogram of an airplane that is turning at 200 m/s in a horizontal circle of radius 10 000 m?

5. What centripetal force is needed to keep a mass of 2.0 kg moving on a smooth floor in a circular path of radius 0.50 m and making one revolution every 4 s?

6. A child having a mass of 30 kg sits 4.0 m from the center of a merry-go-round that is rotating with a period of 10 s. What is the centripetal force acting upon the child?

7. In order to swing a 0.1-kg mass tied to one end of a cord in a horizontal circle of radius 0.5 m so that its tangential speed is 0.2 m/s, what centripetal force must you exert on the mass through the cord?

8. A force of 10 N applied to one end of a cord keeps an object tied to the other end moving at a speed of 2.0 m/s in a horizontal circle of radius 3.0 m. What is the mass of the object?

9. A satellite of mass 100 kg is put into orbit around the earth at a distance at which it weighs 920 N. The period of the satellite is measured and found to be 5.3×10^3 s. (a) What is the centripetal force on the satellite? (b) What is its distance from the earth?

10. Calculate the gravitational force of attraction between a mass of 50 kg and a mass of 60 kg when they are 2.0 m apart.

11. Two similar trucks each having a mass of 2.0×10^5 kg are 40 m apart. (a) What is the gravitational force of attraction between them? (b) How does this force compare with the weight of the trucks?

12. A hydrogen atom consists of a proton having a mass of the order of magnitude of 10^{-27} kg and an electron having a mass of the order of 10^{-30} kg. The distance between them is of the order of 10^{-10} m. What is the order of the gravitational force between them? (For the order of G, use 10^{-10} m³/kg·s².)

13. (a) Compute the value of g at a point in space that is 7.0×10^6 m from the center of the earth. (b) What will be the value of g at a point twice as far from the earth as that in (a)?

14. (a) What is the force exerted by the earth's gravitational field on a body having a mass of 2.0 kg at a point in space where the value of g = 9.0 m/s²? (b) What force does the earth's gravitational field exert on this body at the surface of the earth?

15. What is the value of g at a point in the earth's gravitational field where a 10-kg mass weighs 95 N?

Group 2

16. A student whirls a 0.2-kg stone tied to one end of a cord in a vertical circle at 30 revolutions per minute. The cord is 1.0 m long. (a) What is the period of the motion in s? (b) Neglecting the effect of gravity, what centripetal force must be supplied by the student? (c) Including the effect of gravity, what is the total upward force exerted on the stone by the student when the stone is at the lowest part of its swing? (Note that both the centripetal force and the force needed to balance the weight of the stone must be supplied.)

17. An 80.0-kg body stands on a weighing scale on the earth's equator. Since it moves in a circle as the earth rotates, there is a centripetal force acting upon it. (a) Taking the radius of the earth as approximately 6×10^6 m, and its daily period of rotation as about 9×10^4 s, and π^2 as approximately 10, estimate the centripetal force acting on the body. (b) If the earth were not rotating, what would the body weigh at the equator, assuming g = 9.8 m/s²? (c) How much less than (b) does the scale read because of the effect of the earth's rotation?

18. What does a 1000-kg rocket weigh when it is 7.0×10^6 m from the center of the earth?

19. At what distance from the center of the earth does g = 2.0 m/s²

20. The strength of the earth's gravitational field at the moon is 2.8×10^{-3} m/s² and the mass of the moon is 7.4×10^{22} kg. (a) What is the centripetal force exerted on the moon by the earth? (b) If the period of the moon's motion is 2.3×10^6 s, what is the distance to the moon?

21. The gravitational attraction between the sun and the earth supplies the centripetal force that holds the earth in a nearly circular orbit. Using the following approximate values, estimate the centripetal force exerted by the sun on the earth. Mass of the earth = 6.0×10^{24} kg; radius of the earth's orbit = 1.5×10^{11} m; period of the earth's orbit = 3.2×10^7 s.

22. From the following approximate data, estimate the force of attraction between the earth and the sun. Mass of the sun = 2.0×10^{30} kg; mass of the earth = 6.0×10^{24} kg; radius of the earth's orbit = 1.5×10^{11} m. (This force is the centripetal force computed in problem 21. How closely do these two estimates agree?)

23. The mass of the earth is about 80 times that of the moon. A rocket fired toward the moon has traveled two-thirds of its journey. What is then the ratio of the earth's force of attraction on the rocket to that of the moon?

24. At a point 6.7×10^6 m from the center of the earth, g = 9.0 m/s². What velocity must be given to an earth satellite to send it into a circular orbit at this distance?

25. (a) What is the value of g at a point 400 000 m above the earth? (b) What velocity must be given to an earth satellite to send it into a circular orbit at this altitude?

Applying Physics

1. Centripetal force may be used to determine the weight of a body when we have a known mass at our disposal. To illustrate the method, obtain a length of glass tubing about 15 cm long and fire polish its ends. Wrap the outside of the tubing with adhesive tape, as shown in Fig. 8-9. Pass a piece of string about 1 m long through the tube and fasten a nickel to one end of the string by means of a strip of plastic tape. Tape a pile of 3 nickels together with plastic tape and attach this pile to the opposite end of the string. Make an ink mark at a point on the string 30 cm from the center of the single nickel. The single nickel will serve as a known mass of 5 g. The pile of nickels will be the body whose weight is being measured.

   Hold the glass tube vertically and swing the single nickel over your head in a horizontal circle. Allow the radius of the circle to increase until it is at the 30-cm mark and adjust the speed of rotation until the weight of the 3 nickels is supported by the force exerted on the string by the rotating nickel. Count the number of complete turns made by the nickel in a 10-s interval and determine the period of rotation.

   The weight of the 3 nickels is the centripetal force that keeps the single nickel moving in a circle. Compute it from the relationship

$$f_c = 4\ \pi^2\ mr/T^2.$$

   How accurate is this answer? Assuming the mass of each nickel to be 5 g, compute the weight of 3 nickels from the relationship $w = mg$ and compare your value of w with f_c. What are some sources of error in this method of determining the weight of an object?

2. If there is an elevator in your building and you can obtain a bathroom weighing scale, you can do the following experiment. Put the scale on the elevator floor and stand on it. Note your weight before the elevator starts moving. Note the reading on the scale (a) when the elevator starts moving; (b) when it has attained its normal steady speed; and (c) when it begins to come to a stop. Explain the changes in the scale reading.

3. Fill a glass tumbler about two-thirds full of water. Rotate the water by stirring with a teaspoon. Note that as the water rotates, it climbs up the sides of the glass forming an inverted conical trough. The height to which the water rises is a measure of the centripetal force exerted by the glass on the water. Observe what happens to the height to which the water rises against the glass sides and to the depth of the trough as you gradually increase the rate of stirring. Explain what you observe.

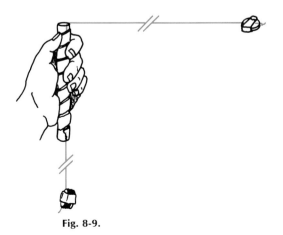

Fig. 8-9.

Acrobatics:
The physics of twist

A figure skater spins effortlessly around an axis through the center of her body; a gymnast executes a full twist during his dismount; and a diver performs a two-and-a-half somersault with two full twists. What do all of these athletic maneuvers have in common? Angular momentum, the amount of momentum an object generates as it spins. Just as linear momentum is a function of a body's mass and velocity, angular momentum depends on a body's mass and angular speed.

For example, the angular momentum of a rock spinning at the end of a string is the product of its mass m, its speed v, and the radial distance r. Application of an external net force is needed to change the linear momentum of a body, and a torque must be applied to change the angular momentum of a rotating body. In addition, angular momentum is conserved, just as linear momentum is.

If you ride a bicycle, you can demonstrate the effects of angular momentum and torque. A bike at rest is very difficult to balance because the wheels have no angular momentum and the slightest torque will tip

the bike. The wheels of a moving bike, however, have considerable angular momentum; a much greater torque is needed to change the magnitude or direction of spin of the wheels.

Divers, gymnasts, and acrobats use two basic rotational motions in their routines: twists and somersaults. All acquire some angular momentum when they launch themselves into the air. It is because this initial angular momentum is conserved that their sometimes astonishing twists and spins are possible. In other words, reducing angular momentum about one axis of rotation induces it in another axis of rotation and vice versa.

For example, suppose a diver somersaulting forward in the layout position (body straight) throws one arm above his head and the other down to his waist. This change in distribution of body mass reduces his somersaulting angular momentum. But angular momentum is conserved. Thus, the loss of somersaulting angular momentum must be compensated for by a gain in twisting angular momentum. The diver's body begins to twist around a vertical axis running from head to toe. It isn't possible to completely convert the somersaulting motion to twist, so the body does both. To eliminate the twist before entering the water, the diver throws his arms back to their original positions.

If your physics lab has a low friction turntable, test one effect of changing angular momentum. Stand on the turntable with a book in each hand and your arms at your sides. Have a classmate start you spinning *slowly* by pulling on one edge of the turntable. To observe the effect of a change in angular momentum, raise your arms so they extend straight out from your body. Because you have redistributed your mass and increased r, your angular velocity v must decrease. You spin more slowly.

Law of Work

9

1. To learn how forces do work.
2. To learn the role of friction in resisting forces.
3. To note the relationship between the work input and the work output in a frictionless machine.
4. To note the relationship between the work input and the work output in a practical machine.
5. To learn how to compute and measure power.

Aims

9-1 Force and Work

We are constantly using forces to move ourselves, vehicles, and bodies of all kinds from one place to another. When a force moves a body on which it acts in the direction of the force, we say it has done *work*. A mover lifting a box from the floor to a table or pushing a piano a few meters across the floor is exerting a force that is doing work. In each case, a force is actually moving the body in the direction in which the force is acting.

The work done by a force is found by multiplying the force by the distance it has pushed the body in the direction of the force.

$$\textbf{Work} = \textbf{Force} \times \textbf{Distance}$$
$$\textbf{W} = \textbf{f} \times \textbf{d}$$

The unit of work is obtained simply by combining the unit of force and the unit of distance. It is the newton-meter, which is also called the *joule* (JOOL), symbol, J. *A newton-meter, or a joule, is the work done by a force of 1 newton in moving an object a distance of 1 meter.*

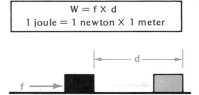
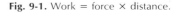

Fig. 9-1. Work = force × distance.

Sample Problems

1. What work is done by a girl who pushes a box along a floor with a force of 80.0 N for a distance of 10.0 m?

   Solution:
   $$f = 80.0 \text{ N} \qquad d = 10.0 \text{ m}$$
   $$W = f \times d$$
   $$W = 80.0 \text{ N} \times 10.0 \text{ m}$$
   $$W = 800 \text{ N·m} = 800 \text{ J}$$

2. A boy raises a 20.0-kg rock 1.50 m. (*a*) What force does the boy use to raise the rock? (*b*) What work does he do?

Solution:
(*a*) The boy must use an upward force equal to the weight of the rock.
$$m = 20.0 \text{ kg} \qquad g = 9.80 \text{ m/s}^2$$
$$w = mg = 20.0 \text{ kg} \times 9.80 \text{ m/s}^2 = 196 \text{ N}$$
(*b*) $f = 196$ N $\qquad d = 1.50$ m
$$W = f \times d$$
$$W = 196 \text{ N} \times 1.50 \text{ m}$$
$$W = 294 \text{ N·m} = 294 \text{ J}$$

Test Yourself Problems

1. What work is done in pushing a cart 3.00 m with a force of 250 N?
2. How much work is done in lifting a 2.50-kg box a distance of 6.00 m?

Fig. 9-2. No matter how much force is exerted, no work is done in supporting the dancer unless there is upward displacement against gravity.

Photo by Kenn Duncan

9-2 Direction of Motion and Work

A force does work only when it actually moves the object it is pushing in the direction of the force. The boy in sample problem 2 does work because his upward force of 196 newtons actually moves the rock upward 1.50 meters. If, at this point, the boy holds the rock motionless, he does no further work, although still exerting an upward force of 196 newtons. Since this force is no longer moving the rock upward, it is no longer doing work.

When a child pulls a sled forward by means of a rope making an angle to the ground, it seems that an upward force acting along the rope is doing work in pulling the sled forward. However, this is not true. Only that component of the child's force that acts parallel to the ground is doing the work of pulling the sled forward. To find the work done here, we must multiply the horizontal component of the child's force by the horizontal distance it moved the sled. (See Sec. 4-20.)

Sample Problem

A child pulls a sled over a level field of snow with a force of 50.0 N by means of a rope making an angle of 30.0° with the ground. How much work is done in pulling the sled forward 20.0 m?

Solution:

The horizontal component of the child's force is f_h = 50.0 N × cos 30.0° = 50.0 N × 0.866 = 43.3 N

f_h = 43.3 N d = 20.0 m

$W = f_h \times d$
W = 43.3 N × 20.0 m = 866 N·m = 866 J

Test Yourself Problem

A laborer is pulling a loaded cart by means of a rope that makes an angle of 20.0° with the horizontal. The laborer exerts a force of 100 N. How much work is done when the cart is moved forward 8.00 m?

9-3 Machines Help Us Do Work

We use machines to help us do work. We need them because the forces we have available are not always suitable for the jobs we must do. In changing a tire of a car, for example, we cannot, with our muscles alone, exert enough force to lift the car. However, by applying our force to a jack, we can change it into a force large enough to lift the car comfortably.

Again, such a simple task as cutting the grass on a lawn would be exhausting and endless if we had to do it without the aid of any device. With a lawn mower, the task is done quickly and easily. The jack and the lawn mower are examples of *machines*. We use machines to make better use of the forces at our command in doing the work of everyday living in the home, on the farm, in industry, and in commerce. We shall see that machines do not save us any work. They simply make it easier or more convenient to do a given job.

9-4 Ideal Machines

In every machine, friction caused by the rubbing of parts against each other interferes with the best operation of the machine. These frictional losses vary from machine to machine and tend to obscure the basic law that machines obey. Hence, for simplicity, we are going to imagine that we are dealing with machines that have no frictional or other losses. Such machines are called perfect or *ideal machines*. Since ideal machines do not exist, we can experiment with them only in thought. But, just as in the case of Galileo's thought experiments on falling bodies, we shall find that our thought experiments and analyses of ideal machines will help us to discover the basic law that underlies the operation of all machines. This is the *law of work*. Later we shall show how the law of work must be modified to apply to real machines.

An ideal machine is free of friction.

9-5 Law of Work

To illustrate the law of work, consider the person in Fig. 9-3 who is lifting a 400-newton box by means of a combination of pulleys. We shall assume that this is an ideal machine and that the pulleys are weightless. Note that the weight of the 400-newton box is supported and shared by four strands that go around the lower two pulleys. Each strand therefore supports one fourth of the weight, or 100 newtons. The operator pulls on only the outermost strand, hence must exert a force of 100 newtons. Thus, the 400-newton box is being raised with a force of only 100 newtons.

It looks as if we are getting something for nothing. However, we soon learn that this is not so, because for every 4.00 meters of rope pulled in, the pulley system raises the box only 1.00 meter. Thus, while we can lift the box with only one-fourth as much force as is needed to lift it without the pulleys, the force must be exerted four times as far as the distance the box is raised.

Now the work put into a machine by the operator is called the *work input*. It is found by multiplying the force applied by the operator, called the *effort, E,* by the distance, D_E, over which that force acts. Thus,

$$\text{Work Input} = E \times D_E$$

In this case, the 100-newton effort acts through 4.00 meters so that the work input = 100 newtons × 4.00 meters = 400 joules.

The work done by a machine in raising a load or overcoming a resistance is called the *work output*. It is found by multiplying the

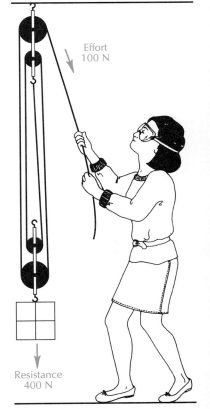

Fig. 9-3. The law of work applied to a pulley system with four supporting strands.

load or *resistance*, R, by the distance, D_R, it is moved. Thus,

$$\text{Work Output} = R \times D_R$$

The pulley system raised the 400-newton resistance 1.00 meter, so the work output = 400 newtons × 1.00 meter = 400 joules.

Notice that the work output of 400 joules is exactly equal to the work input of 400 joules. This illustrates the law of work which is true for all ideal machines. It states:

In an ideal machine, work is neither gained nor lost. The work output is exactly equal to the work input.

$$\text{Work Output} = \text{Work Input}$$
$$R \times D_R = E \times D_E$$

9-6 Application of the Law of Work to the Inclined Plane

The mover in Fig. 9-4 is using a machine called an inclined plane to push an 800-newton safe into a truck whose platform is 1.00 meter above the ground. The inclined plane is simply a 5.00-meter plank that makes a ramp from the ground to the truck platform. Without this plank, the mover would have to lift the safe directly upward a distance of 1.00 meter. This would require the exertion of an upward force equal to the weight of the safe and is more force than the mover can comfortably exert. This is avoided by pushing the safe up the inclined plank. This requires much less force than is needed to lift the safe, but the mover must now push the safe the 5.00-meter length of the plank to make the safe rise 1.00 meter. Assuming this is an ideal machine exactly how much force must be exerted to push the safe up the plank? We can determine this from the law of work as follows.

The work input is the force exerted times the distance over which it acts, which is 5.00 meters.

$$\text{Work Input} = E \times D_E = E \times 5.00 \text{ meters}$$

The work output is the work done in lifting the 800-newton safe 1.00 meter.

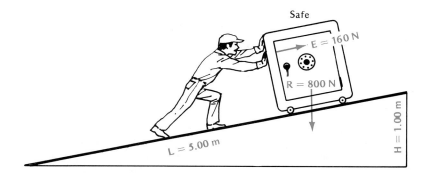

Figure 9-4. The law of work applied to an inclined plane.

Work Output = $R \times D_R$ = 800 newtons × 1.00 meter

Setting the work input equal to the work output, we have:

$$E \times 5.00 \text{ meters} = 800 \text{ newton-meters}$$
and
$$E = 160 \text{ newtons}$$

Thus, we see that one can push the 800-newton safe up the inclined plane with only 160 newtons of force or only one fifth the weight of the safe. However one pays for this advantage by having to push the safe 5.00 meters to make it rise only 1.00 meter.

Sample Problems

1. A 4.00-m plank is used to make an inclined plane for the purpose of rolling an oil barrel weighing 500 N onto a platform 1.20 m above the ground. What force must be applied to the barrel?

    Solution:
    $D_E = 4.00$ m $R = 500$ N $D_R = 1.20$ m

    Work Input = Work Output
    $E \times D_E = R \times D_R$
    $E \times 4.00$ m $= 500$ N $\times 1.20$ m $= 600$ N·m
    $E = 150$ N

2. A certain set of pulleys enables the operator to lift a load weighing 1200 N by applying a 200-N force. How much rope must the operator pull in to make the load rise 2.00 meters?

    Solution:
    $E = 200$ N $R = 1200$ N $D_R = 2.00$ m
    Work Input = Work Output

    $E \times D_E = R \times D_R$
    200 N $\times D_E = 1200$ N $\times 2.00$ m $= 2400$ N·m
    $D_E = 12.0$ m

Test Yourself Problems

Note. Assume that these are ideal machines.

1. An 8.00-m ramp is used as an inclined plane to push a safe onto a truck platform that is 2.00 m above the ground. If the safe weighs 1,800 N, what force is needed to push it up the ramp?
2. In a set of pulleys, the load rises 1.00 m for every

4.00 m of rope that the operator pulls in. What load can be lifted with this machine when the operator applies a force of 200 N?

3. The work put into a certain hoisting machine is 600 J. If the machine is being used to raise a load of 400 N, how high does it raise the load?

9-7 Force-Distance Relationship in Ideal Machines
From the law of work, it follows that:

$$\frac{R}{E} = \frac{D_E}{D_R}$$

This relationship states that what we gain in force in an ideal machine, we lose in distance, and vice versa. In a machine where the resistance is greater than the effort, the distance the effort moves is greater than the distance the resistance is moved in the same ratio. This was demonstrated in the case of the inclined plane (Section 9-6).

In a machine where the effort is greater than the resistance, the effort's distance is smaller than the resistance distance in the same ratio. The bicycle is such a machine. In it, the rider exerts a large

What we gain in force in a machine, we pay for in distance.

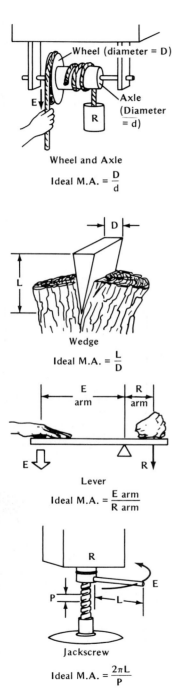

Wheel and Axle

Ideal M.A. $= \dfrac{D}{d}$

Wedge

Ideal M.A. $= \dfrac{L}{D}$

Lever

Ideal M.A. $= \dfrac{E\ arm}{R\ arm}$

Jackscrew

Ideal M.A. $= \dfrac{2\pi L}{P}$

force on the pedal which, acting through a chain and gear system, turns the rear wheel with a much smaller force. By doing this, the rider gains distance because when the pedal is moved a small distance, the rear wheel moves forward a proportionately larger distance.

9-8 Simple Machines and Mechanical Advantage

All machines, however complex, consist of one or more of six simple machines: the pulley, the inclined plane, the wheel and axle, the lever, the wedge, and the jackscrew. Each of these machines is characterized by its *ideal mechanical advantage,* which is the ratio of the distance the effort moves to the distance the resistance is moved. For an ideal machine, this ratio is equal to the ratio of the resistance overcome to the effort applied (as noted in Section 9-7). The ideal mechanical advantage expresses the force-distance relationship in an ideal machine. Thus, for a machine with an ideal mechanical advantage of 10, we know that, in the absence of frictional losses, any effort we apply to the machine will overcome a resistance 10 times as great. To pay for this advantage, however, the effort will have to act through 10 times the distance it moves the resistance. In a real machine, additional effort will be needed to overcome friction. Hence the *real mechanical advantage,* which is defined as the resistance divided by the actual effort, will be less than the ideal mechanical advantage. However, the distance relationships will be the same as for an ideal machine. Thus, in a real inclined plane, such as the one shown in Fig. 9-4, the actual effort needed to push the safe up the plane will be more than 160 N, but the effort will still move five times as far as it raises the safe.

The ideal mechanical advantage of each of the simple machines can be obtained from its appearance or its dimensions. The ideal mechanical advantage of a set of pulleys is equal to the number of strands holding up the resistance. The pulley system in Fig. 9-3 has 4 strands holding up the resistance and therefore has an ideal mechanical advantage of 4. The ideal mechanical advantage of an inclined plane is its length divided by its height. The inclined plane in Fig. 9-4 has an ideal mechanical advantage of 5.00 m ÷ 1.00 m = 5. Fig. 9-5 shows the remaining simple machines and the relationships that give their ideal mechanical advantages.

9-9 Friction in Real Machines

Friction in real machines can be favorable as well as unfavorable. Because friction resists the motion of all moving parts of the machine, some of the work input is wasted in overcoming it. We therefore usually try to reduce friction. However, in some machines, such as auto brakes, friction is useful and it is desirable to increase it.

9-10 Friction and Machine Design

In designing machines, one must consider how friction will affect the operation of the machine. We know that the friction between two rubbing surfaces depends on their roughness. Thus, when it is desirable to reduce friction, surfaces should be made smooth, as on the underside of skis and toboggans. On the other hand, when it is desirable to increase frictional resistance, surfaces that rub against each other should be made rough. For example, both the rubber tires of a car and the road they travel on must be rough enough so that the frictional force between them will provide adequate traction when the car is moving and adequate resistance when it is slowing to a stop. Otherwise, the car will skid, as it can on an icy road.

In some machines, friction is controlled by varying the normal force that presses two sliding surfaces together. This is the case in the brakes of some bicycles. When the rider squeezes a hand lever, a clamp grasps the front wheel, so it begins to slow down. The more pressure the rider exerts, the greater is the normal force pressing the clamp against the wheel. This increases the frictional resistance between the clamp and the wheel and stops the bicycle sooner.

Fig. 9-6. The high speeds achieved by this skater illustrate the relatively low friction between skate blades and ice.

9-11 Reducing Friction in Machines

One of the methods of reducing friction in a machine is to polish and smooth out the surfaces that slide over each other. This is done to the cylinders of an automobile engine so that the pistons will slide through them smoothly. Another method is to use ball bearings so that moving surfaces will roll rather than slide over each other. A third method is to lubricate the parts that move over each other. Here, a film of oil separates the rubbing surfaces so that each body slides over the slippery oil film rather than over the other body.

9-12 Law of Work in Real Machines

To operate a real machine, it is necessary to apply more force than would be needed if the machine were ideal. The extra effort is needed to overcome the frictional resistance of the machine. It follows, therefore, that the actual work input is always greater than the useful work output. If the only losses occurring in the machine are due to friction, the difference between the work input and the work output is the work done in overcoming friction. In this case the law of work becomes: *Work Input = Useful Work Output + Work Used in Overcoming Friction.* Thus, while not all of the work put in is regained as useful work, it is all accounted for and the total work on the right side of the equation is equal to the work put in.

Fig. 9-7. An almost frictionless machine! In this ultracentrifuge, the friction of rubbing parts is eliminated by supporting the rapidly rotating toplike part in midair by a magnet.

9-13 Efficiency

To describe a machine's actual performance, we use the term *efficiency*. The efficiency of a machine is given by the formula:

$$\text{Efficiency} = \frac{\text{Output}}{\text{Input}} \times 100\%$$

The efficiency tells us what percent of the work put into a machine is returned as useful output. The remainder of the work put into the machine is used in overcoming friction and other useless resistance. Since work output and work input are equal in a perfect machine, the efficiency of such a machine would be 100%.

In a set of pulleys that is 60% efficient, 60% of the work put in by the operator is used in lifting the load. The remaining 40% of the input is used in overcoming the friction of the wheels and the weight of the movable pulleys.

Sample Problems

1. A set of pulleys is used to raise a 150-N weight 4.00 m. To do this, the operator applies a force of 40.0 N and pulls in 20.0 m of rope. Find the (a) work output; (b) work input; (c) work used to overcome friction; and (d) efficiency.

Solution:

(a) $R = 150$ N $D_R = 4.00$ m
    Work Output $= R \times D_R$
    Work Output $= 150$ N $\times 4.00$ m
    Work Output $= 600$ N·m $= 600$ J

(b) $E = 40.0$ N $D_E = 20.0$ m
    Work Input $= E \times D_E$
    Work Input $= 40.0$ N $\times 20.0$ m
    Work Input $= 800$ N·m $= 800$ J

(c) The work used to overcome friction is: Work Input − Work Output $= 800$ J − 600 J = 200 J

(d)   Efficiency $= \dfrac{\text{Output}}{\text{Input}} \times 100\%$

     Efficiency $= \dfrac{600 \text{ J}}{800 \text{ J}} \times 100\% = 75\%$

2. A machine having an efficiency of 70.0% lifts a 5.00 kg object 10.0 m. (a) What is the weight of the object? (b) What is the work output? (c) What is the work input? (d) How much work was expended in overcoming friction?

Solution:

(a) $m = 5.00$ kg $g = 9.80$ m/s²
    $w = mg$
    $w = 5.00$ kg $\times 9.80$ m/s²
    $w = 49.0$ N $= R$, the resistance

(b) $R = 49.0$ N $D_R = 10.0$ m
    Work Output $= R \times D_R$
    Work Output $= 49.0$ N $\times 10.0$ m
    Work Output $= 490$ N·m $= 490$ J

(c) Efficiency $= \dfrac{\text{Output}}{\text{Input}} \times 100\%$

    $70.0\% = \dfrac{490 \text{ J}}{\text{Input}} \times 100\%$

    Input $= 490$ J $\times \dfrac{100\%}{70.0\%}$

    Input $= 700$ J

(d) The work used to overcome friction is:
    Work Input − Work Output
    $= 700$ J − 490 J = 210 J

Another way to do this is to note that since the efficiency is 70.0%, 30.0% of the work input is lost in overcoming friction. Hence the work used to overcome friction is:
    $30.0\% \times 700$ J $= 210$ J

Test Yourself Problems

1. A box weighing 400 N is pulled up a 40.0-m incline to a point 1.60 m above its starting level. The force needed to pull the box is 20.0 N. Find (a) the work output; (b) the work input; (c) the work expended on friction; and (d) the efficiency.

2. A 60.0% efficient engine is used to pull a car 1000 m along a track. If the force applied by the engine is 2500 N, what is (a) the work output? (b) the work input; and (c) the work used to overcome friction and other losses?

9-14 Power

We are often interested not only in how much work is done but also in how fast it is done. *The rate at which work is done is called power.* We find the average power of a machine or device by dividing the work it does by the time it takes to do it.

$$\textbf{Average Power} = \frac{\textbf{Work}}{\textbf{Time}}$$

$$\textbf{P} = \frac{\textbf{w}}{\textbf{t}}$$

If a machine does 50 newton-meters of work in 10 seconds, its average power is:

$$P = \frac{50 \text{ N} \cdot \text{m}}{10 \text{ s}}$$

$$P = 5.0 \text{ N} \cdot \text{m/s or } 5.0 \text{ J/s}$$

Fig. 9-8. This huge mechanical shovel illustrates how a lot of power can be harnessed through combinations of simple machines to do useful work. Can you find where the operator sits?

The unit in which power is measured is the *watt,* symbol, W. *The power supplied by an engine or motor that does work at the rate of 1 joule per second is 1 watt.*

The unit of power is the watt (W).

A watt is therefore the rate at which a force of 1 newton does work when it moves a body at the rate of 1 meter per second. The watt is commonly used to measure the power of electrical devices and circuits but it is also a general unit used to measure the power of any engine. Since the watt is a rather small unit, it is often more convenient to measure power in kilowatts. *A kilowatt is 1000 watts.*

A nonmetric unit of power commonly used in commerce and industry is the horsepower. *A horsepower is equal to 746 watts.* Thus a ten horsepower motor may also be identified as a 10 × 746 = 7460-watt motor.

If we know the power of an engine or motor in watts, we can find the work done by it in a given time by simply multiplying its power by the time. Thus a 1-watt engine working for 1 second, does 1 watt-second of work. Since a watt is 1 joule per second, a watt-second is equal to 1 joule. An engine doing work at the rate of 1 kilowatt for 1 hour is said to do 1 kilowatt-hour of work. The kilowatt-hour (symbol, kWh) is the unit used commercially to measure electrical energy. Since there are 3600 seconds in an hour, 1 kilowatt-hour is equal to 1000 watts × 3600 seconds = 3.6 × 10^6 watt-seconds or 1 kilowatt-hour = 3.6 × 10^6 joules.

The power needed to launch the space shuttle is about 10^{10} watts. This is equal to the power output of 10 large electric power plants.

Sample Problems

1. A student pushing a wagon with a force of 40 N moves it 12 m in 10 s. Find (a) the work done; and (b) the power exerted.

Solution:
(a) $f = 40$ N $d = 12$ m $t = 10$ s
$$w = f \times d$$
$$w = 40 \text{ N} \times 12 \text{ m}$$
$$w = 480 \text{ J} = 4.8 \times 10^2 \text{ J}$$

(b) $P = \dfrac{w}{t}$

$$P = \frac{480 \text{ J}}{10 \text{ s}} = 48 \text{ J/s} = 48 \text{ W}$$

2. How much work can a 250-W motor do in 12 s?

Solution:
$P = 250$ W $t = 12$ s

$P = \dfrac{w}{t}$ whence $w = P \times t$

$w = 250$ W $\times 12$ s $= 3000$ W $\cdot$ s $= 3.0 \times 10^3$ J

3. A motor exerting a steady force of 10 N on an object keeps it moving forward at a speed of 2.0 m/s. What is the power of the motor?

Solution:
$f = 10$ N $v = 2.0$ m/s

$$P = \frac{w}{t} = \frac{f \times d}{t} = f \times \frac{d}{t} = f \times v$$

$P = 10$ N $\times 2.0$ m/s $= 20$ N$\cdot$m/s $= 20$ W

4. A 50.0-kg girl climbs 4.0 m up a rope in the school gymnasium in 20 s. (a) How much force must she exert to pull herself up the rope? (b) What is her power?

Solution:
(a) To pull herself up the rope, the girl must exert a force equal to her weight. This force is
$$f = mg$$
$m = 50.0$ kg $g = 9.80$ m/s^2
$f = 50.0$ kg $\times 9.80$ m/s$^2 = 490$ N

(b) $f = 490$ N $d = 4.0$ m $t = 20$ s
$$P = \frac{f \times d}{t}$$
$$P = \frac{490 \text{ N} \times 4.0 \text{ m}}{20 \text{ s}}$$
$P = 98$ J/s $= 98$ W

Test Yourself Problems

1. A student weighing 500 N raises his body 0.300 m on a chinning bar in 1.50 s. (a) What work was done? (b) What was the student's power?
2. How much work can a 125-W motor do in 2.50 s?
3. An 8000-N elevator rises at a speed of 0.30 m/s. What is the power of the motor that operates it?
4. The operator of a pulley system exerts a force of 80.0 N over a distance of 12.0 m for a period of 8.00 s. What is the operator's power?

CHAPTER REVIEW

Summary

Work is done by a force f when it moves an object through a distance d in the direction in which the force is acting. The work done is $\mathbf{W} = \mathbf{f} \times \mathbf{d}$. Work is measured in newton-meters or joules. A **joule** is one newton-meter.

Machines are devices that can change the direction and magnitude of a given force so as to make it more suitable for the work it is to do. The force applied to a machine is called the **effort**. The force that the machine is trying to overcome or the load it is trying to move is called the **resistance.** The work done in operating a machine is called the **work input.** The work done on the resistance in moving or lifting it is called the **work output.** An **ideal machine** is one in which there are no frictional losses. All ideal machines obey the **law of work** which states: *In an ideal machine, the work output is always equal to the work input.*

For any machine, whether real or ideal, the **ideal mechanical advantage** tells us the relationship between the distance the effort moves and the distance the resistance moves. It is defined as the ratio of the effort distance to the resistance distance. It can be obtained from the dimensions of the machine. The **real mechanical advantage** of a machine is defined as the ratio of the resistance to the actual effort needed to operate the machine. The real mechanical advantage is always smaller than the ideal mechanical advantage.

Friction is the force that resists the movement of the surface of one body over the surface of another. In a practical machine, some effort applied to the machine is used to overcome the friction between its many parts. As a result, the work input in a real machine is always greater than the work output. In a real machine, the law of work takes the form: *Work Input = Work Output + Work Used to Overcome Friction.*

The **efficiency** of a machine tells what percent of the work put into the machine is returned as useful output. It is given by:

$$\text{Efficiency} = \frac{\text{Output}}{\text{Input}} \times 100\%$$

The efficiency of a machine may be increased by reducing the friction between its moving parts by such means as lubrication and the use of ball bearings.

The rate at which work is done is called **power** and is found from the relationship:

$$\text{Power} = \frac{\text{Work}}{\text{Time}}$$

The metric unit of power is the **watt** representing a rate of working of one joule per second. A watt is therefore the rate at which a force of one newton does work in moving an object at a speed of one meter per second. A **horsepower** is equal to 746 watts.

Questions

Group 1

1. (a) Under what conditions does a force do work? (b) In what units is work measured?
2. In which of these cases is a youngster doing work in the school gymnasium? (a) Climbing up a rope. (b) Hanging motionless from a chinning bar. (c) Lifting a set of weights from the floor to overhead.
3. A child is pulling a wagon by a rope making an angle of 30° with the ground. (a) Why isn't all of the child's force doing work as the wagon moves forward? (b) What part of it is doing work?

4. (a) What is the law of work for an ideal machine? (b) If an ideal machine enables an effort to overcome a resistance 10 times as large as itself, how does the distance the effort moves compare with the distance the resistance moves?
5. (a) What is friction? (b) What are some causes of friction? (c) How may friction be reduced in a machine?
6. A machine has an ideal mechanical advantage of 5. What does that tell us?
7. (a) How does the work input compare with the work output in a real machine? (b) What does

the difference between input and output represent?

8. (a) Define efficiency. (b) What part of the work put into a machine having an efficiency of 65% is used in overcoming friction? (c) What can be done to increase the efficiency of a machine?

9. (a) Define power. (b) Define a watt. (c) Two motors do the same amount of work but one takes twice as long as the other. Compare the power of the motors.

10. How much work can a 1-kW motor do in 1 s?

Group 2

11. The earth's gravitation provides the centripetal force that keeps a satellite moving in a circular orbit around the earth. Does the centripetal force do any work? Explain.

12. A laborer pushes a cart weighing 1000 N along a level road with a horizontal force of 100 N.

Which of these forces does work as the cart moves forward? Explain.

13. (a) In doing a certain amount of work, how does the actual effort one must apply to a real machine compare with the effort that would be needed if the machine were ideal? (b) Show that the efficiency of a machine may be expressed as: $\dfrac{\text{ideal effort}}{\text{actual effort}} \times 100\%$.

14. A powerful rifle is one that propels a bullet with a high muzzle velocity. To what extent does this use of the word "powerful" agree with the definition of power in physics?

15. What information would you need to compute the power of a motor that operates an elevator in a building?

Problems

Group 1

1. Two motorists exert a 400-N force on a stalled car to move it a distance of 60 m. How much work do they do?

2. How much work is done by a girl weighing 500 N in climbing 5.0 m up a rope in the school gymnasium?

3. A 60.0-kg boy lifts himself on a chinning bar a distance of 0.30 m. (a) What does the boy weigh? (b) What force must he exert to lift himself? (c) How much work does he do?

4. A student pulls a wagon on a level road by means of a rope making an angle of 45° with the ground. The force exerted is 80 N and the wagon moves 20 m. (a) What part of the force is doing work? (b) How much work is done?

5. A gardener pushes a lawnmower 40.0 m across a level lawn with a force of 120 N along the handle which makes an angle of 30.0° with the ground. (a) How much of the applied force does work? (b) How much work is done?

6. A piano weighing 2000 N is pushed up an inclined plane 3.00 m long into a truck whose platform is 1.20 m above the ground. (a) Ignoring frictional losses, what force must be exerted to push the piano? (b) What is the ideal mechanical advantage of the plane?

7. Using a set of pulleys, a mover raises a 1000-N

crate a distance of 5.00 m. This is accomplished by pulling in 20.0 m of cord. (a) If this were an ideal machine, how much effort would have to be applied? (b) If the effort actually applied is 300 N, what force is used to overcome friction? (c) What is the work output? (d) What is the work input? (e) What is the real mechanical advantage?

8. An effort of 120 N is applied to a machine to move a 720-N resistance 4.0 m. Assuming this to be an ideal machine, over what distance must the effort act?

9. By applying a force of 75.0 N over a distance of 8.00 m to a set of pulleys, a student raises a 300-N object to a height of 1.50 m above the floor. What is (a) the work input; (b) the work output; and (c) the efficiency?

10. For the set of pulleys of problem 9, find both the ideal and the real mechanical advantage.

11. In a set of pulleys, 8 cords hold up a resistance of 640 N. (a) What is the ideal effort? (b) How much cord must the effort pull in to raise the resistance 2 m?

12. A certain machine does 5000 J of useful work and expends 1000 J in overcoming friction. Find (a) the work input; and (b) the efficiency of this machine.

13. A 600-N box is raised 6.0 m by a machine that

is 60% efficient. What is (a) the work input; (b) the work output; and (c) the work used in overcoming friction and other losses?

14. What is the power of a motor that does 700 J of work in 35 s?

15. How much work does a 250-W motor do in 20 s?

16. How long will it take a 500-W motor to raise the 1000-N hammer of a pile driver 10 m?

17. A boy applies a steady force of 32 N to a cart and pushes it 12 m in 16 s. What is his power?

18. A motor raises a 5000-N elevator 7.5 m in 15 s. Find (a) the work done; (b) the power of the motor.

19. A 60.0-kg woman climbs a staircase to a level 11.0 m above her starting point in 66.0 s. (a) How much force does she exert in raising herself from step to step? (b) How much work does she do in getting to the higher level? (c) What is her power?

20. A 300-W motor is used to run a small electric cart along a level plane. If it takes 100 N to overcome frictional resistance, at what speed does the motor keep the cart moving?

Group 2

21. A machine having an efficiency of 50% is used to operate a machine having an efficiency of 60%. (a) What is the combined efficiency of the two machines. (b) What work must be put into this combination to obtain a work output of 6.0×10^5 J?

22. A worker is pulling a crate across a warehouse floor by means of a rope making an angle of 25.0° with the floor. A force of 200 N is applied to the rope and pulls the crate 20.0 m in 28.0 s. (a) How much work is done? (b) What is the power?

23. A 20-kilowatt motor operates an elevator weighing 1.0×10^4 N. What is the maximum speed at which the motor can raise the elevator?

24. A worker is rolling a 480-N barrel up a 16-m ramp onto a platform 2.0 m above the floor. The efficiency of this arrangement is 40%. What is the (a) work output; (b) work input; (c) effort applied to the barrel?

25. An airplane weighs 2.0×10^5 N and is designed to gain altitude at the rate of 10 m/s. What must be the minimum power of its engines?

26. A 16 000-N car moves up a hill at 12 m/s. The hill is 360 m long and 15 m high. (a) What work does the car's engine do in getting it to the top of the hill? (b) How long does it take? (c) What power does the engine furnish?

27. An 80-kg gymnast raises himself 0.40 m each time he chins on the gymnasium chinning bar. If he chins ten times in 20 s, what is his average power?

Applying Physics

1. Estimate the power of the motor that operates the elevator in a school, office building, or apartment house as follows. Find out the maximum capacity or weight the elevator is designed to lift. This is generally available on a notice posted in the elevator. Express this weight in newtons. Measure the distance between two adjacent floors and multiply by the number of floors to estimate the total height in meters the elevator is raised when going from the lowest to the highest floor. Now time the elevator as it makes a full run from the bottom to the top floor. From the formula for power, calculate the power of the motor.

2. Determine your power by measuring the time it takes you to run up two flights of stairs.

   CAUTION: Do not do this if your medical condition dictates otherwise.

Find the height you have raised your body by measuring the height of each step and multiplying by the number of steps. Then multiply this height in meters by your weight in newtons and divide the product by the time in seconds. The result will be your power in watts.

3. Tie a string around the lower part of a glass tumbler and fasten a weak rubber band to the string. Pull the empty tumbler over a smooth table top with just enough force to keep it moving at uniform speed. Measure how much the rubber band stretches.

   Now repeat the experiment with the glass tumbler half full of water and again with the tumbler nearly full of water. In each case, measure how much the rubber band stretches. What does this experiment show about the relationship between the force needed to overcome friction and the weight of the object being pulled?

10 Energy and Its Conservation

The SI unit of energy is the joule (J).

10-1 Energy Units

Since energy is the ability to do work, we measure energy in terms of the amount of work it can do. Hence, the unit used to measure energy is the same as that used to measure work, the newton-meter or joule. If all the energy obtainable from a quantity of gasoline can do 1000 joules of work, we say the gasoline contains 1000 joules of energy.

10-2 Conservation of Energy in Ideal Machines

We have seen that in ideal machines the work output is exactly equal to the work input. Another way of stating this principle is to say that the energy we get from an ideal machine is exactly equal to the energy we put in. The machine simply converts the work or energy we put into it to a form more suitable for the job to be done. In a combination of pulleys, for example, the energy input in the form of a small effort acting over a long distance is changed into energy in the form of a large force lifting a heavy load a short distance. However, no matter what energy changes take place in the machine, the energy available to do work as output is exactly equal to the work put in. Thus, in the ideal machine none of the energy put in is lost and no new energy is created.

10-3 Energy Conservation in Real Machines

For a practical machine, in which the only losses are due to friction, some of the energy put in is used to overcome friction. If we add the machine's energy output in the form of useful work to the energy used to overcome friction, we find that the sum is equal to the work or energy put into the machine. Thus, once more we can account for all the energy put into the machine.

Our study of machines gives us a preview of a most important property of energy. That is:

> When energy is transferred from one body to another or from one form to another, no energy is lost and no new energy is created.

10-4 Kinetic Energy

Anyone who has seen the damage produced by floods or raging wind storms knows that moving water and moving air can possess enormous amounts of energy. Unfortunately, when such energy is uncontrolled, it is quite capable of destroying entire communities. However, the energy of moving water and air may also be put to useful purposes. By setting up windmills and waterwheels, this energy can be used to operate mills, generate electricity, and do other useful work.

Kinetic energy is energy associated with motion.

Fig. 10-1. The kinetic energy of a raging wind storm exerts a powerful destructive force.

Moving air and water are examples of bodies that possess a type of energy called *kinetic* (kih·NEH·tik) *energy. By kinetic energy we mean the energy that a body has because of its motion.* Any moving body has kinetic energy because it is able to do work by moving other bodies. Thus, moving cars, trains, ships, airplanes, and projectiles all possess kinetic energy. So do the earth, the moon, the planets, and all other heavenly bodies that are in motion.

10-5 Expression for Kinetic Energy

Let us suppose that a block having mass m resting on a frictionless table is pushed by a constant force f for a distance d. The work done by the force on the block is $f \times d$. This work causes the block to be accelerated and to acquire kinetic energy. If a is the acceleration produced by f, then $f = ma$. It follows that the work done on the block is:

$$fd = mad$$

For a body that starts from rest and is accelerated at a constant rate a, the speed acquired by the body after traveling a distance d is given by $v^2 = 2\,ad$ whence,

$$ad = \frac{v^2}{2}$$

Substituting this value of ad in $fd = mad$, we have:

$$fd = \frac{mv^2}{2}$$

This relationship tells us that the work done on the block by the force in speeding it up from rest to speed v is equal to $mv^2/2$. This quantity is defined as the kinetic energy KE of the body. That is,

$$KE = \tfrac{1}{2}\,mv^2$$

The kinetic energy of a body tells us how much work was done on that body to bring it to its present speed from rest. It also tells us how much work that body can do by moving other bodies until it is brought to rest. Like work, kinetic energy is expressed in newton-meters, or joules.

Sample Problem

(a) What is the kinetic energy of a mass of 5.0 kg moving at 4.0 m/s? (b) If the mass was accelerated from rest for a distance of 10 m, what force was applied to it?

Solution:

(a) $m = 5.0$ kg $v = 4.0$ m/s
$KE = \tfrac{1}{2}\,mv^2$
$KE = \tfrac{1}{2}(5.0 \text{ kg})(4.0 \text{ m/s})^2$
$KE = 40 \text{ kg·m}^2/\text{s}^2 = 40 \text{ N·m} = 40 \text{ J}$

(b) $d = 10$ m $KE = 40 \text{ N·m}$

The work done by the force on the body is equal to the kinetic energy acquired by the body.
$fd = \tfrac{1}{2}\,mv^2 = KE$
$f \times 10\,m = 40 \text{ N·m}$
$f = 4.0 \text{ N}$

Test Yourself Problems

1. What is the kinetic energy of a car having a mass of 1.5×10^3 kg moving at 30 m/s?
2. A force of 10 N is applied to a body on a practically frictionless table over a distance of 0.80 m.

(a) What is the kinetic energy it imparts to the body? (b) If the body starts from rest and has a mass of 4.0 kg, what velocity does the force impart to it?

10-6 Force and Change in Kinetic Energy

Since an unbalanced force changes the velocity of the body on which it acts, it also changes its kinetic energy. If the force acts in the same direction in which the body is already moving, the force increases the body's kinetic energy. If the force acts in a direction opposite to that in which the body is moving, it decreases the body's kinetic energy. The work done by the force on the body is equal to the difference it produces in the kinetic energy of the body. Thus, if a force f acting upon a body of mass m and initial speed v_o, over a distance d, increases the speed of the body to v, then

$$\mathbf{fd} = \tfrac{1}{2}\,\mathbf{mv}^2 - \tfrac{1}{2}\,\mathbf{mv}_o{}^2$$

where fd is the work done by the force f, and the expression on the right is the change in kinetic energy produced by f on the body. The work fd is positive if the force increases the kinetic energy of the body and negative if the force decreases the kinetic energy of the body.

> With maximum braking force, a car needs four times the distance to stop when its speed is doubled.

Sample Problem

When the brake is applied to a car having a mass of 1000 kg, its speed is reduced from 30 m/s to 20 m/s. (a) How much work does the brake do on the car? (b) If the brake is applied for a distance of 25 m, what force does it exert on the car?

Solution:

(a) $m = 1000$ kg $v_o = 30$ m/s $v = 20$ m/s

The initial kinetic energy of the car was

$$\tfrac{1}{2}\,mv_o{}^2 = \tfrac{1}{2}(1000 \text{ kg})(30 \text{ m/s})^2$$
$$= 450\,000 \text{ N·m or J}$$

The final kinetic energy of the car was

$$\tfrac{1}{2}\,mv^2 = \tfrac{1}{2}(1000 \text{ kg})(20 \text{ m/s})^2$$
$$\tfrac{1}{2}\,mv^2 = 200\,000 \text{ N·m or J}$$

The work done by the brake is

$$fd = \tfrac{1}{2}\,mv^2 - \tfrac{1}{2}\,mv_o^2$$
$$fd = 200\,000 \text{ N·m} - 450\,000 \text{ N·m}$$
$$fd = -250\,000 \text{ N·m} = -2.5 \times 10^5 \text{ N·m or J}$$

The work is negative because it is slowing the car down.

(b) $d = 25$ m   Work $= -2.5 \times 10^5$ N·m
$$\text{Work} = fd$$
$$-2.5 \times 10^5 \text{ N·m} = f \times 25 \text{ m}$$
$$f = -1.0 \times 10^4 \text{ N}$$

The force is negative because it is acting opposite to the motion of the car.

Test Yourself Problem

The speed of a truck having a mass of 2.0×10^3 kg increased from 16 m/s to 25 m/s. (a) What work was done by the engine to produce this increase in speed? (b) If the increase in speed took place over a distance of 50 m, what force was acting upon the truck?

10-7 Potential Energy

Often, a body possesses energy in a stored form that is not readily noticeable and is called *potential* (poh·TEN·shul) *energy*. We may define potential energy as *the ability of a body to do work because of the relative position of its parts or because of its position with respect to other bodies.* A wound clock spring has potential energy because the relative position of its parts leaves it in a tense condition. As the spring unwinds, its stored energy is paid out gradually to do the work of running the clock.

An object some distance above the floor or ground has potential energy because of its relative position with respect to the earth. When the body is allowed to fall, the earth's gravitational attraction pulls it downward and thus enables it to move other objects

> Potential energy is associated with the position or state of strain of a body.

Fig. 10-2. Potential energy stored in the tennis racket and ball is changed into kinetic energy as the ball speeds away.

and do work. For this reason, the energy stored in such a body is referred to as *gravitational potential energy*. When the raised weight in a pile driver is allowed to fall, it uses its gravitational potential energy to drive a pile into the ground. Similarly, as the raised weights in a grandfather clock fall to the floor, they do work in running the clock.

On a large scale, the water stored up behind dams also has gravitational potential energy because of its high position with respect to the place into which it empties. This potential energy is used to furnish the work needed in electric powerhouses to run water turbines which operate generators and furnish electricity.

The potential energy of a body is measured by the work it is able to do when its position or the position of its parts is allowed to change. Like all other forms of energy, it is measured in work units: newton-meters or joules.

10-8 Gravitational Potential Energy and Base Level

Gravitational potential energy is particularly important to us because it concerns all bodies on or near the earth. We are especially interested in that part of the gravitational potential energy of a body that can be released to do work in a particular situation. As we have seen in the case of the weights in a grandfather clock or the weight in a pile driver, work is done only when a body is allowed to move or fall from a higher level to a lower level. *The difference between the potential energy of the body at the higher level and its potential energy at the lower level is the energy released to do work.*

For this reason, in any given situation, we compute only the difference between the gravitational potential energy that the body has in its given position and that which it will have when allowed to fall to some selected lower or base level. For objects in a room, for example, we generally select the floor as a convenient base level of potential energy because it is the lowest level to which objects in the room normally descend. The potential energy stored in a 5-newton weight on a table top 1 meter above the floor is the work that it takes to raise the weight from the floor to the top of the table. This is 5 newtons times 1 meter, or 5 newton-meters, or 5 joules. It means that, on the table, the weight has 5 joules of potential energy more than it has on the floor. By letting the weight fall or move down to the floor, we can use this stored energy to do 5 joules of work such as running a clock.

Once the weight has reached the floor, it no longer has gravitational potential energy with respect to the floor as a base level. However, it does have gravitational potential energy with respect to any level lower than the floor. If, for example, a hole were cut in the floor, the weight could then move or fall to the floor of the room below. In doing so, it could do more work. In this case, it

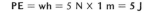

$$PE = wh = 5\ N \times 1\ m = 5\ J$$

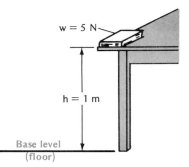

Fig. 10-3. Gravitational potential energy is measured with respect to a base level.

would be convenient to select the floor of the room below as the base level of potential energy. Since this level now determines how far objects in the room can descend, it also determines how much of their potential energy we can put to use in this situation.

10-9 Gravitational Potential Energy Near the Earth's Surface

To compute the gravitational potential energy that a body has over a selected base level, we simply compute the work needed to raise it from the base level to its actual position. For bodies near the surface of the earth, this work is equal to the weight of the body w times the height it was lifted h. Therefore,

$$\text{Potential Energy} = w \times h$$

and since

$$w = mg$$

$$PE = mgh$$

Sample Problems

1. What is the potential energy of an elevator having a mass of 500 kg when it is 10 m above the bottom of its shaft?

   Solution:
   The base level of potential energy here is the bottom of the shaft.

   $m = 500 \text{ kg} \qquad g = 9.8 \text{ m/s}^2 \qquad h = 10 \text{ m}$
   $\qquad PE = mgh$
   $\qquad PE = 500 \text{ kg} \times 9.8 \text{ m/s}^2 \times 10 \text{ m}$
   $\qquad PE = 49\,000 \text{ J} = 4.9 \times 10^4 \text{ J}$

2. A student weighing 500 N walks up a flight of stairs to a level 3.50 m higher than the starting point. How much potential energy is gained?

   Solution:
   $w = 500 \text{ N} \qquad h = 3.50 \text{ m}$
   $\qquad PE = w \times h$
   $\qquad PE = 500 \text{ N} \times 3.50 \text{ m} = 1.75 \times 10^3 \text{ J}$

   Note that the base level from which we measure potential energy here is the level of the starting point.

Test Yourself Problems

1. What potential energy does a 60-kg student gain when climbing a gymnasium rope and raising his body 5.0 m above the floor?
2. How high above the street level must a 5 000-N elevator be to have a potential energy with the respect to the street level of 75 000 J?

3. (a) How much work is done on a 0.10-kg ball when it is thrown upward to a height of 5.0 m above the thrower's hand? (b) What is the ball's gain in potential energy?

10-10 Elastic Potential Energy

When a helical spring is stretched or compressed by a force, work is done because the force acts over a distance. Consider the spring in Fig. 10-4, where a force f has stretched a spring by a distance x. From Hooke's law (Section 3-7), $f = kx$, where k is the spring constant of this particular spring. Note that f is not constant but increases as x increases. It is zero for $x = 0$ and increases to $f = kx$ when the elongation of the spring rises to x. The average value of f

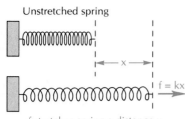

Unstretched spring

f stretches spring a distance x

Fig. 10-4. The stretching force is proportional to the elongation of the spring.

over the distance x is therefore only one-half its final value, or $kx/2$. The work done by this average force as it acts over x is $W =$ average force × distance $= (kx/2)x = \frac{1}{2}kx^2$. This work is stored in the spring as potential energy, called *elastic potential energy* because it is related to the elasticity of the spring. Thus the elastic potential energy PE stored in a spring when stretched a distance x is

$$\text{PE} = \text{W} = \frac{kx^2}{2}$$

This relationship applies to springs that are compressed as well as to those that are stretched.

10-11 Graphical Determination of Elastic Potential Energy

Table 10.1 lists various values of the force stretching a spring versus the elongation of the spring each produces. This data is plotted in Fig. 10-5. Let P be any point on the graph having a force f as its ordinate and the resulting elongation of the spring x as its abscissa. Recall that by Hooke's law, $f = kx$. Now, consider right triangle PQO. Its legs are kx and x. Its area is one-half the product of its legs, or $\frac{1}{2}(kx)x = \frac{1}{2}kx^2$. This is exactly the same expression obtained for the elastic potential energy of a spring in section 10-10. Thus, the area under the graph at any value of x is equal to the elastic potential energy stored in the spring when stretched a distance x.

Let us apply this relationship to point A in Fig. 10-5. The area of the shaded right triangle is the elastic potential energy stored in the spring when it is stretched a distance of 0.20 meter. The area of this right triangle is half the product of its legs, or $\frac{1}{2}(10.0 \text{ N})(0.20 \text{ m})$ $= 1.0 \text{ N} \cdot \text{m}$ or 1.0 J.

The slope of the graph tells us the spring constant. The slope is found by dividing the ordinate of any point on the graph by its abscissa. Thus, dividing the ordinate of point A, 10 N, by the abscissa of point A, 0.20 m, the spring constant $k = 10 \text{ N} \div 0.20 \text{ m}$, or 50 N/m.

Table 10.1 ELONGATION OF A SPRING BY VARIOUS FORCES	
STRETCHING FORCE, f (newtons)	DISTANCE STRETCHED, x (meters)
0.0	0.00
5.0	0.10
10.0	0.20
15.0	0.30
20.0	0.40

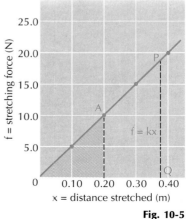

Fig. 10-5

Sample Problem

Fig. 10-6 is the graph of a spring showing the relationship between the force stretching the spring and the distance it is stretched. (a) Find the spring constant. (b) Find the potential energy stored in the spring when stretched 0.020 m.

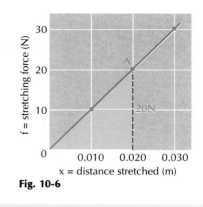

Fig. 10-6

Solution:

(a) The spring constant is the slope of the graph, which is the ordinate of point A divided by its abscissa, or 20 N ÷ 0.020 m = 1000 N/m.

(b) This is the area of the shaded triangle. PE = $\frac{1}{2}fx$ = $\frac{1}{2}$(20 N)(0.020 m) = 0.20 J. The potential energy can also be obtained from PE = $\frac{1}{2}kx^2$ = $\frac{1}{2}$(1000 N/m)(0.020 m)^2 = 0.20 J.

Test Yourself Problems

1. The spring constant of a spring is 200 N/m. What force is needed to compress the spring a distance of 0.025 m?

2. What is the energy stored in the spring in problem 1?

10-12 Interchange of Potential and Kinetic Energy

All about us are bodies whose potential energy is changing to kinetic energy or whose kinetic energy is changing to potential energy.

A ball thrown vertically upward leaves the hand with a certain speed and a corresponding amount of kinetic energy. This kinetic energy is completely converted to gravitational potential energy as the ball rises and comes to a stop at its highest point. Then, as the ball falls back to earth, its potential energy is gradually changed back again to kinetic energy. Since the ball returns to the level from which it started with the same speed with which it left the hand, it has exactly as much kinetic energy at the end of its flight as it had at the beginning. Thus, although its energy changed from kinetic to potential and back to kinetic again, none of its initial energy was lost.

A pendulum bob passes through a similar series of energy changes. At the highest point of the swing, the bob is momentarily at rest and all its energy is potential. As the bob swings downward toward the center position, its potential energy gradually changes to kinetic. Then, as the bob passes the center position and rises to the opposite end of its swing its kinetic energy changes back again into potential energy. If we measure the heights of the bob above the floor at each end of its swing, we note that they are approximately equal. This shows that the bob has just as much gravitational potential energy at the end of a swing as it had at the beginning. Again, in spite of the changes from potential to kinetic energy and then back again to potential energy, no energy was lost.

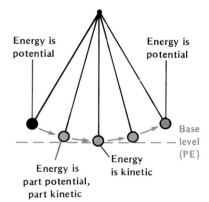

Fig. 10-7. The energy of a swinging pendulum changes continuously from potential to kinetic to potential and back.

10-13 Conservation of Potential and Kinetic Energy

The above examples illustrate the following general principle:

When the only energy changes that take place in a body or a system of bodies are those from potential to kinetic energy or from kinetic to potential energy, no energy is lost. In such a system, the sum of the potential and kinetic energies remains constant at all times. In symbols:

$$PE + KE = \text{total E} = \text{constant}$$

When the only energy changes that take place in a system are from potential to kinetic, or *vice versa*, the total energy of the system remains constant.

Let us verify this relationship in the case of a ball weighing 1 newton that is dropped from a height of 8 meters. Table 10.1 lists its potential and kinetic energies when the ball's height above the ground is 8, 6, 4, 2, and 0 meters. To obtain its potential energy at any height *h,* we use the relationship $PE = wh$. Thus at $h = 6$ m, we have $PE = 1 \text{ N} \times 6 \text{ m} = 6 \text{ J}$.

To obtain the kinetic energy at any height we use the relationship $KE = \frac{1}{2} mv^2$. Now from $w = mg$, we have $m = w/g$. Also, for a falling body, $v^2 = 2\,gd$ where *d* is the distance fallen. Hence $KE = \frac{1}{2}\,(w/g)(2\,gd) = wd$. Thus at $h = 6$ m, the distance fallen *d* is 8 m − 6 m = 2 m, and $KE = wd = 1 \text{ N} \times 2 \text{ m} = 2 \text{ J}$. The kinetic energies at all other heights are obtained in the same manner by multiplying the weight by the distance it has fallen.

Note from Table 10.2 that, as the ball falls, its potential energy decreases from 8 joules to 0 joules, while at the same time, its kinetic energy increases from 0 joules to 8 joules. In the right column, however, we find that for each height, the sum of the potential and kinetic energies is always 8 joules, equal to the energy the ball had at the start. Thus as the ball falls, each joule of potential energy lost is replaced by 1 joule of kinetic energy so that the sum of the potential and kinetic energies remains constant.

Table 10.2

FALLING BALL (1N)	HEIGHT (m)	POTENTIAL ENERGY (J)	KINETIC ENERGY (J)	TOTAL ENERGY (J)
○	8	8	0	8
○	6	6	2	8
○	4	4	4	8
○	2	2	6	8
●	0	0	8	8

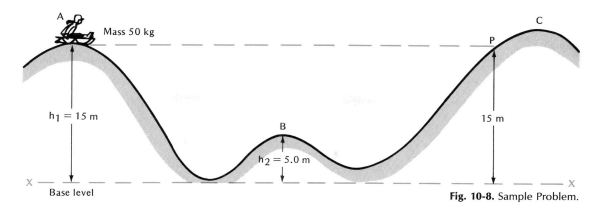

Fig. 10-8. Sample Problem.

Sample Problem

A child on a sled starts from rest and coasts down hill A. (See Fig. 10-8.) (a) If the mass of the child and sled is 50 kg, what kinetic energy will the sled have when it reaches the top of hill B? (b) How far does the sled go up hill C? Assume there are no frictional losses.

Solution:

(a) $m = 50$ kg $g = 9.8$ m/s²
    $h_1 = 15$ m $h_2 = 5.0$ M

The potential energy of the sled at A with respect to base level XX is

$PE_A = mgh_1 = 50$ kg $\times 9.8$ m/s² $\times 15$ m $= 7350$ J

The potential energy of the sled at B is $mgh_2 = 50$ kg $\times 9.8$ m/s² $\times 5.0$ m $= 2450$ J.

In going from A to B, the sled has lost potential energy equal to

$PE_A - PE_B = 7350$ J $- 2450$ J $= 4900$ J

According to the relationship $PE + KE = $ constant, the loss in potential energy must be equal to the gain in kinetic energy. The kinetic energy of the sled at B will therefore be 4.9×10^3 J.

(b) The sled will stop moving up hill C when all of its kinetic energy has been converted into potential energy. This will happen when the sled reaches the point P, 15 m above base level XX. Its potential energy will then be exactly equal to its initial potential energy, which is also the total energy with which it started.

Test Yourself Problem

A 1200-kg car moving at 20.0 m/s turns off its engine as it begins to coast up a hill. At what height above its starting level will the car come to a halt?

10-14 General Expression for Gravitational Potential Energy

The expression $PE = mgh$ for the gravitational potential energy applies only to bodies that are near the surface of the earth where g is practically constant for different values of h. However, as a body is moved far away from the earth, g decreases steadily and this fact must be taken into account in determining the body's potential energy.

Consider a mass m at a distance R from the center of the earth. Since this mass is attracted toward the earth, a force must be applied to it and work must be done to move it away from the earth against the gravitational attraction. It can be shown that the quan-

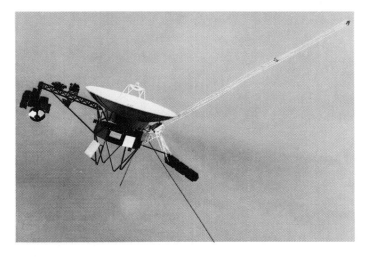

Fig. 10-9. After achieving escape velocity, the unmanned Voyager spacecraft went on to make history by visiting the outer planets of the solar system. Propelled by the gravity of Saturn, Voyager 1 has escaped the solar system itself.

tity of work or energy that must be applied to move a mass m to an infinite distance from the earth is given by GMm/R, where G is the gravitational constant and M is the mass of the earth.

It is customary to define the gravitational potential energy of a body located at a given distance from the center of the earth as the energy or work that must be supplied to the body to move it to infinity from that point. On this basis, the gravitational potential energy of a body located at infinity is given the value zero and represents the base level from which gravitational potential energy is measured. For a body located at any other point, the potential energy is

$$PE = -\frac{GMm}{R}$$

The minus sign indicates that this is the energy the body *lacks* to be able to move from its present position to infinity.

Sample Problem

What is the potential energy of a mass of 1.0 kg situated at the surface of the earth where $R = 6.4 \times 10^6$ m?

Solution:

$G = 6.7 \times 10^{-11}$ m³/kg·s²
$M = 6.0 \times 10^{24}$ kg
$m = 1.0$ kg
$R = 6.4 \times 10^6$ m

$$PE = -\frac{GMm}{R}$$

$$PE = -\frac{6.7 \times 10^{-11}\ \text{m}^3/\text{kg·s}^2 \times 6.0 \times 10^{24}\ \text{kg} \times 1.0\ \text{kg}}{6.4 \times 10^6\ \text{m}}$$

$$PE = -6.3 \times 10^7\ \text{J}$$

Test Yourself Problem

What is the potential energy of a 600-kg satellite moving in a circular orbit 6.5×10^6 m from the center of the earth?

10-15 Binding Energy

We may think of the attractive force between the earth and a body as the force that binds the body to the earth. The work or energy that must be supplied to the body to move it to infinity against the opposition of this gravitational attraction, and thus to enable the body to escape from the earth, is called its *binding energy*. As we have seen, this energy is equal to GMm/R. The sample problem above shows that for a 1-kilogram body located at the earth's surface, the binding energy is 6.3×10^7 joules. This means that 6.3×10^7 joules of work or energy must be transferred to a 1-kilogram mass to enable it to escape from the earth.

10-16 Escape Velocity

To enable a rocket to escape from the earth, it must be given kinetic energy at least equal to the energy that binds it to the earth. If v_e is the speed needed by the rocket to escape and m is the mass of the rocket, the kinetic energy needed for escape is $\frac{1}{2} mv_e^2$. This must be equal to the binding energy GMm/R. Solving for v_e gives

$$\mathbf{v}_e = \sqrt{\frac{2 \, \mathbf{GM}}{\mathbf{R}}}$$

For a rocket fired from the earth's surface, where $R = 6.4 \times 10^6$ meters, v_e comes out to be 1.1×10^4 meters per second or about 40 000 kilometers per hour. It is noteworthy that the escape velocity is the same for all bodies at the same distance from the earth's center. It does not depend upon the mass of the body.

A rocket that is launched at escape velocity will slow down as it moves further and further away from the earth. However, its velocity will never be reduced to zero. It will therefore continue into space and escape from the earth. A rocket that is launched at less than its escape velocity will gradually slow down and come to a halt at some point in space. It will then fall back to the earth.

10-17 Elastic and Inelastic Collisions

A particularly important example of the power and usefulness of the law of conservation of energy is in the study of *elastic collisions*. An elastic collision between two bodies is one in which the sum of the kinetic energies of the bodies before the collision is equal to the sum of the kinetic energies of the bodies after the collision. In an elastic collision, therefore, whatever kinetic energy is lost by one body is gained by the other body.

Most ordinary collisions are inelastic. The colliding bodies deform each other, generate heat, or remove kinetic energy from the system in other ways. As a result, the total final kinetic energy is less than the total initial kinetic energy. However, the law of conservation of momentum continues to hold for inelastic as well as for elastic collisions.

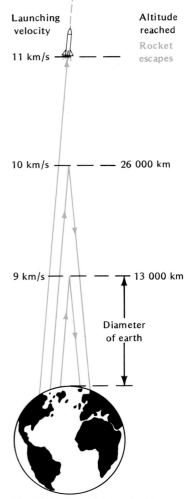

Fig. 10-10. Launching velocities needed to send a rocket to various altitudes.

In an elastic collision, kinetic energy is conserved. In an inelastic collision, it is not conserved.

There are a number of collisions like those between two steel balls or between two billiard balls that are very nearly elastic. In nuclear physics in particular, some of the collisions that we observe between the tiny subatomic and atomic particles of matter are elastic and thus lend themselves to fairly simple analysis. Much basic information about these particles comes from the study of their elastic collisions. Among the simplest of these is the head-on collision.

10-18 Head-on Collision Between Two Particles

In solving elastic collision problems, we generally combine the use of two powerful principles: the principle of conservation of kinetic energy and the principle of conservation of momentum. Let us illustrate the procedure for a head-on collision between an elastic ball of mass m, having initial velocity v_1, and a second elastic ball of mass M that is initially at rest. Let v_2 be the velocity of the first ball after the collision and w_2 be the velocity of the second ball after the collision. Because this is a head-on collision, all the motion after the collision will take place in the line of the original motion of the first ball. (See Fig. 10-11.)

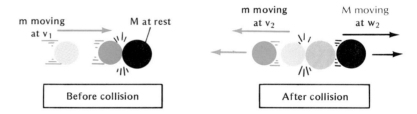

Fig. 10-11. Head-on collision between two masses.

The total kinetic energy before the collision is simply that of the first ball, $\frac{1}{2} mv_1^2$. The total kinetic energy after the collision is the sum of the kinetic energy of the first ball, $\frac{1}{2} mv_2^2$, and the kinetic energy of the second ball, $\frac{1}{2} Mw_2^2$. Since in an elastic collision the kinetic energy before the collision is equal to the kinetic energy after the collision,

$$\tfrac{1}{2} mv_1^2 = \tfrac{1}{2} mv_2^2 + \tfrac{1}{2} Mw_2^2 \tag{1}$$

From the law of conservation of momentum, the total momentum before the collision is equal to the total momentum after the collision. Before the collision, the total momentum is that of the first ball mv_1. After the collision, the total momentum is the sum of the new momentum of the first ball mv_2 and the momentum of the second ball Mw_2. Thus we have the equation

$$mv_1 = mv_2 + Mw_2 \tag{2}$$

Now, we know the masses of the two balls and the initial velocity v_1. The unknown quantities are v_2 and w_2, the velocities of the two balls after collision. Since we have two equations, (1) and (2), relating these velocities, we can solve for them. The algebraic details of the solution are given in Section 10-20. It yields

$$v_2 = \frac{m - M}{m + M} v_1 \qquad (3)$$

$$w_2 = \frac{2m}{m + M} v_1 \qquad (4)$$

10-19 Elastic Collision Between Equal Masses

Equations (3) and (4) give the velocities of the two balls after the collision. In the special case in which both masses are equal, the result is of particular interest. For this case, substituting $m = M$ in (3), we get $v_2 = 0$, and in (4) we get $w_2 = v_1$. This means that the first ball comes to a complete stop after the collision, while the second ball moves off with exactly the same velocity, v_1, that the first ball had before the collision.

10-20 Derivation of Equations (3) and (4)

Optional derivation for honor students

To derive equations (3) and (4) above, we begin with the energy equation (1).

$$\tfrac{1}{2} m v_1^2 = \tfrac{1}{2} m v_2^2 + \tfrac{1}{2} M w_2^2$$

$$\tfrac{1}{2} m v_1^2 - \tfrac{1}{2} m v_2^2 = \tfrac{1}{2} M w_2^2$$

$$m(v_1^2 - v_2^2) = M w_2^2$$

$$m(v_1 - v_2)(v_1 + v_2) = M w_2^2 \qquad (a)$$

Now, from the momentum equation (2),

$$m v_1 = m v_2 + M w_2$$

$$m(v_1 - v_2) = M w_2 \qquad (b)$$

Now, dividing equation (a) by equation (b) (equals divided by equals gives equals), we have,

$$\frac{m(v_1 - v_2)(v_1 + v_2)}{m(v_1 - v_2)} = \frac{M w_2^2}{M w_2}$$

and,

$$v_1 + v_2 = w_2 \qquad (c)$$

Substituting this value for w_2 in (b),

$$m(v_1 - v_2) = M(v_1 + v_2)$$

$$m v_1 - m v_2 = M v_1 + M v_2$$

$$m v_1 - M v_1 = m v_2 + M v_2$$

$$v_1(m - M) = v_2(m + M)$$

Solving for v_2,

$$v_2 = \frac{m - M}{m + M} v_1 \qquad (3)$$

To get w_2, substitute this value of v_2 in (c).

$$v_1 + v_2 = w_2$$

$$v_1 + \left(\frac{m - M}{m + M}\right) v_1 = w_2$$

$$w_2 = \frac{2m}{m + M} v_1 \tag{4}$$

Sample Problem

A neutron moving at 10^5 m/s makes a head-on elastic collision with a helium nucleus initially at rest. Assume that a helium nucleus has a mass 4 times that of a neutron. (a) What is the velocity of the helium nucleus after collision? (b) What is the velocity of the neutron after the collision?

Solution:

(a) If we take the mass of the neutron to be 1 unit of mass, the helium nucleus will therefore have a mass of 4 units. For purposes of this problem it is unimportant to know the actual size of this mass unit since the mass units cancel out in equations (3) and (4).

$m = 1$ unit $M = 4$ units $v_1 = 10^5$ m/s

$$w_2 = \frac{2m}{m + M} v_1$$

$$w_2 = \frac{2 \times 1}{1 + 4} 10^5 \text{ m/s}$$

$$w_2 = 0.4 \times 10^5 = 4 \times 10^4 \text{ m/s}$$

After collision, the helium nucleus will have 2/5 of the velocity of the neutron that struck it.

(b)
$$v_2 = \frac{m - M}{m + M} v_1$$

$$v_2 = \frac{1 - 4}{1 + 4} 10^5 \text{ m/s}$$

$$v_2 = -0.6 \times 10^5 \text{ m/s}$$

$$v_2 = -6 \times 10^4 \text{ m/s}$$

The negative sign shows that the neutron rebounds and reverses direction after the collision with 3/5 of its original velocity.

Test Yourself Problem

A 2.0-kg sphere moving at 2.0 m/s collides head on with a 1.0-kg sphere at rest. Assuming that this is an elastic collision, find the velocity of each sphere after the collision.

10-21 Heat and Conservation of Energy

Our study of simple machines and of potential and kinetic energy has shown that while energy may change its form, all the energy put into a system can be accounted for. We notice, however, that often some of the energy put into a system becomes unavailable to do useful work. As an example, suppose you are pulling a block of wood over a floor. If your force is equal to the frictional resistance between the block and the floor, the block will move along at constant speed and will have constant kinetic energy. When you stop pulling the block, friction soon brings it to a stop. What has happened to the block's kinetic energy? The fact that the block and the floor became a little warmer suggests that this energy was converted into heat.

As another example, consider the case of a weight which is allowed to fall to the ground from a height. We have seen that, as such a body falls, its potential energy is gradually converted into kinetic energy without loss. After the weight hits the ground and stops moving, it no longer has either its original potential energy or its final kinetic energy, but both the weight and the ground on which it fell have become warmer. In losing its kinetic energy as it struck the ground, the weight has produced heat.

These and similar experiences lead us to suspect that *heat is a form of energy* and that the "lost" kinetic energy is replaced by an equal amount of heat energy. We shall see, as we study heat in the next chapters, that this turns out to be true. Thus, the principle of conservation of energy is extended to include heat energy as well as potential and kinetic energy.

Heat is a form of energy.

10-22 Transformation of Energy

We have seen in Chapter 1 that the different forms of energy include electrical energy, sound, light, nuclear energy, chemical energy, and magnetism. Each of these forms of energy can supply forces that can move objects and thus give them kinetic energy which in turn may be changed into potential energy or into heat. Thus a major property of energy is its ability to change its form.

We become keenly aware of this fact when the electric fuse blows out in our homes. The lights go out. The electric range and toaster won't work. The electric motors in the clock and refrigerator stop. The radio and television sets go dead. We suddenly realize that in using electrical energy, we are changing it into other forms of energy. In the lamps and in the television set, electrical energy is being changed into light. In the motor, it is being changed into motion or kinetic energy. In the electric range and toaster, it is being changed to heat. In the electric bell and radio, it is being changed to sound. The ease with which electrical energy can be changed into other forms of energy is the main reason it is so useful.

The behavior of electrical energy illustrates that all forms of energy can be transformed into other forms of energy. In fact, in making the best use of any given form of energy, we frequently invent devices that will transform it into some other more desirable form. In the electric generator, we change the kinetic energy used to rotate the armature into electrical energy. In the gasoline engine, we change the chemical energy of gasoline into heat and then change the heat energy into the kinetic energy of the engine's moving parts.

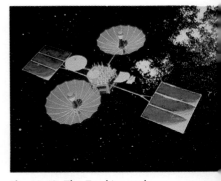

Fig. 10-12. The Tracking and Data Relay Satellite (TDRS) was placed in a synchronous orbit around the earth in 1988 from the space shuttle Discovery. The solar panels on the booms perpendicular to the axis of the two dish antennas will convert the sun's light energy into electrical energy to power the satellite.

10-23 General Principle of Energy Conservation

Does the conservation of energy principle which we have found to be true for kinetic, potential, and heat energy also apply to the other forms of energy? Careful experimentation over the last few centuries has shown that it does. If a gallon of gasoline is burned in a car, its chemical energy changes to an equal quantity of heat energy. When this heat energy is used to run the engine, some of it is converted to the kinetic energy of the moving parts and some remains in the form of heat of the exhaust gases and of the engine itself. If we add up the kinetic energy and the heat remaining, the total is equal to the original quantity of chemical energy in the gasoline. None of the energy has been lost during the transformations, thus illustrating the general principle:

> Energy can be changed from one form to another but cannot be created or destroyed.

10-24 Conservation of Mass

While physicists were discovering the law of conservation of energy, chemists were finding that matter obeys a similar law. They observed that in a chemical reaction, matter may change its form but the quantity of matter left at the end of the reaction always comes out to be the same as the quantity present at the beginning. Thus, when hydrogen and oxygen are allowed to combine, they assume different forms and produce water. However, the mass of the water turns out, within experimental error, to be equal to the combined masses of the oxygen and hydrogen of which it was made.

Chemists summarized these observations in the law of conservation of mass which states:

> Matter can be changed from one form to another but its mass cannot be created or destroyed.

10-25 Conservation of Mass and Energy

For many years, it was a source of wonder that mass and energy should obey such similar laws. Today we know that, under certain circumstances, mass and energy are interchangeable; that mass can be converted into energy and that energy can be converted into mass. These two conservation laws were therefore not exactly correct as stated but had to be amended to include the possibility of mass-energy transformations. They are now parts of the following more general law:

> The combined quantity of mass and energy in the universe remains constant.

This law tells us that neither energy nor mass can be created or destroyed. Whenever energy appears, it can only have been produced from other forms of energy or from mass. Similarly, whenever mass appears it can only have been produced from some form of matter or from energy. In a word, all the mass and energy that

Fig. 10-13. The bluish light given off by this nuclear reactor illustrates the conversion of matter into light as well as heat energy.

was ever in the universe is still here today. All that has been happening through the ages is that the original supply of energy and mass has been steadily changing its form.

The total of all energy and mass in the universe remains the same.

10-26 Equivalence of Mass and Energy

The relativity theory of Einstein enables us to determine how much energy is equivalent to a given quantity of mass and, conversely, how much mass is equivalent to a given quantity of energy. If E is the energy obtainable from a mass m, and c is the velocity of light, the relationship between them is:

$$E = mc^2$$

Since c^2 is a very large number, it follows that, even if m is small, E will be quite large. Thus, the conversion of a small amount of mass can yield a very large amount of energy, as we shall see in the following sample problem.

Mass is converted into energy in nuclear weapons, nuclear reactors, and the sun.

Sample Problem

(a) What energy is obtained when 1.0 kg of mass is completely converted into energy? (b) How much electrical energy in kilowatt-hours can be obtained from this?

Solution:
(a) m = 1.0 kg
c = velocity of light = 3.0×10^8 m/s
$E = mc^2$
$E = 1.0 \text{ kg} \times (3.0 \times 10^8 \text{ m/s})^2$
$E = 9.0 \times 10^{16}$ N·m or J

(b) 1 kWh = 3.6×10^6 J
$$E = \frac{9.0 \times 10^{16}}{3.6 \times 10^6} = 2.5 \times 10^{10} \text{ kWh}$$

This is 25 billion kWh. If we assume that electrical energy costs 4 cents per kWh, its dollar value would be 1 billion dollars.

Test Yourself Problem

How much mass must be converted into energy to produce 1 kWh of energy?

10-27 Importance of Conservation of Energy and Mass

The principle of conservation of energy and mass is one of the most powerful tools the scientist has at his disposal. In a world of continuous change, it is a principle that remains constant and that is a firm starting place for analyzing the complex changes that occur in nature. It applies not only in all the fields of physical science, but in the life sciences as well. It applies to each part of the universe from the largest galaxies to the smallest atomic particles. Throughout the whole field of physics we shall see over and over again how this principle opens the door to new understandings and new knowledge about matter and energy.

CHAPTER REVIEW

Summary

Energy is the ability to do work. It is measured in units of work: *newton-meters,* or *joules.*

Kinetic energy is the energy a body has because of its motion. It is given by the relationship

$$KE = \frac{1}{2}mv^2$$

where m is the mass of a body moving at speed v. When an unbalanced force f acts upon a body over a distance $d,$ it does work equal to fd on that body. It also changes the kinetic energy of that body in accordance with the relationship

$$fd = \frac{1}{2}mv^2 - \frac{1}{2}mv_o^2$$

where m is the mass of the body, v_o is its initial speed, and v is its final speed.

Potential energy is the ability of a body to do work because of the relative position of its parts or because of its position with respect to other bodies.

The gravitational potential energy of a body depends upon its distance from the earth; the greater the distance of a body from the earth, the greater its gravitational potential energy. In describing the gravitational potential energy of a body, it is customary to select some base level and then tell the difference between the gravitational potential energy of the body in the given position and that which it would have if it were permitted to drop to the base level. This difference in gravitational potential energy is the work that can be done by the body or the energy it can transfer to other bodies in falling from the upper to the base level. For a body of mass m near the surface of the earth, the gravitational potential energy is given by

$$PE = mgh$$

where h is the vertical distance of the body from the selected base level.

The more precise expression for the gravitational potential energy of a mass m at a distance R from the center of the earth is

$$PE = -\frac{GMm}{R}$$

where G is the gravitational constant and M is the mass of the earth.

The elastic potential energy stored in a spring when it is stretched or compressed is given by

$$PE = \frac{1}{2}kx^2$$

where k is the spring constant and x is the elongation or compression of the spring. Energy exists in many forms and may change from one form to another. Energy may also change to mass and mass may change into energy according to the relationship:

$$E = mc^2$$

where m is the mass of a body, c is the velocity of light, and E is the energy equivalent to the mass m.

In all these energy and mass changes, the **law of conservation of energy and mass** applies. It states that the total quantity of energy and mass in the universe remains constant.

In collisions between two bodies, energy is generally transferred from one body to the other as well as transformed into other energy forms. There are two kinds of collisions, elastic and inelastic. In an **elastic collision,** the sum of the kinetic energies of the two bodies before the collision is equal to the sum of their kinetic energies after the collision. In an **inelastic collision,** some of the initial kinetic energy of the colliding bodies is converted into heat and other energy forms. As a result, the sum of the kinetic energies of the bodies after the collision is smaller than the sum of their kinetic energies before the collision. However, in both kinds of collisions, the total momentum remains constant in accordance with the law of conservation of momentum.

Questions

Group 1

1. (a) Define energy. (b) In what unit is it measured? (c) What is meant by kinetic energy? (d) Give three illustrations of bodies having kinetic energy.

2. Two toy cars are moving over a table at equal constant speeds. (a) If the first car has twice the mass of the second, how does its kinetic energy compare with that of the second? (b) If a given braking force will stop the first car in a distance of 1 m, in what distance will the same braking force bring the second car to a stop?

3. (a) Define potential energy. (b) Upon what does the quantity of potential energy stored in a clock spring depend? (c) Taking the floor of the laboratory as the base level of energy, what determines the quantity of gravitational potential energy possessed by a book lying on the laboratory table?

4. (a) What is the relationship between the gravitational potential energy of a body and its binding energy? (b) What effect does the mass of a body have on its escape velocity? (c) What effect does the distance of a body from the earth's center have on its escape velocity?

5. How does the elastic potential energy stored in a spring when it is stretched 0.1 m compare with the potential energy stored when the spring is stretched 0.2 m?

6. A rocket is fired into space from the earth at an initial velocity equal to the escape velocity. As the rocket moves farther away from the earth, what happens to (a) its velocity; (b) its kinetic energy; (c) its potential energy? (d) What is the value of the gravitational potential energy of the rocket when it is infinitely distant from the earth?

7. A rocket is fired into space and given an initial kinetic energy smaller than its escape velocity. As the rocket moves farther away from the earth, what happens to (a) its velocity; (b) its kinetic energy? (c) When the rocket comes to a stop in space, what is the relationship between its potential energy and its initial kinetic energy?

8. At what parts of the swing of a pendulum does it have (a) its greatest potential energy; (b) its greatest kinetic energy? (c) How does its greatest potential energy compare with its greatest kinetic energy? (d) How does its total energy at any instant compare with its maximum potential energy?

9. (a) What is meant by an elastic collision? (b) What two conservation laws apply to elastic collisions? (c) Which of these laws does not apply to inelastic collisions?

10. A steel ball falls from a table to the floor and then comes to rest. (a) What changes were taking place in its energy while it was falling? (b) What happened to its initial energy after it came to

rest? (c) State the law of conservation of energy as it applies to this example.

11. (a) What evidence is there that mass and energy are equivalent? (b) What is the relationship that determines how much energy is equivalent to a given mass of matter? (c) Under what conditions is the law of conservation of mass, as used by chemists, true? (d) What is the general law of conservation of mass and energy?

Group 2

12. After being dropped from a height of 1 m, a rubber ball strikes the floor and rebounds to a height of 0.8 m. (a) Neglecting the effects of air friction, explain what happened to the energy lost by the ball. (b) Is the collision between the ball and the floor elastic? Explain.

13. As an empty coal car coasts at constant speed under a coal chute, a load of coal is dropped vertically into it. What effect does the addition of the coal have on (a) the kinetic energy of the car; (b) it velocity?

14. Determine from the expression for kinetic energy the combination of the units kilogram, meter, and second that is equivalent to the joule.

15. A bullet of given mass is fired vertically upward from a gun and the height to which it rises is determined. How can both the kinetic energy and the velocity with which the bullet leaves the gun be calculated from this information?

16. A ball of putty thrown horizontally against a solid cement wall sticks to the wall. (a) Compare the momentum acquired by the wall with that of the ball before the collision. (b) Why isn't the momentum acquired by the wall noticed? (c) What happens to the kinetic energy originally possessed by the ball?

Problems

Group 1

1. (a) What is the kinetic energy of a 4.0-kg mass moving at 5.0 m/s? (b) If the mass was accelerated to this speed from rest by a force of 20 N, over how long a distance did it act?

2. As a block of wood of mass 2.0 kg slides along a floor, its speed decreases from 4.0 m/s to 1.0 m/s. (a) How much kinetic energy does it lose? (b) How much work does the frictional resistance of the floor do upon the block? (c) If the reduction in speed takes place over a distance of 10 m, what is the force of friction acting on the block?

3. A freely falling body starting from rest falls 4.9 m during the first second and acquires a speed of 9.8 m/s. (a) If the body has a mass of 1.0 kg, what is its kinetic energy at the end of the first second? (b) What work does it take to raise the body to its previous position? (c) How much potential energy does the body lose during the first second?

4. A force of 10 N is applied to a small cart for a distance of 2.0 m. (a) What work is done on the cart? (b) Assuming there are no frictional losses, what final kinetic energy does the force transfer to the cart? (c) If the mass of the cart is 10 kg, what is the final speed of the cart?

5. The speed of an automobile of mass 1600 kg increases uniformly from 20 m/s to 30 m/s. (a) What is the increase in its kinetic energy? (b) If the acceleration increases uniformly over a distance of 50 m, what force is accelerating the automobile?

6. What is the kinetic energy of an electron (mass = 9.1×10^{-31} kg) when it is moving at 3.0×10^7 m/s (1/10 the velocity of light)?

7. An archer pulls the middle of the bowstring back a distance of 0.30 m. If the average force applied is 6.0 N, what potential energy is stored in the stretched string of the bow?

8. What potential energy does a 60.0-kg gymnast acquire in climbing 5.00 m up a vertical pole?

9. A 2000-kg truck is raised 15 m in an elevator. What is the increase in its potential energy?

10. A 2.0-kg rock falls from a height of 25 m to a point 20 m above the ground. (a) How much potential energy does it lose? (b) How much kinetic energy does it gain?

11. (a) What is the gravitational potential energy of a 100-kg rocket when it is at a distance of 1.0×10^7 m from the center of the earth? (Assume $G = 6.7 \times 10^{-11}$ m³/kg·s², and the earth's mass $M = 6.0 \times 10^{24}$ kg.) (b) What kinetic energy does the rocket need at this distance to escape from the earth? (c) What is the escape velocity for this value of R?

12. What is the binding energy of a 50-kg youth to the earth?

13. A mass of 2.0 kg is dropped from a height of 5.0 m. Make a table showing its potential and kinetic energies when it is at each of these distances from the ground: (a) 5.0 m; (b) 4.0 m; (c) 2.0 m; (d) 0.0 m.

14. The hammer of a pile driver weighing 1.5×10^5 N is raised 4.0 m. (a) How much potential energy does it acquire? (b) If it is allowed to fall, what will be its kinetic energy when it reaches its starting position?

15. A projectile weighing 25 N is fired directly upward and rises to a height of 400 m. (a) What is its potential energy at the top of its flight? (b) With what kinetic energy does it leave the cannon? (c) What is its kinetic energy at a height of 300 m?

Group 2

16. A cart of mass 500 kg starts from rest at the top of one hill, rolls to the bottom, and then ascends a second hill. The height of the first hill is 30 m and that of the second hill is 10 m. What is the kinetic energy of the cart when it reaches the top of the second hill? (Neglect friction.)

17. A pendulum has a bob whose mass is 1.0 kg. At either of the highest points of its swing, the bob is 0.10 m higher than it is at the lowest point of its swing. (a) How much more potential energy does the bob have at its highest points than it

does at its lowest point? (b) What is the kinetic energy of the bob when it passes through the lowest point of its swing? (c) What is its velocity at that point?

18. A neutron moving at 1.5×10^7 m/s has a head-on elastic collision with a nitrogen nucleus at rest. Assuming that the mass of the nitrogen nucleus is 14 times that of the neutron, find (a) the velocity of the nitrogen nucleus after the collision; (b) the velocity of the neutron after collision.

19. Two gas molecules having equal masses have an elastic head-on collision. At the moment of collision the velocity of the first molecule is 2.0×10^3 m/s while the second molecule is at rest. What are the velocities of the molecules after the collision?

20. How much mass must be converted into energy per second to yield 1.0 watt of power?

21. In a certain reaction, a negative and a positive electron meet and annihilate each other. In their places, there appears energy in the form of gamma radiation. If the mass of an electron (positive or negative) is 9.1×10^{-31} kg, how much gamma radiation energy will appear for each two electrons that annihilate each other?

22. A spring has a constant of 2.0×10^3 N/m. (a) What force does it take to stretch it 0.05 m? (b) How much potential energy is stored in the spring? (c) Make a graph plotting the stretching force against the elongation it produces.

Applying Physics

1. Determine how much potential energy a ball loses on making one bounce. Drop a rubber ball from the heights of 2.5, 2.0, 1.5, 1.0 and 0.5 m above the floor. In each case measure the height to which the ball rises after hitting the floor. In each case obtain the ratio of the height h_2 to which the ball returns with the height h_1 from which it was dropped. The ratio h_2/h_1 is equal to mgh_2/mgh_1 and is therefore the fraction of the original potential energy possessed by the ball which it regains after one bounce. Make a graph showing how this fraction varies with the distance fallen by the ball. What conclusions can you draw from your graph about the part of its energy the ball loses after a bounce?

2. Using a ruler's straightedge as a guide, line up five pennies so that they make contact with each

other. Put a sixth penny in line with the five but about 2 cm from the first penny. Snap the sixth penny with the fingers so that it collides head on with the first of the lined-up pennies. Notice that practically all of the momentum of the incident penny is picked up by the fifth penny in the line while the remaining four hardly move.

Repeat the experiment, but this time put two pennies in line with the five and about 2 cm from the first penny. Snap the two pennies so that they collide with the first of the lined-up pennies. How many of the five lined-up pennies now pick up the momentum of the two incident pennies? Can you explain this in terms of equations (3) and (4) in Section 10-16? Would the same effects be observed if nickels were used for some but not all of the pennies? Try it.

Radar and Aviation

Radar, that is, radio detection and ranging, was first put to practical use early in World War II. The British used it to warn of sea and air attacks in 1939. The United States started using it in 1940 to track aircraft and also to aim antiaircraft guns. Germany had something similar at about the same time.

Radar is an electronic system that detects and locates moving or fixed objects. In the case of aircraft, radar can determine the direction, distance, altitude, and speed of a plane much farther away than the eye can detect. In addition, modern radar sets provide continuous output, and they are equally effective at night and in rain or snow or dense fog.

All radar sets other than optical or laser radars, which send out light signals, transmit electromagnetic waves in the radio portion of the electromagnetic spectrum (see Section 16-31). The most common type of radar is the so-called pulse radar. Pulse radar sends out signals in powerful bursts, or pulses.

As the diagram shows, an oscillator produces an electric signal. This signal is then fed to a transmitter, which is turned on and off rapidly, producing short pulses of waves. These waves are amplified and released through the antenna. The antenna also gathers waves reflected back from an object, in this case a plane. The reflected waves, or echoes, are amplified and displayed on a cathode ray tube similar to the picture tube of a television set. Radar echoes show up on the screen as a bright spot, or blip.

The position of a blip in relation to a compass scale shows the direction of the object detected. The blip's distance from the center of the calibrated screen shows how far

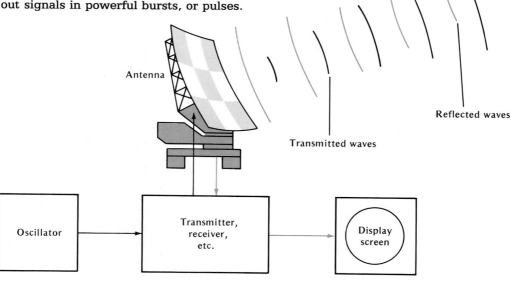

Antenna

Reflected waves

Transmitted waves

Oscillator

Transmitter, receiver, etc.

Display screen

away the object is. And its speed is found by measuring the time it takes the blip to cover a certain distance on the screen.

Radar is the most important aid used by an airport's air traffic controllers. With it, they know the position and velocity of every aircraft within about 80 kilometers of the airport. They are thus able to direct pilots to those altitudes and courses that provide the greatest margin of safety.

Air Traffic Controllers

During peak hours at any major airport, there will be many aircraft on the ground and in the nearby air. As mentioned, air traffic controllers guide the pilots of all of these planes. Controllers keep track of the planes' positions and velocities within an assigned airspace and on the ground, and make certain there is enough space between planes.

Using both radar and visual observation, "airport tower controllers" organize and control traffic into and out of an airport. As a plane approaches the airport, an "arrival controller" directs it to a runway, or, if traffic is heavy, places it in a safe holding pattern above the airport. As a plane nears the runway, a "local controller" monitors its approach and landing, and delays departures that would interfere. When the plane has landed, a "ground controller" takes over. The procedure is reversed for departures.

Between airports commercial air traffic is guided by "enroute controllers." There are 24 enroute control centers around the U.S. These controllers monitor all air traffic within a given region of airspace and transfer planes to adjoining teams of controllers.

Air traffic controllers often direct several planes at one time. They must be able to make quick decisions, with safety their uppermost consideration. In addition to the ability to operate radar, radio, and other equipment, controllers must understand the principles of physics that govern high-speed motion in three dimensions.

Controller trainees are selected by means of a competitive Federal Civil Service examination. A combination of work experience and college is a basic requirement. Trainees receive both on-the-job training and formal instruction. They must work several years in progressively more responsible positions to become fully qualified. Competition for air traffic controller jobs is expected to be keen through the year 2000. During the 1990s the Federal Aviation Administration (FAA) plans to automate many of the air traffic controllers' tasks.

Other occupations that involve air traffic monitoring are airline-radio operators, who control the communications centers of aircraft, and airplane dispatchers, who schedule traffic into and out of airports.

UNIT 3

Heat and the Structure of Matter

(Left) Aerial view of volcanic activity in Hawaii. (Above) Viewing the volcanic activity at ground level. The respirator helps to filter out noxious gases and smoke particles.

In this unit, we pursue our study of matter and energy by investigating the nature of heat. We learn from the work of Count Rumford (1753–1814) and James Prescott Joule (1818–1889) that heat is a form of energy; that it may be converted into work, and that work may be converted into heat; and that the law of conservation of energy applies to heat.

Our study of the changes that occur in bodies when they gain or lose heat leads us to an understanding of the internal structure of matter. We find that all bodies of matter are composed of molecules that are in constant motion. Molecules exert forces varying from attraction to repulsion upon each other. These forces together with the motion of the molecules account for the solid, liquid, and gaseous states of bodies and for the changes of state when bodies gain or lose heat.

We note that heat passes from one body to another in the form of electromagnetic waves known as infrared rays. When heat is absorbed by a body it increases the kinetic and/or the potential energy of its molecules. It becomes part of the internal energy associated with the molecules of the body.

The identification of heat as a form of energy has been applied in numerous heat engines that have been harnessed to do most of our work. They have profoundly changed our methods of producing food and goods and of transporting them and ourselves to all parts of the world. They are responsible for great changes in the power and economies of nations. They have also created enormous environmental pollution problems. Our continuing task is to control the ever increasing use of heat energy and heat engines in the best interests of our own and future generations.

11 Measuring Temperature and Heat

Aims

1. To learn the difference between temperature and heat.
2. To find out how to measure temperature.
3. To find out how to measure heat.

11-1 Temperature

Temperature tells how hot or cold a body is.

A common effect of heat on the bodies to which it is applied is that it makes them warmer. A piece of iron held in a flame for a short time feels warmer to the touch than it did before it was heated. The quantity that tells how warm or cold a body is with respect to some standard body is called its *temperature*. Hot bodies such as molten iron have relatively high temperatures, while cold bodies such as liquid helium have rather low temperatures.

Precise information concerning heat and the effects it produces on the bodies of matter to which it is applied can be obtained by using the measuring device called the *thermometer*. In scientific work, temperatures are usually measured with thermometers having one of two scales that are marked off in the *SI* unit of temperature, the Celsius degree. Although the temperature unit used on both scales is the same, the zero point of each scale is different, so different readings are obtained with the use of each scale. The two scales are called the Celsius (SEL·see·us) scale and the Kelvin scale. We shall deal only with these two scales. When a temperature is given, the scale is indicated after the number of degrees by the letter C for Celsius and the letter K for Kelvin. For example the temperature of a classroom is generally kept at 20°C or 293 K. (The *kelvin* is a unit of temperature now used without the term degree or its symbol.)

11-2 Quantity of Heat

Heat and temperature are different things. When a body at a higher temperature is in contact with a body at a lower temperature, heat flows from the first to the second body. A thermometer measures the temperature of the hot body but it does not tell how much heat that body is able to supply to the colder body. Two bodies may have the same temperature but may be able to supply quite different quantities of heat when put into contact with colder bodies under the same conditions.

Heat is measured in calories and joules.

For example, a liter of boiling water and a half liter of boiling water have the same temperature, about 100°C. However, under

204

the same conditions, the liter can furnish twice as much heat to a colder body as the half liter. Thus, if the boiling water is used to melt ice, the liter can melt twice as much ice as the half liter.

Thus, heat may be defined as the energy that flows from hotter bodies to colder bodies. The SI unit for measuring heat is called the joule (J) (see Section 9-1) or the kilojoule (kJ), which is 10^3 joules. It takes 4.19 kilojoules to raise the temperature of 1 kilogram of water 1 Celsius degree. This is an important quantity of heat and is equal to the older heat unit known as the kilocalorie. The kilocalorie and its subunit, the calorie, are still widely used in reference books, commerce, and industry. In this book, however, heat will be measured exclusively in the SI units of joules and kilojoules. Where heat units are expressed elsewhere in calories or kilocalories, you can convert to joules or kilojoules knowing that 1 calorie is equal to 4.19 joules, or that 1 kilocalorie is equal to 4.19 kilojoules.

The fact that it takes 4.19 kilojoules to raise the temperature of 1 kilogram of water by 1 Celsius degree is used, as we shall see below, in measuring the transfer of heat from one body to another.

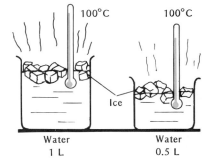

Fig. 11-1. One liter of water can supply twice as much heat as a half liter of water at the same temperature.

The kilocalorie, sometimes called the *large calorie,* is the unit still used by nutritionists in describing the energy obtainable from foods.

11-3 Measuring Quantity of Heat

A quantity of heat is measured by applying it to a known mass of water at a known initial temperature. After the heat has been added to the water, the new temperature of the water is measured. From the rise in temperature and the known value of the mass of the water, the heat absorbed by the water is readily computed as shown in the following examples.

Sample Problems

1. How many joules does a burner supply to 2.00 kg of water in raising its temperature from 20.0°C to 80.0°C?

Solution:
The temperature of the water increases 80.0°C − 20.0°C = 60.0 C°. Since it takes 4.19 kJ to raise the temperature of 1 kg of water 1 C°, it requires 60 × 4.19 kJ or 251 kJ to raise the temperature of 1 kg of water 60.0 C°. To raise the temperature of 2.00 kg of water 60.0 C° therefore requires the addition of 2.00 × 251 = 502 kJ.

2. How much heat is given up by 5.0 kg of water in a fish tank in cooling from 25°C to 10°C?

Solution:
The temperature of the water fell 25°C − 10°C = 15 C°. Since it takes 4.19 kJ of heat to raise the temperature of 1 kg of water 1 C°, 1 kg of water must give up 4.19 kJ of heat to its surroundings to become 1 C° colder than before. In order for 1 kg of water to become 15 C° colder, the water must give up 15 × 4.19 or 63 kJ. For 5.0 kg of water to become 15 C° colder than before, the water must give up 5.0 × 63 kJ = 3.1 × 10² kJ.

Test Yourself Problems

1. A tank containing 2.0 kg of water is heated from 30°C to 50°C. How much heat did the water absorb?

2. How much heat is given up by 0.50 kg of water in cooling from 80.0°C to 55.0°C?

3. How much heat is needed to raise the temperature of 3.0 kg of water from 40°C to 52°C?

Fig. 11-2. A cold object heats a colder one! Here a block of dry ice at about −80°C applies enough heat to the teakettle containing liquid nitrogen at about −200°C to make it boil.

11-4 Flow of Heat and Temperature

When two bodies at different temperatures are brought into contact, *heat will flow out of the body at the higher temperature and into the body at the lower temperature.* The flow will continue until both bodies come to the same temperature. Thus, *the temperatures of two bodies determine the direction in which heat will flow from one body to the other when the bodies are put into contact or are near one another.*

Every body is capable of furnishing energy in the form of heat to a body colder than itself. What is the source of this energy? We shall see in Chapter 12 that it is the sum total of the kinetic and potential energies of the internal parts of the body, namely, its molecules and atoms. We call this its *internal energy.* When a warmer body transmits heat to a colder body, it loses some of its internal energy which is gained by the colder body. While the energy is passing from one body to the other, it is described as heat. Once the energy becomes associated with the molecules and atoms of a body, it is described as internal energy.

When a block of iron, heated to red heat, is dropped into a bucket of cold water, its temperature will decrease while that of the water will increase. Heat is passing steadily from the iron into the water and will continue to do so until both iron and water come to the same temperature. The net flow of heat from the iron to the water then stops. The iron has lost some of its internal energy while the water has gained an equal amount.

11-5 Sense of Temperature and Heat Flow

How good a thermometer are you? Fill three large jars with water as follows: one with ice water; one with tap water at room temperature; and one with hot water. Put the right hand in the ice water and the left hand in the hot water, keeping both hands immersed for about two minutes. Now take the hands out and immerse both hands in the tap water. Does this water feel warm to one hand and cool to the other? Can you explain why?

Fig. 11-3. Simple experiment illustrating the fallibility of the human body as a temperature measuring instrument.

Ice water   Tap water at   Hot water
                   room temperature

Evidently, the hands are not dependable thermometers. Our sense of hot and cold depends not only upon the temperature of the tap water but also on the direction in which heat flows when the hands are immersed in it. The left hand, which has been in the hot water, is at a higher temperature than the tap water. Heat, therefore, flows from that hand into the tap water. Since the left hand is losing heat, it feels cool. The right hand, which has been in the ice water, is at a lower temperature than the tap water. Heat therefore flows from the tap water into the right hand making that hand feel warm.

In general, an object feels cool to the touch when its temperature is lower than that of the hand so that the hand loses heat to it. An object feels warm to the touch when its temperature is higher than that of the hand so that heat passes from it into the hand.

Heat flows from warmer to colder bodies.

11-6 Mercury Thermometer

Thermometers are devices that are sensitive to temperature changes and are equipped with scales from which the temperature can be read. The ordinary mercury thermometer consists of a tube of very thin bore, having at its bottom a bulb filled with mercury. The tube is evacuated and sealed at the top.

The action of this thermometer depends upon the fact that mercury, like most substances, expands when it gets warmer and contracts when it gets colder. When the temperature rises, both the glass tube and the mercury it contains expand. Since the mercury expands more than the glass, it rises in the tube to a level that is a measure of its temperature. This can be read off on the scale marked on the glass tube. When the temperature falls, both the glass tube of the thermometer and the mercury contract. Since the mercury contracts more than the glass tube around it, it falls in the tube to a new level. This level read against the scale is a measure of the temperature.

11-7 Fixed Points of a Celsius Thermometer

Before we can measure any temperature we must select standard temperatures with which to compare it. Two temperatures used as standards on the Celsius scale are the temperature at which water freezes and the temperature at which water boils. These temperatures are selected because they can be readily produced and because water that is either freezing or boiling maintains a constant temperature. However, both of these standard temperatures must be taken when the atmospheric pressure is "normal."

The atmosphere is the ocean of air, several hundred miles high, that covers the earth. The weight of this air causes the atmosphere to press upon or to exert a pressure on all the objects immersed in it. *Pressure is the force exerted on the surface of a body per unit of area.* The pressure exerted by the atmosphere is not constant the world over but varies from day to day and from place to place. To have a standard pressure to which we can refer, it is customary to define as "normal" the average value of the pressure of the atmosphere at sea level taken over a long period of time. *Normal atmospheric pressure* turns out to be about 10^5 *newtons per square meter.* This means that when the atmospheric pressure is normal in a given place, the atmosphere exerts a force of 10^5 newtons on each square meter of surface of any body situated at that place.

Since both the temperature at which water freezes and the temperature at which it boils vary with changes in the atmospheric pressure, it is necessary to select the specific atmospheric pressure for which these temperatures will be used as standards. On the Celsius scale, the temperature of boiling water under normal atmospheric pressure is defined as 100°C, while the temperature of freezing water under normal atmospheric pressure is defined as 0°C. Because these temperatures are used as standards in fixing the temperature scale of the thermometer, they are called *fixed points*.

11-8 Making a Celsius Thermometer

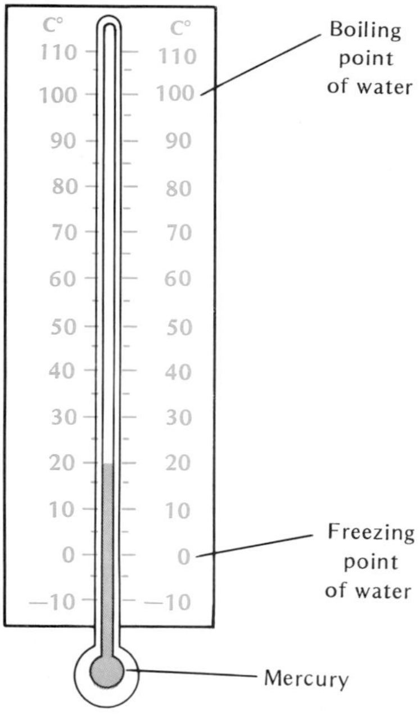

Once the fixed points are selected, it becomes a simple matter to make a temperature scale. Here is how the Celsius scale is put on an unmarked mercury thermometer assuming that the atmospheric pressure is normal.

First, the bulb of the thermometer is put into a mixture of ice and water that has been standing for several minutes. Such a mixture has the same temperature as freezing water or as melting ice. The mercury in the thermometer will contract and fall to a certain level. At this level a mark is made on the glass and labeled 0°C.

Now the bulb of the thermometer is removed from the ice-water mixture. Water is brought to a boil and kept boiling while the bulb of the thermometer is held in it. The mercury will rise as it expands and will stop at a point in the upper part of the thermometer. A mark is made at this level and labeled 100°C.

Now the space between the 100°C mark and the 0°C mark is divided into one hundred equal steps. Each one of these steps is a Celsius degree (C°). The scale is now completed by marking off these 1-degree steps above the 100°C mark as well as below the 0°C mark.

Fig. 11-4. A mercury thermometer reading in Celsius degrees.

11-9 Water Is Unsuitable for Thermometers

Water is not used in thermometers because of its unusual behavior in the range of temperatures between 4°C and 0°C. When water is cooled, it contracts only until its temperature reaches 4°C. On further cooling, from 4°C to 0°C, water expands instead of contracting. On reaching 0°C, water expands further as it turns to ice. After the water has turned to ice, further cooling will cause it to contract.

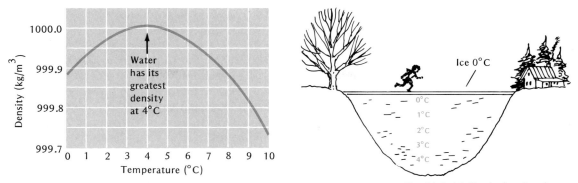

Fig. 11-5. (**a**) Graph showing the variation in density of water near its freezing point. (**b**) The resulting temperature layers in an ice-covered pond.

Thus, water has its smallest volume and its greatest density when its temperature is 4°C.

The extraordinary expansion of water on cooling from 4°C to 0°C is most important to fish and other creatures that live in lakes in the north and south temperate zones. As such lakes lose their heat to their surroundings in winter, only the surface of the lake freezes while the water below it settles into temperature layers as shown in Fig. 11-5(b). Since water is densest at 4°C, the water at the bottom of the lake will have this temperature. Above it will be progressively less dense layers of water with temperatures passing through 3°C, 2°C, and 1°C. Finally, the water in contact with the ice will be at the same temperature as the ice, or 0°C. This layering effect makes it possible for aquatic life to survive the winter in the water below the ice.

11-10 Mercury a Standard Thermometric Substance

The Celsius scale has been defined using a mercury thermometer. Will a thermometer containing a liquid other than mercury give the same temperature scales if it is made in the same way? Suppose, for example, that the bulb of a thermometer containing glycerine is first put into an ice-water mixture to find the 0°C mark and then into the boiling water to find the 100°C mark. If the space between these two marks is divided into one hundred parts, will the readings of this thermometer agree at all temperatures with those of a mercury Celsius thermometer?

The answer is *no*. The glycerine and mercury thermometers will agree in measuring the temperatures of an ice-water mixture and of boiling water because these temperatures were taken as the fixed points for both. They will not agree at intermediate temperatures. For example, when a mercury thermometer shows that the temperature of a given mass of warm water is 50°C, the scale of the glycerine thermometer will read about 47 degrees. The reason for this is that glycerine does not expand with a rise in temperature in the same way as mercury does. If, therefore, we accept the scale obtained with the mercury thermometer as the standard Celsius scale, the readings of the glycerine thermometer cannot be accepted as the correct Celsius temperature values.

It is found that, like glycerine and water, different liquids expand with increasing temperature in quite different ways. In using a liquid thermometer to make a standard Celsius temperature scale, it is therefore necessary to select a particular liquid to be used as a standard measuring material. Historically mercury has served this purpose. It will be seen that this was a fortunate choice. In practice, the expansion of alcohol with rise in temperature is sufficiently similar to that of mercury to make it feasible to use alcohol in many thermometers in place of mercury.

11-11 Need for an Absolute Temperature Scale

Depending upon the particular properties of mercury to define the temperature scale has serious disadvantages. For example, we cannot use a mercury thermometer to measure temperatures below −39°C, at which mercury freezes, or above 360°C, at which mercury boils. It would be desirable to have a means of measuring temperature that does not depend upon the properties of any particular substance. Such a method was first suggested by a study of the expansion of gases upon heating. Out of it finally came the *absolute or Kelvin* temperature scale which is independent of the properties of any particular substance.

11-12 Constructing a Gas Thermometer

The simplified gas thermometer shown in Fig. 11-6 makes use of the expansion of a gas when heated and its contraction when cooled to measure temperature. It consists of a gas-filled tube of thin, uniform bore closed at its upper end by means of a droplet of mercury. The gas is kept from leaving the tube by the atmospheric pressure that pushes downward on the mercury. Any gas, including air, may be used.

When the gas is heated, it expands, pushing the mercury droplet upward. When the gas is cooled, it contracts, permitting the atmospheric pressure to push the mercury downward. The changes in the position of the mercury which correspond to changes in the volume of the gas may therefore be used to measure changes in temperature.

All that is now needed is to put a suitable scale on the thermometer. However, before this can be done, it must be noted that the readings of this gas thermometer will change not only with changes in temperature but also with changes in the atmospheric pressure. An increase in the atmospheric pressure will force the mercury droplet downward and squeeze the gas into a smaller volume. A decrease in the atmospheric pressure will allow the gas to expand into a larger volume and to push the mercury droplet upward. If, therefore, the changes in the volume of the gas are to represent

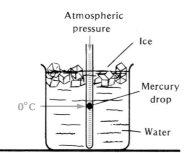

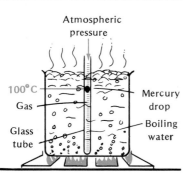

Fig. 11-6. Method of calibrating a gas thermometer.

only changes in temperature, it is necessary that the atmospheric pressure remain constant.

For any fixed value of the atmospheric pressure, we can put a Celsius scale on the gas thermometer just as was done for the mercury thermometer. The scale will then apply as long as the atmospheric pressure remains at that fixed value. When the atmospheric pressure changes, the readings on this scale must be corrected to allow for the change in the pressure.

11-13 Comparing Gas and Mercury Thermometers

To compare the behavior of a gas thermometer and a mercury thermometer, let us put a Celsius scale on a gas thermometer. For simplicity, select a time when the atmospheric pressure is normal. The thermometer is first surrounded by a mixture of ice and water and the level to which the mercury droplet falls after the gas has fully contracted is marked 0°C. The thermometer is then held in boiling water and the level to which the mercury droplet rises after the gas has fully expanded is marked 100°C. The intermediate temperatures are now marked off by subdividing the distance between the 0°C mark and the 100°C mark into one hundred equal parts. Each of these subdivisions represents one Celsius degree (1 C°). Units of this size may now be used to mark off temperatures above 100°C and below 0°C (Fig. 11-6).

If many different temperatures are measured by this gas thermometer and also by the mercury thermometer, the readings of both thermometers always agree no matter what gas is used. This indicates that all gases undergo the same pattern of expansion upon heating as does mercury and that gas thermometers and mercury thermometers may be used interchangeably.

It also means that in setting up the Celsius temperature scale it is no longer necessary to rely on the particular properties of mercury. Any substance in gaseous form will serve just as well. Thus, the scale for measuring temperature no longer depends on the properties of any particular substance but on the common expansion and contraction characteristic of any matter in gaseous form.

11-14 Expansion of Gases

It makes no difference what gas is used in the gas thermometer since *all gases, at constant pressure, expand with rising temperature in almost exactly the same way.* The French scientist, Jacques Charles, found that, *when the pressure on any gas is held constant, its volume increases by about 1/273 of its volume at 0°C for every Celsius degree that its temperature rises, and decreases by the same volume for every Celsius degree that its temperature falls.*

This principle can be observed on a gas thermometer by noting that each subdivision on its scale representing one Celsius degree is about 1/273 as long as the distance from the bottom of the tube

All gases at constant pressure expand with rising temperature at the same rate.

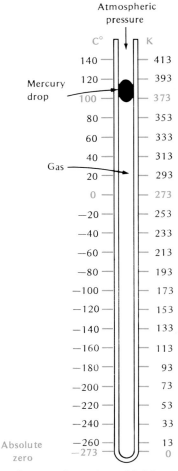

Atmospheric
pressure

C°		K
140		413
120		393
100		373
80		353
60		333
40		313
20		293
0		273
−20		253
−40		233
−60		213
−80		193
−100		173
−120		153
−140		133
−160		113
−180		93
−200		73
−220		53
−240		33
−260		13
−273		0

Mercury drop

Gas

Absolute zero

Fig. 11-7. Comparison of Kelvin and Celsius scales on a gas thermometer.

The lowest possible temperature is −273.15°C.

to the 0°C mark. Since the thermometer tube has a uniform bore, the volume of the gas in any part of it is proportional to the length of that part. It follows that the volume of the gas that occupies an interval of one Celsius degree is very close to 1/273 of the volume occupied by the entire gas at 0°C. Therefore, when the temperature of the gas in the thermometer changes by one Celsius degree, its volume changes by about 1/273 of its volume at 0°C.

Accurate measurements show that different gases vary slightly from each other in their expansion and contraction per degree change in temperature. However, the fact that the variation is so small points up a most important similarity in the behavior of all gases.

11-15 Predicting the Lowest Possible Temperature

The constant rate at which the volume of any gas under constant pressure decreases upon cooling makes it possible to imagine the existence of a lowest possible temperature and to estimate its value. Suppose that a given volume of gas at 0°C and constant pressure is cooled continuously. For each Celsius degree that its temperature falls, it will lose 1/273 of its volume. It may be inferred, therefore, that if the gas is cooled to a temperature 273 degrees below 0°C, or −273°C, it will lose 273/273 of its volume, or its entire volume. In other words, the gas may be expected to shrink to nothing!

This startling prediction suggests that for any substance in gaseous form, *there is a lowest possible temperature located at about −273°C.* This temperature, at which the volume of all gases would, theoretically, shrink to nothing, is called *absolute zero.*

Actually, no gas can be cooled to absolute zero because, before it reaches this low temperature, a gas changes first to a liquid and then to a solid. However, the rate of contraction of gases with decreasing temperature is not the only evidence that suggests the existence of a lowest possible temperature. Much other theoretical and experimental evidence points to the same conclusion and agrees in placing absolute zero at −273.15°C.

More will be said about the meaning of absolute zero in Chapters 12 and 14. Here, it is interesting to note that our present theories of heat and matter predict that it is not possible to cool any object down to absolute zero. However, modern techniques have succeeded in attaining temperatures within a small fraction of a degree of absolute zero.

11-16 Kelvin or Absolute Temperature Scale

Lord Kelvin, the British physicist, conceived the idea of making a temperature scale having its zero value at the absolute zero of temperature. This scale is called the *absolute temperature scale.* It is also known as the *Kelvin scale.*

To achieve greater precision in the measurement of temperature, the Kelvin scale is based on a single new fixed point that can be produced more accurately than the boiling point and freezing point of water. It is the temperature of the triple point of water. *The Triple point of water is that condition of temperature and pressure in which water confined in a closed vessel can exist with all three of its states—ice, water, and vapor—in equilibrium.* That is, the quantities of ice, water, and vapor in the vessel exist together and remain constant with the passage of time. This triple equilibrium occurs only at a temperature of 0.01°C and a pressure of 4.58 millimeters of mercury.

On the Kelvin scale, this triple point temperature of water is used as a fixed point and defined as 273.16 K. The kelvin unit of temperature is chosen to be the same size as the Celsius degree. It follows that, since 0.01°C = 273.16 K, 0°C = 273.15 K and −273.15°C = 0 K. Thus, the lowest possible temperature, −273.15°C, is the zero point on the Kelvin scale. All other Kelvin temperatures are positive numbers.

To change a Celsius temperature to kelvins, we need simply add 273.15° to it. For our purposes, 273° is close enough. Thus,

$$K = °C + 273$$

where K is the temperature in kelvins and C is the temperature in degrees Celsius. Thus, 0°C is 0 + 273 = 273 K; 100°C is 100 + 273 = 373 K; and −100°C is −100 + 273 = 173 K.

To change kelvins to degrees Celsius, simply subtract 273. Thus, 283 K = 283 − 273 = 10°C, and 183 K = 183 − 273 = −90°C.

11-17 Range of Temperatures We Measure

As we have seen, there is a lowest possible temperature, 0 K. There does not seem to be an upper limit of temperature. In the lower part of the scale, we deal with such temperatures as 4 K, the boiling point of helium, and temperatures of only fractional parts of a kelvin. At the upper part of the scale, we encounter temperatures ranging from that at which gold melts, 1336 K, to the temperature at the interior of the sun, about 15 million kelvins. Scientists attempting to obtain energy from nuclear fusion reactions have attained temperatures above 20 million kelvins.

Different kinds of thermometers are needed to measure temperatures at different parts of the scale. Each thermometer depends on some physical change that varies with the temperature. We have seen how the mercury or alcohol thermometer uses the expansion and contraction of a liquid to measure changes in temperature. The range of a liquid thermometer is limited by the liquid's boiling and freezing points. To measure very low temperatures, helium-filled gas thermometers are commonly used. There are also thermometers

Fig. 11-8. Apparatus used to determine the triple point of water.

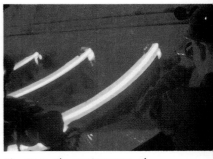

Fig. 11-9. The temperature of hot steel is being determined from its color with the use of an instrument called an optical pyrometer.

that make use of electrical and magnetic properties of certain elements that change in a regular manner with changes in temperature. For measuring high temperatures, there are thermometers that analyze the light emitted by a hot body, that convert heat directly into electricity by means of devices called thermocouples, and that consist of wires whose electrical resistance changes with changes in temperature.

CHAPTER REVIEW

Summary

The **temperature** of a body tells whether it is warmer or colder than other bodies. Temperature is measured by **thermometers** and is expressed in degrees in two temperature scales, the **Celsius** and the **Kelvin** scales.

Heat is a form of energy and is measured in **joules** and **kilojoules.** It takes 4.19 kilojoules to raise the temperature of 1 kilogram of water by one Celsius degree. The flow of heat from one body to another is determined by the temperatures of the two bodies. In general, heat flows from the body at the higher temperature to the one at the lower temperature.

The mercury thermometer is a basic temperature-measuring instrument. It measures temperature by using the fact that mercury expands when warmed and contracts when cooled. Two temperatures commonly used as standards in making a scale for measuring temperature are the melting point of ice and the boiling point of water under normal atmospheric pressure. These temperatures are called the **fixed points** of the temperature scale. On the **Celsius scale,** the melting point of ice is assigned the value 0°C and the boiling point of water is assigned the value 100°C.

The **Kelvin** or **absolute scale** of temperature uses only one fixed point, the temperature of the **triple point of water,** which is given the value **273.16 K.** It chooses the lowest possible temperature, $-273°C$, as its zero point and gives it the value 0 K. Since the kelvin is the same size as the Celsius degree, the relationship between a kelvin and the corresponding Celsius temperature is:

$$K = °C + 273$$

The expansion or contraction of mercury with temperature changes is used to set up the temperature scale, and only substances that expand and contract in a manner similar to that of mercury may replace mercury in thermometers. These include gases. All gases expand and contract with change in temperature at practically the same rate, which is about 1/273 of the volume of the gas at 0°C per Celsius degree of temperature change. The rate at which gases contract upon cooling suggests the existence of a lowest possible temperature at $-273°C$, known as **absolute zero.** This is the temperature taken as 0 on the Kelvin scale.

Questions

Group 1

1. How is heat measured?

2. A block of iron is dropped into a barrel of water having a temperature of 50°C. What flow of heat will take place if the temperature of the iron is (a) 40°C; (b) 50°C; (c) 60°C?

3. A person immerses one hand in a basin of hot water for about 2 min, then immerses it in a basin of lukewarm water. Explain why the lukewarm water feels cool.

4. (a) What are the fixed points of a thermometer? (b) Why is it necessary to specify that the atmospheric pressure be normal in defining the freezing point and the boiling point of water as fixed points? (c) What other temperature is used as a fixed point?

5. (a) Explain how a Celsius scale is put on a mercury thermometer. (b) What changes would you have to make in the scale of this thermometer to make it read kelvins instead of Celsius degrees?

6. Suppose that water is used in a thermometer instead of mercury. Explain what happens to the level of the water in the thermometer as the temperature changes from (a) 5°C to 4°C; (b) 4°C to 3°C; (c) 2°C to 3°C. (d) If a water thermometer is at 4°C and the temperature changes, why cannot the thermometer indicate whether the temperature is rising or falling?

7. (a) Why is it necessary, in making a standard scale for measuring temperature with a liquid thermometer, to specify that the liquid in the thermometer be mercury? (b) What property must a liquid have if it is to be used in place of mercury in a thermometer?

8. (a) Using a diagram, explain how a gas thermometer behaves when the temperature rises while the atmospheric pressure remains constant. (b) How does the thermometer behave when the atmospheric pressure rises while the temperature remains constant? (c) Under what conditions can we be sure that a change in the level of the mercury droplet in the gas thermometer is completely the result of a change in temperature?

9. (a) Why does a gas thermometer behave in about the same way regardless of what gas is used in it? (b) How does the rate at which gases contract when cooled enable us to predict the existence of a lowest possible temperature? (c) What is the predicted lowest temperature? (d) Why can no gas actually be cooled to absolute zero?

10. What temperature on the Kelvin scale corresponds to (a) absolute zero; (b) the melting point of ice; (c) the boiling point of water? (d) Compare the size of a kelvin with the size of a Celsius degree.

Group 2

11. Two thermometers contain equal quantities of mercury but one has a much thinner bore than the other. Which thermometer will be more sensitive to a small change in temperature? Explain.

12. Ample heat can be supplied by the Arctic Ocean, but it is difficult to extract. Explain.

Problems

Group 1

1. A pot containing 0.500 kg of water is heated until the temperature of the water rises from 20.0°C to 60.0°C. How much heat is added to the water?

2. A beaker containing 0.15 kg of water cools from 80°C to 70°C. How much heat is given up by the water?

3. After 419 kJ of heat is supplied to 2.00 kg of water at 20.0°C, what is the new temperature of the water?

4. How many Celsius degrees does the temperature of a mass of 1.00 kg of water fall for each 83.8 kJ of heat it loses?

5. How much heat is required to raise the temperature of 10.0 kg of water a total of 20.0 Celsius degrees?

6. A mass of 150 kg of water cools from 95°C to 35°C. How much heat is given up by the water to its surroundings?

7. A tank contains 2.0×10^3 kg of water. How much heat is needed to raise the temperature of the water from 15°C to 30°C?

8. Convert each of these temperatures from degrees Celsius to kelvins: (a) −270°C; (b) −5°C; (c) 25°C; (d) 95°C; (e) 150°C.

9. Convert each of these temperatures from kelvins to degrees Celsius: (a) 4 K; (b) 173 K; (c) 275 K; (d) 385 K; (e) 1000 K.

10. At 0°C, the volume of a gas maintained at normal atmospheric pressure is 546 cm³. (a) How many cm³ will the volume of this gas increase or decrease for each Celsius degree that its temperature changes? (b) What will be the volume of this gas at −50°C?

Group 2

11. A gas thermometer contains 273 cm³ of air at 0°C and normal atmospheric pressure. On cooling at constant pressure, the volume of the gas shrinks to 158 cm³. (a) What is the temperature of the gas in kelvins? (b) What is its temperature in degrees Celsius?

12. When 1.0 kg of hot water at 90°C is mixed with 2.0 kg of cold water at 0.0°C, the temperature of the mixture becomes 30°C. Show that the heat lost by the hot water is exactly equal to that gained by the cold water.

Applying Physics

1. How does the rate at which a mass of water gives up heat depend upon the difference between its temperature and that of the room? Fill a cup with boiling water and slowly stir the water with a mercury thermometer. Record the temperature of the water every 3 min until the temperature drops to about 25°C. Make a graph plotting the temperature of the water against the elapsed time. The slope of the graph at each temperature point shows the rate at which the water is losing heat. How does this rate change as the temperature decreases?

2. You can check one fixed point of a home mercury or alcohol thermometer as follows. Make a mixture of ice and water and let it stand for several minutes. Dip the thermometer bulb into the mixture and leave it there long enough to permit the mercury in the thermometer to stop falling. Take the reading. If it is not 0°C and the thermometer has been accurately made, what accounts for the difference between its reading and 0°C?

Thermal Imaging: A way to see in the dark

There is a way to see things in the dark without turning on the lights. You will need an infrared detector, a camera, and a display unit for viewing. In other words, a thermographic system.

Thermography is the technique for sensing the temperature of objects by the infrared radiation they emit. A commercial infrared detector can respond to and record temperatures from 15°C to 45°C.

A typical thermal imaging system is similar to a video TV system. First the image of an object is recorded on a screen using only the infrared rays it emits—there is no visible light. Then a special tube, such as the pyroelectric vidicon tube, converts this image into a set of electronic signals. Finally, these signals are transformed on a fluorescent screen into a visible image of the object. The variations in the brightness of this image correspond to the variations in the temperature of different parts of the object.

The infrared imaging technique is finding many useful applications in industry and medicine. For example, it can be used to monitor the process of machine welding of seams and also in the manufacturing of circuit boards. Weak spots can be detected by thermal imaging as the work is being done.

Thermal imaging has become a very successful tool in medicine. A doctor can determine how deep the injury goes in a burn victim or the condition of the circulatory system just below the skin. Breast cancer can also be diagnosed. The reason for these uses is based on the fact that diseased or damaged tissue has a different temperature from surrounding healthy tissue.

The photograph below left shows a thermogram of a normal hand with cold fingers. The thermogram below right shows the hands of a person who has poor circulation.

You may be reading about thermal imaging in the future. It is a relatively new technique for which new uses are sure to be found that will benefit your life.

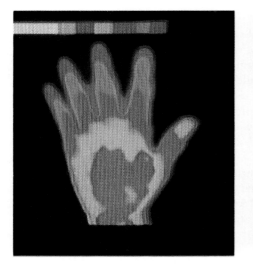

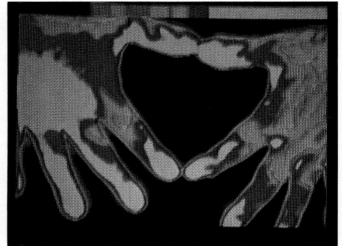

12 Heat Is Energy

Aims

1. To learn that when heat is added to a substance the temperature of the substance rises or the substance changes state.
2. To find that the law of conservation applies to heat, that is, in any transfer of heat from one body to another, heat is neither created nor destroyed.
3. To show that heat can do work and is, therefore, a form of energy.
4. To understand that the three laws of thermodynamics describe the behavior of physical processes involving work and heat.

Heat and Temperature

12-1 Energy Transformation and Heat

In any transfer of heat between two bodies, the heat gained by one body is equal to the heat lost by the other.

Rub your hands together and note that they get warm. You are converting work into heat. The engine of a car converts the heat of burning fuel into the work of moving the car. This ready conversion of work into heat and heat into work indicates that heat is a form of energy. Hence the law of conservation of energy should apply here. To show this, we can begin by showing that heat has its own conservation law: *Whenever heat is transferred from one body to another, no heat is created or destroyed.* The quantity of heat lost by one body is gained by the other.

12-2 Heat Conservation in Water Mixtures

Suppose 1.00 kilogram of cold water at 10.0°C is mixed thoroughly with 1.00 kilogram of warm water at 50.0°C in an insulated container. If the heat gained by the container is negligible, the temperature of the resulting mixture will be 30.0°C, halfway between the two initial temperatures. The warm water has lost heat and has become cooler. The cold water has gained heat and has become warmer. How does the quantity of heat gained by the cold water compare with the quantity of heat lost by the warm water?

To determine the quantity of heat gained by the cold water, it must be remembered that it requires 4.19 kilojoules of heat to raise the temperature of 1 kilogram of water by 1 Celsius degree. In this example, the temperature of the cold water rose from 10.0°C

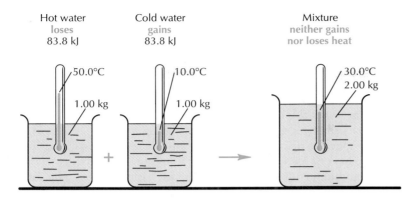

Hot water
loses
83.8 kJ

Cold water
gains
83.8 kJ

Mixture
neither gains
nor loses heat

50.0°C

1.00 kg

10.0°C

1.00 kg

30.0°C
2.00 kg

+

Fig. 12-1. The heat lost by the hot water is gained by the cold water.

to 30.0°C, or 20.0 Celsius degrees. To raise the temperature of 1.00 kilogram of water 20.0 Celsius degrees requires 20.0 × 4.19 kilojoules, or 83.8 kilojoules. Therefore, the cold water gained 83.8 kilojoules of heat.

The quantity of heat lost by the warm water may be determined similarly. When 1 kilogram of water loses 4.19 kilojoules of heat, its temperature falls by 1 Celsius degree. Here, the temperature of the warm water fell from 50.0°C to 30.0°C, or 20.0 Celsius degrees. Hence the warm water lost 20.0 × 4.19 kilojoules, or 83.8 kilojoules of heat. Since the mixing was done in an insulated container with negligible heat loss, we may infer that the cold water gained the 83.8 kilojoules of heat that was lost by the warm water.

This example of the conservation of heat is typical of what happens when two quantities of water at different temperatures are mixed. Heat will continue to pass from the warmer water to the colder water until all the water is at the same temperature. No heat is created or destroyed in the mixing process. Whatever quantity of heat is lost by one part of the mixture is gained by the other.

12-3 Specific Heat

Heat is also conserved when bodies other than water exchange heat. For example, when a hot iron bolt is dropped into a bucket of oil, the heat lost by the bolt as it cools is gained by the oil. Before we can find the heat gained or lost in any mixing process, it is necessary to know how to measure the heat gained or lost by materials other than water. This leads to the concept of *specific heat.*

The specific heat (c) of a substance is the quantity of heat needed to raise the temperature of a unit mass of that substance by 1 Celsius degree.

In SI units, heat is measured in joules (J), mass is measured in kilograms (kg), and temperature difference is measured in Celsius degrees (C°). The units of specific heat are then joules per kilogram-Celsius degree (J/kg · C°). The specific heat of gold, for example, is 129 joules per kilogram-Celsius degree. This means that 129 joules is the amount of heat needed to raise the temperature of 1 kilogram of gold by 1 Celsius degree. The 129 joules also represents the heat lost by 1 kilogram of gold when its temperature falls by 1 Celsius degree.

The specific heats of different substances vary widely. The specific heat of lead is 128 joules per kilogram-Celsius degree; that of aluminum is 900 joules per kilogram-Celsius degree. The ratio of the specific heats of aluminum and lead is 900 J/kg · C° divided by 128 J/kg · C°, or 7.03. Thus, it takes 7.03 times as much heat to raise the temperature of 1 kilogram of aluminum by 1 Celsius degree as it does to raise the temperature of 1 kilogram of lead. It follows that when 1 kilogram of aluminum cools by 1 Celsius degree, it releases to its surroundings 7.03 times as much heat as will 1 kilogram of lead in cooling 1 Celsius degree.

12-4 Specific Heat of Water

The specific heat of water is 4.19×10^3 J/kg · C°.

We have seen that it takes 4.19 kilojoules, or 4.19×10^3 joules, to raise the temperature of 1 kilogram of water by 1 Celsius degree. That is to say, the specific heat of water is 4190 joules per kilogram per Celsius degree. (For convenience, this is often expressed as 4.19 kilojoules per kilogram-Celsius degree.) From Table 12.1 it can be seen that the specific heat of water is very high compared to that of most common substances. This means that to raise the temperature of a given amount of water requires much more heat than is needed to raise the temperature of an equal mass of another substance by the same amount. It also means that water releases more heat on cooling a given amount than does an equal mass of another substance on cooling by the same amount. This property of water is used in the hot-water radiator. The hot water in the radiator gives up to the room 4.19 kilojoules of heat for each kilogram of water that cools by 1 Celsius degree. This is far more heat than would be released on cooling 1 Celsius degree by each kilogram of most other liquids that might be used in the radiator.

The specific heat of water changes sharply with a change of state. In the form of ice or steam, water's specific heat is only about half its value when the water is in liquid form.

Table 12.1 Specific Heats (J/kg · °C)	
alcohol (ethyl)	2430
aluminum	900
copper	387
gold	129
ice	2090
iron	448
lead	128
mercury	138
silver	234
steam	2010
water	4190

12-5 Effect of Specific Heat of Water on Climate

It is well known that the climate of places near lakes and other large bodies of water is cooler in summer and warmer in winter than the climate in other places at the same latitude and altitude.

This is explained by the fact that the specific heat of water is much higher than that of land or air. As a result, for every degree that the temperature rises, the water in a lake absorbs much more heat than does an equal mass of surrounding land or air.

The lake acts as a reservoir of heat in the region. In spring and summer, when the region is getting warmer, the lake absorbs a much larger share of the available heat than do equal masses of the surrounding land and air, thus helping to keep the region cool. In fall and winter, when the region is getting colder, the lake gives up much of the heat it has absorbed, thus helping to keep that region warmer.

12-6 Computing a Gain or Loss of Heat

When the specific heat of a substance is known, the heat gained by a body of that substance as its temperature increases is readily determined. If m is the mass of the body and c is the specific heat of the substance, the heat needed to raise the temperature of that body by 1 Celsius degree is equal to $m \times c$. Now suppose the temperature of the body increases by ΔT Celsius degrees, where ΔT is the difference between its final and initial temperatures ("Δ", the capital Greek letter delta, means "change in"). It follows that the quantity of heat, Q, gained by the body is given by

$$Q = m \times c \times \Delta T$$

By similar reasoning, it is evident that the heat lost by a body when its temperature falls by ΔT Celsius degrees is also given by the above expression.

In computing Q, m is expressed in kilograms, c in joules per kilogram-Celsius degree, and ΔT in Celsius degrees; Q then is expressed in joules. If c is expressed in kilojoules per kilogram-Celsius degree, Q is expressed in kilojoules. (The values of the specific heats listed in Table 12.1 can be converted to kilojoules per kilogram-Celsius degree by dividing by 1000.)

Using this expression for Q, it is possible to measure the heat gained or lost by each of several bodies when heat is passing from some of them to the others. It is found that no heat is permanently lost in transfer. The heat lost by some of the bodies is gained by the others.

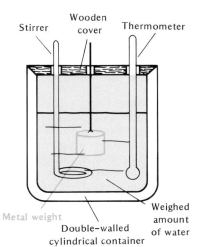

Fig. 12-2. A simple calorimeter used to measure the specific heat of metals and other solid objects.

Sample Problems

1. A 0.500-kg aluminum griddle is heated from 20.0°C to 30.0°C. The specific heat of aluminum is 900 J/kg · C°. How much heat was absorbed by the griddle?

Solution:

$m = 0.500$ kg $c = 900$ J/kg · C°
$\Delta T = 30.0°C - 20.0°C = 10.0$ C°
$Q = m \times c \times \Delta T$
$Q = 0.500$ kg $\times 900$ J/kg · C° $\times 10.0$ C°
$Q = 4.50 \times 10^3$ J, or 4.50 kJ

Note the position of the degree symbols (°). When ° appears *before* the Celsius symbol, C, as in 30.0°C and 20.0°C, it signifies *temperature readings* of 30.0 degrees and 20.0 degrees on the Celsius scale. When ° appears *after* the symbol C, as it does in 10.0 C°, it signifies a *temperature difference* of 10.0 Celsius degrees.

2. A 1.0-kg ingot of iron at 100°C is cooled by being plunged into a vat containing 1.0 kg of water at 0.0°C. What is the temperature of the ingot and water mixture after the mixture has come to equilibrium? The value of the specific heat of iron is 0.448 kJ/kg · C°.

Solution:
Let T°C be the required final temperature. The heat lost by the ingot in going from a temperature of 100°C to T°C is:

$$Q \text{ (lost)} = m \times c \times \Delta T$$

where $m = 1.0$ kg, $c = 0.448$ kJ/kg · C°, and $\Delta T = (100°C - T°C)$.

So,

$Q \text{ (lost)} = (1.0 \text{ kg})(0.448 \text{ kJ/kg} \cdot \text{C°})(100°C - T°C)$
$Q \text{ (lost)} = (44.8 - 0.448T) \text{ kJ}$

The heat gained by the water as it is heated from 0.0°C to T°C is

$$Q \text{ (gained)} = m \times c \times \Delta T$$

where $m = 1.0$ kg, $c = 4.19$ kJ/kg · C°, and $\Delta T = (T°C - 0.0°C)$.

So,

$Q \text{ (gained)} = (1.0 \text{ kg})(4.19 \text{ kJ/kg} \cdot \text{C°})(T°C - 0.0°C)$
$Q \text{ (gained)} = 4.19T \text{ kJ}$

Since no heat is lost as the heat passes from the iron ingot to the water, the heat gained by the water is equal to the heat lost by the ingot.

$$Q \text{ (gained)} = Q \text{ (lost)}$$
$$4.19T \text{ kJ} = (44.8 - 0.448 \, T) \text{ kJ}$$
$$4.64T = 44.8$$

So

$$T = 9.7$$

and

$$T°C = 9.7°C$$

Test Yourself Problems

See Table 12.1 for values of specific heat.
1. How much heat does a 10.0-kg copper ball absorb when its temperature is raised by 10.0 C°?

2. The copper ball above is heated to 500°C and then dropped into a tub containing 20.0 kg of water at 0.00°C. What is the temperature of the mixture when it comes to equilibrium?

Changes of State

12-7 Hidden Heat

Ordinarily when a body is heated, its temperature rises. However, this is not true when the substance being heated is changing its state. Examples are when a heated solid at its melting point is changing to a liquid or when a heated liquid at its boiling point is changing to a gas. In these cases, although heat is added to the solid or liquid, the temperature of the solid or liquid remains the same. Thus the added heat seems to have disappeared into the heated substance without changing its temperature. If heat is conserved, this hidden heat cannot be destroyed. To find out what happened to this heat, consider first the change of state in which a heated solid (such as ice) changes to a liquid (water). This change is called *melting* or *fusion*.

12-8 Melting or Fusion

In a classroom demonstration, a student broke several dozen ice cubes into small pieces and put them in a pan. The student then inserted a mercury thermometer among the pieces of ice and noted that the temperature was a few degrees below 0°C.

The pan was placed on a warm hot plate and the ice was stirred. The temperature rose to about 0°C. After this, the ice began to melt, forming a mixture of water and ice. However, the temperature of the mixture remained at 0°C until all the ice had melted. Then as more heat was added, the temperature of the water slowly rose.

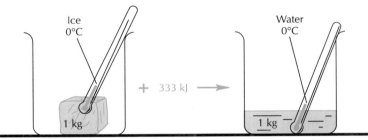

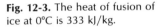

Fig. 12-3. The heat of fusion of ice at 0°C is 333 kJ/kg.

12-9 Heat of Fusion

The above demonstration illustrates two important facts concerning melting or fusion. First, to melt each unit of mass of a solid, such as ice, a definite quantity of heat has to be absorbed. Second, in crystalline solids, such as ice, the temperature remains the same during the melting process although heat has been added to melt the solid. The temperature at which a solid melts is called its *melting point*. The melting point of ice under normal atmospheric conditions is 0°C.

The quantity of heat needed to melt 1 kilogram of a solid substance at its melting point is called the heat of fusion of that substance. The heat of fusion is a fixed quantity for any particular solid crystalline substance but differs from one substance to another. *The heat of fusion of ice is 333 kilojoules per kilogram.* This means that to melt into water at 0°C, 1 kilogram of ice at 0°C must absorb 333 kilojoules of heat.

Ice has a relatively high heat of fusion. This property makes it a good cooling agent. When the ice in a frozen pond melts in the spring, it cools the air above it by absorbing from it 333 kilojoules of heat for each kilogram of ice that melts.

12-10 Freezing or Solidification

Can the 333 kilojoules of heat absorbed by a kilogram of ice on melting be completely recovered? To find out, we must study the heat changes that take place as water is cooled and turned into ice. *The change of a liquid to a solid is called freezing or solidification.* This occurs at a temperature called the *freezing point,* which is the same temperature as the melting point.

When water is cooled and changed to ice, the process will occur in a series of steps, as did the melting of ice. The temperature of the water will first fall to 0°C, its freezing point, and the water will then begin to solidify. However, the temperature of the water and ice mixture will remain at 0°C until all the water has frozen. After this, on further cooling, the temperature of the ice will fall below 0°C.

In the above process, the temperature of the water remains at 0°C although the water is releasing heat. When this quantity of heat is measured, it is found that for each kilogram of water that froze, 333 kilojoules of heat were released by the water. Thus, heat is conserved during the melting of ice or the freezing of water. When 1 kilogram of ice at 0°C absorbs 333 kilojoules of heat, it melts. When the resulting kilogram of water freezes again, it releases 333 kilojoules of heat to its surroundings. The melting and freezing of a substance occur at the same temperature. However, *during melting, a substance in solid form absorbs heat, while during freezing, a substance in liquid form releases heat.*

The freezing of water and the melting of ice in a lake tends to stabilize the local climate. The lake gives up heat to the air on freezing and absorbs heat from the air on melting.

12-11 Solar Heating

Melting and solidification make it possible to store the heat from the sun in a house and use it as needed. One system of solar heating works as follows. See Fig. 12-4(a).

The sun's rays are allowed to enter a glass-covered chamber in the upper part of the house. Here the sun warms the air, which is then circulated by a fan over a series of drums containing Glauber's salt, a solid substance that has a low melting point. As the salt absorbs heat from the hot air, it melts. This melted salt now contains a supply of heat for heating the house.

At night or on cloudy days, a fan pumps cool air from the house over and around the drums of melted salt. Then, as the salt slowly solidifies, it gradually gives up its heat to the air. The air is warmed and is returned to the house by a second fan. Thus, the air in the house circulates continuously between the drums of salt and the house and brings the heat set free by the solidifying salt in the drums into the house. A system like this can store enough solar heat to warm a small house during several sunless days.

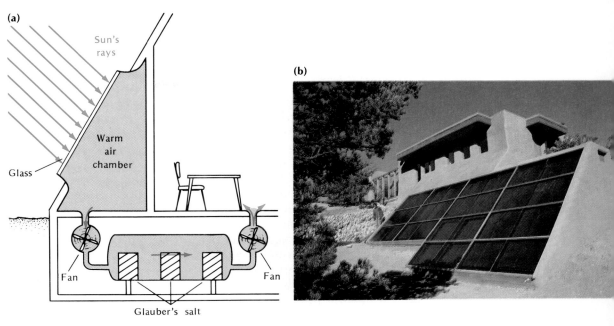

Fig. 12-4. (a) A solar heating system of simple design, in which Glauber's salt melted by the sun is solidified by cool air from the house, which is, in turn, warmed. **(b)** A solar-heated home

Sample Problem

A 2.0-kg piece of ice at 0.0°C melts in a tub and the resulting water comes to room temperature at 20°C. How much heat did the ice absorb from its surroundings?

Solution:

First find the quantity of heat absorbed by the ice in melting. Then add to this the heat absorbed by the resulting water in warming from 0.0°C to 20°C.

To melt 1 kg of ice at 0.0°C requires 333 kJ.
To melt 2.0 kg of ice requires 2.0 × 333 = 666 kJ.

The ice then becomes 2.0 kg of water. The heat needed to raise the temperature of this water from 0.0°C to 20°C is found from the relationship

$$Q = m \times c \times \Delta T$$

where m = 2.0 kg, c = 4.19 kJ/kg · C°, and ΔT = (20°C − 0.0°C) = 20 C°.

Q = 2.0 kg × 4.19 kJ/kg · C° × 20 C°
Q = 168 kJ

The total heat absorbed by the ice is therefore equal to the 666 kJ absorbed in melting plus the 168 kJ absorbed on warming up to 20°C, making a total of 666 + 168 = 834 kJ, or 8.3 × 10² kJ.

Test Yourself Problems

1. How much heat is needed to melt 50.0 kg of ice at 0.0°C?
2. A bucket holding 10.0 kg of water at 10.0°C was cooled until the water froze completely. (a) How much heat did the water lose in being cooled from 10.0°C to 0.0°C? (b) How much heat did the water release upon freezing?

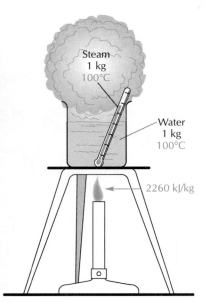

Fig. 12-5. The heat of vaporization of water at 100°C is 2260 kJ/kg.

12-12 Evaporation and Boiling

If the temperature of water is taken at regular intervals while it is being heated, the temperature rises steadily until it reaches about 100°C. At this temperature the water boils and begins to turn into steam. Further heating causes more and more of the water to turn into steam, but this further heating does not change the temperature of the water. Thus, it is seen that it requires a definite quantity of heat to change a given mass of water into steam and that during the process the temperature of the water remains the same.

The change of state in which a liquid changes to a gas (or vapor) is called *evaporation*. Some evaporation takes place at the surface of a liquid at *all* temperatures. You can notice this when a glass of water is left standing for some time. The level of the water slowly goes down as some of the water molecules at the surface escape into the air. As a liquid is heated and its temperature rises, more of the surface molecules escape into the air. At a certain temperature, not only do the surface molecules escape but molecules throughout the liquid also begin to escape into the air. Bubbles of vapor then form throughout the liquid as well as at the surface. The temperature at which this occurs is called the *boiling point*. Different liquids have different boiling points. Under normal atmospheric pressure, the boiling point of mercury is about 360°C, while that of water is 100°C.

The quantity of heat absorbed by 1 kilogram of a liquid in changing into a gas or vapor at its boiling point is called its heat of vaporization. *The heat of vaporization of water is 2260 kilojoules per kilogram.* This means that 2260 kilojoules of heat must be absorbed by 1 kilogram of water at 100°C to change the liquid into 1 kilogram of water vapor, or steam, at the same temperature. Further heating of the resulting steam will simply raise the temperature of the steam.

12-13 Condensation

If steam at 100°C is allowed to cool, it will change into water at 100°C. In this process, although the steam is losing heat, its temperature remains at 100°C. The change in which a gas such as steam or water vapor turns into a liquid is called *condensation*. Condensation and boiling take place at the same temperature. However, *during condensation, heat is given up by the substance, while during boiling, heat is absorbed by the substance.*

Steam behaves as a typical gas. When 1 kilogram of steam turns into liquid water at 100°C, it releases 2260 kilojoules of heat. Note that during condensation, 1 kilogram of steam returns exactly the same quantity of heat that is absorbed by 1 kilogram of liquid water during evaporation. Thus, once again heat is conserved. The heat absorbed by the water at 100°C in changing to steam is stored in the steam and is fully returned when the steam condenses.

A burn from steam at 100°C is more serious than one from water at 100°C because the steam has 2260 kJ per kilogram more heat.

12-14 Steam Heating

Steam heating systems in buildings use steam to carry heat from a central furnace to the radiators in the various rooms. First, the furnace heats water in a boiler to produce steam. Assuming that the boiling takes place at normal pressure, the furnace supplies 2260 kilojoules of heat to change each kilogram of water to steam.

The steam is now piped to every radiator in the house. Here, the steam condenses to liquid water at 100°C. In doing so, each kilogram of steam releases to the radiator, and in turn to the air in the room, the 2260 kilojoules of heat it absorbed when it was vaporized. Each kilogram of water that is formed in the radiators in the house then returns to the boiler. The cycle is repeated so that each kilogram of water absorbs 2260 kilojoules of heat from the furnace and becomes steam. Each kilogram of steam then releases 2260 kilojoules of heat to the radiators in the rooms and becomes water again.

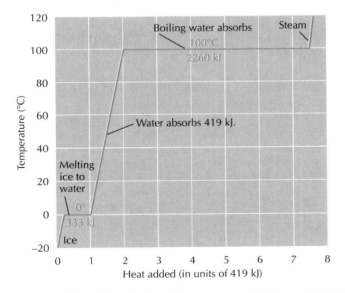

Fig. 12-6. Changes in state as 1 kilogram of ice is heated from −20°C to 100°C.

12-15 Graph of the Changes of State of Water

Fig. 12-6 summarizes the changes of state that take place at normal atmospheric pressure as 1 kilogram of ice at −20°C is steadily heated. At first, the ice rises in temperature as it absorbs heat until it reaches its melting point of 0.0°C. At this point the temperature remains constant and the ice completely melts as it absorbs an additional 333 kilojoules of heat. The resulting water rises in temperature at a rate of 1 C° per 4.19 kilojoules as it absorbs additional heat until it reaches its boiling point of 100°C. Again the temperature remains constant and the water turns to steam as it absorbs an additional 2260 kilojoules of heat. The steam then rises in temperature as it absorbs additional heat.

Fig. 12-7. A pressure cooker raises the boiling point of water and aids cooking at high altitudes, where atmospheric pressure is significantly lower than at sea level.

12-16 Factors Affecting the Boiling and Freezing Points of Water

The values of 100°C for the boiling point of water and 0.0°C for the freezing point are true only when the pressure on the water is equal to normal atmospheric pressure and when the water is pure. Normal atmospheric pressure is the average pressure exerted by the atmosphere at sea level. When the pressure on the water differs from normal atmospheric pressure or when the water has dissolved material in it, both the boiling point and the freezing point of water differ from the values at normal atmospheric pressure.

12-17 Effect of Pressure on the Boiling Point of Water

When the pressure on water is greater than normal atmospheric pressure, water must be heated to a temperature above 100°C to make it boil. The greater the pressure, the greater is the rise in the boiling point of water.

When the pressure on water is less than normal atmospheric pressure, water boils at a temperature below 100°C. The lower the pressure, the lower is the temperature at which water boils. Thus, on a mountaintop where the atmospheric pressure is considerably below normal, water boils at temperatures significantly below 100°C. Cooking food at these lower temperatures can take longer. In fact, the temperature may be too low to cook food properly. In such places pressure cookers are commonly used to raise the boiling point of water to a temperature suitable for cooking.

A pressure cooker is a strong-walled pot fitted with a cover that can be clamped on. When a small quantity of water is heated in the pot, it generates steam. The confined steam and hot air exert a pressure on the water that is much higher than the normal atmospheric pressure. As a result, the boiling point of the water in the pot is raised. At the higher pressure and the consequently higher temperature in the pot, food cooks faster than it would in an open pot.

The same principle is applied to steam boilers where a high pressure on the water may raise the boiling point to over 200°C.

12-18 Effect of Pressure on the Freezing Point of Water

The freezing point of water is also affected by pressure. When the pressure is greater than atmospheric pressure, water freezes and ice melts at a temperature lower than 0°C. Mechanical pressure exerted on ice has the same effect as the pressure exerted by the atmosphere. Thus, the pressure of the blades of an ice skater lowers the melting point of the ice and causes the ice to melt. The skater actually glides along on a thin layer of water. As the skater

Fig. 12-8. A skater glides along on a thin film of water produced by the melting of ice due to the pressure of the skates' thin blades.

passes beyond a given spot, the extra pressure on that spot is eliminated. This return to normal atmospheric pressure raises the freezing point and causes the thin layer of water to refreeze.

12-19 Effects of Dissolved Material on the Boiling and Freezing Points of Water

When a substance is dissolved in water, it raises the boiling point of the water and lowers its freezing point. This is true for all substances that dissolve in water. That is, the boiling point of water containing dissolved matter is higher than the boiling point of pure water. Likewise, the freezing point of water containing dissolved matter is lower than the freezing point of pure water. As an illustration, consider a solution of table salt dissolved in water. At normal atmospheric pressure, salt water must be heated to a temperature above 100°C to make the water boil and must be cooled to a temperature below 0°C to make it freeze. The greater the quantity of salt dissolved in the water, the higher is the boiling point and the lower is the freezing point of the salt solution. In winter, salt sprinkled over a snow-covered road makes the snow melt faster. Can you explain why this happens?

Heat Transfer

In studying the exchange of heat between bodies, it is found that heat may be transferred from one body to another by any of three distinct processes. These processes are called *conduction, convection,* and *radiation.*

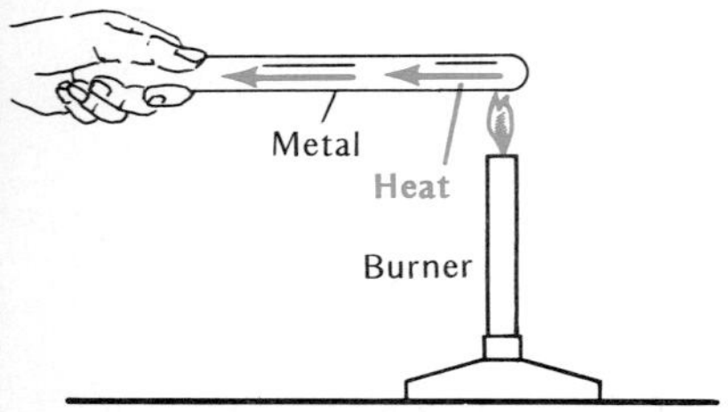

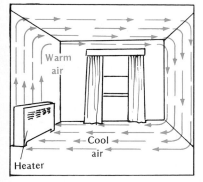

Fig. 12-9. Transfer of heat by conduction.

12-20 Conduction

Many substances, notably metals, have the ability to let heat pass through them. These substances are called *conductors* and the *process whereby heat is transmitted through them is called conduction.* If a conductor such as a bar of copper is heated at one end, the heat travels steadily through the bar until the whole bar is warm. This is characteristic of the behavior of conductors and explains why the handle of a metal spoon with its lower end in hot soup soon becomes too hot to touch.

Substances that do not conduct heat well are called *insulators.* Among the common insulators are wood, paper, plastics, air, and water.

In the laboratory and in industry, both conductors and insulators are used to control the transfer of heat. Conductors are used when it is necessary to transfer heat from place to place. Kitchen pots, for example, are made of good conductors such as aluminum to allow heat from the gas or electric stove to pass through them readily. Insulators are used when it is necessary to prevent or retard the transfer of heat. Water or steam boilers used in buildings are generally coated with insulating material to reduce the escape of heat through the walls of the boilers.

12-21 Convection

To heat a substance that is a poor conductor of heat, it is necessary to bring each part of it to the heater to be heated directly. In liquids and gases, most of which are poor conductors of heat, this is done by the formation of a *current* inside the liquid or gas. The current circulates the liquid or gas continually so that each part of it is brought in direct contact with the heater over and over again. *The process whereby heat is transported through a liquid or gas by means of currents is called convection.* When a liquid or gas is heated from below, these currents form spontaneously and are called *convection currents.*

As an illustration, consider how a steam radiator heats the air in the room by convection currents, which form when the air in contact with the hot radiator becomes warm. The air then expands and becomes less dense than the surrounding air, which moves in under it and forces it to float upward to the top of the room. The new colder air, now in contact with the radiator, is heated in turn and is also displaced upward when surrounding colder air moves under it and takes its place. This process continues and results in the formation of convection currents in which newly heated air continually rises above the radiator while cooler air moves in and takes its place. The convection currents continue to circulate the air, bringing every part of the air in the room to the radiator over and over again.

Fig. 12-10. Transfer of heat by convection.

12-22 Radiation

One has only to stand in the sun to become aware of the great quantity of heat produced by the sun that reaches the earth. Since the sun's heat travels through the near vacuum that exists in space between the sun and the earth, it cannot reach the earth by either conduction or convection. The process by which it does reach the earth resembles that by which the sun's light travels and is called *radiation*.

In radiation, heat is transferred by electromagnetic waves called infrared rays. The manner in which energy is transferred by means of waves will be discussed more fully in Chapter 15. Here it is sufficient to note that while infrared rays are invisible, they resemble light waves in many ways. They belong to the electromagnetic family of waves, all of which travel at 3×10^8 meters per second in a vacuum, the speed of light. They are reflected by mirrors and other polished surfaces. They can be focused, like light, with curved mirrors or lenses, and can even be used to take photographs if film sensitive to infrared waves is used.

All objects, whether warm or cold, are constantly exchanging heat with each other by emitting and receiving heat in radiant form. The higher the temperature of a body with respect to its surroundings, the faster it will radiate heat to its surroundings; the lower the temperature of a body with respect to its surroundings, the faster it will absorb radiant heat. Hotter bodies are therefore net losers of heat while colder bodies are net gainers. Radiant heat is noticed particularly when one is near a hot object, such as a hot iron or an open fire.

Fig. 12-11. Transfer of heat by radiation.

Heat, Energy, and Work

12-23 The Mechanical Equivalent of Heat

Thus far, we have noted that heat is conserved as long as it is being exchanged between bodies of matter. If heat is a form of energy, the energy it represents should also be conserved when heat is changed into work or other forms of energy, or when work is converted into heat. Among the major experimenters who investigated the relationship between heat and work were the American-born Benjamin Thompson (1753–1814), also known as Count Rumford, and the Englishman James Prescott Joule (1818–1889).

Count Rumford noticed that when a cannon was bored out of a piece of solid brass, both the cannon and the rotating boring tool became extremely hot. Apparently, unlimited quantities of heat were being generated by the friction between the boring tool and the solid brass, the quantity depending only upon how long the boring was continued. Rumford inferred that this heat was being

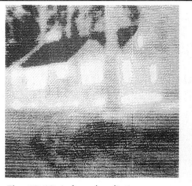

Fig. 12-12. Infrared radiation produced this thermogram, which shows where heat is escaping from this house, resulting in a waste of fuel.

produced at the expense of work done in operating the boring tool. In other words, this work was steadily being converted into heat so that the greater the work input, the greater would be the heat generated.

The specific task of determining the quantitative relationship between work and heat was undertaken by Joule. Using many different kinds of experiments, Joule accurately measured the quantity of work that, when completely converted into heat, produces what was in his day the standard unit of heat, the kilocalorie. You will recall from Section 11-2 that a kilocalorie is the quantity of heat needed to raise the temperature of 1 kilogram of water by 1 C°. Joule found that 4.19 kilojoules of work was equivalent to 1 kilocalorie of heat. This quantitative relationship between work and heat is called *the mechanical equivalent of heat*. As a result of Joule's work, it became evident that heat is a form of energy and that a quantity of work could be converted into an equivalent quantity of heat.

12-24 Joule's Experiment

To determine the relationship between work and heat, Joule used an apparatus like that in Fig. 12-13. Here, water in an insulated container is stirred by a set of rotating paddles attached to a drum. A cord wound around the drum runs over a pulley and is attached to a weight w. When the weight is allowed to fall, it pulls the cord and rotates the paddles. The work done by the paddles in stirring the water is then converted into heat that causes the temperature of the water to rise.

In practice, the apparatus is arranged so that the weight falls very slowly and acquires very little kinetic energy. The work done by the weight in stirring the water is then very nearly equal to the change in its gravitational potential energy as it falls through the height h. This work is equal to the weight times the height through which it has fallen, or wh.

To determine the quantity of heat generated by this work, Joule measured the mass, m, of the water in the container and the increase, ΔT, in its temperature and substituted these values in the relationship $Q = m \times c \times \Delta T$. When he compared the work done by the paddles and the heat they generated, he found that it always took 4.19 kilojoules of work to generate the quantity of heat needed to raise the temperature of 1 kilogram of water by 1 C°. (Recall that this quantity of heat is the kilocalorie.)

Joule did other experiments involving the conversion of chemical, electrical, and other forms of energy into heat. He found that these conversions all take place at the same rate of exchange as that between work and heat. This confirmed that heat is a form of energy and that no energy is lost in converting other forms of energy into heat.

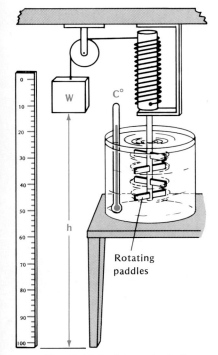

Fig. 12-13. Joule's method of measuring the mechanical equivalent of heat.

12-25 Converting Heat to Work

So far we have seen that a quantity of work can be converted to an equal quantity of heat. It remains to be seen if a quantity of heat can be converted to an equal quantity of work without loss of energy.

A device for converting heat into work is called a *heat engine*. The steam engine, the automobile engine, the jet engine, and the rocket are typical heat engines in everyday use. Essentially, each of these, when supplied with heat, provides a force that moves objects and thus does work on them.

Figure 12-14 illustrates how a very simple engine can exert such a force. It consists of an air-filled cylinder closed at one end and fitted with a smooth-fitting piston at the other. Resting on the piston is a weight, or load, to be lifted by the engine. As the air in the cylinder is heated, it expands and exerts pressure that forces the piston upward. This raises the load and thus does work.

The quantity of work done by this engine is the product of the upward force exerted by the heated air on the piston and the distance the load was lifted. The quantity of heat used by the engine to do this work is equal to the total quantity of heat supplied to the engine minus that part of the heat that was needed to raise the temperature of the confined air. When the part of the heat used to do work is compared with the quantity of work done by it, it is found that there is no loss. Each joule of heat supplied to the engine to do work is converted to 1 joule of work.

Thus, whether heat is converted to work or work is converted to heat, there is no loss of energy in the process.

12-26 Internal Energy

In this chapter, the evidence that heat is conserved during transfer from one body to another and that heat and work are interchangeable led to the conclusion that heat is a form of energy. Just what kind of energy is it?

First of all, it is known that heat is energy that flows naturally from one body to another only when there is a difference in temperature between the two bodies. As heat enters a body, it may raise the temperature of the body or it may cause a change of state of the matter that makes up the body. In either case, it produces a change throughout the body and is stored in the body in some form that we call *internal energy*. As will be discussed in Chapter 13, the *internal energy of an object is the sum of the potential and kinetic energies of the individual molecules of which it is made.* The internal energy of a body reappears as heat when the body is put into contact with a body colder than itself. The warmer body then loses some of its internal energy by transferring it in the form of heat to the cooler body. In the cooler body the heat received is distributed throughout the body as increased internal energy.

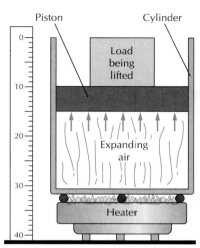

Fig. 12-14. A simple heat engine converts heat into work.

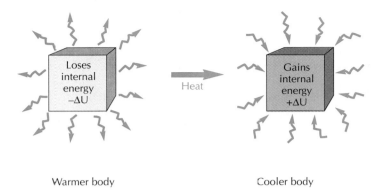

Warmer body   Cooler body

Fig. 12-15. Heat spontaneously flows from a body at higher temperature to one at a lower temperature.

12-27 Thermodynamics and Systems

Thermodynamics is the study of physical processes involving heat, work, and internal energy. A simple example of such a process, one involving only heat and internal energy, is shown in Fig. 12-15. Here a warmer body and a cooler body are placed near each other in a space insulated from all other bodies. Because there is a difference of temperature between them, heat flows from the warmer to the cooler body. The heat absorbed by the cooler body increases its internal energy, which becomes evident by either a rise of temperature or a change of state. At the same time, the loss of heat by the warmer body decreases its internal energy, which becomes evident by a drop in temperature or a change of state.

Thermodynamics is concerned with *systems*. A system is a space singled out for study. It has definite boundaries so that it is possible to define what is inside the system and what is outside the system. A system may contain a combination of substances in various physical states, objects, engines, motors, and other kinds of devices. A system can interact with the environment outside the system either by absorbing heat from the environment or transferring heat to it. A system is also capable of doing work on its environment or having work done on it by its environment.

To illustrate, consider the system shown in Fig. 12-16. It consists of an air-filled cylinder closed at one end and with a freely moving piston at the other. If heat from outside the system is applied to the cylinder, it will cause the air to expand and push the piston outward, thus doing work against the opposing atmospheric pressure. In this case, the system takes in heat and uses it to do work on its environment.

Suppose that instead of heating the cylinder, we push on the piston and thereby compress the air in the cylinder. We, acting as part of the environment, are now doing work on the system. What will be the effect of this work? To answer this question we must turn our attention to the first law of thermodynamics.

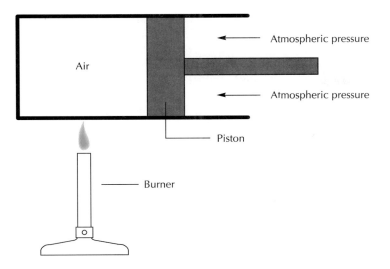

Fig. 12-16. Heating the air in the cylinder causes the air to expand, forcing the piston outward against atmospheric pressure. In this case, the system (the air in the cylinder) does work on the environment (the atmospheric pressure).

12-28 The First Law of Thermodynamics

The first law of thermodynamics states:

Whenever heat is transformed into work or another form of energy, or another form of energy is transformed into heat, there is no loss of energy.

This law is simply the law of conservation of energy restated so that it includes heat, internal energy, work, and all other forms of energy. Expressed as an equation, the first law of thermodynamics is written:

$$Q = \Delta U + W$$

where Q represents the quantity of heat that enters or leaves the system, ΔU is the change in internal energy of the system, and W is the work done on or by the system.

The values of Q, ΔU, and W may be positive or negative depending on the process. The value of Q is positive when heat enters the system and is negative when heat leaves the system. The value of ΔU is positive when the internal energy of the system increases and is negative when the internal energy of the system decreases. The value of W is positive when the system does work on the environment and negative when the environment does work on the system.

12-29 Application of the First Law of Thermodynamics to a Heat Engine

Consider the heat engine in Fig. 12-17. In this engine, the air in the cylinder makes up the system. The cylinder, the piston, the load, the heater, and everything else constitute the environment outside the system. As the heater transfers a quantity Q of heat to the

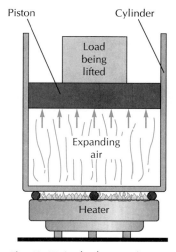

Fig. 12-17. In the *heat engine* shown, a heater transfers heat to the air trapped beneath the piston, causing the air to expand and lift the load.

trapped air in the cylinder, the air temperature rises, signifying an increase in its internal energy, ΔU. At the same time, the heated air expands and exerts an upward force on the piston that raises the piston and the load on it. Thus, the system does work W on the environment. Consequently, when the heater is turned on, Q, ΔU, and W are all positive because heat has been added to the system, its internal energy has increased, and it has done work on the environment.

Suppose the heater transfers 1000 kilojoules of heat to the system and the system then does 400 kilojoules of work on the environment. Then, applying the first law of thermodynamics,

$$Q = \Delta U + W$$
$$1000 \text{ kJ} = \Delta U + 400 \text{ kJ}$$
$$\Delta U = 1000 \text{ kJ} - 400 \text{ kJ} = 600 \text{ kJ}$$

In this operation of the engine, the internal energy increased by 600 kJ.

Now suppose that the heater is turned off so that the air in the cylinder cools. During cooling, heat is transferred from the system to the environment. When the system loses heat, the value of Q is negative. As the air in the cylinder cools, the external force exerted by the environment (the sum of the weights of the piston, the load, and the downward force exerted by the atmosphere) becomes greater than the upward force exerted by the trapped air against the piston. This imbalance in forces causes the piston to move downward and compress the gas into a smaller volume. The environment, therefore, does work on the system, and W has a negative value. Finally, since the temperature of the confined air decreases as it cools, the internal energy also decreases and the value of ΔU is negative. Thus, all three quantities, Q, W, and ΔU are negative when the air in the cylinder is allowed to cool so that heat leaves the system.

When a system does work on the environment, the value of W is positive; when the environment does work on the system, the value of W is negative.

12-30 Other Applications of the First Law of Thermodynamics

In Joule's experiment, illustrated in Figure 12-13, the system consists of the water being stirred by paddles in a container. Everything other than the water is the environment surrounding the system. As the paddles rotate, they exert forces that stir the water and do work on it. Because work is being done on the system, W has a negative value. Because the container is well insulated, no heat can enter or leave the system. (An insulated system, such as this one, in which heat can neither enter or leave the system is called an *adiabatic system*.) Since the value of Q is zero, the first law in this situation reduces to:

$$0 = \Delta U + W$$

Since W has a negative value, ΔU must have a positive value equal in magnitude to W. The equation tells us that all the work done on the system by the paddles is converted into an increase in the internal energy of the system. This increase in internal energy is made evident by an increase in the temperature of the water.

As another example, consider a system consisting of a closed, rigid box filled with air. Since the box is rigid, there is no way in which the enclosed air can do work on the environment or in which the environment can do work on the air. Therefore, the value of W is zero. According to the first law, it follows that when a quantity of heat, Q, is added to the box, the equation of the first law of thermodynamics becomes

$$Q = \Delta U + 0$$

The equation states that all the heat added to the system increases the internal energy of the box and the enclosed air. Thus, the temperature of the box and the enclosed air will rise.

Sample Problem

When 5000 J of heat is added to the engine shown in Fig. 12-16, the engine does 1250 J of work. What is the change in the internal energy of the system?

Solution:

Since heat enters the system and work is done by the system, $Q = +5000$ J and $W = +1250$ J.

$$Q = \Delta U + W$$
$$\Delta U = Q - W$$
$$\Delta U = 5000 \text{ J} - 1250 \text{ J} = 3750 \text{ J}$$

Test Yourself Problems

1. A quantity of hot air is enclosed in an insulated cylinder, which is fitted with a piston. The hot air is allowed to expand and push the piston outward against the pressure of the atmosphere. If the air does 4000 J of work, how much internal energy has it lost?

2. Assume that 10 000 J of heat from the environment is able to enter the cylinder in the problem above. The air in the cylinder does an additional 1500 J of work as a result. What is the change in internal energy of the air?

12-31 The Second Law of Thermodynamics

Let us return to the system shown in Fig. 12-15, where heat is flowing from a warmer body to a cooler one. Assume that the two bodies are isolated so that neither heat nor work can be exchanged with the environment. The first law of thermodynamics states that the energy of the system will be conserved. It does not indicate in which direction the heat will flow. However, from experience we know that heat will spontaneously flow only from the warmer body to the cooler one and not in the reverse direction. This fundamental property of heat is the basis for the second law of thermodynamics, which states:

Heat flows spontaneously only from a body at a higher temperature to one at a lower temperature. To reverse the flow, work must be done.

Fig. 12-18. The exhaust gases from the tailpipe of a vehicle carry away the unused heat.

12-32 Thermodynamics of an Engine

The fact that, according to the second law of thermodynamics, heat travels spontaneously from one body to another only if there is a difference in their temperatures, limits the efficiency with which an engine can convert heat into work.

Consider an automobile engine. For heat to be transferred to the engine, there must be a *heat source* that supplies heat at a high temperature and a *heat sink,* or heat reservoir, that receives heat at a lower temperature. The source of the heat is the fuel that is burned inside the cylinders of the engine. As the fuel burns, it releases heat. This heat raises the temperature of the gases produced during the combustion of the fuel. As the gases get hotter, they expand against the pistons of the engine, causing the pistons to move and do the work that runs the automobile. These gases, while still hot but at a lower temperature, are ejected from the tail pipe of the car into the outside air, which serves as the heat sink. The heat in these exhaust gases is wasted in the sense that it is not converted to useful work by the engine. The part of the heat input that does work in the engine is, therefore, the difference between the quantity of heat provided by the burning fuel and the quantity of heat that is discharged in the exhaust gases.

No engine can completely convert to work all the heat it absorbs from a fixed temperature source.

From the point of view of conversion of heat to work, the second law of thermodynamics can be restated as follows: *It is impossible for any engine to absorb heat from a fixed temperature source and completely convert that heat to work.* There is always some unconverted heat that must be delivered to a heat sink at a lower temperature.

12-33 Limits on the Efficiency of an Engine

Since no engine can convert all the heat supplied to it into work, the efficiency of an engine depends on how much of the heat supplied to it is discharged in the exhaust. The greater the quantity

Fig. 12-19. A jet engine is an efficient heat engine that converts the heat of burning gases into mechanical energy to propel aircraft at high velocities.

of the heat discharged, the lower is the efficiency of the engine. Noting this relationship, the French engineer Sadi Carnot (sah·DEE Kar·NOH/1796–1832) proved that the maximum efficiency of an engine can be computed from the expression

$$\text{Efficiency} = \frac{T_1 - T_2}{T_1} \times 100\%$$

where T_1 is the Kelvin temperature at which the heat enters the engine and T_2 is the Kelvin temperature at which heat leaves the engine in the exhaust gases. Thus, the efficiency of an engine that takes in heat at 600 K and ejects the exhaust gases at 400 K is:

$$\text{Efficiency} = \frac{(600 \text{ K} - 400 \text{ K})}{600 \text{ K}} \times 100\% = 33.3\%$$

This means that only 33.3% of the heat put into the engine is converted to work. The rest is discarded in the exhaust gases.

It is clear that the greater the temperature difference between heat supplied to the engine and the exhaust heat, the greater the efficiency of the engine. No heat engine can be 100% efficient since it would have to eject heat at a temperature of 0 K. This is impossible. Actually, practical engines rarely attain efficiencies higher than 40%.

Typical Engine Efficiencies	
Auto (gasoline)	20%
Steam turbine	30%
Diesel	35%

12-34 Reversing the Direction of Heat Flow

Although heat flows spontaneously only from a warmer to a colder body, it is possible to *make* it flow in the opposite direction. However, to do this, work must be done. This reversal of heat flow occurs, for example, in a refrigerator, which transfers heat from the air and food inside the refrigerator to the air outside. The transfer of heat from the cooler interior of the refrigerator to

Fig. 12-20. In cities, the air-conditioning units for large buildings consume huge quantities of electrical energy and vent huge quantities of waste heat into the atmosphere.

the warmer air in the room requires work. You pay the electric utility company for this work, which is provided by the electrical energy that runs the refrigerator.

Another device that reverses the direction of heat flow is a *heat pump*. It is used to heat a house by removing heat from the cold air outside the house and transferring it as heat at a higher temperature to the interior of the house. Its operation is similar to that of a refrigerator. Like a refrigerator, a heat pump must do work to transfer heat from a heat source (outside air) at a lower temperature to a heat sink (inside air) at a higher temperature.

12-35 Entropy

The natural, spontaneous flow of heat from a body at higher temperature to one at a lower temperature is an irreversible process because *by itself* heat cannot flow in the opposite direction. We have seen that the second law of thermodynamics is an expression of this irreversible process.

The concept of entropy is a quantitative way of expressing this irreversibility. Actually, it is the change in entropy, ΔS, that measures the thermodynamic change that takes place in a system to which heat has been added or from which heat has been removed. The change in entropy is defined as

Fig. 12-21. Rudolf Clausius (1822–1888), a pioneer in the field of thermodynamics, originated the concept of entropy.

$$\Delta S = \frac{\Delta Q}{T}$$

where ΔQ is the heat added to or lost by a system and T is the kelvin temperature. If heat enters the system, ΔQ is positive. If heat leaves the system, ΔQ is negative. Thus ΔS may be positive or negative depending on the sign of ΔQ.

It should be noted that the expression for entropy given above is applicable only to bodies whose temperature remains practically constant when a comparatively small quantity of heat, ΔQ, is added to or taken away from them. Very large bodies fulfill this condition.

Sample Problem

In a system isolated from the outside environment, 10 J of heat passes from a large slab of iron at a temperature of 400 K to another large slab of iron at a temperature of 200 K. What is the change in entropy of the system?

Solution:
For the hotter slab of iron, $\Delta Q_1 = -10$ J and $T_1 = 400$ K. Hence,

$$\Delta S_1 = \frac{\Delta Q_1}{T_1} = \frac{-10J}{400 \text{ K}} = -0.025 \text{ J/K}$$

For the cooler slab of iron, $\Delta Q_2 = +10$ J and $T_2 = 200$ K. Hence,

$$\Delta S_2 = \frac{\Delta Q_2}{T_2} = \frac{+10 \text{ J}}{200 \text{ K}} = +0.050 \text{ J/K}$$
$$\Delta S = \Delta S_1 + \Delta S^2$$
$$\Delta S = (-0.025 \text{ J/K}) + (+0.050 \text{ J/K}) = +0.025 \text{ J/K}$$

The positive value of ΔS indicates that the spontaneous transfer of heat from a warmer to a cooler body is an irreversible process in which the entropy of the system increases.

Test Yourself Problems

1. A piece of ice at 273 K absorbs 819 kJ of heat and melts to water at 273 K. What is the change in entropy of the ice?

2. In an isolated system, 4.00 J of heat passes from a large body at a temperature of 4000 K to a large body at 400 K. If the two bodies retain their initial temperatures, find the change in entropy of this system.

12-36 Irreversible Processes and Entropy

It can be shown that if the thermodynamic change taking place within a system is irreversible, the change in entropy of that system will always increase. To illustrate, consider a system of two large bodies in contact with each other, one at kelvin temperature T_2 and the other at a lower kelvin temperature T_1. Suppose a small quantity of heat passes from the warmer to the cooler body. Let ΔS_2 be the change in entropy of the warmer body and ΔS_1 be the change in entropy of the cooler body. Let ΔQ_2 be the heat lost by the warmer body and ΔQ_1 be the heat gained by the cooler body. Note that the magnitudes of ΔQ_2 and ΔQ_1 are the same (all the heat lost by the warmer body is gained by the cooler body), but their signs are opposite. Heat lost is considered negative, whereas heat gained is considered positive. The changes in entropy of the two bodies are given by the following expressions:

$$\Delta S_2 = \frac{\Delta Q_2}{T_2}$$

$$\Delta S_1 = \frac{\Delta Q_1}{T_1}$$

The change in entropy of the system, ΔS, is the sum of the changes in entropy of the two bodies making up the system

$$\Delta S = \Delta S_1 + \Delta S_2$$

Therefore:

$$\Delta S = \frac{\Delta Q_2}{T_2} + \frac{\Delta Q_1}{T_1}$$

Note that $\Delta Q_2/T_2$ is negative because ΔQ_2 is heat lost, and $\Delta Q_1/T_1$ is positive because ΔQ_1 is heat gained. Noting that the magnitude of ΔQ_2 is equal to that of ΔQ_1 and that T_2 is greater than T_1, it follows that the magnitude of $\Delta Q_1/T_1$ is greater than that of $\Delta Q_2/T_2$. Their sum ΔS is therefore positive, denoting a net increase in the entropy of the system. Thus, in this irreversible transfer of heat from a warmer to a cooler body, the entropy of the system increases.

12-37 Direction of Change of the Entropy of the Universe

Most of the changes that take place in the world, such as the conversion of work to heat, are irreversible. In such situations, the net entropy of the system involving the change and its environment always increases. Thus, the second law of thermodynamics may be restated as follows:

In any system in which irreversible changes are taking place, the net entropy of the system and its environment can only increase. Since most of the changes taking place in the universe are irreversible, the entropy of the universe is increasing.

The entropy of the universe is increasing.

12-38 Entropy and the Convertibility of Heat to Work

Suppose we have a system in which a body at a higher temperature is allowed to remain in contact with a body at a lower temperature until both bodies are at the same temperature. According to the first law of thermodynamics, there is no loss of energy during this transfer of heat. However, as we have seen earlier, there is an increase in the entropy of the system. This increase is related to the loss of opportunities to convert heat into work. Before the transfer, the warmer body could be used as the heat source in an engine while the cooler body could have served as the heat sink. The engine could then have converted some of the heat into work. Now, with both bodies at the same temperature, that opportunity is lost. Thus, the rise in entropy represents a loss of opportunity to convert heat to work. Since heat is continually moving from warmer to cooler bodies throughout the universe, it follows that there is a continual increase in entropy accompanied by an increasing loss of opportunities to convert heat into work.

Fig. 12-22. The vast expanse of snow and ice on Antarctica contains a great quantity of heat that is not available for doing work unless it can flow to a colder body.

12-39 Entropy, Order, and Disorder

When we examine energy on the molecular level, we can recognize two types—energy that is random and energy that is orderly. Consider a projectile in flight. It has internal energy, which is the sum total of the potential and kinetic energies of its molecules. This energy is random and not fully available for work because the molecules move in all directions. Superimposed on this random energy is the kinetic energy of the projectile as a whole, which gives each molecule a velocity in the direction of motion of the projectile. This common molecular motion is organized energy and is fully available to do work.

When the projectile strikes a target and is stopped, this organized energy is converted to random energy in the form of the increased internal energy of the projectile and the target. This is evident in the rise in the temperatures of the projectile and the target.

Projectile's
organized motion

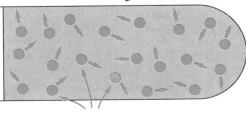

Random
molecular motion

Fig. 12-23. The motion of a projectile in flight is highly organized. On a microscopic level, its molecules exhibit random motion.

This example illustrates a general process in which the order represented by organized energy is replaced by the disorder represented by random energy. *Entropy is a quantitative measure of the disorder of a system.* If the changes taking place in a system are irreversible, so that they occur naturally only in one direction, it can be shown that the entropy of the system will always increase. Because most of the changes that take place in the universe are irreversible, the entropy, or disorder, of the universe is increasing. We see this when perfume escapes from an open bottle into a room, when a glass falls and breaks into pieces, and, on a large scale, when mountains are worn down by wind and running water. In each case the increased disorder means an increase in entropy.

12-40 The Third Law of Thermodynamics

The third law of thermodynamics states:

There is no process by which a body may be cooled to a temperature of 0 K, that is, absolute zero.

We may lower the temperature of a body until it is very close to absolute zero but we can never cause the body to reach absolute zero. This can be understood from the fact that heat travels spontaneously from a body at a higher temperature to one at a lower temperature. To remove heat from a body at a temperature near absolute zero, we must have a heat sink at a still lower temperature to which the heat can flow. It is evident that to cool a body to absolute zero we will need a heat sink at a temperature lower than absolute zero. This is impossible by the definition of absolute zero. So far, by the use of very clever means, physicists have been able to attain temperatures as low as 0.3 K.

CHAPTER REVIEW

Summary

When heat is transferred from one body to another, no part of it is destroyed. The heat lost by one body is gained by the other.

The addition of heat to a substance raises its temperature provided the substance is not at its melting or its boiling temperature. The heat needed to raise the temperature of a unit mass of a substance one degree is called its **specific heat.** Different substances have different specific heats. The specific heat of water is much larger than that of most substances.

When a crystalline solid at its **melting point** is heated, it begins to turn to a liquid without further change in temperature. The heat needed to change a unit mass of a solid at its melting point to a liquid is called its **heat of fusion.** The heat absorbed by a solid in melting is stored in the resulting liquid as **internal energy.** All of it is recovered when the liquid freezes and becomes solid again. For crystalline solids, melting and freezing occur at the same temperature. In melting, a substance in solid form absorbs heat, while in freezing the same substance in liquid form gives off the heat previously absorbed.

When a liquid at its **boiling point** is heated, it begins to turn to a gas or vapor without further change in temperature. The heat needed to change a unit mass of a liquid at its boiling point to a gas or vapor is called its **heat of vaporization.** The heat absorbed by a liquid in boiling is stored in the resulting gas as internal energy. All of it is recovered when the vapor condenses and becomes a liquid again. **Condensation** takes place at the same temperature as boiling or **evaporation.** However, in evaporation, a substance in liquid form absorbs heat; in condensation, the same substance in gas form gives off the heat previously absorbed.

Under normal atmospheric pressure, water boils at 100°C and freezes at 0°C. When the pressure on water is increased, its boiling point is raised and its freezing point is lowered. A decrease in the pressure on water to a point lower than normal atmospheric pressure results in a decrease in its boiling point.

Heat is transferred from one place to another by conduction, convection, and radiation. **Conduction** is the method whereby heat is distributed through certain substances by being passed on from one part of the substance to another. Metals are among the best conductors of heat. **Convection** is the method whereby heat is distributed by the formation of currents inside heated liquids and gases. **Radiation** is the method whereby heat is transferred through space by electromagnetic waves known as heat or **infrared rays.** These waves travel at the speed of light. Although they are invisible, they resemble light waves in many ways. They are reflected from silvered, white, and shiny surfaces and are absorbed by dark, rough surfaces. They can also be used to make photographs when the proper film is used in the camera.

Work can be transformed into heat and heat can be transformed into work. The **mechanical equivalent of heat** is the quantity of work that is converted into one unit of heat. It is 4.19 kilojoules, the quantity of work that when converted to heat will raise the temperature of 1 kilogram of water by 1 Celsius degree. The fact that chemical, electrical, and other forms of energy can also be converted into heat at the same rate of exchange indicates that heat is a form of energy and can be converted into other forms of energy in accordance with the law of conservation of energy.

Thermodynamics is the study of physical processes involving heat, work, and internal energy. Three laws of thermodynamics apply to these processes. The first is the **law of conservation of energy** restated to include heat, namely, in any transformation of energy in which heat is involved, there is no loss of energy. The second law of thermodynamics states that heat naturally travels in only one direction, from a warmer to a cooler body. The third law of thermodynamics states that there is no way to bring a body's temperature to absolute zero.

A **heat engine** is a device for converting heat into work. Because of the second law of thermodynamics, there is a limit to the **efficiency** that any heat engine can attain. The maximum efficiency an engine can attain in converting heat into work is given by

$$\text{Efficiency} = \frac{T_1 - T_2}{T_1} \times 100\%$$

where T_1 is the kelvin temperature of the heat put into the engine and T_2 is the kelvin temperature of the heat in the exhaust gases that leave the engine.

Questions

Group 1

1. (a) Define specific heat. (b) The specific heat of aluminum is about one-fifth that of water. By about how many degrees Celsius will 4.19 kJ of heat raise the temperature of 1 kg of aluminum?

2. Mercury has a much lower specific heat than water. (a) If 1 kg of mercury at 100°C is mixed with 1 kg of water at 0°C, will the temperature of the mixture be 50°C, above 50°C, or below 50°C? Explain your answer. (b) What will happen to the heat lost by the mercury during the mixing of the two liquids?

3. Explain how the high specific heat of water may help to stabilize the climate in a lake region.

4. (a) Describe a situation in which a solid substance absorbs heat but undergoes no change in temperature. (b) What is meant by the heat of fusion of the solid? (c) How do we know that the heat absorbed by the solid has not been destroyed?

5. When a lake that has been frozen in winter melts in the spring, what effect does it have on the temperature of the air above it?

6. Explain how the processes of melting and solidification may be used in a solar heating system to store heat and then to liberate it.

7. A kilogram of water at 100°C is converted to steam at 100°C and then reconverted to water at 100°C. Describe the heat changes that have taken place and show that no heat has been gained or lost.

8. Explain how the condensation of steam in a steam radiator releases heat to the air in the room.

9. (a) Name several substances that are good conductors of heat. (b) Name several other substances that are insulators. (c) In controlling the flow of heat from one place to another, what use is made of conductors and insulators?

10. (a) What substances may be heated by convection? (b) Explain how a steam radiator in a room sets up convection currents of the air in the room.

11. Compare the behavior of the heat radiated from a hot body with the behavior of the light emitted by a lamp with respect to each of the following: (a) speed of travel; (b) ability to pass through a vacuum; (c) ability to be reflected by mirrors; (d) ability to be absorbed; (e) visibility; (f) ability to make photographs.

12. (a) How did Joule's experiment show that work and heat are interchangeable? (b) What is meant by the mechanical equivalent of heat? (c) What properties of heat enable us to identify it as a form of energy?

13. How does the boiling point of water carried in a hot-air balloon change as the balloon rises?

14. How could you speed up the melting of snow in the school yard?

15. If, in your area, water boils at too low a temperature to allow you to cook efficiently, what two methods could you use to raise the boiling point of water?

16. A liter of hot water is mixed with a liter of cold water until the mixture comes to a stable temperature. (a) Is energy lost in this process? (b) Is there a change in the entropy of this system? Explain your answers.

17. A refrigerator transfers heat from food at a lower temperature to the outside air, which is at a higher temperature. Does this contradict the second law of thermodynamics? Explain.

18. When a body loses heat, it suffers a loss in entropy. How can this be consistent with the

second law of thermodynamics, which states that the universe's entropy is increasing?

19. A regular window air conditioner was placed in the middle of a room instead of in the window. Could it cool the room in this position? Explain.
20. Distinguish between heat and internal energy.
21. What limits the ability of an engine to convert all the heat supplied to it into work?
22. (a) What is an irreversible process? (b) Give two examples of irreversible processes.
23. What is the relationship between entropy and disorder?

Group 2
24. An ice cube at −5°C is dropped into a swimming pool full of water at 0°C. What will happen to (a) the temperature of the water in the pool; (b) the mass of the ice cube?
25. The sun radiates heat to stars hotter than itself and absorbs radiant heat from stars colder than itself. Is this fact inconsistent with the statement that heat energy is transferred from bodies at higher temperatures to bodies at lower temperatures. Explain.
26. Air at normal atmospheric pressure is enclosed in a cylindrical tank equipped with a piston. A person pushes the piston until the air is compressed into one-half its former value. (a) What evidence is there that energy has been transferred to the confined air? (b) How can some of this energy be regained and made to do work?
27. Explain why a burn resulting from contact with steam at 100°C may be more damaging than one resulting from contact with the same mass of water at 100°C.
28. Only part of the heat put into an engine is converted into work. What happens to the rest of it?

29. All other forms of energy eventually are converted into heat, which travels to bodies at lower temperatures. In time, all the world's energy can be expected to be in the form of internal energy, all at one temperature. (a) Would it be possible to do work with this store of energy? (b) Would changes of weather occur in such a world? Explain.
30. The temperature at the surface of the ocean is several degrees warmer than the temperature of the water at a depth considerably below the surface. Engineers are planning to use this situation to do work that can be used to make electricity. How is this possible?
31. The insulated rigid box shown in Fig. 12-24 is separated into two sections by a diaphragm. There is a gas in section A and a vacuum in section B. When the diaphragm is removed, the gas expands and fills up section B as well as A. (a) Show, by applying the first law of thermodynamics, that the change in internal energy in this process is zero. (b) What does that tell you about the temperature of the gas after it expands?

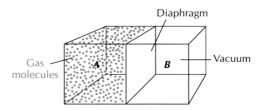

Fig. 12-24.

32. A moving car is brought to a stop by its brakes. (a) What happens to the kinetic energy of the car? (b) How did stopping the car affect the entropy of the universe? Explain.

Problems

Group 1
   See Table 12.1 for the values of specific heat called for in these problems.
1. How much heat does it take to raise the temperature of 2.00 kg of copper from 50°C to 70°C?

2. If it takes 512 J to increase the temperature of 2.00 kg of a substance 2.00 C°, what is the specific heat of this substance?
3. If 8960 J of heat is added to 2.00 kg of iron at 30.0°C, what is the new temperature of the iron?

4. How much heat does 8.00 kg of silver lose on cooling from 100°C to 75°C?

5. If 1.0 kg of water at 20°C is mixed with 2.0 kg of water at 50°C, what will be the final temperature of the mixture?

6. When 2.00 kg of water at 80.0°C is mixed with 4.00 kg of water at 40.0°C, the resulting temperature of the mixture is 53.3°C. Verify that heat was conserved by finding the heat lost by the hotter water and the heat gained by the colder water.

7. A 0.50 kg piece of aluminum at 200°C is dropped into 4.0 kg of water at 0.0°C. What will be the final temperature of the mixture?

8. How much heat must be added to 50.0 kg of ice at 0.0°C to change it to water at the same temperature?

9. A tray containing 0.250 kg of water at 20°C is put into a refrigerator. How much heat must the refrigerator remove from the water to turn it into ice at 0.0°C?

10. A 0.500-kg piece of ice at 0.0°C is dropped into a container of water in which it melts. If the final temperature of the water is 20°C, how much heat did the ice absorb from the water?

11. How much heat is needed to melt 1 kg of ice at 0°C and then raise the temperature of the resulting water to 100°C?

12. A quantity of water is boiling at normal atmospheric pressure in a pot over a gas burner. How much heat must the burner supply for each kilogram of water that leaves the pot as steam?

13. Steam from a tea kettle at 100°C enters a room and condenses. The resulting water then cools to 20°C (room temperature). How much heat does each kilogram of steam give up to the air in the room?

14. In Joule's apparatus, a weight of 840 N falls through a distance of 2.00 m while turning the paddles. (a) How much work is done by the weight in turning the paddles? (b) Into how much heat is this work converted as the paddles stir the water?

15. In Joule's apparatus, a weight of 10 000 N is allowed to fall a distance of 0.838 m. If 1 kg of water is being stirred in the container, how much does its temperature change?

16. An engine receives 5000 kJ of heat and does 2000 kJ of work. How much of the heat supplied to the engine is converted into internal energy?

17. A piece of ice in an ice-and-water mixture at 0°C receives 546 kJ of heat. What change takes place in the entropy of the ice?

18. The burning fuel inside an engine attains a temperature of 800 K. The temperature of the exhaust gases is 400 K. What is the maximum efficiency of this engine?

19. An isolated system consists of two large bodies, one at 100°C and the other at 50°C. If 2.0 J of heat passes from the first body to the second, what is the net change in the entropy of the system?

20. One kilogram of a gas in an insulated cylinder is compressed by a piston that exerts an average force of 4380 N acting over a distance of 0.10m. (a) How much does the internal energy of the gas change? (b) If the specific heat of the gas is 2190 J/kg · C°, what change takes place in the temperature of the gas?

Group 2

21. A block of aluminum weighing 490 N comes to rest after falling 10.0 m. (a) How much of the block's energy is converted to heat? (b) If all the heat is retained by the block, what is the change in the temperature of the block?

22. A 1.00-kg piece of ice at −10.0°C is dropped into a reservoir of water at 0.00°C. (a) How much heat does the ice absorb on reaching a temperature of 0.00°C? (b) How much of the water turns to new ice in this process?

23. The heat sink of an engine is at 290 K. What must be the temperature of the heat source to give the engine a maximum efficiency of 20.0%?

24. A 0.040-kg bullet traveling at 500 m/s becomes embedded in a tree. How much heat is generated from the kinetic energy of the bullet?

25. Suppose that, contrary to the second law of thermodynamics, a quantity of 2.00 J of heat passes spontaneously from a body at 200 K to a warmer body at 400 K. (a) Assuming that there is no change in the temperatures of the bodies during the transfer of this small quantity of heat, compute the change of entropy for this system. (b) How does the answer to (a) show that the original supposition is false?

26. One kilogram of steam at 373 K condenses into 1 kg of liquid water without change of temperature. What is the change in entropy in this process?

Applying Physics

Estimate the efficiency of an electric coffee maker in heating boiling water. Put 1 liter of cold water into the coffee maker and measure its temperature. Turn on the heater and record the temperature of the water at 2-minute intervals until the water reaches a temperature of 90°C. Knowing that 1 liter of water has a mass of 1 kg, find the number of joules absorbed by the water. Divide the number of joules by the total time of heating in seconds to obtain the rate of heating in watts. Compare this value with the wattage rating indicated on the coffee maker. This tells the rate at which electrical energy is furnished to the coffee maker. What percent of the energy put in as electricity is delivered as heat to the water?

Plot the temperature of the water against the elapsed time. Determine from your graph whether the heating takes place at a constant rate.

Global Warming

Human activity has had an influence on the environment for many centuries. However, the most significant alteration in the atmosphere has occurred in the last two centuries since the Industrial Revolution. The burning of fossil fuels such as coal, oil, and natural gas has released huge quantities of carbon dioxide into the earth's atmosphere. This carbon dioxide is believed to be the major cause of a suspected warming of the earth's surface and the lower layers of the atmosphere.

The possibility of such global warming is cause for alarm since a change of only a few Celsius degrees in the mean global temperature would have a major impact on the world's climate. At the peak of the last ice age 18,000 years ago, the mean global temperature was only five Celsius degrees lower than it is today. The expected long-term effects of global warming include an increase in cloud cover, the northward movement of precipitation belts, and an overall increase in global precipitation. The polar regions would experience greater temperature increases than the tropics. The resultant melting of the polar ice would raise

According to some estimates, deforestation may be responsible for as much as 20 percent of the annual global buildup of carbon dioxide. Trees consume carbon dioxide for photosynthesis and in inhabited areas provide cooling shade that cuts energy costs in hot weather.

sea levels, promoting coastal flooding and the mixing of salt water into fresh water sources. At the same time, the interior regions at middle latitudes would be subjected to severe drought.

If the global temperature is indeed increasing, it is probably due to an increase in what is called the "global greenhouse

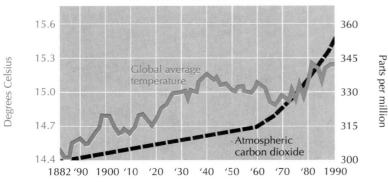

100 Years of Carbon Dioxide

Global average temperature

Atmospheric carbon dioxide

Global warming has accompanied the approximately 17 percent increase in atmospheric carbon dioxide over the last century.

effect.'' The earth's atmosphere readily transmits short-wavelength solar radiation to the earth's surface, which absorbs some of this sunlight and re-radiates it back into space as longer-wavelength infrared rays, or heat. Like the glass of a greenhouse, carbon dioxide and the other so-called ''greenhouse gases'' in the earth's atmosphere absorb some of this heat and re-radiate a portion back toward the earth's surface. Thus heat is trapped between the atmosphere and the earth's surface much like it is trapped inside a greenhouse.

Were it not for this natural global greenhouse effect, the average temperature of the earth would be only $-18°C$, a full $33C°$ cooler than the current $15°C$. However, by disturbing the natural balance through the generation of excess greenhouse gases, human activity may be raising the earth's temperature above its natural level. Evidence that global warming is taking place comes from temperature records compiled for the past 100 years; they show an approximately $0.5C°$ rise in that period. Scientists predict that the global mean temperature could rise another $3C°$ in the next 50 years unless the growth in carbon dioxide production is curbed.

Assessing the impact of human activity on the global climate is extremely difficult because of the number and complexity of the factors involved. Furthermore, the ability of the earth to respond to such stresses and restore equilibrium is not well understood. Uncertainty is introduced into any prediction by many factors, including natural climate variability, the cooling effect of volcanic ash, and changes in the strength of the sun's radiation. But the primary influences are cloud cover and the activity of the oceans.

Clouds cool the earth by reflecting up to 75 percent of incoming sunlight and warm it by trapping re-radiated infrared radiation. In a warmer atmosphere caused by a heightened greenhouse effect, water evaporation and cloud production will increase. Whether the net result would be greater cooling or greater warming is still the subject of scientific investigation.

The activity of the oceans also injects a high level of uncertainty into global warming predictions. The capacity of the oceans to store heat is 1000 times greater than that of the atmosphere. The mixing of the absorbed heat throughout the depths of the world's oceans could delay the warming of the earth for a decade or up to a century.

Any attempt to limit global warming must be undertaken at the international level. Measures that could be beneficial include energy conservation in the areas of heating, air-conditioning, and transportation. Also helpful would be the more efficient use of fossil fuels or the use of alternative energy sources such as solar energy, geothermal energy, wind, or nuclear power.

During the unusually hot summer of 1988, a record drought left barges stranded on the Mississippi River. Such newsworthy events raised public awareness to the threat of global warming.

13 Heat and Molecular Motion

Aims

1. To examine the evidence that suggests the molecular structure of matter and the existence of attractive and repulsive forces among molecules.
2. To study the nature of the forces and molecular motions in solids, liquids, and gases.
3. To understand that heating a body of matter to a higher temperature increases the motion of its molecules, while cooling the body slows down its molecular motion.
4. To examine the evidence that led to the identification of the particles of which matter is composed as molecules and atoms.

Structure of Matter

13-1 Heat and the Structure of Matter

Consider the experience of hammering a nail into a piece of hardwood. The nail and the hammer become quite hot. After the hammering has stopped, the kinetic energy possessed by the moving hammer is gone. In its place, heat has been produced and has been distributed through the hammer and the nail. Just what is heat? How is it associated with the matter in which it is now contained? To answer these questions, we must consider the nature of matter itself, a study that leads to our present idea or theory of the internal structure of matter.

13-2 Theory of Matter

To formulate a theory of the structure of matter, we begin with the facts obtained from our observations and measurements of matter. These facts provide clues that suggest a picture or model of the way in which matter is put together. The model and its properties constitute our theory of matter.

Matter is composed of particles separated from each other and in constant motion.

In this chapter, we shall see that common facts about matter suggest the following model of its structure. Matter appears to be composed of tiny individual particles that are separated from each other and are always in motion. At close distances, these particles seem to be held together by attractive forces while, at still closer distances, they seem to repel each other. The addition of heat to matter generally has the effect of causing its individual particles to move move vigorously. Cooling matter has the effect of making its particles move more slowly.

On the basis of this simple model, we are able to understand many of the common properties of matter and many of the effects of heat upon it. The evidence suggesting this model will now be examined.

13-3 Internal Motion of Matter

Many common experiences suggest that every body of matter consists of tiny particles, each of which is in constant motion. The nature of these particles will be explored later in this chapter. Here, it is sufficient to note that very often they occur in units called molecules. *A molecule is defined as the smallest part of a substance that has all the properties of that substance.*

When a small amount of gas having a distinctive odor is released in one corner of a room, it is gradually detected all over the room. This shows that the molecules of the gas have moved of their own accord to every part of the room. *The self-spreading of the gas is known as diffusion.* A similar effect is noted in a liquid when a substance is allowed to dissolve in it. If a spoonful of sugar is carefully put into a cup of water without stirring and left for a few days, it will be found that every part of the water has been sweetened. This indicates that sugar molecules have moved to every part of the water.

A little water spilled on the table soon disappears into the air. Here, again, is evidence of molecular motion. The movement of the water molecules has enabled them to evaporate and diffuse into the air.

That even the molecules of solids are also in constant motion can be shown by pressing a plate of gold and a plate of lead into close contact. After a long period of time, particles of gold will be found to have entered the lead plate and vice versa.

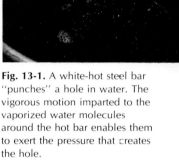

Fig. 13-1. A white-hot steel bar "punches" a hole in water. The vigorous motion imparted to the vaporized water molecules around the hot bar enables them to exert the pressure that creates the hole.

13-4 Space Between Molecules

The last experiment indicates that even in solids such as lead there are empty spaces between molecules. It is through these spaces that a molecule of gold can make its way into the lead. Another clue suggesting the existence of spaces between the molecules of all substances is the fact that substances generally contract into smaller volumes when they are cooled. Such contraction is readily explained by assuming that, during cooling, the molecules of a substance are pulled closer together, thus reducing the empty spaces between them.

An interesting demonstration suggesting the existence of empty spaces among the molecules of a liquid consists of mixing equal volumes of alcohol and water. It will be found that the mixture occupies less space than the sum of the volumes of the two liquids. The molecules of one liquid appear to be filling in spaces between the molecules of the other. See Fig. 13-2.

The spaces among molecules are greatest in gases, less in liquids, and least in solids.

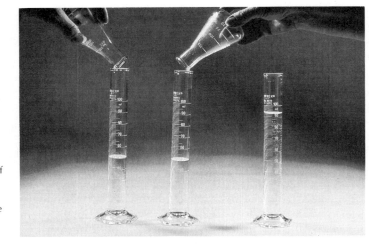

Fig. 13-2. When equal volumes of alcohol and water are mixed, the volume of the mixture (at right) is less than the sum of their separate volumes. Molecules of one liquid fill in spaces between molecules of the other.

In gases, under ordinary conditions, the molecules appear to be very much further apart than those in liquids and solids. This is suggested by the fact that gases can readily be compressed into much smaller volumes while both solids and liquids are almost incompressible even when very large forces are applied to them.

13-5 Attraction Between Molecules

How do molecules band together to form the ordinary objects of everyday experience? Although molecules are constantly moving, it is obvious that they must cling together to form solid objects like iron nails or masses of liquid like drops of water. It seems reasonable to infer from this that, at close distances, molecules attract each other. We can distinguish two kinds of attraction, *adhesion* and *cohesion*.

Adhesion. The force with which unlike molecules attract each other is called adhesion. Adhesion explains why paste sticks to paper, why glue sticks to wood, why water wets a drinking glass, and why chalk adheres to the chalkboard. In each case, molecules of different kinds are attracting each other.

Cohesion. The force with which like molecules attract each other is called cohesion. An iron nail remains in one piece because the cohesive forces among its molecules hold them firmly together. Cohesion is greatest in solid substances, where the attractive forces among the molecules not only hold the solid object together but also keep it in a definite shape.

The cohesive forces in liquids are weaker than those in solids. They are great enough to keep the liquid in one mass but not great enough to give the liquid a shape of its own. The liquid therefore assumes the shape of its container.

In a gas, the molecules are so far apart that the cohesive forces

Fig. 13-3. The smooth flat surfaces of these Johansson measuring blocks used by machinists enable their molecules to come very close together when the blocks are put into contact. The cohesion of the molecules holds the blocks firmly together.

among them are negligible. As a result, a gas literally flies apart unless it is kept together in a tightly closed container. When a gas is unconfined, its molecules move freely in all directions. This lack of cohesion among gas molecules explains why the odor of gas escaping from the kitchen range is soon noted in all parts of the home.

13-6 Repulsion Between Molecules

A large quantity of gas, such as air, is easily compressed into a smaller volume, as is done when you pump air into an automobile tire. However, when an attempt is made to compress a solid such as a piece of brass, or a liquid such as a quantity of oil or water, it is found that this is not possible with forces of ordinary size. Even when very large forces are applied to a solid or a liquid, they produce relatively little compression.

These facts suggest that the molecules of solids and liquids resist being pressed closer together, or that they repel each other. The repulsive force comes into play only when the molecules are very close together, as they are in solids and liquids. In gases, where the molecules are ordinarily far apart compared to their sizes, the repulsive force disappears so that it is easy, under ordinary conditions, to compress a gas into a smaller space.

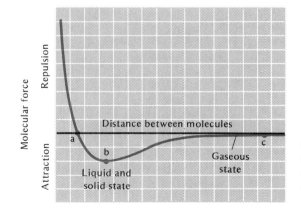

Fig. 13-4. Graph showing how the force between two molecules changes from repulsion to attraction with the distance between them.

13-7 Force and Distance Between Molecules

From the above observations, it appears that the force exerted by one molecule upon another depends upon the distance between them. The relationship is shown graphically in Fig. 13-4. Here the force between two molecules is plotted on the vertical axis and the distance between them on the horizontal axis. Repulsive force is shown above the horizontal axis and attractive force below the horizontal axis.

Going from the extreme right of the graph to the left, notice the following changes in molecular force. When the molecules are far apart, as they are in gases, the force between them is practically zero. This is shown by the typical point *c*. As the molecules come closer together, the attractive force between them increases steadily until, when the distance between them is *b*, the attraction is at a maximum.

As the molecules move still closer together, a second force comes into play, which causes the molecules to repel each other. This repulsive force increases rapidly as the molecules continue to come closer. At first it steadily reduces the attractive force until, when the molecules are the distance *a* apart, the resultant of the attractive and repulsive forces between them is zero. As the molecules come still closer, the resultant force between them becomes one of repulsion and rises sharply.

13-8 Heat and Molecular Motion

As water in a teakettle is continually heated, it begins to change to steam that issues from the spout. The heat seems to have given the individual molecules of water leaving the spout as steam enough additional speed to escape from the main body of the water and spread through the air. A similar effect is noticed when any other liquid is vaporized by continued heating.

These observations suggest that one of the effects of heat on the molecules of a substance is to make them move faster. This conclusion is also suggested by the fact that all bodies of matter, whether solid, liquid, or gas, expand when they are heated. Here, adding heat makes the molecules of the body spread further apart, thus causing the body itself to increase in volume.

If it is assumed that heating causes the molecules to move more vigorously, the reason for their spreading apart becomes clear. It is simply that the cohesive forces pulling the molecules together are not able to hold rapidly moving molecules as close together as more slowly moving molecules. Hence, as additional heat makes the molecules move faster, the distances between the molecules increase and the body which they form continues to increase in volume.

If the application of heat causes the molecules of a body to move faster, it may be expected that the reverse process is also true. That is, increasing the motion of the molecules of a body is equivalent to applying heat to it. One evidence supporting this view is the heating up of an object such as a nail when it is placed on an anvil and struck repeatedly by a hammer. The hammer blows apparently make the molecules of the nail move faster and this increased molecular motion is noticed in the heating of the nail.

13-9 Brownian Movement

Almost direct visual evidence of the motion of molecules and the effect of heat upon them is provided by an observation first reported by the biologist Robert Brown in 1827. It is called *Brownian*

The destructive force of an explosion is attributable to the violent molecular motion induced by the heat released in the explosion.

movement in his honor. Brown noticed that when tiny particles, such as those within pollen grains of plants, are suspended in water, they undergo an unceasing, zigzag motion that is readily observed under the microscope. Furthermore, when the water is heated, the motion of these particles becomes more rapid and more vigorous. Similar motions and effects are noticed in observing smoke or dust particles suspended in air, and other tiny particles suspended in liquids or gases.

You can understand Brownian movement if you consider what happens to a suspended pollen particle when it is hit by the rapidly moving water molecules all around it. The blow of a single water molecule is too small to have a noticeable effect on the particle since the mass of the pollen particle is very much larger than that of the water molecule. However, at each moment, very large numbers of molecules are striking each pollen particle from all directions. From time to time, many more molecules strike the particle on one side than on the other. The resultant of all the molecular blows causes the particle to move a short distance in the direction of this resultant.

As the water molecules continue to bombard the suspended pollen particle, their combined blows produce a resultant that varies in both direction and magnitude. This varying resultant gives the pollen particle the characteristic zigzag motion.

When the water is heated, the increased motion of the particles indicates that the heat has increased the motion of the water molecules. The bombardment by the faster molecules accounts for the move vigorous Brownian movement of the pollen particles.

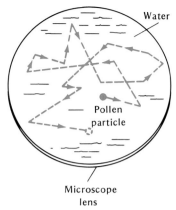

Fig. 13-5. Zigzag Brownian motion of a pollen particle in water.

13-10 States of Matter and Molecular Motion

A substance can generally exist as either a solid, a liquid, or a gas. Water, for example, can be in the form of ice, water, or steam. From the molecular point of view, the state that a substance assumes depends upon two opposing effects. One of these is the never-ending motion of the molecules that tends to make them

Fig. 13-6. Cooling may sharply change the properties of a solid substance. The upper white arcs show the path of a bouncing rubber ball as it moves to the right. The lower white streak is the path of the same ball after being cooled in liquid nitrogen. It has lost its bounce.

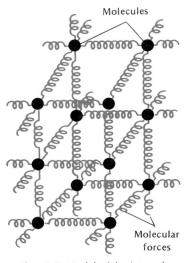

Fig. 13-7. Model of the internal structure of a solid.

separate from each other and move off in all directions. The other is the cohesive or attractive force between molecules that tends to pull them closer together.

The greater the motion of the molecules inside a body, the harder it is for the cohesive forces to hold them together, and the greater the separation between molecules becomes. When the molecules are far enough apart, the cohesive forces between them become zero, as shown in Fig. 13-4. The molecules are then free to move in any direction and the body is then in the gaseous state.

On the other hand, the slower the motion of the molecules of a body, the more tightly the cohesive forces are able to hold them together and the greater is the tendency for the body to assume the compact, solid state. According to theory, heating usually causes molecules to move faster while cooling slows them down. So we should expect bodies heated to high temperatures to be in gaseous form. Bodies cooled to low temperatures should be in solid form. At intermediate temperatures, bodies should be in liquid form. Thus, the theory predicts that the state of a given body of matter depends upon its temperature. This agrees with the facts.

A picture of the internal structure of solids, liquids, and gases can now be imagined in terms of the molecular theory of matter.

13-11 Structure of Solids

In a solid, the attractive forces of the molecules are not only strong enough to hold the body together but are also able to keep the body in a definite and unchanging shape. This means that although each individual molecule moves continually, it is compelled by its neighbors to remain in the neighborhood of a fixed equilibrium position in the solid. The forces acting between neighboring molecules may be pictured by imagining that the molecules are connected by tiny springs, as shown in Fig. 13-7.

As a molecule moves in any given direction, it will stretch or compress the imaginary springs attached to it. The steadily increasing molecular forces represented by these imaginary springs then pull the molecule back toward its equilibrium position. As a result, the motion of a molecule in a solid is generally limited to vibrating ceaselessly about its equilibrium position.

The vibratory motion of molecules can be detected photographically, as shown in Fig. 13-8. Here the vibratory motion of the molecule causes its pictures to appear blurred rather than sharp.

The small distances of the world of molecules are measured in decimal parts of the nanometer (nm), which is one billionth (10^{-9}) of a meter. In a typical solid, the distance between molecules is about 0.35 nanometer. The molecules vibrate about their equilibrium positions with an amplitude of about 0.01 nanometer and a frequency of about 10^{13} hertz.

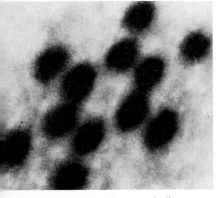

Fig. 13.8. A hexamethylbenzene molecule (magnified about 100 000 000 times its actual size).

In very many solids, the attractive forces not only hold the molecules tightly together but also force them to assume an orderly pattern that repeats itself throughout the solid. This is noticed from the fact that many solids have a regular crystalline structure. In some cases, the crystals are so large that they can be seen with the naked eye or with a simple microscope. In many more cases, the crystalline structure can be noticed only by examining the substance with X rays.

Fig. 13-9. A large crystal of quartz grown artificially.

13-12 Structure of Liquids and Gases

When a substance is in liquid form, its molecules vibrate more vigorously than they do when in the solid form. While the attractive forces are still strong enough to keep the molecules together and at about the same average distance apart, they are not strong enough to keep each molecule in a specific location. Individual molecules can move through the mass of the liquid but cannot readily escape from it. This molecular behavior enables a liquid to flow and to adjust its shape to the shape of its container.

In a gas, the molecules move much faster than they do in the solid or liquid form. Because the molecules in a gas are at least ten times as far apart as they are in the solid or liquid state, the attractive force between them is practically zero. They therefore move freely in straight lines until collisions with other molecules change their direction of motion. Under ordinary room conditions, a molecule of nitrogen in the air has an average speed of about 500 meters per second. Such rapid motion of the molecules, coupled with the lack of attractive force between them, explains why gases maintain neither a definite volume nor a definite shape, and why a gas spreads to all parts of any container into which it is put.

13-13 Gas Pressure and Molecular Motion

The relatively free and rapid motion of the molecules of a gas explains the fact that gases confined in a container exert pressures. When air is pumped into a toy balloon, you can see that the air blows up the balloon. The air is exerting a pressure in all directions and is pushing out the sides of the balloon. Any gas enclosed in a container exerts a similar pressure in all directions.

The fact that only gases exert this particular kind of pressure is explained by the freedom of motion of their molecules. In due time, each of the freely moving molecules of a gas will collide with a wall of the container and give a tiny blow to that wall. Because of the large number of molecules in even a small sample of a gas, billions and billions of molecules bombard each wall of the container at every instant. As a result, the net effect of the molecular blows on the walls of the container is observed as a steady force distributed uniformly over equal areas of the walls of the container rather than as a series of separate blows.

The ceaseless bombardment by fast moving gas molecules on the walls of their container results in a pressure on the walls.

When a gas in a closed container is heated, its pressure increases. The molecular theory explains this by stating that heating increases the speed of the gas molecules. They therefore hit the walls of the container harder and more often than before, thus increasing the pressure on them.

13-14 Heat and Internal Energy

So far we have limited our discussion to the kinetic energy of molecules associated with their ceaseless motion. We shall see in Chapter 14 that the increase in kinetic energy of the molecules of a body brings about a rise in its temperature. However, molecules also have potential energy. We have seen that the gravitational attraction between any two masses results in the storage of gravitational potential energy between them. Similarly, the attraction between two molecules results in the storage of molecular potential energy between them. Like gravitational potential energy, molecular potential energy increases as the separation between the molecules increases beyond the distance at which the attractive force begins.

Since heating a body generally causes its molecules to spread further apart, the addition of heat results in a corresponding increase in its molecular potential energy. Thus, heating a body increases not only the kinetic but also the potential energy of its molecules. *The sum total of the kinetic and potential energies of all the molecules of a body is known as its internal energy, or its thermal energy.* When a body gains heat from other bodies, its internal energy increases, while when a body gives heat to other bodies, it loses some of its internal energy.

Thermal energy may be in potential and kinetic forms other than those we have discussed. For example, molecules may have internal potential energy resulting from the redistribution of the particles of which they are composed. They may also have "internal" kinetic energy associated with motions in which they spin like tops or in which their parts vibrate with respect to each other.

Heat absorbed by a body is stored in it as internal energy in the form of the kinetic and potential energy of its molecules.

13-15 Molecular Explanation of Changes of State

We have seen that when ice and other crystalline solids are heated, their temperatures rise until the melting point is reached. At that point, in spite of the addition of more heat, the temperature stays the same while the solid melts. (See Sec. 12-15.)

This can be explained from the molecular point of view. The heat applied to the solid first increases both kinetic and potential energies of its molecules. The rise in kinetic energy is noticed as a rise in temperature. The increased potential energy results from the separation of the molecules as their motion increases. When the melting point of the solid is reached, all the additional heat goes to separate the molecules from each other sufficiently to let them move freely through the liquid that is forming. None of it goes to increase the kinetic energy of the molecules. Thus, the temperature remains the same until all the solid has turned to liquid.

As the liquid is now heated, the process is repeated. Again, the heat first increases the kinetic energy of the molecules and raises the temperature of the liquid until the boiling point is reached. At that point, all the additional heat goes to free the molecules from each other's attractions completely and change the liquid to gas. Since none of the heat is used to increase the kinetic energy of the molecules, the temperature remains the same until all the liquid has turned to gas. After that, further heating increases the kinetic energy of the molecules of gas and thus further raises the temperature.

Granules of Matter

13-16 Nature of Particles of Matter

The evidence presented thus far has suggested that matter is granular in structure and consists of separate particles, many of which are called molecules. It has not told us very much about the specific nature of the individual molecules and other particles of matter. How large are they? Of what are they made? How great is their mass? How many of them are contained in a given body of matter? How do they combine to form the many different substances of nature?

It is time now to examine these important questions. Many of the answers come from the chemists who, in studying how the different kinds of matter combine with each other, laid the basis for the development of two major discoveries, that of chemical elements and that of atoms.

13-17 Chemical Elements

As early as 1661, the English scientist Robert Boyle suggested that there were in nature only a limited number of different simple substances called *elements*. The major property of an element is that *it cannot be subdivided by chemical means into parts or substances different from itself*. Boyle asserted that all other substances are simply combinations of two or more elements.

Boyle proposed the idea that all matter consists of combinations of a limited number of elements.

As the science of chemistry developed, Boyle's assertion was tested by Lavoisier and many other chemists and was found to be in agreement with the facts. The growing evidence in favor of the existence of elementary chemical substances led to a methodical search for them which resulted in the identification of the more than one hundred elements known today and listed in the Appendix. As Boyle had guessed three centuries ago, all substances are either one of these elements or combinations of two or more of them.

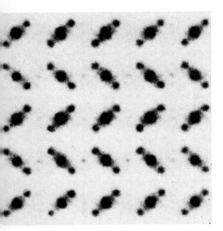

Fig. 13-10. X-ray photograph of iron sulfide molecules in a crystal of pyrite, magnified 2.2 million diameters, illustrates the law of definite proportions. It shows that each molecule consists of a central atom of iron flanked by two sulfur atoms.

13-18 Law of Definite Proportions

As chemists studied how elements combine to form a new substance or compound, they discovered the *law of definite proportions*. This law states that *whenever two or more elements form a particular compound, the elements combine in definite proportions by weight and these proportions are always the same for any sample of that compound.* Water, for example, is a compound of the two elements hydrogen and oxygen combined in the ratio of eight parts of oxygen by weight to one part of hydrogen. This 8 to 1 oxygen-hydrogen ratio is found to be true for all samples of ordinary water regardless of the size of the sample.

Similarly, for carbon dioxide, a common compound of carbon and oxygen, the ratio by weight of carbon to oxygen is 3 to 8.

Why should the elements combine in definite proportions? The attempt to answer this question was one of the factors that led the Englishman John Dalton (1766–1844) to suspect the existence of basic particles of matter called *atoms* and to propose a theory describing their behavior.

13-19 Theory of Atoms

Dalton suggested that each of the chemical elements is composed of tiny indestructible particles or atoms, which, for any given element, are exactly alike. Thus, there are as many different kinds of atoms as there are different kinds of elements. Compounds are made when the atoms of two or more elements combine. In such combinations, there is no loss of mass; that is, the mass of the compound formed is equal to the sum of the masses of the atoms that compose it.

The smallest unit of a compound is a molecule. When different atoms unite to form any given molecule, the atoms always combine in the same definite proportions. Thus, every molecule of a compound consists of exactly the same combination of atoms as every other molecule of that compound. The ratio by weight of any two different atoms contained in a single molecule of a compound is therefore the same constant value for all its molecules. It follows that the ratio by weight of the two elements corresponding to these atoms in any sample of that compound will also be that same constant value. Thus, Dalton's idea of atoms offers a plausible explanation for the law of definite proportions.

Applying the Dalton theory to the case of water, it is noted that each water molecule consists of two atoms of hydrogen and one of oxygen. Since the usual oxygen atom weighs sixteen times as much as the usual hydrogen atom, the ratio by weight of the one oxygen atom to the two hydrogen atoms in each water molecule is 16 to 2, or 8 to 1. It follows that no matter how many or how few molecules there are in a sample of water, there will always be eight times as much oxygen by weight in the sample as there is hydrogen.

Fig. 13-11. John Dalton, English schoolmaster, proposed his atomic theory about 1800.

Dalton's explanation proved substantially correct. It required only minor modification when it was later discovered that the atoms of each element have different forms, called *isotopes*.

13-20 Atomic Mass Units

Once the idea of atoms was established it became possible, by studying how elements combine in chemical reactions, to determine the relative mass of any atom with respect to other atoms. It was found that if the oxygen atom was taken as a standard and assigned a mass of sixteen units, the relative masses of many of the other atoms came out to be nearly whole numbers. These units are called *atomic mass units,* symbol, u.

In atomic mass units, hydrogen atoms have a mass of about 1, helium atoms have a mass of about 4, carbon atoms have a mass of about 12, and so on. The fact that the atomic mass of hydrogen is about one and that the atoms of many of the other elements are approximately whole numbers suggests that all atoms may be made of combinations of two or more hydrogen-like units. This idea will be explored further in Chapter 21.

Later research showed that ordinary oxygen is a mixture of several kinds of oxygen atoms having the same chemical properties but different masses. These are called *isotopes* of oxygen. It was also found that almost *all other elements are also mixtures of isotopes of atoms having the same chemical properties but different masses.*

To set up a more precise way of measuring atomic masses, it was decided to select one of the isotopes of another element as the standard of atomic mass. The standard selected was the most abundant of the isotopes of carbon, which, by agreement, is assigned a mass of exactly 12 atomic mass units. The masses of the isotopes of all other atoms are measured against this standard. On this basis, the mass of the most abundant of the oxygen isotopes comes out to be very close to 16 atomic mass units. In addition, there are two much rarer oxygen isotopes of masses equal to about 17 and 18 atomic mass units.

The standard for measuring atomic masses is the most abundant isotope of carbon, which is assigned a mass of exactly 12 atomic mass units.

13-21 Masses of Molecules

Continuing this study of matter, chemists found that for every element or compound, there is a smallest unit that has all the properties of that element or compound. As we have seen, this unit is called a molecule. Usually, a molecule consists of two or more atoms. There are certain elements, however, such as helium and neon, whose molecules consist of only one atom. For these elements, the molecule and the atom are identical.

The mass of a molecule is simply the sum of the masses of the atoms it contains. An oxygen molecule, for example, consists of two oxygen atoms. Assuming that both of these atoms are the isotope whose mass is 16 atomic mass units, it follows that the mass of this oxygen molecule is 16 + 16, or 32 atomic mass units. The most common type of hydrogen molecule consists of two hydrogen atoms. Since each of these has a mass of 1 atomic mass unit, the mass of this hydrogen molecule is 1 + 1 or 2 atomic mass units. In the case of the helium molecule which consists of only one helium atom, the mass of the molecule is the same as the mass of the helium atom, or 4 atomic mass units.

13-22 The Mole and Avogadro's Number

A *mole* (symbol, mol) is that quantity of a substance whose mass in grams is numerically equal to the mass of one of its molecules in atomic mass units. Since the molecule of oxygen considered in the above paragraph has a mass of 32 atomic mass units, a mole of oxygen is 32 grams of oxygen. Similarly, a mole of the most common form of hydrogen is 2 grams of hydrogen, while a mole of helium is 4 grams of helium.

As a result of investigations begun by the Italian Amedeo Avogadro (1776–1856), it was discovered that a mole of any substance has the same number of molecules as a mole of any other substance. *The number of molecules in a mole is called Avogadro's number and is 6.02×10^{23} molecules.* It follows that there are 6.02×10^{23} oxygen molecules in one mole (or 32 grams) of oxygen, the same number of helium molecules in one mole (or 4 grams) of helium, and the same number of hydrogen molecules in one mole (or 2 grams) of hydrogen.

A principle discovered by Avogadro gives additional information concerning the molecular make-up of a gas. *Avogadro's principle states that equal volumes of different gases at the same temperature and pressure have equal numbers of molecules.* This means that under the same pressure and temperature conditions, there are the same number of molecules in a liter of nitrogen, a liter of oxygen,

> The number of molecules in a mole of a substance is the same for all substances. It is Avogadro's number, 6.02×10^{23} molecules.

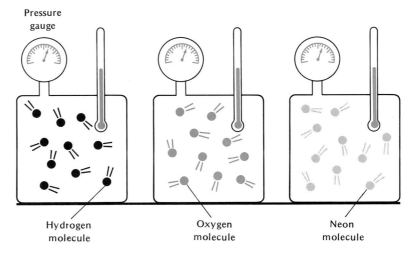

Pressure gauge

Fig. 13-12. Avogadro's principle: equal volumes of all gases at the same temperature and pressure have the same number of molecules.

Hydrogen molecule   Oxygen molecule   Neon molecule

a liter of hydrogen, or a liter of any other gas. It is found experimentally that *one mole of any gas at normal atmospheric pressure and at 0°C has a volume of about 22.4 liters,* or 22.4×10^3 cubic centimeters.

Sample Problems

1. A mole of nitrogen has a mass of 28 g. How many molecules are there in 7.0 g of nitrogen?

   Solution:
   One mole or 28 g of nitrogen contains 6.0×10^{23} molecules. 7.0 g is 7.0/28 or ¼ of a mole of nitrogen. It therefore contains $\frac{1}{4}(6.0 \times 10^{23})$ molecules or 1.5×10^{23} molecules.

2. Equal volumes of oxygen and nitrogen at the same temperature and pressure have masses of 8.0 g and 7.0 g respectively. Given that the mass of an oxygen molecule is 32 u, find the mass of a nitrogen molecule.

Solution:
Since the nitrogen and oxygen have the same volume, they have the same number of molecules. The ratio of the mass of 7.0 g of nitrogen to the mass of 8.0 g of oxygen is therefore equal to the ratio of each nitrogen molecule (m_N) to the mass of each oxygen molecule (32 u).

$$\frac{7.0 \text{ g}}{8.0 \text{ g}} = \frac{m_N}{32 \text{ u}}$$

$$m_N = 28 \text{ u}$$

Test Yourself Problems

1. A mole of carbon dioxide gas has a mass of 44 g. How many molecules are there in 33 g of carbon dioxide?

2. Equal volumes of oxygen and argon have masses of 48 g and 60 g respectively. If the mass of an oxygen molecule is 32 u, what is the mass of an argon molecule?

13-23 Determining Avogadro's Number

There are many ways of determining Avogadro's number. All of them give essentially the same answer. The most direct way is to make an actual count of the number of molecules in 1 mole of substance. Such a count became possible with the discovery of radioactivity at the close of the nineteenth century.

Radioactivity gets its name from the fact that the atoms of certain elements, such as uranium, polonium, and radium, naturally emit rays. Polonium, for example, emits rays known as alpha particles that turn out to be the cores or nuclei of helium atoms. Shortly after they are emitted, alpha particles pick up electrons from their surroundings and become complete atoms of helium gas. The alpha particles emitted by a piece of polonium can be detected and counted by letting them fall upon a screen covered with luminous paint like that on a television-tube screen. On striking this screen, each alpha particle produces a tiny flash of light called a scintillation.

To determine Avogadro's number, a piece of polonium is put into an evacuated glass tube and the number of alpha particles it emits are counted. The helium gas formed by these particles is collected and its mass is determined in grams. The number of particles is now divided by this mass to give the number of helium

Avogadro's number can be obtained by actually counting the molecules in a measured volume of helium.

atoms contained in 1 gram of helium gas. Since a mole of helium has a mass of 4 grams, this number is multiplied by 4 to obtain the number of atoms in a mole of helium. This is Avogadro's number because each atom is also a molecule of helium.

13-24 Actual Masses of Atoms and Molecules

Once Avogadro's number is known, it is a simple matter to obtain the mass of a single molecule of a substance by simply dividing the mass of 1 mole of the substance by Avogadro's number. Thus, 1 mole of oxygen has a mass of 32.0 grams and contains 6.02×10^{23} molecules of oxygen. Hence, one oxygen molecule has a mass of 32.0 grams $\div 6.02 \times 10^{23}$ molecules or 5.32×10^{-23} gram. Since each oxygen molecule consists of 2 oxygen atoms, the mass of an oxygen atom is ½ of 5.32×10^{-23} gram or 2.66×10^{-23} gram.

In the same way, we can show that the common isotope of hydrogen, the lightest of the elements, has a molecule having a mass of 3.34×10^{-24} gram and an atom having a mass of 1.67×10^{-24} gram. At the other end of the scale, the heaviest atoms known have masses about 250 times that of hydrogen.

Today, by means of the instrument called the *mass spectrograph,* it is possible to measure the masses of atoms directly. The values obtained by this direct way agree well with those obtained by the use of Avogadro's number.

13-25 Dimensions of Molecules and Atoms

Until recent times, it was not possible to "see" molecules or atoms. Today, by means of X rays, the electron microscope, and the field ion microscope, it is possible to make intelligible "pictures" of molecules and atoms, to note their arrangement in substances, and to measure their dimensions directly. See Figs. 13-8, 13-10, and 13-13. Various different investigations show that the linear dimensions of different atoms average about 0.2 nanometer. The dimensions of molecules depend on the number of atoms they contain and vary from a few to several hundred nanometers.

Fig. 13-13. The locations of individual atoms of iridium as photographed by the field ion microscope. Magnification is about one million diameters.

CHAPTER REVIEW

Summary

To understand what happens when bodies of matter absorb or emit heat energy, we must study the internal structure of matter. There is convincing evidence that all matter consists of tiny individual particles called **molecules** that are separated from each other and are in ceaseless motion. Molecules attract each other at close distances and band together to form the solid and liquid substances of everyday experience. The force of attraction between like molecules is called **cohesion;** that between unlike molecules is called **adhesion.**

The attractive forces among molecules decrease rapidly as the distance between them increases. In gases, the molecules are usually so far apart that their attractive forces on each other are negligible. For this reason, gases do not have any specific volume. Their molecules spread out or **diffuse** freely to all parts of any container in which the gas is enclosed.

As molecules are pushed closer and closer together, the force of attraction between them gradually turns into a force of repulsion. This explains why solids and liquids are practically incompressible under ordinary conditions.

Heating a substance to higher temperatures increases the motion of its molecules. At high temperatures, the fast-moving molecules of a substance become completely separated from each other and the substance is in a gaseous state. At low temperatures, the molecules of a substance have much less motion and are held together firmly by the attractive forces between them so that the substance is usually in solid form. At intermediate temperatures, a substance is generally in liquid form.

The increased molecular motion caused by heating also explains thermal expansion of substances as well as the increased pressure that results when a confined gas is heated.

The chemist's study of matter shows that all matter is made of combinations of basic substances called **elements.** An element is a substance that cannot be broken down by chemical means into other substances. Each element is composed of tiny particles called **atoms.** All other substances are made by combining the atoms of two or more elements. The smallest particle of a substance is called a **molecule.**

In measuring the masses of atoms, the mass of the most abundant **isotope** of carbon is taken as a standard and is said to have a mass of exactly 12 **atomic mass units (u).** The masses of all other atoms are measured using this standard. A **mole** of a substance is that quantity whose mass in grams is numerically equal to the mass of one molecule of that substance in atomic mass units. A mole of any substance contains 6.02×10^{23} **molecules.** This number is known as **Avogadro's number.**

Avogadro's principle states that *equal volumes of gases at the same temperature and pressure have equal numbers of molecules.* A mole of any gas at normal atmospheric pressure and at 0°C occupies a fixed volume, which is about 22.4 liters, or 22.4×10^3 cubic centimeters.

Questions

Group 1

1. Describe one behavior of solids, one of liquids, and one of gases suggesting that the molecules are in ceaseless motion.

2. Describe one evidence that there is empty space between the molecules of (a) a solid; (b) a liquid; (c) a gas.

3. Explain each of the following in terms of the attractive forces between molecules. (a) Ink adheres to paper. (b) It is much harder to break a steel rod in two than it is to break a glass rod of the same dimensions. (c) The odor of perfume spilled in one corner of a room is gradually noticed in all parts of the room.

4. (a) What evidence is there that molecules repel each other when they are brought close enough together? (b) Why is this repulsive force not usually noticed among the molecules of a gas?

5. How does Brownian motion provide evidence that (a) molecules are in constant motion; (b) the

addition of heat to a substance increases the motion of its molecules? (c) What other evidence suggests that one of the effects of heating a substance is to increase the motion of its molecules?

6. In terms of molecular theory, (a) explain how the continuous addition of heat to a mass of ice will eventually turn it into water and then turn the water into steam; (b) compare the freedom of movement of the molecules in ice, water, and steam.

7. How does the molecular theory explain (a) the pressure exerted by gases on the walls of their containers; (b) the expansion of solids and liquids when heated?

8. (a) From the chemical point of view, what is the major property of an element? (b) What are the smallest parts of elements? (c) How are compounds made?

9. (a) State the law of definite proportions. (b) How does the concept of atoms explain this law?

10. (a) What is an isotope of an element? (b) What isotope is used today as the standard of atomic mass? (c) On the basis of this standard, what is the mass of the most abundant isotope of oxygen?

Group 2

11. Equal volumes of gas A and B are collected at normal atmospheric pressure and at 0°C. The mass of gas A is 4 times as large as that of gas B. Compare (a) the number of molecules in gas A with the number in gas B; (b) the mass of each molecule of gas A with that of each molecule of gas B.

12. (a) Explain in terms of molecular motion why a liquid evaporates more rapidly when it is heated. (b) What effect do the forces among molecules have on the tendency of the liquid to evaporate?

13. Explain why the pressure of a gas increases when the gas is squeezed into a smaller volume, even though the gas is kept at the same temperature.

14. Describe a procedure for determining the mass of one molecule of any gas from Avogadro's principle and from the fact that one mole of oxygen has a mass of 32 g.

15. Using a mass spectrograph, it is possible to measure the actual mass of a single nitrogen molecule. The mass of a mole of nitrogen can be determined from its chemical behavior. How can this information then be used to determine Avogadro's number?

Problems

Group 1

Note: In these problems use 6.0×10^{23} as the value of Avogadro's number, and 32 as the mass of an oxygen molecule in atomic mass units.

1. The isotopes of hydrogen have masses of 1, 2, and 3 u respectively. A hydrogen molecule can be made of any pair of hydrogen atoms. Make a list of the different combinations that will make a hydrogen molecule and determine the mass in u of each.

2. How many molecules are there in 96 g of oxygen?

3. A molecule of the most abundant form of hydrogen has a mass of 2 u. How many molecules are there in 10 g of this form of hydrogen gas?

4. A carbon dioxide molecule is made of 1 atom of carbon of mass 12 u and 2 atoms of oxygen each of mass 16 u. (a) What is the mass of 1 mole of carbon dioxide? (b) How many molecules are there in 11 g of carbon dioxide?

5. A mole of argon has a mass of 40 g. If each molecule of argon consists of one atom, what is the mass of an atom of argon (a) in u; (b) in g?

6. When measured at the same temperature and pressure, equal volumes of carbon monoxide and oxygen have masses of 7.0 g and 8.0 g respectively. (a) What is the mass in u of 1 molecule of carbon monoxide? (b) How many molecules are there in this sample of carbon monoxide?

Group 2

7. A mole of nitrogen has a mass of 28 g. (a) What is the mass in g of 1 molecule of nitrogen? (b) From the fact that each nitrogen molecule consists of 2 atoms, find the mass in grams of a nitrogen atom.

8. Equal volumes of fluorine and oxygen at the same pressure and temperature have masses of 9.5 g and 8.0 g respectively. (a) What is the mass of 1 mole of fluorine? (*b*) If each molecule of fluorine is composed of 2 atoms, what is the mass of 1 atom of fluorine in u; (*c*) in g?

9. A mole of aluminum atoms has a volume of 10 cm³. (*a*) If each atom is considered to be a small cube whose faces make contact with those of each neighboring atom, what is the volume occupied by each atom? (*b*) Estimate the linear dimensions of an atom of aluminum by determining the order of magnitude of the edge of the cube containing each atom.

10. Equal volumes of bromine gas and oxygen measured at the same temperature and pressure have masses of 320 g and 64 g respectively. What is the mass in u of a bromine molecule?

Applying Physics

1. When alcohol and water are mixed, the volume of the mixture is smaller than the sum of the volumes of the two liquids before mixing. Some of the molecules of one liquid appear to occupy the empty spaces among the molecules of the other. How much of such empty intermolecular space is available to be filled when equal volumes of alcohol and water are mixed?

   Mix 200 mL of alcohol and 200 mL of water and measure the volume of the mixture. The difference between this volume and 400 mL is the volume of the formerly empty intermolecular space now being filled. What percent is this of the total volume occupied by the two liquids separately?

2. Repeat this experiment by mixing alcohol and water in different proportions by volume. For example, 1 to 3, 1 to 2, 2 to 1, and 3 to 1. In each case, determine the percentage of the volume occupied by the two liquids separately that is lost when the liquids are mixed. Do your results suggest any relationship between the percentage of total volume lost and the composition of the mixture?

14

The Kinetic Theory of Gases

Aims

1. To investigate the general gas laws and to learn how to apply them.
2. To develop a theory of molecular motion that explains the behavior of gases both qualitatively and quantitatively.
3. To apply this theory to understanding the varying conditions of a gas such as its temperature and pressure.
4. To evaluate the theory in terms of its successes and limitations.

Properties of Gases

14-1 Toward a Quantitative Theory of Gases

To be entirely satisfactory, a theory relating matter and heat must go beyond a qualitative explanation of the facts with which it deals. It must also be able to explain the quantitative relationships among them. For example, a complete molecular theory of heat should do more than explain in a general way why matter expands upon heating. It should be able to predict exactly how much a given body of matter will expand for a given rise in temperature. In this respect, the molecular theory of heat is far from complete.

We don't know enough about the molecular structure of matter and the forces that molecules exert upon each other to make accurate quantitative predictions of this kind concerning solids and liquids. In dealing with gases, however, the molecular theory has achieved much success in explaining and predicting gas behavior in quantitative terms.

A simplified molecular theory of gases, known also as the kinetic theory, will be introduced in this chapter. To lay the groundwork for this theory, the experimental laws that gases obey will first be presented. We shall then show how the kinetic theory of gases explains these laws and gives a deeper insight into the inner nature of gases.

14-2 Describing the State of a Gas

The condition of a constant mass of a gas may change in three ways: in pressure, in volume, or in temperature. Thus, when the air inside a bicycle tire pump is compressed by the piston, the pressure of the air in the pump increases, its volume decreases, and its temperature increases.

Fig. 14-1. When you compress the air in a tire pump, the volume decreases while both the pressure and temperature increase.

It follows that the state or condition of a gas at any moment is fully described when its pressure, volume, and temperature are given. The laws that describe the changes that take place in gases under different conditions are therefore relationships among these three quantities. Three such laws will be described in this chapter. Since gas pressure plays a most important part in these laws, we must first review its meaning and the method of its measurement.

The state of a gas is given by its volume, pressure, and temperature.

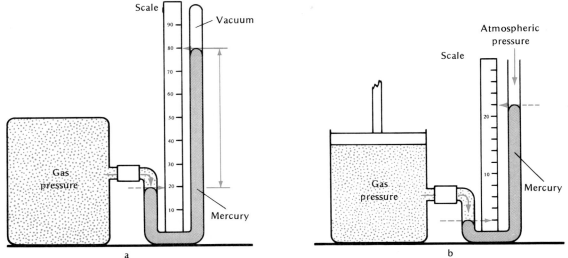

Fig. 14-2. The manometer in (**a**) measures the actual pressure of the confined gas, while that in (**b**) measures the difference between atmospheric pressure and that of the confined gas.

14-3 Measuring Gas Pressure

The principle underlying the measurement of the pressure exerted by a confined gas is illustrated by the mercury *manometer* (muh·NAH·muh·ter) or pressure gauge consisting of a glass U-tube containing mercury, as shown in Fig. 14-2 (a). The right arm of the U-tube is sealed at the top and has a vacuum above the mercury. The left arm of the U-tube is connected to the container of the gas whose pressure is to be measured. A meterstick mounted between the arms of the manometer serves to measure the difference between the levels of mercury in the two arms. This difference is the length of that part of the mercury column in the right arm that is being supported by the pressure of the gas in the container.

The pressure of the gas is usually expressed by giving the length of this column of mercury. Thus, the pressure exerted by the air in the earth's atmosphere under normal conditions may be expressed as 76 centimeters or *760 millimeters of mercury*. This means that *normal atmospheric pressure* is able to support a column of mercury 76 centimeters high. Or, it means that a gas exerting a pressure equal to normal atmospheric pressure will push the mercury 76 centimeters higher in the right arm of the manometer than it is in the left.

A practical manometer often used in the laboratory to measure gas pressure is illustrated in Fig. 14-2 (b). Here, the right arm of the U-tube is open to the pressure of the atmosphere. The difference between the mercury levels in the arms of the U-tube now represents the difference between the pressure of the gas being measured and the atmospheric pressure. Since the pressure of the atmosphere is easily measured by means of a barometer, the actual pressure exerted by the gas can readily be determined.

14-4 Meaning of Pressure

The reading of gas pressure in centimeters or millimeters of mercury is readily converted into the usual unit of pressure, namely the *pascal* (PAS·kul). *The pascal is defined as a pressure of one newton per square meter.* First, however we must understand the meaning of pressure. *Pressure is defined as the force exerted per unit area of a surface.* Consider a rectangular vat containing 10 000 newtons of water and having a bottom area of exactly 10 square meters. The 10 000-newton weight of the water is distributed over the 10 square meters of the bottom of the vat. The pressure, or force exerted on each square meter of the bottom, is therefore 10 000 newtons divided by 10 square meters, or 1000 newtons per square meter. This equals 1000 pascals (symbol, Pa) or 1 kilopascal (KIL·oh·pas·kul, symbol, kPa).

In describing large pressures, normal atmospheric pressure is sometimes used as a pressure unit. It is called *one atmosphere* and is a pressure of 1.01×10^5 newtons per square meter, or 101 kilopascals. As we have learned, this is the pressure that will hold up a column of mercury 760 millimeters high. Two atmospheres is a pressure of 202 kilopascals; three atmospheres is a pressure of 303 kilopascals; and so forth.

14-5 Computing Gas Pressure

The pressure exerted at the bottom of a column of mercury in a tube having a uniform cross section is simply the weight of the mercury divided by the base area of the tube. In the case of the pressure gauge in Fig. 14-2 (a), the pressure to be found is that exerted by the excess of mercury in the right arm of the gauge. The weight of this excess mercury column is equal to its mass m times the acceleration of gravity g, or mg. The mass of the mercury column is equal to its volume times its density.

Now the volume of the mercury column is its base area A times its height h, and the mass of this column is $A \times h \times d$, where d is the density of the mercury. Substituting this expression for the mass in mg, the weight of the excess mercury in the right arm of the gauge becomes $Ahdg$. This weight is distributed over the base area A. Hence, the pressure p at the bottom of the mercury column is $Ahdg$ divided by A, or:

$$p = hdg$$

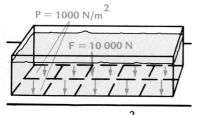
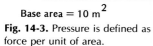

P = 1000 N/m^2

F = 10 000 N

Base area = 10 m^2

Fig. 14-3. Pressure is defined as force per unit of area.

where h is the height of the column, d is the density of mercury, and g is the acceleration of gravity.

In applying this relationship, h is expressed in meters, d in kilograms per cubic meter, and g in meters per second per second. The pressure p then comes out in newtons per square meter, or pascals.

Sample Problem

Convert a pressure of 20 cm of mercury into pascals. The density of mercury is 13 600 kg/m³.

Solution:

$h = 20$ cm $= 0.20$ m $d = 13\ 600$ kg/m³
$g = 9.8$ m/s²

$p = hdg$
$p = 0.20$ m $\times$ 13 600 kg/m³ $\times$ 9.8 m/s²
$p = 2.7 \times 10^4$ N/m² or Pa $= 27$ kPa
(to two significant figures).

Test Yourself Problem

Convert the following pressures into pascals: (*a*) 750 mm of mercury, (*b*) 50 cm of mercury, (*c*) 0.90 m of mercury.

14-6 Boyle's Law

Consider the closed cylinder in Fig. 14-4 which is fitted at the upper end with a gas-tight piston and is filled with a gas. The volume of the gas depends upon how much pressure is exerted by the piston. As the piston is pushed down, it exerts more pressure on the gas and squeezes it into a smaller volume. The gas, in turn, resists being compressed by exerting a counterpressure on the piston. The compression stops when the counterpressure of the gas is equal to the pressure exerted by the piston.

The Englishman Robert Boyle (1627–1691) discovered that, to a close approximation, *the volume of a gas kept at constant temperature varies inversely with the pressure exerted upon it.* That is, doubling the pressure applied by the piston in Fig. 14-4 will squeeze the gas into about half its former volume. Trebling the pressure

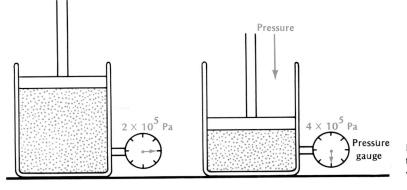

Fig. 14-4. Boyle's law: at constant temperature, the volume of a gas varies inversely with its pressure.

will squeeze the gas into about one-third of its former volume, and so forth. On the other hand, decreasing the pressure applied by the piston to one-third of its initial value will permit the gas to push the piston up until it expands to about three times its initial volume.

This relationship, known as *Boyle's law*, may be written: *at constant temperature,*

$$\frac{V_1}{V_2} = \frac{P_2}{P_1}$$

where V_1 and P_1 are the initial volume and pressure of the gas and V_2 and P_2 are its final volume and pressure. Experiment shows that all gases obey Boyle's law rather closely except when they are too highly compressed or at too low a temperature.

14-7 Graph of Boyle's Law

From $V_1/V_2 = P_2/P_1$ it follows that

$$P_1V_1 = P_2V_2 = \text{a constant}$$

Thus, no matter what changes in volume and pressure take place in a given mass of any gas at constant temperature, the product of its pressure and its volume remains constant.

This relationship is shown in the graph in Fig. 14-5 in which the volume of a gas is plotted against its pressure with the temperature kept constant. The curve shown is called a *hyperbola* (hy·PER·-buh·luh). Note that at A, $P = \frac{1}{2}$ unit, $V = 3$ units, and PV is therefore $\frac{1}{2} \times 3 = 1\frac{1}{2}$ units. At B, $P = 1$ unit, $V = 1\frac{1}{2}$ units, and PV is again $1 \times 1\frac{1}{2} = 1\frac{1}{2}$ units. At C, $P = 1\frac{1}{2}$ units, $V = 1$ unit, and PV is again $1\frac{1}{2} \times 1 = 1\frac{1}{2}$ units.

Fig. 14-5. PV = a constant.

Sample Problem

A volume of 100 m³ of hydrogen is compressed at constant temperature until its pressure rises from 140 kPa to 200 kPa. What is the final volume of the hydrogen?

Solution:

$V_1 = 100 \text{ m}^3 \qquad P_1 = 140 \text{ kPa}$
$V_2 = ? \qquad P_2 = 200 \text{ kPa}$

$$\frac{V_1}{V_2} = \frac{P_2}{P_1}$$

$$V_2 = \frac{P_1V_1}{P_2}$$

$$V_2 = \frac{140 \text{ kPa} \times 100 \text{ m}^3}{200 \text{ kPa}} = 70.0 \text{ m}^3$$

Test Yourself Problems

1. A volume of 0.20 m³ of a gas is compressed at constant temperature until its pressure rises from 150 kPa to 200 kPa. What is its final volume?

2. A volume of 0.10 m³ of a gas expands at constant temperature to a volume of 0.25 m³. If the initial pressure was 100 kPa, what is the final pressure?

14-8 Charles' Law

This law was discovered by Jacques Charles (1746–1823) and independently by Joseph Louis Gay-Lussac (1778–1850). It expresses the fact already noted in Chapter 11 that all gases, when kept at the same constant pressure, expand or contract with changes in temperature at approximately the same rate.

Specifically, any gas at constant pressure expands or contracts very nearly 1/273 of its volume at 0°C for every degree that its temperature changes. Table 14.1 shows the approximate volumes of a sample of a gas kept at constant pressure for the corresponding Celsius and Kelvin temperatures. For simplicity, we have selected a sample of gas whose volume is 273 cubic meters at 0°C.

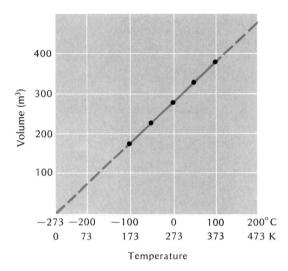

Table 14.1		
TEMPERATURE (°C)	TEMPERATURE (K)	VOLUME (m³)
−100	173	173
−50	223	223
0	273	273
50	323	323
100	373	373

Fig. 14-6. Charles' law: at constant pressure, the volume of a gas is proportional to its Kelvin temperature.

The straight-line graph in Fig. 14-6 illustrates the law of Charles and Gay-Lussac which states: *When a gas is kept under constant pressure, its volume varies directly with its Kelvin temperature.* That is, **at constant pressure:**

$$\frac{V_1}{V_2} = \frac{T_1}{T_2}$$

where V_1 and V_2 are the initial and final volumes and T_1 and T_2 are the corresponding initial and final Kelvin temperatures of the gas.

Note that as the graph is extended to the left, as shown by the dotted line, it cuts the temperature axis at −273°C where the volume of the gas shrinks to zero. Recall from Section 11-15 that this was one of the early clues suggesting that −273°C is the lowest possible temperature.

Charles' law suggests that at −273°C the volume of a gas would shrink to zero.

Sample Problem

A volume of 0.20 m³ of nitrogen at 20°C is heated at constant pressure to a temperature of 313°C. What is the new volume of the nitrogen?

Solution:
$V_1 = 0.20 \text{ m}^3$
$V_2 = ?$
$T_1 = 20°C = 20 + 273 = 293 \text{ K}$
$T_2 = 313°C = 313 + 273 = 586 \text{ K}$

$$\frac{V_1}{V_2} = \frac{T_1}{T_2}$$

$$V_2 = \frac{V_1 T_2}{T_1}$$

$$V_2 = \frac{0.20 \text{ m}^3 \times 586 \text{ K}}{293 \text{ K}}$$

$$V_2 = 0.40 \text{ m}^3$$

Test Yourself Problems

1. A volume of 1.00 m³ of a gas at 900 K is cooled at constant pressure to 300 K. What is the new volume?

2. To what Kelvin temperature must 0.050 m³ of helium at 273 K be heated at constant pressure to raise its volume to 0.150 m³?

14-9 General Gas Law

A general gas law relating the changes in pressure, volume, and Kelvin temperature of a gas is obtained by combining Boyle's law and Charles' law. Let a gas whose initial pressure, volume, and temperature are P_1 V_1 and T_1 undergo changes that give it a final pressure, volume, and temperature of P_2, V_2, and T_2. Then the *general gas law* is given by the formula:

$$\frac{P_1 V_1}{T_1} = \frac{P_2 V_2}{T_2}$$

Note that when the temperature of the gas is held constant so that $T_1 = T_2$, this law reduces to $P_1 V_1 = P_2 V_2$, which is Boyle's law. Again, when the pressure of the gas is kept constant so that $P_1 = P_2$, this law reduces to $V_1/T_1 = V_2/T_2$, which is Charles' law.

Sample Problem

A volume of 0.020 m³ of air at 27°C and a pressure of 100 kPa is compressed until its pressure becomes 250 kPa and its temperature is 102°C. What is its new volume?

Solution:
$V_1 = 0.020 \text{ m}^3 \qquad P_1 = 100 \text{ kPa}$
$T_1 = 27°C = 273 + 27 = 300 \text{ K}$
$V_2 = ? \qquad\qquad P_2 = 250 \text{ kPa}$
$T_2 = 102°C = 273 + 102 = 375 \text{ K}$

$$\frac{P_1 V_1}{T_1} = \frac{P_2 V_2}{T_2}$$

$$V_2 = \frac{P_1 V_1 T_2}{P_2 T_1}$$

$$V_2 = \frac{100 \text{ kPa} \times 0.020 \text{ m}^3 \times 375 \text{ K}}{250 \text{ kPa} \times 300 \text{ K}}$$

$$V_2 = 0.010 \text{ m}^3$$

Test Yourself Problems

1. A volume of 0.060 m³ of a gas at 300 K and a pressure of 75 kPa expands until its volume is 0.100 m³ and its temperature is 200 K. What is the final pressure of the gas?

2. A gas having a volume of 2.00 m³, a pressure of 200 kPa, and a temperature of 127°C is compressed and cooled until its pressure is 250 kPa and its volume is 1.00 m³. What is its final temperature?

14-10 Universal Gas Constant

The general gas law states that no matter what changes take place in a given mass of gas, the product of its pressure and volume divided by its Kelvin temperature remains constant. Thus, if P, V, and T are the pressure, volume, and Kelvin temperature of a gas at any moment, the gas law states that $PV/T = c$, a constant, or $PV = cT$.

The value of c is usually different for different gas samples. However, when the equation is rewritten for samples each consisting of 1 mole of gas, it is found that the value of c becomes the same constant for all gases. This constant, called R, is therefore a *universal gas constant*. Thus, for 1 mole of any gas, the *general gas law* becomes:

$$PV = RT$$

The value of R can be calculated from the fact that 1 mole of any gas at normal atmospheric pressure and at 0°C has a volume of 22.4 liters, or 0.0224 cubic meter. Normal atmospheric pressure is 101.3 kilopascals, or 1.013×10^5 newtons per square meter; and 0°C is equal to $0 + 273$ or 273 K. Substituting these values into $PV = RT$ and solving for R gives $R = 8.31$ joules per mole-kelvin. *In using $PV = RT$ with this value of R, P must be expressed in pascals, V in cubic meters, and T in kelvins.*

$R = 8.31$ J/mol·K

Sample Problem

What is the volume of 1 mole of air at a pressure of 1.00×10^5 Pa and at a temperature of 1000 K?

Solution:
$P = 1.00 \times 10^5$ Pa
$R = 8.31$ J/mol·K $T = 1000$ K

$PV = RT; V = RT/P$

$$V = \frac{8.31 \text{ J/mol·K} \times 1000 \text{ K}}{1.00 \times 10^5 \text{ Pa}}$$

$V = 8.31 \times 10^{-2}$ m³

Test Yourself Problem

What is the pressure on 1 mole of nitrogen when it is at a temperature of 600 K and its volume is 6.00×10^{-2} m³?

14-11 Relating Temperature and Pressure

Solving for the pressure in $PV = RT$ gives $P = RT/V$. If the volume of the gas is held constant, R/V is constant and $P = $ (a constant) $\times T$. Hence, *when the volume of a gas is kept constant, its pressure varies in direct proportion to its Kelvin temperature;* **or, at constant volume,**

$$\frac{P_1}{P_2} = \frac{T_1}{T_2}$$

where P_1 and T_1 are the initial values of the pressure and temperature of a gas and P_2 and T_2 are the final values.

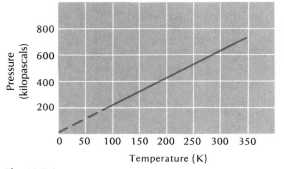

Fig. 14-7. In a gas at constant volume, the pressure is proportional to the Kelvin temperature.

Table 14.2	
TEMPERATURE (kelvins)	PRESSURE (kilopascals)
100	200
150	300
200	400
250	500
300	600
350	700

Table 14.2 lists the pressures exerted by a sample of gas when heated at constant volume over a range of temperatures. These values are plotted in the straight-line graph of Fig. 14-7. Note that the extended graph cuts through the origin and thus predicts that at absolute zero, or 0 K, the pressure of the gas is also zero.

14-12 Constant-Volume Gas Thermometer

The straight-line relationship between gas pressure and temperature when the gas is held at constant volume is used in the constant-volume gas thermometer. In this instrument a gas is kept at constant volume and the changes in its temperature are measured by the changes in its pressure. Each value of the pressure corresponds to a specific Kelvin temperature. The constant-volume gas thermometer is sensitive and reliable and is therefore favored in scientific work.

14-13 Limitations of the Gas Laws

Boyle's law, Charles' law, and the general gas law are experimental relationships based upon the measurements of many gases. They are not exact descriptions of the behavior of any particular gas but, within certain limits, are good approximations of the behavior of all gases. In general, all gases obey these laws rather closely provided they are not too near the high-pressure and low-temperature conditions at which they condense into liquids.

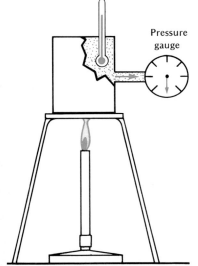

Fig. 14-8. Heating a gas at constant volume increases both its pressure and temperature.

Pressure gauge

Theory of the Ideal Gas

14-14 The Ideal Gas

The fact that all gases, in spite of the differences among them, resemble each other in their conformity to the general gas law and its special cases, suggests that there are basic similarities in their internal structure. We are now prepared to form a theory of the internal structure of gases that can account for the observed similarities in their behavior.

Just as it was found useful in the study of machines to imagine an ideal machine (Section 9-4) so it will now be useful to imagine an *ideal gas*. An ideal gas is defined as one that obeys the general gas law $PV = RT$ exactly and under all conditions of temperature and pressure. A *real gas* differs from an ideal gas in that it obeys the general gas law only approximately and only within certain limits of temperature and pressure.

An ideal gas is defined as one that obeys $PV = RT$ for all pressures and temperatures.

14-15 Molecular Model of an Ideal Gas

We have seen that, under ordinary conditions, gas molecules are believed to be relatively distant from one another. Accordingly, they are acting almost free of the attractive and repulsive forces between molecules at the close distances at which they are found in solids and liquids. Each gas molecule is therefore thought to move in a straight line uninfluenced by other molecules until it collides with one of them. Over a period of time, each molecule will take a zigzag path consisting of many short straight-line segments.

Assuming this model of a gas, what properties must its molecules have to enable them to exert pressure and occupy volume in such a way that the gas always obeys the general gas law $PV = RT$? We begin by imagining a simple model of an ideal gas whose molecules have these characteristics:

The model of an ideal gas explains all the gas laws as the result of the motions and elastic collisions of gas molecules.

(1) *Small size.* The sizes of molecules are so small that molecules may be treated as points rather than as small bodies with definite volumes.

(2) *Lack of forces.* No forces act between any two molecules except when they collide. Between collisions, each molecule moves in a straight line at constant speed in accordance with the law of inertia.

(3) *Elastic collisions.* All collisions involving molecules are elastic and obey both the law of conservation of momentum and the law of conservation of energy. Therefore, a molecule striking the wall of the gas container at right angles with a given speed will rebound from that wall with the same speed in the opposite direction. Furthermore, each collision is assumed to occur over a time interval that is negligibly small.

(4) *Large number.* The total number of molecules is very large so that collisions among them, and between them and the walls of the container, cause the molecules to keep changing their directions and their speeds. As a result, the number of molecules having each particular speed and direction remains essentially unchanged from moment to moment.

(5) All laws obeyed by this gas model are entirely the result of the motions and collisions of its molecules. The theory of the behavior of this ideal gas model is called the *kinetic theory of gases.*

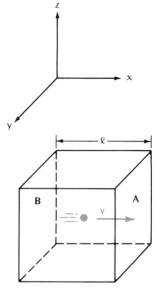

Fig. 14-9. A single molecule moves back and forth, colliding successively with walls A and B.

14-16 Deriving Pressure of an Ideal Gas

We shall now derive an expression for the pressure exerted on the walls of its container by an ideal gas. In terms of our picture of an ideal gas, the pressure on each wall of a container of gas is the result of the never-ending collisions of the molecules of the gas with that wall. We can therefore derive an expression for the pressure exerted by the gas on the walls of its container by adding up the effects of these molecular collisions.

In Fig. 14-9 is shown a gas-filled cubical box having edges of length l. Since the molecules are moving in all possible directions, they keep colliding elastically with each other and with the six walls of the cube. The repeated reboundings of these randomly moving molecules from the sides of the box have the same average effect on those sides as though $\frac{1}{3}$ of the molecules in the box were moving directly back and forth in the x direction, $\frac{1}{3}$ were moving directly back and forth in the y direction, and $\frac{1}{3}$ were moving directly back and forth in the z direction. Thus, if there are N randomly moving molecules in the box, their effect on side A is the same as though $N/3$ of them were shuttling back and forth between A and B in a direction always perpendicular to these surfaces.

We shall now use this simplified picture to compute the pressure exerted by the $N/3$ molecules colliding with side A. First, we shall follow a typical single molecule shuttling between A and B and determine how many collisions it makes with wall A during a time interval t. Second, we shall determine the change in momentum suffered by this molecule during time t as a result of its collisions with wall A. This change of momentum determines the force this molecule exerts on wall A. Finally, we shall determine the total change of momentum suffered by all $N/3$ of the molecules bombarding side A during time t. This will enable us to compute the average force and the pressure exerted by all the molecules striking side A.

14-17 Derivation of the Pressure Formula

(1) Number of Collisions Per Molecule. Consider a single molecule moving at speed v at right angles to wall A. On collision with A, the molecule reverses direction and moves toward the opposite wall B. Here, on collision with B, the molecule reverses direction and again moves toward A. Thus, successive collisions with A and B keep the molecule moving back and forth across the box.

Since the walls of the box are elastic, the speed of the molecule after each collision with A or B is the same as it was before the collision. Thus, assuming the time taken by each collision to be negligibly small, the molecule maintains its original speed v. The distance traveled by the molecule during a small interval of time t is therefore vt. Since $2\,l$ is the round trip distance from A to B and back, $vt/2\,l$ is the number of round trips made by the molecule during time t. The number of collisions each molecule makes with side A during time t is therefore also $vt/2\ l$.

(2) *Momentum Change Per Molecule.* As the molecule strikes wall A with speed v, it rebounds from it with the same speed but in the opposite direction. If m is the mass of the molecule, its momentum before the collision with A is mv. After the collision, its momentum is $-mv$. The change of momentum suffered by the molecule on collision with A is therefore $mv - (-mv) = 2\ mv$.

The molecule suffers a similar change of momentum during each of the $vt/2\ l$ collisions it makes with wall A during the time t. The total change of momentum suffered by the molecule during t is therefore $vt/2\ l \times 2\ mv = mv^2t/l$.

(3) *Pressure Exerted by the Ideal Gas.* Now, consider the $N/3$ molecules that are shuttling back and forth between A and B. They have many different speeds ranging from zero to some maximum value. If $\bar{v}^2$ is the average value of the squares of the speeds of all the molecules, the average change of momentum per molecule during time t is $m\bar{v}^2t/l$. The change of momentum during time t suffered by all $N/3$ molecules in their collisions with A is therefore:

$$\frac{N}{3}\frac{m\bar{v}^2t}{l}$$

On collision, each molecule exerts a tiny separate blow on A. However, because the number of the molecules is very large, the net effect of their individual blows is to exert a steady average force F on A. Wall A therefore reacts to the bombardment of the molecules by exerting an equal and opposite force on the molecules. According to the momentum form of the second law of motion in Sec. 7-25, this force times the time it acts is equal to the change of momentum it produces in the molecules that strike it during that time. Thus,

$$Ft = \frac{N}{3}\frac{m\bar{v}^2t}{l}$$

and,

$$F = \frac{N}{3}\frac{m\bar{v}^2}{l}$$

Now the pressure P on wall A is equal to the force F divided by the area of A which is $l \times l$, or l^2. Dividing both sides of the above equation by l^2, we have:

$$\frac{F}{l^2} = P = \frac{N}{3}\frac{m\bar{v}^2}{l^3}$$

Since l^3 is the volume V of the cubical box, this can be written:

$$\mathbf{P} = \frac{1}{3}\frac{\mathbf{N}}{\mathbf{V}}\mathbf{m\bar{v}^2}$$

Gas pressure is the result of the impulses resulting from the collision of gas molecules with the walls of their containers.

The kinetic theory thus gives us an expression for the pressure exerted by a confined gas that depends only upon the number of molecules in the container, the volume of the container, the masses of the molecules, and their speeds. Let us now examine further the meaning of this pressure formula and see how well it agrees with the experimental facts.

14-18 Pressure and Average Molecular Kinetic Energy

The expression $\frac{1}{2}\, m\bar{v}^2$ is the average kinetic energy per molecule of the ideal gas. Letting E_{ke} represent $\frac{1}{2}\, m\bar{v}^2$ in:

$$P = \frac{1}{3}\frac{N}{V}\, m\bar{v}^2$$

gives,

$$P = \frac{1}{3}\frac{N}{V}\, (2\, E_{ke})$$

and,

$$\mathbf{P} = \frac{2}{3}\frac{\mathbf{N}}{\mathbf{V}}\, \mathbf{E}_{ke}$$

Since N/V is the number of molecules per unit volume and E_{ke} is the average kinetic energy per molecule, the pressure turns out to be numerically equal to two-thirds of the total kinetic energy of the molecules in a unit volume of the gas.

14-19 Kinetic Interpretation of Temperature

If the volume of a given mass of gas is kept constant, V and N are constant and so is $2N/3V$. The pressure of the ideal gas then becomes simply $P = (\text{a constant}) \times (E_{ke})$ and the pressure depends only upon the average kinetic energy of the molecules.

However, it was seen in Section 14-11 that, for a gas kept at constant volume, the pressure is directly proportional to the Kelvin temperature. This suggests that the Kelvin temperature of a gas is proportional to the average kinetic energy E_{ke} of its molecules. Thus according to the kinetic theory of gases, *the temperature of a gas is simply a measure of the average kinetic energy of its molecules*. It follows that *any two gases whose molecules have the same average kinetic energy will have the same temperature and vice versa*.

The Kelvin temperature is a measure of the average kinetic energy of the molecules of a gas.

14-20 Kinetic Theory Predicts the General Gas Law

Again making use of the kinetic theory interpretation of temperature, it can be shown that the molecular model of the ideal gas predicts the general gas law. First the relationship $P = (2N/3V)E_{ke}$ is rewritten in the form $PV = \frac{2}{3}\, NE_{ke}$. Then, since E_{ke} is proportional to the Kelvin temperature, and since N is a constant for a given sample of a gas, $\frac{2}{3}\, NE_{ke}$ is equal to some constant c times the Kelvin temperature T. The equation $PV = \frac{2}{3} N_{Ke}$ becomes, therefore, $PV = cT$. This is the general gas law which takes the form $PV = RT$ when we are dealing with 1 mole of a gas.

14-21 Average Kinetic Energy Per Molecule

It is now possible to determine the constant of proportionality that relates the Kelvin temperature and the average molecular kinetic energy. Consider 1 mole of an ideal gas. The number of molecules it contains is Avogadro's number N_a. The gas therefore obeys the following two equivalent relationships:

$$PV = \frac{2}{3} N_a E_{ke}$$

and,

$$PV = RT$$

Note that the left sides of these equations are the same. It follows that $\frac{2}{3} N_a E_{ke} = RT$, whence,

$$E_{ke} = \frac{3}{2} \frac{R}{N_a} T$$

R is the universal gas constant, 8.31 joules per mole-kelvin; N_a is Avogadro's number, 6.02×10^{23}, which is also a universal constant. Hence, R/N_a is a universal constant. It is given the letter k and called *Boltzmann's constant*. If we substitute k for R/N_a, the expression for E_{ke} becomes:

$$\mathbf{E}_{ke} = \frac{3}{2} \mathbf{kT}$$

where $k = 1.38 \times 10^{-23}$ joule per molecule-kelvin.

According to this relationship, the proportionality factor that links the average kinetic energy of the molecules of a gas and its Kelvin temperature is 3/2 times the universal constant k. Since this proportionality factor is the same for all gases, it follows that *different gases at the same temperature have the same average kinetic energy per molecule.*

14-22 Molecular Speed

The speed of a molecule having the average kinetic energy E_{ke} can be obtained by substituting for E_{ke} its value, $\frac{1}{2} m\bar{v}^2$. Then,

$$E_{ke} = \frac{1}{2} m\bar{v}^2 = \frac{3}{2} kT$$

and,

$$\bar{v}^2 = \frac{3\,kT}{m}$$

The square root of $\bar{v}^2$ is the speed of a molecule having the

Fig. 14-10. An inflated balloon will expand when exposed to a strong heat source and will contract when subjected to extreme cold.

$k = 1.38 \times 10^{-23}$ J/molecule·K

average kinetic energy E_{ke}. It is called the *root-mean-square speed* of the molecules and is given the symbol v_{rms}.

$$\mathbf{v}_{rms} = \sqrt{\overline{\mathbf{v}^2}} = \sqrt{\frac{3\,kT}{m}}$$

For a given gas, this relationship shows that the molecular speed v_{rms} varies directly as the square root of the Kelvin temperature. Thus, when the temperature increases by a factor of 4, say from 100 K to 400 K, v_{rms} increases by a factor of $\sqrt{4}$, or 2, and becomes twice as great as it was at the lower temperature.

For different gases at the same temperature, v_{rms} varies inversely as the square root of the masses of their molecules. Consider two gases such that the mass of each molecule of the first gas is four times as great as the mass of each molecule of the second gas. If the temperature is the same, the molecular speed v_{rms} is $1/\sqrt{4}$ or ½ as great in the first gas as it is in the second.

14-23 Interpretation of Absolute Zero

From the relationships between E_{ke} and T and between v_{rms} and T, it is seen that when T is zero, both the average kinetic energy per molecule and the molecular speed v_{rms} are zero. This means that according to the kinetic theory, the molecules of an ideal gas at the temperature of absolute zero are at rest. However, when a real gas nears temperatures in the region of absolute zero, its state changes first to liquid and then to solid. Since the kinetic theory does not apply at temperatures at which a substance is no longer a gas, the correct interpretation of absolute zero is more complicated than the theory of the ideal gas suggests.

Sample Problems

1. What is the average kinetic energy per molecule of a gas at the temperature of melting ice?

   Solution:
   $k = 1.38 \times 10^{-23}$ J/molecule·K
   $T = 0°C = 273$ K

   $E_{ke} = \dfrac{3}{2} kT$

   $E_{ke} = \dfrac{3}{2}(1.38 \times 10^{-23}\ \text{J/molecule·K})(273\ \text{K})$

   $E_{ke} = 5.65 \times 10^{-21}$ J

2. Assuming that the above gas is hydrogen, whose molecule has a mass of 3.34×10^{-27} kg, what is the root-mean-square speed of its molecules?

Solution:

$v_{rms} = \sqrt{\dfrac{3\,kT}{m}}$

$v_{rms} = \sqrt{\dfrac{3(1.38 \times 10^{-23})(273)}{3.34 \times 10^{-27}}}$

$v_{rms} = 1.84 \times 10^3$ m/s

This is nearly 2 kilometers per second. It indicates that even at temperatures as low as that of melting ice, molecules move at very high speeds. Actual measurement of molecular speeds shows that the speeds predicted by kinetic theory are of the right order of magnitude.

Test Yourself Problems

1. (a) What is the average kinetic energy per molecule of oxygen at a temperature of 100 K? (b) What is the average kinetic energy per molecule of hydrogen at this same temperature?

2. The mass of a molecule of oxygen is 5.34×10^{-26} kg. What is the root-mean-square speed of its molecules at 100 K?

14-24 Success of Kinetic Theory Model

The kinetic theory began by defining an ideal gas as one that conforms to the general gas law not only in the range of pressures and temperatures at which all gases conform but under all conditions of temperature and pressure. A model of the molecular makeup of an ideal gas was then described, and the behavior of this model predicted in terms of the motions and collisions of its molecules.

The model suggested that the Kelvin temperature of an ideal gas is proportional to the average kinetic energy of its molecules. Assuming this interpretation of temperature, it was then shown that the model of an ideal gas predicts the general gas law as a consequence of the motions and collisions of its molecules. Since a real gas behaves approximately like an ideal gas over a limited range of temperatures and pressures, the molecular behavior of the model gas gives us an insight into the molecular behavior of real gases under these conditions.

The model was able to explain the general gas law in terms of molecular motion.

14-25 Limitations of the Ideal Gas Model

The ideal gas model is too simple to explain some of the facts about the behavior of real gases, particularly when they are under high pressure or at very low temperature. It also does not explain why real gases differ in the closeness with which they obey the general gas law. The model must therefore be revised to make its behavior correspond more closely to that of real gases. It is beyond the purpose of this book to look into all the revisions needed. Here, three will be mentioned briefly.

The ideal gas model ignores the sizes of molecules as negligibly small. In real gases, the sizes of the molecules have an effect on the behavior of the gases and must be taken into account.

The ideal gas model assumes that molecules move in short, straight lines between collisions and have no other motion. There is evidence that the molecules of real gases can have two other motions. They may rotate and their parts may vibrate with respect to each other. A complete gas theory must take these additional motions into account.

Ideal gas molecules are assumed to have no energy other than the kinetic energy associated with their straight-line motions and to have no forces act upon them except when they collide. In real gases, it is known that molecules attract each other at close distances. Work is therefore needed to pull them apart and work done

The model failed to explain why the general gas law applies only within certain limits of temperature and pressure.

in this way is stored in the gas as potential energy. Potential energy may also be stored in the atoms or parts of atoms of which molecules are made. Thus, the molecules of a real gas possess both kinetic and potential energy, and both must be taken into account in understanding its behavior.

CHAPTER REVIEW

Summary

Pressure is defined as force exerted per unit of area. The *SI* metric unit of pressure is the **pascal,** which equals one newton per square meter. Pressure is also commonly expressed in standard **atmospheres** and in **millimeters of mercury** when measured by a mercury **manometer.**

All gases closely follow the **general gas law:**

$$\frac{P_1V_1}{T_1} = \frac{P_2V_2}{T_2}$$

provided that the gases are not too highly compressed or at too low a temperature. Here, P_1, V_1, and T_1 are the initial values of the pressure, volume, and Kelvin temperature of a gas, and P_2, V_2, and T_2 are the final values. This law is also written $PV = RT$, where R is the universal gas constant, 8.31 joules per mole-kelvin.

For a gas at constant temperature, the general gas law reduces to **Boyle's law, $P_1V_1 = P_2V_2$.** For a gas at constant pressure, the general gas law reduces to **Charles' law, $V_1/V_2 = T_1/T_2$.** For a gas at constant volume, the general gas law reduces to **$P_1/P_2 = T_1/T_2$.**

An **ideal gas** is defined as one that conforms exactly to the general gas law under all conditions. The **kinetic theory of gases** proposes a model of an ideal gas that attempts to explain the behavior of the gas in terms of the motions and collisions of its molecules. This theory is able to show that the **pressure P exerted by an ideal gas** is given by $P = (N/3\ V)m\bar{v}^2$ where N is the number of gas molecules, V is the volume of their container, m is the mass of each molecule, and $\bar{v}^2$ is the average value of the squares of the molecular speeds.

This relationship leads to the conclusion that the Kelvin temperature of an ideal gas is a measure of the average kinetic energy of its molecules. Increasing the temperature of a gas increases both the kinetic energy and the speed of its molecules according to the relationship $E_{ke} = {}^3/_2\ kT$ where E_{ke} is the **average kinetic energy** of a gas molecule, T is the Kelvin temperature, and k is Boltzmann's constant, 1.38×10^{-23} joule per molecule-kelvin. This expression indicates further that when T is absolute zero, the kinetic energy of the molecules of a gas is also zero and molecular motion stops.

The pressure formula also yields the relationship $PV = cT$ which is a form of the general gas law.

Although the kinetic theory of the ideal gas claims some remarkable successes, it has some serious limitations. These arise from the fact that the theory ignores the sizes of the molecules, the forces that they exert upon each other, their internal potential energy, and their internal vibratory motion. A complete theory must take these and other characteristics of molecules into account.

Questions

Group 1

1. (a) Explain how to use a U-tube containing mercury as a manometer for measuring gas pressure. (b) What is the meaning of a pressure described as 30 cm of mercury?

2. A gas is compressed at constant temperature into 0.1 of its original volume. What change has taken place in its pressure?

3. The temperature of a gas enclosed in a steel tank rises from 100 K to 300 K. What change has taken place in the pressure inside the tank?

4. (a) Under what condition does the general gas law reduce to Boyle's law? (b) If the pressure on an enclosed gas kept at constant temperature is steadily increased, will there ever come a time when the gas is squeezed into zero volume? (See Fig. 14-5.)

5. (a) Explain how Charles' law predicts the value of absolute zero to be at $-273°C$. (b) If the volume of a gas is kept constant, what prediction does the general gas law make about the value of the pressure when the gas is cooled to 0 K? (Assume the gas does not change its state.)

6. (a) How is an ideal gas defined? (b) Under what conditions does the behavior of a real gas fail to agree with that of an ideal gas?

7. What assumptions are made in the kinetic theory model of an ideal gas concerning (a) the size of its molecules; (b) the forces they exert upon each other; (c) the types of collisions which they undergo.

8. (a) What molecular quantity is being measured when the temperature of an ideal gas is taken? (b) Can a space that is a perfect vacuum have a temperature? Explain.

9. Explain in terms of the kinetic theory of gases why allowing the volume of a gas kept at constant temperature to increase results in a decrease in the pressure it exerts on the walls of its container.

10. (a) What are some of the successes achieved by the kinetic theory of the ideal gas? (b) What are some of its limitations? (c) Why should the behavior of a real gas that is highly compressed deviate seriously from that of an ideal gas?

Group 2

11. An automobile tire hardens after it has been riding over a bumpy, rocky road for some time. Explain in terms of the kinetic theory why the pressure in the tire increases?

12. A rigid container has n molecules of gas A in it. An equal number of molecules of gas B is then introduced into it. If the average kinetic energies of the molecules of the two gases were equal before the gases were mixed, what change, if any, has taken place in (a) the temperature; (b) the pressure inside the container?

13. A gas is enclosed in a cylinder fitted with a piston. When the piston is moved so as to compress the gas, the temperature of the gas rises. Where do the molecules of the gas get the additional kinetic energy corresponding to this rise in temperature?

14. Two different gases have the same volume, temperature, and pressure but their molecules have unequal masses. Compare (a) the total kinetic energy; (b) the root-mean-square speeds of the molecules of the two gases.

15. A mole of gas A and a mole of gas B have the same temperature and volume but gas B has 16 times the mass of gas A. Compare (a) the pressures; (b) the numbers of molecules; (c) the masses of the molecules; (d) the root-mean-square speeds of the molecules of the two gases.

Problems

Group 1

1. Convert the following pressures into pascals: (a) 0.10 m of mercury; (b) 500 mm of mercury; (c) 60 cm of mercury; (d) 4.0 atmospheres.

2. A mass of air is compressed at constant temperature from a volume of 0.500 m³ to a volume of 0.300 m³. If the initial pressure of the gas was 100 kPa, what was the final pressure of the gas?

3. A constant mass of helium in a balloon expands at constant temperature as it rises from the ground where the pressure is 990 kPa to an altitude where the pressure is 660 kPa. If the initial volume of the gas was 300 m³, what is the final volume?

4. An initial volume of 0.566 m³ of air is heated at constant pressure from 10°C to 100°C. What is the final volume?

5. To what temperature must 1.00 m³ of a gas at 50°C be heated at constant pressure in order for its volume to double?

6. A rigid steel tank contains carbon dioxide at 30°C and 505 kPa pressure. If the temperature of the tank and its contents is reduced to −33°C, what is the final pressure in the tank?

7. The temperature of the air in a tire rises from 27°C to 47°C. If the initial pressure of the air in the tire was 2.0×10^5 Pa, what was the final pressure?

8. A volume of 4.0×10^{-4} m³ of oxygen at a temperature of 7.0°C and a pressure of 1.4 kPa is compressed until its volume is reduced to 1.0×10^{-4} m³ and its temperature rises to 57°C. What is the final pressure of the oxygen?

9. A mass of gas has a volume of 0.12 m³ at a temperature of −23°C and a pressure of 1.6 kPa. What will be the temperature of the gas when its volume decreases to 0.080 m³ and its pressure rises to 3.2 kPa?

10. What is the volume of 1 mole of a gas when at a temperature of −223°C and a pressure of 1.00×10^5 Pa?

11. What is the pressure of 1 mole of hydrogen whose volume is 0.020 m³ and whose temperature is 21°C?

12. What is the average value of the kinetic energy of the molecules of any gas at (a) 50°C; (b) 100°C? (c) What is the total kinetic energy of a mole of gas at each of these temperatures?

Group 2

13. A molecule of hydrogen has a mass of 3.3×10^{-27} kg. What is its root-mean-square speed in hydrogen gas at 27°C?

14. A molecule of oxygen has 8 times the mass of a helium molecule. The root-mean-square speed of oxygen molecules at 0°C is 460 m/s. What is the root-mean-square speed of a molecule of helium at 0°C?

15. The mass of an oxygen molecule is 16 times that of a hydrogen molecule. At a certain temperature, the root-mean-square speed of the hydrogen molecule is 2000 m/s. What is the root-mean-square speed of an oxygen molecule at the same temperature?

16. The temperature of a gas is −50°C. (a) To what temperature must it be heated to double the average kinetic energy of its molecules? (b) To what temperature must it be heated to double the root-mean-square speed of its molecules?

17. The mass of a hydrogen molecule is 3.3×10^{-27} kg. (a) If it strikes the wall of its container at right angles with a speed of 2000 m/s, what change of momentum does it undergo? (b) If 10^{28} such molecules strike 1 m² of the container wall per second, what is the pressure on that wall in pascals? (Note: The pressure is the total change of momentum per second per square meter of the wall.)

Applying Physics

1. At what speed do molecules of ammonia gas diffuse through the air? Open a bottle of concentrated ammonia water at one end of a room. Then stand several meters from the bottle and measure the time it takes the ammonia gas escaping from the bottle to reach you and be detected by its odor. Estimate the rate at which ammonia molecules diffuse through the air by dividing the distance traveled by the ammonia gas from the bottle to you by the time of travel.

2. Compare the relative rates of diffusion of two gases through air as follows. Soak two wads of cotton, one with hydrochloric acid and one with ammonia water. Put one of the wads at each end of a 1-meter glass tube about 1.5 cm in diameter and close both ends with stoppers.

Hydrogen chloride gas issuing from the acid and ammonia gas issuing from the ammonia water diffuse toward each other. After a while, the molecules of these two gases meet and react chemically to form a ring of white powder (ammonium chloride) on the inside of the tube. Measure the distances between the white ring and each end of the tube. The ratio of these distances is the ratio of the speeds of diffusion of the two gases through the air. Which gas diffuses faster? Can you explain why one gas diffuses faster than the other?

How Snowflakes Form: Clues to pattern formation in nature

Think of the billions upon billions of snowflakes that fall in a typical winter's snowstorms. You've known since childhood that each snowflake, although similar to others in form, is unique with respect to its detailed structure. Physicists are interested in how snowflakes form, for they think this process will throw light on the mysterious mechanism of pattern formation in nature. From the shapes of galaxies to crystal structures in alloys, to bilateral symmetry in mammals, to the shapes of individual cells, all inanimate and animate structures come into being by pattern formation. Thus, the lowly snowflake may help us understand the formation of something as vast as a galaxy of stars.

Physicist James Langer of the Institute for Theoretical Physics in Santa Barbara, California has developed a theory of snowflake formation. Langer knew that heat is released when the water surrounding a speck of dust—the seed of a snowflake—freezes. He knew also that when a snowflake grows it acquires a hexagonal shape, and that the sharp-tipped structure unique to snowflakes is particularly stable, while other shapes are not. Moreover, physicists have known the mathematical equations of growth for snowflakes for years. They just haven't been able to solve them, not even on a computer.

Langer's task was to use the physical laws that govern the change of a liquid to a solid crystal in equations that describe the growth of a hypothetical snowflake, and to use the equations in a computer simulation. He did this, and got results similar to those shown in the diagram. As the program ran, pictures of snowflake-like shapes emerged from virtually nothing. The circle at the center represents the original tiny blob of ice frozen around a speck of dust.

Langer also has a tentative explanation for why each snowflake is unique. As the spherical piece of ice that is the start of a snowflake grows, it crystallizes in the hexagonal shapes of ice crystals, and bulges out at six points. But the snowflake is tumbling about in the atmosphere, experiencing a countless variety of different temperatures and humidities. This constantly changing set of growth conditions is different for each snowflake, thus the growth of the six original branches and all side branches will be unique.

Langer believes that the mechanism of pattern formation throughout nature may be similar to the process involved in snowflake growth. As he refines his theory and applies it, he expects to learn more about how patterns develop, a phenomenon that until now has been a mystery. Watch for further developments. We may soon have valuable new clues to the formation of the myriad shapes found in the natural world.

The New Automobile Aerodynamics

Look closely at the shape of almost any new automobile. You'll see right away that there are differences from the shapes of cars of just a few years ago. There are good reasons for these changes. Fuel prices shot up during the 1970's. Also, Federal regulations now require increased fuel efficiency. As a result, designers are producing car shapes that do a better job of overcoming wind resistance.

The principal tool in this effort is the wind tunnel. A proposed body model is mounted in a wind tunnel. Then all the forces and pressures the wind imposes on the body shape are measured. It's not unusual for wind speeds of 60 miles per hour to be used. All of the data collected is then fed into a complex formula that yields a quantity called the drag coefficient, or Cd. The drag coefficient of a vehicle is a direct measure of its fuel efficiency. The lower the Cd, the greater the fuel efficiency.

In a parachute, a device designed for maximum wind resistance, the Cd is as high as 1.35. By contrast, airplane wings have a Cd of about 0.05. A car shape with a Cd of 0.30 would be quite good. Sixty years ago cars had an average Cd of about 0.70. Today it is in the vicinity of 0.40. Experts estimate that dropping the Cd from 0.40 to 0.30 would save about 10 billion gallons of gasoline a year.

Engineers have discovered that a car's frontal area and design, the smoothness of its overall surface, and how the wind behaves as it passes over the body are the major factors in producing a low Cd. As efforts to achieve lower Cd's continue, you'll see interesting changes taking place in automobile body shapes.

Automotive Engineers

The design and production of automobiles requires the input of many different kinds of engineers. You have already seen that

aerodynamic engineers work to produce car shapes with the lowest possible coefficient of drag. Other mechanical engineers are involved in all phases of the automobile industry, from designing body structures, engines, suspensions, and exhaust emission reduction devices, to testing, maintenance, and production operations.

The work of the automotive engineer can range from the greasy and noisy environment of an engine shop to the quiet and cleanliness of an office. Engineers must be able to work as part of a team. They should have creative and analytical minds, and a capacity for detail. Engineers often use calculators and computers to solve mathematical problems. They must be able to consult with other engineers and communicate well, both orally and in writing.

Beginning automotive engineering jobs usually require a bachelor's degree in engineering, although some jobs are open to experienced technicians or people with degrees in mathematics or the physical sciences. High school students interested in engineering should take all the mathematics and physical science courses offered in their high school. Beginning automotive engineers usually perform routine tasks at first, then gradually work their way up to more responsible positions. Many engineers must continue their education throughout their careers.

There are many different types of engineers, from those who work to prolong human life to those who design chemical molecules to meet a specific purpose. In the future, engineers will be needed to develop sources of energy as well as to design energy saving systems for automobiles, homes, and other buildings. In addition, engineers will be needed to solve environmental problems. Today engineering is the second largest profession, behind teaching.

Opportunities for mechanical engineers are expected to increase about as fast as the average for all occupations through the year 2000. Because of foreign competition and uncertain economic conditions, however, growth in job opportunities for automotive engineers is less certain.

UNIT 4

Wave Motion, Sound, and Light

(Left) Small wavelets, or ripples, distort the image of the boat in the water. (Above) The behavior of ripples in water illustrates the behavior of transverse waves.

In this unit we study wave motion as a means of transmitting energy in sound and light. We learn that sound is transmitted by longitudinal waves. Through the work of Christian Huygens (1629–1695) and Thomas Young (1773–1829), we are shown that the behavior of light on reflection, refraction, interference, and diffraction can also be explained by assuming that light is transmitted by means of waves. The further discovery that light can be polarized leads to the conclusion that light waves must be transverse. The wave theory reaches its climax in the work of James Clerk Maxwell (1831–1879) who identified light as a member of a large family of electromagnetic waves of different wavelengths that travel at the same speed in a vacuum or in air and that conform to the same laws.

The wave theory proves eminently successful in explaining the behavior of light in transit; that is, moving through space or through transparent substances. However, it is unable to account for the manner in which light is emitted and absorbed by matter. The work of Max Planck (1858–1947), Albert Einstein (1879–1955), and Neils Bohr (1885–1962) established that in its emission and absorption, light acts as if it consists of particles each of which is a discrete packet of energy called a photon. This leads finally to the quantum theory of light which provides an explanation of both its wavelike and particle-like properties.

Understanding of the behavior of light and the larger electromagnetic family of waves has vastly expanded our ability to observe and communicate with the world about us. Today, radio communications keep us in instant touch with every part of the earth. Optical and radio telescopes enable us to probe into the far reaches of space invisible to the naked eye. Optical microscopes and X-ray machines extend our vision into the interior of bodies and into the world of the very small. Finally, the camera enables us to make permanent records of events while they are happening and to store them for future study and reference. Thus the application of our understanding of the nature of light and electromagnetic waves has given us powerful tools for advancing the search for more knowledge.

15

Energy Transfer by Waves

Aims

1. To study the nature of wave motion as a means of transferring energy.
2. To study some of the phenomena characteristic of wave motion such as reflection, refraction, interference, and diffraction.
3. To understand the nature and transmission of sound as an example of wave motion.

Properties of Waves

15-1 Wave Motions

Probably the most familiar waves are those seen on the surface of bodies of water. Less obvious waves are those that transmit sounds through the air and through liquid and solid bodies. As we study the radiation of heat and light, we shall see that this process also has the characteristics of wave motion. An understanding of wave motion will therefore give us deeper insight into the nature of heat and light.

15-2 Waves in Water

Touch your finger several times to the surface of water in the center of a basin. Circular ripples will spread out in all directions from the place where the water was disturbed. If you float small pieces of cork on the water, you will see each piece of cork bob up and down once as each ripple passes under it. Since it takes work or energy to raise the pieces of cork, this means that each ripple is transferring some of the energy supplied by the finger in disturbing the water to each piece of cork. The ripples are a method of transferring energy supplied at the center of disturbance over the surface of the water.

If you repeatedly touch the water and remove your finger at a regular rate, evenly spaced concentric circular ripples will form. These successive ripples move outward from the center of disturbance, growing larger and larger as they do so. As each ripple passes under one of the floating pieces of cork, it causes that piece to bob up and down. With the passage of successive ripples, each piece of cork vibrates with a regular up and down motion. Ripples following each other at regular intervals are called *periodic waves*, or simply *waves*. A single ripple or disturbance is called a *pulse*.

294

15-3 Waves Transfer Energy

Exactly what is it that moves forward as successive waves pass over water? It is not the water. The water in any particular place alternately rises and falls, but it remains in approximately the same place. Only the energy that makes successive masses of water rise and fall moves forward with the waves. This energy is supplied by the agent that is disturbing the water and producing the waves.

Water waves illustrate characteristics common to all waves. *They are the means whereby energy is transmitted through a substance or medium when the medium is disturbed.* As periodic waves pass through the medium, they cause the particles of the medium to vibrate.

15-4 Types of Waves

There are two types of waves, *transverse* and *longitudinal. Transverse waves are those that cause the particles over which they pass to vibrate at right angles to the direction in which the waves are moving.* We shall see that light and heat appear to be transmitted by transverse waves. Water waves are not true transverse waves but they approximate this type of wave, as is shown when waves move over the surface of a lake. They cause small boats and other floating objects to bob up and down, or to vibrate at right angles to the direction in which the waves are moving.

Longitudinal waves cause the particles over which they pass to vibrate in a direction parallel to the direction in which the waves are moving. Sounds are transmitted through the air by longitudinal waves. As the sound of a bell moves northward, it can be shown that it causes the molecules of the air through which it passes to vibrate in a north-and-south direction.

To better understand the properties of waves, we shall first study the behavior of water waves as an approximation of transverse waves, and then the behavior of sound waves as an example of longitudinal waves.

Fig. 15-1. The surfer is using the energy of the wave to speed him to the shore.

15-5 Parts of Waves

A series of water waves consists of crests and troughs following each other in regular succession, as shown in Fig. 15-2. Each crest and its adjacent trough make up one wave. *The distance between two successive crests or two successive troughs is called one wavelength.* The dotted horizontal line shows the surface of the water before the waves disturbed it. As the crest of a wave passes under a floating piece of cork, the piece of cork is raised above the dotted line and then returned to it. As the adjacent trough passes under the piece of cork, it is lowered below the dotted line and then returned to it. In this manner, each wavelength causes the piece of cork to make one complete up-and-down vibration as it passes by.

The height of the crest of a wave above the original undisturbed

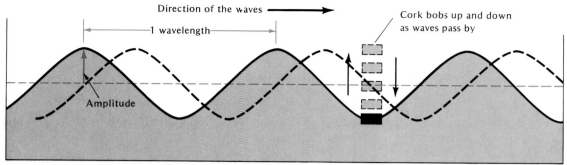

Fig. 15-2. The waves transmit energy to the cork by making it vibrate up and down.

surface represented by the dotted line is called the *amplitude* of that wave. Since the amplitude of a water wave determines how high it will raise the water over which it passes, the amplitude is a measure of the work the wave can do and therefore of its energy. The greater the amplitude of water waves, the more energy they transmit to the water over which they pass.

15-6 Wave Frequency

The number of waves per second that leave the point where they are being formed is called the *frequency* of the waves. This is also *the number of waves that pass any given point each second.* Each wave passing over a mass of water causes it to make one complete up-and-down movement or vibration. As a result, waves of a given frequency cause the particles of water over which they pass to vibrate at the same frequency.

The unit of frequency is the hertz (Hz). 1 Hz = 1 cycle/s

The frequencies of waves are expressed in *cycles per second*. A cycle is one complete wave or vibration. A cycle per second is known as a *hertz*, symbol Hz. A *kilohertz* is 1000 cycles per second, symbol, kHz. A *megahertz* is 1 000 000 cycles per second, symbol, MHz. The dimensions in which hertz are expressed are reciprocal seconds, s^{-1}.

15-7 Frequency, Wavelength, and Speed

Water waves move forward at a definite speed which depends upon the depth of the water. The speed of water waves can be measured by noting the time taken by the crest of a wave to travel a measured distance and then dividing the distance by the time.

If v is the speed of a series of waves, f their frequency, and λ (the Greek letter *lambda*) is their wavelength, it is easily seen that:

$$v = f\lambda$$

For example, consider a set of circular water waves of wavelength 10 centimeters that leave a given point with a frequency of 2 waves per second. At the end of the first second, exactly 2 complete waves, consisting of 2 troughs and 2 crests, have been formed. Since each wavelength is 10 centimeters long, the 2 waves will extend over a distance of 2 × 10, or 20 centimeters. This means that the front of the first wave traveled 20 centimeters from the source in 1 second or at a speed of 20 centimeters per second. It illustrates the fact that the speed of the waves is equal to their frequency of 2 waves per second times their wavelength of 10 centimeters, or 20 centimeters per second.

The relationship $v = f\lambda$ is true not only for water waves but for waves of all kinds.

Sample Problem

The speed of radio waves in air is 3×10^8 m/s. What is the wavelength of radio waves having a frequency of 5×10^5 Hz?

Solution:

$v = 3 \times 10^8$ m/s $f = 5 \times 10^5$ Hz

$v = f\lambda; \lambda = v/f$

$$\lambda = \frac{3 \times 10^8 \text{ m/s}}{5 \times 10^5 \text{ s}^{-1^*}}$$

$\lambda = 6 \times 10^2$ m

*(See Section 15-6).

Test Yourself Problems

1. What is the frequency of a sound whose speed in air is 330 m/s and whose wavelength is 1.65 m?

2. The frequency of certain light waves is 6.0×10^{14} Hz. Taking 3.0×10^8 m/s as the speed of light, find the wavelength of these waves.

(a) (b)

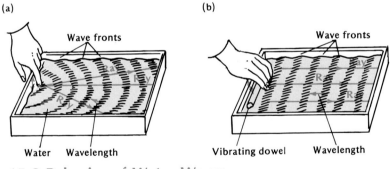

Water   Wavelength   Vibrating dowel   Wavelength

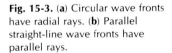

Fig. 15-3. (a) Circular wave fronts have radial rays. (b) Parallel straight-line wave fronts have parallel rays.

15-8 Behavior of Water Waves

Some of the properties of waves may be observed in an ordinary rectangular baking pan 25 centimeters wide by 40 centimeters long filled with water to a depth of about one centimeter. The pan is put about a meter below a strong concentrated source of light. By dipping the finger into the water surface at the middle of the pan and moving the finger up and down at a regular rate, a series of wave crests in the form of concentric circles is generated. The circular boundary of each wave crest is called a *wave front*. Lines perpendicular to the wave fronts are called *rays*. For a series of concentric circular wave fronts, the rays are simply the common radii of the circles. The ray at each point of a wave front shows the direction in which that particular point of the waves is moving. See Fig. 15-3 (a).

Fig. 15-4. A circular ripple tank showing the shadows of parallel wave fronts below the tank.

A particularly simple series of waves may be formed by laying a straight piece of wooden dowel about 20 centimeters long into the water and moving it up and down at a regular rate. The wave fronts are now nearly straight lines instead of arcs of circles and they are parallel to each other. The rays associated with these parallel straight-line wave fronts are perpendicular to the wave fronts and therefore parallel to each other. A bundle of such parallel rays is called a *parallel beam*. See Fig. 15-3 (b).

15-9 Ripple Tank

A special laboratory arrangement for studying the behavior of waves is the ripple tank, which consists essentially of a framed window mounted horizontally about 0.5 meter above the floor, filled with water to a depth of about one centimeter. At a point on one side of the tank, a small bead attached to an electrically driven mechanical arm dips into the water and is made to vibrate at a regular rate. This vibrating bead is called the wave generator because, as it vibrates, it generates a series of regularly spaced circular waves that spread out over the tank.

To make the waves visible, a strong point source of light is mounted about 0.5 meter above the tank and a sheet of white paper is spread on the floor below the tank. As the light passes through the tank and falls on the sheet of paper, it forms a series of bright concentric circles wherever it passes through wave crests. Thus, the movement of these crests is easily seen.

By making a straight dowel vibrate and serve as the wave generator instead of a bead, a series of straight waves is produced as shown in Fig. 15-4.

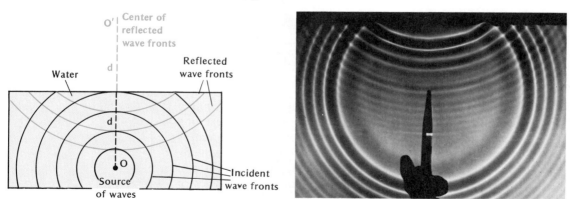

Fig. 15-5. The reflection of circular waves is shown in the drawing (left) and photo (right).

15-10 Reflection

When a series of water waves strikes an obstacle, the waves are turned back, or reflected. To observe this, hold the finger about 8 centimeters from the middle of one of the long sides of a rectangular pan of water. Form a series of circular waves by dipping the finger into the water. Each circular wave spreads out until it hits the wall of the container, as shown in Fig. 15-5. Here, it turns "inside out" as it is reflected and reverses its direction of motion.

After reflection, the wave fronts remain circular, continuing to increase in radius. By extending the radii of the reflected waves backward through the wall of the container, it is found that their center is a point as far behind the reflecting wall as the point at which the waves originated is in front of it.

15-11 Law of Reflection

When the reflecting wall is very large compared to the wavelengths of the water waves falling upon it, the waves obey a simple law of reflection. This law can readily be observed by means of a series of parallel straight-line wave fronts.

A series of parallel straight-line wave fronts is generated in a ripple tank and directed so that the waves fall obliquely upon a straight barrier. Note that the waves are reflected after they strike the barrier, moving in the direction shown in Fig. 15-6. The waves falling upon the barrier are called *incident waves*. Those that are reflected from the barrier are called *reflected waves*.

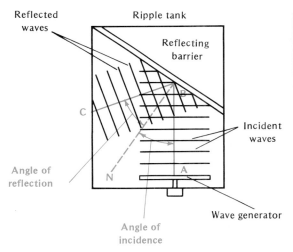

Fig. 15-6. The reflection of straight-line parallel wave fronts illustrates the law of reflection: the angle of reflection is equal to the angle of incidence.

Consider the path of a typical ray such as *AB*, which is perpendicular to the incident wave fronts and is called an *incident ray*. After reflection, *AB* turns in the direction *BC* perpendicular to the reflected wave fronts. *BC* is called a *reflected ray*. Let *NB* be the perpendicular to the reflecting wall at *B*. *NB* is called a *normal*. The angle between the incident ray *AB* and the normal is called the *angle of incidence*. The angle between the reflected ray *BC* and the normal is called the *angle of reflection*. It is seen that, on reflection, a ray obeys the following law: *the angle of reflection is equal to the angle of incidence.*

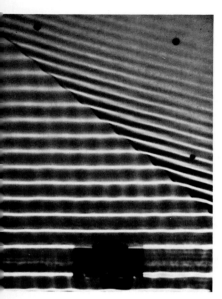

This law is general and holds true for all possible angles of incidence and for waves of all wavelengths.

15-12 Refraction of Water Waves

The speed of water waves depends upon the depth of the water and decreases as the water becomes less deep. When water waves pass from deep water to shallower water or vice versa, the change in their speed usually causes them to change direction sharply. The sudden changing of the direction of the waves is called *refraction.*

Refraction may be illustrated in a ripple tank by lining part of the bottom of the tank with a triangular piece of glass as shown in Fig. 15-7. The water in the tank is now less deep above the glass triangle than it is everywhere else.

When a series of parallel straight waves is now generated at the deep end of the tank, they advance, break, and change direction as they pass into the less deep water. The typical ray *AB* changes direction sharply on crossing the boundary between the deeper and less deep water and takes the new direction *BC*. The ray is said to be refracted as it crosses the boundary between the two water depths. *AB* is called the incident ray and *BC* is called the refracted ray.

Fig. 15-7. Parallel straight-line waves moving upward are refracted toward the right as they pass into shallower water in the upper half of the photo.

15-13 Superposition and Interference

Different waves can travel over a water surface at the same time. If two sets of circular waves are generated at two different points in the pan of water at the same time, each set appears simply to pass through the other. Actually, the effects of the two sets of waves on any small mass of water over which they pass at the same time combine by a process called *superposition.*

Consider a point on the water over which both sets of waves are passing. If a crest of the first set of waves arrives at the given point at the same time as a crest of the second set of waves, the waves will assist each other to produce an especially high crest. At that point, the amplitude of the new crest is equal to the sum of the amplitudes of the two superimposed crests.

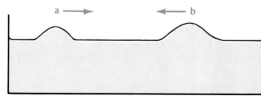

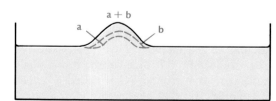

Fig. 15-8. Constructive interference: as two wave crests are superimposed, they form one crest equal to their sum.

If troughs of both sets of waves arrive at the given point at the same time, they will again assist each other to produce an especially large trough equal in amplitude to the sum of the amplitudes of the two superimposed troughs. In these two instances, the two waves are exactly "in step" at the given point and are said to be *in phase.* The net effect of such waves in phase is to assist each other. Such waves are said to *interfere constructively* with each other. (See Fig. 15-8).

If a crest of either set of waves arrives at the given point at the same time as a trough of the other set of waves, the waves will cancel out all or part of each other's effects. The net amplitude produced by the two waves will be equal to the difference between their individual amplitudes and in the direction of the larger one. If the amplitudes of the waves are equal, their net effect is to cancel each other out. Waves that are "in opposite step" with each other in this manner are said to be *in opposite phase* and are said to *interfere destructively* with each other. (See Fig. 15-9).

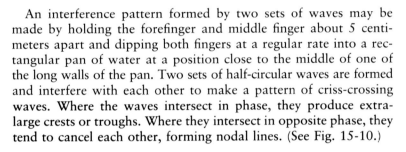

An interference pattern formed by two sets of waves may be made by holding the forefinger and middle finger about 5 centimeters apart and dipping both fingers at a regular rate into a rectangular pan of water at a position close to the middle of one of the long walls of the pan. Two sets of half-circular waves are formed and interfere with each other to make a pattern of criss-crossing waves. Where the waves intersect in phase, they produce extra-large crests or troughs. Where they intersect in opposite phase, they tend to cancel each other, forming nodal lines. (See Fig. 15-10.)

Fig. 15-9. Destructive interference: as the crest and trough of two waves are superimposed, they form a crest or trough equal to their difference.

15-14 Diffraction

Waves have the ability to travel around corners and obstacles in their paths. This phenomenon is called *diffraction.*

To illustrate diffraction in water waves, block off the middle of a ripple tank with a pair of straight barriers, leaving a small vertical opening or slit about 1 centimeter wide between them. Make a series of straight-line wave fronts at one end of the tank. As each wave reaches the barriers, all of it is reflected and turned back except the part that passes through the opening. That part spreads out on leaving the opening and assumes a nearly circular form, showing that the oncoming waves bend around or are diffracted by the edges of the slit as they pass through.

By changing the width of the opening through which the waves pass, it can be shown that *the degree of diffraction that the waves undergo depends upon the ratio of their wavelength to the width of the opening through which they pass.* When the wavelength is about the same size as the opening, or larger than the opening, the waves are sharply diffracted around the edges of the opening. See

Fig. 15-10. Interference pattern formed by water waves made by two point sources vibrating in phase. The points of destructive interference are clearly seen as nodal lines radiating out from the wave sources.

Fig. 15-11 (a). When the wavelength is small compared to the opening, there is very little diffraction around the edges of the opening. See Fig. 15-11 (b).

Diffraction also enables waves to pass *around* obstacles, provided their wavelength is about the same size as the obstacle or larger. In a harbor, you can see that the water waves whose wavelength is generally equal to or larger than the length of small boats pass around these boats quite readily. The same waves are not diffracted sufficiently to get around a large ship whose dimensions are much larger than their wavelength. The water near the side of the ship facing away from the waves is protected by the ship and remains undisturbed.

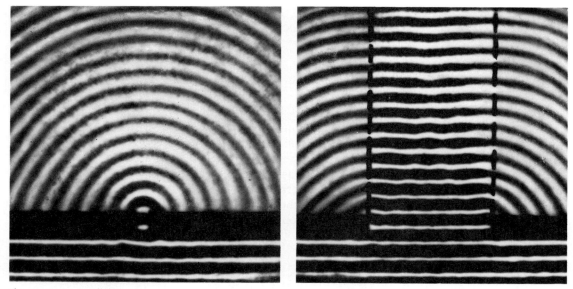

Fig. 15-11. (**a**) Parallel water waves generated at bottom are strongly diffracted as they pass through an opening about the same size as their wavelength. (**b**) Diffraction effects are small as waves pass through an opening much larger than their wavelength. Emerging wave fronts are only slightly curved.

15-15 Mechanism of Diffraction

To understand how water waves are diffracted, notice that the waves make each particle of water over which they pass vibrate up and down at the same frequency as that of the waves. Each particle can therefore be considered a tiny generator of new wavelets having the same frequency and wavelength as the waves that set it into vibration. Utilizing this idea, the Dutch physicist Christian Huygens suggested that each point of a moving wave front acts like a generator of new secondary wavelets. These are superimposed upon each other. By interfering constructively and destructively with each other, they form a single new wave front that turns out to be their common tangent.

In Fig. 15-12, several points on the straight wave front *AB* are shown generating secondary wavelets. The common tangent of the secondary wavelets is the new wave front *A'B'*, which is also a straight wave front. Now each point on $A'B'$ also becomes a vibrating source of secondary wavelets which combine to produce the new straight wave front $A''B''$. In this manner, a straight wave remains straight as it moves forward.

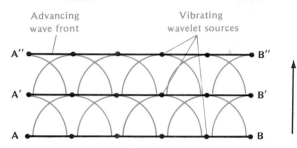

Advancing wave front

Vibrating wavelet sources

Fig. 15-12. The secondary wavelets combine to form the advancing wave front.

Fig. 15-11 (a) shows what happens when several straight-line waves pass through an opening about the same size as their wavelength. As the water inside the opening is set vibrating by the oncoming waves, each of its particles becomes a generator of secondary wavelets. Because the opening is about the same size as the wavelength of the oncoming waves, all the particles are very close together and the wavelets they generate very nearly coincide. Their net effect is therefore approximately the same as that which would be produced by a large single particle vibrating inside the opening. Such a single vibrating particle would produce the nearly circular wave fronts shown. The smaller the width of the opening, the more nearly circular will be the emerging wave fronts.

On the other hand, when the opening width is large compared to the wavelength of the incident waves, the wavelets produced by the water particles inside the opening will overlap over a sizable distance. As shown in Fig. 15-11 (b), they will then combine to produce new wave fronts that are nearly straight. Diffraction effects may then be too small to be noticed. It will be seen that this is often the case for light waves.

Nature of Sound

15-16 Sound Travels by Waves

You have no doubt noticed that a loud sound, such as the blare of a loudspeaker, sets certain objects in a room vibrating. The sound is moving across the room and transmitting energy to the bodies over which it passes. This behavior is characteristic of waves. It illustrates the fact that sound is transmitted through air and other substances by means of waves. As the sound waves generated by the loudspeaker pass through the room, they cause all particles of substances in their path to vibrate. The passage of sound waves over the ear causes the ear drum to vibrate and thus makes a person aware of the sound.

15-17 Sound Waves from Vibrating Bodies

Just as a body vibrating on the surface of water generates water waves, so a vibrating body in air generates sound waves in air. The wave fronts radiate out in all directions from the vibrating source. The sound waves produced by a violin are made by the vibrations of its strings. The sounds coming from a radio loudspeaker are produced by the vibrations of its diaphragm, while the sound of the human voice is made by the vibrations of the vocal cords.

The ear is not able to hear all the sounds made by bodies that vibrate. People with very good hearing can hear sounds made by bodies that make from 20 to 20 000 vibrations per second. In general, bodies that vibrate at high frequency produce high, screeching sounds. These are described as sounds of *high pitch*. Bodies that vibrate slowly produce low sounds, or sounds of *low pitch*.

15-18 How Sound Waves Form

To understand how a vibrating body generates sound waves in air, consider the vibrating string shown in Fig. 15-13. As the center of the string moves to the right, it compresses the air directly in front of it. This slight pressure is passed on to the air a little further on, which, in turn, passes it on to the air still further on, and so forth. As a result, the pressure exerted by the string as it moves to the right travels steadily outward and causes each tiny mass of air through which it passes to be compressed.

Fig. 15-13. A vibrating stretched string produces alternate compressions and expansions of air.

When the string completes its motion to the right, it begins to move back to the left. The pressure on the air immediately to the right of the string is now removed, giving that air room to expand. This air, in turn, removes the pressure on the air further on so that it also can expand. As this process continues, the relaxation of the pressure on the air is transmitted outward through successive masses of air, and they expand one after another.

Each time the string moves forward and back once, first a pressure or compression and then a relaxation of the pressure or an expansion travels outward from the string. *One compression and one expansion make up one complete wave.* The number of waves produced per second is equal to the *frequency* of vibration of the string. As a result, each mass of air through which the waves pass contracts and expands at the same regular frequency as that of the vibrating string.

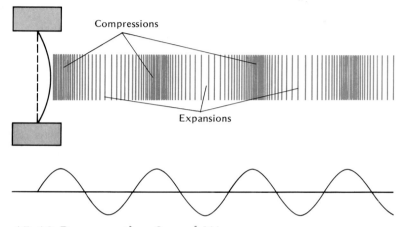

Fig. 15-14. Representation of sound waves by a sine curve.

15-19 Representing Sound Waves

A series of sound waves may be represented by a curve resembling the cross section of water waves and known mathematically as a *sine curve.* The crests represent contractions of the air while the troughs represent expansions. A *wavelength* is represented by the distance occupied by one crest and its adjacent trough. The number of crests and troughs produced per second is the *frequency* of the waves. The amplitude or height of the crests above the base line represents the *amplitude of vibration* of the particles of air over which the waves are passing. *The amplitude of sound waves determines the loudness of the sound;* the greater the amplitude the louder the sound.

15-20 Speed of Sound in Air

The speed of sound waves in air may be determined by measuring the time taken by a loud sound, like that from a cannon, to travel a known distance. When the cannon is fired into the air, observers will see its flash immediately because light travels at 3×10^8 meters per second and reaches them almost at once. However, the sound of the shot, which travels much more slowly than light, takes several seconds to reach them. The observers can measure the time that elapses between seeing the flash and hearing the shot. This is the time taken for the sound waves to travel the known distance between the cannon and the observers. Dividing this distance by the time gives the speed of sound.

It turns out that at normal atmospheric pressure and 0°C, the

All sounds travel at the same speed in air.

speed of sound in air is about 330 meters per second. The speed of sound in air increases with rising temperatures at about 0.6 meter per second per Celsius degree, and decreases at the same rate when the temperature falls.

In air, all sound waves travel at the same speed regardless of their frequency. This is evident in the fact that the sounds of different frequencies coming from all the instruments in a band reach distant listeners in the audience at the same time.

15-21 Reflection of Sound Waves and Echoes

When sound waves fall upon a hard surface, they are reflected. The sounds made in a room are reflected many times as they rebound from the walls, ceiling, and floor. This multiple reflection is often noticed in empty rooms and is called *reverberation*.

Echoes and reverberation are produced by the reflection of sound waves.

An *echo* is another of the effects of *reflection of sound waves.* Ordinarily we do not hear the reflected sounds in a room as echoes. This is explained by the fact that the sensation of any sound remains in our consciousness for about one-tenth of a second after the sound is gone. In the usual small room, a reflected sound returns less than one-tenth of a second after the original sound was heard. The sensation of the original sound is still with us and blends with the newly received reflected sound. When a reflected sound returns more than one-tenth of a second after the original sound was heard, the sensation of the original sound is over. The reflected sound is therefore heard as a separate sound, or echo.

To produce an echo, the reflecting surface must be about 17 meters or more from the observer. A sound going to such a surface and back to the observer must travel at least $2 \times 17 = 34$ meters. This is the distance sound travels in air at normal room temperature in 0.1 second. The reflected sound is therefore received by the observer 0.1 second or more after the original sound was made and is heard as a distinct echo.

15-22 Absorption of Sound

In halls and auditoriums, it is desirable to eliminate echoes and excessive reverberation of sound waves. This is done by covering or replacing smooth, hard reflecting surfaces in the room with irregular surfaces and soft materials such as cork, curtains, draperies, and sound-absorbing wallboard. Soft materials do not vibrate freely and therefore absorb the energy of the sound waves that fall upon them. They pass the absorbed sound energy on to their molecules, causing them to vibrate more vigorously. The energy of the sound is thus converted into additional molecular energy or heat.

15-23 Speed of Sound in Other Media

Sound can travel through any elastic substance. In general, it moves through liquids and solids faster than it does through air. The speed

Fig. 15-15. This modern concert hall in Berlin employs specially designed reflecting panels that break up the sounds produced on stage while sound-absorbing materials in ceilings and walls inhibit echoes.

of sound in water, for example, is about four times that in air, while its speed in steel is about fifteen times as fast as in air.

Since sound waves consist of the vibrations of the matter through which they pass, they cannot exist in or travel through a vacuum. This can be shown by suspending a ringing electric bell under a bell jar and exhausting the air from the jar. As the air is removed, the sound of the bell becomes fainter and fainter. When a high vacuum has been attained, the bell can no longer be heard.

15-24 Length of Sound Waves

When the speed and frequency of sound waves are known, their wavelengths can be computed from the relationship noted in Sec. 15-7.

$$Velocity = Frequency \times Wavelength$$

For example, let us compute the wavelength of the sound at the lower end of the audible range whose frequency is 20 hertz. Assuming that the speed of sound is 340 meters per second (at 17°C) and substituting in $v = f\lambda$, we have 340 meters per second = 20 hertz $\times \lambda$, whence $\lambda = 340 \div 20 = 17$ meters.

Since the speed of sound in air is the same value for all frequencies, it follows that the greater the frequency of a sound, the shorter must be its wavelength. The frequencies of the highest audible sounds are in the neighborhood of 20 000 hertz or 1000 times as great as the frequency of 20 hertz whose wavelength we have computed. Their wavelengths are therefore about one-thousandth of those of the lowest audible frequencies or about 17 millimeters.

The higher the frequency of a sound, the shorter its wavelength.

15-25 Diffraction of Sound Waves

Most sounds travel freely around corners and obstacles. That is why a car horn is readily heard around a street corner although the car itself cannot be seen. This is an example of the diffraction of sound waves.

As noted for water waves, the degree to which waves are diffracted depends upon the ratio of their wavelength to the size of the opening through which they pass or the obstacle around which they pass. When the length of the sound waves is larger than or about the same size as the opening or the obstacle, the waves are strongly diffracted and bend sharply around its edges. When the length of the waves is very much smaller than the opening or obstacle, they undergo very little bending and pass its edges in very nearly straight paths. This explains why low-pitched sounds like the car horn which have long wavelengths are heard much better around the edges of openings and obstacles than are high-pitched sounds which have very short wavelengths.

Sounds of longer wavelength are diffracted more than those of shorter wavelength.

15-26 Sympathetic Vibrations or Resonance

The sound waves produced by any vibrating body set up vibrations in the objects over which they pass. In most objects, these induced vibrations are small. However, if an object happens to have a natural frequency of vibration equal to that of the body producing the sound waves, that object will be set into vigorous vibration. It is then said to be in *sympathetic vibration* or in *resonance* with the original vibrating body.

Fig. 15-16. Tuning forks demonstrating sympathetic vibrations.

Sympathetic vibrations can take place only between two bodies that have the same natural frequency of vibration. To demonstrate this in the laboratory, two tuning forks having the same frequency are mounted on resonator boxes designed to increase the loudness of their sounds. The forks are placed several feet apart and one of them is struck and allowed to vibrate for several seconds. The vibrations of its prongs are then stopped with the hand and the second fork can be heard vibrating sympathetically.

15-27 Explanation of Sympathetic Vibrations

When pushing a child in a swing, we time our pushes to match the natural frequency of vibration of the swing. Our series of light pushes is able to cause the swing to vibrate through a wide arc.

A similar effect takes place in sympathetic vibration. When the first body is vibrating, it sends out sound waves that cause the second body to vibrate by exerting a series of tiny pressures on it. If these pressures happen to be timed properly, they build up a large vibration in the second body, just as happens in the case of the swing. This condition is fulfilled only when the two bodies have the same natural frequency of vibration.

Fig. 15-17. This collapse of the Tacoma Narrows Bridge in Washington was caused by the disastrous resonance effect of successive gusts of wind timed to the natural vibration frequency of the bridge.

15-28 Interference of Sound Waves

Many sounds can pass through the same air and be heard at the same time. Just as for superimposed water waves, the amplitudes of such superimposed sound waves combine to interfere constructively and destructively.

Interference of sound waves may be demonstrated using any two vibrating bodies, such as two tuning forks, that have slightly different frequencies of vibration. When both tuning forks are set vibrating at the same time, their sound waves interfere with each other to produce a throbbing effect called *beats*. Between beats, the two sound waves virtually cancel each other to produce near silences.

15-29 Explanation of Beats

To understand the production of beats, consider what happens to the mass of air that is next to the ear when sound waves from the two tuning forks pass through it. When these sound waves arrive at this mass of air in phase, both waves will either make the air expand or make it contract at the same time. As a result, the amplitudes of the two waves will add up so that this mass of air undergoes an extra-large expansion or contraction. The air then causes the eardrum to vibrate vigorously and the ear hears this as an extra-loud sound or beat.

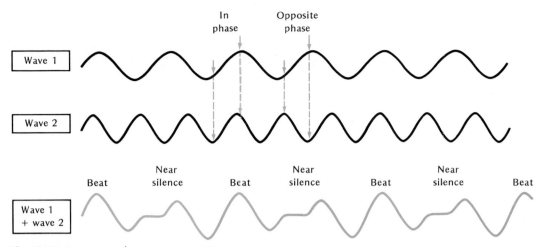

Fig. 15-18. As two sound waves of slightly different frequency alternately come into phase and into opposite phase, they produce audible beats.

The superposition of two sound waves of slightly different frequencies is heard as beats.

When the two sound waves arrive at the mass of air in opposite phase, one set of waves tends to make the air expand while the other tends to make the same air contract. In this case, the amplitudes of the two waves act opposite to each other. If the amplitudes are nearly equal, the waves cancel each other's effects. The air then momentarily stops vibrating and the ear becomes aware of a temporary silence.

Since the two sets of sounds have slightly different frequencies, they also have slightly different wavelengths. As these waves pass through a given mass of air, they will be alternately in phase and in opposite phase at a regular rate. When in phase, the two waves produce a beat. When in opposite phase, they produce a near silence. *The number of beats produced per second is equal to the difference between the frequencies of the two sets of waves.* For example, a tone of 200 hertz and one of 205 hertz, when played together, produce two sets of waves that are alternately in phase and in opposite phase at any point through which they pass 5 times a second. As a result, we hear 5 beats per second.

The top two curves of Fig. 15-18 show graphically how two sets of waves of slightly different wavelength (or frequency) get into phase and into opposite phase at a regular rate as they move along. The third curve is obtained by superimposing the first two and combining their amplitudes. You can see that, when the two sets of waves are in phase, they produce an extra-large amplitude which is heard as a beat. When they are in opposite phase, they nearly cancel each other, producing a momentary near silence.

15-30 Doppler Effect

When a rapidly moving car is coming toward you, the pitch of its horn sounds higher to you than it does when the car is at rest. When the car passes by and speeds away from you, the pitch of its horn becomes markedly lower. Similar changes in pitch take place when the car stands still and you move rapidly toward or away from it. Since the pitch of the horn depends upon the frequency of the sound waves which you receive, these observations show that both your motion and that of the car are able to change the frequency of the waves you receive. This change of frequency is called the *Doppler effect* after the man who first explained it.

The Doppler effect is the change in the frequency of the waves received by an observer whenever the wave source and the observer are in relative motion toward or away from each other. It applies to all kinds of waves—water waves and light waves, as well as sound waves. Whenever the source of the waves and the observer are in relative motion toward each other, the frequency of the waves received by the observer is higher than the frequency of the source. Whenever the source of the waves and the observer are in relative motion away from each other, the frequency of the waves received by the observer is lower than that of the source. The greater the relative speed of the source of the waves with respect to the observer, the greater is the Doppler shift, or change of frequency, observed.

The Doppler shift in the frequencies of the light emitted by a source moving with respect to the observer is noticed as a change in the color of the light. This fact is very useful to astronomers. By measuring the Doppler shift in the light emitted by a moving star or other heavenly body, astronomers can determine not only whether the body is moving toward or away from the earth but also how fast it is moving. Astrophysicists interpret the Doppler effect, or "red shift," in radiation arriving from distant galaxies and "quasars" as indicating that they are moving outward on the fringes of the observable universe at close to the speed of light.

The Doppler shift in the frequencies of light coming from all the stars and heavenly bodies indicates that they are all moving away from each other and that the universe is expanding.

15-31 Doppler Shift for a Moving Source

Let us see how the Doppler shift takes place when a source of light or sound is moving toward or away from a stationary observer. Consider Fig. 15-19 (a) in which a source of sound vibrating at frequency f is moving to the right at speed v_s. The widening circles represent the successive compressions of the sound waves leaving the source.

The motion of the source causes the sound waves to be crowded together in front of the source and to be spread further apart behind the source. The effect of this is to decrease the wavelength of the waves in front of the source and to increase the wavelength of the waves behind the source. Since all the waves travel at the same speed, an observer in front of the moving source will receive more than f waves per second while an observer behind the source will receive fewer than f waves per second.

Consider the case of an observer who is in front of the source. If the speed of sound in air is v, all the waves in front of the

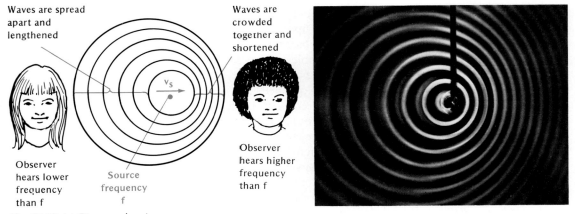

Waves are spread apart and lengthened

Waves are crowded together and shortened

v_s

Observer hears lower frequency than f

Source frequency f

Observer hears higher frequency than f

Fig. 15-19. (a) Diagram showing the Doppler effect for a moving source of sound. (b) The same effect produced by a vibrating source moving to the right through water.

observer that are within a distance equal to v will arrive in one second. This number is the frequency f' of the waves and is equal to v/λ', where λ' is the wavelength of the shortened waves in front of the source.

Now, during any second, the source moves forward a distance v_s and emits f waves. At the end of the second, the first of the f waves emitted has traveled a distance v from the starting point of the source. The last of the f waves emitted is just leaving the source that is now a distance v_s from its starting point. The f waves generated during that second are therefore compressed into a length equal to $v - v_s$, so that λ' is equal to $(v - v_s)/f$. Substituting this value for λ' in $f' = v/\lambda'$ gives:

$$f' = f\left(\frac{v}{v - v_s}\right)$$

In a similar way it can be seen that, for an observer stationed behind the moving source, the observed frequency is:

$$f' = f\left(\frac{v}{v + v_s}\right)$$

Sample Problem

A train moving at 20 m/s sounds a horn of frequency 200 Hz. Assuming that the speed of sound is 340 m/s, what is the frequency of the sound heard by (a) a person toward whom the train is moving, and (b) someone from whom the train is moving away?

Solution:
(a) $f = 200$ Hz $v = 340$ m/s.
    $v_s = 20$ m/s

$$f' = f\left(\frac{v}{v - v_s}\right)$$

$$f' = 200 \text{ Hz}\left(\frac{340 \text{ m/s}}{340 \text{ m/s} - 20 \text{ m/s}}\right)$$

$$f' = 212 \text{ Hz}$$

(b) $$f' = f\left(\frac{v}{v + v_s}\right)$$

$$f' = 200 \text{ Hz}\left(\frac{340 \text{ m/s}}{340 \text{ m/s} + 20 \text{ m/s}}\right)$$

$$f' = 189 \text{ Hz}$$

Test Yourself Problem

A train engineer blows a whistle of 400-Hz frequency as the train approaches a station. If the speed of the train is 25 m/s, what will be the frequency of the sound of the whistle heard by an observer in the station? Assume the speed of sound is 340 m/s.

15-32 General Doppler Formula

If both the source of sound and the observer are moving, both motions contribute to the Doppler effect. For an observer and a source of sound moving directly *toward* each other, the observed frequency is:

$$f' = f\left(\frac{v + v_o}{v - v_s}\right)$$

where v_o is the speed of the observer. When the observer and the source of sound are moving directly *away* from each other, the observed frequency is:

$$f' = f\left(\frac{v - v_o}{v + v_s}\right)$$

CHAPTER REVIEW

Summary

Waves are means whereby energy is transferred through a medium. As **periodic waves** pass through a medium, they cause the particles of the medium to vibrate at a regular rate. In **transverse waves,** these vibrations are at right angles to the direction in which the waves are moving. Light and other electromagnetic waves are examples of transverse waves. In **longitudinal waves,** these vibrations are parallel to the direction in which the waves are moving. Sound is transmitted by longitudinal waves.

Waves are described by their *amplitudes, wavelengths, frequencies, and speeds.* In water waves, the **amplitude** is the height of the crest of the wave above the original undisturbed surface. The **wavelength** is the distance between two successive crests or troughs. The **frequency** is the number of waves that leave the point of origin of the waves each second.

For any set of periodic waves, the speed v is equal to the product of the frequency f and the wavelength λ; that is, $v = f\lambda.$

Water and other waves are **reflected** upon encountering a wall or obstacle in such a manner that *the angle of reflection is equal to the angle of incidence.* Water and other waves often change direction when, in passing from one medium to another, their speed changes. The changing of direction of waves resulting from a change in their speed is called **refraction.**

Two or more sets of water and other waves may pass through the same space at the same time. Their effects are then combined by a process called **superposition.** If the waves arrive in phase at a given place, they assist each other to produce an especially high crest or an especially low trough at that place. They are then said to **interfere constructively.** If the waves arrive in opposite phase at a given place, they cancel out all or part of each other's effects. They are then said to **interfere destructively.**

Water and other waves are able to bend around the edges of obstacles and narrow openings by a process called **diffraction.** The degree of diffraction undergone by waves depends upon the ratio of their wavelength to the width of the opening through which they pass. The larger this ratio, the sharper is the diffraction around the edges of the opening.

Sound waves are generated by vibrating bodies. They are longitudinal waves that cause each mass of air or other medium through which they pass to vibrate at a regular rate equal to the frequency of the body producing the waves. In a mass of air, this vibration consists of alternate expansions and contractions. The **frequency** of sound waves determines the **pitch** of the sound produced; the higher the frequency the higher the pitch. The **amplitude** of sound waves determines the **loudness,** or **volume,** of the sound produced; the greater the amplitude the louder the sound.

Sound waves do not pass through a vacuum but travel readily through gases, liquids, and elastic solids. Sound waves travel at 330 meters per second in air at 0°C and increase in speed as the temperature of the air rises. Like all other waves, they undergo **reflection, refraction, diffraction,** and **interference.** When sound waves pass over a body having the same natural frequency of vibration as the frequency of the sound waves, they cause it to vibrate especially vigorously. This effect is known as **resonance,** or **sympathetic vibration.**

When any source of sound, light, or other wave phenomenon is in relative motion toward or away from an observer, the motion causes the frequency of the waves received by the observer to be higher or lower than the frequency of the source. This shift in frequency is called the **Doppler effect.**

Questions

Group 1

1. (a) What evidence is there that energy is being transferred as waves pass over the surface of a lake? (b) What form does this energy take after the water has settled and become calm again?

2. (a) Distinguish between transverse waves and longitudinal waves. (b) Referring to water waves, define wavelength, amplitude, and frequency.

3. Waves are made in a pan of water by dipping a finger at a regular rate into the middle of the surface of the water. (a) Make a drawing showing three successive wave fronts. (b) Indicate the wavelength of this series of waves. (c) Draw four rays.

4. (a) What is the law of reflection? (b) Explain how to use a ripple tank or some other arrangement to illustrate the law of reflection.

5. (a) What is meant by refraction? (b) What causes refraction in water waves?

6. (a) What is meant by diffraction? (b) How may diffraction of water waves be illustrated?

7. (a) How did Huygens explain the mechanism of diffraction? (b) What effect does the ratio of the wavelength to the width of the opening have on the degree of diffraction undergone by a series of waves on passing through the opening?

8. (a) How are sound waves generated? (b) What determines the pitch of a sound? (c) In what direction do sound waves traveling north cause the air particles over which they pass to vibrate?

9. (a) Explain how a vibrating string with a frequency of 400 Hz generates sound waves in air. (b) How many expansions and compressions will a mass of air undergo per second as these waves pass through it?

10. (a) How can it be demonstrated that sound does not pass through a vacuum? (b) Describe how to measure the speed of sound in air.

11. (a) How does the speed of a sound vary with the temperature of the air? (b) Compare the speed in air of two sounds, one of frequency 200 Hz and the other of frequency 400 Hz (c) Compare the speed of sound in air with its speed in water and in steel.

12. (a) What is the cause of echoes and reverberations? (b) What can be done in halls to reduce undesirable reverberations and echoes?

13. (a) What evidence is there in daily experience that sound waves are strongly diffracted? (b) Why is a low-pitched sound more likely to pass around an obstacle than a high-pitched sound?

14. Describe an experiment illustrating sympathetic vibration.

15. (a) Describe an experiment illustrating interference between two sets of sound waves. (b) Explain how the two sets of waves produce a beat. (c) How do they produce a silence or near silence?

16. A train sounding a horn of constant frequency slows down as it nears a station. What change in pitch is heard by a commuter standing in the station?

17. How may the Doppler effect be used to measure the speed of a source of sound of known frequency that is moving toward or away from the observer?

Group 2

18. Two sets of water waves are being generated at two different points in a tank of water. The waves have equal amplitudes, equal wavelengths, and equal velocities. A piece of cork is deposited at a point on the surface where the waves from both sources always arrive in opposite phase. What will be the combined effect of both waves on the piece of cork? Explain your answer.

19. (a) Referring to question 18, the piece of cork is now deposited at a point on the surface where the waves from both sources always arrive exactly in phase. What will be the combined effect of both waves on the piece of cork? Explain your answer. (b) How will the motion of the cork in (a) differ from the motion it has when only one of the wave generators is in operation instead of both?

20. Suppose the two sets of water waves in question 18 have equal wavelengths and velocities but unequal amplitudes. What will then be the answers to questions 18, 19 (a) and 19 (b)?

21. At a factory, a whistle of fixed frequency is blown at constant loudness. Compare the sound waves of the whistle as heard by an employee 100 m from the factory with those heard by an employee 1 km from the factory with respect to (a) frequency and (b) loudness. (c) When workers in a car ride toward the factory at high speed, what change takes place in the number of waves reaching them per second? (d) What effect does their motion have on the pitch of the sound they hear?

22. A piano tuner is trying to adjust the tension of a string so that it has the same frequency as a certain standard tuning fork. What tests can be made to determine when the string and the tuning fork agree exactly in frequency?

Problems

Group 1

1. What is the wavelength of radio waves whose frequency is 6.0×10^6 Hz? The speed of radio waves is 3.0×10^8 m/s.

2. What is the frequency of light whose wavelength is 5.0×10^{-7} m?

3. At 0°C, sound travels in air at 330 m/s. What is the frequency of a sound whose wavelength is 2.00 m?

4. A soldier hears the sound of the firing of a distant cannon 6.00 seconds after seeing the flash. If the temperature is 20°C, how far is the soldier from the cannon?

5. A girl hears her echo return from a wall 10.0 s after she makes a sound. Assuming the speed of sound is 330 m/s, how far is the girl from the wall?

6. The frequency of a certain sound is 440 Hz. What is the wavelength of this sound when the temperature of the air is (a) 20°C; (b) 30°C?

7. Two piano strings are played together producing 3 beats per second. The pitch of the first string is then raised by tightening it until its frequency is 280 Hz. When the strings are now played together they produce 4 beats per second. What is the frequency of vibration of the second string?

8. A source of sound having a frequency of 300 Hz is moving at 25 m/s. If the speed of sound in air is 340 m/s, what is the frequency of the sound heard by (a) someone in front of the moving source; (b) someone behind the moving source?

Group 2

9. A bullet having a velocity of 680 m/s hits a target 510 m away. If the speed of sound is 340 m/s, how soon after the bullet hits the target does the sound of the shot reach the target?

10. A miner drops a stone into a mine shaft 122.5 m deep. If the temperature is 20.0°C, how soon afterward does he hear the stone hit the bottom of the shaft?

11. The frequency of a vibrating string having a fixed tension is found to vary experimentally with the length of the string as shown in Table 15.1. Make a graph of these results. What numerical relationship can you infer between the length of this string and its frequency of vibration?

12. A sounding body vibrating at 400 Hz is moving away from someone who hears a pitch corresponding to 380 Hz. If the speed of sound in air is 330 m/s, what is the speed of the sounding body?

13. A source of sound and an observer are moving directly toward each other. The source has a frequency of 500 Hz and a speed of 30 m/s. The speed of the observer is 20 m/s. If the speed of sound is 330 m/s, what is the frequency of the sound heard by the observer?

14. If the directions of both source and observer in problem 13 are reversed, while the speeds remain the same, what frequency would be heard by the observer?

Table 15.1

LENGTH (cm)	FREQUENCY (Hz)
100	152
90	170
80	190
70	215
60	250
50	300

Applying Physics

1. Here is an experiment that illustrates the principle of sympathetic vibration, or resonance. Stretch a horizontal string about 60 cm long tightly between vertical supports. Suspend 2 pendulums having equal lengths and bobs of equal mass from it at a distance of about 30 cm from each other. Set one of the pendulums into vibration. Notice what happens to both pendulums over a period of time. By what means does each pendulum transmit some of its energy to the other? How long does it take each pendulum to transfer its energy to the other? Is this a periodic process?

   Repeat the experiment with pendulums of equal length one of which has a bob of much larger mass than the other. Explain any differences between the behavior of these pendulums and those used above.

   Repeat the experiment with pendulums of different length and having bobs of equal masses. Explain your observations.

2. If a strong concentrated light source is located about 2 m above a bathtub, the tub, filled with about 20 cm of water, makes an effective ripple tank. Disturbing the water with the end of a stick produces circular waves visible as bright bands on the bottom of the tub. By vibrating the end of the stick vertically at a regular rate, periodic waves are produced. Reflections may be observed when the waves hit the flat side of the tub. Periodic straight waves may be obtained by laying a straight dowel, about 30 centimeters long, on the water and moving it up and down at a regular rate. The superposition of waves may be observed by making circular waves in two different parts of the tub and seeing what happens where they meet. See if you can estimate the speed at which the waves travel by timing their motion over a fixed distance.

3. Beats can be demonstrated on the home piano. Simply strike two adjacent keys at the left end of the keyboard at the same time. The sound waves of the two tones will alternately reinforce and destroy each other and you will hear beats. Repeat the experiment with pairs of adjacent keys in different parts of the keyboard. Note where on the keyboard beats are heard best. Can you explain why?

The Nature of Light

16

Aims

1. To identify light as a form of energy.
2. To study the properties and behavior of light.
3. To investigate and evaluate the different theories that have been advanced to explain the properties and behavior of light.

16-1 Light and Understanding

Most of our knowledge of the world about us comes through the light reaching our eyes. For this reason, we associate light with understanding. For example, we say we *see the light,* or *are enlightened,* when we mean we understand something. We also speak of *being in the dark* when we mean the opposite. This unique attitude toward light gives us a particular interest in its nature and properties.

16-2 Light as a Form of Energy

That light is a form of energy is confirmed by the fact that it can do work on matter. Delicate experiments show that light exerts a tiny pressure capable of moving matter. In the photoelectric cell or "electric eye," light does the work needed to eject electrons from the light-sensitive surface on which it falls. In this process, light is converted into electrical energy. Light can also be converted into other forms of energy. The sunlight brought to a focus on a piece of paper with a magnifying glass is converted into enough heat to ignite the paper. In the green plant, light energy is converted into the chemical energy needed by the plant for growth.

16-3 Light and Vision

Without light there can be no vision. We see a body only when light coming from that body enters the eye.

A visible body may be either *luminous* or *illuminated.* A luminous body produces and emits its own light. Examples include the sun, stars, and various artificial electric and other light sources used in the home and industry.

An illuminated body is one that can be seen only when light from a light source falls upon it. Some of this light is turned back or reflected by the surface of the body. When this reflected light enters the eye of an observer, the body from which it came can be

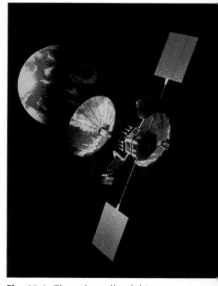

Fig. 16-1. The solar cells of this communications satellite use the sun's energy to power the electronic components of the satellite which relays telephone and television signals across oceans and continents of the earth.

317

seen. Most of the things in everyday experience are illuminated and are seen by the light they reflect. Typical illuminated bodies are the desks, chairs, and chalkboards of a classroom. The moon and the planets are also illuminated objects made visible by the sunlight that they reflect.

16-4 Transmission, Absorption, and Reflection

When light from an object travels toward the eye, the light must usually pass through the air or other matter between the object and the eye. If the intervening matter happens to be like air or a glass window, a small part of the light is absorbed, but most of it passes through without undergoing much distortion. On receiving this light, the eye sees clearly the object from which it came. Materials such as glass, air, cellophane, and water are said to be *transparent*. They let light pass through them in such a way that we can see through them.

Some substances let light pass through them but distort the light so that objects cannot be seen through them. Such substances are said to be *translucent*. Examples of translucent substances are ground glass, thin sheets of paper, and the frosted glass of an electric light bulb.

Certain objects do not let light pass through them at all and are said to be *opaque*. Light falling on an opaque body is usually partly reflected and partly absorbed. Examples of opaque bodies are a sheet of metal and the walls of a room.

16-5 Speed of Light

The fact that the light from the sun passes through 150 billion meters (150 Gm) of empty space on its way to the earth shows that light can travel through a vacuum. The speed with which light travels in a vacuum is approximately 300 million meters per second. It was first determined in 1676 by the Danish astronomer Ole Roemer, during his study of the motions of the moons of the planet Jupiter. Roemer discovered that the light coming from these satellites takes about 1000 seconds to travel the 300-billion-meter distance across the earth's orbit. Dividing the distance of 3×10^{11} meters by the travel time of 1×10^3 seconds gives 3×10^8 meters per second as the speed of light in a vacuum. This is equivalent to 300 000 kilometers per second.

The speed of light is a fundamental universal constant.

The speed of light in a vacuum plays a most important part in the theories of matter and motion because it appears to be the greatest speed any form of matter or energy can have. Light travels at only a little less than this maximum speed in air, but travels at considerably slower speeds in other transparent substances. The speed of light in water, for example, is about three-fourths as fast as its speed in a vacuum. In a diamond, light travels less than half as fast as it does through a vacuum.

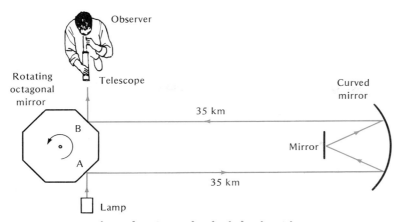

Fig. 16-2. Michelson's method of measuring the speed of light.

16-6 Measuring the Speed of Light in Air

A very precise method for measuring the speed of light in air was developed in 1926 by Albert Michelson, an American physicist. His apparatus was similar in principle to that shown in Fig. 16-2.

A beam of light is reflected from one of the eight faces of a rotating octagonal mirror at position A and made to travel toward a combination of a curved mirror and a plane mirror located 35 kilometers away. Here, the beam is made to turn back by reflection from the two mirrors and returns to the face of the rotating octagonal mirror at position B. If the rotating octagonal mirror happens to be in the exact position shown in the figure at the instant that the beam returns to it, the beam will be reflected by the mirror face at B into the observer's telescope. The observer will then see its light. If the rotating octagonal mirror is in any other position at that instant, the returning beam will be reflected by the face at B in some other direction. It will therefore not enter the telescope and will not be seen by the observer.

The rotating octagonal mirror passes through a position exactly like the one shown in the figure every time it makes one-eighth of a turn. The observer will therefore see the beam when the mirror rotates at such a rate that it makes exactly one-eighth of a turn in the time that it takes the beam to travel the 70-kilometer distance from the rotating mirror to the distant curved mirror and back again. In practice, the observer measures the rate at which the mirror must turn to see the beam in the telescope. From this rate, the time taken by the rotating mirror to make one-eighth of a turn may be computed. This is the time taken by the light to make its 70-kilometer round trip. When the 70-kilometer distance is divided by this measured time, the velocity of the light beam in air is found.

16-7 Colors of Light

Light has many different colors. Samples of colored light may be obtained by passing a beam of white light through a colored transparent material such as a piece of colored glass. When the white

Albert Michelson was the first American physicist to win the Nobel prize. He won it in 1907.

The color of a transparent substance is the color of the light it transmits.

light of a searchlight beam or of the sun is passed into a plate of red glass, the emerging light is red. When the same beam of white light is passed through a plate of green glass, the emerging light is green. Again, when the beam is passed through a plate of blue glass, the emerging light is blue. In general, the color of light that passes through a piece of colored glass is always the color of that glass.

These observations suggest two conclusions. The first is that *white light is a mixture of all colors of light.* The second is that a *piece of colored glass transmits principally its own color of light and absorbs all other colors.* For example, when white light passes through a red plate of glass, the glass transmits mainly the red part of the white light and absorbs most of the rest of the light. Similarly, a green plate of glass transmits mainly green light and absorbs all other colors. From these observations, it can be predicted that if red light is passed into green glass, it will be absorbed so that practically no light gets through. Experiment shows that this is the case.

Summarizing, we may conclude that the color of glass and other transparent substances is the color of the light that they transmit. The ability of a transparent substance to transmit one color of light and absorb others is called *selective transmission.*

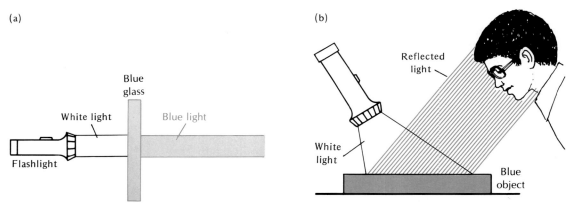

(a)   (b)

Fig. 16-3. Diagrams illustrating (**a**) selective transmission and (**b**) selective reflection of light.

16-8 Color of Opaque Objects

The color of an opaque object is the color of the light it reflects to the eye. When white light shines upon a red piece of cloth, the cloth reflects the red part of the white light and absorbs practically all the other colors of light. The red cloth therefore owes its color to the fact that it reflects red light but absorbs all other colors of light.

The color of an opaque object is the color of the light it reflects.

The ability of an opaque body to reflect one color of light and absorb others is called *selective reflection.* In general, the colors of opaque bodies are the results of selective reflection. A blue book appears blue when illuminated by white light because it reflects the blue part of the white light and absorbs the other colors of light. A blue book illuminated by blue light will reflect that light and therefore again appear blue. However, when it is illuminated by some other color of light, the blue book absorbs nearly all the light falling upon it and appears to be black. For example, when red light is shone upon the blue book, the book absorbs the red light and no light is reflected to the eye. *The eye judges a body from which it receives no light to be black.*

A white object is one whose surface reflects all colors of light that fall upon it. Such an object will appear to be white in white light, green in green light, red in red light, and so on. In each case, it reflects the color of the light falling upon it and therefore appears to be that color.

16-9 Light Appears to Travel in Straight Lines

The path of a beam of sunlight passing through an opening in the clouds or of a powerful beam of light coming from a motion picture projector in a darkened theatre is made visible by the numberless dust particles suspended in the air. As these particles reflect the light of the beam to the eye, they show its path through the air. It is seen that the edges of the beam are straight lines. Such a beam may be imagined to be composed of an infinite number of straight lines of light bundled together. Imaginary straight lines of light are called *rays*.

Fig. 16-4. Searchlight beams travel in straight lines.

Since a single ray has no thickness, it cannot be produced in the laboratory. However, the behavior of a ray can be approximated and studied by observing the behavior of a very narrow parallel-sided beam of light. The paths of such narrow beams of light in air, water, and other transparent substances show that, to a high degree of approximation, light travels in straight lines. That is why we usually see objects only if they are in our line of sight and do not see around the corners of opaque objects.

The usual straightness of its path is one of the most important characteristics of light and explains many common observations of its behavior. Three examples selected for discussion here are the formation of shadows, the formation of images in the pinhole camera, and the manner in which light spreads out from a point source.

16-10 Shadows from a Point Source of Light

When an opaque object cuts off the light rays coming from a source of light, the space behind the object is in darkness and is called a *shadow*. In the simplest case, the light source is so small or so distant from the opaque object that all its light rays may be thought of as diverging from a single point. Such a source is called a point source of light. If the object is not too small, the shadow that is formed by a point source of light lies between the rays that just pass over the extreme edges of the object, as shown in Fig. 16-5.

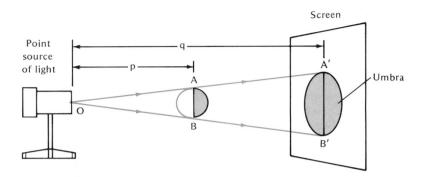

Fig. 16-5. A point source of light produces a sharp shadow.

On a wall or screen, the shadow is seen as a totally dark area called an *umbra.* The umbra has a rather sharp boundary that separates it from the fully lighted area around it.

From the similar triangles OAB and $OA'B'$, it follows that:

$$\frac{AB}{A'B'} = \frac{p}{q}$$

where AB is the size of the object, $A'B'$ is the size of the shadow, p is the object's distance from the light source, and q is the shadow's distance from the light source. For example, if q is three times as large as p, it can be predicted that the shadow $A'B'$ will be three times as large as the object AB. Actual measurement shows that this relationship is true if the object is not too small.

16-11 Shadows from Large Sources of Light

When the light source is larger, as in the case of an electric light bulb, the shadow formed generally consists of a dark central part called the *umbra* surrounded by a lighter, less distinct shadow called the *penumbra* that gradually blends with the illuminated area around it. If the source of light and the opaque object are both spheres, the boundaries of the shadow can be determined by drawing straight lines representing light rays from the top and the bottom points of the source to the top and bottom of the object, as shown in Fig. 16-7.

The umbra lies between the two inner rays. It is a region of total darkness because every light ray traveling in a straight line toward it from the source is intercepted by the opaque object before it can reach this area. The penumbra, occupying the remaining shaded area, is only in partial darkness because light rays can reach it from some parts of the light source but not from others.

Fig. 16-6. The sun casts a sharp shadow of the gate. Because the sun is so distant, it acts like a point source of light.

16-12 Eclipses

Eclipses are caused by the shadows of heavenly bodies. An *eclipse of the sun* occurs when the moon moves into a position directly between the sun and the earth. The moon's shadow then falls upon

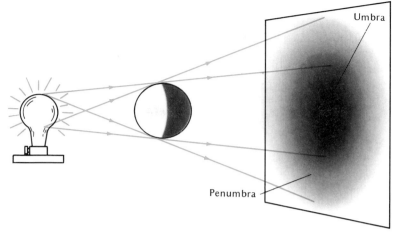

Fig. 16-7. The shadow formed by a large source of light consists of a dark umbra and a diffuse penumbra.

the earth, forming an umbra and a penumbra. People living in the region where the umbra falls find that their view of the sun is completely cut off by the moon and that they are in darkness. In this region, the eclipse is said to be total. People living in regions where the penumbra falls find that only part of the sun's disc is covered by the moon. In these regions, there is only partial darkness and the eclipse is said to be partial.

An *eclipse of the moon* occurs when the earth moves into a position directly between the sun and the moon. The umbra of the earth's shadow then encloses the moon, leaving it temporarily in near darkness. Since the moon is thus deprived of most the sun's light, it is only dimly seen until it moves out of the umbra of the earth's shadow.

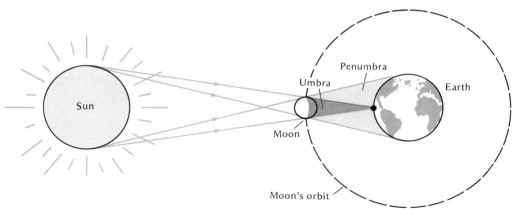

Fig. 16-8. An eclipse of the sun can occur only at new moon.

16-13 Pinhole Camera

The pinhole camera is simply a rectangular lightproof box with a pinhole in the center of its front side and a waxed paper screen or photographic film at the opposite side. It will form an image on its screen of any luminous or sufficiently illuminated object placed in front of it. The luminous object sends its own light into the camera. An illuminated object will send its reflected light into the camera.

An image forms because light rays leaving each point of the object pass through the pinhole and form a spot of light on the screen representing the image of that point. A bright point of the object sends more light through the pinhole and makes a brighter image of itself than does a dimmer point of the object. The screen of the camera thus has on it a collection of lighter and darker spots of light that are in the same arrangement as the corresponding points on the object from which they came. These spots of light form an image of the object on the screen.

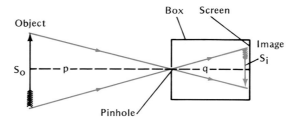

Fig. 16-9. A pinhole camera forms an inverted real image.

16-14 Characteristics of Pinhole Camera Image

An image formed by light rays that fall on a screen is called a *real image*. A photograph of this image can be obtained if the pinhole camera is equipped with a suitable shutter and if a photographic film is substituted for the screen.

Because light rays travel in straight lines, the image in the pinhole camera is inverted. From Fig. 16-9 you can see that a ray of light coming from the top of the object and passing through the pinhole hits the bottom of the screen. Similarly, a ray of light coming from the bottom of the object and passing through the pinhole hits the top of the screen. The image of the top of the object therefore appears at the bottom part of the screen while the image of the bottom of the object appears on the upper part of the screen.

From the similar triangles formed by the object, the image, and the two light rays, it can be predicted that:

$$\frac{S_o}{S_i} = \frac{p}{q}$$

where S_o is the size of the object, S_i is the size of the image, p is the object's distance from the pinhole, and q is the image's distance from the pinhole. This relationship is found to be experimentally correct, provided that the diameter of the pinhole is not too small.

16-15 Strength of a Light Source

The relationship between the intensity of the illumination cast upon a surface by a point source of light and the distance between the light source and the surface is another consequence of the straight-line path of light. To explore this relationship, it is useful to define the units in which light sources and the intensity of illumination of a surface are measured.

The strength of a light source, called its *luminous intensity,* is measured by comparing it with an internationally accepted standard source of light. Until quite recently, this light standard was simply a candle made according to a specific formula. However, this proved to be too crude and unsteady for modern needs.

Today, the standard light source is a cavity lined with thorium oxide, and kept white hot at the solidification temperature of platinum. *The unit of luminous intensity is called the candela* (kan·DEL·uh, symbol, cd). It is the rate at which 1/60 of a square centimeter of this source emits light. A lamp emitting light at the rate of one candela is said to have a luminous intensity of one candela. A lamp emitting light at twice this rate has a luminous intensity of 2 candelas, and so forth.

A typical 40-watt tungsten-filament lamp has an intensity of about 35 candelas. Fluorescent lamps are much more efficient light producers. Thus a 40-watt fluorescent lamp has a luminous intensity of about 200 candelas.

The *SI* unit of luminous intensity is the candela.

16-16 Intensity of Illumination

The rate at which light falls upon a unit area of a surface is called the *intensity of illumination and is measured in lumens* (LOO·menz). To define a lumen (lm), imagine a 1-candela point source of light at the center of a hollow sphere 1 meter in radius. The rate at which light energy falls upon one square meter of the inner surface of this sphere is 1 lumen. It follows that if a 5-candela light source were placed at the center of this sphere, each square meter of it would receive 5 times as much illumination as this, or 5 lumens. Thus, the intensity of illumination is directly proportional to the strength of the light source providing the light.

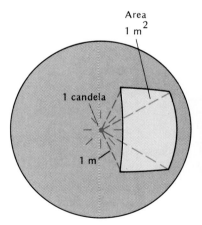

Fig. 16-10. The intensity of illumination of a 1 candela light source falling on 1 square meter of surface 1 meter away is equal to 1 lumen.

16-17 Inverse Square Law of Illumination

Fig. 16-11 shows how the light rays diverge as they leave a point source of light such as a carbon arc. The light that falls upon an area of one square meter at a distance of one meter from the light source spreads out over an area of 4 square meters when it has traveled 2 meters from the source. The same light has spread over an area of 9 square meters when it has traveled 3 meters from the source, and over an area of 16 square meters when it has traveled 4 meters from the source. Thus, as the surface moves further away from the light source, the intensity of illumination on the surface decreases from its value at one meter to $\frac{1}{2^2}$ or $\frac{1}{4}$ that value at 2 meters, to $\frac{1}{3^2}$ or $\frac{1}{9}$ that value at 3 meters, and to $\frac{1}{4^2}$ or $\frac{1}{16}$ that value at 4 meters.

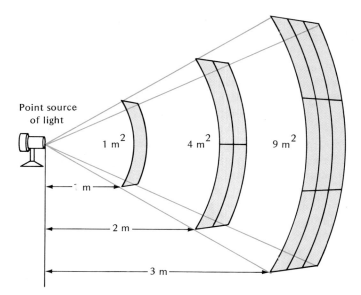

Fig. 16-11. The intensity of illumination from a point source of light is inversely proportional to the square of the distance from the source.

It is evident that *the intensity of illumination cast upon a surface by a point source of light is inversely proportional to the square of the distance between the surface and the light source.* Remember that the intensity of illumination upon a surface is also directly proportional to the strength of the light source. Then we have the relationship:

$$I = \frac{cd}{r^2}$$

where I is the intensity of illumination in lumens, cd is the strength of the light source in candelas, and r is the distance in meters between the light source and the illuminated surface.

Sample Problem

What is the intensity of illumination cast on a gymnasium floor by a 100-cd lamp hanging 5 m above the floor?

Solution:

$cd = 100$ cd $r = 5$ m

$$I = \frac{cd}{r^2}$$

$$I = \frac{100 \text{ cd}}{(5 \text{ m})^2} = 4 \text{ lm}$$

Test Yourself Problems

1. A 20-cd lamp is 0.50 m above a desk. What is the illumination on the desk?

2. How strong a point light source is needed to cast 10 lumens of light on a surface 2.0 meters from the source?

16-18 Diffraction of Light

The evidence furnished thus far by the formation of shadows, the operation of the pinhole camera, and the inverse square law of illumination, supports the conclusion that light travels in straight lines. However, closer examination of these very effects shows that this conclusion is not strictly true. Although the effect is usually very small, it is observed that light actually bends around the edges of bodies, and thus departs from a straight-line path. Light, like water and sound waves, is diffracted around the edges of obstacles in its path.

As early as the middle of the seventeenth century, the Italian Francesco Maria Grimaldi made an important observation. When an object placed between a point source of light and a screen is very small, its shadow is slightly smaller than it should be if the light rays from the source actually pass the edges of the object in straight lines. This observation suggested that light was bending around the edges of the object, thus slightly reducing the size of the shadow.

Grimaldi confirmed this evidence of the diffraction of light by a second experiment. He allowed light from a point source of light to pass through a tiny pinhole and fall upon a screen. He noticed that the size of the bright spot on the screen is larger than is to be expected if light travels precisely in straight lines. Here, the light is bending around the edges of the hole and increasing slightly the size of the spot of light on the screen.

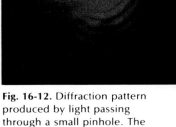

Fig. 16-12. Diffraction pattern produced by light passing through a small pinhole. The source of the light was a laser.

16-19 Demonstration of Diffraction

A modification of Grimaldi's second experiment is simply performed as follows. Cut a razor-thin slit in the middle of an index card. Holding the slit vertically in front of the eye, look at a lighted electric bulb surrounded by a cylindrical paper shade that just covers the bulb. Such a shade can be made from a sheet of white paper held around the bulb by means of a rubber band. It will be noted that the shade looks wider than it is and that its vertical edges consist of fuzzy, colored bands.

Next, hold the slit horizontally in front of the eye. The shade appears elongated and its top and bottom edges are fuzzy, colored bands.

In each case, the apparent increase in the dimensions of the shade is explained by the diffraction of the light from the shade as it passes around the edges of the slit into the eye. The colored bands are the result of the interference of light to be explained in more detail in Chapter 19.

16-20 What Determines the Path of Light

How can we reconcile the fact that light is diffracted around the edges of an object with the fact that it is also observed to travel in straight lines? The answer lies in the size of the opening through which, or the size of the obstacle around which, light passes.

Light is always diffracted on passing the edge of an object. However, when a beam of light passes through a large opening and falls upon a screen, the effects of diffraction in slightly increasing the

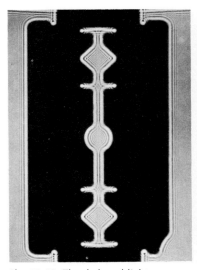

Fig. 16-13. The dark and light bands outlining the shadow of the razor blade result from the diffraction and interference of light when passed around the edges of the blade.

size of the spot on the screen and in blurring its boundaries are too small to be noticed against the brightness and size of the spot. As a result, light is observed to travel in straight lines in this situation.

Straight-line paths of light are also noted when a point source of light forms the shadow of a relatively large object on a screen. Again, the effects of the diffraction of light around the edges of the object are negligible compared to the size of the shadow and the brightness of the area around it.

It is only when light passes through a very small opening or around a very small object that the effects of diffraction become important. In these cases, the bending of the light around the edges of the opening or of the object produces an effect that is a significant factor in determining the pattern of light and dark areas formed by the light when it falls on a screen.

Historically, two opposing theories were advanced to explain the straight-line motion of light on the one hand and its diffraction around the edges of bodies or openings on the other. One was the particle theory of light proposed by Sir Isaac Newton. The other was the wave theory of light proposed by the Dutch physicist Christian Huygens (HY·gunz).

16-21 Newton's Particle Theory

According to Newton, light consists of tiny *particles of matter* that are ejected from a luminous body. As long as no force acts upon them, the particles obey the law of inertia by moving in straight lines at constant speed. This is the situation under most ordinary conditions. It explains why light is usually observed to travel in straight lines.

To explain the diffraction of light, Newton suggested that when light particles come very close to matter, they are attracted by it. The particles of light passing close to the edge of an opening or of a body will be attracted by the edge. As a result, they will be deflected from their straight-line path and be made to "bend" around the edge.

16-22 Wave Theory

This theory suggests that light consists of a series of waves that radiate from a light source just as water waves radiate from a disturbed point on the surface of a pond. In the case of water waves, we saw in Section 15-15 that the rays of the wave fronts travel in straight lines as long as they are passing through openings or around obstacles whose dimensions are considerably longer than the lengths of the water waves. In passing through small openings or around small objects, water waves are diffracted by an amount that depends on the ratio of the wavelength to the size of the opening or object.

The behavior of light is similar to that of water waves. From the fact that the diffraction of light is observed only when it passes through a very narrow slit and not when the slit is wide, it can be concluded that light waves must have very small wavelengths. This would explain why the paths of light rays through wide slits are practically straight lines.

16-23 Interference of Light

There was not enough experimental evidence about the behavior of light in the seventeenth century times of Newton and Huygens to enable scientists to choose between the particle and the wave theories. In the next two centuries, new evidence was obtained that decided overwhelmingly in favor of the wave theory. It was found, for example, that just as the wavelength of a sound determines its pitch, so does the wavelength of light determine its color. Among the most significant evidence of the wavelike nature of light was the demonstration of *interference* in light by Thomas Young, an English scientist, in 1802.

We have seen that two sets of sound waves can interfere to produce alternate beats and silences. Young showed that two sets of light waves can also interfere to produce alternate areas of brightness and darkness. This will happen when two sets of light waves having the same wavelength are superimposed upon each other. In those places where the crests and troughs of one set of waves always coincide with the crests and troughs of the other set of waves, the waves combine to produce an especially bright area. In those places where the crests of one set of waves always fall upon the troughs of the other, the waves neutralize each other so that the place will be dark.

16-24 Demonstration of Interference

To demonstrate the interference of light waves, dip the open end of a glass tumbler into a saucer of slightly soapy water and then remove the glass tumbler. A soap film forms across the mouth of the glass. Stand a few feet from a window and tip the tumbler so that the light of the sky can be seen in the soap film. Hold the tumbler in this tipped position for a minute or two. Horizontal bands of different colors of light will form at the upper part of the soap film and gradually broaden and move downward. The color effects seen are caused by the interference of the many different wavelengths or colors present in white light. Fig. 16-14 shows interference patterns produced in soap films by white light.

To simplify the effect, now permit only one color of light, for example, blue light, to fall upon the film. This can be done by holding a blue piece of cellophane above the soap film so that only the blue part of the daylight passes through it and falls on the film. The pattern now seen consists of alternate bands of blue light and darkness.

The fact that light can form interference patterns is major evidence of its wavelike nature.

Fig. 16-14. Interference patterns produced in soap films by white light: (top) in a thin film; (bottom) in a partially settled film.

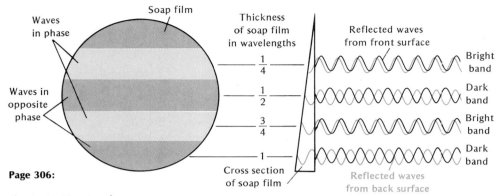

Page 306:

Fig. 16-15. How interference bands are formed in a soap film by blue light.

16-25 Explaining Interference in Soap Films

To understand how the blue and black interference pattern forms, you must understand that the soap film has a front and a back surface as shown in Fig. 16-15. It is also wedge-shaped because tipping the tumbler causes the liquid to flow downward so that the film is thin at the top and thickens progressively toward the bottom.

When the blue light waves fall upon the front surface of the film, some of the waves are reflected at the front surface and some enter the film and are reflected at the back surface. The two sets of reflected waves are then reunited and interfere with each other as they leave the film. At those parts of the film where the two sets of waves leave in phase, they reinforce each other to produce a bright band. At those parts of the film where the two sets of waves leave the film in opposite phase, they interfere destructively with each other to produce a dark band.

Two factors determine whether the two sets of reflected waves will be in or out of phase. The first is that the waves reflected from the front surface of the film always snap into opposite phase on reflection. That is, a crest is reflected as a trough and a trough is reflected as a crest. This sudden shift of phase takes place whenever waves of any kind, traveling in one substance, are reflected at the surface of a second substance, in which their speed is slower than in the first. Since light travels more slowly in the soap film than it does in air, light waves reflected from the front surface of the soap film return in opposite phase.

The second factor affecting the phase relationship between the two sets of reflected waves is the thickness of the film. The waves that enter the film and are reflected from its back surface travel a distance equal to twice the thickness of the film before rejoining the waves reflected from the front surface. Where the film is ¼, ¾, ⁵⁄₄, or any odd number of quarter wavelengths thick, this extra distance is ½, ³⁄₂, ⁵⁄₂, or an odd number of half wavelengths. The waves that have traveled it therefore come out of the film in a phase opposite to that with which they entered the film. A crest emerges as a trough, and vice versa. These emerging waves are therefore in phase with those reflected from the front surface where crests are also reflected as troughs, and vice versa. The two sets of waves therefore reinforce each other to produce bright bands at those parts of the film that are an odd quarter wavelength thick, as shown in Fig. 16-15.

Where the film is ½, ²⁄₂, ³⁄₂, or any number of half wavelengths thick, the extra distance traveled by the waves that enter the film is 1, 2, 3, or any whole number of wavelengths. The waves that are reflected from the back surface of the film therefore emerge in the same phase with which they entered. As a result they are in opposite phase to that of the waves reflected from the front surface. These two sets of waves therefore interfere destructively to produce dark bands at those parts of the film that are a whole number of half wavelengths thick, as shown in Fig. 16-15.

16-26 Polarization

While the phenomenon of interference confirms the wavelike nature of light, it gives no clue as to whether light waves are transverse or longitudinal. Evidence that light consists of transverse waves was first supplied by the discovery of the phenomenon of *polarization*.

When waves are passing over a water surface, the particles of water near the surface vibrate approximately in the single up and down direction at right angles to the direction of the waves. The waves forming a beam of light differ from this in that they contain vibrations in all possible directions at right angles to the direction in which the beam is traveling. By passing a beam of light through certain crystalline substances we can remove all vibrations from the beam except those parallel to a certain axis of the crystal. The emerging beam then contains vibrations only in that one direction and is said to be *plane polarized*.

The crystal that polarized the beam is called a *polarizer*. Crystal substances commonly used to polarize light are tourmaline and the artificially made "Polaroid." A polarizing crystal acts as though it consists of many slitlike openings parallel to its axis. Vibrations of light parallel to these slits pass through them. Vibrations perpendicular to the slits are blocked.

The polarization of light indicates that light waves are transverse.

16-27 Detection of Polarized Light

Since plane-polarized light does not appear different to the eye from ordinary light, special means are needed to detect it. This is done by placing a second polarizing crystal, called the *analyzer*, in the path of the polarized beam. If the analyzer is held so that its axis is parallel to the vibrations present in the polarized beam, it will permit the polarized beam to pass through it.

If the analyzer is turned through 90 degrees so that its axis is

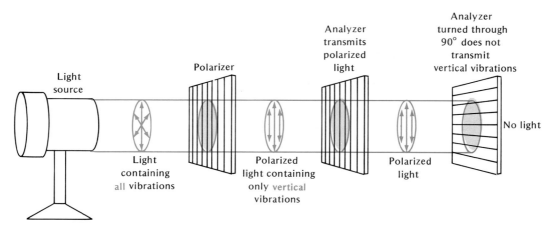

Fig. 16-16. Polarization of light.

perpendicular to the vibrations of the polarized beam, the analyzer will not permit the beam to pass through it. Thus, by adjusting the position of the analyzer, the polarized beam can be blocked or transmitted at will. (See Fig. 16-6.)

16-28 Polarization by Reflection

Passing light through crystals is not the only means of polarizing light. Light is also polarized when it is reflected from smooth surfaces at certain angles. In fact, much of the light reflected by water surfaces, by glass, and other highly reflecting surfaces, is plane

Fig. 16-17. Polarization by reflection. At left, a view through ordinary sunglasses that do not remove the glare. At right, the same view through polarized sunglasses.

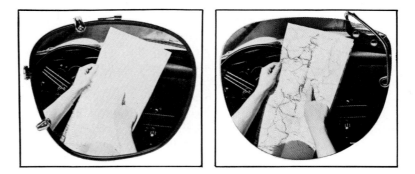

polarized. This fact is commonly used in manufacturing sun glasses designed to eliminate the glare from strongly reflecting surfaces. Many sun glasses are made of Polaroid, a manufactured material that effectively polarizes light. The Polaroid glass or plastic acts like an analyzer and filters out that part of the reflected light that is polarized at right angles to its axis. This reduces the quantity of light coming from the reflecting surface to the eye and thus eliminates glare.

16-29 Demonstration of Polarization

With a pair of Polaroid discs from sun glasses, the polarization of light can be simply demonstrated. Hold one of the Polaroid discs in front of an electric light so that it polarizes the light that passes through it. Examine this light using the second Polaroid disc as an analyzer by holding it in front of the eye and looking at the first Polaroid disc, or polarizer. Turn the analyzer slowly until the polarizer looks black. In this position, the analyzer does not permit the polarized light to pass through it.

Now slowly turn the analyzer through 180 degrees. Notice that the polarizer gradually brightens, reaching a maximum brightness at 90 degrees. It then gradually dims to become black again at 180 degrees. This pattern is completed when the analyzer is turned through another 180 degrees. The analyzer completely blocks the polarized light in the 0- and 180-degree positions and transmits it fully in the two positions at right angles to them. At intermediate positions of the analyzer, the polarized light is only partially blocked.

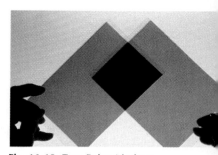

Fig. 16-18. Two Polaroid plates are held with their transmitting axes at right angles. Light polarized by the first plate does not pass through the second.

16-30 Further Development of the Wave Theory of Light

By the middle of the nineteenth century, the evidence suggesting that light behaves like a regularly spaced series of transverse waves was overwhelming. Such waves accounted not only for the diffraction, interference, and polarization of light, but also for the reflection and refraction of light to be considered in the next chapter. Two questions, however, remained unanswered. First, what is the medium that transmits the vibrations caused by these waves? Second, what is the nature of the vibrations themselves?

In all other regular wave motions, the vibrations are transmitted by some medium. Water surfaces transmit water waves. Air and other substances transmit sound waves. In each case, the particles of a material medium are set into vibration by the passing waves.

Light, however, travels through a vacuum. Here, there is no material medium to vibrate and transmit the waves. To overcome this difficulty, it was proposed that there exists a hitherto undetected substance called the *ether* that pervades all space and is the medium that transmits the transverse vibrations of light. Many experiments were tried to detect the ether but all failed. When, finally, an especially precise experiment by Michelson and Morley also failed to detect the ether, Albert Einstein suggested that the idea of the ether be abandoned and that it be assumed instead that light needs no material medium in which to travel.

The answer to the second question, as to the nature of the vibrations transmitted by light, was given by the English scientist James Clerk Maxwell (1831–1879). In his *electromagnetic theory of light,* Maxwell showed that when electric charges vibrate, they

Maxwell's theory established that light waves are electromagnetic and predicted the existence of radio waves many years before they were actually observed and produced by Hertz.

generate transverse waves which travel through a vacuum with the speed of light. These waves consist of vibrating electric and magnetic forces acting at right angles to each other and to the direction of motion of the waves. They are called *electromagnetic waves.* They vary widely in wavelength, from the gamma rays emitted by radium with wavelengths of a few trillionths (10^{-12}) of a meter (or picometers), to radio waves with wavelengths of several kilometers.

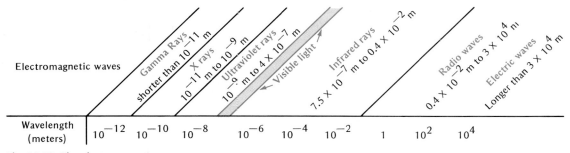

Fig. 16-19. The electromagnetic wave family.

16-31 Electromagnetic Wave Family

Only a very small part of the very large family of electromagnetic waves is visible. This is the part called light. Fig. 16-19 lists the various kinds of waves in the electromagnetic family. Differences in wavelength among the visible light waves are seen by us as different colors.

Somewhat longer in wavelength than the longest visible waves, which are red in color, are the invisible *infrared rays.* These are commonly experienced by us as heat and are the means whereby heat is radiated through space. Still longer in wavelength are the *radio waves* used in radio communication.

Somewhat shorter in wavelength than the shortest visible waves, which are violet in color, are the so-called *ultraviolet rays.* These invisible rays are important in making pictures by photography because they strongly affect photographic film. They are also responsible for producing the summer tan acquired by exposure to the sun.

Shorter in wavelength than the ultraviolet rays are the penetrating *X rays* used by the doctor and the dentist to examine interior parts of the body opaque to ordinary light. The most penetrating and shortest of the electromagnetic waves are the *gamma rays* emitted by radioactive substances.

16-32 Granular Nature of Light Energy

As long as light is moving through a substance or through space, its behavior can be explained in terms of the properties of electromagnetic waves. When, however, light is absorbed or emitted by matter, it seems to act like a stream of individual particles or packets of energy rather than like waves.

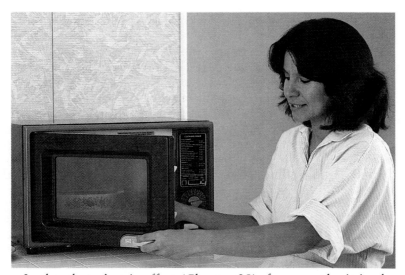

Fig. 16-20. A microwave oven uses electromagnetic radiation with a wavelength intermediate between that of infrared rays and radio waves—from several millimeters to several meters.

In the photoelectric effect (Chapter 29), for example, it is observed that when light falls upon certain bodies, it ejects electrons from them. Here the light acts as if it consists of individual particles of energy, called *quanta* or *photons,* each of which acts upon one electron and knocks it out of the illuminated body.

Again, when light is emitted by the atoms of a gas in an electrical discharge tube, such as a commercial neon light, it appears to consist of streams of individual photons. Each photon retains all its energy while it is passing through space. It gives up energy only upon interacting with a particle of matter such as an electron or an atom.

These observations are typical of many which illustrate that *the energy associated with light is bundled in tiny individual packets.* Thus, it appears that the energy of light and other electromagnetic radiations is *granular* in structure.

16-33 Quantum Theory of Light

The quantum theory of light arose out of the need to reconcile the wavelike behavior of light with its particle-like behavior. It was developed by Planck, Bohr, Einstein, and others during the early part of this century. According to this theory, light is emitted from luminous bodies in tiny packets of energy called photons. *Each photon is associated with light waves of a specific frequency.* It is therefore identified with the color associated with that frequency and *has a fixed, definite quantity of energy proportional to that frequency.*

Photons can "collide" and interact with electrons and other tiny particles of matter. In these interactions, their behavior resembles that of particles. The wavelike behavior of light is explained by assuming that every photon is guided in its motion through space

Light energy is emitted and absorbed in discrete packets called photons.

The motion of photons is guided by their probability waves.

by its own special "probability wave." The probability wave determines the chance that, at a particular instant, all the energy of a photon will be found at a given point of the wave front.

The superposition and interaction of the probability waves associated with the photons account for interference, diffraction, and the other wavelike properties of light. In these situations, the quantum theory gives the same results as the wave theory. Therefore in the next chapters we can continue to apply the wave theory correctly to those situations in which light exhibits wavelike properties.

When light and matter interact, only the quantum theory satisfactorily explains the behavior of light. This we shall see in Chapter 29.

CHAPTER REVIEW

Summary

In this chapter, some of the properties of light are examined in order to investigate the nature of light. We have seen that light is a form of energy because it can do work and can be changed into other forms of energy. We have also observed that in such situations as the **transmission of light** through a vacuum or through air, the formation of **shadows,** the formation of the **real image** in a **pinhole camera,** and the spreading out of light energy from a point source, light behaves as though it consists of **rays** that travel in straight lines. This behavior suggested to Newton that a light **beam** consists of a stream of high-speed particles moving in a straight line.

However, the **diffraction of light** in passing through very narrow openings or around very small bodies suggests the theory that light energy is transferred through space or matter by waves rather than particles. These waves have extremely small wavelengths. Their diffraction is therefore noticed only when the opening through which they pass, or the obstacle around which they bend, is very small. In all other situations, the paths in which light rays travel are essentially straight lines. Thus, the **wave theory** accounts for the diffraction of light as well as its apparent straight-line motion.

The discovery of the **interference** and of the **polarization** of light further confirmed the wave theory of light and suggested that light consists of transverse rather than longitudinal waves. The theory was then extended by Maxwell in his **electromagnetic theory of light** which stated that light waves consist of vibrating electric and magnetic forces acting at right angles to each other and to the direction of motion of the waves. According to the electromagnetic theory, these waves are generated by vibrating electric charges. Light is only one of the members of a large **spectrum** of electromagnetic waves which differ from each other principally in wavelength. All electromagnetic waves travel at the **speed of light, 3×10^8 meters per second,** and have wavelike properties similar to those of light.

The electromagnetic theory of light was highly successful in explaining the wavelike behaviors that are associated with light while it is in transit from one place to another. These include *reflection, refraction, diffraction, interference,* and *polarization.* However, it could not explain why light energy behaves as though it were packaged in bundles, called **photons,** when the light is absorbed or is emitted by matter. This failure made necessary the formulation of the quantum theory of light in which the wavelike and particle-like behaviors of light are reconciled.

According to the **quantum theory of light,** light is emitted from luminous bodies in definite packets of energy called **quanta,** or **photons.** Each photon is guided in its motion through space by a "probability wave" with which it is associated. The superposition and interference of the probability waves of photons account for all the wavelike behaviors of light, such as interference and diffraction, and give exactly the same results as the wave theory. Each photon is associated with the particular frequency assigned to it by the wave theory. The energy contained by a photon is proportional to its associated frequency and, for visible light, determines the color of that photon.

Questions

Group 1

1. (a) State one evidence that light can do work. (b) State one evidence that light can be transformed to other forms of energy.

2. Explain how we see ordinary nonluminous objects, such as classroom desks and chairs.

3. (a) What three things may happen to the sun's light energy after it falls upon the surface of the ocean? (b) As divers descend into the ocean on a sunny day, they notice that it becomes darker the deeper they go. What happens to the light that enters the water?

4. (a) How do we know that light can travel through a vacuum? (b) How may the speed of light in air or in a vacuum be measured? (c) How does the speed of light in water, glass, and air compare with its speed in a vacuum? (d) Why is the speed of light in a vacuum a particularly important number in physics?

5. You are given two red and two green light filters and a spotlight that provides a beam of white light. (a) How can you demonstrate that white light contains red and green components? (b) How can you demonstrate the selective transmission property of the red filter?

6. (a) What determines the color of an opaque object? (b) An American flag is illuminated only by red light. What colors will its red, white, and blue parts appear to be? Explain your answer.

7. An object is midway between a point source of light and a screen. (a) Make a prediction concerning the relative height of the object and its shadow on the screen. (b) On what property of light is your prediction based?

8. (a) Show, by drawing the appropriate light rays, the limits of the umbra of the moon's shadow on the earth during an eclipse of the sun. (b) If the moon and the earth remain the same distance apart while both move directly away from the sun, what happens to the size of the umbra? (c) Can the umbra of the moon's shadow ever have a diameter greater than that of the moon? Explain.

9. (a) Explain how a pinhole camera forms an image of an object on its screen. (b) What determines the size of the image? (c) Why is the image inverted?

10. Since the distance from the sun to the earth is known, as is also the intensity of illumination it casts upon the earth's surface, a rough estimate of the luminous intensity of the sun can be obtained from the relationship $I = cd/r^2$. Why is it inaccurate to apply this relationship to the sun in this way?

11. (a) What evidence is there that light is diffracted around the edges of obstacles? (b) The bending of light is noticed only when light passes through very narrow openings or around very small objects. What does this suggest about the size of the wavelengths of light?

12. (a) Describe an experiment demonstrating interference between two sets of light waves. (b) What will be seen at those places where the two sets of waves are in phase? Out of phase?

13. (a) Describe how a beam of light from a flash-light may be polarized? (b) How does the composition of the polarized beam differ from the composition of the beam before it was polarized? (c) Why is the intensity of the polarized beam less than that of the beam from which it was obtained?

14. (a) How may the fact be detected that much light reflected from water, glass, and polished surfaces is polarized? (b) How do Polaroid eyeglasses reduce glare?

15. (a) What does the fact that light can be polarized suggest about the nature of light waves? (b) According to Maxwell's electromagnetic theory, what do the vibrations of light waves consist of?

16. (a) List the various types of electromagnetic radiation in order of increasing wavelength. (b) State one use or property characteristic of each type.

17. (a) What is a quantum or photon of light? (b) How does it differ from the particles of Newton's theory of light?

Group 2

18. In which of these situations does light show its photon nature and in which does it show its wave nature: (a) light passing through a narrow slit; (b) light falling on a metal and ejecting electrons from it; (c) light emitted by the atoms in a neon lamp; (d) light passing through a polarizer?

19. It has been shown that both matter and light are granular in structure. (a) In what way do the elementary granules of matter differ from the elementary granules of light? (b) In what way may photons of visible light differ from each other? (c) How is this difference seen by an observer?

20. (a) What is the purpose of the probability waves associated with photons? (b) How do these waves account for such wavelike behaviors of light as interference and diffraction?

21. The distance traveled by light in a year is called one light-year. If an explosion took place tonight on a star that is ten light-years away, when would observers on the earth be able to see it? Is it possible for observers on the earth to see explosions or other changes in stars at the moment they happen? Explain.

22. If the hole of a pinhole camera is too large, the image of each point of the object is a blurred area on the screen instead of a point. Show by a ray diagram like that of Fig. 16-9 why this happens.

23. Explain why we cannot keep improving the sharpness of the image formed by a pinhole camera by making this hole smaller and smaller.

24. Explain why radio waves are much more readily diffracted on passing through narrow openings than are the other electromagnetic waves.

25. Can two plane-polarized light waves interfere if the plane of one is perpendicular to the plane of the other? Explain.

Problems

Group 1

1. The distance from the earth to the moon is about 3.8×10^8 m. A radar impulse traveling at the speed of light is sent to the moon and, after reflection, returns to the earth. About how long does it take to make the round trip?

2. A baseball fan in a ball park is 100 m from the batter's box when the batter hits the ball. (a) How long after the batter actually hits the ball does the fan see it occur? (b) Assuming the speed of sound to be 333 m/s., how long after the batter actually hits the ball does the fan hear the crack of the bat?

3. It takes 4 years for light from certain stars to reach the earth. How far away are these stars from the earth?

4. A pencil 10 cm long is 20 cm in front of a pinhole camera. The screen of the camera is 15 cm from the pinhole. What is the length of the image of the pencil?

5. Where should the screen of the pinhole camera of problem 4 be put if it is desired to have an image of the pencil that is 4 cm tall?

6. A 32-cd point source of light is hanging 2 m above a desk. What is the intensity of illumination on the desk?

7. What is the intensity of illumination cast by a 1000-cd light from a lighthouse on a ship at a distance of 2000 m?

Group 2

8. Two lamps throw the same intensity of illumination on a wall when one lamp is 5.0 m from the wall and the other is 8.0 m from the wall. If the nearer lamp has a rating of 25 cd, what is the luminous intensity of the other?

9. The intensity of illumination received from a distant lamp by an observer is 1.0×10^{-2} lm. If the lamp is known to have a strength of 64 cd, what is the distance between the lamp and the observer?

10. A 10-cd lamp and a 40-cd lamp (both point sources) cast equal intensities of illumination on a wall. If the 10-cd lamp is 6.0 m from the wall, how far is the 40-cd lamp from the wall?

11. The screen of a pinhole camera is 12 cm from the pinhole. An illuminated rectangular card having dimensions of 3.0 cm by 4.0 cm is placed 18 cm from the pinhole. (a) What is the area of the image of the card formed by the camera? (b) If the distance between the screen and the pinhole is decreased to 4.0 cm, how many times brighter does the image appear than before?

12. (a) How long does it take light to travel the 70-km distance in the Michelson method of measuring the speed of light described in Section 16-6? (b) What is the smallest number of turns per second that the octagonal mirror must make for the observer to receive the light after its 70-km round trip?

Applying Physics

1. Measure the height of an outdoor object such as a flagpole or tree by the length of its shadow in the late afternoon sun. Hold a meterstick vertically with one end resting on the ground. Measure the length of the shadow of the stick cast by the sun. Now, measure the length of the shadow of the flagpole or tree. The height of the object is found from the fact that the ratio of the object's height to the length of its shadow is equal to the ratio of the meterstick's height to the length of its shadow.

2. Make a comparison of the intensities of two lamps by means of a piece of white paper in the middle of which is a circular grease spot. The lamps to be compared are put in a room having no other source of light and set about 4 m apart. The paper with the grease spot is now placed between the lamps and is held perpendicular to the line joining them. If one side of the paper gets more light than the other, that side will look brighter except at the middle. Here, more light passes through the grease spot than is replaced by the light coming from the other side of the paper. As a result, the grease spot looks dark. When both sides of the paper are equally illuminated, the grease spot will be just as bright as the rest of the paper and will seem to disappear.

The paper with the grease spot is moved closer to one lamp or the other until the grease spot seems to disappear. For this position, the intensity of illumination cast by one lamp on its side of the paper is equal to the intensity of illumination cast by the second lamp on its side of the paper. Measure the distance of each lamp to the grease spot. Then, if cd_1 and cd_2 are the luminous intensities of the two lamps and r_1 and r_2 are their corresponding distances to the grease spot, $cd_1/r_1^2 = cd/r_2^2$, or $cd_1/cd_2 = r_1^2/r_2^2$. Determine the ratio of the luminous intensities of the two lamps from this relationship.

PHYSICS✛PLUS

Fiber Optics

The transmission of light through hair-thin, transparent fibers—a process known as fiber optics—has revolutionized the field of communication.

The so-called optical fibers have many advantages over the copper wire traditionally used to transmit electrical signals for communication systems. A single optical fiber has the capacity to transmit 2.5 trillion pages of data per second. Optical fibers for communication are made from glass and so are much less expensive than copper wiring. Unlike metal wire, fiber optic cable never corrodes and is immune to interference from electrical or magnetic fields. In addition, it can be made much smaller in diameter than

A bundle of 660 single glass fibers passing through the eye of a needle.

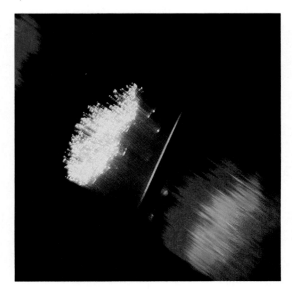

metal wire and therefore is also much lighter in weight than metal wire.

Because of these features, the major long-distance carrier in the United States began in 1988 to replace 2 billion miles of phone lines with fiber optic cable. There are also plans to run a second set of optical cables across the Atlantic and Pacific oceans so that 40,000 telephone conversations can be carried simultaneously.

Although the cost of fiber optic cable dropped from $7 per meter in 1977 to $0.23 per meter in 1988, it is still considered too expensive to be strung directly to consumers' homes ("distribution" lines) and is limited to "transmission" lines. However, it is predicted that within the next 30 years almost every home will be "fiber-networked" as manufacturing, installation, and maintenance costs drop. Then telephone company fiber optic wiring could also be used to carry cable television at a lower cost and with an improvement in signal quality.

Fiber optics has also found application in the field of medicine. Inserted through natural openings or small cuts and then passed along natural passageways in the body, optical fibers can be used to inspect the internal organs without surgery. The fibers can also be used as the vehicle for transmitting the laser light used in laser surgery. In the bloodstream, optical fibers can be used as "sensors" to determine blood velocity and oxygen content.

Fiber optics is already playing an impressive role in our lives. However, today's applications are only the forerunners of the opportunities that lie ahead.

Reflection and Refraction

17

Aims

1. To observe and study how light is reflected and refracted.
2. To see how the wave theory of light accounts for its reflection and refraction.
3. To learn how the refraction of light through a prism can be used to separate white light into its component colors.

17-1 Law of Reflection

When a parallel beam of sunlight or a beam from a searchlight falls upon a flat, shiny surface like that of a mirror, the beam leaves the mirror at an angle equal to the one at which it struck the mirror. This results from the reflection of the individual rays composing the beam.

To study how one ray of the beam acts on being reflected, consider the arrangement in Fig. 17-2 (a). Here, a ray of light is approximated by letting a beam from a spotlight pass through two small holes. The ray is then allowed to fall on a plane mirror and the direction in which the ray is reflected is noted. The ray falling on the mirror is called an *incident ray*. After it is reflected, it is called a *reflected ray*.

To measure angles made by the light rays we usually erect a perpendicular called the *normal* at the point where the reflection occurs. The angle between the incident ray and the normal is called the *angle of incidence*. The angle between the reflected ray and the normal is called the *angle of reflection*. For all angles of incidence you will find that the following *law of reflection* holds true:

Fig. 17-1. The raylike behavior of light is shown here as a laser beam is reflected successively by mirrors.

A ray of light is reflected from a plane surface so that the angle of reflection is exactly equal to the angle of incidence. Furthermore, the incident ray, the normal, and the reflected ray all lie in the same plane.

17-2 Wave Theory Explains Reflection

The wave theory readily explains the law of reflection. In Fig. 17-2 (b) we have a point source of light S sending spherical wave fronts to a plane mirror through a slit. S is the common center of all these wave fronts. After reflection, these wave fronts remain spherical but reverse their direction so that their common center

(a)

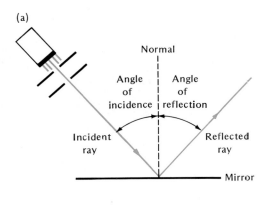

(b)

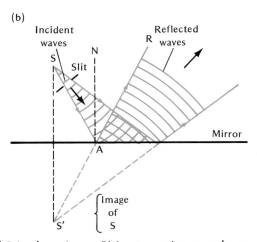

Fig. 17-2. (a) The law of reflection. (b) Wave theory explanation of the law of reflection.

is now S', the image of S in the mirror. S' is at a point exactly as far behind the mirror as S is in front of it.

Note that rays of spherical or circular waves are simply their radii; from this we see that SA is an incident ray and AR its reflected ray. AN is the normal at A. Since line SS' is parallel to AN, it follows from elementary geometry that the angle of incidence SAN and the angle of reflection NAR are each equal to angle $SS'A$, and are therefore equal to each other. Thus, the ray SA obeys the law of reflection.

17-3 Regular and Diffuse Reflection

Everyday experience reveals the differences between the way in which light is reflected from a plane mirror or the surface of still water and the way in which it is reflected from a page of this book. One of these differences is that reflected images can be seen in the mirror and in the water but not in the page of the book. The type of reflection that takes place at a smooth flat surface like a mirror is called *regular reflection*. The type of reflection that occurs at a relatively rough surface like paper is called *diffuse reflection*.

Fig. 17-3 (a) shows a beam of parallel rays being reflected regularly from a smooth plane surface. Each individual ray is reflected so that its angle of reflection is equal to its angle of incidence. Since all rays are reflected from the same flat surface, the rays remain parallel after reflection.

Fig. 17-3 (b) shows a beam of parallel rays being reflected diffusely from a rough surface. Note that, to light waves, a surface like this page is not a flat plane but consists of millions of tiny flat surfaces facing in all directions. On reaching the page, every ray of the beam hits a different one of these tiny flat surfaces and is therefore reflected in a different direction. As a result, the light reflected from the page is scattered in all directions.

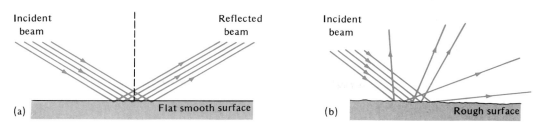

Fig. 17-3. (a) Regular reflection. (b) Diffuse reflection.

The scattering that is caused by diffuse reflection is very desirable. It enables us to see, from any direction or position, the object that is diffusing the light.

17-4 Refraction

A stick half immersed in water appears to be broken. This can be shown by immersing the lower half of a pencil in a basin of water at an angle of about 45 degrees to the surface of the water. When viewed from above, the pencil appears to be broken at the point where it enters the water. This is a typical effect produced by the refraction of light.

Refraction is the sudden change in the direction of a light ray that takes place at the moment it passes obliquely from one substance into another. In general, there are two types of refraction. One is illustrated when a ray of light passes from air into glass. The second is illustrated when a ray of light passes from glass into air.

Refraction is the bending of light rays as they pass obliquely from one medium into another.

17-5 Passage of Light from Air into Glass

The refraction of light on passing from air into glass may be studied by means of the optical-disc apparatus shown in Fig. 17-4. A narrow beam of light representing a ray is obtained by passing a parallel-sided beam of light from a spotlight through a narrow slit. The narrow beam is then allowed to enter a half circular plate of glass at different angles to the diameter. The path of the ray is observed against a white background on which there is an angular scale.

The ray entering the glass is called the incident ray. After it has entered the glass, it is called the *refracted ray*. In the first case of Fig. 17-4, the incident ray passes from air into glass at right angles to the glass surface. In this case, the ray continues into the glass without changing direction.

In the second case of Fig. 17-4, the incident ray enters the glass obliquely. To measure the angle at which the ray strikes the glass, the normal line NN' is drawn perpendicular to the glass surface at the point at which the light ray strikes it. It is seen that the ray divides into two parts. One is the reflected ray that returns to the air from the glass surface in accordance with the law of reflection. The second is the refracted ray which changes direction by bending toward the normal as it enters the glass. As before, the angle between the incident ray and the normal is called the *angle of incidence*. The angle between the refracted ray and the normal is called the *angle of refraction*. Since the ray bends toward the normal on passing from air into glass, the angle of incidence is greater than the angle of refraction.

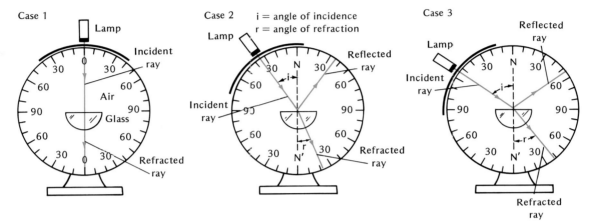

Fig. 17-4. Refraction of a ray on passing from air into glass.

In the third case of Fig. 17-4, the angle of incidence has been increased. It is seen that the angle of refraction also has increased as the ray is again refracted toward the normal.

From these observations we may note the following five facts concerning refraction of light on passing from air into glass: (1) As long as a ray of light remains in either air or glass, its path is a straight line. (2) When the ray passes at right angles from air into glass, its direction remains unchanged. (3) When the ray passes obliquely from air into glass, it is refracted at the glass surface and bends toward the normal. (4) As the angle of incidence increases so does the angle of refraction. (5) For all angles of incidence, the incident ray, the normal line, and the refracted ray are in the same plane.

17-6 Snell's Law and the Index of Refraction

The type of refraction observed when a light ray passes from air into glass occurs whenever a light ray passes from air or a vacuum into any denser medium such as water, oil, or diamond. The seventeenth-century Dutch scientist Willebrord Snell discovered that each substance refracts light rays entering it from air or a vacuum according to a very definite rule now known as *Snell's law*. The law states that *on passing from air (or a vacuum) into another substance, a light ray is always refracted toward the normal in such a direction that the ratio of the sine of its angle of incidence to the sine of its angle of refraction is a constant.* This constant usually has a different value for each substance and is called the *index of refraction* of that substance. In symbols,

$$\frac{\sin i}{\sin r} = n_s$$

where i is the angle of incidence, r is the angle of refraction, and n_s is the index of refraction of the substance.

17-7 Measuring the Index of Refraction

The optical disc shown in Fig. 17-4 can be used to measure the index of refraction of a substance. A ray of light is allowed to enter a flat surface of the substance obliquely and its angles of incidence and refraction are then measured. The sines of both angles are then looked up in a table like the one on page 703. Dividing the sine of the angle of incidence by the sine of the angle of refraction gives the index of refraction of the substance.

Table 17.1 gives the index of refraction of each of several substances. Note that they are all greater than 1. This must be so because, in passing obliquely from air or a vacuum into another substance, a light ray is always refracted toward the normal. The sine of its angle of incidence is therefore always greater than the sine of its angle of refraction.

Strictly speaking, the index of refraction of a substance is defined as the ratio of the sine of the angle of incidence to the sine of the angle of refraction for a ray of light passing from a vacuum into the substance rather than from air into the substance. However, the difference between these two ratios is so small that, for most purposes, it can be neglected. In this book, the index of refraction of a substance will refer to the value obtained by observing a ray entering the substance either from air or from a vacuum.

Table 17.1 Indices of Refraction	
MEDIUM	n
Air	1.00
Water	1.33
Carbon disulfide	1.63
Quartz	1.46
Crown glass	1.52
Flint glass	1.66
Polyethylene	1.50
Diamond	2.42

Sample Problem

Find to the nearest degree the angle of refraction of a ray of light passing from air into water at an incident angle of 30°.

Solution:
From Table 17.1, the index of refraction of water n_w = 1.33.

The angle of incidence is $i = 30°$.

From the table on page 703, $\sin 30° = 0.500$.

By Snell's law:

$$\frac{\sin i}{\sin r} = n_w$$

$$\sin r = \frac{\sin i}{n_w}$$

$$\sin r = \frac{0.500}{1.33} = 0.375$$

In the table on page 673, it is found that $\sin 22° = 0.375$. Hence, r is 22°.

Test Yourself Problem

What is the angle of refraction when a ray of light passes from air into diamond at an incident angle of (a) 30°; (b) 40°; (c) 50°?

Case 1

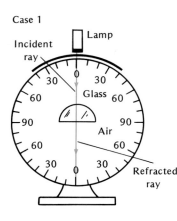

Case 2   I = angle of incidence
R = angle of refraction

Case 3

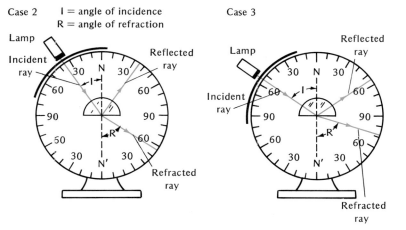

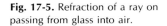

Fig. 17-5. Refraction of a ray on passing from glass into air.

17-8 Passage of Light from Glass into Air

A different type of refraction occurs when light passes from glass into air instead of from air into glass. In Fig. 17-5, a light ray enters the semicircular glass plate along a radius and emerges into the air across the diameter of the plate. In the first case, the incident ray suffers no change of direction as it passes from the glass to the air along the normal to the glass surface.

In the second case, the incident ray inside the glass falls obliquely upon the flat surface. The ray divides into two parts. One is reflected by the surface in the usual manner. The second part is the refracted ray that passes out of the glass into the air. As it crosses the glass surface, the ray is bent away from the normal. The angle of refraction is therefore greater than the angle of incidence.

In the third case, the angle of incidence has been increased. Now you find that the angle of refraction has also increased.

17-9 Reversibility of Light Rays

The manner in which a ray is refracted on passing from glass into air is typical of the manner in which a ray is refracted in passing from any substance into air or a vacuum. If the refracted ray in the above demonstrations falls perpendicularly upon a plane mirror, it will reverse its direction and retrace its entire path as it re-enters the glass. This illustrates the fact that the path of a light ray is reversible. That is to say, if the direction of a light ray is reversed, it will return over exactly the same path on which it came.

From Snell's law, a ray passing from air into a substance conforms to the relationship $\sin i / \sin r = n_s$. If the path of the ray is reversed, the angles i and r reverse roles; r now becomes the new angle of incidence I, while i becomes the new angle of refraction R. Hence, for a light ray passing from a substance into air, Snell's law takes the form:

$$\frac{\sin I}{\sin R} = \frac{\sin r}{\sin i} = \frac{1}{n_s}$$

Thus, Snell's law takes two forms. When a ray passes from air or a vacuum into a denser substance, the law is:

$$\frac{\sin i}{\sin r} = n_s$$

When a ray passes from a denser substance into air or a vacuum, the law is:

$$\frac{\sin I}{\sin R} = \frac{1}{n_s}$$

17-10 General Law of Refraction

From the two forms of Snell's law, it follows that a ray of light passing from any substance into any other substance obeys the relationship:

$$\frac{\sin i}{\sin r} = n_{1.2}$$

Here, $n_{1.2}$ (read: *n-one-two*) is called the *relative index of refraction* of the second substance with respect to the first. It is given by:

$$\mathbf{n}_{1.2} = \frac{\mathbf{n}_2}{\mathbf{n}_1}$$

where n_2 and n_1 are the indices of refraction of the second and first substances respectively.

You can see from this relationship that a ray of light passing obliquely from a substance having a lower index of refraction n_1 into one having a higher index of refraction n_2 is bent toward the normal. In this case, $n_{1.2} = n_2/n_1$ is greater than 1. This means that $\sin i$ is greater than $\sin r$ and therefore that angle i is greater than angle r.

Similarly, you can see that a ray of light passing obliquely from a substance having a higher index of refraction into one having a lower index of refraction is bent away from the normal.

A light ray bends toward the normal if $n_{1.2} > 1$ and away from the normal if $n_{1.2} < 1$.

Sample Problem

A ray of light is passing from glass ($n_g = 1.50$) into water ($n_w = 1.33$) at an angle of incidence of 30°. What is the angle of refraction?

Solution:
The relative index of refraction of water with respect to glass is:

$$n_{g,w} = \frac{n_w}{n_g} = \frac{1.33}{1.50} = 0.889$$

Also,

$$\sin i = \sin 30° = 0.500$$

By Snell's law:

$$\frac{\sin i}{\sin r} = n_{g,w}$$

$$\sin r = \frac{\sin i}{n_{g,w}}$$

$$\sin r = \frac{0.500}{0.889} = 0.562$$

If you now look up the angle whose sine is 0.562 in the table on page 703, you will find that, to the nearest degree,

$$r = 34°$$

Note that the ray of light is refracted away from the normal.

Test Yourself Problem

A ray of light passes from carbon disulfide into water. What is the angle of refraction when the ray is incident at (a) 30°; (b) 40°; (c) 50°?

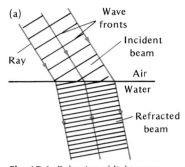

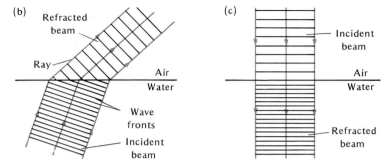

Fig. 17-6. Behavior of light waves on passing (a) obliquely from air into water; (b) obliquely from water into air; (c) normally from air into water.

17-11 Wave Theory Explanation of Refraction

We saw in Section 15-12 that water waves are refracted because they change speed on passing from shallow into deeper water or vice versa. The wave theory of light states that light is refracted because its waves also change speed when they pass from one substance to another.

To explain why light is refracted toward the normal on passing obliquely from air into a substance, the theory assumes that the speed of light waves is decreased on entering the substance. In Fig. 17-6 (a), a beam of parallel rays of light is passing obliquely from air into water. The wave fronts of such a beam are a series of parallel plane surfaces at right angles to the beam. As each wave front passes into the water, it slows down. However, since the left part of each wave front enters the water first, that part slows down first. The rest of the wave front continues to travel in air at the original speed until it too enters the water and is slowed down. The effect of this successive slowing down of each part of the wave front from left to right is to cause the wave front to pivot about its left end and to move into a new direction closer to the normal than before. The rays which are perpendicular to the new wave front are therefore refracted toward the normal.

To explain why light is refracted away from the normal on passing obliquely from a substance into air, the wave theory assumes that the speed of light waves is increased as they pass out of the substance into the air. In Fig. 17-6 (b), the wave fronts of a beam of light are passing obliquely from water into air and speed up as they do so. Since the left part of each wave front leaves the water first, there is a successive speeding up of the emerging wave front from left to right. As a result, the wave front pivots about its right end as it passes into the air taking a direction further from the normal than before. This change of direction of the wave fronts is noticed as the refraction of the rays away from the normal.

Light waves passing obliquely from one medium into another are refracted because the waves slow down or speed up on entering the new medium.

The failure of the beam to change direction on passing into or out of a substance along the normal is explained with the aid of Fig. 17-6 (c). Here, all parts of each wave front are slowed down at the same time. The direction of the wave fronts therefore does not change and no refraction takes place.

17-12 Wave Theory Interpretation of Snell's Law

The wave theory is able to do more than merely explain how refraction takes place. It can give the exact relationship that should be followed by a light ray in passing from one substance to another. As will be shown below, the theory predicts that if v_a is the speed of light in air and v_s is its speed in a given substance, then light rays passing from air into that substance conform to the relationship:

$$\frac{\sin i}{\sin r} = \frac{v_a}{v_s}$$

Since, by Snell's law sin i/sin r is also equal to n_s, this indicates that $n_s = v_a/v_s$. Thus, according to the wave theory, *the index of refraction of a substance is the ratio of the speed of light in air to its speed in that substance.* This prediction can be checked experimentally by measuring n_s, v_a, and v_s for many different substances. It is found to be true.

The index of refraction of a substance is equal to the velocity of light in air divided by its velocity in the substance.

17-13 Derivation of Snell's Law

We shall now show how the relationship:

$$\frac{\sin i}{\sin r} = \frac{v_a}{v_s}$$

is derived from the wave theory of light.

Consider Fig. 17-7 in which an incident light beam consisting of plane parallel wave fronts is passing from air into a substance s, such as water. BE is the wave front whose left end is about to enter

Optional derivation for honor students

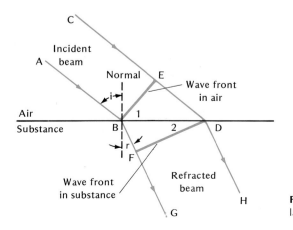

Fig. 17-7. Derivation of Snell's law.

the substance s. AB and CD are the light rays forming the opposite edges of the incident beam and BG and DH are their respective refracted rays. Finally, FD is the position taken by BE at the moment it completely enters the substance s.

As the wave front moves from position BE into position FD, the light at E travels the distance ED in air while, at the same time, the light at B travels the distance BF in the substance. The ratio of ED to BF is therefore the ratio of the speed of light in air v_a to that in the substance v_s or:

$$\frac{ED}{BF} = \frac{v_a}{v_s}$$

Now, in right triangle BED, angle 1 is equal to the angle of incidence i, because its corresponding sides are perpendicular to those of angle i. For a similar reason, angle 2 in right triangle BFD is equal to the angle of refraction r.

It was shown in Section 4-17 that the sine of an acute angle in a right triangle is the ratio of the opposite side to the hypotenuse. It follows in triangle BED that:

$$\sin 1 = \sin i = \frac{ED}{BD}$$

and in triangle BFD, that:

$$\sin 2 = \sin r = \frac{BF}{BD}$$

Dividing the first equation by the second,

$$\frac{\sin i}{\sin r} = \frac{ED}{BD} \div \frac{BF}{BD} = \frac{ED}{BF}$$

and since,

$$\frac{ED}{BF} = \frac{v_a}{v_s}$$

it follows that:

$$\frac{\sin i}{\sin r} = \frac{v_a}{v_s}$$

17-14 Measuring the Speed of Light in a Substance

The relationship $n_s = v_a/v_s$ enables us to make a simple determination of the speed of light in a substance when we know the speed of light in air. We need only to measure n_s by the method described in Sec. 17-7 and substitute in the above relationship. Thus, knowing that for water $n_s = 1.33$ and that $v_a = 3.0 \times 10^8$ meters per second, it follows that $1.33 = 3.0 \times 10^8$ meters per second $\div v_s$. Thus, v_s, the velocity of light in water, comes out to be 2.3×10^8 meters per second.

17-15 Relative Index of Refraction

The relative index of refraction of substance 2 with respect to substance 1 is $n_{1,2} = n_2/n_1$. But $n_2 = v_a/v_2$ where v_2 is the speed of light in substance 2 and $n_1 = v_a/v_1$ where v_1 is the speed of light in substance 1. Hence,

$$n_{1,2} = \frac{v_a}{v_2} \div \frac{v_a}{v_1} = \frac{v_1}{v_2}$$

Thus, the relative index of refraction of substance 2 with respect to substance 1 is the ratio of the velocity of light in substance 1 to its velocity in substance 2.

17-16 Effects Caused by Refraction

Many common optical illusions are the result of refraction. As shown in Fig. 17-8, the refraction of light on passing from water into air causes objects under the water to appear to a viewer to be nearer the surface of the water than they actually are. Objects seen obliquely through thick plate glass windows or through prisms appear to be displaced from their true positions. In each of these cases, light is refracted twice, once on passing from air into the glass and again on leaving the glass to enter the air. Fig. 17-9 shows the paths of typical light rays in these cases.

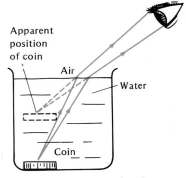

Fig. 17-8. Refraction makes the coin appear to be nearer the surface of the water than it is.

(a)   (b)

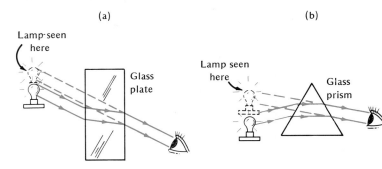

Fig. 17-9. The refraction of light by (**a**) a glass plate, and (**b**) a triangular prism, causes an object viewed through the plate or prism to appear to be displaced from its true position.

Probably the most important effect caused by refraction is the lengthening of the day. Although the difference between the speed of light in a vacuum and in air is small enough to be neglected in most ordinary situations, it causes a significant degree of refraction in the light rays coming from the sun when they enter and pass through the earth's atmosphere. As a light ray passes from nearly vacuous outer space into the comparatively dense atmosphere, it is refracted toward the earth. As a result, the sun appears to be displaced from its true position and we see it and receive some of its light for a short time before it actually comes over the horizon in the morning. The same effect is repeated in the evening when refraction brings us sunlight for a short time after the sun has actually set. Thus, the net result of the refraction of the sun's light by our atmosphere is to lengthen the day.

Fig. 17-10. Apparent flattening of the setting sun is caused by refraction of its rays as they slant through the earth's atmosphere.

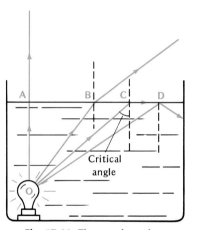

Fig. 17-11. The rays from the lamp pass out of the water into air when they are incident at angles less than the critical angle. Rays incident at angles greater than the critical angle are totally reflected.

17-17 Total Reflection

Light rays passing from a substance of higher refractive index to one of lower refractive index bend away from the normal. For them therefore, there is a certain angle of incidence for which the refracted ray makes a right angle with the normal and comes out along the surface between the two media. The angle of incidence for which this happens is called the *critical angle*. A ray in the more refractive substance that is incident upon the surface at an angle greater than the critical angle is unable to pass through the surface. Instead it is reflected back into the original substance. Such a ray is said to be *totally reflected*.

In Fig. 17-11, a number of rays are shown passing from water into air. The ray *OA* strikes the surface normally and the emerging ray undergoes no change of direction. The ray *OB* is incident at an angle smaller than the critical angle and the refracted ray is bent away from the normal. The ray *OC* is incident at the critical angle and the refracted ray coincides with the surface of the water. The ray *OD* is incident at an angle greater than the critical angle and is totally reflected.

17-18 Computing the Critical Angle

The critical angle can be computed from Snell's law: $\sin i / \sin r = n_2/n_1$ where n_1 is greater than n_2. When i is the critical angle, r is exactly 90 degrees and $\sin r = \sin 90° = 1$. Here, $\sin i / 1 = n_2/n_1$ and the sine of the critical angle is determined by the ratio n_2/n_1.

Sample Problem

Find the critical angle for light rays passing from water into air.

Solution:

$$n_2 = n_{air} = 1.00 \qquad n_1 = n_{water} = 1.33$$

$$\sin i = \frac{n_2}{n_1}$$

$$\sin i = \frac{1.00}{1.33} = 0.75$$

From the table on page 703, the angle whose sine is 0.75 is 49°.

Test Yourself Problem

What is the critical angle for light rays passing from (*a*) quartz into water; (*b*) diamond into air; (*c*) carbon disulfide into water?

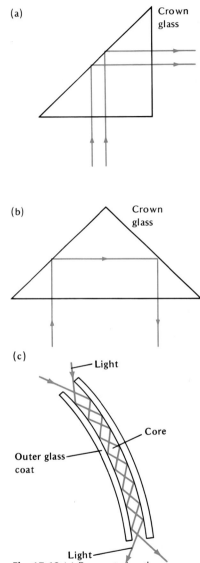

(a) Crown glass

(b) Crown glass

(c) Light

Core

Outer glass coat

Light

Fig. 17-12 (a) Rays entering the short side of a right-angled prism are totally reflected through 90°. **(b)** Rays entering the long side of the prism reverse their direction as a result of two internal reflections. **(c)** An optical fiber "pipes" light around curves by total reflections.

17-19 Totally Reflecting Devices

Total reflection is important in designing prisms to act as reflectors in binoculars and other optical instruments. A right triangular prism with 45-degree angles made of crown glass makes an excellent device for bending light rays through an angle of 90 degrees. Light rays entering one of the short sides of the prism as shown in Fig. 17-12 (a) are incident on the long side of the prism at an angle of 45 degrees. This is greater than the critical angle of crown glass which is only 42 degrees. The rays are therefore totally reflected and emerge at right angles to their original direction. Similarly, light rays entering the prism perpendicular to the long side reverse their direction as shown in Fig. 17-12 (b). Optical instrument designers prefer prisms to plane mirrors for the task of changing the direction of light rays by reflection because prisms lose much less light by absorption than do mirrors.

Total reflection is also used in specially coated optical glass fibers. Light entering an optical fiber is conducted through it almost without loss by successive total reflections from the outer coat of the fiber. Such fibers act like light pipes. They are used in medical instruments to guide light into interior parts of the body and in communications. (See *Physics+Plus* after Chapter 16.)

17-20 Refraction of White Light by a Prism

Isaac Newton was the first to show that all light is colored and that what is called white light is merely a combination of all the colors of light. He did this by passing a narrow beam of sunlight into a triangular glass prism and catching the light that emerged on a screen. He saw that the light on the screen was arranged in a rainbow-like assembly of colors in the order: red, orange, yellow, green, blue, indigo, and violet. In some manner, the refraction of the white light through the prism had separated the light into the colors of which it is composed. (See Fig. 17-15.)

Newton reasoned that if white light is actually a combination of the colored lights seen on the screen, then, bringing the colors together again should produce white light. He showed that this is actually what happens, by passing the colored lights emerging from the first prism into a second inverted prism. The colors were recombined in the second prism into a beam of white light.

17-21 Dispersion of Light

Sunlight is not the only source of light that consists of a mixture of many colors. The lights given by candles, incandescent electric lights, fluorescent lamps, and nearly all other sources of light are

Fig. 17-13. An artist's representation of Newton using a prism to disperse sunlight.

also mixtures of different colors of light. You can show this by passing the light from these sources through a triangular glass prism, as Newton did with sunlight, and catching the emerging colored beams on a screen. The process of separating light into its component colors is called *dispersion*. The assembly of colors that appears on the screen is called the *spectrum* of the light that entered the prism. Different light sources produce different spectra.

17-22 How Prisms Disperse Light

The ability of a triangular prism to disperse light passing through it into a spectrum depends upon the fact that different colors of light are refracted different amounts by the glass. This means that *the glass has a slightly different index of refraction for each color of light.* For violet light, the glass has the greatest index of refraction. Violet light is therefore *refracted more* by the glass prism than all the other colors. Next in order come indigo, blue, green, yellow, and orange light. For red light, the glass has the smallest index of refraction. Red light is therefore *refracted less* by the prism than all other colors of light.

17-23 Speed of Colored Lights in Glass

In air or in a vacuum, all the colors of light travel at the same speed. On entering glass, however, all colors of light slow down, each by a different amount. The index of refraction of any color of light is the ratio of its speed in air to its speed in glass. Since the index of refraction of glass for violet light is greater than that for all other colors, the *speed* of violet light through the glass must be *slower* than that of the other colors of light. From the degree of refraction undergone by each of the other colors of light, it follows that their speeds in glass increase gradually in the order: indigo, blue, green, yellow, and orange. Red light, with the lowest index of refraction, *travels faster* through the glass than all the other colors.

17-24 Dispersion by Diamonds and Water Drops

Glass is not the only substance in which different colors of light travel at slightly different speeds. Other substances having this property are water, carbon disulfide, lucite, and diamond. Since they can disperse light into a spectrum, they are called dispersive substances.

Fig. 17-14. Dispersion of sunlight by raindrops produces a rainbow.

The sparkling colors seen in cut diamonds and the brilliant colors of the rainbow are produced in the same way. The colors of the diamond are the result of the dispersion by the diamond of the white light that enters it. The colors of the rainbow are the result of the dispersion by raindrops of sunlight. (See Fig. 17-14.)

Fig. 17-15. Dispersion of white light into its spectrum by a glass prism.

17-25 Wavelength and Color of Light

We shall see in Chapter 19 that *each color of light has a different frequency and therefore a different wavelength*. Wavelengths of light are frequently measured in units of 10^{-10} meters called *angstroms* (symbol, A). In *SI* metric measure the corresponding units are *nanometers* (symbol, nm), equal to 10^{-9} meters. *One nanometer equals ten angstroms*. In air, the wavelengths of light range from 400 nanometers, which we see as violet, to 750 nanometers, which we see as red.

Arranged according to increasing wavelength, the colors of light take the order: violet, indigo, blue, green, yellow, orange, and red. Referring to Fig. 17-15, you can see from this order that the shorter the wavelength of a color of light, the more it is refracted by the prism. This means that the shorter the wavelength of a color, the slower is its speed in glass. In theory, each wavelength of light should differ in color from other wavelengths. Actually the eye cannot tell the difference in color between two wavelengths that differ by only small amounts.

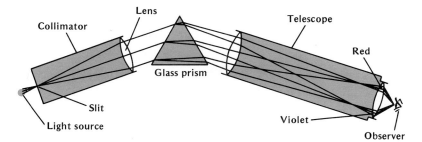

Fig. 17-16. Diagram of a prism spectroscope.

17-26 Spectroscope

The triangular glass prism gives us a rough means of separating and analyzing the light coming from any source into its component colors. To do this more precisely, we use the *prism spectroscope,* a device having a triangular glass prism as its main part. The light to be analyzed is passed into the spectroscope through a tube called the *collimator*. The collimator has a slit at one end and a converging lens at the other. Its job is to sort out a beam of parallel rays and to direct them into the prism. The prism then refracts and disperses this beam into a spectrum which is viewed through the observer's telescope. High-grade spectroscopes usually have a scale built into the instrument that makes it possible to measure the actual wavelengths of the colors seen in the telescope.

A very important use of the spectroscope is the identification of substances by the composition of the light that they emit when they are stimulated electrically and by other means. By analyzing the emitted light, the spectroscope is able to detect traces of substances too minute to be examined chemically and to give information not only of the presence of the substance but also of its quantity.

17-27 Emission Spectra

The spectrum produced by a source of light is called an *emission spectrum*. When different light sources are viewed through the spectroscope, they are found to have two main types of spectra. One of them is the *continuous emission spectrum* in which all the colors merge into each other in a continuous band from the red light at one end to the violet light at the other. Continuous spectra are emitted by hot incandescent solids and liquids as well as by incandescent gases under high pressure. Examples of continuous spectra are those produced by incandescent electric lights and carbon arc lamps. (See Fig. 17-17a.)

The second type of spectrum emitted by a light source is the *line emission spectrum* which consists of several individual lines of color separated by dark spaces. Line emission spectra are produced by gases or vapors under low or normal pressures when they are stimulated to emit light by heat, electric discharges, and other means. Examples of light sources that emit line spectra are gas discharge tubes, such as the neon tube used in advertising, and electric sparks. (See Fig. 17-17, b & c.)

17-28 Fingerprints of the Elements

Elements can be identified by their line spectra.

Each of the more than one hundred chemical elements can be vaporized and made to emit a line spectrum. Because the line spectrum emitted by each element is different from that emitted by any other element, line emission spectra can serve as "fingerprints" to identify the atoms of the elements that produce them. When scientists see in a spectroscope the highly colored lines that belong to the line emission spectrum of a particular element, they know that atoms of that element must be present because no other atoms emit those identical spectral lines.

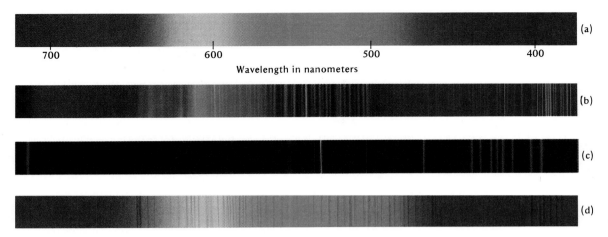

Fig. 17-17. Types of spectra: (a) continuous spectrum; (b) bright-line spectrum of iron; (c) line spectrum of atomic hydrogen; (d) Fraunhofer lines (absorption spectrum).

17-29 Absorption Spectra

When white light is passed through various transparent substances and then examined in the spectroscope, it is often found that some of the colors usually present in the spectrum of white light are missing. In their places we have dark absorption lines or bands surrounded by the remaining colors. Such a spectrum is called an *absorption spectrum.*

When the absorbing substance is a solid or a liquid, we obtain an absorption spectrum consisting of one or more wide black bands in an otherwise continuous spectrum. When the absorbing substance is a gas or vapor, we obtain a line absorption spectrum consisting of individual black lines interspersed throughout the continuous spectrum. (See Fig. 17-17d.)

17-30 The Sun's Spectrum

A study of the sun's spectrum illustrates how the spectroscope gives us information about the sun's composition. The main body of the sun consists of incandescent gases at high temperature and pressure and apparently produces a continuous spectrum as expected. However, when we examine the spectrum of the sun in the spectroscope, we find that it is actually a line absorption spectrum with many black lines distributed throughout it. These are called *Fraunhofer lines* in honor of the Bavarian optician who first observed and mapped them accurately.

The Fraunhofer lines indicate that the sun is surrounded by an atmosphere of cooler gases that are absorbing some of the wavelengths of light coming from the main body of the sun. Since a gas absorbs the very same wavelengths of light that it emits in its own spectrum when properly stimulated, the dark absorption lines tell us what gases absorbed them. Thus, two Fraunhofer lines called the D-lines appear in the yellow part of the spectrum where two

Before it was found on earth helium was discovered on the sun by Lockyer in 1868 when he identified its Fraunhofer lines in the sun's spectrum.

lines in the emission spectrum of sodium are normally found. The absorption of these lines by the sun's atmosphere indicates that sodium vapor is present in the sun's atmosphere. In a similar way, the other Fraunhofer lines show the presence of other chemical elements in the sun's atmosphere that absorbed them and thus reveal to us the composition of the sun's atmosphere.

CHAPTER REVIEW

Summary

In the processes of reflection and refraction, the behavior of light is best described in terms of rays. **Reflection** is the rebounding of a ray of light when it falls upon the surface of a substance. The **law of reflection** states that in every reflection of a ray of light from a surface, *the angle of reflection is equal to the angle of incidence;* also, *the incident ray, the normal, and the reflected ray lie in the same plane.*

Refraction is the sudden change in direction that a ray of light undergoes in passing obliquely from one substance into another. The **wave theory** explains that light is refracted on passing from one medium into another because its speed changes. Each substance has a characteristic **index of refraction** n_s, which is the ratio of the speed of light in air or in a vacuum v_a to the speed of light in that substance v_s; that is, $\mathbf{n_s} = \mathbf{v_a/v_s}$.

A light ray is refracted *toward the normal* when it slows down on passing obliquely from a substance of lower index of refraction to one of higher index of refraction. A light ray is refracted *away from the normal* when it speeds up on passing obliquely from a substance of higher index of refraction to one of lower index of refraction. For this kind of refraction, there is an angle of incidence called the **critical angle** for which the angle of refraction is 90 degrees. Any ray passing from a substance of higher index of refraction to one of lower index of refraction at an angle greater than the critical angle is **totally reflected** and returns to the substance from which it came.

Rays obey the following quantitative relationships known as **Snell's law:**
For a ray passing from air or a vacuum into a substance:

$$\frac{\sin i}{\sin r} = n_s$$

where i is the angle of incidence, r is the angle of refraction, and n_s is the index of refraction of the substance.

For a ray passing from any substance into air or a vacuum:

$$\frac{\sin I}{\sin R} = \frac{1}{n_s}$$

where I is the angle of incidence and R is the angle of refraction.

For a ray passing from any substance into any other substance:

$$\frac{\sin i}{\sin r} = n_{1.2}$$

where $n_{1.2}$ is the **relative index of refraction** of substance 2 with respect to substance 1. If n_1 and n_2 are the indices of refraction of the substances 1 and 2, $n_{1.2} = n_2/n_1$. Also, $n_{1.2} = v_1/v_2$ where v_1 and v_2 are the velocities of light in substances 1 and 2 respectively.

The speeds of the different colors or wavelengths of light are slowed down by different amounts when they enter glass and other refractive substances. The shorter the wavelength, the more its speed is reduced. For this reason, glass and other refractive materials have slightly different indices of refraction for the different colors or wavelengths of light. Such substances can therefore be used to **disperse light** from any source into its spectrum or component colors by passing the light through triangular prisms made of them. When this is done to white light from the sun or an incandescent lamp, a **continuous spectrum** consisting of all colors of light is the result.

The **spectroscope** is a precise instrument for studying the spectra associated with different sources of light. In general, very hot solids and liquids and very hot gases under pressure emit **continuous spectra.** Gases under lower pressure emit **line spectra** when properly excited by heat or by electrical means. Each chemical element has its own characteristic line spectrum by which it can be identified.

When a continuous spectrum is passed through a cool gas, the gas absorbs from it the lines that it normally emits when excited, thus forming a line **absorption spectrum.** The series of black **Fraunhofer lines** in the sun's continuous spectrum are formed in this way.

Questions

Group 1

1. Draw a horizontal and a vertical line on your paper to represent the edges of two plane mirrors at right angles to each other. Draw any ray that falls on the horizontal mirror at such an angle that it is also reflected from the vertical mirror. Show by construction that, after being reflected from both mirrors, the ray takes a direction opposite and parallel to its initial direction. Repeat this construction for a second ray having a different initial direction.

2. (a) Distinguish between regular and diffuse reflection. (b) Why is it desirable that the pages of a book be rough rather than smooth and glossy?

3. (a) What is refraction? (b) In which direction is light refracted on passing obliquely from air or a vacuum into a denser substance? (c) What happens to the angle of refraction as the angle of incidence decreases from 45° to 0°?

4. (a) State Snell's law. (b) How can this law be used to measure the index of refraction of a substance?

5. A small lamp is at the bottom of a fish tank full of water. Make a drawing showing (a) the path of a light ray passing along the normal to the surface from the water into the air; (b) the path of a light ray passing obliquely at an angle smaller than the critical angle from the water into the air. (c) In (a) and (b) show the paths of the reflected rays as well.

6. (a) What is meant by the reversibility of light rays? (b) Make a diagram showing a ray of light being reflected from a horizontal plane mirror. Insert a second plane mirror in the path of the reflected ray in such a way that the ray will reverse its direction and return to its starting point. (c) Make a diagram showing the path of a ray of light entering a water tank at an angle of incidence of about 45°. Put a plane mirror in the path of the refracted ray in such a way that the ray will be caused to reverse its direction and will retrace its path.

7. Explain, according to the wave theory: (a) why light is not refracted on passing along the normal to the surface between two substances; (b) why light is refracted toward the normal on passing obliquely from air or a vacuum into a denser substance; (c) why light is refracted away from the normal on passing obliquely from a denser substance into air or a vacuum.

8. (a) According to the wave theory, what does the index of refraction of a substance with respect

to air or a vacuum represent? (b) How can the speed of light in a substance be determined from its index of refraction? (c) How does the speed of light in all other substances compare with its speed in air?

9. Show by a diagram why objects viewed obliquely through a thick glass plate appear to be displaced from their true positions.

10. (a) Explain how refraction results in the lengthening of the day. (b) Show by a diagram how we can see the sun at sunset even though it is actually slightly below the horizon.

11. (a) How may a triangular glass prism be used to separate the light from an incandescent spotlight into the colors of light of which it is composed? (b) Why does the prism refract the different colors of light different amounts?

12. (a) Glass has a slightly higher refractive index for blue light than it does for red light. What does this fact tell about the relative speeds with which these colors of light travel in glass? (b) What evidence is there that diamond has a slightly different index of refraction for each of the colors of light?

13. Arrange the different colors of light in (a) the order of increasing wavelength; (b) the order of increasing frequency. (c) When a given color of light passes from air into a denser substance, its frequency remains constant. What happens to its wavelength?

14. (a) Describe a prism spectroscope. (b) Explain how it is used to analyze a source of light.

15. (a) Distinguish between a continuous emission spectrum and a line emission spectrum. (b) What kinds of light sources produce each of them?

16. Why may line spectra be used as an accurate means of identifying each of the different elements?

17. The sun's spectrum with its black Fraunhofer lines is an absorption spectrum. How do the Fraunhofer lines tell us what elements are present in the sun's atmosphere?

Group 2

18. (a) Why may a light ray be totally reflected only on passing from a substance of higher refractive index to one of lower refractive index? (b) With the aid of a diagram, define the critical angle.

19. Crown glass has a critical angle of 42°. Diamond has a critical angle of 24°. Explain why a diamond sparkles more than a piece of crown glass cut to the same design.

20. Two similar basins are filled to the same level with water ($n = 1.33$) and carbon disulfide ($n = 1.63$). A coin is put on the bottom of each basin. Which coin will appear to be nearer to the eye of an observer looking down into each of the basins? Explain your answer with the aid of diagrams similar to that in Fig. 17-8.

21. Although the light from the sun is refracted on passing from the vacuum of space into the atmosphere, it is not separated into its spectrum. What does this tell us about the speeds with which the different colors of light travel in air?

22. To a man swimming at the bottom of an indoor swimming pool, a lamp in the ceiling above the pool appears to be further away than it actually is. Explain.

Problems

Group 1

1. A ray of light falls upon the center of a horizontal rectangular mirror at an angle of 30°. (a) Construct a diagram showing the direction of the reflected ray. (b) Keeping the incident ray in the same direction, now rotate the mirror about a horizontal axis passing through its center so that the right half of the mirror is 20° below its initial position. Construct a second diagram showing the new direction of the reflected ray. (c) How does the angle between the reflected rays in (b) and (a) compare with the angle through which the mirror was rotated?

2. Referring to the procedure in problem 1, prove that when an incident ray having a fixed direction falls upon a slowly rotating plane mirror, the reflected ray always turns through an angle twice as great as that through which the mirror turns.

3. A ray passes from air into water ($n = 1.33$) at an angle of incidence of 50°. Find the angle of refraction to the nearest degree.

4. A ray of light is incident on a block of polyethylene ($n = 1.50$) at angles of 0°, 30°, 45°, 60°, and 80° in succession. (a) In each case, determine the angle of refraction and list it in a table

opposite its corresponding angle of incidence. (b) Plot the angles of refraction against the corresponding angles of incidence on a graph. Are they in direct proportion to each other?

5. Two rays of light are incident at an angle of 45° on a piece of quartz (n = 1.46) and a diamond (n = 2.42) respectively. Compare their angles of refraction.

6. A certain color of light is incident upon a rectangular block of quartz at an angle of 30.0°. It is refracted by the quartz so that the angle of refraction is 20.0°. What is the index of refraction of quartz for this particular light?

7. (a) A ray passes from water (n = 1.33) into air at an angle of incidence of 20°. Find the angle of refraction to the nearest degree. (b) Find the angles of refraction for angles of incidence of 30° and 40°. (c) What change takes place in the angle of refraction as the angle of incidence is increased?

8. Assuming 3.0×10^8 m/s as the speed of light in a vacuum, find the speed of light in (a) flint glass (n = 1.66); (b) carbon disulfide (n = 1.63); (c) quartz (n = 1.46).

9. The speed of light in a certain clear plastic is 2.0×10^8 m/s. What is the index of refraction of the plastic?

10. Find the critical angle for light rays passing from (a) crown glass (n = 1.52) into air; (b) crown glass into water (n = 1.33).

Group 2

11. (a) What is the relative index of refraction of water (n = 1.33) with respect to carbon disulfide (n = 1.63)? (b) What is the relative index of refraction of carbon disulfide with respect to water? (c) How much larger is the speed of light in water than it is in carbon disulfide?

12. A layer of water is supported on top of a layer of carbon disulfide. A ray of light passes from the water into the carbon disulfide at an angle of incidence of 45°. Determine the angle of refraction to the nearest degree.

13. Determine the ratio of the velocity of light in polyethylene (n = 1.50) to that in diamond (n = 2.62).

14. The index of refraction of crown glass for violet light is 1.53, while for red light it is 1.51. (a) Assuming the speed of light in a vacuum is 3.00×10^8 m/s, what is the speed of violet light in crown glass? (b) Similarly, what is the speed of red light in crown glass?

15. Referring to problem 14, suppose both violet and red light enter a block of crown glass from the air at an angle of incidence of 30.0°. (a) What will be the difference in the sines of their angles of refraction? (b) What will be the difference in their angles of refraction?

Applying Physics

We can find the approximate value of the index of refraction of a liquid by comparing its real depth with its apparent depth. Fill a long cylindrical glass jar with water or any other liquid whose index of refraction is to be measured. Hold a ruler vertically in the liquid and note its depth. Hold a second ruler parallel to the first outside the jar. Sighting down both rulers as vertically as possible, raise the outside ruler until its lower end seems to be at the same level as the other. The position of the liquid surface on the outside ruler shows the apparent depth of the liquid. Divide the real depth of the liquid by the apparent depth in order to obtain the index of refraction of the liquid.

18

Mirrors, Lenses, and Optical Instruments

Aims

1. To learn how mirrors and lenses form images.
2. To learn how mirrors and lenses are used in optical instruments to increase the range and the accuracy of human vision.

Mirrors

18-1 Types of Mirrors

The most familiar type of mirror is the plane mirror, consisting of a flat piece of glass silvered on one side. Curved mirrors may be either concave or convex. In each case the reflecting surface is a small portion of a hollow sphere. The center of the sphere is called the *center of curvature* of the mirror. The line joining the center of curvature and the middle point of the mirror is called the *principal axis*. When the inside spherical surface of the mirror is silvered and used as the reflecting surface, we have a *concave mirror*. When the outside spherical surface is silvered and used as the reflecting surface, we have a *convex mirror*.

If the size of a concave or convex mirror is small compared to the radius of the sphere of which it is a part, the mirror reflects light rays in a relatively simple manner. It is then capable of forming clear images of objects. We shall limit our study to mirrors of this sort.

18-2 Concave Mirrors Are Convergent

When a beam of rays parallel to the principal axis of a concave mirror falls upon the mirror, the rays are reflected so that they converge and meet in a point called the *principal focus* of the mirror. *This point is located on the principal axis at a distance equal to approximately half the radius of the mirror.* The distance from the principal focus to the mirror is called the *focal distance* of the concave mirror.

The converging action of a concave mirror can be understood by approximating its shape with many small plane mirrors as shown in Fig. 18-2. Each ray parallel to the principal axis of the mirror is incident on one of these tiny mirrors and is reflected at any angle to the normal equal to the angle of incidence. Since different rays

Fig. 18-1. In this solar cooker the sun's rays are brought to a focus by the concave mirror. The pot, which is located at the focus, is heated as a result. (Note that the concave mirror forms an image that is upside down. See Section 18-11.)

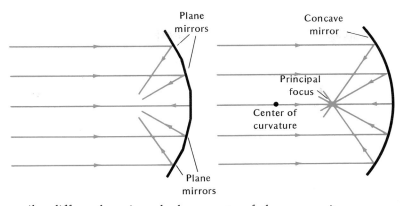

Fig. 18-2 Converging action of a concave mirror.

strike differently oriented plane parts of the composite concave mirror, their reflected rays converge into the principal focus.

Strictly speaking, a spherical concave mirror brings parallel rays to a focus only approximately and only if its radius of curvature is relatively large. Where a mirror capable of precise focusing of parallel rays is desired, curved mirrors of parabolic shape are used. (See Section 18-18.)

18-3 Solar Furnace

The ability of a concave mirror to gather parallel light rays into a point is used in one type of solar furnace. It consists of a large concave mirror mounted so that it can be turned to receive the rays of the sun and bring them to a focus. Temperatures as high as 5000°C can be produced at the focus in this way.

18-4 Searchlight Reflectors

If a point source of light is put at the focus of a concave mirror that part of its light that goes toward the mirror and is reflected from it emerges as a beam of rays parallel to the principal axis. This behavior, shown in Fig. 18-3, is another example of the reversibility of light rays. Each ray leaving the principal focus is reflected from the concave mirror in a direction paralle to the principal axis. This is exactly the reverse of the path taken by the rays in Fig. 18-2. There, each incident ray is parallel to the principal axis and passes into the principal focus after reflection.

Thus, the concave mirror has the ability to change divergent rays into parallel rays. Where a highly concentrated beam of light is needed, as in powerful searchlights, large concave reflectors are often placed behind a strong *point source* of light to produce such a beam. If a beam of parallel rays is desired, the light source is put at the focus of the concave reflector. A small concave reflector may be put in front of the light source to reflect back to the main mirror those light rays that move directly forward from the source. In automobile headlights and in flashlights, where a slowly diverging

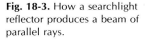

Fig. 18-3. How a searchlight reflector produces a beam of parallel rays.

beam is preferred, the source of light is placed a little closer to the mirror than the principal focus.

18-5 Convex Mirrors Are Divergent

When a beam of rays parallel to the principal axis of a convex mirror falls upon the mirror, the rays are reflected so that they spread apart or diverge. If the reflected rays are extended backward through the mirror, one can see that they pass through a single point behind the mirror. This point is approximately at a distance of half the radius behind the mirror and is the principal focus of a convex mirror. Because the reflected rays only seem to come from this point but do not actually pass through it, it is called a *virtual focus*.

The diverging action of a convex mirror on parallel light rays can be understood by making an approximation of such a mirror by putting together very many small plane mirrors. Fig. 18-4 shows how the reflection of the individual light rays in the incident beam by the differently oriented plane parts of the composite convex mirror causes the rays to diverge.

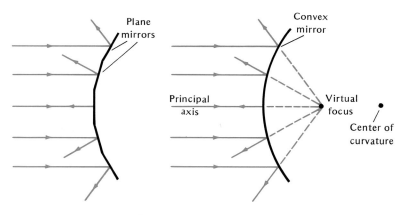

Fig. 18-4. Diverging action of a convex mirror.

18-6 Two Types of Images

Mirrors and lenses form two types of images, *real* and *virtual*. A *real image* is one that can be put on a screen like the image seen on a moving picture screen. This image is composed of light that is cast upon the screen by the motion picture projector. It is seen by looking at the screen.

A *virtual image* is the kind of image seen in a plane mirror. Here, the image seems to be located behind the mirror. However, no light actually falls at the place where the image seems to be. The image therefore cannot be put on a screen. It can only be seen by looking into the mirror.

In general, real images are usually seen on screens on which they are projected. Virtual images can only be seen by looking into the mirror or lens producing them.

18-7 How Real Images Form

Every object can be thought of as a collection of many points. To simplify matters, let us see how an image of just one of these points is formed.

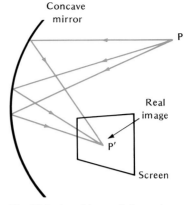

Look at the period at the end of a sentence from several different directions. Notice that you can see it from any direction. Because this period is reflecting light in all directions, it acts like a point source of light. In the same way, every point of an object acts like a point source of light from which rays of light diverge in all directions.

To see how a real image forms, put a point P several focal distances in front of a concave mirror as shown in Fig. 18-5. Now follow several diverging rays of light leaving P. As they fall upon the mirror, they are reflected in such a manner that they converge and meet again at point P'. If a screen is placed at P', a real image of P will be seen there.

Fig. 18-5. A real image is formed when the diverging rays leaving P actually meet again in P' after reflection.

This illustrates how a real image of a point is formed. A concave mirror or other optical device is used to cause the diverging rays of light leaving the point to converge. Where the converging rays meet in a new point, they form a real image of the point from which they came.

18-8 How Virtual Images Form

To see how a virtual image forms, put a point P in front of a plane mirror. Again, follow several rays diverging from P. After reflection from the mirror, the rays continue to diverge as shown in Fig. 18-6. However, on extending the reflected rays backward through the mirror, you find that they meet at P', a point behind the mirror. The eye seeing these reflected rays therefore judges them to be coming not from P but from P'. Thus an image of P is seen at P'. Since no light actually comes from P', the image seen there is a virtual image.

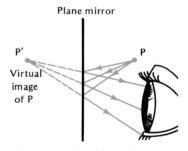

Fig. 18-6. A virtual image is formed when the diverging rays leaving P continue to diverge after reflection in such a way that they seem to come from P'.

This illustrates how a virtual image of a point is formed. An optical device such as a plane mirror is used to change the direction of the diverging rays of light leaving the point so that they seem to come from another point. A virtual image is then seen at the second point.

18-9 Images Formed by Plane Mirrors

From our knowledge of how rays of light act, we can predict the location and size of the image of an object formed by a mirror, a lens, or any other optical device. This is done by constructing a ray diagram in which the position of the image of any selected point of the object is obtained by following two light rays leaving that point. We usually represent the object by an arrow and obtain the images of its two end points. A line joining the images of the two end points represents the image of the entire object.

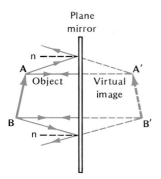

Fig. 18-7. Ray diagram showing how a virtual image is formed by a plane mirror.

To illustrate the method, we apply it to the location of the image of the object *AB* formed by the plane mirror in Fig. 18-7. To find the image of point *A* of the object, two diverging rays are drawn and followed. For simplicity, we select as one of these rays the one normal to the mirror. This ray is reflected back upon itself. The second ray is any other one that falls upon the mirror. To find the direction after reflection, the normal is drawn and a protractor is used to make the angle of reflection equal to the angle of incidence.

The diagram now shows two reflected rays that came originally from point *A*. Since these rays diverge, they do not intersect and do not form a real image. However, when they are extended backward by the dotted lines, their extensions intersect at the point *A'*. Because the reflected rays appear to be coming from this point, a virtual image of *A* is seen at *A'*.

This procedure is now repeated for the point *B* to obtain its virtual image *B'*. Then *A'* and *B'* are joined by line *A'B'*, which is the virtual image of *AB*.

18-10 Characteristics of Plane Mirror Images

The ray diagram confirms the observed characteristics of the image formed by a plane mirror. *The image is virtual, erect, and the same size as the object.* It is also exactly *as far behind the mirror as the object is in front of it.* Finally, it differs from the object in that it is *a symmetrical counterpart of it.* This can be observed by looking at the image of a right hand in a plane mirror. The image is not that of a right hand but of a left hand.

18-11 Images Formed by Concave Mirrors

A concave mirror can form both real and virtual images. *Virtual images are always erect. Real images are always inverted.* Which type of image is formed depends upon how far the object is from the mirror. *When the object is further from the mirror than the focal distance, the mirror makes a real image of it. When the object is nearer than the focal distance, the mirror makes a virtual image of it. When the object is located exactly at the focus of the mirror, no image is formed.* In this case, as we found in Section 18-4, the reflected rays are parallel and will not meet to form an image no matter how far they are extended in either direction.

The size of the image is also determined by the position of the object. *When the object is farther away than the center of curvature or two focal distances from the concave mirror, the image formed is smaller than the object. When the object is located exactly at two focal distances from the mirror, the image and object are the same size. When the object is closer to the mirror than two focal distances, the image is larger than the object.*

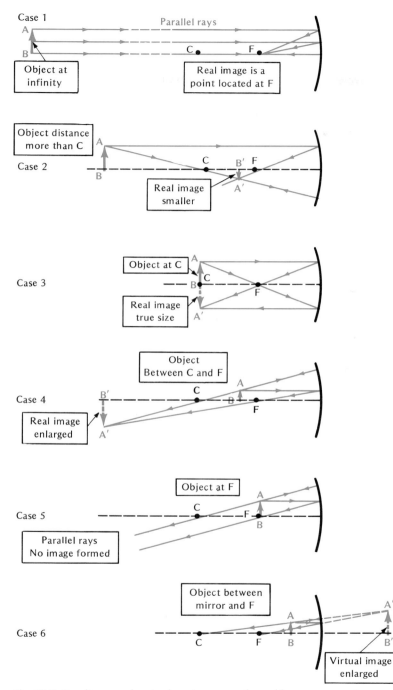

Fig. 18-8. Ray diagrams showing how images are formed by a concave mirror as the object distance changes.

18-12 Ray Diagrams for Concave Mirrors

As in the case of the plane mirror, the image of each point of an object in a concave mirror is found by following two rays of light that leave that point. For simplicity, these rays are chosen as follows. One of them is the ray that is parallel to the principal axis of the mirror. As was seen in Fig. 18-2, such rays pass through the focus of the mirror after reflection. The second ray is the one that passes through the center of curvature of the mirror. Since this ray coincides with a radius of the mirror, it hits the surface of the mirror at right angles and is therefore reflected back upon itself. If these two rays converge after reflection, the point where they meet is a real image of the given object point. If these rays diverge after reflection, they are extended backward by dotted lines until they meet in a point behind the mirror. That point is a virtual image of the given point of the object.

Fig. 18-8 shows ray diagrams for six typical positions of the object as the object starts from infinity and moves toward the mirror. To simplify the diagrams, the object is taken as an arrow whose bottom point is on the principal axis of the mirror. It is then necessary to find only the image of the top of the arrow A. Two rays are drawn from A to the mirror and their real or virtual intersection after reflection gives A′, the image of A. Since the image of B, the bottom of the arrow, must lie on the principal axis, and since the entire image comes out vertical, a perpendicular from A′ to the principal axis represents the complete image, A′B′.

18-13 Applications of Concave Mirrors

In addition to their use in the solar furnace, and as reflectors for searchlights and automobile headlamps, concave mirrors are sometimes used as shaving mirrors to give a magnified image of the face. The shaving mirror operates as in case 6 where the object, the face, is nearer to the mirror than the focal length. A magnified, virtual, erect image of the face is seen in the mirror.

Most of the great astronomical telescopes in the world are reflecting telescopes in which the main optical part is a large concave mirror. This telescope was invented by Sir Isaac Newton. In its simplest form, a *reflecting telescope* consists of a tube at the bottom of which is a concave mirror of long focal length. Since light rays entering the telescope from a star are essentially parallel, they are brought to a point by the concave mirror. This point of light is the image of the star and can be viewed conveniently by means of a small plane mirror and eyepiece positioned in the upper part of the tube, as shown in Fig. 18-9. By this arrangement, the observer views the star images from the side of the telescope with minimum blocking of the light entering the telescope.

Because stars are at very great distances from the earth, their images in the telescope are always points of light and show no

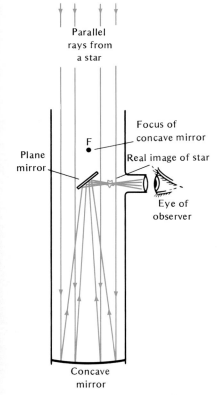

Fig. 18-9. Diagram of a reflecting telescope arranged for viewing the star images from one side.

Parallel rays from a star

Plane mirror

F

Focus of concave mirror

Real image of star

Eye of observer

Concave mirror

details of the structure of the stars themselves. A major advantage of using large concave mirrors in telescopes is their large light-gathering capacity, which makes it possible for them to reveal the presence of very faint stars. The large reflecting telescope at Mt. Palomar, California, has a concave mirror more than 5 meters in diameter.

18-14 Image Formed by Convex Mirrors

A convex mirror produces only virtual images regardless of the position of the object. These images are seen behind the reflecting surface of the mirror. *They are always erect and smaller than the object.* Fig. 18-11 shows how the image is constructed. As before, two rays leaving the top of the object *A* are traced. You can see that after reflection from the mirror, the rays diverge. The reflected rays are therefore extended backward through the mirror until they meet at *A'*. From *A'* a perpendicular is dropped to the principal axis to give *A'B'*, the virtual image of *AB*.

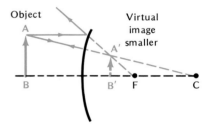

Fig. 18-11. Ray diagram showing how an image is formed by a convex mirror.

The ability of the convex mirror to give reduced images of objects makes it particularly useful as a rearview mirror for the automobile. Such a mirror gives a wider view of the roadway than is possible with a plane mirror.

18-15 Object-Image Relationships in Spherical Mirrors

It is evident from the ray diagrams of Figs. 18-8 and 18-11 that the position and size of the image in each case depends upon the focal length of the mirror and the position and size of the object. A geometrical analysis of any of these diagrams, like that in the next section, shows that the following two general relationships apply to all spherical mirrors, whether concave or convex.

Fig. 18-12. A convex rearview mirror on a car gives a wider field of viewing but the virtual image is reduced in size.

First, for all positions of the object and image:

$$\frac{S_o}{S_i} = \frac{p}{q}$$

where S_o is the size of the object, S_i is the size of the image, p is the distance of the object from the mirror, and q is the distance of its image from the mirror.

Second, if f is the focal distance of the spherical mirror, it turns out that:

$$\frac{1}{p} + \frac{1}{q} = \frac{1}{f}$$

Remember that for a large spherical mirror, f is equal to one-half its radius of curvature R; that is, $f = R/2$.

Optional derivation

18-16 Deriving Spherical Mirror Relationships

Consider the ray diagram of Fig. 18-13. Here, to simplify the derivation, the ray from A that is parallel to the principal axis has been omitted. In its place is drawn the ray that goes from A to the center of the mirror, O. Since the principal axis is the normal to the mirror at O, this ray and its reflected ray make equal angles with the principal axis. A second ray from A is drawn through the center of curvature of the mirror as before. The two reflected rays intersect at A' thus giving the image of A. The image, $A'B'$, is then obtained by dropping a perpendicular from A' to the principal axis.

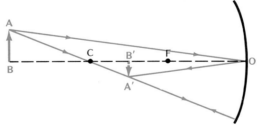

Fig. 18-13. Derivation of the object-image relationships in a concave mirror.

Now, right triangle ABO is similar to right triangle $A'B'O$. Hence,

$$\frac{AB}{A'B'} = \frac{BO}{B'O}$$

We now substitute the following:

$$AB = S_o, \text{ the size of the object,}$$
$$A'B' = S_i, \text{ the size of the image,}$$
$$BO = p, \text{ the object's distance from the mirror,}$$
$$B'O = q, \text{ the image's distance from the mirror,}$$

This gives the first relationship:

$$\frac{S_o}{S_i} = \frac{p}{q}$$

Next, note that right triangle, ABC is similar to right triangle $A'B'C$. Hence,

$$\frac{BC}{B'C} = \frac{AB}{A'B'} \quad \text{or} \quad \frac{BC}{B'C} = \frac{S_o}{S_i} = \frac{p}{q}$$

Now, $BC = BO - CO$, where CO is the radius of the mirror and is equal to twice its focal length, or $2f$. Hence $BC = p - 2f$. Again, $B'C = CO - B'O = 2f - q$. Substituting these values for BC and $B'C$ in:

$$\frac{BC}{B'C} = \frac{p}{q} \quad \text{gives} \quad \frac{p - 2f}{2f - q} = \frac{p}{q}$$

whence:

$$pq - 2fq = 2fp - pq$$
$$2fq + 2fp = 2pq$$

Dividing every term on both sides of the equation by $2fpq$ gives the second relationship:

$$\frac{1}{p} + \frac{1}{q} = \frac{1}{f}$$

18-17 Problems Involving Spherical Mirrors

In applying the above two object-image relationships for spherical mirrors, you will find that q comes out negative when the image is virtual. A negative value of q means that the image is behind the mirror instead of in front of it.

For convex mirrors, the focal length f is negative because it is behind the reflecting surface of the mirror and is a virtual focus. It will be seen for this mirror that the image distance q always turns out to be negative. This is in agreement with the observation that the images made by convex mirrors are always seen behind the mirror surface and are virtual.

Sample Problems

1. An object is 30 cm from a concave mirror of radius 10 cm. (a) At what distance from the mirror will the image be formed? (b) If the object is 5 cm tall, how tall is its image?

Solution:

(a) $p = 30$ cm $f = R/2 = 10$ cm$/2 = 5$ cm

$$\frac{1}{p} + \frac{1}{q} = \frac{1}{f}$$

To clear fractions, multiply each term by pqf and cancel.

$$\frac{pqf}{p} + \frac{pqf}{q} = \frac{pqf}{f}; \quad qf + pf = pq$$

$$\text{whence } q = \frac{pf}{p - f} = \frac{30 \text{ cm} \times 5 \text{ cm}}{30 \text{ cm} - 5 \text{ cm}} = 6 \text{ cm}$$

Since q is positive, the image is real.

(b) $S_o = 5$ cm $p = 30$ cm $q = 6$ cm

$$\frac{S_o}{S_i} = \frac{p}{q}, \text{ whence } S_i = \frac{qS_o}{p}$$

$$S_i = \frac{6 \text{ cm} \times 5 \text{ cm}}{30 \text{ cm}} = 1 \text{ cm}$$

2. An object is 5.0 cm from a concave mirror having a focal length of 10 cm. Find the image distance.

Solution:
$p = 5.0$ cm $f = 10$ cm

$$\frac{1}{p} + \frac{1}{q} = \frac{1}{f}$$

Solve for q as in problem 1.

$$q = \frac{pf}{p - f}$$

$$q = \frac{5.0 \text{ cm} \times 10 \text{ cm}}{5.0 \text{ cm} - 10 \text{ cm}} = -10 \text{ cm}$$

The negative value of q indicates that the image is virtual and 10 cm behind the mirror. The image is also twice as large as the object since q is twice as large as p.

3. A convex mirror of focal length 4 cm makes an image of an object that is 12 cm from the mirror. What is the distance of the image from the mirror?

Solution:
$p = 12$ cm $f = -4$ cm

(f is negative because the mirror is convex.)

$$\frac{1}{p} + \frac{1}{q} = \frac{1}{f}$$

$$q = \frac{pf}{p - f}$$

$$q = \frac{12 \text{ cm} \times (-4 \text{ cm})}{12 \text{ cm} - (-4 \text{ cm})}$$

$$q = -3 \text{ cm}$$

The image is virtual and is seen 3 cm behind the mirror.

Test Yourself Problems

1. An object is 30 cm from a concave mirror having a radius of 20 cm. (*a*) At what distance from the mirror will the image be formed? (*b*) What are the characteristics of the image? (*c*) If the object is 10 cm tall, how tall is the image?
2. The object in the above example is moved toward the mirror to a point 8.0 cm from the mirror. (*a*) Find the position of the image. (*b*) What are its characteristics? (*c*) What is its size?

3. A convex mirror has a focal length of 6 cm and makes an image of an object 6 cm in front of it. (*a*) Find the position of the image. (*b*) What are its characteristics? (*c*) If the size of the object is 10 cm, what is the size of the image?

18-18 Defects of Spherical Mirrors

A mirror with a spherical surface does not actually have exactly the right shape to bring all the incident rays parallel to its principal axis into its principal focus. Rays that fall upon the outer edges of such a mirror come to a focus sooner than those that fall upon the middle part of the mirror. For this reason, mirrors with spherical surfaces cannot produce perfectly clear images. Since this defect is caused by the spherical shape of the mirror, it is called *spherical aberration.*

If the diameter of a spherical concave mirror is small compared to its radius of curvature, the aberration is usually small enough to be neglected. In other cases, spherical aberration is eliminated by giving curved mirrors the shape of a *paraboloid,* a surface that has the proper curvature to produce perfect focusing. Such mirrors are commonly built into automobile headlights to improve their focusing qualities.

Questions

Group 1

1. A person 1.8 m tall stands 2.0 m in front of a full-length plane mirror. (a) How tall is the image? (b) How far behind the mirror is it located? (c) If the person now walks toward the mirror, what happens to the size of the image?

2. A 10-cm ruler is held vertically in front of a vertical plane mirror 5 cm high. The top of the mirror and the top of the ruler are on the same horizontal level. The eye of an observer is directly behind the top of the ruler. Show, by drawing the complete paths of one ray of light leaving the bottom of the object, one leaving the middle of the object, and one leaving the top of the object, that the observer can see the image of the entire ruler in the mirror.

3. An object represented by an arrow 3 cm tall is 6 cm in front of a long vertical plane mirror. (a) Make a ray diagram to scale showing how the image is formed. (b) Describe the image.

4. (a) How would you use the rays of the sun to determine the focal length of a concave mirror? (b) How may a concave mirror use the sun's heat and light to develop high temperatures?

5. (a) Where should a small electric light be put with respect to a concave mirror of focal length 10 cm in order that the rays of light falling on it emerge parallel to the principal axis? (b) What device uses a concave mirror in this manner?

6. An object consisting of a vertical arrow 2 cm high is 8 cm from a concave mirror having a radius of 6 cm. The base of the object is on the principal axis. (a) Make a ray diagram to locate the image of the object. (b) Now move the object to a position 2 cm from the mirror and again locate its image by means of a ray diagram. (c) Which of these images is virtual? (d) What is the difference between a virtual and a real image?

7. Where should an object be placed with respect to a concave mirror to form (a) an enlarged virtual image; (b) an enlarged real image? (c) What purpose is served by having the concave mirror in a reflecting telescope as large as possible?

Group 2

8. Two vertical plane mirrors hinged at an edge are at right angles to each other. A point P is 2 cm from one of the mirrors and 3 cm from the other. Each mirror makes a direct image of P as well as an image of the image of P in the other mirror. (a) Using the fact that a plane mirror image is as far behind the mirror as is the object in front of it, make a diagram showing the location of all the images formed by the mirrors. (b) How many images are there? (c) Repeat (a) and (b) when the angle between the mirrors is changed to 60°.

9. (a) Show by means of ray diagrams that a convex rear-view mirror gives the driver of the car a wider view of the roadway behind him than does a plane mirror of the same size. (b) What is one disadvantage of the convex rear-view mirror as compared to a plane rear-view mirror?

10. (a) Five parallel rays spaced 0.5 cm apart fall on a concave mirror having a 12.0-cm radius. Show by constructing the angles of reflection of the reflected rays that the rays come to a focus at a point about 6 cm from the mirror. (b) Repeat the construction with the same parallel rays but with a mirror having a radius of 4 cm. Do the reflected rays now come to a sharp focus? Explain.

Problems

Group 1

1. An object is 1 m in front of a plane mirror. A photographer standing 3 m from the mirror wishes to take a picture of the image of the object in the mirror. For what distance should the camera be focused?

2. A student runs toward a plane mirror at the rate of 2 m/s. At what rate is the distance between the student and the image decreasing?

3. Each line of Table 18.1 gives some information about a concave or convex mirror. Supply the missing information.

Table 18.1

	OBJECT DISTANCE (cm)	IMAGE DISTANCE (cm)	RADIUS (cm)	FOCAL LENGTH (cm)
Mirror A	10	30		
Mirror B	20			10
Mirror C	40	10		
Mirror D		36	24	
Mirror E	10			−10
Mirror F		−4.0		−12

4. (a) For each mirror in problem 3, state whether the mirror is concave or convex and whether the image is real and inverted, or virtual and erect. (b) For each mirror in problem 3, determine the size of the image if the object is 10 cm high.

5. Make a ray diagram to check your answers for mirror B and mirror E in problem 3.

6. A motorist sees the image of a car in the convex rear-view mirror whose focal length is 1.0 m. If the car is 1.6 m tall and 7.0 m away, what is the size of its image?

7. A shaver's face is 10 cm from a concave shaving mirror whose focal length is 20 cm. (a) At what distance from the mirror is the image? (b) How much does the mirror magnify the shaver's face?

Group 2

8. Show by means of a derivation similar to that in Section 18-16 that the relationship $S_o/S_i = p/q$ also applies to the convex mirror.

9. Two reflecting telescopes have concave mirrors of equal focal length. The diameter of one of them is 15 cm, and that of the other is 10 cm. How much brighter is the image of a star as seen in the larger telescope than the image of the same star as seen in the smaller telescope?

10. A concave mirror has a focal length of 20 cm. Where should an object be placed in order that its virtual image should be twice as tall as the object?

Lenses

18-19 Types of Lenses

Lenses are key parts in such devices as eyeglasses, cameras, and magnifying glasses. A lens is usually a piece of glass or other transparent substance whose opposite sides are very smooth spherical surfaces. The line joining the centers of the spheres of which each surface is a part is called the *principal axis* of the lens. There are two main types of lenses; *converging or convex lenses* and *diverging or concave lenses*. A converging lens is one that has its thickest part in the middle and grows thinner toward the edges. A diverging lens is one that has its thinnest part in the middle and grows thicker toward the edges. Cross sections of some usual types of converging and diverging lenses are shown in Fig. 18-14.

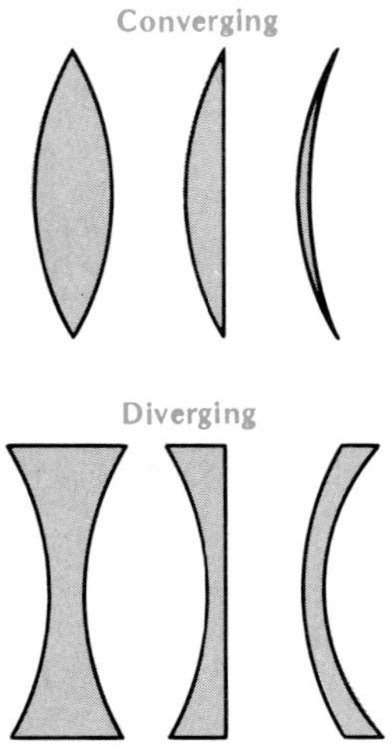

18-20 Convex Lenses Are Convergent

As its name implies, a converging lens makes light rays passing through it converge. When rays parallel to the principal axis of a converging lens pass through the lens, the rays are refracted by the lens so that they meet in a point. This point is called the *principal focus* of the converging lens. The distance from the focus to the center of the lens is called the *focal length*. In general, the thicker a converging lens is, the shorter its focal length and the faster it makes light rays converge.

The ability of the converging lens to bring light rays to a focus is explained by Fig. 18-15. Here, a glass block and four prisms are put together so that they approximate the shape of a converging lens. Light rays passing through the upper prisms are refracted downward. Those passing through the lower prisms are refracted upward, while those going through the center block pass through without change of direction.

Converging

Diverging

Fig. 18-14. Various types of convex and concave lenses.

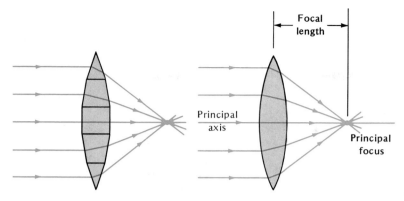

18-21 Concave Lenses Are Divergent

When rays parallel to the principal axis of a diverging lens pass through the lens, the rays are refracted by the lens so that they diverge and therefore do not meet. However, when the diverging rays are extended back through the lens, their extensions meet in a point. This is the principal focus of the lens and its distance from the lens is called the focal length. It is a *virtual focus* because the rays of light emerging from the lens seem to come from this point but never actually pass through it. Fig. 18-16 shows how a set of prisms that approximates a diverging lens refracts light rays so that they diverge.

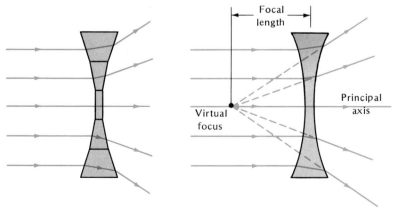

18-22 Images Formed by Converging Lenses

A converging lens brings a beam of parallel rays to a focus. In this respect, its function resembles that of the concave mirror. We shall see that the converging lens produces the same kinds of images as does the concave mirror for similar positions of the object.

A converging lens can produce both real and virtual images, depending upon the position of the object. The real images are always inverted, while the virtual images are always erect. *If the object is further from the lens than one focal distance, the lens makes a real image of it. If the object is nearer than one focal*

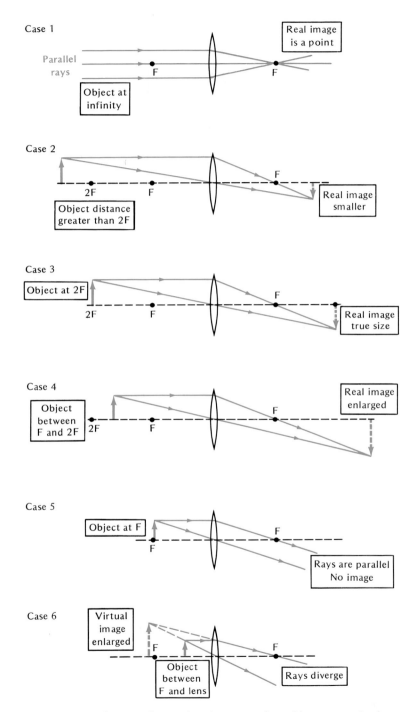

Fig. 18-17. Ray diagrams showing how images are formed by a converging lens as the object distance changes.

distance, the lens makes a virtual image of it. When the object is exactly at the principal focus, the lens refracts the light coming from each point of it so that the rays are parallel on leaving the lens and do not meet to form an image.

The size of the image is also determined by the position of the object. As in the concave mirror, *when the object is more than two focal distances from the lens, the image formed is smaller than the object. When the object is exactly at two focal distances from the lens, the image is the same size as the object. When the object is closer to the lens than two focal distances, the image is larger than the object.*

18-23 Ray Diagrams for Converging Lenses

These diagrams apply only to converging lenses that are relatively thin. Thick lenses must be treated as combinations of thin lenses.

To construct the image formed by a converging lens, two rays of light from each point of the object are followed, just as was done for the concave mirror. One of these is the ray parallel to the principal axis of the lens. This ray is refracted by the lens so that it passes into the principal focus of the lens. The second ray is the one that passes through the center of the lens. In a thin lens, the middle of the lens is very nearly like a thin plate of parallel-sided glass. The second ray therefore passes through without significant change of direction. If these rays meet after passing through the lens, they form a real image at their meeting point. If these rays diverge after leaving the lens, they are extended backward through the lens by dotted lines until their extensions meet in a point. That point is a virtual image of the given point of the object.

Fig. 18-17 shows ray diagrams for the same six positions of the object used with the concave mirror in Fig. 18-8. To simplify matters, as was done for mirrors, the object is made an arrow with its base on the principal axis of the lens. Thus we need only to find the image of the top of the object. A perpendicular dropped to the principal axis from that image point then represents the whole image. Compare each case with the corresponding case of the concave mirror.

18-24 Images Formed by Diverging Lenses

No matter where the object is put with respect to a diverging lens, the image formed is always virtual, erect, and smaller than the object. It is also on the same side of the lens as the object.

In Fig. 18-18 the usual two rays are drawn from the top of the object A through the lens. The rays diverge on leaving the lens and therefore do not form a real image. The diverging rays are therefore extended backward through the lens until they meet at A'. Point A' is the virtual image of A. The perpendicular $A'B'$ dropped from A' to the principal axis of the lens is the virtual image of AB.

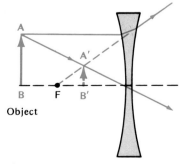

Fig. 18-18. Ray diagram showing how a virtual image is formed by a concave lens.

18-25 Object-Image Relationships in Lenses

The mathematical form of the relationships between the positions and sizes of an object and its image as formed by either converging or diverging lenses is exactly the same as the form of the relationships that apply to curved mirrors.

$$\frac{S_o}{S_i} = \frac{p}{q}$$

and

$$\frac{1}{p} + \frac{1}{q} = \frac{1}{f}$$

where S_o is the size of the object, S_i is the size of the image, p is the distance of the object from the center of the lens, q is the distance of the image from the center of the lens, and f is the focal length of the lens.

Optional derivation

18-26 Derivation of Lens Relationships

As in the case of the spherical mirror, these relationships can readily be derived from the ray diagram in Fig. 18-19. Here, right triangles ABO and $A'B'O'$ are similar so that:

$$\frac{AB}{A'B'} = \frac{BO}{B'O} \quad \text{or} \quad \frac{S_o}{S_i} = \frac{p}{q}$$

This is the first relationship.

Now draw OP to form right triangle POF which is similar to right triangle $FB'A'$. In these triangles,

$$\frac{PO}{A'B'} = \frac{OF}{FB'}$$

Now $PO = AB$, $OF =$ focal length f, and $FB' = B'O - OF = q - f$. Substituting in the above equation, we have:

$$\frac{AB}{A'B'} = \frac{f}{q-f}$$

But,

$$\frac{AB}{A'B'} = \frac{S_o}{S_i} = \frac{p}{q}$$

Hence,

$$\frac{p}{q} = \frac{f}{q-f}$$

and,

$$pq - pf = qf \quad \text{or,} \quad qf + pf = pq$$

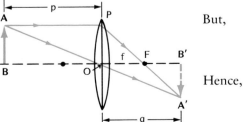

Fig. 18-19. Derivation of the object-image relationships for a convex lens.

Dividing each item on both sides of the equation by pqf gives the second relationship:

$$\frac{1}{p} + \frac{1}{q} = \frac{1}{f}$$

18-27 Problems Involving Lenses

In applying the lens relationships, it will be seen that q comes out negative when the image formed is virtual. A negative value of q means that the image is on the same side of the lens as the object. A positive value of q means that the image is real and that it and the object are on opposite sides of the lens.

For diverging lenses, the focal length f is negative because it is virtual and on the same side of the lens as the object. For diverging lenses, the image distance q always turns out to be negative. This confirms the fact that all images made by a diverging lens are virtual and seen on the same side of the lens as the object.

Sample Problems

1. An object 10 cm high is 15 cm from a converging lens whose focal length is 10 cm. (a) How far is the image from the lens? (b) How tall is it?

   Solution:
   (a) $p = 15$ cm $f = 10$ cm

   $$\frac{1}{p} + \frac{1}{q} = \frac{1}{f}$$

   To clear fractions, multiply by pqf and cancel.

   $$\frac{pqf}{p} + \frac{pqf}{q} = \frac{pqf}{f}; \quad qf + pf = pq$$

   whence $q = \dfrac{pf}{p - f}$

   $$q = \frac{15 \text{ cm} \times 10 \text{ cm}}{15 \text{ cm} - 10 \text{ cm}} = 30\text{cm}$$

   (b) $S_o = 10$ cm $p = 15$ cm $q = 30$ cm

   $$\frac{S_o}{S_i} = \frac{p}{q}; \quad S_i = \frac{qS_o}{p}$$

   $$S_i = \frac{30 \text{ cm} \times 10 \text{ cm}}{15 \text{ cm}} = 20 \text{ cm}$$

2. An object is 2 cm from a converging lens whose focal length is 3 cm. (a) How far is the image from the lens? (b) What kind of image is it?

   Solution:
   (a) $p = 2$ cm $f = 3$ cm

   $$\frac{1}{p} + \frac{1}{q} = \frac{1}{f}$$

   Solve for q as in problem 1.

   $$q = \frac{pf}{p - f}$$

   $$q = \frac{2 \text{ cm} \times 3 \text{ cm}}{2 \text{ cm} - 3 \text{ cm}} = -6 \text{ cm}$$

   (b) Since q is negative, the image is virtual and on the same side of the lens as the object. Since q is 3 times as large as p, the image is 3 times as large as the object.

3. An object is 24 cm from a diverging lens whose focal length is 8 cm. How far from the lens is the image?

   Solution:
   $p = 24$ cm $f = -8$ cm (f is negative for a diverging lens.)

   $$\frac{1}{p} + \frac{1}{q} = \frac{1}{f}$$

   Solve for q as in problem 1.

   $$q = \frac{pf}{p - f}$$

   $$q = \frac{24 \text{ cm} \times (-8 \text{ cm})}{24 \text{ cm} - (-8 \text{ cm})} = -6 \text{ cm}$$

   Since q is negative, the image is virtual.

Test Yourself Problems

1. An object 10 cm tall is 18 cm from a converging lens whose focal length is 12 cm. (*a*) How far from the lens is the image? (*b*) How large is the image? (*c*) What kind of image is it?
2. The object in problem 1 is moved up to 8 cm from the lens. (*a*) How far from the lens is the image now? (*b*) How large is the image? (*c*) What kind of image is it?
3. An object 10 cm tall is 20 cm from a diverging lens whose focal length is 20 cm. (*a*) How far from the lens is the image? (*b*) How large is the image? (*c*) What kind of image is it?

18-28 Defects of Lenses

No single converging lens having spherical surfaces can actually produce a perfectly sharp image. This failure of a converging lens to focus the light passing through it sharply is caused by two major defects: *spherical aberration* and *chromatic aberration*.

18-29 Spherical Aberration

Spherical aberration is caused by the fact that light rays passing through the edges of a converging lens are made to converge faster than light rays passing through the middle part of the lens. As a result, those light rays that come from a given point of an object and pass through the edges of the lens come to a focus sooner than those light rays from the same point that pass through the middle of the lens. Hence, the image of that point is blurred. Because the lens makes a blurred image of each of the points of the object in this same manner, it follows that the image of the entire object will be somewhat blurred.

A common method of avoiding spherical aberration is to put a circular stop or diaphragm in front of a lens so that it cuts off the rays passing through the edges of the lens. The remaining rays which pass through the middle part of the lens focus sharply to form a clear image. This has the disadvantage, of course, of limiting the amount of light that passes through the lens.

Spherical aberration is reduced by putting a circular stop in front of the lens, thus using only the central part of the lens.

18-30 Chromatic Aberration

Chromatic aberration is caused by the fact that transparent substances like glass refract the different colors of light by different amounts. We found in Section 17-22 that glass refracts violet light most and red light least. Therefore, when a beam of white light passes through a glass lens, the violet rays will converge into a focus first, then the blue, green, yellow, orange, and red. The image that is formed by these rays will not only be blurred but it will also be colored around the edges.

A single lens cannot be corrected for chromatic aberration, but a combination of two lenses properly chosen can eliminate it. In practice, a converging lens of crown glass is combined with a diverging lens of flint glass. The lenses in the combination are so

Chromatic aberration is corrected by using a combination of a converging and a diverging lens.

designed that the dispersion of white light into its spectrum caused by one of them is cancelled by the other. This lens combination therefore permits all the colors of light passing through the lens to come to a focus in the same point. This is called an *achromatic lens.*

Optical Instruments

18-31 Applications of Lenses

The eye is certainly the application of the lens that is most important to us. However, its capabilities are limited. It cannot see a very small object or a very distant object. It cannot record and store for future reference a sight we would like to remember. To assist the eye and extend its range, various optical instruments utilizing the properties of lenses have been devised.

18-32 The Eye

The eye consists of an opaque eyeball that is filled with a clear liquid. At the front of the eyeball is a circular transparent window called the cornea (KOR·nee·uh). Behind the cornea is the iris, which is a diaphragm having a hole in its middle called the pupil. The iris can open or close to increase or decrease the size of the pupil. It regulates the amount of light entering the eye.

Next is a crystalline converging lens held in position by the ciliary (SIL·ee·ur·ee) muscle. This lens casts tiny real images of the objects we see on a light-sensitive tissue at the back of the eye called the retina (RET·in·uh). Attached to the retina is the optic nerve that connects the eye to the brain.

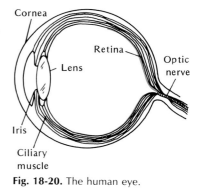

Fig. 18-20. The human eye.

18-33 Operation of Seeing

The eye itself does only part of the job of seeing. The rest is done by the brain. First the cornea and lens of the eye cast on the retina an image of the object at which we are looking. An object can be seen clearly only if it is many focal distances from the lens. Its image on the retina is formed as in case 2 of Fig. 18-17 and is inverted and generally much smaller than the object itself.

Seeing involves both the eye and the brain.

However, if we are to "see" accurately, this image must be properly interpreted. The interpretation is done by the brain. Nerve signals representing the image formed on the retina are sent over the optic nerve from the retina to the brain. The brain then interprets these signals to show us the true erect position and the true size of the object. What we "see" is therefore not the image that forms on the retina of the eye but the interpretation that the brain makes of that image.

18-34 Accommodation

The eye automatically adapts itself to make clear images of distant objects or nearby objects as the situation requires. This process, called *accommodation,* is accomplished by the ciliary muscle, which has the ability to change the focal length of the eye's lens by changing its thickness. For seeing nearby objects, the lens of the eye is made thicker so that its focal length decreases. For seeing more distant objects, the lens is made thinner so that its focal length increases. In the normal eye these changes in the focal length of the lens make it possible for the image of an object to be focused upon the retina whether the object is near or far.

18-35 Eyeglasses

The defective eye is generally farsighted or nearsighted. In the farsighted eye, the lens can become thin enough to focus distant objects clearly on the retina. However, the lens is not able to become thick enough to bring the images of nearby objects to a focus on the retina. In this case, the lens of the eye cannot cause the light from a nearby object to converge rapidly enough to focus an image upon the retina. The lens must therefore be assisted. This is done with a converging lens that is used as an eyeglass or contact lens in front of the eye. The two lenses together cause the light entering the eye to converge just rapidly enough to focus the image of the object exactly on the retina. In this way, converging lenses are used to correct farsightedness.

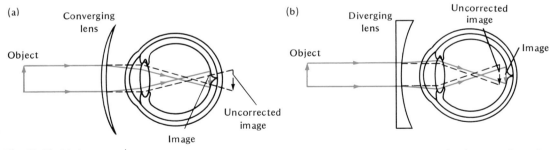

Fig. 18-21. (a) A convex lens corrects farsightedness. **(b)** A concave lens corrects nearsightedness.

In the nearsighted eye, the lens can become thick enough to focus nearby objects clearly on the retina but cannot become thin enough to focus distant objects on the retina. In this case, the light from the distant object converges too rapidly and comes to a focus before it reaches the retina. To make the light entering the eye converge less rapidly, a diverging contact lens or eyeglass is placed in front of the eye. The combination of the diverging eyeglass and the converging lens of the eye gives just the right amount of converging action to bring the image of distant objects squarely on the retina. In this way, diverging lenses are used to correct nearsightedness.

18-36 Camera

The camera is essentially a lightproof box having a converging lens at one end and a light-sensitive film at the opposite end. In the simple box camera, the lens is fixed in position with respect to the film. Theoretically, this camera can produce a clearly focused image for only one position of the object. Practically, however, the light rays from any object whose distance from the lens is large compared to the focal length are brought to a sufficiently approximate focus to give an acceptable picture.

In more precise cameras, the lens is mounted in a bellows or other arrangement so that it can be moved closer to or further from the film. The mobility of the lens makes it possible to focus the image on the film sharply for different positions of the object and to get equally clear pictures in each case. Light is normally excluded from the camera by a shutter that covers the opening where the lens is mounted.

When a picture is to be taken, the shutter is snapped open for a short time and then shut. During this time, light from the object being photographed enters the camera and forms a real, inverted image on the film. The quantity of light that enters while the shutter is open is regulated by a diaphragm that can be adjusted to admit more or less light into the camera as needed. After the image has fallen on it, the film is said to be exposed. It is then ready for the chemical processing that develops the picture and fixes it permanently.

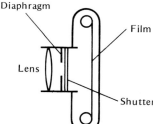

Fig. 18-22. A simple camera.

18-37 Types of Cameras

In the camera, the object is always more than one focal distance from the lens so that the image on the film always comes out real and inverted. Depending upon the purpose for which the camera is used, the image may be smaller than, equal to, or larger than the object.

In the snapshot camera, the object is generally many focal lengths from the lens. The arrangement of object, lens, and image is like that in case 2, Fig. 18-17, and a real, inverted image smaller than the object is formed on the film. The motion picture camera is another example of case 2.

In the copying or duplicating camera, the picture or document to be reproduced is at twice the focal distance from the camera lens, as shown in case 3, Fig. 18-17. The image formed on the film is a true-size copy of the object.

In the enlarging camera, the object or picture to be enlarged is put between one and two focal distances from the lens as in case 4, Fig. 18-17. In this case the image formed is real, inverted, and enlarged.

Fig. 18-23. A strip of motion picture film showing successive pictures of a running giraffe. When these pictures are seen in rapid succession, the viewer gets the illusion of continuous motion.

18-38 Projector

The projector used to show slides or motion picture films on a screen is, in principle, nothing more than a converging lens. The slide or film to be projected is placed between one and two focal distances from the converging lens as in case 4, Fig. 18-17. The lens then casts a real, inverted, and enlarged image of the slide on the screen. Since the lens inverts the image, the slide is deliberately held upside down. The image on the screen then comes out right side up.

In a practical projector, the slide to be projected is brightly illuminated by a strong source of light such as an arc lamp or a powerful incandescent lamp. The illumination is concentrated and evenly distributed over the slide by a combination of two converging lenses, called a condenser, placed between the light source and the slide. The slide is placed a little more than one focal distance from the projecting lens whose position can be adjusted to focus the image clearly upon the screen.

18-39 Motion Pictures and Persistence of Vision

Motion pictures are made possible by the fact that every image formed on the retina of the eye remains for about a sixteenth of a second after the object that produced the image has been removed. The ability of the eye to retain an image on its retina for this short period of time is called persistence of vision.

A motion picture film is simply a series of individual frames or pictures of a changing scene taken by a camera in rapid succession. Each picture in the series is therefore only slightly different from the one before it. When these pictures are projected on a screen in rapid succession, the persistence of vision makes them blend into each other so that the viewer gets the illusion of continuous motion. A rotating shutter interrupts the light from the projector to the screen while the film moves from frame to frame, thus preventing blurring of the images.

18-40 Simple Microscope

The simplest microscope or magnifying glass consists of a single converging lens of short focal length. The object to be magnified is put a little closer than one focal length to the lens which then forms a virtual, erect, and magnified image of it. The image is seen by looking through the lens as shown in Fig. 18-24.

When the lens is used to best advantage, the object is placed just inside the focus. The image then is formed at the distance of most distinct vision which is about 25 centimeters from the lens. The magnification produced by the lens is the ratio of the size of the image to the size of the object. This ratio is equal to the ratio of the image distance from the lens, which is 25 centimeters, to the object distance from the lens, which is *f,* the focal length of the lens. Hence, the magnifying power of the simple microscope is equal to 25/*f* if *f* is given in centimeters.

A converging lens with a focal length of 5 centimeters will be able to magnify an object 25 ÷ 5, or 5 times. A lens with a focal length of 2.5 centimeters will have a magnifying power of 25 ÷ 2.5, or 10. In general, the shorter the focal length of the converging lens the higher is its magnification.

18-41 Compound Microscope

Higher magnifying power than is obtainable with a single converging lens is achieved by using a compound microscope composed of two converging lenses. These lenses are set in the opposite ends of a tube mounted in a stand so that it can be lowered or raised for focusing. The object to be magnified is placed under the lower lens which has a very short focal length and is called the objective lens. The final magnified image is viewed through the upper lens which is called the eyepiece. The instrument generally has a horizontal platform under the objective on which to put the object. There is also either a bright light source or a converging mirror-and-lens combination to gather enough light to illuminate the object well.

The main optical system of the compound microscope is shown in Fig. 18-25. The object to be magnified is placed between one and two focal distances below the objective lens. For this position of the object, the objective lens forms a real, inverted, and magnified image. The eyepiece is so situated that this real image of the object falls just inside of its focus. The eyepiece therefore acts like a magnifying glass and makes a virtual, magnified image of the real image produced by the objective. This virtual image is the one the eye sees. Note that each lens contributes to the magnification of the image.

18-42 Refracting Telescope

The refracting astronomical telescope consists of two converging lenses set at the opposite ends of a tube: an objective lens of long focal length, and an eyepiece of short focal length. As in the compound microscope, the objective lens forms a real, inverted image of a star or some other distant object at a point just within the focus of the eyepiece. The eyepiece then acts like a magnifying glass and makes an enlarged, virtual image of the real image produced by the objective. This virtual image is the one seen by the eye. As Fig. 18-26 shows, it is inverted with respect to the object being viewed through the telescope.

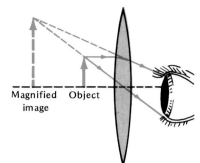

Fig. 18-24. The simple mircroscope or magnifying glass.

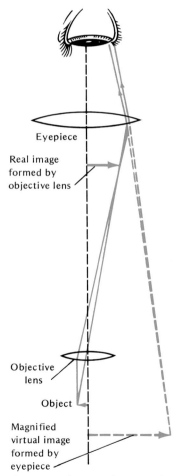

Fig. 18-25. The optical system of the compound microscope.

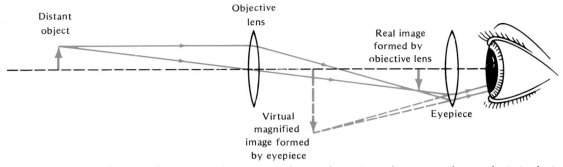

Distant
object

Objective
lens

Real image
formed by
objective lens

Virtual
magnified
image formed
by eyepiece

Eyepiece

Fig. 18-26. The optical system of
the refracting telescope.

In telescopes to be used to view objects on the earth, it is desirable to have an image that is erect. This is done by putting an
intermediate converging lens between the objective lens and the
eyepiece to invert the image made by the objective lens. The final
image made by the objective lens then comes out erect.

CHAPTER REVIEW

Summary

Mirrors and lenses are able to make two types of images of objects: *real
and virtual*. A **real image** of a point of an object forms when the diverging
rays of light from that point are brought together into another point after
falling upon a mirror or lens. The image of an object is the sum total of the
images of all its points. Real images are generally seen on a screen and
are always *inverted*. They can be formed only by **concave mirrors** and
converging lenses.

A **virtual image** of a point of an object forms when the rays of light from
that point continue to diverge after falling upon a mirror or lens but do so
in such a way that they seem to originate at a point other than the original
point. This imaginary point of origin of the reflected or refracted rays is the
virtual image of the original point. Virtual images of objects are seen only
by looking into the mirror or lens that forms them. Virtual images of objects
are always *erect* and can be formed by *all types of mirrors and lenses*.

The position and size of the image formed by a mirror or lens may be
determined by means of a ray diagram. Ray diagrams apply closely only
to mirrors having large radii of curvature and to thin lenses. Ray diagrams
are based on the assumptions that light rays travel in straight lines while
in the same medium, and that they follow the laws of reflection and refraction. They do not take into account the effects of the dispersion, interference, and diffraction of light. The geometry of ray diagrams shows that
the following two relationships apply to all spherical mirrors and lenses.

$$\frac{S_o}{S_i} = \frac{p}{q} \quad \text{and} \quad \frac{1}{p} + \frac{1}{q} = \frac{1}{f}$$

where S_o is the size of the object, S_i is the size of the image, p is the
distance of the object from the mirror or lens, q is the distance of the image
from the mirror or lens, and f is the **focal length** of the mirror or lens.

Concave mirrors and converging lenses produce exactly the same types of images. Convex mirrors and diverging lenses also produce the same types of images which are always virtual, erect, and smaller than the object.

Spherical concave mirrors and convex diverging lenses are both subject to **spherical aberration.** In addition, lenses are subject to **chromatic aberration** due to variations in the speed of light of different colors or wavelengths through them.

Mirrors and lenses are used in many optical instruments to concentrate light, to form magnified images of small objects, to bring distant objects into view, and to make various kinds of photographs.

Questions

Group 1

1. (a) Explain how to measure the focal length of a thin converging lens. (b) What happens to the focal length of a converging lens as the lens is made thicker? (c) What is meant by chromatic aberration? (d) How is it corrected?

2. A thin converging lens is 5 cm in diameter and has a focal length of 2 cm. Draw ray diagrams to find the image of an object consisting of a vertical arrow 1.5 cm tall that stands on the principal axis at a distance from the lens of (a) 3 cm; (b) 5 cm; (c) 8 cm. (d) What happens to the image size as the object moves away from the lens?

3. (a) Using the same lens and object as in question 2, draw a ray diagram to locate the image formed when the object is 1 cm from the lens. (b) Can a picture of this image be made by putting a photographic plate at the place where the image is located? Explain.

4. Show by one or more ray diagrams that a diverging lens forms only virtual images, smaller than the object.

5. Compare the camera and the eye as to (a) the type of lens that produces the image; (b) the manner in which the image is focused; (c) the kind of image that is formed on the light-sensitive screen; (d) the manner in which the image is "interpreted." (e) How are lenses used in correcting for nearsightedness and farsightedness?

6. (a) What is meant by spherical aberration of a converging lens? (b) How is it overcome in many camera lenses?

7. (a) When a converging lens is to be used as a motion picture projector to throw an enlarged image of each frame of the film on a screen, where must the film be placed with respect to the lens? (b) Why must the film be upside down? (c) How is the illusion of motion obtained when the frames appear on the screen in rapid succession?

8. (a) Where should the object be placed for highest magnification by a simple microscope? (b) What does the magnification depend upon?

9. Explain how each of the two converging lenses of the compound microscope contributes to the magnification of the object being viewed.

10. In the terrestrial telescope, the intermediate lens is inserted to invert the image formed by the objective lens. Should the intermediate lens be positioned so as to make a real or a virtual image of the image formed by the objective lens? Explain.

Group 2

11. A converging lens of diamond ($n = 2.42$) and a lens of glass ($n = 1.52$) have the same shapes. Which will make a more powerful magnifying glass? Explain.

12. A glass converging lens ($n = 1.52$) is put into a fish tank full of water ($n = 1.33$) and a beam of parallel rays of light is passed through it. Do the rays come to a focus at the same distance from the lens as they do when the lens is in the air? Explain.

13. Why is there chromatic aberration in lenses but none in plane or spherical mirrors?

14. A parallel-sided beam of violet light is passed through a convex lens and brought to a focus. The same lens is then used to bring a parallel-sided beam of red light to a focus. (a) Explain why the focal length of the lens is different for violet light than it is for red light. (b) How would the focal length of the lens for yellow light compare with its focal length for red light?

Problems

Group 1

1. A camera lens having a 6.0-cm focal length is 6.6 cm from the film. At what distance from the lens should the object be to obtain a clear image?

2. An object 5.0 cm tall is 20 cm in front of a diverging lens of 4.0-cm focal length. (a) At what distance from the lens is the image? (b) What is its size?

3. Each line of Table 18.2 gives some information about a converging or diverging lens. Supply the missing information.

Table 18.2

	OBJECT DISTANCE (cm)	IMAGE DISTANCE (cm)	FOCAL LENGTH (cm)
Lens A	24	24	
Lens B	6.0		8.0
Lens C	5.0		−20
Lens D		25	5.0
Lens E	12		−24
Lens F	50	40	

4. Construct a ray diagram to verify your answers to Lens B and Lens C in problem 3.

5. (a) For each lens in problem 3, state whether the lens is a converging or diverging lens and whether the image is real and inverted, or virtual and erect. (b) Also, determine the size of the image in each case if the object is 12 cm tall.

6. It is desired to use a converging lens as a projector to make a real image 100 cm tall of an object that is 10 cm tall. If the object is to be put 8.8 cm from the lens, what is the focal length of the lens needed for this task?

7. A converging lens is used in a copying camera to make true-sized copies of documents. If the documents are always at a fixed distance of 30 cm from the lens, what is the focal length of the lens?

8. A stamp that is a square 2 cm on a side is magnified by a simple microscope having a focal length of 5 cm. How long is the side of the magnified image of the stamp seen under the microscope?

Group 2

9. The real image of a slide put 0.4 cm from the objective lens of a compound microscope is formed at a distance of 16 cm from the lens. The eyepiece of focal length 2.5 cm then magnifies this image. (a) How many times was the slide magnified by the objective? (b) What is the magnifying power of the eyepiece? (c) By how many times do the two lenses of the microscope magnify the slide?

10. An object is 20 cm in front of a converging lens of 4.0 cm focal length. A second converging lens of 2.5 cm focal length is 8 cm behind the first lens. (a) Find the position of the image made by the first lens. (b) Using this image as the object, find the position of its image in the second lens. (c) What magnification does this pair of lenses give?

Applying Physics

1. To observe how the magnifying power of a converging lens depends upon its thickness, put a pencil into a glass of water and hold it vertically against the side of the glass nearest you. Now slowly move the pencil away from you and note the increasing magnification. The cylindrical shape of the glass and of the water it contains causes that part of the water and glass between you and the pencil to act like a cylindrical lens which thickens as the pencil is moved further away. Measure the width of the image of the pencil when it is furthest from you and compare this with the actual width of the pencil to obtain the magnifying power of this arrangement.

2. To measure the focal length of a converging lens such as a magnifying glass, use the lens to cast the image of a lighted candle on a vertical wall or screen. Measure the object and image distances and compute the focal length from the relationship $1/p + 1/q = 1/f$. Check your value by focusing the sun's rays with the lens and measuring the distance to the focus.

The Hubble Space Telescope

Nearly 20 years after its conception, the most expensive and technologically advanced astronomical observatory was launched in April 1990 by the space shuttle Discovery. Placed in orbit above the earth's distorting atmosphere, the $1.5-billion Hubble Space Telescope (HST) is capable of producing images 10 times sharper than any instrument on the ground. This is comparable to being able to distinguish a car's left headlight from its right at a distance of 2 500 miles, the distance between Washington, D.C., and Los Angeles.

The HST was named in honor of Edwin Hubble, the scientist who confirmed that the universe is expanding and determined the rate of that expansion. During its expected 15-year lifetime, scientists will use the HST to look 14 billion years into the past to measure the rate of expansion of the universe and determine its age and density, a prerequisite for understanding the fate of the universe. In addition, astronomers hope to see galaxies in the process of formation. The HST will also be used to search for planets orbiting stars within 15 light-years of Earth (approximately 9×10^{13} miles).

The HST consists of three major systems—the optical telescope assembly, an assortment of light-detecting instruments and guidance sensors, and a support systems module. Shortly after launch it was discovered that the curvature of one of the telescope's mirrors was off by about 4 percent of the diameter of a human hair. Consequently, the telescope's ability to distinguish close objects and to detect faint light from the early universe will be impaired until astronauts can make

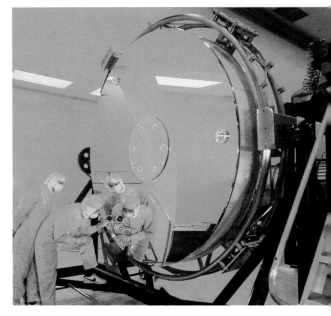

adjustments. However, in the interim the HST will still be in constant use, as observation time will be reassigned to utilize other instruments. Among these instruments is a wide-field planetary camera, which can image large, faint objects, such as galaxies, as well as planets. This camera is expected to produce most of the images. There is also a faint-object camera with a smaller field of view and greater sensitivity to ultraviolet light. The HST's Goddard high-resolution spectograph is sensitive only to the ultraviolet emissions of stars. It is designed to detect objects about 1000 times dimmer than any seen so far. The simplest instrument is the high-speed photometer designed to measure fleeting fluctuations in brightness.

For some time to come the HST should provide astronomers with exciting discoveries and intriguing new mysteries to unravel.

19 Interference and Diffraction

Aims

1. To study the interference and diffraction of light as further evidence of its wave nature.
2. To learn how we can make a direct measurement of the wavelengths of light of different colors.

Interference and diffraction support the wave theory of light.

19-1 Interference, Diffraction, and Image Formation

We have seen that light generally travels in straight lines through a given medium. Observations of interference and diffraction phenomena require us to revise this concept of the straight-line propagation of light. We can actually observe that light bends significantly around the edges of objects. An important effect of this bending is a limit to the sharpness of the images obtainable with mirrors and lenses, and hence instruments in which they are used.

19-2 Interference Pattern of Two Sets of Water Waves

As the waves from two light sources pass through the same space, their crests and troughs combine at each point and form an interference pattern. Such a pattern can be visualized by examining a similar interference pattern in water waves. (See Fig. 15-10.)

Two sets of water waves of equal wavelengths and amplitudes are generated in a ripple tank by two sources, S_1 and S_2, separated by a small distance. The sources consist of a pair of beads dipping into the surface of the water and made to vibrate in unison at a fixed frequency. As a result, the sources are in phase, simultaneously producing crests, then troughs, then crests, and so on.

Fig. 19-1 shows the instantaneous pattern of semicircular crests and troughs that have left each wave source. The crests are shown by solid lines and the troughs by dotted lines. The distance from one crest to the next, or from one trough to the next, is one wavelength. The distance between a crest and the adjacent trough is one-half wavelength.

19-3 Lines of Reinforcement

Consider the points labeled 1, each of which is equidistant from the wave sources S_1 and S_2. At every alternate point, a crest of the waves coming from S_1 intersects a corresponding crest of the waves coming from S_2. At the remaining points, a trough of the waves coming from S_1 intersects a trough of the waves coming from S_2. The waves from S_1 and S_2 therefore arrive in phase at each of these

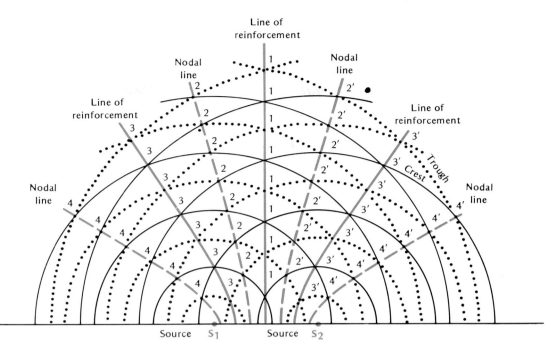

Fig. 19-1. Interference pattern formed by the waves generated by two sources.

points and reinforce each other. Extra-large crests are produced at points where two crests intersect and extra-large troughs are formed wherever two troughs intersect.

As time goes on, every crest and trough spreads out and moves further away from its source. The points of intersection of any two crests or of any two troughs then move away from the sources of the waves. The path they follow is the line drawn through the points labeled 1 and is called a *line of reinforcement.*

This line of reinforcement passing through the points labeled 1 is the perpendicular bisector of the line joining S_1 and S_2. Since any point on it is equidistant from S_1 and S_2, the waves from the two sources always arrive in phase at that point. The two sets of waves therefore reinforce each other *at all points* of the line of reinforcement. This is observed on the surface of the water over which the waves are passing where waves of extra-large crests and troughs are seen moving continuously away from the sources in the direction of this line of reinforcement.

19-4 Condition for Lines of Reinforcement

Now consider the points labeled 3. At each of these points, the waves from the two sources also meet in phase. At alternate points, their crests arrive together; at the intervening points, their troughs arrive together. The line passing through the points labeled 3 is therefore a line of reinforcement similar to the line passing through the points labeled 1. It differs from the first line of reinforcement, however, in that every point on it is not equidistant from S_1 and S_2, but is one wavelength further from S_2 than it is from S_1. As seen on the surface of the water, this line resembles the first line of reinforcement. Waves of extra-large crests and troughs move continuously away from the sources in the direction of this second line of reinforcement.

(a)

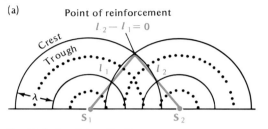

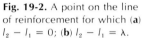

Point of reinforcement
$l_2 - l_1 = 0$

(b)

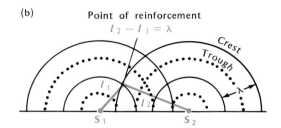

Point of reinforcement
$l_2 - l_1 = \lambda$

Fig. 19-2. A point on the line of reinforcement for which (**a**) $l_2 - l_1 = 0$; (**b**) $l_2 - l_1 = \lambda$.

Additional lines of reinforcement may also be formed. They are easily located by means of the relationship which the points on them must obey. In Fig. 19-2, let l_1 and l_2 be the distances of any point from the wave sources S_1 and S_2 respectively. Then for all points on the line of reinforcement passing through the points labeled 1 in Fig. 19-1, $l_2 - l_1 = 0$. For all points on the line of reinforcement passing through the points labeled 3, $l_2 - l_1 = \lambda$, where λ is the Greek letter *lambda* and represents one wavelength of the interfering waves. It can be seen that the next lines of reinforcement formed consist of points for which $l_2 - l_1 = 2\lambda$, $l_2 - l_1 = 3\lambda$, $l_2 - l_1 = 4\lambda$, and so forth. In general, all lines of reinforcement consist of points for which $l_2 - l_1 = n\lambda$ where n is zero or a whole number. The maximum number of lines of reinforcement that can be formed is limited by the ratio of the distance between the sources of the waves and their wavelength.

A symmetrical set of lines of reinforcement forms on the right half of the pattern. One such line passes through the points marked 3′ in Fig. 19-1.

19-5 Nodal lines

Intersecting waves in opposite phase cancel each other to produce nodal lines.

Next, consider the points labeled 2, each of which is one-half wavelength further from S_2 than from S_1. At each of these points, the crest of one set of waves intersects a trough of the other set. Since the waves are in opposite phase, they cancel each other at these points of intersection. As time passes by and both sets of waves spread out, the points of intersection of their crests and troughs also move outward along the line passing through the points labeled 2. This line is called a *nodal line*. It is observed on the surface of the water as a line on which the two sets of waves cancel each other. On this line, the water neither rises nor falls but remains at its original undisturbed level.

(a)   Nodal point

$$l_2 - l_1 = \frac{\lambda}{2}$$

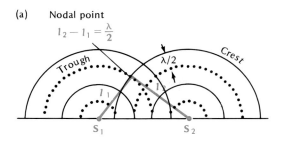

(b)   Nodal point

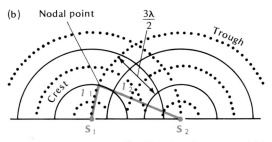

Fig. 19-3. A point on a nodal line for which (a) $l_2 - l_1 = \frac{1}{2}\lambda$; (b) $l_2 - l_1 = 1\frac{1}{2}\lambda$.

19-6 Condition for Nodal Lines

The line passing through the points labeled 4 is also a nodal line. At every point of this line, the waves from the two sources arrive in opposite phase and cancel each other. This can be seen from the fact that each of the points labeled 4 is one and one-half wavelengths further from S_2 than from S_1 and is the intersection between a crest of one set of waves and a trough of the other. On the surface of the water, this nodal line, too, is observed as one on which the cancellation of the waves leaves the water undisturbed.

The points on these two nodal lines obey the relationships $l_2 - l_1 = \lambda/2$ and $l_2 - l_1 = 1\frac{1}{2}\lambda$ respectively, as shown in Figs. 19-3 (a) and (b). The next nodal lines that are formed will consist of points for which $l_2 - l_1 = 2\frac{1}{2}\lambda$, $l_2 - l_1 = 3\frac{1}{2}\lambda$, and so on. In general, nodal lines consist of points for which $l_2 - l_1 = (n + \frac{1}{2})\lambda$ when n is zero or a whole number. Each different value of n gives another nodal line and the number of nodal lines is limited by the ratio of the distance between the sources and their wavelength.

A symmetrical set of nodal lines forms on the right half of the pattern. Two such lines pass through the points labeled 2′ and 4′ in Fig. 19-1.

The photograph in Fig. 19-4 shows a typical pattern of nodal lines and lines of reinforcement made by two sources of water waves. Note that the water is at rest on the nodal lines while waves of extra-large amplitude pass outward from the two sources between the nodal lines. Midway between the nodal lines are the lines of reinforcement. If these were light waves instead of water waves, the nodal lines would be lines of darkness. Midway between them would be lines of maximum brightness corresponding to the lines of reinforcement.

19-7 Interference Pattern with Sources out of Phase

In the preceding experimental arrangement, the sources of the water waves vibrate in unison and are in phase. If the vibrations of the wave sources are adjusted so that one of them lags a little behind the other, it will generate its crests and troughs a little later than does the first source. These out-of-phase sources will also produce an interference pattern consisting of nodal lines and lines

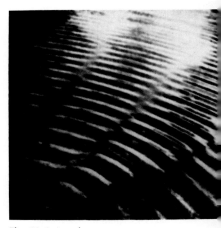

Fig. 19-4. Interference pattern made by the waves generated by two sources in a ripple tank. Nodal lines are clearly seen.

Fixed interference
patterns are produced by
two light sources that
maintain a constant
phase difference with
respect to each other.

(a)

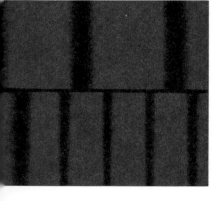

(b)

Fig. 19-5. Interference patterns produced by (a) red, (b) blue light passed through a pair of narrow slits placed close together (top) and slightly farther apart (bottom).

of reinforcement, but the pattern will no longer be a symmetrical one like that obtained when the sources were in phase. Instead, the nodal lines and lines of reinforcement will shift toward the delayed source.

Suppose the sources keep changing their phase relationship toward each other rapidly so that now, the first produces its crests and troughs before the second, and then, the second produces its crests and troughs before the first. The interference pattern will then keep shifting rapidly, first toward one of the sources and then toward the other. Where there is a nodal line at one instant there will be a line of reinforcement the next and vice versa. The average effect will be that no fixed interference pattern is formed.

19-8 Condition for Fixed Interference Pattern

To produce a fixed interference pattern, the sources must either remain in phase or maintain a constant phase difference with respect to each other. This requirement is particularly difficult to meet in producing an interference pattern between light waves because the parts of a light source that emit light waves are its very large number of atoms.

Each atom does not radiate light continuously, but in bursts lasting only about 10^{-8} second. Depending upon the time when different atoms began radiating, different pairs of the millions of atoms of a light source emitting light at the same moment may be in all possible phase relationships. The interference pattern made by a particular pair of atoms will therefore usually have its nodal lines and lines of reinforcement in different places from those occupied by the nodal lines and lines of reinforcement made by other pairs of atoms radiating light at the same time. As these many overlapping interference patterns are superimposed, they obliterate each other.

To obtain a fixed interference pattern, we must have two light sources that maintain the same phase relationship with each other. Such sources are said to be *coherent*. A method for obtaining them was first discovered by the Englishman Thomas Young.

19-9 Double-Slit Interference Pattern

Thomas Young duplicated for light the conditions for producing a fixed interference pattern of nodal lines and lines of reinforcement like the one observed with water waves. To obtain two sources of light that would remain in phase, like the two sources of water waves, he used diffraction to divide the light from a single source into two separate sources.

A modification of his experimental arrangement is shown in Fig. 19-6. Since it is difficult to obtain a source of light that radiates waves of only one wavelength, we will be content to work with a light source that radiates a narrow range of wavelengths. Such a

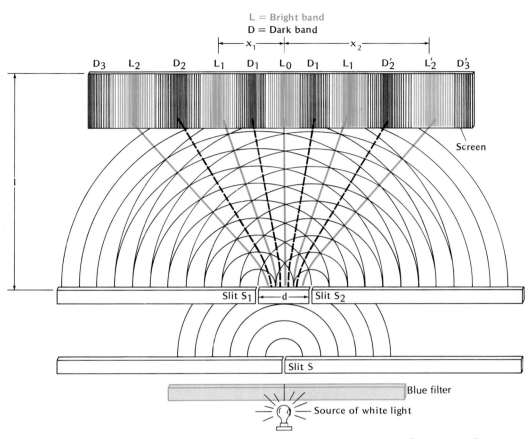

L = Bright band
D = Dark band

Fig. 19-6. Interference pattern produced by Young's double-slit experiment.

source may be obtained by passing white light through a blue glass filter so that only a blue band of waves passes through.

The blue light is now passed through a very narrow slit S, by which it is diffracted. The slit S acts like a new source of waves which then pass through two distant, very narrow, and closely adjacent slits, S_1 and S_2. Each of these secondary slits now acts like a separate source of light waves which are allowed to fall on the screen where they form an interference pattern.

Since the same crest or trough coming from S arrives at slits S_1 and S_2 at the same time, the waves that leave at any instant from S_1 and S_2 start in the same phase. That is, when a crest from S arrives at S_1 and S_2, each of the twin slits sends out a crest, and when a trough from S arrives at S_1 and S_2, each of the twin slits sends out a trough. The effect of this arrangement is that S_1 and S_2 act like two generators that remain in phase at all times.

The interference pattern made by the light waves from S_1 and S_2 consists of nodal lines and lines of reinforcement similar to those observed earlier with the water waves. In Fig. 19-6, the crests of the waves and the lines of reinforcement are shown as full lines.

The troughs are not shown, but the nodal lines, which are situated midway between the lines of reinforcement, are shown by dotted lines. On lines of reinforcement, the light waves from S_1 and S_2 combine to produce maximum brightness. Wherever these lines intersect the screen, there is a bright band of blue light. On nodal lines, the light waves from S_1 and S_2 interfere to produce darkness. Where the nodal lines fall upon the screen there is a band of darkness. The pattern seen on the screen therefore consists of regularly spaced bright bands of blue light separated by bands of darkness.

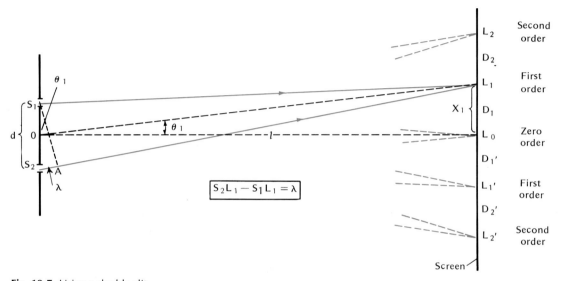

Fig. 19-7. Using a double-slit interference pattern to measure the wavelength of a light source.

19-10 Measuring Wavelength

If a color of light other than blue is used in Young's double-slit arrangement, the interference pattern formed on the screen is essentialiy the same, except that the distance between the centers of the bright bands is different than it is for blue light. Since the disance between the bright bands depends upon the wavelength of the light that forms them, it is a means of measuring that wavelength.

To measure the wavelength of the light by means of the double-slit interference pattern, the distance of the screen from the double slit should be very much greater than the distance between the slits. In Fig. 19-7, the screen is at a great distance l from two slits separated by the distance d. L_0, L_1, L_1', L_2, and L_2' are the positions of the centers of the bright bands. L_0 is called a band of first order because it is one wavelength further from S_2 than from S_1; L_2 is called a band of second order because it is two wavelengths further from S_2 than from S_1, and so on. L_1' and L_2' are symmetrical first- and second-order bands on the other side of L. Between these bright bands are the first- and second-order dark bands, D_1, D_1', D_2, D_2'.

We shall show in the next section that if x_1 is the distance on the screen between L_0 and L_1, and λ is the wavelength of the light passing through the slits to form the interference pattern, then:

$$\lambda = \frac{x_1 d}{l}$$

Since x_1, d, and l can be measured directly, the wavelength λ can be obtained by substituting the measured values in this relationship.

Sample Problem

Yellow light from a sodium vapor lamp passes through two slits that are 0.0200 cm apart and falls on a screen 100 cm away. The first-order bright band is 0.295 cm from the middle of the central band. What is the wavelength of the light?

Solution:

$x_1 = 0.295$ cm $d = 0.0200$ cm $l = 100$ cm

$$\lambda = \frac{x_1 d}{l}$$

$$\lambda = \frac{0.295 \times 0.0200}{100} \text{ cm}$$

$$\lambda = 5.90 \times 10^{-5} \text{ cm}$$

Test Yourself Problems

1. A certain wavelength of light passes through two slits that are 0.0167 cm apart and falls on a screen 100 cm away. The first order bright band is 0.300 cm from the middle of the central band. What is the wavelength of the light?

2. If the screen in problem 1 is moved to a position 200 cm away from the slits, what effect would this have on the position of the first order bright band with respect to the central band?

19-11 Derivation of Wavelength Formula

Optional derivation for honor students

Consider only the zero and first-order bands L_0 and L_1 which have been isolated in Fig. 19-7 and are a distance x_1 apart. With L_1 as a center and S_1L_1 as a radius, draw arc S_1A. Since S_1L_1 is very large compared to d, the arc S_1A is very nearly a straight line. No important error is incurred, therefore, in treating S_1A as a straight line instead of the arc of a circle and considering S_1A perpendicular to both OL_1 and S_1L_1. Also, since L_1 is a first-order bright band, $S_2L_1 - S_1L_1 = \lambda$, and since S_1L_1 is equal to AL_1, S_2A is equal to λ.

Now, angles S_2S_1A and L_1OL_0 are equal because their sides are perpendicular to each other. Calling these angles θ_1 (θ is the Greek letter *theta*), we have from right triangle S_2S_1A,

$$\sin \theta_1 = \frac{\lambda}{d}$$

and from right triangle L_1OL_0,

$$\sin \theta_1 = \frac{x_1}{OL_1}$$

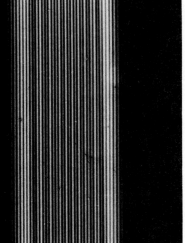

Fig. 19-8. Interference pattern equivalent to that produced by a pair of very narrow slits. Note the equal spacing between the bright bands.

whence:

$$\frac{\lambda}{d} = \frac{x_1}{OL_1}$$

Since l is very large compared to x_1, OL_1 and l are very nearly equal. Substituting l for OL_1 gives:

$$\frac{\lambda}{d} = \frac{x_1}{l}$$

whence:

$$\lambda = \frac{x_1 d}{l}$$

19-12 Measuring Wavelength by Higher-Order Bands

By similar reasoning, λ can be obtained by measuring the distance x_2 between L_0 and the second-order band L_2. For this band, the difference between S_1L_2 and S_2L_2 is 2λ so that:

$$2\lambda = \frac{x_2 d}{l} \qquad \text{or,} \qquad \lambda = \frac{x_2 d}{2l}$$

Similarly, from the position of the third-order band L_3, we have the relationship:

$$\lambda = \frac{x_3 d}{3l}$$

where x_3 is the distance between L_0 and L_3. In general, to measure λ for the nth-order band L_n, we have the relationship:

$$\lambda = \frac{x_n d}{nl}$$

Since x_n is larger than x_1, x_2, x_3, etc., it can be measured more accurately than they can. A more precise value for λ can therefore be made by working with the band of highest order that is seen clearly.

19-13 Predicted Spacing Between Bands

From:

$$\lambda = \frac{x_1 d}{l} = \frac{x_2 d}{2l} = \frac{x_3 d}{3l} = \frac{x_n d}{nl}$$

it follows that:

$$x_1 = \frac{x_2}{2} = \frac{x_3}{3} = \frac{x_n}{n}$$

whence:

$$x_2 = 2x_1, \qquad x_3 = 3x_1, \qquad \text{and} \qquad x_n = nx_1$$

Our expression for λ therefore predicts that the bright bands will be spaced regularly at a distance x_1 apart. Observation shows that this is the case. (See Fig. 19-8).

19-14 Separation of Light into Spectra

From:

$$\lambda = \frac{x_1 d}{l} \quad \text{we have:} \quad x_1 = \frac{l}{d}\lambda$$

Since l and d are constants, x_1 is directly proportional to λ. This means that the greater the wavelength of the light falling on the double slit, the greater is the spacing between its bands, and the greater is the distance of its bands from the central band L_0. This relationship explains how a double slit can separate white light or any other light mixture into the various wavelengths of which it is composed.

When white light is passed through a double slit, only the central band L_0 is white. Here, all the component colors or wavelengths present in the white light have the central bright bands of their interference patterns superimposed on each other. The resulting mixture is white light.

However, each higher-order band on both sides of the central white band L_0 is dispersed into a spectrum. Here the component colors of white light appear in the following order of nearness to L_0: first violet, then indigo, blue, green, yellow, orange, and red. There are thus first-order, second-order, and higher-order spectra corresponding to each band. Since the violet component of white light is displaced least from L_0, it follows from $\lambda = x_1 d/l$ that it has the shortest wavelength. The red component which is displaced most from L_0 has the longest wavelength. The wavelengths of the other colors lie between those of violet and red in the order of their appearance on the screen.

The action of the double slit confirms that the color of light waves depends upon their wavelength. Theoretically, every different wavelength of light represents a different color, and light consisting of only a single wavelength is said to be monochromatic. Actually the eye is not able to distinguish between wavelengths of light that are too close together. Instead, it recognizes groups of wavelengths as having certain colors. A range of the shortest visible wavelengths is recognized as violet light; ranges of longer ones as blue, green, yellow, and orange lights; and a range of the longest wavelengths as red light. The range of wavelengths associated with each color of the spectrum when light travels in a vacuum is given in Table 19.1. The wavelengths are expressed in units of 10^{-9} meters or nanometers.

Fig. 19-9. Interference patterns produced by white light passed through double slits placed close together (top) and slightly farther apart (bottom).

Table 19.1

COLOR	WAVELENGTH (10⁻⁹ m)
Violet	380-450
Blue	450-500
Green	500-570
Yellow	570-590
Orange	590-610
Red	610-750

Fig. 19-10. Diffraction of white light by a diffraction grating.

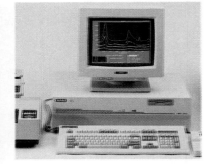

Fig. 19-11. Instruments used to measure light frequencies, like the spectrometer shown above, commonly include diffraction gratings to disperse the light that is being analyzed.

19-15 Grating Spectroscope

The double slit can be used like a prism to disperse the light from any source into a spectrum and analyze it into its component wavelengths. However, because so little light passes through the two narrow slits, and because the light is then spread over the central band and several higher-order spectra, each spectrum may be very faint and hard to observe. To overcome this difficulty, a diffraction grating consisting of a very large number of closely spaced parallel slits is used. Each of the many pairs of adjacent slits in the grating disperses the light passing through it into the usual double-slit pattern seen on the screen. Furthermore, all of these double-slit patterns are superimposed upon each other, making the pattern that appears on the screen essentially the same as that for a double slit but very much brighter.

The amount of dispersion that takes place when light passes through a double slit depends upon the distance between the two slits. The smaller this distance, the more spread out each of the spectra will be. This is also true of a diffraction grating which may have as many as 10 000 parallel slits per centimeter. With the distance between their slits only 10^{-4} centimeter, such gratings can provide wide dispersion of the light passing through them and make possible very precise measurements of its component wavelengths.

In practice, *diffraction gratings* are made by a ruling machine that scratches many thousands of parallel, equidistant lines on a plate of glass with a sharp diamond point. The narrow unscratched glass strips between the lines act like parallel slits through which the light passes on entering the grating.

19-16 Frequency of Light Waves

Since both the velocity v and the wavelength λ of light waves can now be measured, their frequency f can be determined from the relationship $v = \lambda f$. A knowledge of the frequency of light waves is particularly useful because it is a constant property of the waves. In contrast, the wavelength of waves of constant frequency varies with the medium in which the waves are traveling. The higher the index of refraction of the medium, the more slowly the waves travel in it, and the shorter is their wavelength.

Sample Problems

1. What is the frequency of the yellow light from a sodium vapor lamp whose wavelength in air is 5.9 $\times$ 10^{-7} m?

Solution:
The velocity of light in air = 3.0 $\times$ 10^8 m/s

$\lambda = 5.9 \times 10^{-7}$ m
Since $v = \lambda f$,

$$f = \frac{v}{\lambda}$$

$$f = \frac{3.0 \times 10^8 \text{ m/s}}{5.9 \times 10^{-7} \text{ m}} = 5.1 \times 10^{14} \text{ Hz}$$

2. What is the wavelength of this yellow light in glass having an index of refraction $n_g = 1.5$?

Solution:
The index of refraction of glass is:

$$n_g = \frac{v_a}{v_g}$$

where v_a is the velocity of light in air and v_g is its velocity in glass.

Now, for air $v_a = f\lambda_a$ where λ_a = wavelength in air; for glass $v_g = f\lambda_g$ where λ_g = wavelength in glass.
Hence,

$$n_g = \frac{v_a}{v_g} = \frac{f\lambda_a}{f\lambda_g} = \frac{\lambda_a}{\lambda_g}$$

Substitute $n_g = 1.5$ and $\lambda_a = 5.9 \times 10^{-7}$ m in

$$n_g = \frac{\lambda_a}{\lambda_g}$$

$$1.5 = \frac{5.9 \times 10^{-7} \text{ m}}{\lambda_g}$$

$$\lambda_g = 3.9 \times 10^{-7} \text{ m}$$

Notice from this problem that the index of refraction of a substance n_x for a particular frequency of light can be expressed as the ratio of its wavelength in air (or vacuum) λ_a to its wavelength in that substance λ_x; that is,

$$\mathbf{n}_x = \frac{\mathbf{v}_a}{\mathbf{v}_x} = \frac{\lambda_a}{\lambda_x}$$

Test Yourself Problems

1. What is the frequency of a line in the spectrum of hydrogen whose wavelength is 6.6×10^{-7} m?

2. What is the wavelength of the light in problem 1 when it is passing through glass having an index of refraction of 1.5?

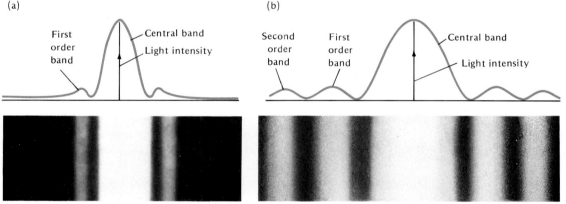

(a) First order band / Central band / Light intensity

(b) Second order band / First order band / Central band / Light intensity

Fig. 19-12. The diffraction pattern from a single narrow slit (a) broadens when the slit is made narrower (b). The curves above each photo show the variation in light intensity with increasing distance from the center of the patterns.

19-17 Single-Slit Diffraction Pattern

We can now study in more detail how light is diffracted on passing through a single narrow opening. It is observed that the effect of the diffraction of light is negligible when the light passes through a slit that is very large compared to its wavelength. As the width of the slit is made smaller and smaller, the amount of diffraction of the light around the edges of the slit becomes larger and larger. When the light passing through a very narrow slit is allowed to fall on a distant screen, it forms a pattern of bright and dark bands. These bands suggest that interference has taken place among the light waves that leave the slit in a manner similar to the interference that takes place between the light waves leaving a double slit.

The pattern formed by the single slit can be simplified for study purposes by passing a single color of light, consisting of a narrow band of wavelengths, through the slit instead of white light. Figs. 19-12 (a) and (b) show that the *diffraction pattern* then consists of a central bright band that is much larger and brighter than the higher order bands on both sides of it.

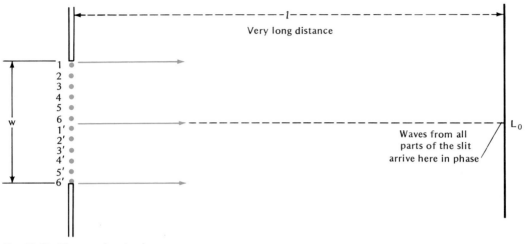

Fig. 19-13. Diagram showing how the central bright band forms in a single-slit interference pattern.

19-18 Explaining the Single-Slit Diffraction Pattern

In Sec. 15-15, the diffraction of water waves was explained by observing that each point of a wave front is, in effect, a new source of wavelets. The diffraction pattern formed by light passing through a narrow slit can be explained in a similar manner.

Consider a slit of width w through which plane waves of monochromatic light pass and fall on a very distant screen. As each wave passes through the slit, each point of it acts like a source from which light waves radiate in all directions. An evenly spaced sampling of twelve of these points is shown in Fig. 19-13. Since all of these light sources are always part of the same original wave fronts, they are in phase.

For simplicity, let the screen be so distant from the slit that the lines drawn from the twelve points in the slit to a point on the screen can be considered parallel for all practical purposes. To this degree of approximation, the waves that reach L_0 in Fig. 19-13 travel the same distance from each of the sources. They therefore always arrive at L_0 in phase, making this a position of maximum brightness.

Now consider the point D_1 in Fig. 19-14, whose direction from the slit is given by the angle θ_1, selected so that the waves leaving the lower edge of the slit travel exactly one wavelength more to reach D_1 than do the waves leaving the upper edge of the slit. At point D_1, the light waves from the slit will cancel each other's effects so that D_1 will be a position of darkness.

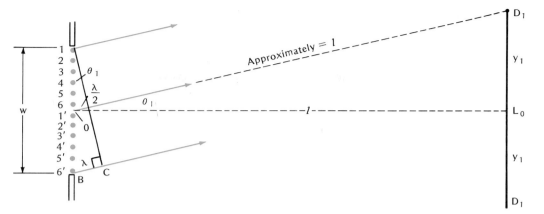

Fig. 19-14. Diagram showing how first-order dark bands form in the single-slit interference pattern.

The manner in which this cancellation takes place can be understood by considering the effects of the matched pairs of sources 1 and 1', 2 and 2', 3 and 3', and so forth. The light waves leaving point 1' travel one-half wavelength more in getting to D_1 than those leaving point 1. These pairs of waves therefore arrive at D_1 in opposite phase. As the crest of one is superimposed upon the trough of the other, the waves cancel each other. The light waves from points 2' and 2 also arrive in opposite phase at D_1 and cancel each other's effects for the same reason. So also do the light waves from points 3' and 3, 4' and 4, 5' and 5, and 6' and 6. In this manner, all the light waves arriving at D_1 pair off to cancel each other, making D_1 a position of darkness.

A symmetrical position of darkness is formed at D_1', at an equal distance below L_0. Between D_1 and D_1' will be a bright band of light having its maximum intensity at L_0. From the right triangle ABC in Fig. 19-14, you can see that D_1 and D_1', the first-order dark bands, form in the directions for which $\sin \theta_1 = \lambda/w$. In a similar manner, one can show that second-order dark bands will form in a direction given by the angle θ_2 such that $\sin \theta_2 = 2\,\lambda/w$. Third-order dark bands form in a direction given by the angle θ_3 such that $\sin \theta_3 = 3\,\lambda/w$, and so on.

Between the adjacent pairs of dark bands there will continue to be partial cancellation of the light waves arriving from the slit, but many of the waves will also arrive nearly enough in phase to produce a band of light. Since only part of the waves leaving the slit contribute to the brightness of the side bands, they are much less bright than the central band where all the waves reinforce each other.

19-19 Effect of Slit Width on Diffraction

As the width of the slit is decreased, $\sin \theta_1$ which is equal to λ/w becomes larger, and therefore so does the angle θ_1. The first-order dark bands D_1 and D_1' then move further from the center of the pattern, and the size of the central bright band increases, as shown

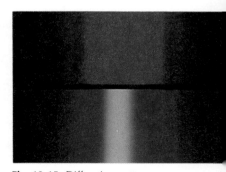

Fig. 19-15. Diffraction patterns produced by red (top) and blue (bottom) light passing through a single slit.

The narrower the slit, the greater is the diffraction of light passing through it.

Fig. 19-16. Diffraction pattern produced by white light passing through a single slit.

in Figs. 19-12 (a) and (b). Thus as the width of the slit decreases, the light passing through it is diffracted more than before and makes a broader diffraction pattern.

When the width of the slit is equal to one wavelength, $\sin \theta_1 = \lambda/w = 1$, and $\theta_1 = 90°$. For a slit of this size, the bright central band extends over a 180-degree arc around the slit. Light leaving the slit then goes in all possible directions and there are no dark bands in the pattern.

When two adjacent very narrow slits are used as in the Young double-slit arrangement, each slit makes its own pattern containing a wide central bright band and much less prominent side bands. As the two patterns are superimposed on the screen, the central bands interfere to give the characteristic double-slit pattern.

19-20 Measuring Wavelength with a Single Slit

The single slit may also be used to measure wavelengths of light. Making the same approximations used in the case of the double slit, we have in right triangle D_1OL_0, $\sin \theta_1 = y_1/l$, where y_1 is the distance between the center of the pattern and the first-order dark band D_1 (Fig. 19-14). Since $\sin \theta_1$ is also equal to λ/w, $\lambda/w = y_1/l$, whence,

$$\lambda = \frac{wy_1}{l}$$

A wavelength may now be determined by measuring w, y_1, and l, and substituting their values in this relationship. When a given wavelength is measured by means of a double slit as well as by means of a single slit, both values are found to agree with each other. This fact further confirms the wave theory interpretation of the behavior of light.

Sample Problem

A monochromatic source of light of wavelength 4.0×10^{-5} cm passes through a single slit and falls on a screen 90 cm away. If the distance of the first-order dark band is 0.30 cm from the center of the pattern, what is the width of the slit?

Solution:

$\lambda = 4.0 \times 10^{-5}$ cm

$l = 90$ cm $y = 0.30$ cm

$\lambda = \dfrac{wy_1}{l}$ $w = \dfrac{\lambda l}{y_1}$

$w = \dfrac{4.0 \times 10^{-5} \text{ cm} \times 90 \text{ cm}}{0.30 \text{ cm}} = 0.012$ cm

---

Test Yourself Problems

1. Monochromatic light passes through a single slit of width 0.010 cm and falls on a screen 100 cm away. If the distance from the center of the pattern to the first dark band is 0.60 cm, what is the wavelength of the light?

2. Light of wavelength 4.5×10^{-5} cm passes through a single slit and falls on a screen 100 cm away. If the slit is 0.015 cm wide, what is the distance from the center of the pattern of the first dark band?

19-21 Single-Slit Pattern with White Light

When white light is passed through a single slit, each of its component colors makes similar patterns of diffraction bands. However, the bands made by the longer wavelengths of light are spread out by the slit more than those made by the shorter wavelengths. The pattern seen on the screen is the result of the superposition of the patterns made by each of the colors. The large composite central band is a mixture of all the colors falling on it and is white. Positioned symmetrically on both sides of it are bands of many colors resulting from the overlapping of the patterns of each of the colors. (See Fig. 19-16.)

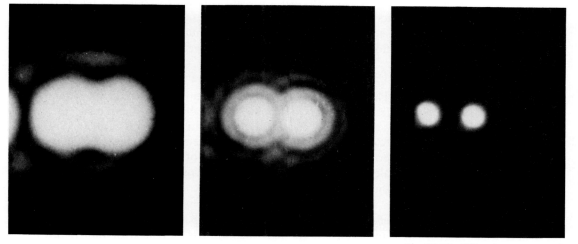

19-22 Resolving Power of a Lens

It can now be seen how diffraction limits the magnifying power of lenses and of optical instruments containing lenses. When a converging lens makes an image of a distant point source of light, the image is not a sharp point as indicated by the ray diagram for the lens but a diffraction pattern consisting of a circular disc surrounded by alternate rings of light and darkness.

The lens acts like a circular opening and light waves are diffracted on passing through such an opening just as they are diffracted on passing through a slit. Furthermore, just as the ratio of the width of a slit to the wavelength of the light passing through it determines the size of the diffraction pattern for a slit, so the ratio of the diameter of a lens to the wavelength of the light passing through it determines the size of the diffraction pattern for a lens. The smaller this ratio, the larger and more spread out will be the diffraction pattern.

When the lens makes images of two neighboring distant point sources of light, each image is a circular diffraction pattern. (See Fig. 19-17.) As long as the two point sources are separated by a

Fig. 19-17. Diffraction patterns of the images of two point sources of light made by a lens of small diameter (left); a lens of larger diameter (center); and a lens of still larger diameter (right). The patterns decrease in size as the diameter of the lens increases. At right, the patterns are small enough to permit clear separation of the images.

large enough distance, the diffraction patterns of their images are completely separate and distinct. However, when the two point sources are close enough together, the diffraction patterns of their images overlap and fuse. The two images can then no longer be distinguished from each other. Thus, the size of the diffraction patterns made by the lens determines how close two points can be and yet have their images clearly distinguishable from each other.

The ability of a lens to produce clearly distinguishable images of objects that are very close together is called its *resolving power*. By determining the smallest distance apart that two neighboring points can be and yet have their images distinguishable from each other, the resolving power of a lens of a microscope or a telescope limits its *magnifying power*. It is useless for a lens to magnify points separated by less than this minimum distance because the diffraction patterns of their images merge indistinguishably.

19-23 Increasing Resolving Power

The resolving power of a lens can be increased by reducing diffraction effects. This can be done by increasing the size of the lens (see Fig. 19-17) and by using light of shorter wavelengths. Large telescopes, both reflecting and refracting, achieve high resolving power by using lenses or mirrors of large diameters. This is equivalent to increasing the opening through which the light passes, and thus results in decreasing the effects of diffraction.

In microscopes, the circumference of the objective lens is generally a circle of small diameter so that diffraction becomes very important when high magnification is needed. Here, diffraction effects may be reduced by viewing the object with visible light of the shorter wavelengths such as blue or violet light, or with ultraviolet light whose wavelengths are considerably shorter than those of visible light. When ultraviolet light is used, the image formed must be thrown upon a fluorescent screen to become visible. Such a screen glows when ultraviolet light falls upon it, and reveals the image.

19-24 Coherent Light

Light from an ordinary source such as an incandescent or fluorescent lamp is basically disorganized. It consists of waves of many different wavelengths emitted in short random bursts by the atoms of the source. Furthermore, as was explained in Section 19-8, even those groups of waves that happen to have the same wavelength are in many different phase relationships to each other. As a result, there is very little reinforcement among them and their energy becomes diffused rather than concentrated.

To produce better organized light, it is desirable to have a light source all of whose waves have the same wavelength and are exactly in phase with each other. Light having these special properties

In a coherent light source, all the waves have the same wavelength and are in phase.

would be perfectly coherent. Only in recent years has a light source capable of producing such highly *coherent light* been developed. It is called a *laser*. The name stands for Light Amplification by the Stimulated Emission of Radiation, the process whereby laser light is produced. The success of the laser in producing coherent light is another confirmation of the wave theory of the reinforcement and interference of light.

Fig. 19-18. A converging lens brings a laser beam to a focus where the high concentration of energy makes the air glow brightly.

19-25 The Laser

In the laser, the source of light generally consists of the atoms of a solid, such as a ruby crystal, or of the atoms of a gas contained in a tube. The atoms are artificially stimulated to emit light waves of only a single wavelength. This is done in such a way that the waves automatically fall into phase with each other the moment they are emitted. The laser then gathers together the contributions of enormous numbers of atoms to produce a beam of coherent light. (See Sections 32-24 to 32-26 for further details.)

Lasers are powerful beams of coherent light.

Laser light is generally emitted in the form of a beam of parallel rays. The perfect mutual reinforcement of the waves in the beam keeps the rays parallel. The beam therefore does not spread out as it moves through space, and it loses very little energy on its journey. This property makes it ideal for sending signals over very long distances. Such signals have been successfully detected after making the round trip between the earth and the moon.

Another outstanding characteristic of a laser beam is its relatively high concentration of energy. This characteristic arises from the fact that the beam is produced by the cooperative action of billions of atoms emitting their bursts of light energy in perfect phase with each other. This energy can be further greatly concentrated by bringing the beam to a focus with a converging lens. The concentration of energy at the focus may then be so great that air in its vicinity is immediately brought to incandescence. (See Fig. 19-18.) Focused laser beams are also capable of vaporizing and penetrating heavy steel plates and of punching holes in diamonds.

Fig. 19-19. An eye doctor treats a patient's diseased retina with the blue-green coherent light of an argon gas laser to prevent possible blindness.

19-26 Modern Applications of Lasers

The laser beam is finding a wide range of uses. As a research instrument, it is proving invaluable in the precise measurement of length. Among its unusual practical applications is its use by surgeons for the painless removal of eye tumors.

Laser beams are replacing cables and radio waves as carriers of signals and telephone calls. They have opened up the new field of *holography*, a remarkably realistic method of projecting three-dimensional images. (See *Physics+Plus* after this chapter.)

Finally, lasers are providing a powerful tool to the nuclear physicist seeking to develop nuclear fusion as a source of energy. To obtain the temperatures of billions of degrees Celsius needed to bring about the fusion of atomic nuclei, physicists concentrate a battery of high-energy laser beams on them.

CHAPTER REVIEW

Summary

In this chapter, we have seen that the effects of diffraction and interference can be accounted for both qualitatively and quantitatively by the wave theory of light. The theory explains correctly the appearance of the **interference pattern** of Young's double slit as a series of regularly spread alternating bright and dark bands. This pattern also provides a method for measuring wavelengths of light by means of the relationship $\lambda = x_1 d/l$ where x_1 is the distance between the center of the interference pattern and the first-order bright band, d is the distance between the slits, and l the distance between the slits and the screen.

The wave theory is further applied successfully in explaining the **diffraction pattern** produced by the single slit. This pattern, consisting of an extra-wide central bright band with fainter higher-order bands on both sides of it, also provides a means of measuring wavelengths of light. Agreement between the values for wavelengths measured with double slits and those measured with a single slit strongly supports the wave theory. Finally, we have seen how diffraction and interference determine the **resolving power** of a lens and therefore limit its **magnifying power.**

The wave theory of light has so far proved to be remarkably successful in explaining how the path of light is changed by reflection, refraction, and diffraction, and how its energy is redistributed through interference. The production of **coherent light** by **lasers** is still another success of the theory.

However, a basic question remains to be answered. *Of what do the vibrations of light waves consist?* Maxwell found the answer to this question in the study of electricity and magnetism, to be undertaken in the next chapters.

Questions

Group 1

1. S_1 and S_2 are two sources of waves in a water tank, each generating waves of wavelength λ. Small pieces of cork are floating at points A, B, C, D, and E of the surface. The differences between the distances of each of these points from S_1 and S_2 are as follows: A, zero; B, 2λ; C, 4.5λ; D, 2.7λ; E, 2.5λ. Assuming that the sources S_1 and S_2 are in phase, (a) which of these points is on a line of reinforcement; (b) which is on a nodal line; (c) which is on neither? (d) Describe the motion, if any, a piece of cork on a line of reinforcement undergoes. (e) What motion, if any, does a piece of cork that is on a nodal line undergo?

2. Using compass and ruler, make a diagram of the interference pattern made by two sources of waves, S_1 and S_2, that are in phase and are 2 cm apart. Assume that the wavelengths are 1 cm long and draw six crests that have left each source; omit the troughs. Draw the lines of reinforcement. Then draw a similar diagram, this time with the sources of the waves, S_1 and S_2, 3 cm apart.

   In each of the two diagrams, draw dotted lines midway between the lines of reinforcement which show the approximate positions of the nodal lines. (a) What happened to the spacing

between the nodal lines when the distance between S_1 and S_2 was increased? (b) What happened to the number of nodal lines? (c) What two quantities determine how many nodal lines will form?

3. Referring to Young's double-slit experiment, (a) how are two sources of light that remain in phase obtained? (b) How is light of one color obtained? (c) What is seen on the screen where it is intersected by a nodal line? (d) By a line of reinforcement? (e) Describe the spacing of the bands in the interference pattern that is formed on the screen.

4. Referring to Young's double-slit experiments and assuming the light source is monochromatic light, (a) what happens to the distance between the centers of the bright bands as the distance between the two slits is decreased? (b) How does the distance between the bright bands in the interference pattern made by monochromatic red light compare with the distance between the bright bands made by monochromatic violet light? (Look up the range of wavelengths of these two colors of light in Table 19.1, Sec. 19-14.)

5. In measuring a wavelength of light by means of a double slit, it is desirable to make a pattern in which the bright and dark bands are as widely spaced as possible. Referring to the relationship

$\lambda = x_1 d/l$, what two changes in the experimental setup may be made to obtain a wider spacing of the pattern made by a given source of light?

6. (a) Describe the arrangement of the different colors of light in a typical bright band formed by a double slit when white light is passed through the slits. (b) Which color of light is deviated most from its original path by the double slit? (c) Which color is deviated least from its original path?

7. Why are no interference bands seen on a screen illuminated by two separate sources of light such as two candles?

8. Compare the single-slit diffraction pattern made by monochromatic light with the double-slit diffraction pattern made by the same light as to: (a) the size of the central bright band; (b) the spacing of the bright bands.

9. Given a monochromatic source of light of known wavelength, how can the single-slit experiment be used to measure the width of a very narrow slit?

10. (a) What effect does decreasing the width of a single slit have on the size of the central bright band made when monochromatic light passes through it? (b) What is the relationship between the width of the slit and the wavelength of the light when the central bright band just extends over a 180°-arc around the slit?

Group 2

11. (a) What determines the size of the diffraction pattern made by a lens when light of given wavelength passes through it? (b) How do the diffraction patterns made by a lens of two very close point sources of light determine if the images of those points can be distinguished from each other? (c) What is meant by the resolving power of a lens?

12. (a) Why does a lens of larger diameter have a higher resolving power than a lens of smaller diameter? (b) Why may the resolving power of a lens or microscope be increased by illuminating the object with ultraviolet light rather than with visible light? (c) How is an object illuminated by ultraviolet light made visible?

13. Stars close together in the sky are sometimes photographed through a telescopic camera having a filter that transmits only the blue and violet parts of the stars' spectra. Why does this result in a better picture of the stars than could be made without the filter?

14. The ability of a diffraction grating to separate light from a light source into its spectrum depends upon the way in which it is ruled. What effect does decreasing the distance between the rulings (or slits) have on the intensity and on the spreading out of the spectra produced?

15. In the spectrum of white light formed by a diffraction grating, violet light is deviated the least and red light is deviated the most. In the spectrum of white light formed by a prism, red light is deviated the least and violet light the most. Explain the difference between the action of the grating and the action of the prism.

16. (a) In what ways is the light energy in a coherent or laser beam better organized than the light energy in a beam of ordinary light? (b) A beam of laser light is passed through a double slit and allowed to fall on a screen. What kind of pattern is made upon the screen?

Problems

Note: Convert all lengths in these problems to the same units.

Group 1

1. Monochromatic light passes through two slits 0.030 cm apart and falls on a screen 120 cm away. The first-order bright band is 0.16 cm from the middle of the center band. What is the wavelength of the light?

2. In problem 1, how far from the center band are the second- and third-order bright bands?

3. Yellow light of wavelength 6.0×10^{-5} cm passes through two narrow slits 0.020 cm apart and forms an interference pattern on a screen 180 cm away. How far apart are the central bright band and the first-order bright band?

4. Violet light of average wavelength 4.0×10^{-5} cm is passed through two narrow slits 0.010 cm apart. At what distance must the screen be put so that the distance between the bright bands is 0.30 cm?

5. Green light passes through a double slit for which $d = 0.2$ mm and falls on a screen 2 m away. The distance between the fourth-order

bright band and the central band is 2 cm. (a) What is the spacing between bright bands? (b) What is the wavelength of the light?

6. A photograph of the interference pattern made at 200 cm from a double slit through which monochromatic light is passing shows that the bright bands are 0.40 cm apart. The distance between the slits is 0.020 cm. (a) What is the wavelength of the light? (b) What is its color?

7. (a) What is the frequency of blue-violet light whose wavelength is 4.5×10^{-5} cm in air? Take the speed of light as 3.0×10^{10} cm/sec. (b) What is the frequency of this light in water?

8. What is the wavelength of the light in problem 7 in glass whose index of refraction is 1.5?

9. What is the frequency of light whose wavelength in a vacuum is 600 nm?

10. A monochromatic source of light passing through a single slit of width 0.02 cm forms an interference pattern on a photographic film 200 cm from the slit. (a) If the distance between the dark bands on opposite sides of the central band is 1.0 cm, what is the distance between the center of the pattern and the first dark band? (b) What is the wavelength of the light?

Group 2

11. Light of wavelength 5.5×10^{-5} cm is passed through a single slit and falls on a screen 160 cm away. What must be the width of the slit so that the distance from the center of the first-order dark band to the center of the central band will be 0.40 cm?

12. What is the distance between the two first-order dark bands of a single-slit interference pattern made with a slit 0.025 cm wide with light of wavelength 5.4×10^{-5} cm on a screen 80 cm away?

13. In the single-slit relationship $\sin \theta_1 = \lambda/w$, the angle θ_1 measures how wide the bright central band will be. The wider the central band, the more the light in the beam "bends around the edges of the slit" and the greater is the diffraction effect. (a) Compare the values of $\sin \theta_1$ for yellow light of wavelength 600 nm that is passed through one slit of width 0.010 cm and one of 0.10 cm. (b) Noting that for small angles $\sin \theta_1$ is proportional to θ_1, by what factor is the bending of the light around the edges of the narrower slit larger than that around the edges of the wider slit?

14. Show by the method followed in Section 19-10 that, for the double slit, the angle θ_2 at which the bright band of the second order occurs is given by $\sin \theta_2 = 2 \lambda/d$, the angle θ_3 at which the band of third order appears is given by $\sin \theta_3 = 3 \lambda/d$, and so on.

15. Radio waves that are emitted by two radio transmitters that are in phase behave like light waves coming from a double slit. Just as for light waves, the angle θ_1 at which the first-order reinforcement takes place is given by $\sin \theta_1 = \lambda/d$. (a) If two such transmitters 1000 m apart each send out waves of length 100 m and equal amplitude, what is the $\sin \theta_1$? (b) What is θ_1 to the nearest degree? (c) In what two directions would the two sets of waves cancel each other to produce the equivalent of the first-order nodal lines?

Applying Physics

The single-slit pattern can be seen in the following way. Make a narrow slit by cutting a rectangular opening 25 mm by 5 mm in the middle of a piece of cardboard. Over this opening, adjust the positions of two double-edged razor blades so that their edges make a very narrow slit.

CAUTION: Cover the edges of the blades with clear cellophane tape to avoid being cut by them.

Secure the razor blades in position with cellophane tape.

Now, make a second slit about 25 mm long and 1 mm wide in a piece of cardboard and paste a piece of red cellophane over it. This slit, held vertically in front of a bright source of light, such as a 100-watt electric light bulb, provides a narrow source of red light. Look at this red light through the very narrow slit between the razor blades to see the interference pattern made by it.

Replace the red cellophane on the second slit with a blue or violet piece of cellophane. Compare the band pattern for blue light with that for red light.

Reset the razor blades to make the slit narrower and repeat the observation. What happens to the pattern as the slit is made narrower? Now see what happens to the pattern when the slit is made wider.

Laser Technology

It's not only highly trained specialists in technical fields such as astronomy, medicine, telecommunications, electronic publishing, and manufacturing who have been affected by the development of sophisticated laser technology. Of course, eye surgeons, astronomers, typesetters, and so on do use lasers. But the impact of laser technology extends into our everyday lives more than we may have realized.

Three applications of lasers with which many of us have had personal experience are compact disks (CDs), cash registers that read Universal Product Code (UPC), and three-dimensional images produced by a process called *holography.*

Compact Disks.   Due to their superior sound quality, CDs have become the

preferred means of recording and reproducing music. A 13-centimeter diameter CD can hold more music than a 33⅓-rpm record. Unlike magnetic tapes or records, CDs are not subject to noise (record or tape "hiss") due to tiny imperfections in the disk's surface or the presence of dust or fingerprints. You can play a CD as often as you like and it will never wear out because, unlike a tape or a record, no needle or "head" is ever in direct contact with the CD.

It is the grooves in the plastic of a CD that store information. A low-power diode laser in a CD player is used to decode that information. It is very important that the laser light be accurately focused on the disk. To achieve this, an electromagnet in the player is used to keep a light-focusing lens at a constant distance from the disk. The laser light is reflected with varying intensity from the different grooves. These variations in intensity are read by a light-sensitive detector and decoded to produce sound.

Universal Product Code.   Just about everything you buy has a UPC or "bar code" on the label or price tag. The ten-digit code, composed of parallel bars of variable width and spacing, identifies the manufacturer and specific item. The benefits of UPC for the customer include quicker service and more accurate pricing. Businesses benefit from increased productivity of store cashiers and the automation of inventory control.

At the store's checkout counter, a cashier passes the UPC over a window through which a laser beam is directed. (The low-powered helium-neon laser beam used is not harmful to the eyes.) The dark areas of the UPC absorb the laser light, whereas the light areas reflect it. The variations in the intensity of this reflected light are read by a

light-sensitive detector. This information is sent to a central computer, which identifies the product and its current cost at that store and relays this information to the cash register. The customer receives a printed receipt with a detailed description of the sale, including the date, time, item name, and price.

Holography. Holograms, three-dimensional images created by holography, are a commonplace today in the world of advertising, marketing, and entertainment. Think of that image on a credit card, badge, or other flat surface that seems to move as you shift your view. You may even have had the experience of walking around, and viewing from all angles, a three-dimensional image of an object or a person.

Holography is a system of lensless photography that creates three-dimensional images—in space, or on a surface—that seem real. To make a hologram, an object is illuminated with the coherent light from a laser. The interference patterns of the light waves reflected from the object are recorded on a piece of film.

If you look at the exposed film you will not see a recognizable image. However, the complete three-dimensional shape and appearance of the object have been encoded in the film's coating. This is because the path of each light wave, reflected from a different point on the object's surface, is unique. When the exposed film is illuminated with a laser beam or, in some cases, ordinary white light, the object is reproduced as a three-dimensional image.

Acoustics of Concert Halls

Think of the last time you heard a band playing in a large gymnasium, and you will understand the need for acoustical engineering. Concert halls, in particular, must be designed so that audiences hear the music as it is played, with little or no distortion.

Architectural acoustics analyzes the pathways of sound waves in a hall, together with the shape of the hall, and determines the best shapes for different sounds. Music consists of a rapidly changing complex of tones. The enemy of music is reverberation, or echoing.

Acoustic engineers use the concept of reverberation time to evaluate a hall. This is the time required for the energy of a sound wave to drop to one-millionth of its original value. When reverberation time is very low, reflected sounds reach the listener at almost the same time as the original sound. When reverberation time is high, reflected sounds reach the listener at the same time as new sounds from the source. In this latter case,

confusion can result. A reverberation time of one second is best for speech, and one to three seconds is best for music.

Four qualities are important in establishing the musical character of a hall: definition, or clarity; fullness of tone; balance; and blend.

Definition has to do with the audience's ability to recognize the various instruments and the notes being played. Fullness of tone is the richness, body, or warmth of the music. These two qualities are dependent on reverberation time. A short reverberation time favors definition, a longer one fullness of tone.

Balance and blend are more under the control of the musicians. The arrangement of the instruments, stage or pit design, and the presence of reflecting surfaces near the musicians affect how the sound has been "mixed" by the time it reaches the audience. Audience members too affect the quality of music. To the acoustic engineer, one person in an audience is equivalent to 0.4 m² of open window!

How ceiling shape affects the paths of sound waves. In (b) the redesigned ceiling distributes sound more uniformly throughout the hall.

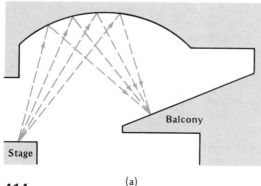

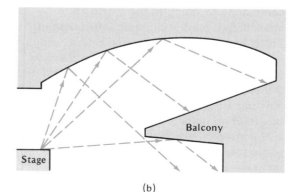

(a)

(b)

Musicians and Related Occupations

Can you imagine a world without music or musicians? It would be a dull and lifeless place indeed. Much of the pleasure people derive from their leisure time depends on music.

Professional musicians earn their livelihood performing for others. They may play in rock groups, jazz combos, dance bands, or classical ensembles, from chamber music groups to symphony orchestras. Many well-known musicians appear as soloists and make recordings.

Musicians spend much of their time practicing and rehearsing. They usually perform at night and on weekends. Often they must travel great distances to perform, especially in the case of extended tours. Many musicians experience unemployment between engagements or "gigs." Others can find only part-time work and must supplement their incomes with other types of jobs.

Competition in the music world is keen. Many people have performing skills, and wish to become professionals. Only the most talented, however, find employment. Although openings are expected to grow about as fast as the average for all occupations, the high level of competition is expected to continue.

Often people who become professional musicians begin studying an instrument at a very early age. Long and intensive training is needed to acquire the necessary performing skill, knowledge of music, and ability to interpret music. Training may be accomplished by means of private study with a professional, in a college or university program, at a music conservatory, or through practice with a group. An audition is usually needed to qualify for admission to an institutional program.

Most professional musicians do not have a college degree, but those with degrees are more likely to work full time.

People considering careers in music should have musical talent, creative ability, versatility, and poise. Self-discipline is a must, for constant study and practice are needed to maintain skills. Long hours, night work, and frequent travel require physical stamina too.

Aside from performers, including singers, some related occupations are music teachers, arrangers, composers, instrument repairers and tuners, and many jobs having to do with the business side of music.

UNIT 5

Electricity

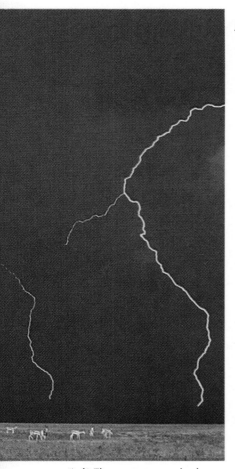

As we return to the study of matter, we are led to the discovery that all matter is electrical in nature. It consists of atoms containing positive charges in the form of protons and negative charges in the form of electrons. The charges exert electrical forces upon each other; like charges repel each other and unlike charges attract each other.

From the work of Michael Faraday (1791–1867) and Robert A. Millikan (1868–1953) we learn that all electric charges are multiples of a natural unit charge that is equal to the quantity of charge on an electron or a proton. This natural unit is the smallest quantity in which positive or negative charges usually occur.

Charges can be made to flow in a circuit, thus constituting an electric current. Such currents are sources of energy that can be readily converted into other forms of energy and do work.

Electrical energy is the most versatile and widely used of all the forms of energy. It is readily applied in motors to do mechanical work, in incandescent light bulbs to give light, in electric heaters to provide heat, in computers to perform calculations and process data, and in telephones, radios, and television sets to provide communications. Its use has changed every aspect of living in the home, factory, office, farm, and mine.

However, there are dark clouds on the horizon. We are reaching the limit of the electrical energy we can expect from the primary sources of energy; namely, waterpower, coal, oil, and gas. In the face of the rising demand for electrical energy we must find other primary energy sources. That is one of the major challenges of modern times and one that we must meet successfully if we are to maintain our way of life.

(Left) The most spectacular form of lightning is produced by a large discharge of static electricity from a cloud to the earth. Here the discharge is over water. (Above) Lightning over the East African savanna, in Tanzania.

20 Electricity and the Nature of Matter

Aims

1. To study the evidence that all bodies of matter contain positive and negative electric charges.
2. To learn how bodies acquire electric charges and how to detect and identify those charges.
3. To propose a theory to explain how bodies gain or lose electric charge and to test the theory.
4. To learn how to control the movement of electrons in charging and discharging bodies and in setting up a continuous electric current.

20-1 Amber and Electricity

As early as 600 B.C., it was known that amber that has been rubbed by cloth attracts small bits of material of all kinds. Amber's ability to attract resembles that of the natural magnet, known as lodestone. However, the attraction of lodestones and magnets is limited mainly to objects made of iron, whereas amber which has been rubbed with cloth attracts all substances.

It was not until the sixteenth century that a methodical study of this curious property of amber was made by William Gilbert, physician to Elizabeth I, Queen of England. He found that many other substances act like amber in acquiring the ability to attract light objects after they have been rubbed against other substances. Gilbert called such substances "electrics," which comes from the Greek word *electron,* meaning amber.

Subsequently, it was shown that every body acquires the ability to attract small bits of matter after it has been rubbed by another body. Bodies showing this attracting property are said to have an *electric charge* or to be *electrified*. Since the electric charges on electrified bodies are usually at rest, they are referred to as stationary or *static electricity*. In the familiar household electric circuits, electric charges are in motion. Streams of moving charges are known as *electric currents*.

20-2 Like Charges Repel; Unlike Charges Attract

A typical charged body is made by rubbing a rod of hard rubber with fur. The charge in the rubber rod is revealed by the fact that it attracts small uncharged pieces of wood, metal, paper, and other materials. In general, an electrically charged body attracts all objects that are uncharged. However, it may either attract or repel other charged bodies.

> Bodies with like charges repel each other; those with unlike charges attract each other.

To demonstrate this, two rubber rods are charged by rubbing each of them with fur. Since both rods are charged in the same way, they can be assumed to have similar charges. One of the charged rubber rods is suspended at its center by means of a silk thread so that it can turn freely in the horizontal plane. The second charged rod is then brought near one end of the suspended rod. You can see that the suspended rod turns away, showing that the two similarly charged rods repel each other.

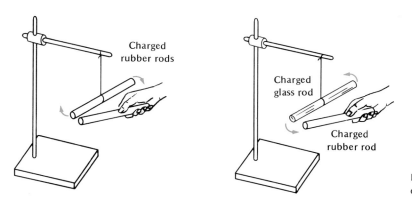

Fig. 20-1. Like charges repel each other; unlike charges attract.

This demonstration is now repeated using two glass rods, each of which has been rubbed with silk. Again you can see that the similar charges on the glass rods repel each other.

However, when one of the charged rubber rods is brought near the suspended charged glass rod, the rods attract each other. Obviously the electric charge on the rubber rod is different from the electric charge on the glass rod.

20-3 Two Kinds of Charges

When any other charged body is brought near the suspended charged glass rod and then near the suspended charged rubber rod, you can see that it repels one of the charged rods and attracts the other. Thus, there are only two kinds of charged bodies: those that repel the charged glass rod, and therefore have a charge like that on the glass, and those that repel the charged rubber rod, and therefore have a charge like that on the rubber.

Following the suggestion of Benjamin Franklin, *the kind of electric charge acquired by a glass rod that has been rubbed with silk*

Benjamin Franklin named the two kinds of electric charge.

is called a positive charge. The kind of electric charge acquired by a rubber rod that has been rubbed with fur is named a negative charge. All positively charged bodies repel each other and all negatively charged bodies repel each other. However, positively charged bodies attract negatively charged bodies.

20-4 Rubbing Produces Equal Opposite Charges

Experiment shows that when two different bodies are rubbed together, both bodies become charged. One acquires a positive charge and the other acquires an equal negative charge. For example, when a glass rod rubbed with silk acquires a positive charge, the silk acquires an equal negative charge. This can be shown by bringing the silk near a suspended negatively charged rubber rod and noting that the rod is repelled. When the charged silk and the charged glass rod are put in close contact, the combination shows no net charge, indicating that the opposite charges are equal and neutralize each other.

Similarly, it can be shown that when a rubber rod rubbed with fur acquires a negative charge, the fur acquires an equal positive charge.

20-5 Theory of Static Electricity

How can we account for the charging of bodies by rubbing? The fact that any two bodies can be charged in this way suggests that matter itself contains large numbers of both positive and negative charges. In a neutral body, these opposite charges are not noticed because they are present in equal quantities. However, when the balance of positive and negative charges is destroyed so that a body has more of one charge than of the other, the body shows a noticeable net electric charge. If there are more positive charges than negative charges, the body is positively charged. If there are more negative charges than positive charges, the body is negatively charged.

Only electrons move from body to body. A neutral body that gains electrons becomes negatively charged; one that loses electrons becomes positively charged.

The charging of two bodies when rubbed together can be explained by assuming that, during contact, charges move from one body to the other, thus upsetting the balance of charges in each of them. There are three possible ways in which this can happen. The first is that both positive and negative charges move between the bodies during contact, leaving one with an excess of positive charge and the other with an excess of negative charge.

The second possibility is that only positive charges move from one body to the other during contact while the negative charges remain fixed in both bodies. The body receiving the additional positive charges then becomes positively charged; the other body, having lost some of its positive charges, is left with an excess of negative charges.

The third possibility is that, during contact, only negative charges move from one body to the other while the positive charges remain fixed. The body receiving the negative charges becomes negatively charged. The other body, having lost some of its negative charges, is left with a net positive charge.

Experimental evidence, some of which will be mentioned in this chapter, indicates that the third possibility best fits the facts. It is found that the negative charges of solid bodies, and of metals in particular, are much more easily detached from these bodies than their positive charges. These readily movable negative charges occur in the form of the tiny particles called *electrons*. Regardless of their sources, *all electrons are alike*. They have two distinguishing characteristics. Electrons have *the smallest negative charge a body can have*. They also have *the smallest rest mass possessed by any particle of matter*.

The positive charges in bodies are also observed to occur in the form of tiny particles known as *protons*. Individual protons are not free to move within solid bodies. Each proton has a positive charge equal to the negative charge on the electron but a mass 1840 times larger than that of the electron.

This outline of a theory relating matter and electricity will be referred to from now on as the *electron theory*. To test this theory, we must now see how well it explains the observed facts and also examine some of the evidence supporting its assumptions.

20-6 Explanation of Charging by Rubbing

It follows from the electron theory that neutral bodies usually acquire electric charges either by gaining electrons from other bodies or by losing them to other bodies. They become negatively charged when they gain electrons and positively charged when they lose electrons.

The equal and opposite charging of two different neutral bodies when rubbed together is explained by assuming that one body holds its electrons less firmly than the second. During contact, electrons rub off the first body and attach themselves to the second. The first body, having lost some of its electrons, is left with an excess of protons, or positive charges. The second body, having gained these electrons, acquires an equal excess of negative charges.

In the charging of glass by rubbing with silk, it is the glass that holds its electrons less firmly and loses some of them to the silk. In the charging of rubber by rubbing with fur, it is the fur that holds its electrons less firmly and loses some of them to the rubber.

20-7 Conductors and Insulators

Electric charges pass readily through certain substances called *conductors*, and pass only with great difficulty through other substances called *insulators*. In general, among the solid substances, all the metals and carbon are good conductors. That is why these

Fig. 20-2. Lightning following the path of least resistance often strikes the tops of tall buildings, as shown here, where it is conducted safely to the ground through conducting cables.

substances are frequently used to conduct electricity in the home and factory. Some liquids are also good conductors, others very poor. But all of the gases, when dry, are insulators. Other common insulators are glass, rubber, mica, silk, cork, plastics, and air.

To demonstrate the behavior of solid conductors, place a metal rod on a glass stand, as shown in Fig. 20-3. Suspend a light metal-coated ball by means of a silk thread so that the ball is in contact with one end of the rod. Now touch the other end of the rod with a negatively charged body and then withdraw the charged body. You can see that the metal rod and the ball repel each other, showing that both have received like charges. Test the charge on the ball by bringing the original negatively charged body near it. The ball is repelled, showing that negative charge has passed from the negatively charged body into both the metal rod and the ball.

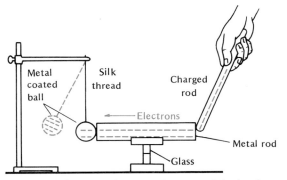

Fig. 20-3. Demonstration of conduction in a metal rod.

Repeat the experiment using a positively charged body to furnish the original charge instead of a negatively charged body. The ball is again charged and repelled. This time, however, it and the rod have received positive charges of electricity.

When a rubber rod or any other insulator is substituted for the metal rod and the above two experiments are repeated, no charge is transferred through the rubber rod to the ball.

20-8 Free Electrons in Metals

According to the electron theory, metals can conduct electricity because they contain large numbers of electrons that are free to move through them. In insulators, there are practically no such *free electrons*.

In Fig. 20-3, as the negatively charged body touches the metal rod, some of its excess electrons pass into the rod and join the free electrons there. At the same time, the repulsion of the negative body forces some of the free electrons in the metal rod to move toward the opposite end of the rod and into the ball. Both rod and ball thus acquire excess electrons and repel each other.

When the positively charged body touches the end of the metal rod, it attracts the free electrons in the rod and ball toward itself. Some of the electrons therefore flow out of the ball and the rod and into the positively charged body. The rod and ball, having lost electrons, now have an excess of positive charge and again repel each other. Note that although only electrons moved, the effect is the same as if the positive charges had moved in the opposite direction.

20-9 Controlling Movement of Electrons

The conduction of metals draws attention to the conditions under which electrons move within solid matter. First there must be a force acting upon them. Since electrons are negative charges, this force can be supplied by other charged bodies. A nearby negatively charged body repels electrons while a nearby positively charged body attracts them.

Secondly, the electrons must be free to move. In metals there are many free electrons that can move readily under the slightest force. For this reason, when it is desired to move electrons from one place to another, a conducting path of metal is usually provided. In insulators, electrons are generally held firmly by the fixed positive charges and cannot move freely. As a result, insulators can lose or gain electrons on touching other bodies only at the points of contact. The charge on one part of the insulator is not conducted to other parts. When it is desired to prevent electrons from moving into or out of a body, the body is surrounded by insulators.

Electrons move more or less freely through conductors, but with great difficulty through insulators.

20-10 Charging by Contact

Rubbing a body with another body is not the only way to charge it. When a neutral body merely touches a charged body, it acquires a charge like that on the body. The neutral body is then said to have been *charged by contact*. When the neutral body touches a negatively charged body, some of the excess electrons are repelled out of the negatively charged body and pass into the neutral body. The neutral body, having gained electrons, becomes negatively charged. When the neutral body touches a positively charged body, it loses some of its electrons to the positively charged body and becomes positively charged.

20-11 Electroscope

To study electric charges more carefully, an instrument sensitive enough to detect and measure small electric charges is needed. The *gold leaf electroscope,* shown in Fig. 20-4, is such an instrument. It consists of a metal rod having a metal ball or knob at its upper end and a pair of gold leaves suspended from its lower end. The rod is mounted in a stopper made of an insulator such as rubber. The stopper fits into an airtight glass container so that the gold leaves are enclosed and protected from stray air currents. Gold is used for the leaves because it can be hammered very thin and makes a very light leaf.

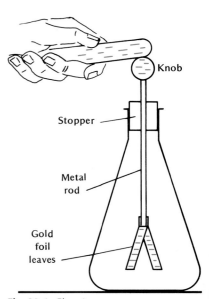

Fig. 20-4. Charging an electroscope negatively by conduction.

To detect a small charge on a body, part of its charge is given to the knob of the electroscope by contact. Because the knob, rod, and leaves of the electroscope are conductors, the charge put on the knob promptly spreads to the leaves. Both leaves thus acquire like charges, repel each other, and spread apart. The greater the charges on the leaves, the more they spread apart. The angle through which the leaves spread apart is therefore a measure of the charge on the leaves. Once a charge is on the electroscope, it is prevented from escaping by the insulating rubber support of its rod and the insulating property of the surrounding air.

20-12 Use of Ground

To remove the charge from an electroscope, we simply connect a metal wire from its knob to the *ground*. This is called *grounding* the electroscope. The leaves of the electroscope then collapse, showing that the electroscope has lost its charge.

In grounding, we make use of two properties of the earth. The first is that the earth is a good conductor. The second is that, because of its huge size it can accept or give up almost unlimited numbers of electrons without appreciably changing its own net charge.

When a negatively charged conductor, such as a negatively charged electroscope, is connected to the ground, electrons are repelled out of the electroscope into the ground until it has effectively lost all of its charge. When a positively charged conductor, such as a positively charged electroscope, is grounded, the flow of electrons is reversed. Electrons from the ground are attracted into the positively charged electroscope until its positive charge is completely neutralized. The earth is thus always gaining or losing electrons, but these changes are far too small to have any significant effect on its net charge.

Practically, in discharging an electroscope, it is sufficient to touch its knob with the finger instead of connecting a wire conductor between the knob and the ground. In so doing, we make use of the fact that the human body is also a conductor and provides a suitable pathway for electrons between the electroscope and the ground.

Grounding the outside metal frames of radios, television sets, and other household appliances is a commonly used safety measure. It prevents the accumulation of electric charges on the outside parts of these appliances by promptly conducting them to the ground. Thus it protects the consumer from the electric shock that might result from making contact with a highly charged conductor.

20-13 Photoelectric Effect in Metals

Part of the evidence for the assumption that only the free electrons, and not the positive charges, move in metals comes from the observation that electrons are rather easily ejected from metals. One can demonstrate with an electroscope that the mere falling of light on a metal is sufficient to eject electrons from it.

In Fig. 20-5, an insulated zinc plate is mounted on the knob of an electroscope by a metal rod, and the zinc and the electroscope are negatively charged. As light from a carbon arc lamp is directed on the zinc plate, the electroscope gradually loses its charge. If the experiment is repeated with the zinc charged positively instead of negatively, the electroscope does not lose its charge when the light is directed upon it. Apparently, the light is able to eject negative charges from the zinc plate, but not positive ones. This experiment repeated with other metals points to the same conclusion: namely, that only negative charges are ejected from the metals by light.

When the experiment is repeated in a vacuum, it is possible to collect and identify these negative charges, which turn out to be electrons.

The ejection of electrons from metals when light falls upon them is called the *photoelectric effect*. The relative ease with which electrons can leave a metal is also evident in the observation that simply heating a metal causes large numbers of electrons to boil off and escape from it. This observation suggests that many electrons are very loosely held by the atoms of a metal. These are the free electrons believed to be responsible for the ability of metals to conduct electricity.

20-14 Equal and Opposite Charges in Metals

The electron theory makes the assumption that neutral bodies actually contain positive and negative charges in equal number. With the electroscope, we can demonstrate that this is true for neutral metal bodies. First, it is noted that the knob, rod, and leaves of an uncharged electroscope constitute a neutral metal body.

We begin by bringing a negatively charged body near the knob of an electroscope but not touching it. The leaves acquire a charge and spread apart. We then withdraw the negatively charged body and notice that the leaves lose their charge and fall together again. Next, we repeat the experiment and test the charge on the leaves by bringing a second negatively charged body near them. The leaves are repelled, showing that they have a negative charge.

We now repeat the demonstration but bring a positively charged body instead of a negatively charged one near the knob of the electroscope but not touching it. Again, the leaves spread apart as they acquire a temporary charge that turns out to be positive. Again, the leaves lose their charge on removal of the charged body.

Since at no time is the electroscope touched by any other body and since the electroscope is well insulated, it neither gains nor loses any electric charges during these experiments. The positive and negative charges observed on its leaves must therefore have

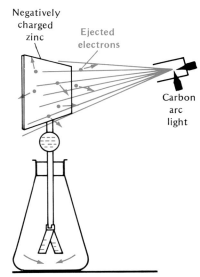

Fig. 20-5. As light ejects electrons from the zinc plate, the electroscope loses its negative charge.

Fig. 20-6. Photographers use light meters that make use of the photoelectric effect in metals.

been in the electroscope all the time. The fact that the electroscope remains neutral both before and after each experiment shows that the opposite charges in it must be present in equal quantities.

It will become increasingly evident that not only neutral metal bodies but all neutral bodies contain equal quantities of positive and negative charges. This suggests that *the molecules and atoms of which all bodies are composed also contain equal and opposite electric charges.*

20-15 Electron Behavior in the Uncharged Electroscope

On the assumption that only the negative charges in the form of free electrons move in metals, the behavior of the uncharged electroscope in the above experiments is simply explained. When the negatively charged body is brought near the knob of the electroscope, it repels some of the free electrons from the knob into the leaves. The leaves, having gained electrons, become negatively charged and repel each other. The knob, having lost these electrons, is equally positively charged. On removing the negatively charged body from the vicinity of the knob, the excess positive charge in the knob attracts the electrons in the leaves back to the knob again. The leaves become neutral again and collapse.

When the positively charged body is brought near the knob of the electroscope, it attracts free electrons from the leaves up into the knob. This gives the knob a temporary negative charge while the leaves get an equal positive charge and spread apart. As soon as the positively charged body is taken away from the knob, the free electrons in the knob are attracted back into the leaves. The leaves become neutral again and collapse.

20-16 Identifying Charges with an Electroscope

An electroscope can be used not only to detect charges of electricity but also to tell whether they are positive or negative. To do this, a known positive or negative charge is put on the electroscope, causing the leaves to spread apart. Then the unknown charge is brought near the knob of the electroscope. If the unknown charge is like the known charge on the electroscope, it will cause the leaves to spread further apart. If the unknown charge is unlike that already on the electroscope, the leaves will come together.

For example, if a negatively charged body is brought near a negatively charged electroscope, it will repel many of the extra electrons on the knob into the leaves. The leaves will thus gain more negative charge and will repel each other more than before.

If a positively charged body is brought near the knob of a negatively charged electroscope, it will attract some of the electrons from the leaves up into the knob. The leaves will therefore become less negative than before and will come closer together.

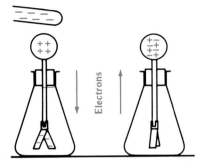

Fig. 20-7. The negatively charged rod repels some of the free electrons from the knob to the leaves.

20-17 Charging by Induction

The ease with which electrons can be moved into and out of a metal conductor makes it possible to give a permanent charge to an insulated conductor by bringing a charged body near but not touching it. This method of charging a conductor is called *charging by induction.* The original charged body always induces a charge opposite to its own in a nearby conductor. Thus, a positively charged body induces a negative charge in a nearby conductor and a negatively charged body induces a positive charge in a nearby conductor.

A person standing near a highly charged body can be in great danger because the charged body induces an equal and opposite charge on the person. A spark can then jump the intervening space and send a large current through the person.

1. Bring negative charge near knob
2. Ground the electroscope
3. Remove ground connection — then the charged body

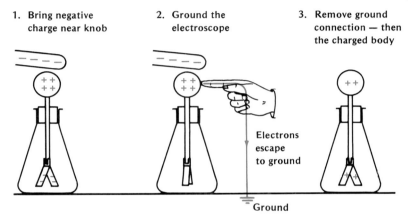

Electrons escape to ground

Ground

Fig. 20-8. Charging an electroscope positively by induction.

20-18 Charging Electroscope Positively by Induction

To charge an electroscope positively by induction, we must use a negatively charged body such as a rubber rod that has been rubbed with fur. The method involves three steps as shown in Fig. 20-8.

1. The negatively charged rubber rod is brought near but not touching the knob of the electroscope. The gold leaves spread apart.

2. While the rubber rod is kept near the knob of the electroscope, the electroscope is grounded by touching its knob with a finger. The leaves come together.

3. The ground connection is removed by withdrawing the finger from the knob; then, the negatively charged body is removed. The leaves of the electroscope now spread apart again, showing that they are permanently charged. The charge on the electroscope is positive. This can be shown by bringing a positively charged body near the knob and observing that the leaves spread further apart.

20-19 Electron Movement During Induction

What happens to the electrons during the process of induction? When the negatively charged rubber rod is brought near the knob of the electroscope in step one, it repels some of the free electrons from the knob to the leaves. The knob thus temporarily becomes positively charged and the leaves temporarily become negatively charged. The leaves show their charge by spreading apart.

When the knob of the electroscope is grounded, some of the remaining free electrons in the knob are repelled into the ground by the nearby rubber rod. The excess free electrons previously driven into the leaves are then attracted into the knob to replace those that have left. The leaves show the loss of their charge by coming together again.

The electroscope has now been made to transfer some of its electrons to the ground. To prevent the return of those electrons, first the ground connection is removed and then the charged body is removed. Since the electroscope lost some of its electrons, it has a shortage of electrons, or a net positive charge. The leaves show this positive charge by spreading apart.

1. **Bring positive charge near knob**

2. **Ground the electroscope**

3. **Remove ground connection — then the charged body**

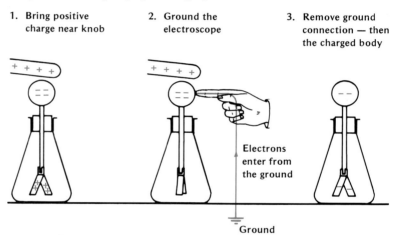

Electrons enter from the ground

Ground

Fig. 20-9. Charging an electroscope negatively by induction.

20-20 Charging Electroscope Negatively by Induction

To charge an electroscope negatively by induction, a positively charged body is brought near its knob as shown in Fig. 20-9 and the same three steps described above are repeated. In this case, the process of induction causes electrons from the ground to flow into the electroscope. The electroscope therefore gains electrons and becomes negatively charged.

20-21 Charged Bodies Attract Neutral Bodies

Induction readily explains why charged bodies attract any neutral conductors. A charged body simply induces an opposite charge in the near end of the conductor. The resulting attraction between the two bodies is then simply the attraction of these opposite charges. For example, a positively charged rod brought near a neutral piece of metal attracts free electrons toward the end of the metal nearest to it. That end of the metal therefore becomes negatively charged and is attracted by the positively charged body.

If the neutral body is an insulator, the explanation is not as simple as it is for a conductor because neither the electrons nor the positive charges move freely in an insulator. The molecules or atoms of an insulator may be imagined to be tiny combinations of equal quantities of positive charges and electrons held together by their mutual attraction, as shown in Fig. 20-10.

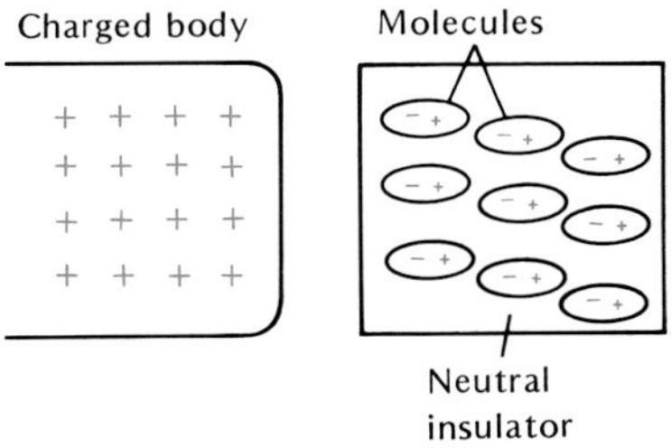

When a positively charged body is brought near an insulator, it cannot pull the electrons away from the positively charged partners. However, by attracting the electrons and repelling the positive charges, it can distort each molecule or atom slightly so that its negative charge is a little nearer to the charged body than its positive charge. Since the positively charged body attracts the nearer negative charges of the molecules or atoms more than it repels the further positive charges, the net result is attraction.

The attraction between a negatively charged body and a neutral insulator is explained in a similar way.

Fig. 20-10. The positively charged body exerts a net attraction on each molecule of the insulator.

20-22 Difference of Potential

We have seen that free electrons move whenever they are repelled by other electrons and negative charges or attracted by positive charges. If a negatively charged body is connected to a positively charged body by a conducting wire, electrons are repelled by the negatively charged body and attracted to the positively charged body. As a result, electrons flow through the wire from the negatively charged body to the positively charged body. This flow of electrons is called a current.

If two negatively charged bodies are connected by a wire conductor, free electrons will be repelled out of the one having the higher concentration of free electrons more strongly than they are repelled out of the one having the lower concentration of electrons. There will again be the flow of electrons, this time from the one with the higher concentration of electrons to the one with the lower concentration.

Bodies with different concentrations of electrons are said to be at *different potentials* or to have a *difference of potential between them.* When such bodies are connected by a wire conductor, electrons flow from one to the other until both bodies have the same concentration of electrons. The bodies are then said to be at the same potential or to have no difference of potential between them. Potential and differences of potential will be studied more precisely in Chapter 22. For present purposes, it is useful to know that differences of potential are measured in volts and are sometimes called voltages.

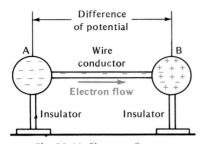

Fig. 20-11. Electrons flow as long as there is a difference of potential between *A* and *B*.

20-23 Steady Direct Current

In Fig. 20-11, the negatively charged metal ball *A* and the positively charged metal ball *B* are connected by a wire conductor. Electrons therefore flow from ball *A* to ball *B* until the difference of potential between them is zero.

Now suppose that as each electron from *A* enters *B*, a way is found to remove one electron from *B* and put it back on *A*. The quantity of negative charge on *A* and the quantity of positive charge on *B* would then remain the same and so would the difference of potential between them. As a result, electrons would keep flowing continuously in the wire conductor from *A* to *B*. A current that flows continuously in one direction is called a *steady direct current* and is abbreviated *DC*.

To maintain a steady direct current in a wire conductor, a device is needed that can maintain a constant voltage or difference of potential between the ends of the wire. *Chemical cells* and *electric generators* are such devices. The generator will be studied later. Here, brief attention will be given to the chemical cell, which was invented by the Italian Alessandro Volta in the eighteenth century, and named the *voltaic cell* in his honor. The so-called *dry cell* used in flashlights and portable radios is the commercial form of the voltaic cell. Usually groups of cells, called a *battery,* are used rather than single cells.

The electric battery was invented by Alessandro Volta (1745–1827).

20-24 Voltaic Cells

In its simplest form, a voltaic cell consists of two dissimilar conductors, called *electrodes,* immersed in a solution of an acid, base, or salt called the *electrolyte.* The typical cell of Fig. 20-12 has an electrode of copper and an electrode of zinc immersed in the electrolyte consisting of sulfuric acid. As chemical action takes place in the cell, electrons accumulate on the zinc electrode, giving it a negative charge. The copper electrode loses electrons and acquires an equal positive charge. Thus a difference of potential is set up between the two electrodes.

When the zinc and copper electrodes are connected by a wire conductor, electrons flow through the wire from the zinc to the copper. For each electron that leaves the zinc and goes through the connecting wire to the copper electrode the chemical action inside the cell removes one electron from the copper electrode and transports it to the zinc electrode. The zinc and copper electrodes therefore maintain their original difference of potential and current continues to flow in the wire.

The task of the cell is *to maintain a constant difference of potential between the electrodes* by continuing to move electrons inside the cell from the copper to the zinc. However, the movement of electrons in this direction is resisted by the attractive force of

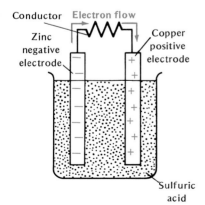

Fig. 20-12. A simple voltaic cell.

the positively charged copper electrode and the repulsive force of the negatively charged zinc electrode. Hence, to move electrons from the copper to the zinc electrode against these two opposing forces, the cell must continually do work. The energy needed to do this work is supplied by the *chemical energy* stored in the cell. When the chemical energy in the cell is consumed, chemical action stops. The electrodes then lose their difference of potential and current stops flowing through the wire.

The total work done by the voltaic cell in moving each electron determines the maximum difference of potential or voltage it can maintain between its electrodes. A simple commercial dry cell furnishes about 1.5 volts, but batteries of them are easily connected together to furnish higher voltages as needed.

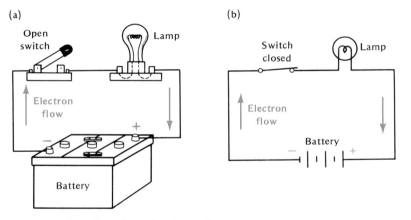

Fig. 20-13. (a) A simple electric circuit. (b) Schematic diagram of the circuit.

20-25 Simple Electric Circuit

A commercial battery has two terminals marked *plus* (+) and *minus* (−) to identify its positive and negative electrodes. Fig. 20-13 (a) shows how a battery is used to light an incandescent lamp. The lamp is essentially a length of thin tungsten wire sealed in a glass bulb. Because the tungsten wire is a much poorer conductor than the copper conducting wires used in a circuit, it gets white hot and emits light when sufficient current is flowing in the circuit.

In the circuit, the electrons leave the negative terminal of the battery along the copper conducting wire, move through the lamp, and return to the positive terminal of the battery. Inside the battery, chemical energy is now used to move the electrons back to the negative electrode to start their trip through the lamp all over again. Thus, the electrons continue to travel over and over again around this closed path, which is known as a *circuit*.

To start or stop the flow of current in the circuit, a *switch* consisting of a metal strap that connects two points of the circuit is used. The switch acts like a little drawbridge for electrons. When

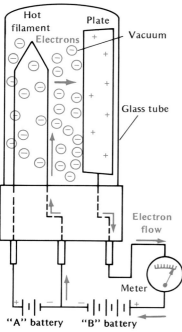

Hot filament

Plate

Electrons

Vacuum

Glass tube

Electron flow

Meter

"A" battery   "B" battery

Fig. 20-14. Electrons ejected by the hot filament in a vacuum tube flow toward the positive plate.

it is closed, electrons travel over it and current flows steadily in the circuit. The circuit is then said to be *closed*. When the switch is open, it is replaced by an airgap in the circuit over which the electrons cannot travel. Current then stops flowing in the circuit and the circuit is said to be *open*.

Fig. 20-13 (b) shows a schematic diagram of the closed circuit. Notice the symbols used to represent a battery, a switch, and a lamp. Note that each cell in the battery is represented by a short line for its minus terminal and a long line for its plus terminal. Three pairs of these lines in the diagram indicate that this battery consists of three cells.

20-26 Demonstrating a Current of Electrons

There is strong evidence that current in the metal conductors of a circuit actually consists of a flow of free electrons. It can be shown that electrons will boil out of a metal and can then be used to conduct a current across a vacuum. In their random movement inside a metal, free electrons continually collide with the vibrating molecules of the metal and acquire a share of their thermal kinetic energy. Raising the temperature of the metal increases the kinetic energy of its molecules, thereby also increasing the kinetic energy of its free electrons. As a result, more and more electrons acquire speeds high enough to break through the surface of the metal and escape. Such ejection of electrons from a heated metal is called *thermionic emission*. It is readily observed by means of the arrangement shown in Fig. 20-14.

The metal to be heated is a tungsten wire called a *filament*. It is enclosed in one end of an evacuated glass tube having a flat piece of metal called a *plate* at the other. The filament is connected to the negative terminal of the battery *B*, while the plate is connected to the positive terminal. This gives the filament a negative charge and the plate a positive charge and sets up a difference of potential between them. An electric *meter* is inserted in the circuit to measure any current that flows through the vacuum in the tube.

First, you find that when the filament is cold, there is no current in the tube. The filament is then heated by letting the battery *A* send a current through it. The meter now shows that a current flows through the vacuum in the tube and increases as the filament gets hotter. This points to the conclusion that negatively charged particles are being ejected from the hot filament and that they then flow across the tube to the plate because they are attracted by the positively charged plate.

To check this conclusion, the connections of the filament and plate to battery *B* are reversed so that the plate becomes negatively, instead of positively, charged. The meter reading now drops to zero. As expected, the negatively charged plate repels the negatively charged particles emitted by the filament and prevents their flow across the tube.

How do we know that the negatively charged particles ejected from the filament are electrons? By means outlined in Chapter 22, these particles can be formed into beams and their individual charges and masses can be determined. Measurement shows that all the negative particles ejected by heating different metals have the same mass and the same quantity of charge as the particle identified as the electron.

Thus it is evident that in the part of the circuit between the filament and the plate, electrons are continually being ejected from the filament, crossing the tube, and entering the plate. This suggests that the electrons continue to move through the closed circuit, passing from the plate through the meter and the battery, and then back to the filament to repeat the circuit.

CHAPTER REVIEW

Summary

A study of the **static electric charges** that appear on bodies when they are rubbed by other bodies shows that there are two kinds of electric charges, **positive** and **negative.** *Like charges repel each other while unlike charges attract each other.*

Normally, a body contains equal quantities of elementary positively and negatively charged particles. It becomes charged when it gains or loses some of its charged particles and this balance is disturbed. Usually, only the negatively charged particles can move from body to body. These are in the form of tiny units called **electrons.** More massive positively charged particles are called **protons.** Neutral bodies acquire a charge when they gain or lose electrons. They become negatively charged on gaining electrons and positively charged on losing some of them. Neutral bodies may become **charged by rubbing,** by contact with charged bodies, and by induction. When **charged by contact,** a body acquires a charge like that of the body that touched it. When **charged by induction,** a body acquires a charge opposite to that of the body that induced the charge. **Electroscopes** are used to detect and identify electric charges.

Substances that transmit electric charges readily are called **conductors;** those that do not are called **insulators.** Most solid conductors are metals and owe their ability to conduct to the presence of many **free electrons** inside them. The slightest electrical forces can make these electrons move readily through the conductor. The continuous flow of electrons through a conductor constitutes a steady **electric current.** Such a flow may be maintained in a conducting circuit by maintaining a constant **difference of potential** between two points in the circuit. The **voltaic cell** is one device that can supply such a constant difference of potential.

The free electrons in a metal are loosely bound to it. This can be demonstrated by shining light upon the metal. Electrons are then ejected from it. This is known as the **photoelectric effect.** Electrons may also be ejected from a metal simply by heating it.

Questions

Group 1

1. Explain how, with a negatively charged rubber rod and a positively charged glass rod, you can determine the sign of the charge on a piece of Lucite that has been rubbed with fur.

2. How does the attraction shown by charged bodies for all neutral bodies differ from the attraction shown by magnets?

3. How does the electron theory explain (a) why a rubber rod becomes negatively charged when rubbed with fur; (b) why a glass rod becomes positively charged when rubbed with silk?

4. A metal ball *A* on an insulated stand and a similar metal ball *B* are 1 m apart. Ball *A* has a negative charge and ball *B* has an equal positive charge. (a) Explain what happens to the electrons on ball *A* when it is connected to ball *B* by thread made of an insulator such as silk. (b) What happens to the electrons on ball *A* when it is connected to ball *B* by a copper wire?

5. (a) Explain how to charge an electroscope negatively by contact. (b) What causes some of the excess electrons that are put on the knob of the electroscope to move down into the leaves? (c) How does the electroscope indicate the presence of the excess electrons? (d) How does the electroscope indicate the relative number of excess electrons present?

6. (a) What is meant by "grounding" a charged body? (b) Explain why grounding a positively charged electroscope causes it to lose its charge. (c) Why are the cabinets and outer metal parts of household electric appliances grounded? (d) Why does the earth remain electrically neutral?

7. A positively charged body attracts a suspended cork ball and repels a suspended metal ball. What conclusions can be drawn about the charge present on each ball from these observations?

8. (a) What is the photoelectric effect in metals? (b) What does this effect suggest about electrons in a metal?

9. A positively charged body is brought near the knob of an uncharged electroscope. (a) Explain why the leaves diverge. (b) How can it be shown that the leaves have a positive charge? (c) Explain why the leaves collapse when the positively charged body is removed.

10. Explain how a charged electroscope may be used to determine if the charge on a charged body is positive or negative.

11. (a) How can it be shown that the metal parts of an uncharged electroscope contain both positive and negative charges? (b) What evidence is there that these opposite charges are present in equal quantities?

12. (a) Explain how to charge an electroscope negatively by induction. (b) Explain how the electroscope acquires excess electrons in this process.

13. A person acquires a dangerously high charge when standing near a highly charged body even though no contact is made. Explain.

14. (a) In what units are differences of potential measured? (b) If two bodies having a difference of potential between them are connected by a conducting wire, in which direction will the electrons flow?

15. (a) What is a steady direct current? (b) If two oppositely charged insulated metal balls are connected by a wire, how long will current flow in the wire? (c) What must be done to obtain a steady direct current flow in the wire?

16. (a) Describe a voltaic cell. (b) How does such a cell maintain a steady direct current in a conducting wire joining its electrodes?

17. In the voltaic cell, where does the energy needed to maintain a constant difference of potential between its electrodes come from?

18. A circuit consists of a lamp connected by copper wire to a battery and a switch. (a) Explain why no current flows when the switch is open. (b) Describe the complete path and direction taken by the electrons when the switch is closed.

Group 2

19. (a) Explain why heating a metal wire to incandescence causes it to eject some of its electrons. (b) On the basis of your explanation, what effect should increasing the temperature of the wire have on the number of electrons it emits per unit of time?

20. A vacuum tube contains a tungsten filament connected to a battery at one end and a metal plate sealed in at the other. The filament is kept at incandescence by the battery. The metal plate is connected to the knob of an electroscope. Explain what happens to the electroscope when

(a) it is given a strong positive charge; (b) it is given a strong negative charge. (c) What does the behavior of the electroscope tell about the charges flowing between the filament and the plate?

21. Two exactly similar metal balls A and B are mounted on insulated stands. A is given a negative charge. (a) Explain how the charge on A can now be shared with B so that each ball has exactly half the original charge. (b) Explain how a charge equal to ¼ of that originally on A can be left on both A and B without the aid of any additional apparatus. (c) Can the process used in (a) and (b) be used to obtain a charge equal to ⅓ the charge on A? Explain.

22. Two similar metal balls A and B are mounted on insulated stands. Ball A has a positive charge. (a) Describe how ball A can be used to give ball B a negative charge. (b) What is the largest negative charge ball B can acquire in this way?

23. When a person standing on an insulated floor touches an insulated charged body, is all the charge thereby removed from it? Explain.

Applying Physics

This experiment with static electricity should be done on a dry day. Since moist air is a conductor, the static charge on a body quickly escapes into the air around the body when the day is humid.

A rubber balloon suspended from a silk thread can serve as a detector of electric charge. The balloon is charged by rubbing it with flannel. Other charged bodies can now be identified by bringing them near the balloon. If the charged body repels the balloon, it has a like charge; if it attracts, it has an opposite charge.

Charge ball point pen cases, plastic objects, and rods of glass or Lucite by rubbing them with flannel. Test each charge and record your result. Now repeat the experiment but, this time, rub the various objects with silk instead of flannel. Does it make a difference in the charge acquired by a body whether it is rubbed by flannel or silk?

Instead of the suspended rubber balloon, a shellacked meterstick supported at its center so that it can rotate freely in a horizontal plane, can be used as a detector of charge. The stick can be supported by a bent wire cradle hanging from a silk thread. One end of the stick is wrapped in metal foil and charged. When a charged object is brought near the charged end of the meterstick, it will attract it if the charges are opposite and repel it if the charges are the same. The meterstick will then rotate about its support. How could this setup be used to tell whether the charge on an object is positive or negative?

21 Natural Unit of Electricity

Aims

1. To investigate how electricity is conducted through liquids and gases.
2. To examine the evidence leading to the discovery that there is a natural unit of electric charge equal to the quantity of charge on the electron.
3. To study the evidence that supports the law of conservation of electric charge and to learn to apply that law.

21-1 Conduction in Liquids

Liquids vary in the ability to conduct electricity. A crude device for testing this ability is the *electrolytic* (ih·lek·troh·LIT·ik) *cell* shown in Fig. 21-1. Here, two conductors called electrodes are immersed in the liquid to be tested and are connected to the terminals of a battery. Usually carbon or platinum electrodes are used because they do not react chemically with the liquids tested. The electrode connected to the positive battery terminal is called the *anode*. The electrode connected to the negative battery terminal is called the *cathode*. A lamp is included in the circuit to show in a qualitative way how much current is flowing.

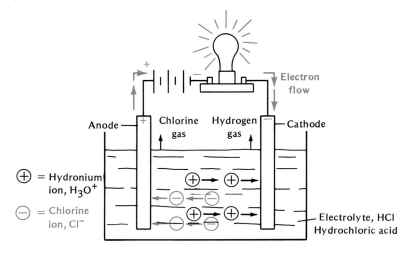

Fig. 21-1. As current passes through the electrolytic cell, the positive ions are attracted toward the cathode and the negative ions are attracted toward the anode.

When different liquids are tested in this circuit, the lamp glows more or less brightly for those that conduct electricity well, and not at all for those that conduct poorly. Two kinds of liquids are found to be conductors, those that are liquid metals like mercury

436

or molten silver, and those that belong to the class of solutions called electrolytes that was mentioned in connection with the voltaic cell. The more important electrolytes are the water solutions of acids, bases, and salts. However, pure water itself is a very poor conductor.

21-2 Ionization in Solutions

We have seen that the current passing through a metal is a flow of electrons. What flows through an electrolytic solution when a current passes through it? To answer this question, suppose we put a solution of hydrogen chloride, called hydrochloric acid, into the electrolytic cell shown in Fig. 21-1. Each molecule of hydrogen chloride contains one atom of hydrogen and one atom of chlorine. As current flows through the solution, hydrogen gas collects at the negative electrode, or cathode, and chlorine gas collects at the positive electrode, or anode. The passage of the electric current appears to be separating the hydrochloric acid into the hydrogen and chlorine atoms of which it is made.

The separating action taking place in the solution can be explained by assuming that on dissolving in water, many of the molecules of hydrogen chloride break up into two oppositely charged parts. One of them is a hydrogen atom having a positive charge. The other is a chlorine atom having an equal negative charge. Such charged atoms and charged groups of atoms are called *ions*. The process by which they form is called *ionization*.

The carriers of current in electrolytic solutions are negatively and positively charged ions.

21-3 Action of the Ions

The hydrogen ions immediately attach themselves to water molecules to form combinations called *hydronium* (hy·DROH·nee·um) *ions*. The chlorine ions also associate with water molecules and become hydrated. Ordinarily, the positive hydronium ions and the negative chlorine ions wander about freely in the solution in all directions. However, when the electrolytic cell is connected to a battery, the anode becomes positively charged and the cathode becomes negatively charged. The positively charged hydronium ions are then attracted toward the cathode and the negatively charged chlorine ions are attracted toward the anode. The flow of current inside the cell therefore consists of positive hydronium ions flowing in one direction and negative chlorine ions flowing in the opposite direction.

When the hydronium ions reach the cathode, which has an excess of electrons, each takes one electron from it and thus neutralizes the positively charged hydrogen ion attached to it. The hydrogen ions thus become hydrogen atoms and are released into the solution. Here they pair up to form hydrogen molecules which gradually come out of the solution as bubbles of hydrogen gas. When the chlorine ions reach the anode, which has a shortage of electrons, they give up their extra electrons and become neutral chlorine atoms. These pair up to form chlorine molecules which gradually come out of the solution as bubbles of chlorine gas.

The behavior of hydrochloric acid solution is typical of all electrolytes. In general, when acids, bases, and salts are dissolved in water, many of their molecules break up into positively and negatively charged ions which are free to move in the solution. When a difference of potential is set up by immersing two oppositely charged electrodes in the solution, the positive ions are pulled toward the cathode and the negative ions are pulled toward the anode. Thus, the current in an electrolyte consists of positive and negative charges flowing in opposite directions.

The very low conductivity of pure water is explained by the fact that water molecules do not readily break up into ions. As a result, very few current carriers are present in pure water.

21-4 Electroplating

The metal atoms in electrolytic solutions form positive ions. Thus, in a copper sulfate solution, the crystals of copper sulfate break up into positively charged copper ions and negatively charged combinations of atoms known as sulfate ions. During the passage of current through a copper sulfate solution, the copper ions move toward the cathode and the sulfate ions move toward the anode. At the cathode, the positively charged copper ions obtain enough electrons to turn them into neutral copper atoms. These atoms stick to the cathode as a thin film of pure copper.

In a similar way, when current passes through a solution of silver nitrate, positively charged silver ions move toward the cathode and are deposited on it as a pure silver film. The negatively charged nitrate ions move to the anode.

The fact that metal ions in solutions form thin coatings of pure metal on the cathode when current is passing through the solution is the basis for the commercial process of *electroplating* illustrated in Fig. 21-2. To electroplate a brass spoon with silver, the spoon is made the cathode in an electrolyte containing silver ions such as a silver nitrate solution. The anode is a bar of silver. As current passes through the cell, silver is gradually deposited on the spoon. At the same time, the nitrate ions go to the silver anode where they gradually combine chemically with the silver anode, producing more silver nitrate. This promptly dissolves in the solution and breaks up into silver and nitrate ions. Thus, the anode, by gradually dissolving in the solution, replaces the silver ions removed by the spoon being plated at the cathode.

To electroplate the spoon with copper instead of silver, the electrolyte used must be a solution of a copper salt such as copper sulfate, while the anode must be a bar of copper.

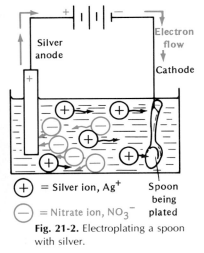

Silver anode

Electron flow

Cathode

$\oplus$ = Silver ion, Ag^+ Spoon being plated

$\ominus$ = Nitrate ion, NO_3^-

Fig. 21-2. Electroplating a spoon with silver.

21-5 Electrolytic Measurement of Current

The English scientist Michael Faraday discovered that the quantity of silver or copper or any other metal deposited at the cathode of an electrolytic cell is proportional to the size of the current and to the length of time that it flows. The quantity of silver, or any other metal, deposited in an electrolytic cell per unit of time can therefore be used to measure the current that deposits it. Current is measured in a unit called the *ampere*, symbol, A. The official *SI* definition of this unit is given in Section 26-6. For present purposes, it is sufficient to know that an ampere is the quantity of current that deposits silver in an electrolytic cell at the rate of 1.118×10^{-6} kilogram per second. Given this fact, a current in a circuit can be measured by simply weighing the amount of silver it deposits per second in an electrolytic cell in kilograms and dividing this number by 1.118×10^{-6} kilogram. A current of 2 amperes is one that deposits $2 \times 1.118 \times 10^{-6}$ kilogram of silver per second; a current of 3 amperes is one that deposits $3 \times 1.118 \times 10^{-6}$ kilogram of silver per second, and so on.

While this method of measuring current gives very accurate results, it is inconvenient. For most practical purposes, direct-reading *ammeters* are used. These are described in Section 26-9.

21-6 Unit of Electric Charge

The flow of water passing through a pipe may be described in terms of the quantity of water that passes through a given cross section of the pipe per unit of time. For example, if 600 liters of water pass out of the end of a water hose in 10 minutes, the flow is expressed as 60 liters per minute, or 1 liter per second. Electric current may also be expressed as the quantity of charge that flows through any given point of a circuit per unit of time. The *SI* unit used to measure quantity of electric charge is called a *coulomb* and is given the symbol C. *The coulomb is defined as the quantity of charge that flows through an electric circuit in one second when the current in the circuit is one ampere.* A current of one ampere is therefore one in which one coulomb of electric charge passes through the circuit per second. A current of 2 amperes is one in which 2 coulombs of electric charge pass through the circuit per second, and so on.

The coulomb represents a rather large unit of electric charge that was developed in the early days of electricity. Later we shall see that in dealing with the very small charges on electrons and other tiny parts of atoms, it is more useful to express the size of electric charges by comparing them with the unit that nature itself has created; namely, the charge on an electron.

21-7 Natural Unit of Electric Charge

The existence of a *natural unit of electric charge* was first discovered by Faraday in his study of the ability of electric currents to

Fig. 21-3. The photograph shows industrial equipment used for tin electroplating.

remove ions from electrolytic solutions. The principle underlying his discovery can be understood from the arrangement in Fig. 21-4, where a battery sends a current through a series of electrolytic cells containing different solutions.

The first cell contains hydrochloric acid; the second contains silver nitrate; and the third contains copper sulfate. All the electrodes are made of platinum, which does not react chemically with these solutions. By inserting a lamp or, better still, an ammeter in different parts of the circuit, we can show that the same quantity of current is passing through all the cells. This means that in a given time the same quantity of charge passes through each cell.

For simplicity, let us consider only what happens at the cathode of each cell. As the current flows, hydrogen collects at the cathode of the first cell, silver deposits on the cathode of the second cell, and copper deposits on the cathode of the third cell.

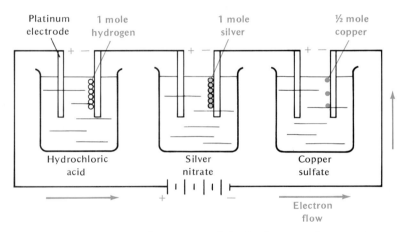

Fig. 21-4. Current flowing through a series of electrolytic cells.

Now let us measure the masses of the hydrogen, silver, and copper delivered by the cells in a given time. For the hydrogen, this can be done by collecting the gas and measuring its mass. The mass of the silver deposited on the cathode of the second cell can be determined by weighing that cathode before and after the given time interval. The mass of the copper deposited at the cathode of the third cell can be measured in the same way. Since we know the mass of a mole of the atoms of each of these elements, we can readily determine how many moles of each element were deposited in the given time.

It is found that for every mole of hydrogen atoms that is delivered in the first cell, one mole of silver atoms is deposited in the second cell and one-half mole of copper atoms is deposited in the third cell. Since a mole of every element contains the same number of atoms, and since the same quantity of electric charge has passed through all the cells during the given time, it follows that the charge carried by each ion of silver is equal to that carried by each ion of hydrogen. Also, the charge carried by each ion of copper is twice that carried by each ion of hydrogen or silver.

When this experiment is repeated with many different solutions, it is found that the quantity of charge carried by all positive ions in solution is either 1, 2, 3, or some whole number of times as large as the positive charge on a hydrogen ion.

When the products formed at the anode are also collected and weighed, it is found that the quantity of charge carried by all negative ions in solutions is also always 1, 2, 3, or some whole number of times as large as the positive charge on the hydrogen ion.

It appears then that the quantity of charge on a hydrogen ion is a natural indivisible unit of electric charge. All other charges, positive or negative, are whole-number multiples of this unit. The hydrogen ion is the proton. The elementary body having the same natural unit of negative charge turns out to be the electron.

21-8 Elementary Unit Charge

Measurement shows that one mole of hydrogen or silver ions carries 96 500 coulombs of charge through the circuit. Since there are 6.02×10^{23} ions in a mole, the charge associated with each hydrogen or silver ion is 96 500 coulombs divided by 6.02×10^{23}, or *1.60×10^{-19} coulomb*. Thus, a study of the ionization of solutions gives a quantitative value for the natural unit of charge. This value is confirmed by measurements made by Millikan and many others and is recognized as the quantity of negative charge on the electron and the quantity of positive charge on the proton.

Using this unit, we find that the number of electrons or protons in a coulomb of charge is $1 \div 1.60 \times 10^{-19}$, or 6.25×10^{18}. *Since one ampere is a flow of one coulomb of charge per second* (Section 21-6), *it is a current in which 6.25×10^{18} electrons pass through the circuit per second.*

The elementary unit of charge is that of the electron or proton, which is 1.60×10^{19} coulomb. No free charge smaller than this has ever been found, although bound particles called quarks are assumed to have only fractional parts of the elementary charge. (See Section 34-25.)

21-9 Conductivity of Gases

Some of the best opportunities to isolate the proton and the electron and to learn how they combine to form the various atoms of matter arose out of the study of the electrical properties of gases. All gases are ordinarily very poor conductors of electricity. This is illustrated by the fact that a charged electroscope surrounded by dry air does not lose its charge for a very long time. Apparently there is no significant number of freely moving electric particles like the electrons in metals or the ions in electrolytes to carry current through the air.

Gases can be made to conduct electricity under conditions that give valuable information about the composition of their atoms. This can be done in a *gas discharge tube* consisting of a glass tube having two metal electrodes sealed into its opposite ends. The air in the tube is removed by a vacuum pump and is replaced by a

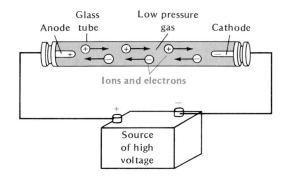

Glass
Anode tube

Low pressure
gas   Cathode

Ions and electrons

Source
of high
voltage

Fig. 21-5. A gas discharge tube.

small amount of the gas to be studied. The electrodes of the tube are then connected to a battery or other source of high difference of potential. This causes one electrode, called the *anode,* to become highly positively charged. The other electrode, called the *cathode* becomes highly negatively charged. Current now passes through the tube and the gas emits light.

Such discharge tubes containing neon and other gases are commonly used as light sources in so-called "neon" advertising signs (see Fig. 21-6) and in tubes used for fluorescent lights. Different gases glow with different colors.

21-10 Ionization in Gases

The usual carriers of the current in the gas discharge tube are found to be positively and negatively charged ions as well as free electrons. However, if the quantity of the gas in the tube is reduced so that there is a near vacuum, one can show that the primary carriers are electrons and positively charged ions of the gas in the tube.

The arrangement shown in Fig. 21-7 consists of a gas discharge tube whose electrodes have a tiny hole in their centers. The gas is between the electrodes which are in the middle part of the tube. At

Fig. 21-6. Bright, colorful, flashing neon signs have long been used in high-density commercial areas by businesses competing for the attention of busy passersby.

each end of the tube is a screen of fluorescent material that has the property of glowing when electric charges strike it. As current flows through the tube, it is observed that a beam of charged particles passes through the hole in the anode and a second beam of charged particles passes in the opposite direction through the hole in the cathode. Each beam causes the fluorescent screen on which it falls to glow.

Now the beam coming from the hole in the anode turns out to consist of negative charges. This is shown by the fact that the beam is attracted toward a positively charged body and repelled from a negatively charged body. What is more, regardless of the gas used, the individual negative charges in this beam turn out to be exactly the same. All of them have a single unit of elementary charge equal to that discovered by Faraday. All of them also have exactly the same mass, which turns out to be only 1/1840 of the mass of a hydrogen atom. These are the particles we have called *electrons*.

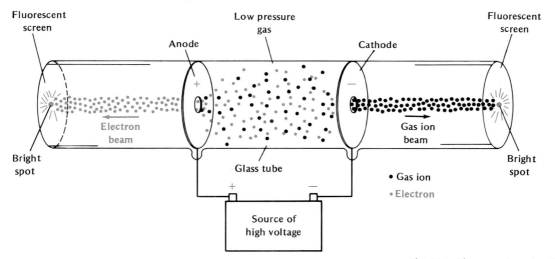

Fluorescent screen

Low pressure gas

Fluorescent screen

Anode

Cathode

Electron beam

Gas ion beam

Bright spot

Glass tube

Bright spot

• Gas ion

• Electron

Source of high voltage

Fig. 21-7. The current carriers in a gas discharge tube are positive ions and electrons that move in opposite directions.

The beam coming from the hole in the cathode turns out to consist of positively charged ions of the gas in the tube. They have different masses depending on the gas in the tube. They may also have different quantities of positive charge but, as was found in electrolysis, the quantity is always a whole number of times as large as the elementary unit of charge.

21-11 Electrical Composition of Atoms

The positive and negative beams in the discharge tube can be explained as follows. Each atom of the gas in the discharge tube consists of a number of elementary negative charges in the form of electrons and an equal number of elementary positive charges. One or more electrons are readily detachable from each atom. As a result, the application of a high difference of potential to the tube detaches one or more electrons from each atom leaving the remainder of the atom positively charged. The gas is thus ionized into positively charged ions of the gas and electrons. The positively charged gas ions are attracted to the oppositely charged cathode; some of them pass through the hole to form the positively charged beam of gas ions. The electrons are attracted to the anode; some of these pass through the hole to form the observed beam of electrons.

A hint that the positive charges of all atoms are made up of protons, the elementary particles of positive charge, goes all the way back to an observation first made by William Prout, a nineteenth-century English physician. Prout noticed that the atomic masses of the different elements are generally very nearly a whole number of times as large as the atomic mass of hydrogen. He drew from this the idea, known as *Prout's hypothesis,* that all other atoms are essentially combinations of hydrogen atoms.

The ionization of hydrogen gas in the discharge tube lends some support to Prout's hypothesis. Here, the beam of positively charged hydrogen ions shows a marked difference from the ions of other gases. While the individual atoms of other gases may carry a number of elementary positive charges, hydrogen atoms always carry only one unit of elementary positive charge. Furthermore, the ion of hydrogen is found to have a mass practically equal to that of a hydrogen atom. This indicates that a hydrogen atom consists of two indivisible charged particles, a hydrogen ion and an electron. The ion of the hydrogen atom is the basic positively charged particle we have called the *proton.*

Since all gas ions in discharge tubes always carry positive charges that are exact multiples of the proton's charge, it is inferred that the positive charges present in atoms are in the form of protons. It follows that the basic charged parts of all atoms consist of equal numbers of the same two particles, protons and electrons.

Prout's hypothesis has proved to be almost correct. All other atoms contain combinations of the same pair of charged particles that form the hydrogen atom. However, we shall see that all atoms except hydrogen contain still a third particle called the neutron. This particle is not usually noticed in studies of the electrical properties of matter because it is electrically neutral. In Chapter 30 we shall see how the three basic particles, the proton, the electron, and the neutron fit together to make all of the different atoms of matter.

21-12 Conservation of Charge

In all cases in which matter becomes electrically charged, charge is acquired by the transfer of positive charges, negative charges, or both from one body to another. When a glass rod is rubbed with silk, the glass acquires a positive charge and the silk acquires an equal negative charge. Rubbing did not create these charges. It simply transferred charged particles, which in this case happen to be electrons, from the glass to the silk. The sum total of positive and negative charges originally present in the neutral glass and silk was zero. After rubbing, the sum of the separate equal and opposite charges produced on the glass and silk remains zero.

A similar relationship is found when molecules ionize. The sum of the charges on the positive ions formed is exactly equal to the sum of the charges on the negative ions formed at the same time. The net charge of the molecules was zero before ionization and the sum of the positive and negative charges formed is still zero after ionization.

These observations are typical of many that suggest the principle of the *conservation of electric charge.* According to this principle:

> In all exchanges of charge between bodies of matter, the algebraic sum of the positive and the negative charges remains constant.

Usually, in the transfer of charged particles from one body to another, no particles are created or destroyed. The actual number of positively charged particles and the actual number of negatively charged particles in both bodies remain constant.

However, there are reactions in subatomic physics in which two equal and oppositely charged particles destroy each other on collision and convert their energy into gamma rays. The reverse process also occurs, resulting in the creation of two equal and oppositely charged particles from gamma rays. Here too, in spite of the disappparance or creation of charged particles, the law of conservation of charge holds. In these reactions, the charged particles are always destroyed or created in pairs of equal and opposite charges. The algebraic sum of their charges before and after the reactions remains zero.

The principle of conservation of electric charge, like the principle of conservation of energy, is a universal principle. No exceptions to it have been observed.

Electric charges can be created or destroyed. However, they are always created or destroyed in equal and opposite pairs. This leaves the algebraic sum of the charges in the universe unchanged.

CHAPTER REVIEW

Summary

The study of the electrical behavior of solids, liquids, and gases in **electrolytic cells** and in **gas discharge tubes** provides evidence that the atoms of all matter normally contain equal quantities of positive and negative electrical charges. These charges are composed of indivisible **natural units of charge,** each of which is *1.60 $\times$ 10^{-19} coulomb.* The basic particles that carry these unit charges are the **electron** that carries a unit negative charge and the **proton** that carries a unit positive charge. The electrical parts of all atoms consist of equal numbers of protons and electrons.

All changes that occur in the electrical charges of bodies of matter result from the transfer of whole numbers of unit charges from one body to another. In all these transfers, the principle of **conservation of charge** holds. *No net charge is lost or gained.*

Free electric charges move under the repulsive forces of like charges and the attractive forces of unlike charges. A stream of charges constitutes an electric current. In currents passing through metals, only the negative charges in the form of **free electrons** flow. In currents passing through electrolytes and ionized gases, both positive and negative particles called **ions** flow. The positive charges move in one direction and the negative charges move in the opposite direction.

Questions

Group 1

1. (a) What three classes of water solutions are electrolytes? (b) Describe an experimental procedure using an electrolytic cell, a meter for measuring current (ammeter), and a battery, to determine which of several electrolytes conducts electricity best.

2. (a) What are the carriers of current in a solution of hydrogen chloride? (b) When an electrolytic cell containing hydrochloric acid is connected to a battery, which of these carriers go to the anode and which go to the cathode? (c) What external evidence is there that the motion referred to in (b) is actually taking place?

3. (a) Explain with the aid of a diagram how a brass object can be plated with silver. (b) What determines how much silver will be deposited on the object?

4. (a) How can an electrolytic cell be used to measure an electric current? (b) In what units is current expressed? (c) How is the coulomb defined?

5. When aluminum oxide is dissolved in molten cryolite, the aluminum oxide ionizes into positive aluminum ions and negative oxygen ions. An electrolytic cell containing a solution of aluminum oxide in molten cryolite is connected in series with the three cells shown in Fig. 21-4 and current is sent through all four cells. It is found that when the current has deposited 1 mole of silver atoms, it has deposited only ⅓ of a mole of aluminum atoms. (a) How many elementary charges are present in each aluminum ion? (b) What is the actual charge in coulombs on 1 aluminum ion?

6. (a) How can it be shown that a gas such as air under normal atmospheric pressure is a very poor conductor of electricity? (b) Under what conditions may the ability of a gas to conduct electricity be greatly increased? (c) How does the "neon sign" make use of the answer to (b)?

7. (a) What are the carriers of current in a discharge tube containing argon gas under low pressure? (b) When a source of high voltage is applied to the tube, which current carriers go toward the anode and which toward the cathode?

8. Current is passing through a low pressure discharge tube such as the one shown in Fig. 21-7 containing argon, and a second similar tube containing neon. The mass of an argon atom is 40 u and that of a neon atom is 20 u. (a) Compare the particles that pass through the holes in the anodes of the two tubes with respect to mass and type of charge. (b) Do the same for the particles that pass through the holes in the cathodes of the two tubes.

9. (a) What does the evidence obtained by passing current through many different elements in gas discharge tubes suggest concerning the electrical particles that are contained in all atoms? (b) What uncharged particle is also known to be a part of all atoms except the common form of hydrogen?

10. (a) How may a gas discharge tube of the type shown in Fig. 21-7 be used to obtain a beam of hydrogen ions? (b) Compare the proton and the electron with respect to mass and charge.

11. Apply the principle of conservation of charge to each of the following: (a) After a body with a positive charge of q_1 coulombs touches a neutral body, its positive charge is reduced to q_2 coulombs. What is the charge on the neutral body? (b) When an electroscope is charged positively by induction, n electrons leave the electroscope and enter the ground. What is the quantity of

positive charge induced in the electroscope? (c) In a water solution of iron chloride, each neutral molecule breaks up into 1 positively charged ion of iron and 3 negatively charged ions of chlorine. How does the quantity of positive charge on each ion of iron compare with the quantity of negative charge on each ion of chlorine?

12. The flame of a candle is first held near the knob of a positively charged electroscope and then near the knob of a negatively charged electroscope. In each case, the electroscope loses its charge. What conclusion is suggested by these observations about the electrical composition of the gases in and around the flame?

Group 2

13. Faraday obtained an accurate value for the electrical charge associated with 1 mole of hydrogen atoms. Millikan obtained an accurate value for the elementary charge of electricity. How may Avogadro's number (Section 13-22) be computed from these two results?

14. Homes are usually supplied with an alternating current that reverses its direction at the rate of 120 times a second. Explain why such a current would not be suitable for electroplating.

15. Discuss the truth of the statement: Any change in the number of elementary charges present in the universe must be an even number.

Problems

Group 1

1. A current of 5.00 A flowed in a copper wire for 20.0 s. (a) How many coulombs of charge passed through the wire during this time? (b) How many electrons flowed through the wire during this time?

2. How many amperes of current flow in a wire through which 10^{18} electrons pass per second?

3. A certain current flowing through a silver electroplating apparatus deposited 1.118×10^{-5} kg of silver in 100 s. How large was the current?

4. A magnesium ion carries 2 elementary units of charge. How many magnesium atoms will be deposited on the cathode of an electroplating cell containing a solution of a magnesium salt by a current of 2 A in 8 s?

5. It takes 289 500 coulombs of charge flowing through an electrolytic cell containing aluminum ions to deposit 1 mole of aluminum atoms. How much charge is associated with each aluminum ion in (a) coulombs; (b) elementary units of charge?

Group 2

6. In an electrolytic cell containing hydrochloric acid, it takes 48 250 C of charge to liberate ½ mole of hydrogen atoms. Assuming that the charge on a hydrogen ion or proton is known to be 1.60×10^{-19} C, determine Avogadro's number.

7. A calcium ion carries two elementary charges. How long will a current of 2.0 A take to deposit 0.010 mole of calcium in an electrolytic cell?

8. A spoon is silver plated for 30 min in an electrolytic cell through which 0.10 A of current is flowing. What mass of silver is deposited on the spoon?

Applying Physics

An electrolytic cell apparatus like that shown in Fig. 21-1 can be made by using two dry cells, a flashlight bulb, and a pair of electrodes consisting of equal lengths of bare copper wire. A glass tumbler can serve as the container for the electrolyte. To keep the electrodes a constant distance apart, they can be mounted so that they hang vertically downward from a small square piece of plywood to which they are fastened with thumbtacks. The plywood square is then put on top of the tumbler so that the electrodes are immersed in the liquid being tested.

Using this apparatus, the relative ability of different liquids and solutions to conduct electricity can be noted qualitatively by the brightness of the bulb. Test a salt solution, a vinegar solution, and a sugar solution. In each case, be sure that the amounts of solution being compared are the same by filling the tumbler to the same level.

Make a graded series of solutions of salt from very dilute to concentrated. Test each and note the effect of the concentration on the conductivity of the solutions.

Artificial Intelligence

Computers can perform so many difficult tasks—guide a spacecraft, analyze the global climate, maintain financial records for nationwide banking systems, and so on—that we tend to label them "intelligent." But their "intelligence" is actually quite limited. There are so many simple things they *cannot* do, for example, understand the meaning of anything, recognize a word regardless of the handwriting or the speaker's accent, guide a robot around an unfamiliar room, or just use plain common sense. These are the capabilities that researchers in the field of artificial intelligence (AI) are trying to build into computers. If they succeed, computers may finally be the equal of the human mind.

Today's computers are programmed with facts and with rules for manipulating them. They process data and do arithmetic much faster than we can. But they lack the "common sense" that we develop through experience in the world. They also lack the pattern recognition skills that we use to understand language and perceive the world through our senses. To remedy this situation many AI researchers are trying a new approach called a *neural net*, or *connectionist*, model. Unlike today's computers, which are logic machines whose structure and organization bear little or no resemblance to the human brain, computers based on a neural net model *would* resemble the brain. As in the brain, information would be stored in a collection of "unit processors" similar to individual nerve cells that communicate via electric signals. Knowledge would reside in the pattern of activity of a collection of such individually "dumb" processors. Memory, logic, and control would not be separate but would be interconnected functions.

While neural net research goes on, progress is being made in developing isolated aspects of AI that have practical application today. So-called "expert systems" make decisions based on "if-then" rules derived from observing human experts at work, such as doctors in a field like medical diagnosis. In the area of language, computer programs are being developed for extracting information from computer databases. Advances in pattern-recognition ability have led to computer enhancement of astronomical photos and the construction of robots that inspect or sort manufactured items.

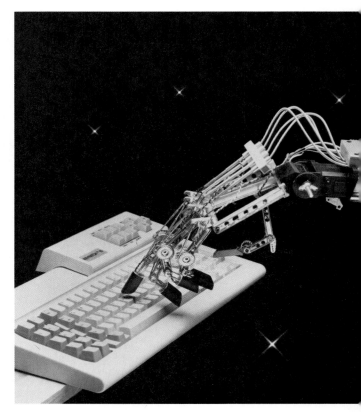

Coulomb's Law and Electric Fields

22

1. To study quantitatively the nature of the forces between electric charges.
2. To develop the concept of the electric field as a transmitter of electric force.
3. To learn how work is done and potential energy is stored in an electric field.
4. To learn how to use electric fields to control and direct the motion of electric charges.

Electric Forces and Fields

22-1 Electric Forces and Properties of Matter

The forces that electrical charges in matter exert upon each other play a major role in determining the properties and structure of matter. They account for such phenomena as cohesion, adhesion, and friction. They also provide the forces that bind electrons and protons together in different kinds of atoms, and that bind atoms together to form molecules of elements and compounds.

22-2 Measuring Electrical Forces

The exact nature of the forces exerted upon each other by charged bodies was determined in France by Charles Coulomb in 1785. He used an apparatus like that shown schematically in Fig. 22-1. A light horizontal rod made of an insulator and having a small conducting sphere at each end is suspended by a fine metal wire. A second conducting sphere on an insulated stand is fixed near one of the movable spheres. Both spheres are given like charges. The force of repulsion acting on the movable sphere then causes the horizontal rod to turn. In doing so, the rod twists the supporting wire until the wire's increasing opposition brings the rod to a stop.

Now it is known from experiment that the angle through which the wire is twisted is directly proportional to the force that is causing the twisting. By preliminary measurement of the angles of twist produced by a series of known forces, the force needed to produce each degree of twist is determined. Thereafter, any angle of twist is easily converted into the force that produced it.

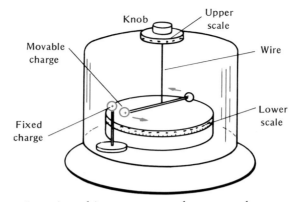

Fig. 22-1. Coulomb's apparatus for measuring the force between two charges.

In using this apparatus, the two spheres are first given like charges and the angle of twist of the suspension wire caused by the repulsion is noted on the lower scale. The movable sphere is then moved toward the fixed sphere until the distance between the spheres is a selected value. This is done by twisting the knob attached to the upper end of the suspension wire. It causes additional twist of the wire which is measured on the upper scale. The total angle of twist of the wire is the sum of the readings on the upper and lower scales and measures the force between the charges at the selected distance.

By a similar technique, this apparatus can also be used to measure the force of attraction between the two spheres when they are given opposite charges.

22-3 Coulomb's Law

Coulomb proceeded in two steps. First, he gave the fixed and movable spheres like charges and measured the force they exerted upon each other as the distance between them was changed. He found that the *force varied inversely as the square of the distance.* Thus, when the distance between the charged spheres was doubled, the force between them decreased to $\frac{1}{2}^2$, or $\frac{1}{4}$, of its previous value. When the distance was tripled, the force decreased to $\frac{1}{3}^2$, or $\frac{1}{9}$, of its previous value, and so on. The same relationship was observed when the spheres were given opposite charges and attracted, instead of repelled, each other.

Then Coulomb put a series of known different charges on the spheres and measured the force between them when they were at a constant distance apart. He found that *the force exerted by the bodies upon each other,* whether attraction or repulsion, *is directly proportional to the quantity of charge on each of the spheres and therefore to their product.* Doubling the quantity of charge on one sphere caused the force between them to double. Doubling the quantity of charge on both spheres caused the force between them to increase to 2×2, or 4 times its former value.

These experimental findings resulted in the law generally called *Coulomb's inverse square law.* It states:

> The force of attraction or repulsion between two electric charges is proportional to the product of the charges and inversely proportional to the square of the distance between them.

22-4 Mathematical Expression for Coulomb's Law

Coulomb's law may be written:

$$F \propto \frac{q_1 q_2}{r^2}$$

where q_1 and q_2 are the quantities of each of the charges, r is their distance apart, and F is the force they exert upon each other. Changing the proportion to an equality gives:

$$\mathbf{F} = \mathbf{K} \frac{q_1\ q_2}{r^2}$$

Note the similarity between Coulomb's law and Newton's law of universal gravitation,

$$f = G\frac{m_1 m_2}{d^2}.$$

(See Section 8-8.)

where K is a constant whose value depends upon the units in which q_1, q_2, r, and F are measured, and upon the medium surrounding the charges.

To determine K, the forces between a known pair of charges at a fixed distance apart is measured. The known values of q_1, q_2, r and F are expressed in *SI* units and are substituted into Coulomb's law. Solving for K gives 9.0×10^9 newton-meters²/coulomb² when the charges are in air or a vacuum. Therefore, for charges in air or a vacuum, Coulomb's law takes the form:

$$F = 9.0 \times 10^9 \frac{\text{N} \cdot \text{m}^2}{\text{C}^2} \cdot \frac{q_1 q_2}{r^2}$$

In using this relationship, q_1 and q_2 must be expressed in coulombs, and r in meters. F will then be in newtons. Positive charges are given a positive sign and negative charges a negative sign. A force of repulsion then has a positive sign while a force of attraction has a negative sign.

Sample Problem

Find the force between two positive 1.0-C charges when they are 1000 m apart.

Solution:

$q_1 = q_2 = 1.0$ C $r = 1000$ m

$$F = 9.0 \times 10^9 \frac{\text{N} \cdot \text{m}^2}{\text{C}^2} \cdot \frac{q_1 q_2}{r^2}$$

$$F = 9.0 \times 10^9 \frac{\text{N} \cdot \text{m}^2}{\text{C}^2} \frac{(1.0\ \text{C})(1.0\ \text{C})}{(1000\ \text{m})^2}$$

$$F = 9.0 \times 10^3 \text{ N}$$

The force is positive, indicating repulsion. Note that all the units cancel except N which is the proper unit for F.

This problem shows that a coulomb is an enormous quantity of charge. Here, two coulomb charges a kilometer apart repel each other with a force of 9000 N! Static charges are usually far smaller than a coulomb and are often expressed in millionths of a coulomb, or *microcoulombs,* symbol μC.

Test Yourself Problems

1. A positive charge of 2.0×10^{-6} C and a negative charge of 5.0×10^{-6} C are 3.0×10^{-2} m apart. What is the force between them?

2. What is the force between a $1.0\ \mu$C positive charge and a $1.0\ \mu$C negative charge that are 1.0 m apart?

22-5 Coulomb's Law Applies to Point Charges

Coulomb's law applies only to charged bodies that are very small compared to the distance between them. The charge on each of such bodies may then be treated as though it is concentrated at a single point. Such a charge is called a *point charge.*

Experiment shows that electrons, protons, ions, and other atomic and subatomic electrically charged particles generally behave like point charges and follow Coulomb's law very precisely. The force between two large charged bodies that are too near each other to be treated as point charges is found by obtaining the vector sum of the forces that all the individual charges in the first body exert upon all the individual charges in the second body.

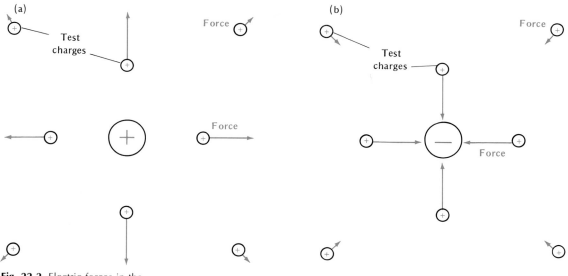

Fig. 22-2. Electric forces in the field of **(a)** a positively charged sphere; and **(b)** a negatively charged sphere.

22-6 Electric Field of Force

A charged body exerts a force on all other charged bodies in its neighborhood. To describe this fact, a charged body is considered to be surrounded by a field of force called its *electric field.* The chief property of an electric field is that it *exerts a force on any charged body placed in it.* To explore the electric field around a given charged body, a small positively charged body called a test charge is imagined placed at each point around it and the force exerted upon the test charge is noted. In this way, the distribution of the forces acting on the test charge at all points in the electric field can be determined.

In Fig. 22-2 (a), several points in the electric field around a small positively charged sphere are explored. At each point, a vector shows the direction and magnitude of the force acting on a small positive test charge put at that point. It is found that the test charge is always repelled radially outward and that, in accordance with Coulomb's law, it is repelled more strongly at points near the charged sphere than at more distant points.

In Fig. 22-2 (b), several typical points in the field around a small negatively charged sphere are explored. In this case, the small positive test charge is attracted by the negatively charged sphere. All the vectors therefore point toward the negatively charged sphere.

In general, electric fields are much more complicated than these because they are produced by many neighboring charged particles and bodies instead of just one. All electric fields are imagined to be explored in the same way, always using a small positive test charge. The resultant electric force exerted on the test charge by the field of a group of charged bodies is the vector sum of the forces exerted by all the bodies separately.

22-7 Electric Field Intensity

The force exerted per unit positive charge on a small test charge placed at any point in an electric field is called the *electric field intensity* at that point. If a positive test charge q, located at a given point in an electric field, is acted upon by a force F, the electric field intensity E at that point is:

$$E = \frac{F}{q}$$

Here F is in newtons and q is in coulombs. Therefore, E is expressed in newtons per coulomb. *The direction of E at any point,* by definition, *is the one in which the field pushes a positive test charge placed at that point.* A negative charge placed at the same point is pushed in a direction opposite to that of E.

Sample Problems

1. A positive charge of 10^{-5} C experiences a force of 0.2 N when located at a certain point in an electric field. What is the electric field strength at that point?

Solution:

$F = 0.2$ N $q = +10^{-5}$ C

$E = \dfrac{F}{q} = \dfrac{0.2 \text{ N}}{10^{-5} \text{ C}} = 2 \times 10^{4}$ N/C

2. The positive charge in problem 1 is replaced by a negative charge ½ as large. What force does the field exert upon it?

Solution:

The force is ½ as great as it was on the greater positive charge, or ½ of 0.2 N = 0.1 N, and is in the opposite direction.

Test Yourself Problems

1. A positive charge of 2×10^{-6} C experiences a force of 0.1 N when at a certain point in the electric field of another positive charge. What is the electric field strength at that point?

2. The electric field intensity at a point is 4.0×10^4 N/C. What is the force on a charge of 0.50 μC placed at that point? (Remember to change μC to C.)

22-8 Electric Field as Transmitter of Force

According to the present view, an electric charge does not exert a force on a second charge directly but through its electric field. The first charge modifies the space around it by setting up an electric field in it. The electric field then exerts the force that acts upon the second charge. In a similar way, the second charge sets up an electric field about itself that then exerts an equal and opposite force on the first charge.

(a)   (b)

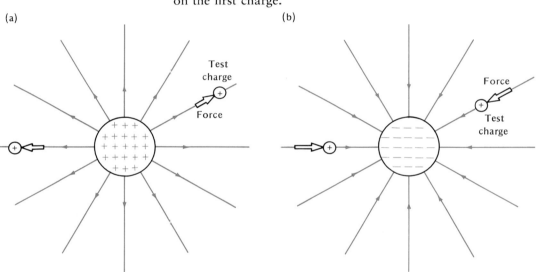

Fig. 22-3. Lines of force around **(a)** a positively charged sphere; and **(b)** a negatively charged sphere.

22-9 Lines of Force

A helpful way of visualizing electric fields around charges is by drawing *lines of force* around them. A line of force is the path which a small positive test charge would take when released in the electric field and allowed to move without acceleration. For example, consider the positively charged sphere in Fig. 22-3 (a). If a tiny positive test charge is released at various points near the surface of the sphere, the force of repulsion will move it radially outward away from the charged sphere on the paths shown. These paths are the lines of force of the electric field of the sphere.

Fig. 22-3 (b) shows the lines of force of the electric field around a negatively charged sphere. A positive test charge placed near this sphere would be attracted toward the sphere along the extended radii as shown by the direction of the lines of force.

Lines of force are drawn so that they show two things about the strength of an electric field at each point. First, the direction of the tangent to the line of force passing through any point shows the direction of the electric field strength at that point. This is the direction in which a positive charge put at that point would be pushed. Second, the electric lines of force are spread apart so that the concentration of the lines of force at any place in the field is proportional to the electric field strength at that place. Where the lines are close together, the field strength is large; where the lines are far apart, the field strength is small.

In Figs. 22-3 (a) and (b), the lines are close together near the charged sphere where the electric field strength is great. As the distance from the charged spheres increases, the lines spread further apart, indicating that the electric field strength is decreasing. This illustrates the fact that the electric force on the positive test charge used to explore the field decreases as its distance from the charged body increases.

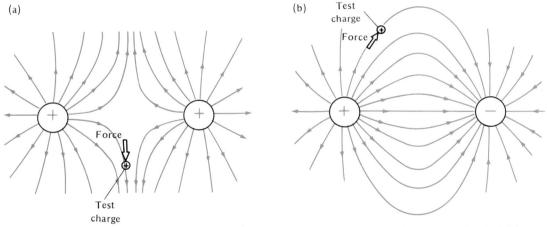

22-10 Fields of Force Between Two Charged Bodies

Figs. 22-4 (a) and (b) show the lines of force in the electric field around two equal positive charges and an equal positive and negative charge. Each line of force shows the path that would be followed by a small positive charge put into the field and allowed to move slowly at constant speed. Note that in the field of the two positive charges, the positive test charge is repelled by both charges and all lines of force lead away from both charges.

In the electric field of two bodies having unlike charges, the positive test charge is repelled by the positively charged body and attracted by the negatively charged body. All lines of force therefore lead away from the positively charged body toward the negatively charged body. In general, it is noteworthy that lines of force begin on positively charged bodies and end on negatively charged bodies.

Fig. 22-4. (a) Electric field around two positive charges. **(b)** Electric field around two opposite charges.

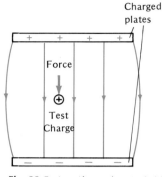

Fig. 22-5. A uniform electric field is set up between two oppositely charged parallel plates.

22-11 Uniform Electric Field Between Parallel Plates

The strength of the electric field around two or more charged bodies is generally not constant. It changes from point to point, becoming smaller as the positive test charge is moved further away from the charged bodies. A very useful electric field is one in which the electric field strength is constant at all points in the field. Such a field is formed between two long parallel plates charged oppositely, as shown in Fig. 22-5. A small positive test charge placed anywhere between these plates is repelled downward by the positively charged plate and attracted downward by the negatively charged plate. If free to move, it would follow a downward path perpendicular to both plates, as shown by the lines of force. The equal spacing of the lines of force shows that the force acting on the test charge per unit of positive charge is constant in all parts of the field, except at points very near the edges of the plates.

22-12 Electric Fields Around Conductors

The electric field inside and around a conductor having either a static charge of electricity or no charge has two special characteristics. First, *the electric field strength inside a conductor is always zero.* Second, *the lines of force of an electric field immediately outside the conductor are always perpendicular to the surface of the conductor.*

These characteristics result from the fact that conductors contain large numbers of free electrons. We can understand why the electric fields of conductors have these two characteristics by supposing that each of them were not true and examining the consequences.

First, suppose that an electric field other than zero appears in a conductor with a static charge or a net zero charge. This field will immediately set the free electrons in the conductor moving. The charge in the conductor cannot therefore remain static as long as there is an electric field inside of it.

Second, suppose the lines of force of the electric field of a conductor with a static charge or a net zero charge were not perpendicular to the outside surface of the conductor. The force exerted by the field at any point on the surface of the conductor would then have a component parallel to the surface of the conductor. This component would cause the free electrons in the conductor to move along the surface, contradicting the condition that the charge on the conductor is static.

22-13 Electric Shielding

The property of a conductor whereby the electric field inside of it is zero makes it possible to shield objects and instruments against the forces exerted by the electrical fields of nearby charged bodies.

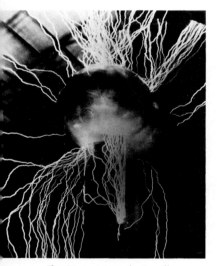

Fig. 22-6. A Van de Graaff generator in which charges carried by a belt inside the cylinder accumulate on the large metal cap. The large difference in potential built up accelerates charged particles used to smash atoms.

By surrounding an electroscope with a metal case, we assure that the electrical field around this instrument remains zero regardless of the presence of charged bodies in the vicinity.

Associated with the absence of an electric field inside a conductor is the fact that any excess charge on a conductor lies wholly on its surface. This must be so because if there were an unbalanced charge inside a conductor, it would set up an electric field around itself. This contradicts the experimental fact that there is no electric field inside a conductor bearing a static charge.

22-14 Van de Graaff Generator

The fact that all the static charge on a conductor lies on its outside surface is used in the *Van de Graaff generator* to pile up an enormous charge on a large, hollow, insulated sphere. This charged sphere induces an equal and opposite charge on the ground so that a difference of potential of many millions of volts is set up between the sphere and the ground. Such large differences of potential are useful in accelerating ions and other charged particles used as projectiles for penetrating the nuclei of atoms.

A very simple model of the Van de Graaff generator is illustrated in Fig. 22-7. A large, hollow, metal sphere is mounted on an insulated stand consisting of a hollow cylinder. Inside the cylinder is a motor-driven silk belt that runs over insulated rollers. Each part of the belt acquires a charge of electrons as it passes and rubs against a glass cylinder on its way up to the sphere. It carries these electrons into the sphere where a metal brush takes them off the silk and conducts them to the inside of the sphere. Since the electrons promptly go to the outside of the sphere, the inside of the sphere remains uncharged. Hence it is able to continue to receive more and more electrons as they are brought up by the belt. The continuous transfer of the electrons to the outside surface of the sphere soon builds up a giant charge on it. The result is a difference of potential of millions of volts between the sphere and the ground.

In practical Van De Graaff generators, the charge is not put on the belt by friction. It is usually put on by a high-voltage machine that either removes electrons from the belt continuously or sprays electrons onto it. In the first case, the ball accumulates a positive charge; in the second, it accumulates a negative charge.

22-15 Work and Potential Energy in an Electric Field

A charge at any point in an electric field possesses potential energy. This becomes evident when the charge is permitted to move with absolute freedom. The electric field then pushes it, steadily increasing its velocity and converting some of its potential energy into kinetic energy. Consider the electric field of the positively charged

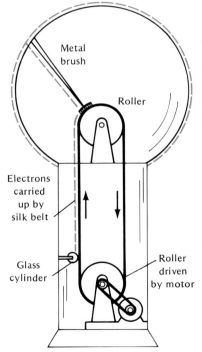

Fig. 22-7. Principle of the Van de Graaff generator.

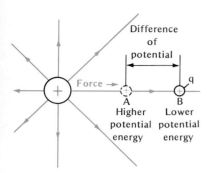

Fig. 22-8. The difference in potential between A and B is the work needed to move a unit positive charge from B to A.

sphere in Fig. 22-8. If a positive test charge q is put at any point in it, such as A, and then set free, it is repelled and accelerated toward the right by the electric field and acquires increasing kinetic energy. This kinetic energy is obtained at the expense of the potential energy possessed by q when it was at A. In accordance with the law of the conservation of energy, the gain of kinetic energy by q is exactly equal to its loss of potential energy.

If the positive test charge is put at B and released, the same behavior indicates that it also has potential energy at this position. However, to move the charge from B to A, an outside force must be exerted upon it. The fact that an outside force must do work to move q from B to A indicates that q has more potential energy when it is at point A than it has when it is at point B. Point A is said to have a *higher electrical potential* than point B and a *difference of potential* is said to exist between points A and B. Again, in accordance with the law of conservation of energy, the work done in moving q from B to A is equal to the difference between q's potential energy at A and B.

22-16 Difference of Potential

The difference of potential between two points is defined as the work it takes to move a unit positive charge from the point at lower potential to that at higher potential. If it takes W units of work to move a positive charge q from point B to point A, it follows from this definition that the difference of potential V between B and A is:

$$V = \frac{W}{q}$$

In *SI* units, W is measured in joules, q is measured in coulombs, and V is expressed in joules per coulomb. Thus, *the SI unit of difference of potential is 1 joule per coulomb.* This unit is called a *volt*, symbol V. Differences of potential are often referred to as voltages. They may be measured by modified forms of electroscopes known as *electrometers* and by *voltmeters*.

The relationship $V = W/q$ may be rewritten in the form:

$$W = Vq$$

We use this form when V and q are given and we wish to find W.

Sample Problem

It takes 5.0×10^{-3} J of work to move a positive charge of 2.5×10^{-4} C from point X to point Y of an electric field. What is the difference of potential between X and Y?

Solution:

$W = 5.0 \times 10^{-3}$ J
$q = +2.5 \times 10^{-4}$ C

$V = \dfrac{W}{q} = \dfrac{5.0 \times 10^{-3} \text{ J}}{2.5 \times 10^{-4} \text{ C}} = 20$ J/C = 20 volts

Test Yourself Problems

1. It takes 8.0×10^{-3} J to move a charge of 4.0×10^{-6} C from one point to another in an electric field. What is the difference of potential between the two points?

2. How much work does it take to move a positive charge of 5.0 μC from a point of lower to a point of higher potential when the voltage between the points is 60 V?

22-17 Electronvolt

The joule is too large a unit for measuring the work done on moving elementary charges, such as electrons or protons, or the small charges on ions. For this purpose, the *electronvolt* is a more convenient unit of energy or work. *An electronvolt is the work done in moving an electron or other body having a unit of elementary charge through a difference of potential of one volt.* Substituting the charge of the electron, 1.60×10^{-19} coulomb, for q and 1 volt for V in the relationship $W = Vq$, we have $W = 1$ volt $\times 1.60 \times 10^{-19}$ coulomb $= 1.60 \times 10^{-19}$ joule as the energy corresponding to 1 electronvolt (symbol, eV). Thus:

The electronvolt is a unit of work particularly useful in dealing with atomic and subatomic particles.

$$1 \text{ eV} = 1.60 \times 10^{-19} \text{ J}$$

Sample Problems

1. The difference of potential between the points A and B in Fig. 22-8 is 100 volts. (*a*) How much work in electronvolts is done by the electric field in moving a free proton from A to B? (*b*) What happens to this work? (*c*) What is this work in joules?

Solution:

(*a*) $V = 100$ volts $q = +1$ electron's charge

$W = Vq$
$Vq = 100$ volts $\times 1$ electron's charge $= 100$ eV

(*b*) This work is used to accelerate the proton. It is converted into the kinetic energy gained by the proton on moving from A to B.

(*c*) $1 \text{ eV} = 1.60 \times 10^{-19}$ J
           $100 \text{ eV} = 1.60 \times 10^{-17}$ J

2. If an electron were put at A instead of a proton, how much work in electronvolts is done as it is moved from A to B?

Solution:
Since an electron and proton have equal quantities of charge, the work done would be numerically the same as for a proton, that is, 100 eV. However, since an electron is a negative charge, this work must be supplied by an outside force which would pull the electron from A to B against the attraction of the positively charged sphere.

Test Yourself Problems

1. Under the force exerted by an electric field an electron is moved from one point to another. The difference of potential between the points is 50 V. How much work is done on the electron (*a*) in eV; (*b*) in J?

2. To move an ion having a charge of 2 electrons from one point to another in an electric field requires 10 eV of energy. (*a*) What is the difference of potential between the two points? (*b*) How many joules of work is done on the ion?

22-18 Electrical Potential Energy

We have seen that the potential energy of an electric charge varies from point to point in an electric field. So far, we have shown how to compute the difference of potential associated with two different points in an electric field from the relationship $V = W/q$. We shall now see how to compute the value of the potential energy itself for the special case of a charge situated at a given point in the electric field around a point charge. This case is particularly important in the study of the structure of atoms. You will find in Chapter 30 that all the positive charges of any atom are concentrated in a central core called its nucleus, while all the electrons of an atom are distributed outside of its nucleus. The nucleus therefore acts like a positive point charge having an electric field around it. Each electron in this field has an amount of potential energy that depends upon its position in the field.

In expressing the gravitational potential energy of a body, we learned in Section 10-8 that a base level such as the surface of the earth must be arbitrarily selected as the level corresponding to zero potential energy. Similarly, in expressing the potential energy of a charge in an electric field, a position of the charge corresponding to the base or zero level of potential energy must first be selected.

A charge in the electric field of a point charge is said to have zero potential energy when it is at an infinite distance from the point charge.

The potential energy of a point charge q_2 when located at a point P in the field of another point charge q_1 is then defined as the work needed to move q_2 from infinity up to the position P. It is shown below that the potential energy U of q_2 when P is at a distance r from q_1 is given by:

$$U = K \frac{q_1 q_2}{r}$$

where K is 9.0×10^9 newton-meters^2/coulomb^2.

Note that the potential energy of q_2 is positive if it represents the work an outside force must do to bring the charge q_2 up to P from an infinite distance. This is the case when q_1 and q_2 are like charges and repel each other. The potential energy of q_2 is negative if it represents the work done by the electric field in pulling the charge q_2 up to P from an infinite distance. This is the case when q_1 and q_2 are unlike charges and attract each other.

Strictly speaking, *the potential energy is associated with both charges.* Each charge has potential energy in the field of the other, and the above expression for the potential energy of q_2 applies to either charge.

Optional derivation for honor students

22-19 Derivation of Potential Energy Formula

The derivation consists of assuming that this is the correct expression for the potential energy, and then showing that the correct expression for the force between two charges, or Coulomb's law, is derived from it.

Consider q_1 to be the fixed charge and q_2 to be in the field of q_1. When the distance between the charges is r, the potential energy of q_2 is Kq_1q_2/r. Now let q_2 be pushed toward q_1 a very small distance Δr. If F is the average force needed to push q_2, the work done by F on q_2 is $F \times \Delta r$.

Since q_2 is now at the closer distance $r - \Delta r$ to q_1, its new potential energy is greater and is equal to $Kq_1q_2/(r - \Delta r)$. Since the energy is conserved, the work done by F in moving q_2 must be equal to the potential energy gained by q_2. That is,

$$F \times \Delta r = \frac{Kq_1q_2}{r - \Delta r} - \frac{Kq_1q_2}{r}$$

Clearing fractions on the right of the equation,

$$F \, \Delta r = Kq_1q_2 \left[\frac{\Delta r}{r(r - \Delta r)} \right]$$

whence,

$$F = \frac{Kq_1q_2}{r^2 - r \, \Delta r}$$

F is the average force exerted on q_2 in moving it over the distance Δr. As Δr is taken closer and closer to zero, F comes closer to the force exerted upon q_2 when it is at the distance r from q_1. When Δr becomes zero, the above expression becomes $F = Kq_1q_2/r^2$ which is Coulomb's law.

A similar proof may be given to show that $U = Kq_1q_2/r$ is the correct expression for the potential energy when q_1 and q_2 are both negative as well as when q_1 and q_2 have opposite charges.

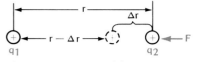

Fig. 22-9. Derivation of the formula for electrical potential energy associated with two charges.

Sample Problem

Assuming that a hydrogen atom consists of one electron and one proton separated by a distance of 5.3×10^{-11} m, what is the potential energy of the electron in the field of the proton?

Solution:

$$q_{\text{proton}} = q_1 = 1.6 \times 10^{-19} \text{ C}$$
$$r = 5.3 \times 10^{-11} \text{ m}$$
$$q_{\text{electron}} = q_2 = -1.6 \times 10^{-19} \text{ C}$$

$$U = \frac{Kq_1q_2}{r} =$$

$$9.0 \times 10^9 \frac{\text{N} \cdot \text{m}^2}{\text{C}^2} \times \frac{(1.6 \times 10^{-19} \text{ C})(-1.6 \times 10^{-19} \text{ C})}{5.3 \times 10^{-11} \text{ m}}$$

$$U = -4.3 \times 10^{-18} \text{ N} \cdot \text{m, or J}$$

Note that the potential energy is negative because the charges are opposite and attract each other.

Test Yourself Problems

1. Referring to the above sample problem, find how much work it takes to move the electron an infinite distance from the proton.

2. A positive charge of 1.5×10^{-7} C and a positive charge of 2.0×10^{-8} C are 3.0×10^{-2} m apart. What is the potential energy associated with these charges?

Measuring and Controlling Electrons and Ions

22-20 Electric Field Intensity Between Charged Parallel Plates

In the remainder of this chapter, we shall see how the forces exerted by electric fields are used to move, control, and measure various types of charged particles. A specially important electric field is that between oppositely charged parallel conducting plates because the electric field intensity E at every point between the plates has the same constant value. This means that a charge put anywhere between the plates has the same constant force acting upon it.

To charge two parallel conducting plates, they are connected to the terminals of a battery which sets up a constant difference of potential V between them. If d is the distance between the plates, it can be shown with the aid of Fig. 22-10 (a) that

$$E = \frac{V}{d}$$

Consider a small positive test charge q located just above the negatively charged plate. The work done in moving this charge up to the positively charged plate depends only on the difference of potential between the plates and is equal to Vq. This work, in turn,

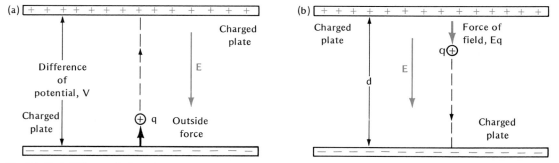

Fig. 22-10. (a) The work done in moving q in opposition to the field E from the lower to the upper plate is Vq. (b) The work done by the field E in moving q back to the lower plate is Eqd.

must be equal to that done by the field when q is released and the field is allowed to push it back to the negative plate, as shown in Fig. 22-10 (b). The downward force exerted by the field on q is Eq. The work done in moving q the distance d between the plates is the force Eq times the distance d, or Eqd. Since $Eqd = Vq$, it follows that $E = V/d$.

Since V is easily measured with a voltmeter and d with a ruler, E is readily determined. When V is measured in volts and d in meters, E is obtained in *volts per meter*. However, as we saw in Section 22-7, E may also be expressed in newtons per coulomb. *A field strength of one volt per meter is equivalent to one of one newton per coulomb.*

Sample Problems

1. What is the electric field strength between two charged metal plates that are 0.050 m apart when a voltmeter shows that they have a difference of potential of 100 volts between them?

Solution:
$d = 0.050$ m $V = 100$ volts

$$E = \frac{V}{d} = \frac{100 \text{ V}}{0.050 \text{ m}} = 2.0 \times 10^3 \text{ V/m}$$

or, since 1 V/m = 1 N/C, $E = 2.0 \times 10^3$ N/C

2. What is the force on a positive charge of 4.0 μC when in this field?

Solution:
$E = 2.0 \times 10^3$ N/C $q = 4.0 \times 10^{-6}$ C

$F = Eq$
$F = (2.0 \times 10^3 \text{ N/C})(4.0 \times 10^{-6} \text{ C})$
$F = 8.0 \times 10^{-3}$ N

This force is in the same direction as E.

3. What is the force on a negative charge of 4.0 μC when in this field?

Solution:
F has the same magnitude as in problem 2 but acts in the direction opposite to E.

Test Yourself Problem

Two oppositely charged parallel plates are 0.20 m apart and have a difference of potential of 500 V between them. (*a*) What is the electric field intensity between the plates? (*b*) What force does the field exert upon an electron in it? (*c*) What force does the field exert on a proton in it?

22-21 Millikan's Measurement of Natural Unit of Charge

The uniform electric field between two charged parallel plates can be used to measure the quantity of charge on a small body placed between them. Consider the tiny positively charged oil drop between the two oppositely charged parallel plates in Fig. 22-11. Two

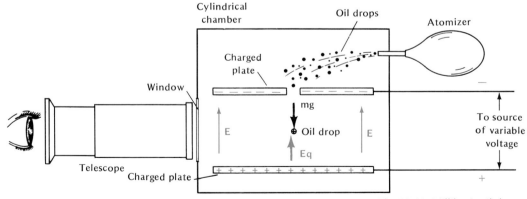

Fig. 22-11. Millikan's oil drop experiment. The drop remains suspended in the lower chamber when $Eq = mg$.

forces are acting upon it. The weight of the oil drop is pulling it downward and the electric field between the plates is pushing it upward.

Now if m is the mass of the oil drop, and g the acceleration of gravity, the downward force on the drop is mg. If q is the positive charge on the oil drop, and E the electric field strength between the

Millikan was awarded a
Nobel prize in 1923 for
his work on measuring the
charge on the electron.

plates, the upward force on the oil drop is Eq. If we now increase
or decrease E until $Eq = mg$, the resultant of the upward and
downward forces on the oil drop is zero, and the drop remains
suspended in mid-air. The value of q can then be determined by
measuring E and mg and substituting in $q = mg/E$.

This is the principle of the method used by Robert A. Millikan,
an American physicist, to measure the charge of the electron. Millikan's early apparatus is shown schematically in Fig. 22-11. As
very tiny drops of oil are sprayed by an atomizer into the upper
part of the chamber, some of them acquire positive or negative
charges by friction as they leave the nozzle. All the drops fall downward under the influence of gravity, but the resistance of the air
prevents their acceleration and soon causes them to move downward at a slow constant speed.

Occasionally a charged drop passes through the hole in the upper
plate of the pair of charged parallel plates and enters the space
between them where its rate of fall can be observed through a
microscope. By adjusting the difference of potential between the
plates by means of the voltage regulator, the falling oil drop can
be stopped and held in mid-air. The electric field strength between
the plates E is then such that $Eq = mg$, or $q = mg/E$, where m is
the mass of the oil drop and q is its charge.

To obtain the value of q, E and mg must be measured. E is
obtained by measuring the difference of potential V between the
plates with a voltmeter and dividing V by the distance d between
the plates. The weight mg of each oil drop is obtained by measuring
the speed with which it falls under the pull of gravity when the
electric field is turned off. Because of the resistance of air friction,
a falling oil drop is not accelerated but falls at a constant speed
that is in a known ratio to its weight.

From the values of mg/E for many oil drops, Millikan found that
the charges on oil drops, whether positive or negative, always turn
out to be a whole number of times as great as a certain indivisible
unit of charge equal to 1.60×10^{-19} coulomb. This minimum unit
of charge is equal to the natural unit of charge observed in Faraday's electrolysis experiments. From much other experimental
work it is known to be the charge on a single electron. Since the
oil drops in Millikan's experiment acquire their charges by gaining
or losing electrons, net charges always turn out to be whole number
multiples of the charge on one electron.

Sample Problem

A negatively charged oil drop weighing 1.59×10^{-14}
N is balanced in the electric field between the two
oppositely charged plates in the Millikan apparatus.
The difference of potential between the plates is 100
volts, and the distance between them is 0.0050 m.
(a) What is the field strength between the plates? (b)
What is the charge on the oil drop? (c) How many
excess electrons are on the oil drop?

Solution:

(a) $V = 100$ volts $d = 0.0050$ m $= 5.0 \times 10^{-3}$ m

$$E = \frac{V}{d} = \frac{100 \text{ V}}{5.0 \times 10^{-3} \text{ m}} = 2.0 \times 10^4 \text{ V/m, or N/C}$$

(b) $mg = 1.59 \times 10^{-14}$ N $E = 2.0 \times 10^4$ N/C

$$q = \frac{mg}{E} = \frac{1.59 \times 10^{-14} \text{ N}}{2.0 \times 10^4 \text{ N/C}} = 8.0 \times 10^{-19} \text{ C}$$

(c) Dividing q by the charge on 1 electron, which is 1.6×10^{-19} C, gives:

$$\frac{8.0 \times 10^{-19}}{1.6 \times 10^{-19}} = 5 \text{ electrons}$$

Test Yourself Problem

A positively charged oil drop weighing 9.6×10^{-14} N is balanced between two oppositely charged plates 1.0×10^{-3} m apart and having a difference of potential of 200 V between them. (a) What is the charge on the oil drop? (b) How many unit charges equal to that on the proton are on it?

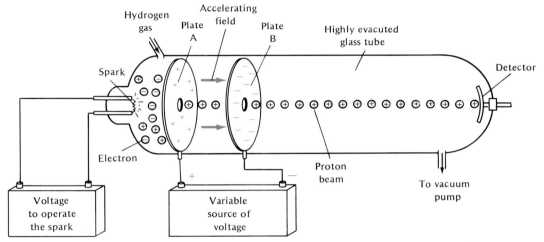

Fig. 22-12. Schematic arrangement for producing a beam of protons.

22-22 Proton and Ion Beams

Much information concerning atoms is obtained by making and studying beams of their ions. Fig. 22-12 illustrates how a beam of positive hydrogen ions or protons is made in a highly evacuated glass tube using the electric field of a pair of parallel charged plates A and B. The protons are supplied by passing an electric spark through a little hydrogen gas allowed to enter at the left end of the tube. This knocks electrons out of many of the hydrogen atoms, leaving many bare protons which pass through the hole in plate A and enter the electric field between A and B. Here the protons, repelled by positively charged plate A and attracted by negatively charged plate B, are accelerated toward plate B. Large numbers of them pass through the hole in plate B and emerge as a beam of protons.

The velocity with which the protons in the beam travel can be controlled by changing the difference of potential applied to the plates. If the potential difference is V and if the charge on each proton is e, the work done by the electric field in moving each proton from A to B is Ve. The electric force that does this work accelerates e from practically zero velocity at A to the velocity v at B. If m_p is the mass of the proton, the kinetic energy it acquires on reaching B is $\frac{1}{2}\,m_p v^2$. Since energy is conserved, this kinetic energy must be equal to the work done on the proton by the field, so that $Ve = \frac{1}{2}\,m_p v^2$. This relationship shows that by increasing V, we can increase the speed of the beam as required.

This proton beam can be used to determine the mass of the proton when V, e, and v are known. The value of e is the same as the charge on the electron. The value of V is measured with a voltmeter. The value of the velocity of the proton beam v can be measured directly by actually timing the particles in the beam as it travels a known distance inside the tube or by a method described in Section 26-17. The mass of the proton is then obtained by solving for m_p in $Ve = \frac{1}{2}m_p v^2$. It turns out to be about 1.7×10^{-27} kilogram.

The masses of the atoms of other gases may be measured in a similar way using beams of their ions. These beams are made in a manner similar to that used to make a proton beam, by replacing the hydrogen in the tube by the gas whose ions are to form the beam.

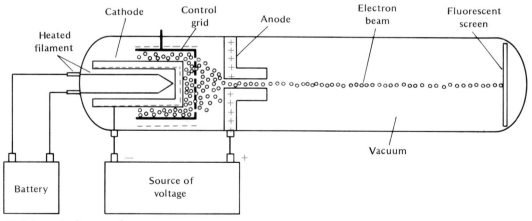

Fig. 22-13. Schematic diagram of an electron gun.

22-23 Electron Gun

Electron beams find a wide variety of uses in research as well as in practical applications. They can be made with an *electron gun* enclosed in a highly evacuated tube. The electron gun consists essentially of a source of electrons called a cathode and one or more positively charged hollow cylinders called anodes which are designed to produce a narrow electron beam.

A simplified electron gun having a cathode and a single anode, consisting of a metal disc with a hole in its center, is shown in Fig. 22-13. A difference of potential applied between the cathode and the anode keeps the anode positively charged. The cathode consists of a cylindrical sleeve of metallic compounds that emit large numbers of electrons when heated. It encloses a tungsten filament which is heated to a high temperature by means of an electric current and heats the cathode in turn. The electrons emitted by the cathode are accelerated toward the anode and pass through the hole in it as an electron beam. Such beams are often called *cathode rays*. The velocity of the electron beam depends upon the difference of potential applied between the cathode and anode. It can be increased or decreased by increasing or decreasing this difference of potential.

An electron gun also contains a device for increasing or decreasing the number of electrons fed to the beam. This is a wire mesh called a *control grid* that surrounds the cathode. Ordinarily, because electrons are so tiny, they can move through the coarse mesh of the control grid freely. However, when the grid is given a negative charge, it repels the electrons and reduces the number that can pass through the wire mesh and go on to the anode. The larger the negative charge put on the grid, the smaller is the number of electrons it permits to pass through it. Thus, by changing the negative charge put on the control grid, it is possible to increase or decrease the strength of the electron beam as desired.

22-24 Determining Mass of the Electron

As in the case of a proton beam, the work done in accelerating each electron in an electron beam is Ve, where V is the difference of potential between the cathode and anode and e is the charge of the electron. This work is converted into the kinetic energy of the electron which is equal to $\frac{1}{2} m_e v^2$, where m_e is the mass of an electron and v is its final velocity. Therefore, as for the proton beam, $Ve = \frac{1}{2} m_e v^2$. Since V, e, and v are known or can be measured, the mass of the electron, m_e, can be determined from this relationship.

The mass of the electron turns out to be 9.1×10^{-31} kilogram. This is nearly 2000 times smaller than the mass of the proton. Because the electron has so small a mass, cathode rays are easily controlled by applying relatively small electric forces to them. This control is used in the cathode-ray tube, which is the picture tube of the television receiver and has many applications as a research tool.

Sample Problem

The difference of potential between the cathode and anode in a simple electron gun is 100 volts. (*a*) What kinetic energy is imparted to each electron by the gun? (*b*) What is the velocity of the beam?

Solution:

(a) The work done by the gun in moving each electron through a difference of potential of 100 V is 100 eV = 100 $(1.6 \times 10^{-19}$ J$)$ = 1.6×10^{-17} J. This work is converted into the kinetic energy imparted to each electron.

(b) $m_e = 9.1 \times 10^{-31}$ kg $KE = 1.6 \times 10^{-17}$ J

$$KE = \tfrac{1}{2} \, m_e v^2, \quad \text{whence} \quad v^2 = \frac{2\,KE}{m_e}$$

Substituting gives:

$$v^2 = \frac{2 \times 1.6 \times 10^{-17} \text{ J}}{9.1 \times 10^{-31} \text{ kg}}$$
$$v = 5.9 \times 10^6 \text{ m/s}$$

Note that the small difference of potential of 100 V gives an electron the very high velocity of nearly six million m/s or about 0.02 times the velocity of light.

Test Yourself Problem

(a) What is the kinetic energy of an electron moving at one-tenth the speed of light? (b) What difference of potential between the cathode and anode of an electron gun is needed to give an electron this energy?

22-25 Cathode-Ray Tube

The cathode-ray tube has found ever-increasing uses in computers, television, radar, testing instruments, and research.

The cathode-ray tube consists of an evacuated glass tube containing an *electron gun* at one end, a *fluorescent screen* painted on the inside surface of the glass at the opposite end, and *two pairs of deflecting plates in between.* The electron gun keeps a narrow beam of electrons moving toward the screen. On its way, the beam passes through the two pairs of oppositely charged deflecting plates whose uniform electric fields are used to control the direction of the beam. Finally, the beam falls upon the screen and produces a fluorescent spot of light where it strikes. The brightness of this spot of light varies with the intensity of the beam.

The action of the deflecting plates is illustrated in Fig. 22-14. When the beam passes between the first pair of plates, its electrons are attracted by the upper positively charged plate and repelled by the lower negatively charged plate. The electron beam is therefore deflected upward in the direction of the upper part of the screen. The amount of deflection produced by the plates increases as the opposite charges upon them are increased. By reversing the charge upon the plates so that the upper plate is negatively charged and the lower plate is positively charged, the beam is deflected downward instead of upward.

In a similar way, charges on the second set of plates control the deflection of the beam horizontally. By varying the charges on the two pairs of deflecting plates, the electron beam may be made to move about from point to point of the screen at will. This process is called *scanning.*

The movement of the beam is shown by the movement of the spot of light it forms on the screen. In the television receiver, this rapidly moving spot of light draws the image seen on the fluorescent screen by the viewer. Because the electron has such a small

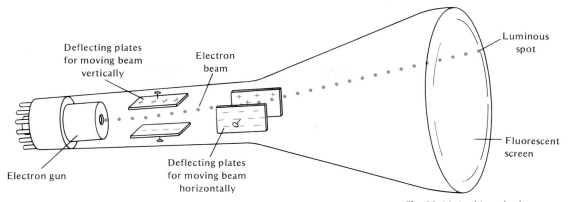

Deflecting plates for moving beam vertically

Electron beam

Luminous spot

Electron gun

Deflecting plates for moving beam horizontally

Fluorescent screen

mass, the electron beam responds extremely rapidly to the forces acting upon it. In the cathode tubes of television receivers, the deflecting plates are usually replaced by electromagnets. These also have the ability to exert deflecting forces on the beam of electrons. See Section 26-13.

Fig. 22-14. In this cathode-ray tube, the charged deflecting plates control the path of the electron beam furnished by the gun.

CHAPTER REVIEW

Summary

The forces of attraction or repulsion exerted upon each other by two point electric charges q_1 and q_2 separated by a distance r is given by **Coulomb's law:**

$$F = \frac{Kq_1q_2}{r^2}$$

According to the modern view, these electric charges do not act upon each other directly but through their electric fields.

An **electric field** is described by giving its strength E at each point in the field. At any point in a field, E is defined as the force exerted on a unit of positive charge situated at that point. If E is the strength of an electric field at a given point, the force F acting on a charge q situated at that point is **F = Eq.**

The **potential energy of a charge** at a given point in an electric field is defined as the work that is needed to move that charge from infinity to its present location. If a positive test charge has more potential energy when located at one point than when located at a second point in a field, the first point is said to be at a higher potential than the second. A **difference of potential** exists between the first and second points. This difference of potential is equal to the work done in moving a unit positive charge from the point at lower potential to the point at higher potential. If W is the work needed to move a positive charge q from one point to another in an electric field, the difference of potential V between those points is given by **V = W/q,** where W is expressed in joules, q in coulombs, and V in volts. A **volt** is equal to 1 joule per coulomb.

When a positive electric charge q, at rest in an electric field, is permitted to move freely, the force of the electric field will push it from its starting point to a point at lower potential. If V is the difference of potential between the initial and the final points, the work done by the field in pushing a positive charge q from the higher potential to the lower potential is $W = Vq$. This work is used in accelerating q and in giving it kinetic energy' equal to $\frac{1}{2} mv^2$ where m is the mass of q and v is its final velocity. By the law of conservation of energy, it follows that $W = Vq = \frac{1}{2} mv^2$. This relationship makes it possible to use electric fields to speed up charged particles as is done in making electron and ion beams.

The work done in moving an electric charge may also be expressed in **electronvolts (eV).** An electronvolt is the work done in moving an elementary positive charge through a difference of potential of 1 volt. One electronvolt is equal to 1.60×10^{-19} joule.

Questions

Group 1

1. (a) Assuming that atoms contain equal numbers of protons and electrons, list the electrical forces that they exert upon each other. (b) Which of these electrical forces tend to hold the atom together? (c) Which tend to make it fly apart?

2. (a) Two like charges exert a certain force upon each other when they are 1 m apart. What happens to the force between them when the distance between them is doubled? Halved? Tripled? (b) What happens to the force between them when the charge on one of them is doubled? Halved? Tripled?

3. A positive test charge is located on the mid-point of the line joining two equally charged positive spheres. If the force exerted by one of the spheres on the test charge is F, what is the direction and magnitude of the resultant force exerted by both spheres on the positive test charge?

4. (a) A positive test charge is on the line joining two equally charged positive spheres and is located at a point $\frac{1}{3}$ of the distance between them. If the force exerted by the nearer sphere on the test charge is F, what is the force exerted by the more distant sphere? (b) What is the direction and magnitude of the resultant force exerted by both spheres on the test charge?

5. If the charge on the sphere nearer to the test charge in problem 4 is replaced by an equal negative charge, what will now be the direction and magnitude of the resultant force exerted on the test charge by both spheres?

6. A positive charge is located at the mid-point of the line joining two positively charged spheres, the first of which has 3 times as much charge as the second. (a) If the force exerted on the test charge by the sphere with the smaller charge is F, what is the force exerted on the test charge by the other sphere? (b) What is the magnitude and direction of the resultant force exerted on the test charge by both spheres?

7. Why may Coulomb's law not be applied to two large charged bodies that are very close to each other?

8. (a) What is meant by the electric field intensity at a point of an electric field? (b) What two things do the lines of force show about the electric field intensity at a point in an electric field?

9. (a) Draw the lines of force of the electric field between two horizontal oppositely charged metal plates, assuming that the upper one is positively charged and the lower one negatively charged. (b) How does the electric field intensity at a point midway between the plates compare with that at a point near either plate?

10. A positively charged dust particle is midway between the charged plates in question 9. (a) What two forces are acting upon it? (b) How will it move?

11. (a) What will determine whether a negatively charged dust particle midway between the charged plates in question 9 will move upward or downward? (b) How can this particle be kept suspended between the charged plates?

12. A solid metal sphere and a hollow metal sphere are given equal negative charges. (a) How do the electrons distribute themselves in each sphere? (b) Compare the electric fields inside each of the spheres. (c) Compare the electric fields outside of each sphere.

13. The leaves of an electroscope are often enclosed in a cylindrical metal container having glass windows at opposite ends. How does the metal case protect the electroscope from the effects of nearby electric charges?

14. With respect to the Van de Graaff generator, explain: (a) how a large charge is built upon the ball; (b) why it is possible to keep adding more and more charge to the inside surface of the ball. (c) What is the electric field intensity inside the charged ball? (d) Where outside the ball is its electric field intensity greatest?

15. A and B are two points in the electric field around a small positively charged sphere. A is nearer to the sphere than B. (a) Which point is at higher potential? (b) As B is moved away from the sphere, what happens to the difference of potential between A and B? (c) Answer (a) and (b) assuming that the sphere is charged negatively instead of positively.

16. (a) Explain how a beam of protons can be obtained. (b) How can the velocity of the protons be increased or decreased.

17. With reference to the electron gun, explain: (a) the source of electrons; (b) how the electrons are accelerated; (c) how their velocity is deter-

mined before they leave the gun; (d) how a grid may be used to control the flow of electrons leaving a gun.

18. With reference to a cathode-ray tube, explain the purpose of (a) the electron gun; (b) the deflecting plates; (c) the screen. (d) What part does this tube play in a television receiver.

Group 2

19. A relatively small electric force can have a great effect on the motion of an electron. Why is this not true when the same force is acting upon a proton?

20. (a) To determine the mass of an electron in an electron beam coming out of a gun, what three quantities must be known or measured? (b) Explain how the electron's mass can be determined from these quantities.

21. Explain the statement: Electric charges act upon each other not directly but through their electric fields.

22. What happens to the potential energy of two like charges that are permitted to separate under their own repulsion when the charges are both (a) positive; (b) negative?

23. A proton shot horizontally toward a vertical positively charged metal plate moves forward until the repulsion by the plate brings it to a halt. (a) What has happened to the initial kinetic energy of the proton at the moment it stops? (b) What will happen to the proton after it has been stopped?

Problems

Group 1

1. What is the force between two equal positive charges of 2.0×10^{-4} C that are 2.0 m apart?

2. What is the force between a positive charge of 0.00060 C and a negative charge of 0.00030 C separated by a distance of 0.30 m?

3. Find the force between a positive charge of 1.0 μC and a positive charge of 2.0 μC when they are 0.030 m apart.

4. Find the force exerted by a negative charge of 5.0 μC on a positive charge of 2.0 μC when the charges are 9.0 cm apart.

5. A hydrogen atom is composed of a proton and an electron at a distance of 5.3×10^{-11} m from

each other. Taking 1.6×10^{-19} C as the elementary unit of charge, compute the force pulling the proton and the electron together.

6. A positive charge of 4.0×10^{-6} C exerts a force of repulsion of 7.2 N on a second charge 0.050 m away. What is the sign and magnitude of the second charge?

7. A positive charge of 8.0×10^{-7} C experiences a force of 0.04 N when located at a certain point in an electric field. What is the electric field strength at that point?

8. What is the magnitude of the force exerted on an electron in a uniform electric field whose intensity is 1000 N/C?

9. It takes 4×10^{-3} J of work to move a positive charge of 2×10^{-5} C from one point of an electric field to another. What is the difference of potential between the two points in (a) J/C? (b) V?

10. The difference of potential between the terminals of a battery is 6 V. How much work is done when a series of negative charges totaling 0.5 C are moved from one terminal to the other?

11. Two parallel metal plates have opposite charges and a difference of potential of 1000 V between them. (a) How much work in electronvolts is done in moving an electron from the negatively charged plate to the positively charged plate? (b) What is this work in joules?

12. An electron beginning at rest is accelerated as it moves in an electric field between two points having a difference of potential of 500 V between them. (a) How much kinetic energy in electronvolts is imparted to the electron? (b) What is the value of the energy in joules?

13. Two positive charges of 3.0×10^{-7} C each are 0.2 m apart. (a) What is the potential energy of the charges? (b) If one of the charges is kept fixed while the other is allowed to move under their mutual repulsion, how much kinetic energy will be imparted to the moving charge when it is infinitely far from the fixed charge?

14. What is the electric field strength between two oppositely charged metal plates that are 0.080 m apart and have a difference of potential of 1600 V between them?

15. What is the force on a negative charge of 9.0 μC in the electric field of problem 14?

16. A negatively charged oil drop weighing 9.6×10^{-14} N is balanced between two oppositely charged plates in the Millikan apparatus when the difference of potential between the plates is 300 V and the distance between them 0.010 m. (a) What is the field strength between the plates? (b) What is the charge on the drop? (c) How many excess electrons are on the oil drop?

17. (a) How much work in electronvolts is done on a proton that is accelerated over a difference of potential of 50 000 V? (b) Express this in joules.

Group 2

18. (a) How much kinetic energy does the proton in problem 17 acquire? (b) Assuming that the mass of the proton is 1.7×10^{-27} kg, what velocity does it acquire?

19. An electron in an electron gun is accelerated by a voltage of 20 000 V. (a) What kinetic energy in joules does it acquire? (b) Assuming that the mass of the electron is 9.1×10^{-31} kg, what is its velocity?

20. (a) What kinetic energy in joules does an electron moving at 3.0×10^6 m/s possess? (b) Convert (a) to electronvolts. (c) What voltage must be applied to the electron to give it this velocity?

21. The charged horizontal deflecting plates in a cathode tube have an electric field of 3.0×10^4 N/C between them. What is the vertical force on each electron in the beam passing between the plates?

22. (a) At what distance from a proton must an electron be to have a potential energy of -5.0 eV? (b) If the distance between the electron and proton is doubled, what is its new potential energy?

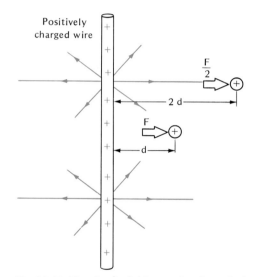

Fig. 22-15. The electric field around a charged wire.

23. The electric field around a long uniformly charged positive wire consists of lines of force extending radially in all directions at right angles to the wire, as shown in Fig. 22-15. The electric field strength around such a charged wire varies inversely as the distance from the wire; that is, $E \propto 1/d$ (unlike the electric field around a point charge which varies as $1/d^2$). Assuming that the force on a unit charge at 1 cm from the wire is 0.012 N, make a table showing how this force

changes as the charge is moved to 2.0, 4.0, 6.0, 8.0, 10.0, and 12.0 cm from the wire. Plot the results on a graph.

24. A second wire similarly charged is placed parallel to the wire in question 23 at a distance of 10 cm from it. What is the force on a unit charge at a point (a) midway between the wires; (b) 2 cm from one wire and 8 cm from the other?

Applying Physics

An idea of the manner in which the electric field acts in the space around an electric charge may be obtained from the following experiment, which requires the air to be quite dry. Cut 2 strips of newspaper about 75 cm long and 5 cm wide. Charge each strip by laying it on a wooden or plastic table top and rubbing it briskly with a piece of fur or wool. Put one end of each of the two charged strips of paper together and hold the common ends so that the strips hang freely. They will repel each other like the leaves of an electroscope. The electric field in the space between the paper strips is pushing each of them away from the vertical.

Bring the hand near but not touching the outside of one of the strips. The charged strip induces an opposite charge on the hand. The new electric field between the hand and the charged strip now causes the strip to move toward the hand.

Bring the hand between but not touching the two charged paper strips. Explain what happens to the electric field between the strips that now causes them to move together.

Charge the paper strips again and let them repel each other as before. Hold a lighted match a short distance to one side of one of the strips. Note that the strips come together. The flame ionizes the air around it. Explain how these ions destroy the electric fields around and between the charged strips.

Electric Circuits

Aims

1. To learn how a steady direct current may be set up and maintained in an electric circuit.
2. To study the factors that control the flow of current in an electric circuit.
3. To study the role of energy in the electric circuit—how it is received, transferred, and transformed.

23-1 Electric Circuits Transfer Energy

We use electric circuits to transfer electrical energy supplied by a battery or generator to such devices as lamps, heaters, radios, and motors. These devices convert electrical energy into light, heat, sound, mechanical work, and other forms. To understand how this transfer occurs, consider first a current made up only of metal conductors carrying a current flowing steadily in one direction. This is called a *steady direct current*. In such a circuit, the current consists of a flow of free electrons.

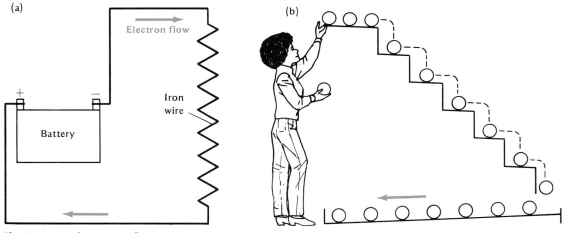

Fig. 23-1. (a) A battery supplies the energy which moves electrons through the circuit. **(b)** Analogy of an electrical circuit in which human muscles provide the energy to move the balls through the circuit.

23-2 Mechanical Analogy of a Circuit

The circuit in Fig. 23-1 (a) consists of an iron wire connecting the negative and positive terminals of a battery. The battery sets up a constant difference of potential between its terminals by piling up and maintaining a constant negative charge on one terminal and a constant positive charge on the other. Under the influence of this difference of potential, electrons flow through the wire from the negative to the positive terminal. The battery now moves the electrons inside itself from the positive to the negative terminal to begin their round trip through the circuit again.

The operation of this electrical circuit is similar in many ways to the operation of the mechanical circuit in Fig. 23-1 (b). The steel balls represent electrons. The upper level at the top of the staircase represents the negative terminal of the battery and the lower level at the bottom of the staircase represents the positive terminal. The difference of electrical potential energy between the battery terminals is represented by the difference in gravitational potential energy resulting from the difference in height between the top and bottom of the staircase. The iron wire is represented by the staircase, and the battery by the figure whose job is to raise each ball arriving at the lower level to the higher level. The balls then roll down the steps and return on the lower level, to be raised again and started on another round trip through the circuit.

23-3 Function of the Battery

Notice that the person operating the mechanical circuit must continually supply energy by raising each ball from the lower level to the upper level. Similarly, *the battery in the electrical circuit must continually supply energy to move each electron inside the battery from the positive terminal to the negative terminal.* Since each ball is lifted through the same height by someone, it is given the same additional potential energy. This is also true of each electron, which receives the same additional electrical potential energy from the battery when it is moved through the constant difference of potential between the positive and negative terminals. The increased potential energy acquired by the electrons then enables them to "fall" through the iron wire circuit just as the increased potential energy acquired by the balls lifted to the top of the stairs enables them to fall down the stairs.

The battery is therefore the source and continuous supplier of the energy moving the electrons around the circuit. It does not supply the electrons that flow in the circuit. They are already present as free electrons in the conductors of the circuit. The battery merely gives them the kinetic energy they need to move through the circuit. It obtains this energy from the chemical energy stored in the materials of which it is made, just as the person in the mechanical circuit obtains mechanical energy from the chemical energy stored up from the oxidation of food consumed.

Fig. 23-2. Electric wheelchairs of advanced design, operated by storage batteries, provide greater independence for the disabled.

23-4 Electric Field Inside Conducting Wire

The electrons in the circuit are repelled by the negative charge piled up on the negative terminal of the battery and attracted by the positive charge piled up on the positive terminal. However, these charges do not act upon the electrons directly but *through their electric field.* Just as the difference of potential between two oppositely charged parallel plates sets up an electric field in the space between them, so does the difference of potential between the two oppositely charged battery terminals set up an electric field inside the wire joining them. Under the forces exerted on them by this field, the free electrons in the metal move continuously from the negative toward the positive terminal of the battery.

Note that the presence of the battery's electric field inside the conductor is not contrary to the observation made in the last chapter that there is no electrical field inside a conductor. That observation applies only when the charge on the conductor remains static and not when currents are flowing in it as they are here.

23-5 Electron Motion in Conductors

As we saw in Section 20-26, the free electrons in a conductor share the collisions and the thermal kinetic energy of the atoms of the metal. Although at room temperature the average speed of the electrons inside a metal is over a million meters per second, they form no current because they are moving in all possible directions. When the battery sets up an electric field inside the conductor, the electrons continue their random motion but are also accelerated by the field in the direction of the positive terminal.

However, before they have speeded up very much they collide with the metallic atoms in their paths and lose some of their kinetic energy to them. These collisions continually reverse the motion of the electrons, as shown in Fig. 23-3, so that their net average speed in the direction in which they are pushed by the field is extremely slow. Also, since the kinetic energy given up by the electrons in the collision is gained by the atoms of the conductor, the flow of electrons generates heat in the wire and raises its temperature.

The motion of the electrons in the mechanical circuit of Fig. 23-1 (b) is illustrated by the movement of the balls down the steps representing the conductor. The steps represent the atoms with which the electrons collide as they drift along through the wire. The balls pick up speed and kinetic energy on falling from each step to the lower one only to lose much of it on colliding with the lower step. The energy lost by the balls is absorbed by the molecules of the steps and generates heat in the steps.

23-6 Electromotive Force

The simple circuit of Fig. 23-1 (a) reveals two conditions needed to keep a current flowing through a conductor. First, *the conductor must be part of a completely closed circuit.* Second, *there must be a source of energy steadily giving the electrons in the conductor the kinetic energy needed by them to move through the circuit.* Such a source of energy is said to provide an *electromotive force,* or an *EMF.* The most widely used sources of EMF are chemical batteries and electric generators.

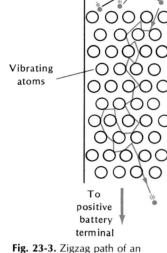

Electrons

Vibrating atoms

To positive battery terminal

Fig. 23-3. Zigzag path of an electron through a conductor.

It should be noted that the EMF supplied by a battery or generator is not a force. It is the energy furnished by the battery or generator to each coulomb of electrons that it sends through the circuit. Each source of EMF has a store of energy which it gradually converts into electrical energy and supplies to the circuit as needed. In the chemical battery, stored chemical energy is converted into electrical energy. In the generator, a supply of mechanical energy is converted into electrical energy.

Batteries or generators change their stores of energy into electrical energy by piling up positive charge on one of their terminals and negative charge on the other. The electrical potential energy stored in the oppositely charged terminals provides the difference of potential responsible for the flow of the electrons in the circuit joining the terminals.

The EMF is the energy supplied to each coulomb of charge that it drives through a circuit.

23-7 EMF and Terminal Voltage

Like the difference of potential that it provides, the EMF of a battery or a generator is measured in *volts*. It should be noted that the EMF of a battery or generator is usually a little larger than the difference of potential it sets up between its terminals when it is in actual use. The reason for this is that some of the energy furnished by the battery or generator to move electrons through the circuit is used in moving the electrons through the battery or generator itself. The remainder of the energy furnished by the battery or generator moves electrons through outside wires or devices connecting its terminals. This remainder determines the *terminal voltage* of the battery or generator.

23-8 Direction of Current and Electron Flow

We have seen that the current in a conductor consists only of negative charges in the form of electrons flowing from the negative to the positive terminal of the battery or generator to which it is connected. In ionized gases or liquids, the current consists of both positively and negatively charged ions flowing in opposite directions. In a proton beam, the current consists of positive charges flowing from the positively charged anode to the negatively charged cathode.

In the 18th century, before anyone knew what actually flows in an electric current, Benjamin Franklin assumed that current consists of positive charges flowing from the positive to the negative terminal of a battery. Now we know that most of the currents with which we deal take place in conductors where electrons flow rather than positive charges. Many books find it simple therefore to take the direction of current as that in which electrons flow; that is, from the negative to the positive terminal of a source of EMF. We shall adopt this direction of current throughout this book. The

Electrons in a conductor flow from the negative to the positive terminal of the source of EMF to which the conductor is connected.

student will find, however, that some other books retain the older idea of current as a flow of positive charges from a positive to a negative terminal.

23-9 Electrical Power

Let us now see how the energy supplied to maintain an electric current in a wire conductor or electrical device is calculated. Suppose q coulombs of charge pass through a wire conductor connected to a battery whose terminals have a difference of potential of V volts between them. The work done by the battery in moving these charges is Vq joules. If these charges pass through the wire in t seconds, the rate at which work is done in moving them is Vq/t joules per second. Since q/t is the number of coulombs passing through the wire per second, it is the current I in the wire. It follows that the rate at which work is done in making the current I flow through the wire is VI. From our study of mechanics, we know that the power P is defined as the rate at which work is done, so:

$$P = VI$$

Electrical power is expressed in watts.

In *SI* units, V is expressed in volts and I in amperes. P then comes out in joules per second which are the power units called *watts*. Therefore, *a watt is the rate at which a battery applying a difference of potential of one volt across a device must supply energy to keep a current of one ampere flowing through it.* That is, *1 watt = 1 volt × 1 ampere* and represents a rate of working of *1 joule per second*.

Practical electrical appliances usually carry a label indicating the rate at which they use electrical energy in watts when supplied with the voltage for which they were designed. A lamp labeled 200 watts is one that requires twice as much electrical energy to operate it per second as does a 100-watt lamp designed for use at the same voltage.

23-10 Electrical Work and Energy

The power tells how much energy is supplied to maintain a current in a conductor or a circuit each second. Hence, the total work or energy W furnished to maintain the current for a given time t is the power multiplied by the time, or Pt. Since $P = VI$,

$$W = VIt$$

Electrical work or energy is expressed in joules or watt-seconds.

In *SI* units, V is in volts, I in amperes, and t in seconds, so that the work or energy W comes out in joules or *watt-seconds*.

The joule or watt-second is too small a unit of energy for many purposes. To measure the large quantities of electrical energy used commercially, the *watt-hour* and *kilowatt-hour* are used as units. A watt-hour is the energy supplied to a circuit or conductor in one hour by one watt of power. Since there are 3600 seconds in an hour, a watt-hour is 3600 watt-seconds or 3600 joules. A *kilowatt-hour* (symbol, kWh) is 1000 watt-hours or 3 600 000 joules. The kilowatt-hours of electrical energy supplied to a device or a circuit may be computed from the relationship, **kilowatt-hours = watts × hours/1000.**

Sample Problem

(a) What power is supplied to a lamp that is operated by a battery having a 12-V difference of potential across its terminals when the current through the lamp is 2.0 A? (b) How much electrical energy has been supplied to the lamp after it has been operating for 4.0 s?

Solution:

(a) $V = 12$ V $I = 2.0$ A

$P = VI = 12$ V $\times 2.0$ A $= 24$ W or 24 J/s

(b) $t = 4.0$ s

$W = VIt = Pt = 24$ J/s $\times 4.0$ s $= 96$ J

Test Yourself Problems

1. What power is supplied to a motor that operates on a 120-V line and draws 1.50 A of current?

2. A lamp is labeled 6.0 V—12 W. (a) What current flows through it when it is operating? (b) How much energy is supplied to it in 1000 s?

23-11 Definition of Resistance

The current in a conductor is the quantity of charge that passes through the conductor per second. We have seen in Section 21-8 that a current in a conductor of one ampere is a flow through it of 6.25×10^{18} electrons per second. When a metal wire is connected to a battery supplying a constant difference of potential, the current flowing through it depends on the degree of opposition which the electrons encounter in the conductor as they collide with the atoms in their paths. This opposition is called the *resistance* of the wire. It is defined as the ratio of the potential difference applied to the ends of the wire to the current flowing through it, or $R = V/I$.

The resistance of a conductor is the opposition it gives to the flow of current through it.

23-12 Ohm's Law.

If a series of increasing potential differences is applied to a given metal wire whose temperature is kept constant, the current in the wire increases in direct proportion to the difference of potential. Plotting the currents I against the corresponding differences of potential V gives the straight line shown in Fig. 23-4 (a). This means that the ratio V/I is constant, and therefore that the resistance R of the metal wire also remains constant in spite of the change in V and I.

Georg Simon Ohm, a German scientist, discovered that this constancy of resistance is true for all metal conductors provided their temperatures are also kept constant. His discovery, known as *Ohm's law*, is written:

$$\frac{V}{I} = R = constant$$

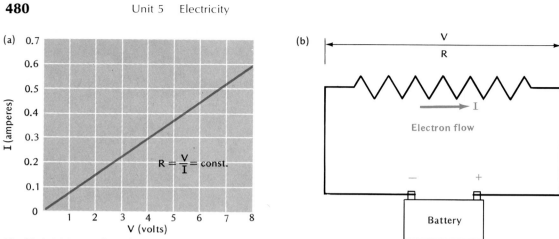

Fig. 23-4. (a) In metal conductors at constant temperature, the current *I* is directly proportional to the applied difference of potential *V*. The conductor's resistance *R* is therefore a constant. **(b)** Ohm's law applies only to conductors for which $V/I = R$ is constant.

where V is the potential difference applied to a conductor of constant resistance R, and I is the current flowing through it.

When V is expressed in volts and I is expressed in amperes, R is expressed in the unit called the *ohm*. You can see from this relationship that one ohm is the resistance of a conductor such that a potential difference of one volt between its ends causes a current of one ampere to flow through it. The symbol for the ohm is the greek letter omega, Ω.

Ohm's law can be rewritten in the forms $V = IR$ and $I = V/R$. In the latter form it expresses the fact that *the current passing through a metal conductor, kept at constant temperature, is proportional to the difference of potential applied to its ends and inversely proportional to its resistance.*

Sample Problems

1. (*a*) What current flows through an 80-Ω coil of wire when it is connected to the terminal of a generator supplying a potential difference of 120 V? (*b*) How many electrons pass through the coil per second?

Solution:
(*a*) $R = 80 \ \Omega$ $V = 120$ V

$$I = \frac{V}{R} = \frac{120 \text{ V}}{80 \ \Omega} = 1.5 \text{ A}$$

(*b*) 1 A = 1 C/s = 6.25×10^{18} electrons/s
$I = 1.5$ A = $1.5 \times 6.25 \times 10^{18}$ electrons/s
$I = 9.4 \times 10^{18}$ electrons/s

2. What is the resistance of an electric iron in which the current is 5.0 A when the difference of potential is the 110 V usually supplied in the home?

Solution:
$I = 5.0$ A $V = 110$ V

$$R = \frac{V}{I} = \frac{110 \text{ V}}{5.0 \text{ A}} = 22 \ \Omega$$

3. An incandescent lamp designed for use on a 120-V household circuit is labeled 60.0 W. What is the operating resistance of the lamp?

Solution:
To obtain R we must know V and I. V is given. I can be found from $P = VI$

$P = 60.0$ W $V = 120$ V

$$P = VI \text{ or } I = \frac{P}{V} \qquad I = \frac{60.0 \text{ W}}{120 \text{ V}} = 0.500 \text{ A}$$

R can now be obtained from

$$R = \frac{V}{I} = \frac{120 \text{ V}}{0.500 \text{ A}} = 240 \ \Omega$$

Test Yourself Problems

1. What is the resistance of an electrical device connected to a 110-V line that draws a current of 10 A?

2. What current flows through a toaster having a resistance of 20 Ω and operating on a 120-V line?

3. A lamp is labeled 6 V—12 W. What is its operating resistance?

23-13 Resistance of Metal Conductors

Many electrical devices such as electric lamps and toasters are simply metal wire conductors whose resistance is constant when their temperatures are kept constant. Such devices obey Ohm's law and are called *resistors*. The resistance of a wire conductor depends upon four factors: *the substance of which it is made, its length, its cross-section, and its temperature.*

Variation with Substance. Conductors made of certain metals such as copper, silver, and aluminum have relatively low resistance compared to similar conductors made of iron, nichrome, and other metals. A copper wire, for example, has only one-seventh the resistance of a similar iron wire. For this reason, the wires that lead the current from a battery or generator to an electrical device such as a lamp or motor and then back again are usually made of copper. On the other hand, a device such as a toaster, in which it is desired to generate heat, is made of a metal alloy such as nichrome that has a relatively high resistance.

Variation with Length and Cross Section. The resistance of a wire made of a specific material varies directly with its length and inversely with its cross section. These relationships can be understood by picturing the flow of electrons in wires of different lengths and cross sections. In a longer wire, the electrons encounter and collide with more atoms in their paths than they do in a shorter wire. They therefore meet more resistance in passing through the longer wire than they do in passing through the shorter wire.

Turning to wires of different cross sections, it is clear that the greater the cross section of a given length of wire, the greater will be the number of spaces between the atoms through which the electrons can pass through that cross section in a given time. So the greater the cross section of a wire, the smaller is its resistance.

Variation with Temperature. Increasing the temperature of a metal conductor increases its resistance, while decreasing the temperature decreases the resistance of the conductor. The tungsten filament of an ordinary incandescent light bulb, for example, has a resistance at its highest operating temperature more than ten times as large as it has when the filament is cold. Again, the behavior of the free electrons in the metal explains this relationship. Heating the metal increases the vibratory motion of its atoms as well as the random speeds of its electrons. As a result, the number of collisions made by the electrons with the atoms in their paths also increases. This retards the forward motion of the electrons and is observed as the increased resistance of the wire.

> **The resistance of a metal wire varies directly with its length and inversely with its cross-section.**

> **The resistance of a metal wire rises as its temperature increases.**

Fig. 23-5. High-speed magnetic trains are both supported and propelled by powerful magnetic fields that are produced by superconductive coils.

23-14 Superconductivity

As the temperature of certain metals approaches absolute zero, their resistance suddenly drops to zero. The property whereby a metal loses its electrical resistance when cooled to a sufficiently low temperature is called *superconductivity*. The temperature at which a metal becomes superconductive is called its *transition temperature*. Different metals have different transition temperatures that range from a fraction of a degree above 0 K to about 25 K. The metals known to be superconductive include aluminum, lead, titanium, thallium, niobium, vanadium, and technecium.

The electrical resistance of certain superconductive metals drops to zero at temperatures near zero Kelvin.

The property of superconductivity is widely used in producing the powerful magnetic fields needed in modern nuclear research instruments and machines. In these, a coil of wire made of a superconductive metal is cooled below its transition point. Since its resistance then drops to zero, even a small electromotive force can send a large current through it. Once started, the current continues to flow without further application of the electromotive force. As we will see in Section 25-14, such a current-carrying coil sets up a magnetic field around itself whose strength is proportional to the current. Since the superconductive current is very strong, so is the corresponding magnetic field.

23-15 Energy Conservation in Resistors

As current passes through a resistor, the enormous numbers of collisions of the moving electrons with the atoms in their paths increases the kinetic energy of vibration of the atoms. As a result, the resistor becomes heated and its temperature rises. Joule showed experimentally that *all the electrical energy furnished to a resistor is converted into an equal quantity of heat energy*. This confirms that the law of conservation of energy applies to the transformation of electrical energy into heat in the resistors of an electrical circuit.

The principle of the experimental method used by Joule is illustrated in Fig. 23-6. A resistor R is immersed in water whose mass and temperature are measured before the current is turned on. A current I is now passed through R which generates heat and warms the water. The current I and the difference of potential across R are measured. After the current has flowed for a measured time t, it is turned off and the temperature of the water is measured again.

The electrical energy supplied to the resistor and converted to heat is obtained in joules by substituting the values of V, I, and t in $W = VIt$. The heat generated by the resistor and used in heating up the water is also obtained in joules from the change in the water's temperature, the specific heat of water, and the known mass of the water. Actually, the heat retained by the container of the water and by R itself is also taken into account.

Joule found that the energy delivered as heat to the water, its container, and R is equal to the electrical energy supplied to the resistor.

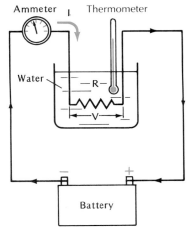

Fig. 23-6. Joule's experiment showing that the electrical energy furnished to R is converted into an equal quantity of heat energy acquired by the water and its container.

23-16 Heat Generated in Resistors

It follows from Joule's experiment that the heat Q generated in a resistor in t seconds is equal to the electrical energy W furnished to the resistor during that time, or $Q = W = VIt$. Substituting $V = IR$ gives $Q = (IR)It$ or:

$$Q = I^2Rt$$

If I is expressed in amperes, R in ohms, and t in seconds, Q comes out in *joules* or *watt-seconds*.

The rate at which heat is generated in a resistor is therefore proportional to the *square* of the current flowing through it and to its resistance. A resistor carrying 2 amperes of current generates heat 2^2 or 4 times as fast as when it is carrying one ampere of current. Again, a resistor of 100 ohms carrying the same current as one of 50 ohms generates heat at twice the rate of the smaller resistance.

In such devices as electric heaters, it is desirable to convert all the electrical energy supplied to the device into heat. However, in all other electrical appliances and parts of a circuit having resistance, the heat generated is an undesirable waste of the electrical energy supplied. Hence every effort is made to reduce the generation of heat to a minimum.

The role of resistance in generating heat is particularly important in the distribution of electric power over power lines. Because such lines are often very long, they have considerable resistance and therefore turn some of the electrical energy passing through them into heat. This energy is wasted, and therefore much effort is being made to reduce the loss by reducing the resistance of the lines. In recent years new materials have been found that are superconductive at temperatures as high as 125 K. They raise the

Fig. 23-7. In electric welding, much heat is generated since the low resistance of the arc permits a very high current.

hope of finding materials that are superconductive at room temperature. Using such zero resistance materials for power lines would eliminate heat losses altogether.

Sample Problem

(a) What current flows through a 20-Ω electric heater operating on a house circuit furnishing a difference of potential of 120 V? (b) How much electrical energy is furnished to it in 10 s? (c) How much heat does it develop in that time?

Solution:

(a) $V = 120$ V $R = 20$ Ω $t = 10$ s

$$I = \frac{V}{R} = \frac{120 \text{ V}}{20 \text{ Ω}} = 6.0 \text{ A}$$

(b) $W = VIt$
     $W = 120 \text{ V} \times 6.0 \text{ A} \times 10 \text{ s} = 7.2 \times 10^3 \text{ J}$

(c) $Q = I^2Rt$
     $Q = (6.0 \text{ A})^2(20 \text{ Ω})(10 \text{ s})$
     $Q = 7.2 \times 10^3 \text{ J}$

Thus, $W = Q$. This merely states that the electrical energy furnished to the heater is wholly converted into heat.

Test Yourself Problem

1. A heater on a 110-V line draws 10 A of current. How long will it take the heater to generate 2.2×10^5 J of heat?

2. How many joules of electrical energy are converted into heat in a device that has a resistance of 10 Ω and draws 6.0 A for 5.0 s?

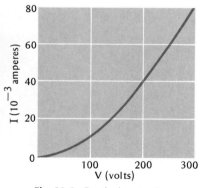

Fig. 23-8. Graph showing how the current *I* in a certain vacuum tube varies with the applied potential difference *V*. The resistance of the tube, R = V/I, is not constant and the tube does not follow Ohm's law.

23-17 Resistance of Conductors That Do Not Obey Ohm's Law

In general, only metal conductors have constant resistance at constant temperature and obey Ohm's law. Other conducting devices, such as electrolytic cells and gas discharge tubes, also offer resistance to the flow of current through them but their resistance is not constant. Instead, their resistance varies with changing difference of potential. Such devices therefore do not obey Ohm's law.

Fig. 23-8 shows how the current *I* passing through a certain radio vacuum tube changes as the applied voltage is increased. Since the graph is not a straight line, V/I is not constant. The resistance of the tube therefore varies as the voltage changes.

23-18 Energy Conversion in Electric Circuits

We have seen that in a metallic resistor, all the energy given up by a current of electrons passing through it is converted into Joule heat. This is not true of resistors that are not metals and do not obey Ohm's law. In an electrolytic cell, for example, only a small part of the electrical energy furnished to the cell is converted into heat. The major share of it is converted into chemical energy and stored in the cell. Again, in an operating electric motor, only some of the electrical energy is converted to heat while the remainder is converted into the kinetic energy of the rotating armature. In this form it is available to move objects and do mechanical work.

Regardless of what transformation is undergone by the electrical energy furnished to an electric circuit, the energy is conserved. The energy supplied by the battery or generator is equal to the sum of the energies transferred to the wires or devices in the circuit as heat, chemical energy, mechanical energy, or other forms of energy.

CHAPTER REVIEW

Summary

The electric circuit is a means of transferring electrical energy from a source such as a battery or generator to the devices in the circuit. In devices that are pure **resistors**, this energy is converted into heat. In other devices, it is converted partly into heat and partly into other energy forms such as chemical and mechanical energy. In all such conversions of electrical energy, the law of conservation of energy holds true: *No energy is created or destroyed.*

The energy supplied by a battery or generator per coulomb of charge that it drives through a circuit is called its **electromotive force** or **EMF.** The battery or generator transmits this energy to the charges moving in the circuit by setting up a **difference of potential** between its terminals. In turn, this sets up an electric field inside the conductors forming the circuit. The field then exerts the force that drives the electrons or electric charges through the circuit. EMF and difference of potential are both measured in **volts, V.**

The opposition offered to the flow of current by the conductors or devices in a circuit is called **resistance** and is measured in **ohms Ω.** The resistance R of a device is defined as the ratio of the difference of potential V applied to it divided by the current I passing through it. The resistance of metal conductors kept at constant temperature remains constant when varying differences of potential are applied to them. Such conductors, called **resistors,** obey **Ohm's law** which may be written $V/I = R = a$ *constant.* According to this law, the current flowing through a resistor is directly proportional to the difference of potential applied to it. The resistance of a metal wire depends upon the substance of which it is made and increases when its temperature is raised. It also varies directly with the length of the wire and inversely with its cross section.

When certain metals are cooled to within a few degrees of absolute zero, they lose all electrical resistance and become **superconductors.**

The resistance of many conductors and devices is not constant when varying differences of potential are applied to them. Such conductors do not obey Ohm's law.

In all electric circuits, the electric power P furnished to a device is given by $P = VI$. **Power** is expressed in **watts** or **joules per second.** The total **electrical energy** furnished to a device is given by $W = VIt$ and is expressed in **watt-seconds** or **joules.** The electrical energy furnished to a resistor R is entirely converted into heat. The number of joules of heat Q generated in a resistor R in t seconds by a current I is $Q = I^2Rt$.

Questions

Group 1

1. Consider a simple circuit consisting of a length of metal wire connected to a chemical battery. (a) What part does the battery play in maintaining a current in the circuit? (b) Why is the terminal voltage of the battery smaller than the total EMF supplied by the battery?

2. (a) Describe the motions of the electrons as they are pushed through the circuit of question 1. (b) What sets up the electric field inside the wire that acts upon the electrons? (c) How does the movement of the electrons through the wire generate heat?

3. (a) In which direction with respect to the positive and negative terminals of the battery do the electrons in the circuit of question 1 move? (b) Describe two kinds of currents that do not consist solely of electrons. (c) In what directions do they flow?

4. (a) Where does the energy that drives a current through an electric circuit come from? (b) In what units is electrical power measured? (c) In what units is electrical energy measured?

5. Two lamps rated 50 watts and 100 watts are designed for use at the same voltage. When both lamps are operating, compare (a) the current passing through each; (b) the rate at which electrical energy is expended in each.

6. The difference of potential applied to the ends of a fixed length of conducting wire is increased steadily. (a) If the wire is kept at constant temperature, what change takes place in the current flowing through it? (b) What ratio remains constant as the difference of potential increases? (c) What is this ratio called? (d) State Ohm's law.

7. (a) What metals have relatively low resistance? (b) For what purpose is it desirable to use such metals in electric circuits or devices? (c) Name a metal alloy that has a relatively high resistance. (d) For what purpose is it used?

8. (a) Compare the resistances of two iron wires of equal cross section, one of which is twice as long as the other. (b) Compare the resistances of two iron wires of equal length, one of which has twice the cross section of the other.

9. Explain in terms of the motions of the free electrons and the atoms inside a metal conductor why the resistance of the conductor rises as the temperature of the conductor increases.

10. (a) What happens to the electrical energy furnished by a battery to move a current through a resistor? (b) How did Joule demonstrate that the energy furnished to the resistor is conserved?

Group 2

11. (a) Explain the statement, "Only certain kinds of conductors follow Ohm's law." (b) Name some electrical devices that do not follow Ohm's law.

12. Into what other forms of energy is electrical energy transformed when a current is sent through (a) the resistor of an electric toaster; (b) an electrolytic cell; (c) a motor. (d) What is the relationship between the electrical energy put into an electric circuit and the various other forms of energy developed in the circuit?

Problems

Group 1

1. A current of 0.20 A is kept flowing in a wire by a 10-V battery for 5.0 s. (a) What power is supplied to the wire? (b) How much electrical energy has been delivered to the wire?

2. How much electrical energy is supplied to a 100-W lamp in 50 s?

3. A 60-W lamp is operating on a 120-V line. (a) How much current passes through the lamp? (b) How many joules of energy are furnished to the lamp per second? (c) How many electrons pass through the lamp in 4.0 s?

4. A 200-V generator is maintaining a steady current in an electric heater. (a) How much work is done on each coulomb of charge moved through the heater? (b) How much work in joules is done on each electron moved through the heater?

5. A 110-V motor draws 0.50 A of current. (a) What power is being supplied to the motor? Assuming

that the motor converts all the electrical energy supplied to it to mechanical work, how many joules of work are done by it in (b) 10 s; (c) 10 min?

6. What is the resistance of a metal conductor through which a 6.0-V battery sends a current of 0.20 A?

7. A difference of potential of 3.0 V is applied to a coil of tungsten wire having a resistance of 2.5 Ω. (a) What is the current in the coil? (b) If the temperature is kept constant while the applied voltage is raised to 5.0 V, what will be the new value of the current?

8. A current of 0.40 A is flowing in a length of nichrome wire of resistance 25 Ω. What difference of potential is being applied to the wire?

9. (a) How much electrical energy is supplied to a 500-W electric heater in 20 min?

10. A 110-V electric iron draws 3.0 A of current. (a) How much electrical energy is supplied to it per hour? (b) How much heat does it develop per hour?

11. A current of 1.2 A flows through a 50-Ω resistor for 5.0 min. How much heat is generated in the resistor?

Group 2

12. Suppose that, in the Joule apparatus of Fig. 23-6, $V = 100$ V and $R = 20$ Ω. (a) Assuming that R remains practically constant with increasing temperature, what will the ammeter read? (b) How much heat will be generated in R in

100 s? (c) Assume that the water in the tank absorbs practically all of this heat. If the water has a mass of 1.0 kg and an initial temperature of 20°C, by how many degrees will its temperature rise?

13. Suppose that in the Joule apparatus of Fig. 23-6 the ammeter reads 2.0 A when the voltage is 120 V. How much heat is generated in R per second?

14. In the V-I curve of Fig. 23-8, you can find the resistance of the tube at any voltage by dividing the voltage by the corresponding current. (a) Find the resistance of the tube when the applied voltage is 100, 200, and 300 V. (b) What happens to the resistance as the voltage rises?

15. The terminals of a battery having an EMF of 1.5 V are connected by a resistor of 7.0 Ω. A voltmeter shows that the actual difference of potential across the battery terminals is 1.4 V. (a) What part of the battery EMF is being used to move the current through the battery itself? (b) What current flows through the resistor? (c) At what rate in watts is heat developed in the resistor? (d) What is the total power in watts furnished to the circuit?

16. When 10 V is applied to a lamp, a current of 0.40 A flows through the filament. When 110 V is applied to the same lamp, 1.20 A flows through the filament. Compare (a) the resistances of the lamp; (b) the power consumed for each of the applied voltages.

Applying Physics

Measure the water-heating efficiency of an electric heater for making tea or coffee. Put a liter (1 kg) of water into the heater and take its temperature. Now heat the water and measure the time it takes to raise the temperature to 90°C. Compute the heat gained by the water from the relationship $Q = ms\Delta T$. (See Section 12-6.)

Now compute the electrical energy supplied to the heater by multiplying the wattage rating shown on the heater by the number of seconds it was in operation.

The efficiency of the heater in heating the water is the ratio of the heat gained by the water to the electrical energy furnished to the heater.

Series and Parallel Circuits

Aims

1. To learn how to connect electrical devices for most effective use of electrical energy.
2. To learn the characteristics of series connections in circuits.
3. To learn the characteristics of parallel connections in circuits.

24-1 Two Types of Connections

Electrical devices in a circuit may be connected in series or in parallel. Devices connected *in series* must be designed so that they operate on the same current. For example, in some trains the lights are connected in series of five similar lamps to an appropriate source of voltage. The same current passes through each of the lamps and causes them to be equally bright. However, all of the lamps go on or off together when the current is turned on or off.

Devices connected *in parallel* must be designed so that they operate on the same voltage. Most household appliances are designed to operate on the voltage that is available in the home and are connected in parallel. Each appliance can be plugged into any electrical outlet and can operate independently of any other appliance. Thus, a lamp in the kitchen may be turned on or off without affecting lamps or other devices operating anywhere else in the house. Each type of circuit has its own characteristics and uses. First we take up series connections.

24-2 Resistors in Series

Figs. 24-2 (a) and (b) show three resistors R_1, R_2, and R_3 connected in series with a generator. Electrons leave the negative terminal of the generator, pass through each of the resistors in turn, and then return to the positive terminal of the generator. This is the only pathway over which the electrons can travel through the circuit. A break anywhere in this pathway results in an open circuit and stops the flow of current.

This circuit illustrates the following major characteristics of series connections:

1. Since the electrons must follow the single pathway through the circuit, the current passing through R_1, R_2, R_3 and all other parts of the circuit is the same.

2. This common current is resisted by R_1, R_2, and R_3 in turn so that the total resistance R to the flow of current in the circuit is:

$$R = R_1 + R_2 + R_3$$

Fig. 24-1. The distribution of alternating current at high voltage is accomplished by series and parallel circuits.

(a)

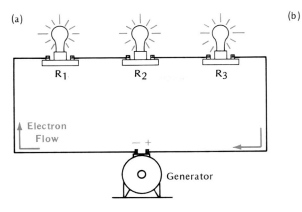

(b)

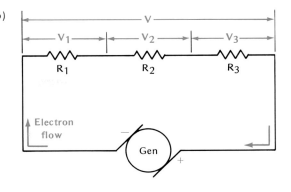

Fig. 24-2. (a) Three lamps connected in series with a generator. **(b)** Schematic drawing using conventional symbols.

3. The current I flowing in the circuit is:

$$I = \frac{V}{R}$$

where V is the difference of potential supplied by the generator to the whole circuit.

4. The total difference of potential applied to the circuit divides among the individual resistors so that the sum of the differences of potential across each of the individual resistors is equal to the total difference of potential across the circuit. That is, if V_1, V_2, and V_3 are the differences of potential across R_1, R_2, and R_3 respectively,

$$V = V_1 + V_2 + V_3$$

This relationship expresses the fact that the energy used to move each coulomb of charge through the entire circuit is simply the sum of the energies used to move that coulomb of charge through each of the resistors in turn.

5. Ohm's law applies to each resistor separately so that $V_1 = IR_1$, $V_2 = IR_2$, and $V_3 = IR_3$. Since I is the same for the whole circuit, the difference of potential across each resistor is proportional to its resistance. V_1, V_2, and V_3 are commonly called the *voltage drops* across their respective resistances.

Sample Problem

A 10-Ω resistor and a 20-Ω resistor connected in series have a 15-V difference of potential applied to them. (a) What is their combined resistance? (b) What current flows through the circuit? (c) What is the voltage drop across each resistor? (d) What power is supplied to each resistor? (e) What power is supplied to the circuit?

Solution:

(a) $R_1 = 10\ \Omega$ $R_2 = 20\ \Omega$ $V = 15$ V

$R = R_1 + R_2 = 10\ \Omega + 20\ \Omega = 30\ \Omega$

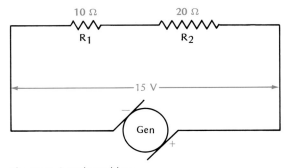

Fig. 24-3. Sample problem.

(b) $I = \dfrac{V}{R} = \dfrac{15 \text{ V}}{30 \text{ }\Omega} = 0.50 \text{ A}$

(c) The voltage drop across R_1 is:

$$V_1 = IR_1 = 0.50 \text{ A} \times 10 \text{ }\Omega = 5.0 \text{ V}$$

The voltage drop across R_2 is:

$$V_2 = IR_2 = 0.50 \text{ A} \times 20 \text{ }\Omega = 10 \text{ V}$$

Thus R_2 which is twice as large as R_1 has a difference of potential across it twice as great as that across R_1.

(d) The power supplied to R_1 is:

$$P = V_1I = 5.0 \text{ V} \times 0.50 \text{ A} = 2.5 \text{ W}$$

The power supplied to R_2 is:

$$P_2 = V_2I = 10 \text{ V} \times 0.50 \text{ A} = 5.0 \text{ W}$$

(e) The power supplied to the circuit is:

$$P = VI = 15 \text{ V} \times 0.50 \text{ A} = 7.5 \text{ W}$$

Note also that:

$$P = P_1 + P_2$$
$$7.5 \text{ W} = 2.5 \text{ W} + 5.0 \text{ W}$$

Test Yourself Problems

1. A 100-Ω resistor and a 20.0-Ω resistor connected in series have a difference of potential of 240 V applied to them. (a) What current flows through the circuit? (b) What is the voltage drop across each resistor? (c) What is the power supplied to each resistor? (d) What is the power supplied to the circuit?

2. A current of 0.50 A flows through a 10-Ω resistor and a 15-Ω resistor connected in series. What is the difference of potential across (a) Each of the resistors; (b) both resistors?

3. What resistance must be put in series with a 240-Ω lamp to reduce the current passing through it to 0.40 A when connected to a 120-V source?

24-3 Internal Resistance of a Cell

Consider the circuit of Fig. 24-4 in which a single chemical cell is connected to a lamp. The EMF of the cell supplies not only the energy that drives the current through the lamp but also the energy that drives the current through the cell itself. Thus, there are two separate resistances in series in this circuit. One is the resistance offered by the lamp and its connecting wires which make up the *external resistance* of the circuit. The other is the *internal resistance* which the cell itself offers to the flow of current through it.

In driving the current through the entire circuit, the EMF divides proportionally between the external and the internal resistance of the circuit as illustrated in the sample problem below. That part of the EMF that drives the current across the external resistance is called the terminal voltage. The remaining part of the EMF drives the current through the cell itself. It follows that the terminal voltage of an operating cell is always less than its EMF. Furthermore, the greater the internal resistance of a cell, the greater is the part of its EMF needed to drive current through it and the smaller will be the terminal voltage available to the external circuit.

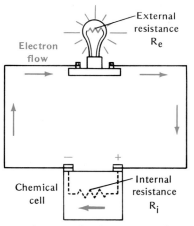

Fig. 24-4. The electrons must be driven through the internal resistance of the cell as well as through the external circuit.

Sample Problem

A battery having an EMF of 6.0 V and an internal resistance of 0.2 Ω is connected to a 2.8-Ω lamp. (a) What is the total resistance of the circuit? (b) What is the current? (c) What is the terminal voltage of the battery? (d) What part of the EMF drives the current inside the battery?

Solution:

(a) The internal resistance, $R_i = 0.2 \ \Omega$
The external resistance, $R_e = 2.8 \ \Omega$

$$R = R_e + R_i = 2.8 + 0.2 = 3.0 \ \Omega$$

(b) EMF = 6.0 V

$$I = \frac{V}{R} = \frac{6.0 \ V}{3.0 \ \Omega} = 2.0 \ A$$

(c) The terminal voltage V_t is simply the voltage drop across the external resistance R_e.

$$V_t = IR_e = 2.0 \ A \times 2.8 \ \Omega = 5.6 \ V$$

(d) The part of the EMF used inside the battery is the voltage drop inside the battery.

$$V_b = IR_i = 2.0 \ A \times 0.2 \ \Omega = 0.4 \ V$$

Another way to obtain V_b is to note that it must be the difference between the battery EMF and its terminal voltage; that is,

$$6.0 \ V - 5.6 \ V = 0.4 \ V$$

Test Yourself Problem

A battery having an EMF of 1.50 V and an internal resistance of 0.05 Ω is connected to a 5.95-Ω lamp. (a) What is the total resistance of the circuit? (b) What is the current? (c) What is the terminal voltage?

(a)

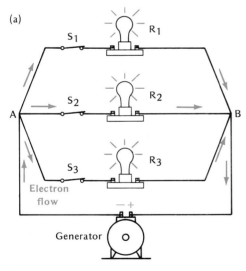

(b)

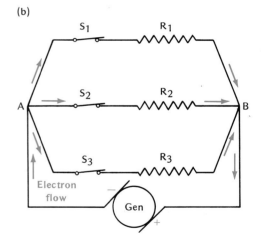

Fig. 24-5. (a) Three lamps, each with its own switch, connected in parallel with a generator. **(b)** Schematic diagram using conventional symbols.

24-4 Resistors in Parallel

In a series circuit, all the devices must be operating at the same time, for if any one device is turned off, the circuit is broken and all current stops flowing. In the home and factory, there are circuits that make it possible to operate any device independently of any other device. For example, in the home, a lamp can be turned on or off without affecting the operation of any other lamp or electrical device. Parallel circuits make such independent operation possible. Resistors are said to be connected in parallel when they connect the same two points of an electric circuit.

Figs. 24-5 (a) and (b) show a generator sending a current through three resistors in parallel connection. Here R_1, R_2, and R_3 connect the points A and B of a circuit. When electrons leaving the negative terminal of the generator arrive at A, they separate into three different streams. If R_1 is the smallest of the three resistors, the largest number of electrons pass through it to B. If R_3 is the largest of the three resistors, the smallest number of electrons pass through it to B. An intermediate number of electrons pass to B through R_2. At B, the electrons leaving R_1, R_2, and R_3 come together again and return along the common lead to the positive terminal of the generator.

Note that alongside of each resistor is a switch that controls the flow of current through it. If S_1 is opened, current stops flowing through R_1 but continues to flow through R_2 and R_3 as before. Similarly, the current in each of the resistors can be turned on or off without affecting the others.

24-5 Characteristics of Parallel Circuits

In a parallel circuit, each device can operate independently of any other device in parallel with it.

The general characteristics of parallel connections may be noted by referring to the circuit of Fig. 24-5 (b).

1. Since the resistors connect the same two points A and B of the circuit, *they have the same voltage or difference of potential across them.*

2. If V is the difference of potential between A and B, it follows from Ohm's law that the current flowing in R_1 is $I_1 = V/R_1$; the current in R_2 is $I_2 = V/R_2$; and the current in R_3 is $I_3 = V/R_3$. These relationships show that *in parallel connection, the smaller the resistor the more current it carries.*

3. Since no electrons are gained or lost in flowing through the circuit, the number of electrons leaving A in a given time must be equal to the number of electrons arriving at B. This means that I, the total current entering or leaving resistors connected in parallel, is equal to the sum of the currents flowing in all the resistors, or:

$$\mathbf{I = I_1 + I_2 + I_3}$$

4. The combined resistance of resistors in parallel to the flow of current between the two points of the circuit that they connect is less than the resistance of any one of them. If, for example, all the resistors were equal, the total resistance encountered by electrons in going from A to B across all three resistors would be only one-third as great as it would be if all the electrons had to cross from A to B on only a single one of the resistors.

The combined resistance R of the three resistors in parallel in Fig. 24-5 (b) may be found from the values of R_1, R_2, and R_3 as follows. From Ohm's law, the total current flowing through all three resistors is $I = V/R$. That part of the total current flowing through R_1 is $I_1 = V/R_1$; that flowing in R_2 is $I_2 = V/R_2$; and that flowing in R_3 is $I_3 = V/R_3$. Substituting the values for I, I_1, I_2, and

I_3 in the relationship $I = I_1 + I_2 + I_3$, gives $V/R = V/R_1 + V/R_2 + V/R_3$. On cancelling V, we obtain

$$\frac{1}{R} = \frac{1}{R_1} + \frac{1}{R_2} + \frac{1}{R_3}$$

Although the above relationships are stated for three parallel resistors, they are quite general and can be extended to any number of parallel resistors.

Sample Problems

1. A heater, a toaster, and an iron are connected in parallel to a generator furnishing a potential difference of 120 V. The resistance of the heater is 12 Ω, that of the toaster is 20 Ω, and that of the iron is 24 Ω. Find (a) the current flowing in each device; (b) the total current leaving the generator; (c) the combined resistance of all three devices. (See Fig. 24-6.)

Solution:

(a) $V = 120$ V

heater $= R_1 = 12$ Ω
toaster $= R_2 = 20$ Ω
iron $= R_3 = 24$ Ω

Heater Current $= I_1 = \dfrac{V}{R_1} = \dfrac{120\ \text{V}}{12\ \Omega} = 10$ A

Toaster Current $= I_2 = \dfrac{V}{R_2} = \dfrac{120\ \text{V}}{20\ \Omega} = 6.0$ A

Iron Current $= I_3 = \dfrac{V}{R_3} = \dfrac{120\ \text{V}}{24\ \Omega} = 5.0$ A

(b) Total current $I = I_1 + I_2 + I_3$
$I = 10$ A $+ 6.0$ A $+ 5.0$ A $= 21$ A

(c) $\quad R = \dfrac{V}{I} = \dfrac{120\ \text{V}}{21\ \text{A}} = 5.7$ Ω

R may also be obtained from the relationship

$$\frac{1}{R} = \frac{1}{R_1} + \frac{1}{R_2} + \frac{1}{R_3}$$
$$\frac{1}{R} = \frac{1}{12} + \frac{1}{20} + \frac{1}{24}$$
$$\frac{1}{R} = \frac{10}{120} + \frac{6}{120} + \frac{5}{120} = \frac{21}{120}$$
$$R = \frac{120}{21} = 5.7\ \Omega$$

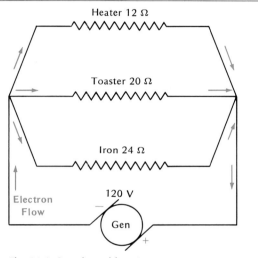

Fig. 24-6. Sample problem 1.

2. Four 18-W lamps are connected in parallel to a 6-V battery. Find (a) the current in each lamp; (b) the resistance of one lamp; (c) the total resistance of the four lamps.

Solution:

(a) For each lamp: $P = 18$ W and $V = 6$ V

From $P = VI$, $I = \dfrac{P}{V} = \dfrac{18\ \text{W}}{6\ \text{V}} = 3$ A

(b) For each lamp: $R = \dfrac{V}{I} = \dfrac{6\ \text{V}}{3\ \text{A}} = 2$ Ω

(c) Since the lamps all have equal resistances, the combined resistance of four of them in parallel is ¼ that of one of them, or ¼ of 2 Ω = 0.5 Ω.

Test Yourself Problems

1. A 240-W lamp and a 12-Ω heater are connected in parallel with a 120-V household supply. (a) What current flows through each? (b) What is the total current furnished by the household supply? (c) What is the combined resistance of the lamp and the heater?

2. Ten 100-W lamps are connected in parallel to a 120-V generator. (a) What current flows through each lamp? (b) What is the resistance of each lamp? (c) What is their combined resistance?

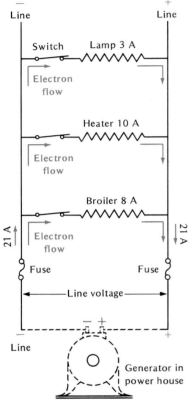

Fig. 24-7. Power supply lines showing parallel connections of appliances, series connections of fuses, and electron flow.

24-6 Parallel Circuits and Overloading

Electricity in the home is usually supplied by two lead wires called *lines* that enter each room. A voltage of about 110 to 120 volts is maintained across these lines by the generators in the power company. Household electrical appliances are supplied with current by connecting one end of each appliance to each of the lines as in Fig. 24-7. They are then connected in parallel. Here, a broiler, a heater, and a lamp, each having its own switch, are connected in this way. When only the broiler's switch is turned on, the line supplies 8 amperes to it. When the heater's switch is also turned on, the line supplies 18 amperes, of which 8 amperes go to the broiler and 10 amperes go to the heater. When the lamp is also turned on, the line now supplies a total of 21 amperes—8 amperes to the broiler, 10 amperes to the heater, and 3 amperes to the lamp. Thus, as additional appliances are connected and turned on in this circuit, the line current becomes dangerously high, which overheats the line and may start a fire. At this point the line is said to be *overloaded*.

24-7 Fuses and Circuit Breakers

To protect circuits or parts of circuits against overloading, *fuses* are inserted in them as shown in Fig. 24-7. A fuse is simply a piece of metal wire or ribbon that has a low melting point. It is generally enclosed in a fireproof case and is connected in series with the line it protects. Thus the entire line current flows through the fuse. When the current becomes larger than the safe value, both line and fuse become hot. The fuse then melts or "blows out" and breaks the circuit. The fuse should not be replaced until the cause of the overload has been determined and corrected.

Fuses sometimes blow out on home circuits because of the development of a *short circuit*. This condition occurs when the two lead lines are accidentally brought into direct contact with each other or are connected by a conductor of very low resistance. In either case, a short circuit of extremely low resistance is formed and a very large current surges through the lines and blows the fuses protecting them.

In many buildings, *circuit breakers* are used instead of fuses to protect electrical circuits. A circuit breaker is essentially an electromagnet (see Section 25-19). When line current flows through it, a magnetic field is produced. The strength of the field is proportional to the size of the current flowing through the circuit breaker. When the current becomes too large, the magnetic field becomes strong enough to pull open a switch and thus turn off the current.

Fig. 24-8. Two types of fuses in common use in homes.

24-8 Resistors in Series-Parallel Combinations

Most circuits include both series and parallel connections. In determining how currents or differences of potential are distributed in the different parts of these circuits, Ohm's law is applied to each of the parts of the circuit as shown in the problem below.

Sample Problem

Consider the circuit of Fig. 24-9 in which the battery supplies a difference of potential of 48 V. (*a*) What is the total resistance of the circuit? (*b*) What is the current flowing through the 36-Ω resistor? (*c*) What is the difference of potential across the 48-Ω resistor? (*d*) What current flows through the 16-Ω resistor? (*e*) What current flows through the 48-Ω resistor?

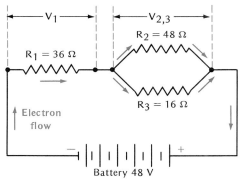

Fig. 24-9. Sample problem.

Solution:

(*a*) $R_1 = 36\ \Omega$ $R_2 = 48\ \Omega$ $R_3 = 16\ \Omega$

Since R_2 and R_3 are in parallel, their combined resistance $R_{2,3}$ is found from $\dfrac{1}{R_{2,3}} = \dfrac{1}{R_2} + \dfrac{1}{R_3}$. Thus

$$R_{2,3} = \frac{R_2 R_3}{R_2 + R_3} = \frac{48\Omega \times 16\Omega}{48\Omega + 16\Omega} = 12\Omega$$

The total resistance of the circuit is:

$$R = R_1 + R_{2,3} = 36\ \Omega + 12\ \Omega = 48\ \Omega$$

(*b*) The current I_1 flowing through the 36-Ω resistor is equal to the total current I leaving the battery. By Ohm's law, this is:

$$I_1 = I = \frac{V}{R} = \frac{48\ \text{V}}{48\ \Omega} = 1.0\ \text{A}$$

(*c*) Let V_1 be the difference of potential across R_1 and $V_{2,3}$ be the difference of potential across both R_2 and R_3.

$$V_1 = IR_1 = 1.0\ \text{A} \times 36\ \Omega = 36\ \text{V}$$

Since 36 of the 48 volts applied to the circuit are used to drive current across the 36-Ω resistor, $48 - 36$, or 12 volts, remain to drive the current across the parallel-connected 48-Ω and 16-Ω resistors. Hence $V_{2,3} = 12$ V

(*d*) The current flowing in the 16-Ω resistor is:

$$I_3' = \frac{V_{2,3}}{R_3} = \frac{12\ \text{V}}{16\ \Omega} = 0.75\ \text{A}$$

(*e*) The current flowing through the 48-Ω resistor is:

$$I_2 = \frac{V_{2,3}}{R_2} = \frac{12\ \text{V}}{48\ \Omega} = 0.25\ \text{A}$$

Note that $I_1 = I_2 + I_3$. The current of 1.0 A leaving R_1 divides into one branch of 0.25 A that passes through R_2 and a second branch of 0.75 A that passes through R_3.

Test Yourself Problem

A 75-Ω resistor is connected in series with three resistors of 120 Ω, 150 Ω, 200 Ω respectively that are connected in parallel with each other. The network is connected to a 100-V generator. (*a*) What is the combined resistance of the three resistors that are connected in parallel? (*b*) What is the total resistance of the circuit? (*c*) How much current is furnished by the generator? (*d*) What is the voltage across the 75-Ω resistor? (*e*) What is the voltage across the 150-Ω resistor? (*f*) What is the current in the 150-Ω resistor?

24-9 Kirchhoff's First Law

All the relationships that we have used in series and parallel circuits are included in two general laws discovered by the German physicist, G. R. Kirchhoff. These laws apply to any electric circuit in which a steady current is flowing. They are especially useful in analyzing complex circuits containing many series and parallel connections.

Kirchhoff's first law states: *The sum of the currents flowing into any point that joins three or more conductors in a circuit is equal to the sum of the currents leaving that point.* This law follows from the principle of conservation of electric charge. In effect, it asserts that none of the electric charges flowing through a circuit are lost or destroyed.

As an example, consider the point in Fig. 24-9 at which R_1 joins R_2 and R_3. The current I flowing into this point from R_1 is one ampere. The currents I_2 and I_3 which flow out of this common point are 0.25 ampere and 0.75 ampere respectively, making a total of one ampere. Thus, the current flowing into the common point is equal to the sum of the currents flowing out of the common point.

24-10 Kirchhoff's Second Law

The second law applies to any closed path made by the conductors in a circuit. Such a path is called a closed loop. The law states: *In any closed loop of a circuit, the algebraic sum of the EMF's is equal to the algebraic sum of all the differences of potential around the loop.* This law follows from the law of conservation of energy. It expresses the fact that the total energy supplied by the sources of EMF to each charge flowing in a loop is equal to the total energy given up by each charge as it moves around the loop.

Before applying Kirchhoff's second law to a loop, we must give each EMF and difference of potential in it a positive or negative sign. To do this, let us imagine transporting a test electron around the loop in a single direction. Differences of potential across resistors and EMF's of batteries or generators in the loop are given a positive sign if the test electron moves through them in the same direction as the current. They are given a negative sign if the test electron moves through them in the direction opposite to that of the current.

To illustrate, consider the closed loop in the sample problem of Fig. 24-9 connecting the battery, R_1, and R_3. Taking a test electron around this loop in the same direction as the current (clockwise), we find that the total EMF is that of the battery. This is 48 volts and positive. The differences of potential across R_1 and R_3 are 36 volts and 12 volts respectively. They are also positive and add up to a sum of 48 volts. Thus, the total EMF is equal to the sum of the differences of potential in the loop, in agreement with Kirchhoff's second law.

As another example, consider the closed loop formed by R_2 and R_3. Take a test electron over R_2 in the direction of the current and back to the starting point over R_3. The difference of potential across R_2 is 12 volts and positive. That across R_3 is also 12 volts but negative because the test electron is moving opposite to the current. The algebraic sum of the differences of potential in the loop is therefore 12 volts − 12 volts = 0. Since there is no source of EMF in this loop, the EMF is also zero thus satisfying Kirchhoff's second law.

24-11 Connecting Ammeters and Voltmeters

Currents flowing in circuits are generally measured by *ammeters*. Electromotive forces and differences of potential are generally measured by *voltmeters*. The mechanisms of one type of ammeter and of one type of voltmeter are described in Sections 26-9 and 26-10. Here, we shall discuss only the method of connecting these instruments in a circuit.

An ammeter is connected in series with the device in which the current is to be measured. This connection assures that the entire current that passes through the device also goes through the ammeter. The insertion of an ammeter into a circuit must not change the resistance of the circuit, for then it would also change the current it is trying to measure. For this reason, ammeters are designed to have *negligible resistance* compared to that of the circuit or of the device whose current they measure.

A voltmeter is connected in parallel with the device across which the voltage or difference of potential is to be measured. In order that it may not change the current or voltage of the circuit into which it is introduced, the voltmeter must have a *very high resistance* compared to that of the device across which it is connected. The combined resistance of the voltmeter and the device will then be practically the same as that of the device by itself. The introduction of the voltmeter will therefore have a negligible effect on the flow of current or on the difference of potential between the two points of the circuit which it connects.

An ammeter is connected in series with a device. A voltmeter is connected in parallel with it.

24-12 Measuring Resistance and Power

A voltmeter and an ammeter can be used to measure the resistance of a device such as a lamp, as shown in Fig. 24-10. The voltmeter is connected in parallel with the device and reads the potential difference across it. The ammeter is connected in series with the device and measures the current passing through it. The meters are read and the resistance of the device is obtained from Ohm's law by dividing the voltmeter reading V by the ammeter reading I; that

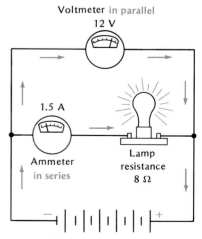

Fig. 24-10. Connections for measuring the resistance of a lamp with a voltmeter and an ammeter.

is, $R = V/I$. Thus, if the voltmeter in Fig. 24-10 reads 12 volts and the ammeter reads 1.5 amperes, the resistance of the lamp is 12 volts ÷ 1.5 amperes = 8.0 ohms.

The readings obtained from the voltmeter and ammeter can also be used to determine the electric power furnished to the conductor or device. Since the electric power $P = VI$, we need only multiply the voltmeter reading in volts by the ammeter reading in amperes to obtain the power in watts. Thus the power furnished to the lamp is 12 volts × 1.5 amperes = 18 watts.

24-13 Measuring EMF and Terminal Voltage of Cells

A voltmeter is used to measure both the EMF of a cell or battery of cells and its terminal voltage. For both measurements, the terminals of the voltmeter are connected to the terminals of the cell. When the cell is on open circuit and not driving a current through an external resistor or other device, the voltmeter reads its EMF. When the cell is on a closed circuit and is driving current through an external resistor or other device, the voltmeter reads its terminal voltage. This is smaller than its EMF by an amount that depends upon the size of the internal resistance of the cell.

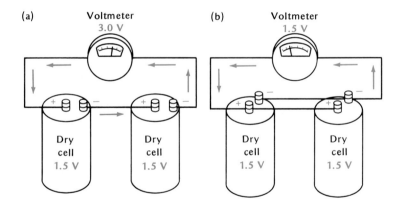

Fig. 24-11. (a) Two cells in series supply twice the EMF of one cell, but twice the internal resistance. **(b)** Two cells in parallel supply the same EMF as one cell, but half the internal resistance.

24-14 Connecting Cells in Series and Parallel

In a flashlight there are often two dry cells. In an automobile storage battery there are usually three, six, or twelve cells. Like all electrical devices, cells may be connected in series or in parallel.

Cells are connected in series as shown in Fig. 24-11 (a) by arranging them in a line so that the positive terminal of one cell is connected to the negative terminal of the next cell in line. The total internal resistance of a series of cells is the sum of the internal resistances of the individual cells. However, the advantage of connecting cells in series is that their EMF's add up. Thus, the EMF of a single dry cell is 1.5 volts. The total EMF of a battery of two dry cells connected in series is 1.5 + 1.5 or 3.0 volts. By connecting increasing numbers of cells in series, it is possible to obtain higher and higher EMF's.

Cells are connected in parallel as shown in Fig. 24-11 (b) by joining all their positive terminals together and joining all their negative terminals together. Only cells having equal EMF's should be connected in this way. The total EMF of a battery of cells connected in parallel is the same as the EMF of one of them. However, the advantage of connecting cells in parallel is that their combined internal resistance is only a fraction of the internal resistance of one of them. For example, when two cells are connected in parallel, their combined resistance is only half the resistance of one of them. This decrease in their total internal resistance makes it possible for the two cells to send a larger current through an external circuit than one of them could send. Cells are usually connected in parallel when they are connected to an external circuit having a very low resistance. In this case, the resistance of the battery is an important part of the total resistance of the circuit and it is desirable to keep it as low as possible.

CHAPTER REVIEW

Summary

Electrical devices may be connected together in two major ways: series and parallel. A circuit consisting of devices connected **in series** has the following characteristics: (1) The current in all parts of the circuit is the same. (2) The total resistance of the circuit is the sum of the individual resistances of each of the devices in the circuit. (3) The difference of potential applied to the circuit divides among the devices so that the voltage drop across each device is proportional to its resistance.

A circuit consisting of devices connected together **in parallel** has these characteristics: (1) All devices joining the same two points of the circuit have the same difference of potential across them. (2) The total current divides among the parallel branches so that the current in each branch is inversely proportional to its resistance. (3) The total current flowing in the circuit is equal to the sum of the currents flowing in its parallel branches. (4) The combined resistance of all the devices connected in parallel is found from the relationship:

$$\frac{1}{R} = \frac{1}{R_1} + \frac{1}{R_2} + \frac{1}{R_3} + \text{etc.}$$

All circuits obey **Kirchhoff's** two general laws. The **first law** states: *The sum of the currents flowing into any point of a circuit is equal to the sum of the currents flowing out of that point.* **Kirchhoff's second law** states: *The algebraic sum of the EMF's in a closed loop of a circuit is equal to the algebraic sum of the differences of potential around the loop.*

Ammeters are connected in series with the devices whose current they are measuring. **Voltmeters** are connected in parallel with the devices across which they are measuring the difference of potential. The resistance

R of a device may be measured by measuring the voltage drop V across it and the current I passing through it from the relationship:

$$R = \frac{V}{I}$$

The power furnished may also be obtained from the relationship:

$$P = VI$$

Chemical cells may be connected in series or in parallel. For cells connected in series, the total EMF is equal to the sum of the individual EMF's of the individual cells. Also, the total **internal resistance** of the cells is equal to the sum of their individual internal resistances.

For cells connected in parallel, the total EMF is equal to that of a single cell. However, the total internal resistance of the cells is a fraction of the internal resistance of each cell. If the cells have equal resistances, the fraction is 1 divided by the number of cells.

Questions

Group 1

1. A 200-Ω lamp and a 100-Ω lamp are connected in series to a generator. Each lamp is in series with its own switch. (a) Explain what happens to the current in each lamp when either of the switches is open. (b) What is the combined resistance of the lamps?

2. If both switches in question 1 are closed, compare (a) the currents flowing through the two lamps; (b) the voltage drops across each of the lamps; (c) the electric power expended in each of the lamps; (d) the rate at which heat is generated in each of the lamps.

3. A cell is operating a flashlight bulb. (a) What is meant by the internal resistance of the cell? (b) Distinguish between the EMF of the cell and its terminal voltage. (c) Why is the terminal voltage of the operating cell smaller than its EMF?

4. (a) If the dry cell of question 3 is replaced by a second cell having the same EMF but having a smaller internal resistance, will its terminal voltage be greater or smaller than that of the first cell? (b) Assuming that the resistance of the bulb remains the same, will the current through it be greater or smaller when it is connected to the second cell than it was when it was connected to the first cell? Explain.

5. Three lamps connected in parallel to a generator have resistances of 10, 20, and 30 ohms respectively. Each lamp has its own switch in series with it. Explain what happens to the flow of current in each lamp when (a) only the switch of the 10-Ω lamp is on; (b) only the switches of the 10-Ω and the 20-Ω lamps are on; (c) all the switches are on.

6. Compare the current flowing in the 10-Ω lamp of question 5 (a) with the currents flowing in this same lamp in 5 (b) and 5 (c).

7. Referring to question 5 (c), compare (a) the differences of potential across each lamp; (b) the current in the 10-Ω lamp with that in the 20-Ω lamp and that in the 30-Ω lamp; (c) the power supplied to the 10-Ω lamp with that supplied to each of the other two lamps. (d) What is the relationship between the total current leaving the generator and the currents passing through the three lamps?

8. (a) Explain why the main lines supplying current to many devices in parallel may become overloaded as more devices are switched into the circuit. (b) Why is overloading dangerous?

9. (a) What is a fuse? (b) How does it protect a circuit or device to which it is connected?

10. (a) How is an ammeter connected to a device whose current it is to measure? (b) Why should an ammeter have a very low resistance?

11. (a) How is a voltmeter connected to measure the difference of potential across a device in a circuit? (b) Why should a voltmeter have a very high resistance?

12. How are an ammeter and a voltmeter used to measure (a) the resistance of an electrical device; (b) the power expended in operating the device?

13. Two cells of equal EMF and equal internal resistance are connected in series. (a) What is their total EMF? (b) What is their combined internal resistance?
14. If the two cells of question 13 are now connected in parallel, what is (a) their combined EMF and (b) their combined internal resistance?
15. A battery is operating an electric heater equipped with its own switch. Explain how a voltmeter can be used to measure (a) the EMF of the battery; (b) the terminal voltage of the battery.

Group 2
16. If you have a supply of 2-Ω resistors, show by diagrams how two or more of these resistors may be connected to give a total resistance of (a) 0.5 Ω; (b) 1 Ω; (c) 3 Ω; (d) 5 Ω.
17. Several lamps and a voltmeter are connected in parallel with a battery. Each lamp has its own switch. You observe that as more lamps are turned on in the circuit, the reading of the voltmeter decreases. Explain.
18. As a battery gets older, its internal resistance increases. (a) What effect does this have on the maximum current the battery can furnish? (b) What effect does this have on its terminal voltage?
19. A lamp is operated by a battery. A boy makes the mistake of connecting a voltmeter in series with the lamp instead of in parallel with it. (a) What effect does this have on the brightness of the lamp? (b) To what voltage does the reading of the voltmeter correspond?
20. Show by diagrams five different series, parallel, or series-parallel combinations in which three different resistors R_1, R_2, and R_3 can be connected in a circuit.

Problems

Group 1
1. Three resistors of 5.00 Ω, 7.00 Ω, and 12.0 Ω are connected in series to a 12.0-V battery of negligible internal resistance. (a) What is the total resistance of the circuit? (b) What current flows through the circuit? (c) What is the voltage drop across each of the resistors? (d) What power is used to drive current through each resistor?
2. Five similar lamps are connected in series to a source providing a difference of potential of 550 V. (a) What is the difference of potential across each lamp? (b) If the current flowing in the circuit is 1.10 A, what is the resistance of each lamp? (c) What is the resistance of all five lamps?
3. A current of 0.25 A flows through two resistors of 12 Ω and 16 Ω which are connected in series with a battery. (a) What is the combined resistance of the two resistors? (b) What is the difference of potential across each resistor? (c) What difference of potential is provided by the battery?
4. (a) What electric power is supplied to each of the resistors of problem 3? (b) What is the total power supplied by the battery?
5. A cell with an EMF of 1.50 V and an internal resistance of 0.20 Ω is connected to a flashlight bulb having a resistance of 1.80 Ω. (a) What is the total resistance of the circuit? (b) What is the current? (c) What is the terminal voltage of the cell? (d) What part of the EMF drives the current through the cell?
6. A variable resistor is connected in series with a 20-Ω heater on a 220-V line and is adjusted until an ammeter shows that exactly 5.0 A flow through the heater. (a) For this setting of the resistor, what is the combined resistance of the resistor and the heater? (b) What is the resistance corresponding to this setting of the resistor? (c) For what value of the variable resistor will the heater current be reduced to 4.0 A?
7. A 50.0-Ω heater and a 250-Ω lamp are connected in parallel to a generator supplying a difference of potential of 125 V. (a) What current flows through each device? (b) What current leaves the generator? (c) What is the combined resistance of the heater and the lamp?
8. Ten 110-Ω lamps are connected in parallel to a 110-V house line. (a) What is the combined resistance of the ten lamps? (b) What current flows in the line when all ten lamps are operating? (c) What power is then being supplied to the circuit by the line?
9. Five 60.0-W electric light bulbs are connected in parallel to a 120-V line. Find (a) the current in each bulb; (b) the resistance of each bulb; (c) the total resistance of the five bulbs.
10. A toaster and a heater are connected in parallel to a 110-V household line. The current in the toaster is 2.0 A and that in the heater is 5.0 A. Find (a) the resistance of the toaster; (b) the resistance of the heater; (c) the line current.

11. Compare the total resistance of a 5-Ω resistor and a 20-Ω resistor connected in parallel with their total resistance connected in series.

12. Five 440-W heaters are connected to a 110-V line. Each heater can be turned on or off independently. (a) What is the increase in the line current as each heater is turned on? (b) If the fuses inserted in the line blow out when the line current exceeds 15 amps, how many heaters can be operated at the same time?

13. An 80-Ω resistor and a 20-Ω resistor are connected in parallel. This combination is then connected in series with a 32-Ω resistor and a 24-V battery. (a) What is the total resistance of the circuit? (b) What current flows through the 32-Ω resistor? (c) What is the difference of potential across the 32-Ω resistor? (d) Across the 80-Ω resistor?

14. Two 50-Ω resistors are connected in parallel and the combination is then connected in series with a similar parallel combination of two 50-Ω resistors. (a) What is the total resistance of the four resistors? (b) If this arrangement of the resistors draws a total current of 2.0 A from a battery, what is the current in each of the resistors?

15. Each of four cells has an EMF of 2.0 V and an internal resistance of 0.050 Ω. Find the EMF of a battery of these cells when connected (a) in series and (b) in parallel. What is the internal resistance of a battery of these cells when connected (c) in series and (d) in parallel?

16. The four cells of problem 15 are connected in series to a motor having a resistance of 5.8 Ω. (a) What is the total resistance of the circuit? (b) What current flows through the motor?

Group 2

17. Two cells, each of EMF 1.5 V and of internal resistance 0.20 Ω, are connected in parallel. The cells are used to send a current through a wire whose resistance is 0.90 Ω. (a) What is the combined EMF of the two cells? (b) What is their internal resistance? (c) What is the total resistance? (d) What current flows through the wire?

18. A voltmeter reads 90 V when connected across the terminals of a battery on open circuit. When the battery is then used to operate several electrical devices, the voltmeter reads 88 V. (a) What is the EMF of the battery? (b) What part of the EMF is used to drive the current through the battery itself? (c) If the current passing through the battery is 5.0 A, what is its internal resistance?

19. Four dry cells, each having an EMF of 1.5 V and an internal resistance of 0.40 Ω are to be used to pass a current through a 0.40-Ω resistor. The cells may all be connected in series or in parallel. They may also be connected in pairs in a series-parallel combination. For which of these three connections of the cells will the current through the resistor be largest?

20. Referring to Fig. 24-12, the ammeter reads 3.00 A. Find (a) the current in R_1; (b) the current in R_2. (c) What is the difference of potential furnished by the generator to the circuit?

21. Three wires in a circuit meet in a common point. The currents flowing in two of them are 2 A and 3 A respectively. What are the possible values of the current in the third wire? Explain your answer.

22. A loop in a circuit consists of three resistors that form a triangle ABC. The difference of potential across AB is 10 V and that across BC is 0 V. If the resistance of AC is 4.0 Ω, what is the current in AC?

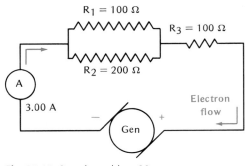

Fig. 24-12. Sample problem 20.

Applying Physics

If your home has circuit breakers, make a diagram of the electrical wiring. Show the incoming lines and the various outlets. Determine the position of each circuit breaker by switching one at a time and noting which parts of the system are then without power.

Note the ampere rating of the circuit breakers for the kitchen. Is there any combination of electrical appliances, which, if turned on simultaneously, would draw enough current to trip the circuit breakers?

"High-Temperature" Superconductors

Over 80 years ago the Dutch physicist Heike Kamerlingh Onnes discovered that mercury, tin, and lead lose all electrical resistance when cooled to a temperature of 4 K. At the time, scientists found this phenomenon, called *superconductivity*, fascinating but of little practical use. The liquid helium needed to produce the extremely cold temperature required is expensive and difficult to work with. In addition, superconducting tin and lead cannot carry large enough currents.

However, all of this was to change. In 1986, Johannes Bednorz and Karl Müller discovered a ceramic material that became superconducting above 30 K. A year later Ching-Wu Chu reported superconductivity at 95 K, a temperature well above that of liquid nitrogen. This was a significant breakthrough, because the cost of liquid nitrogen is only about 2 percent that of a comparable volume of liquid helium. Since Chu's discovery, materials have been found that are superconductive at 125 K. Of course the ultimate goal of the ongoing research is to find a material that is a superconductor at room temperature.

In the meantime, researchers are contending with the inability of many of the so-called "high-temperature" superconductors to remain superconductive in a strong magnetic field. This property could be a major obstacle to the development of large-scale applications such as motors and power transmission, which either require or create magnetic fields. In addition, it must be determined whether the new superconducting materials will be strong and flexible enough to be made into wire or other useful forms. If the cost of the materials makes it practical, superconductors could have a tremendous impact on technology.

There are many possible applications of superconductors. They could be used to construct the windings of electric motors and generators. Equipment built with the superconductors would be lighter, more powerful, and more efficient than conventional machines. Computers could be made more compact, faster, and more efficient by using superconducting materials. Using superconductors in place of conventional wire would eliminate the need to provide internal space in the computer for the cooling components used to control the heat produced by resistance in conventional wiring.

The world's first ship to be powered by superconductivity

Avoiding Mid-Air Collisions

Air travel is by far the safest method of mass transportation. Indeed, mid-air collisions are quite rare. Yet, when one does occur there is usually great loss of life. For this reason, and also because the skies are growing more crowded, electronics engineers have been developing a new type of airborne radar that will help avoid mid-air collisions.

Current air traffic control systems (see "Applications and Careers: Force and Motion" after Chapter 10) determine a plane's course and altitude from ground stations and display this information on the air traffic controller's radar screen. The new system displays a nearby plane's location and altitude on an aircraft's color weather radar screen. Thus, instant warning of a possible collision is available to the pilot of an aircraft in crowded airspace.

The new system, called Traffic Alert and Collision Avoidance System (TCAS), uses a radar transponder. In a transponder, an incoming signal activates transmission of another signal. TCAS spots all aircraft within the vicinity and determines each plane's distance and altitude. TCAS can track several planes at a time.

Suppose a small plane is climbing and is on a collision course with a descending airliner. At 60 seconds before collision, the small plane is 1400 feet below the airliner and climbing. This distance might be shown as $-14 \uparrow$ on the radar screen of the airliner just outside a circle that marks a horizontal distance of two miles from the airliner.

Look at the diagram below. At 40 seconds

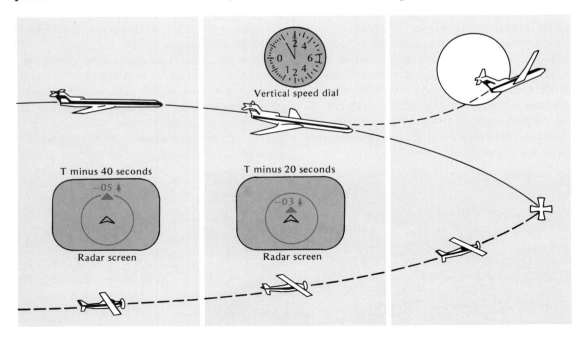

Vertical speed dial

T minus 40 seconds

−05 ↟

Radar screen

T minus 20 seconds

−03 ↟

Radar screen

before collision, the small plane is 500 feet below the airliner ($-05\uparrow$) and less than two miles ahead. This might be shown on the screen in yellow and an alarm sounded. Then, with 20 seconds remaining before collision, a "resolution advisory" could sound a second alarm and the signal ($-03\uparrow$) be shown in red. At the same time a red arrow might appear on the airliner's vertical speed dial, directing the pilot to climb.

TCAS is being developed as a joint project of the aviation industry and the Federal Aviation Administration (FAA).

Electrical and Electronics Technicians

In a very real way, the modern industrialized world runs on electrical and electronic equipment. There are electrical appliances of all sorts, computers in the home, office, and factory, and many different electronics devices in the civilian, medical, and military worlds. Research physicists explore basic phenomena, engineers use such findings to create new devices, and electrical and electronics technicians develop, manufacture, operate, and service the devices.

Because this is a very broad field, many technicians specialize. In general, however, they perform such tasks as following the directions of engineers and scientists, conducting experiments, and testing, adjusting, and repairing equipment. The work is carried out under a wide variety of conditions, in laboratories, shops, factories, and on location—as when repairing or

servicing radar equipment in the cockpit of an airliner, for example.

Job opportunities for electrical and electronics technicians are expected to grow more rapidly than the average for all occupations through the year 2000. Increased demand for computers, communications equipment, military electronics, and consumer goods accounts for this projected growth. Opportunities will be best for graduates of post-secondary technical training programs, particularly those in which trainees acquire practical work experience.

It is possible to qualify for some electrical and electronics technician jobs by means of a combination of education and work experience, but most employers prefer technical training. Such training is available at technical institutes, junior and community colleges, some 4-year colleges and universities, and vocational and technical schools.

Computer technicians at work.

UNIT 6
Electromagnetism

(Above) Harvard University radio telescope in Harvard, Massachusetts. This telescope is being used to search outer space for signals that might have been originated by another civilization somewhere in the Universe. (Left) Radio telescope at Socorro, New Mexico. This telescope has 27 antennas and extends for 21 km (13 mi).

Further investigation of the nature of electrical energy and its relationship to matter leads to the work of Hans Christian Oersted (1777–1851). In 1819, he demonstrated that every electric current generates a magnetic field around itself. This was the first recognition that magnetism is associated with and produced by moving electric charges. Since all matter is composed of electric charges in motion, it follows that all matter has magnetic properties.

The counterpart of Oersted's discovery was made by Michael Faraday (1791–1867). He showed that moving magnets or magnetic fields can set electric charges in motion and thus generate currents. Just as the effects of static charges are visualized as acting upon each other through an electric field, so the magnetic effects of magnets and moving electric charges can be visualized as acting upon each other through a magnetic field.

Understanding of the interrelationships between electricity and magnetism and their fields was achieved by James Clerk Maxwell (1831–1879) in his great theory of electromagnetism. This theory predicted that when electric charges or magnets move, they radiate energy in the form of electromagnetic waves. These include radio waves, heat, light, and X rays which travel through a vacuum at the speed of light and have properties similar to those of light. Maxwell's prediction was confirmed by Heinrich Hertz (1857–1894) who first showed experimentally how radio waves could be produced and detected.

The discovery of electromagnetism introduced three major contributors to change in human life and society. They were the generator, the motor, and radio communication. The generator made it possible to supply electrical energy in the quantities needed by a growing industrial society. The motor provided the means of putting that energy to work. The discovery of electromagnetic waves opened up the era of radio and television that give us practically instant communication with any part of the earth. This has enabled us to exchange information of all kinds with peoples near and far, creating a true "global village."

25

Magnetism and Magnetic Fields

Aims

1. To become familiar with the properties of magnets and their magnetic fields.
2. To understand the interrelationships between electric currents and magnetic fields.
3. To develop a theory of magnetism and to study the evidence supporting that theory.
4. To recognize magnetism as a property of all matter.

25-1 Magnetism, Electricity, and Matter

Further exploration of the nature of electricity reveals that it is closely related to magnetism. We shall see that all magnetic forces and fields are produced by moving electric charges. This relationship between electricity and magnetism gives us new insights into the properties of matter. Every body of matter is composed of electric charges that are always in motion. It follows that they will produce magnetic effects in and around the body and influence its properties.

25-2 Magnetic Poles and the Compass

The first human experience with magnetism involved a magnet found in nature, the iron ore known as magnetite or lodestone. The Greeks knew of it as early as 800 B.C.

It is helpful to begin the study of magnetic forces with an examination of the properties of the familiar magnet. This is generally made of steel and has the ability to exert relatively strong attractive forces on pieces of iron and certain other materials placed near its ends. When a bar magnet is suspended so that it can turn freely in a horizontal plane, it usually comes to rest so that one end of it points in the general northerly direction and the other end points in the general southerly direction. For this reason the end of the magnet that points north is called its *north pole* and the opposite end is called its *south pole*. Every magnet has both a north and a south pole.

The ability of a freely suspended magnet to align itself in the general north-south direction is the property underlying the operation of the magnetic compass long used by mariners as a means of determining direction. A compass is essentially a light bar magnet supported horizontally at its center by a sharp point on which it can turn freely. It is a particularly useful device for exploring magnetic effects in the neighborhood of a magnet.

Fig. 25-1. Magnetic compasses used in the laboratory.

25-3 Forces Between Magnetic Poles

We have seen that electric charges exert attractive and repulsive forces upon each other. The poles of magnets also exert attractive and repulsive forces upon each other.

> Like magnetic poles repel each other; unlike poles attract each other.

This is shown by bringing each pole of a bar magnet near the north and south poles of a compass in turn. The north pole of the magnet repels the north pole of the compass and attracts its south pole. The south pole of the magnet repels the south pole of the compass and attracts its north pole.

(a)   (b)

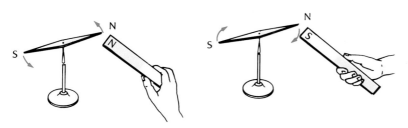

Fig. 25-2. (a) Like magnetic poles repel. **(b)** Unlike magnetic poles attract.

25-4 The Earth Is a Magnet

The forces of attraction and repulsion between the poles of magnets and the north-seeking behavior of one end of the compass needle led William Gilbert in the year 1600 to the conclusion that *the*

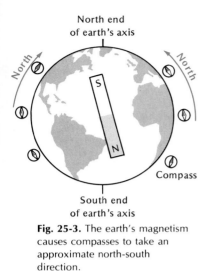

North end
of earth's axis

South end
of earth's axis

Compass

Fig. 25-3. The earth's magnetism causes compasses to take an approximate north-south direction.

All fixed iron objects and structures are magnetized by induction by the earth's magnet.

earth itself is a huge magnet. The south pole of the earth's magnet is near the northern end of its axis of rotation, while the north pole of the earth's magnet is near the southern end of its axis. Since unlike poles attract each other, the north poles of compasses are attracted toward the northern part of the world where the south pole of the earth's magnet is located. This explains why the north pole of a compass points in the northerly direction. (See Fig. 25-3.)

Compasses do not generally point true north because the magnetic poles of the earth are located at considerable distances from the north and south poles of the axis on which it rotates. In the northern hemisphere, the earth's magnetic pole is about 1800 kilometers from the north pole of its axis. Since the north pole of the compass points toward the earth's magnetic pole rather than toward its actual north pole, the compass direction generally deviates from the true north. Mariners and others who use compasses must allow for this deviation in determining the true north.

25-5 Magnetization by Induction

A piece of unmagnetized iron becomes magnetized when placed in the neighborhood of a magnet. Like all magnets, it acquires two opposite poles. This can be shown by sprinkling iron filings around it and noticing that they are attracted to two opposite ends of the piece of iron. The iron is said to be magnetized by *induction*.

Magnetic induction accounts for the ability of the pole of a magnet to attract a nearby piece of unmagnetized iron. The piece of iron becomes magnetized so that the end of it nearer to the pole of the magnet becomes an opposite pole. The attraction that is observed between the pole of the magnet and the piece of iron is therefore simply the attraction of opposite poles.

Iron steam radiators, iron fences, iron structures of buildings, and other iron objects that remain fixed for a long time become magnetized by induction because of the influence of the earth's magnet. Since the south pole of the earth's magnet is in the northern part of the world, the bottoms of these iron objects in the northern hemisphere become magnetic north poles by induction. Their tops become south poles.

25-6 Permanent and Temporary Magnets

A piece of steel magnetized by induction retains its magnetism after it is removed from the neighborhood of the magnet that magnetized it. Steel therefore makes a *permanent magnet.* A piece of soft iron, on the other hand, can only be a *temporary magnet.* After being magnetized by being brought near a permanent magnet, it loses its magnetism as soon as it is moved out of the neighborhood of the permanent magnet.

25-7 Magnetic Fields

Let us return to the iron radiator in the classroom that is magnetized by induction by the earth's magnet. The fact that the radiator is many thousands of miles from either pole of the earth's magnet shows that a magnet can act upon other magnets and pieces of iron even though they are at considerable distances from its poles. We have previously noted that electric charges do not act upon each other directly but through their electric fields. According to the present view, *magnets also do not act upon each other directly but through magnetic fields*. Each magnet sets up a magnetic field in the space around itself through which it acts upon other magnets.

25-8 Mapping Magnetic Lines of Force

To explore the magnetic field at any point near a magnet, a tiny compass is put at that point. The poles of the magnet, acting through its magnetic field on the poles of the compass, force the compass to assume a definite direction. If the space around the magnet is filled with many tiny compasses, each of them is forced to take a definite direction by the magnetic field. One can see that the compasses arrange themselves so as to form curved lines that run from one end of the magnet to the other. These are called *magnetic lines of force*. Like the lines of force of electric fields, they are used to represent the forces in a magnetic field.

It is not practical to use many compasses to map the lines of force in a magnetic field, but there is a simple way to demonstrate the general pattern made by the lines by using iron filings. To illustrate, let us apply it to show the magnetic field of a bar magnet.

The magnet is put on a table and a large sheet of white paper is put over it. Iron filings are sprinkled evenly over the paper. Each iron filing now becomes a little magnet by induction because it is in the field of the bar magnet. If the paper is then tapped gently, the iron filings are temporarily loosened and enabled to turn as compasses would do. The filings arrange themselves along the lines of force as shown in Fig. 25-4.

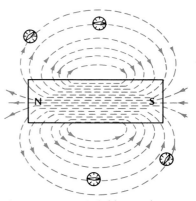

Fig. 25-4. Magnetic field around a bar magnet represented by lines of force.

25-9 Properties of Lines of Force

The field of a bar magnet illustrates the following general properties of magnetic lines of force:

1. The north pole of a compass placed at any point of a magnetic field points in the direction of the tangent to the line of force passing through that point. Since a compass can take only one direction at any point in a field, it follows that lines of force cannot cross each other.

2. The concentration of the lines of force at any place shows the strength of the magnetic field at that place. The largest concentration of the lines of force near the poles of the magnet shows that

The concept of lines of force was introduced by Michael Faraday.

the field of the magnet is strongest there. The lines of force become less concentrated as the distance from the magnet increases, showing that the forces exerted by the magnet decrease as the distance increases.

3. Each magnetic line of force is a closed curve. Outside the magnet, it goes from the north to the south pole. It completes its path inside the magnet where it goes from the south pole to the north pole.

(a)

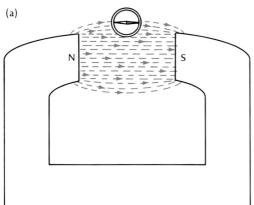

(b)

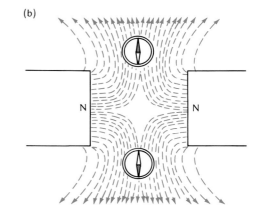

Fig. 25-5. (a) Magnetic field between unlike poles. **(b)** Magnetic field between like poles.

25-10 Magnetic Field Between Unlike Poles

Fig. 25-5 (a) shows the magnetic field between the two unlike poles of a horseshoe magnet or of neighboring magnets. The great concentration of the lines of force between the poles shows that there are strong magnetic forces at and between them. Magnets are often bent into horseshoe shapes to take advantage of the concentrated magnetic fields between their poles.

25-11 Magnetic Field Between Like Poles

Lines of force from like poles repel each other. This is indicated in the map of the field between two nearby north poles shown in Fig. 25-5 (b). Here, the lines of force from the two like poles run toward each other but repel one another so that they never cross.

25-12 Magnetic Field Around a Current

In 1819, Hans Oersted, a Danish scientist, noticed that an electric current flowing in a wire caused a nearby compass to change its direction. This was the first evidence of a connection between magnetism and electricity. Investigating it further, Oersted discovered this important principle:

An electric current flowing in a wire sets up a magnetic field around itself.

The force exerted by this magnetic field is no different from the forces exerted by the magnetic fields of permanent magnets.

The nature of the magnetic field around a current-bearing wire can be investigated by the arrangement shown in Fig. 25-6. A wire passing vertically through a cardboard platform is connected to a battery which sends a direct current through it. Iron filings are sprinkled on the platform which is tapped gently to enable the filings to arrange themselves about the magnetic lines of force around the wire. Notice that there are no poles associated with this field. The lines of force are a set of concentric circles having the wire as a common center. Several compasses placed on the platform show the direction of the lines of force.

If the direction of the current in the wire is reversed by changing the connections to the battery, and the magnetic field is again examined with iron filings and compass, the circular shape of the lines of force is the same as before. However, the direction of the lines of force is reversed. This shows that the direction of the current in a wire determines the direction of the lines of force around it. The strength of the magnetic field around the wire can be varied. It increases with increasing current and decreases with decreasing current.

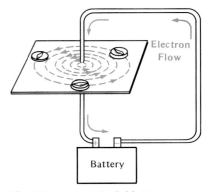

Fig. 25-6. A magnetic field exists around a wire carrying current.

25-13 Finding the Direction of the Magnetic Field Around a Wire

The following rule makes it easy to determine the direction of the lines of force around a wire from the direction in which the electrons flow through it. (See Fig. 25-7.)

Grasp the wire with the left hand so that the extended thumb points in the direction of the electron flow. The curled fingers then point in the direction of the lines of force.

25-14 Magnetic Field Around a Circular Wire and a Coil

Fig. 25-8 (a) shows the lines of force around a current-bearing wire forming a circular loop. Notice that the field is strongest at the center of the loop. A compass placed at this point takes the position of the lines of force at this point and is perpendicular to the plane of the loop.

Fig. 25-8 (b) shows the lines of force around a coil or solenoid of wire in which a current is flowing. Notice that the field resembles that of a permanent bar magnet. The lines of force at one end of it are concentrated like those around the north pole of a permanent magnet. At the opposite end, they are concentrated like those around the south pole of a permanent magnet. The ends of the coil behave as poles do and repel the like poles of other magnets or coils.

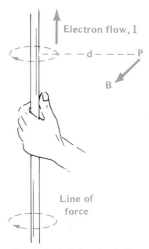

Fig. 25-7. Left-hand rule for finding the direction of the magnetic field around a straight wire.

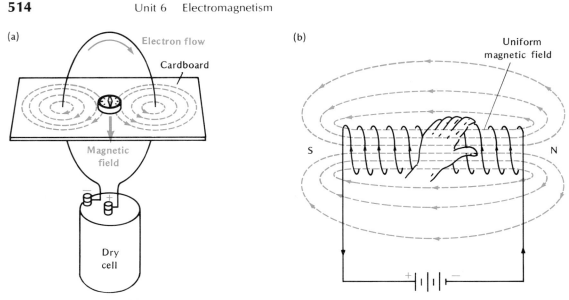

(a)

(b)

Fig. 25-8. (a) Magnetic field at the center of a circular loop. **(b)** Magnetic field around a coil, illustrating the left-hand rule.

The magnetic field inside a solenoid or coil of wire carrying a current is particularly useful because, except near the ends of the coil, it is very nearly uniform. Thus, it provides a way to set up a uniform magnetic field just as a pair of charged parallel plates gives us a way to set up a uniform electric field. The magnetic field inside a coil of wire increases directly with the current through the coil and also with the number of turns per unit length of the coil.

25-15 Finding the Direction of the Magnetic Field of a Coil

Reversing the direction of the current in a circular loop of wire or in a coil of wire reverses the direction of the magnetic field of the loop or coil. Again, the direction of the magnetic field may be determined from a left-hand rule.

Grasp the coil or loop with the left hand so that the fingers curl around it in the same direction as the electrons are flowing. The extended thumb indicates the direction in which the lines of force are going.

25-16 Units of Magnetic Flux and Induction

The representation of a magnetic field by lines of force suggests an imaginary flow or *flux* of magnetic force through the space surrounding a permanent magnet or a wire carrying a current. The total magnetic flux passing through any given area of a magnetic field is measured by means of a unit called the *weber*, symbol, Wb.

The density of magnetic flux at a point of a magnetic field is defined as the flux per unit of area taken perpendicular to the flux at that point. The greater the flux density at a point in a magnetic field, the greater is the strength of the magnetic field at that point. In the International System (SI), *magnetic flux density* is measured in *webers per square meter*, a unit also known as the *tesla (T)*. It is designated by the letter B and is called the *magnetic induction*. The direction of B at a point in a magnetic field is the direction of the lines of force at that point.

A little further on, we shall see that a magnetic field exerts a force on a current-bearing wire placed in it at right angles to its lines of force. The force exerted by a magnetic field on every meter of a wire carrying one ampere of current is used as a measure of the strength of that field and therefore of its magnetic induction B. When B is measured in this way, it is expressed in units of *newtons per ampere-meter*. A magnetic field having a magnetic induction of 1 newton per ampere-meter has a flux density of 1 weber per square meter, or 1 tesla.

Since we are frequently interested in the forces exerted by magnetic fields, we shall often express magnetic inductions in newtons per ampere-meter instead of the equivalent units of webers per square meter, or teslas.

25-17 Magnetic Induction Around a Straight Wire

The strength of the magnetic field at any point P around the long, straight wire carrying a current shown in Fig. 25-7 depends upon only two factors: the distance d of P from the wire and the current I flowing in the wire. It is found experimentally that the magnetic induction B of this field at a point P is directly proportional to the strength of the current I and that it is inversely proportional to the distance d. The value of B at P is therefore proportional to I/d and equal to a constant times I/d. This constant is determined by the units used to express B, I, and d. In *SI* units, B is expressed in newtons per ampere-meter (or webers per square meter), I is expressed in amperes, and d is expressed in meters. The constant then turns out to be 2×10^{-7} and is expressed in newtons per (ampere)^2 or webers per ampere-meter.

$$B_{\text{straight wire}} = \left(2 \times 10^{-7} \frac{N}{A^2} \right) \frac{I}{d}$$

Sample Problem

What is the magnetic induction at a point 1 m from a straight wire carrying a current of 5 A?

Solution:

$I = 5$ A $d = 1$ m

$$B_{\text{straight wire}} = \left(2 \times 10^{-7} \frac{N}{A^2} \right) \frac{I}{d}$$

$$B_{\text{straight wire}} = \left(2 \times 10^{-7} \frac{N}{A^2} \right) \frac{5\ A}{1\ m}$$

$$B_{\text{straight wire}} = 10^{-6}\ \text{N/A} \cdot \text{m, or Wb/m}^2 \text{, or T}$$

Test Yourself Problems

1. What is the magnetic induction at a point 0.5 m from a straight wire carrying 1 A of current?
2. What current should be passed through a straight wire to produce a magnetic induction of 2×10^{-8} N/A·m (or Wb/m^2) at a distance of 0.01 m from the wire?

25-18 Magnetic Induction at the Center of a Circular Loop

The strength of the magnetic field at the center of a circular loop of wire carrying a current I is found experimentally to be proportional to the current passing through the loop and inversely proportional to the radius of the loop r. The value of B at the center of the loop can therefore be expressed as a constant times I/r. In SI units, this constant turns out to be $2\pi \times 10^{-7}$ newtons per (ampere)^2. Thus

$$B_{\text{center of a loop}} = \left(2\pi \times 10^{-7}\ \frac{\text{N}}{\text{A}^2}\right)\frac{I}{r}$$

25-19 Electromagnet

Shortly after Oersted's discovery, it was learned that a powerful magnet could be made by putting a core of soft iron inside a coil carrying current. A combination of a coil of wire wound on a soft iron core is called an *electromagnet*. When a current passes through the coil of an electromagnet, it sets up a magnetic field inside the coil. This magnetic field magnetizes the soft iron core by induction, making one end of it a north pole and the other end a south pole. The magnetic field of the magnetized soft iron core now joins the magnetic field of the coil to produce a very high concentration of lines of force inside the core and consequently strong magnetic poles at its ends.

The ability of a material like the soft iron core to strengthen and concentrate the magnetic field passing through it is called its permeability. When a soft iron core is inside a coil carrying a constant current, the magnetic field in the coil is several hundred times greater than it is when the empty coil carries the same current.

When the current in the electromagnet is turned off, the magnetic field of the coil is destroyed, and the soft iron core also loses practically all of its magnetism. Thus, the electromagnet can be turned on or off at will. The strength of an electromagnet can also be controlled. It can be increased by (1) *increasing the current,* (2) *increasing the number of turns in the coil,* and (3) *replacing its core with one of higher permeability.* Permalloy, an alloy of iron and nickel having a particularly high permeability, makes a very effective core for electromagnets.

The ready control of the electromagnet makes possible a wide variety of applications, ranging from electric bells to cyclotrons.

25-20 Making Permanent Magnets

If a piece of steel is used as the core of an electromagnet, the steel retains its magnetism after the current is turned off. This is a common way to make a permanent steel magnet. The strength of the steel magnet depends upon the strength of the magnetic field inside

Fig. 25-10. A commercial electromagnet in use. The load of iron is released when the current is turned off.

the coil. This can be controlled by varying the current sent through the coil. Up to a point, increasing the current in the coil increases the strength of the magnetization undergone by the steel.

25-21 Permanent Magnetism and Electromagnetism

The study of electromagnetism has shown that magnetic forces and fields exist around electric currents or electric charges in motion. It has also been seen that magnetic forces and fields exist around permanent magnets. We turn now to a theory of permanent magnetism that explains the relationship between these two seemingly independent ways of setting up a magnetic field.

25-22 Internal Structure of Permanent Magnets

Only a small number of materials other than iron have the ability to become strong permanent magnets or to be affected strongly by a magnetic field. These include nickel and cobalt and certain iron alloys which are commonly known as *magnetic substances*. Because this property is strongly characteristic of iron, it is called *ferromagnetism*. (*Ferro-* comes from the Latin word *ferrum* meaning iron.) The origin of ferromagnetism sheds light on the magnetic properties of atoms and their parts.

When a bar magnet is broken into many pieces, each piece, no matter how small, is found to be a complete bar magnet having its own north and south poles. This suggests that if we continued to break a magnet into ever smaller pieces, we would find ultimately that each atom of the original magnet is also a tiny bar magnet.

25-23 Theory of Atomic Magnetism

How do these atomic magnets acquire their poles and magnetic fields? We know that atoms contain equal numbers of electrons and protons. If either of these electric charges moves inside the atom, it produces an atomic current and therefore sets up a magnetic field. The evidence indicates that it is the motion of the electrons in atoms that is responsible for the magnetic behavior of the atom. Each electron acts as though it has two motions; it appears to spin on its axis and to circulate around the core or nucleus of the atom to which it belongs. Each of these motions of the electron is equivalent to a tiny electric current flowing in a circular circuit and sets up a magnetic field of its own.

In most atoms, the tiny magnetic fields set up by the electrons overlap in such a manner as to produce a resultant magnetic field of zero or near zero. In iron and other magnetic substances, the magnetic fields of the electrons do not cancel each other out and each atom has a resultant field like that of a tiny bar magnet. When enough of the atoms in a piece of iron are *lined up* (Fig. 25-11 (a)) so that their magnetic fields combine to produce a large resultant field, the piece of iron as a whole is a magnet with a definite north

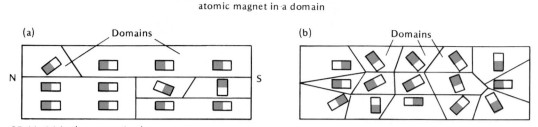

☐☐ Represents a typical
atomic magnet in a domain

Fig. 25-11. (a) In the magnetized bar, domains are pretty well lined up, as are the individual atomic magnets. **(b)** In the unmagnetized bar, the domains are not aligned and their atomic magnets face in many directions.

pole at one end and a definite south pole at the other. When the atomic magnets in a piece of iron *are facing in different directions* (Fig. 25-11 (b)), the net result of their combined magnetic fields is zero. The piece of iron is then unmagnetized and has no resultant magnetic poles.

Thus, the ultimate sources of the magnetic fields of permanent magnets are electric currents—the tiny currents flowing inside their atoms.

25-24 Domains

More detailed research indicates that the atoms of a ferromagnetic substance do not act individually but combine to form groups called *domains*. All atomic magnets in each domain are lined up so that *each domain is a microscopic bar magnet* with its own north and south poles. In unmagnetized substances, the domains face in all directions and neutralize each other's magnetic fields. In magnetized substances, many of the domains are lined up so that they produce a strong resultant magnetic field in one direction through the substance. It is the domains rather than the individual atomic magnets that rearrange themselves when a ferromagnetic substance is put into a magnetic field.

25-25 Checks on the Theory of Ferromagnetism

How well does the theory of atomic magnets and domains explain the known facts about ferromagnetic substances? According to the theory, a bar of iron or steel ordinarily has its magnetic domains facing in all possible directions. It should become magnetized when something is done to line up many of its atomic domains, thus arranging their atomic magnets as in Fig. 25-11 (a). This prediction can be checked by putting the bar in a magnetic field. The atomic magnets in the domains should then tend to line themselves up along the lines of force of the field in two ways. First, the domains already aligned with the field should grow larger as atoms in neighboring domains are forced to line up with the field. Second, some of the other domains should tend to turn as a whole and thus line up their atoms directly with the field. As a result, the bar should become magnetized and show magnetic poles. We have seen that

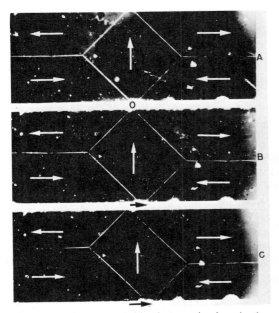

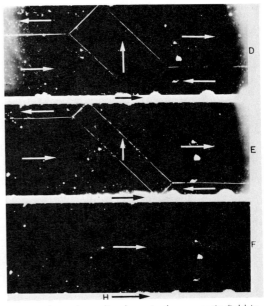

Fig. 25-12. These remarkable photographs show in six stages how a "whisker" of iron, placed in a magnetic field H, acting toward the right, becomes magnetized by induction. The sections of the whisker are domains and the arrows show the direction in which the atomic magnets in each are lined up. As the magnetic field is steadily increased, the domains whose atomic magnets are aligned with the field become larger, while the others become smaller. In the final stage, all the domains and their atomic magnets are lined up with the field.

this actually happens in magnetization by induction. Fig. 25-12 shows the behavior of the domain in a "whisker" of iron.

If the alignment of atomic magnets in a piece of iron is responsible for its magnetization, anything that is done to destroy that alignment will destroy the magnetization. This is verified by the fact that heating or hammering a magnet vigorously demagnetizes it. The effect of hammering or heating is to increase the thermal vibratory motion of the domains and the atomic magnets to such an extent that they cannot retain their alignment.

According to the theory of atomic magnets, a permanent magnet cannot be magnetized beyond a certain maximum strength. That maximum is reached when all of its atomic magnets are aligned with their north poles in one direction and with their south poles in the opposite direction. A magnet in which complete alignment exists is said to be *saturated*. The prediction that magnets have a saturation level agrees with the facts. A piece of iron can be magnetized more and more strongly by being put in stronger and stronger magnetic fields. However, a time is reached when the iron attains its maximum magnetization and becomes saturated. Putting it into stronger magnetic fields will then have no further effect in increasing its magnetic strength.

25-26 Magnetism and the Nature of Matter

All atoms have magnetic properties.

The theory of atomic magnets indicates that since all electrons set up magnetic fields inside of atoms, *all atoms have combinations of magnetic fields around them.* It has already been noted that in the atoms of iron and the ferromagnetic substances, the electronic magnetic fields combine to make each atom a tiny magnet. In these substances, the atomic magnets exert special forces upon one another that cause them to line up and form the domains responsible for their strong magnetic properties.

Many other substances also have atoms that are tiny magnets. However, their atoms do not band together to form domains, and any temporary alignment of them is soon destroyed by their ordinary thermal motion. Such substances are only weakly attracted to magnetic poles and are called *paramagnetic* substances.

Finally, there are also atoms like bismuth and copper in which the magnetic fields of the electrons balance one another so as to add up to zero. When an external magnetic field is brought near such atoms, it changes the orbital motion of their electrons thus increasing the magnetic fields of some of them and decreasing the magnetic fields of others. This destroys the previous magnetic balance and leaves each atom with a net magnetic field opposing the external field. As a result, substances composed of such atoms are weakly repelled by magnetic poles and are called *diamagnetic* substances.

The fact that all atoms are either ferromagnetic, paramagnetic, or diamagnetic indicates that:

> Magnetic fields and their properties are not limited to the atoms of ferromagnetic substances but are associated with all atoms and therefore with matter in general.

CHAPTER REVIEW

Summary

A **permanent magnet** has two poles, a **north** and a **south pole,** that strongly attract iron and certain other substances. *Like magnetic poles repel each other; unlike magnetic poles attract each other.*

The space around a magnet in which it exerts forces on other magnets is called its **magnetic field.** The field of a magnet may be mapped by drawing **magnetic lines of force.**

The unit of **magnetic flux** is the **weber** (Wb). The **magnetic flux density,** also called the **magnetic induction,** is measured in webers per square meter, or its equivalents, newtons per ampere-meter, called **teslas** (T).

Oersted discovered that every electric current produces a magnetic field around itself. When the current is flowing in a straight wire, the magnetic field around it consists of lines of force that are concentric circles having the wire as their center. The magnetic induction or field strength B at a distance d from a wire carrying a current I is given by the formula:

$$B_{\text{straight wire}} = (2 \times 10^{-7} \text{ N/A}^2)\frac{I}{d}$$

The magnetic induction at the center of a circular loop carrying a current I and having a radius r is:

$$B_{\text{center of a loop}} = \left(2 \pi \times 10^{-7} \frac{\text{N}}{\text{A}^2} \right) \frac{I}{r}$$

The magnetic field around a coil of wire carrying current resembles that around a bar magnet. The direction of the magnetic field depends upon the direction of the current and the strength of the magnetic field depends upon the strength of the current.

According to theory, the atoms of strongly magnetic or ferromagnetic substances are tiny bar magnets themselves. These atoms act together in groups called **domains** which are responsible for the magnetic properties of the material. In an unmagnetized bar of iron, the domains face in all possible directions and cancel out each other's effects. In a magnetized bar of iron, huge numbers of the domains are lined up with their north poles facing one end of the bar and their south poles facing the opposite end of the bar. The net result is to form a permanent north magnetic pole at one end and a permanent south magnetic pole at the opposite end.

The origin of the magnetism of the atoms of substances is in the motions of their electrons which are equivalent to tiny currents. In the atoms of most substances, the magnetic fields produced by the motions of their electrons virtually cancel each other. However, in the atoms of iron and certain other substances, the magnetic fields of the atoms add up to produce a resultant field like that of a bar magnet. Each of these atoms therefore acts like a tiny bar magnet.

Ferromagnetic substances are the only ones affected strongly by magnets and magnetic fields, but all other substances are also affected by magnetic fields. They may be **paramagnetic** substances which are weakly attracted by magnetic poles, or they may be **diamagnetic** substances which are repelled by magnetic poles. The observation that all substances have magnetic properties supports the theory that all the magnetic properties of substances originate in the currents associated with the electric charges of their atoms.

Questions

Group 1

1. (a) Given only a bar magnet and a piece of string, explain how to determine which is the north pole of the magnet. (b) How does the property of the bar magnet used in (a) explain the behavior of a compass?

2. (a) What forces do magnetic poles exert upon each other? (b) How can a compass be used to determine which end of a bar magnet is a north pole?

3. (a) How does the fact that the earth itself is a magnet explain why the north pole of a compass points in the general northerly direction? (b) Why doesn't a compass needle generally show the true north?

4. One end of an unmagnetized iron bar is brought near the south pole of a magnet. (a) How does the process of magnetic induction explain why the south pole of the magnet attracts the nearer end of the iron bar? (b) What change takes place in the opposite end of the iron bar? (c) If the bar is made of soft iron, what happens to its magnetism when it is taken away from the neighborhood of the magnet?

5. (a) Explain how iron filings are used to map the lines of force of the magnetic field of a bar magnet. (b) What position does a compass take when put at any point in the magnetic field? (c) What does the concentration of the lines of force in any part of the field show about the field?

6. Draw six complete lines of force around a bar magnet. Show the direction of the lines of force by arrows. Show the direction a compass takes when put at two different points in the field of the magnet.

7. Draw eight lines of force between two nearby south poles showing their directions by arrows. How do the lines of force represent the fact that the poles repel each other?

8. (a) Explain how to demonstrate that a current-bearing wire has a magnetic field around it. (b) What is the shape of the lines of force of this field? (c) What effect does changing the current flowing in a wire have on the magnetic field around it? (d) How does the strength of the magnetic field change as one moves away from the wire?

9. (a) State the left-hand rule for determining the direction of the magnetic lines of force around a current-bearing wire. (b) What effect does reversing the current in the wire have on the magnetic field around it?

10. (a) State the left-hand rule for finding the direction of the magnetic field inside a coil of wire carrying a current. (b) Draw a coil of wire connected to a battery and show six complete lines of force of its magnetic field. (c) At which end of the coil is its north pole?

11. What are the units for measuring (a) magnetic flux; (b) magnetic flux density or magnetic induction?

12. Explain how to use a battery, a switch, and a coil of wire to magnetize a piece of steel.

13. (a) Draw a bar electromagnet connected to a battery. Show the current direction and the poles of the magnet. (b) How does changing the current affect the strength of the electromagnet? (c) If the current is kept constant, while the number of turns per unit length of the coil is increased, what will happen to the strength of the electromagnet?

14. (a) According to the theory of magnetism, how does the arrangement of the atomic magnets in an unmagnetized piece of iron differ from their arrangement when the same piece of iron is magnetized? (b) How are "electron currents" responsible for making the atoms of certain elements into tiny magnets? (c) Why aren't the atoms of all elements tiny magnets?

15. How does the theory of ferromagnetism explain: (a) why breaking a bar magnet produces two smaller magnets each with its own north and south poles; (b) why hammering or heating a permanent magnet destroys its magnetism; (c) Why a piece of iron cannot be magnetized beyond a certain maximum strength?

Group 2

16. An alternating current in a straight wire reverses itself 120 times a second. Describe the changes that take place in the magnetic field around the wire as the current continues to reverse itself.

17. Two straight wires carrying equal but opposite direct currents are held adjacent to each other. Explain why this arrangement has the effect of eliminating the magnetic field around the pair of wires.

18. An electron is moving clockwise at constant speed in a tiny circular orbit. (a) Show by a diagram the direction of the magnetic field it generates. (b) What effect does a change in the speed of the electron have on its magnetic field?

Problems

Group 1

1. What is the magnetic induction (or flux density) at a point 0.5 m from a straight wire carrying a current of 2 A?

2. How large should a current in a wire be so that the magnetic induction at a distance of 0.10 m from the wire is 5.0×10^{-6} Wb/m², or N/A·m?

3. At what distance from a wire carrying a 10-A current is the magnetic induction equal to that of the earth's magnetic field, namely 5×10^{-5} Wb/m²?

4. (a) What is the magnetic induction at the center of a circular loop of wire carrying a current of 2.0 A and having a radius of 0.16 m? (b) If the current is kept constant while the radius of the loop is doubled, what effect will this have on the magnetic induction at its center?

5. What current flowing in a circular loop of radius

1.0 m will produce a magnetic induction at its center of 4.0×10^{-7} Wb/m²?

Group 2

6. Two parallel straight wires are 1.0 m apart. Each wire carries 2.0 A of current in the same direction. What is the resultant magnetic induction produced by both wires at a point (a) midway between them; (b) ¼ of the distance between them?

7. If the two wires of problem 6 carry the same current but in opposite directions, what is the resultant magnetic induction at a point (a) midway between them; (b) ¼ of the distance between them?

8. Two parallel straight wires, 1 m apart, carry currents of 2 amps and 4 amps respectively. At what points between the two wires is the resultant magnetic induction zero when both currents are in the same direction?

Applying Physics

1. Lines of force of the earth's magnetic field usually make an angle with the horizontal in all parts of the earth except the magnetic equator. The force exerted by the earth's magnetic field can therefore be resolved into two components, one vertical and the other horizontal. The horizontal component is the one that causes a compass to turn.

Here is a way to compare the strength of the horizontal component of the earth's magnetic field with the magnetic field in the neighborhood of the pole of a given bar magnet.

Put a small compass equipped with an angular scale on a table and let the needle come to rest. Note on the scale the direction of the N-pole of the needle. This is the direction of the horizontal component of the earth's magnetic field. About 25 cm from the compass, put the N-pole of a bar magnet and line the magnet up so that it points toward the center of the compass and makes a right angle to the direction of the horizontal component of the earth's magnetic field. The compass needle is now being acted upon by two magnetic fields at right angles to each other, the horizontal component of the field of the earth and the field of the bar magnet. (See Fig. 25-13.) Move the N-pole of the bar magnet toward the center of the compass until the compass needle assumes a position 45° to the direction of the horizontal component of the earth's magnetic field. At this distance from the N-pole of the bar magnet, the strength of the horizontal component of the earth's field and that of the bar magnet are equal.

Repeat the experiment with several other bar magnets. The distances from their N-poles at which the compass experiences a force equal to

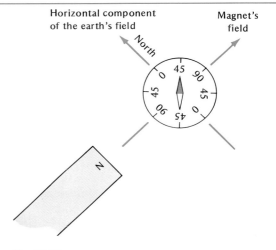

Fig. 25-13

that exerted by the horizontal component of the earth's magnetic field is a qualitative measure of the relative strengths of the magnets' poles.

2. The paths of lines of force may be demonstrated with a small permanent bar magnet. First use the magnet to magnetize a steel sewing needle by stroking it several times in the same direction with one pole of the bar magnet. Then lay the magnet in the middle of a large deep dish and cover the magnet with about an inch of water. Thrust the magnetized needle half way through a cork wafer and float the cork on the water so that the N-pole of the needle points downward. The cork wafer will now move along the lines of force until it comes to rest at the S-pole of the bar magnet or at the edge of the dish. Put the wafer down in different places around the magnet and observe the paths it follows.

26

Forces Exerted by Magnetic Fields

Aims

1. To investigate the nature of the force exerted by a magnetic field on electric currents and moving charges.
2. To learn how magnetic forces are used in meters to measure electric currents.
3. To understand how magnetic forces enable us to convert electrical to mechanical energy in such devices as the motor.
4. To learn how electric and magnetic fields can be used to measure the velocities and masses of charged atomic and subatomic particles.

26-1 Force on a Current in a Magnetic Field

Oersted showed that a current in a conductor acting through its magnetic field exerts forces on a nearby magnet. From the law of action and reaction, one would expect that the magnet acting through its magnetic field exerts equal and opposite forces on the conductor. Michael Faraday showed that this is true by means of the experimental arrangement shown in Fig. 26-1. A straight wire, connected to a switch and battery, is held at right angles to the magnetic field between the poles of a horseshoe magnet. When the switch is closed and current flows through the wire as shown, the magnetic field exerts a force upon the wire that pushes it upward

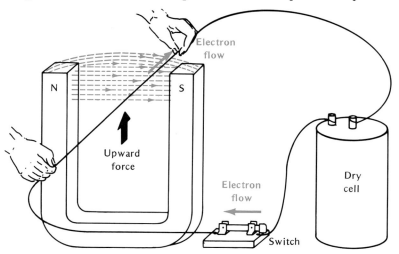

Fig. 26-1. A magnetic field exerts a force on a current-bearing wire.

out of the magnetic field. When the switch is open, there is no force upon the wire. When the direction of the current in the wire is reversed, the force on the wire pushes it downward instead of upward.

These observations suggest that the magnetic field exerts a force on the charges flowing in the wire rather than on the wire itself. In this case, the charges in the wire are free electrons. When these electrons are moving in a current, the magnetic field exerts a force upon them that is transferred to the wire in which the electrons are confined. When the electrons have no net forward motion and are therefore not forming a current, the net force of the magnetic field upon them and upon the wire is zero.

We shall see that this experiment illustrates the more general observation that a magnetic field exerts a force upon an electric charge only when the charge is moving in the field and not when it is at rest.

26-2 Direction of Force

The force exerted by the magnetic field upon the current in a wire is remarkable because it acts neither in the direction of the magnetic field nor in the direction of the current. *It acts in a direction perpendicular to both the current direction and the direction of the magnetic field.* The relationship between the directions of the current, the magnetic field, and the force acting on the wire is remembered by means of the following rule. (See Fig. 26-2.)

Hold the outstretched left hand so that the extended fingers point in the direction of the magnetic field. Turn the hand so that the extended thumb points in the direction of the flow of electrons. The palm will then face in the direction of the force which the magnetic field is exerting upon the wire.

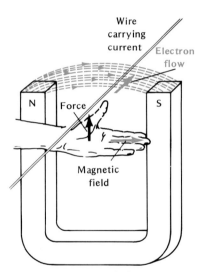

Fig. 26-2. Left-hand rule for finding the direction of the force on a current-bearing wire.

26-3 Magnitude of Force

Faraday found that when the wire and the magnetic field are at right angles to each other, the force exerted upon the wire by the field is proportional to each of three factors. It increases with (1) *the magnetic induction B of the field,* (2) *the current I* flowing in the wire, and (3) *the length l* of the part of the wire that is in the magnetic field. Using appropriate units, the force F exerted on the wire is therefore given by the formula:

$$F = IlB$$

In this expression, F is expressed in newtons, I in amperes, and l in meters. The units used to express B must therefore be newtons per ampere-meter. A uniform magnetic field having a magnetic induction B of 1 newton per ampere-meter is one that exerts a force of 1 newton on each meter of a straight wire carrying a current of 1 ampere and placed at right angles to B. As we have seen in Sec. 25-16, B is also called the magnetic flux density and may be expressed in units of webers per square meter. Recall that one weber per square meter is equal to one newton per ampere-meter, or tesla.

26-4 Effect of Angle Between B and I

The force upon a wire in a uniform field in which the magnetic induction is B is greatest when the direction of the current I in the wire and the direction of B are at right angles to each other. As the angle between B and I decreases from 90 degrees to 0 degrees, the force on the current decreases from a maximum to zero. It is found that *only the component of B perpendicular to I is effective in exerting force upon I*. In this book, we shall limit ourselves to those situations in which B is perpendicular to I. The reader will understand that in situations where B and I are not perpendicular to each other, the component of B perpendicular to I is the effective magnetic induction responsible for the force acting upon I.

Sample Problems

1. A wire carrying a current of 2.0 A is at right angles to a uniform magnetic field in which the magnetic induction B is 0.20 Wb/m². If the length of wire in the field is 10 cm, what is the force on the wire?

Solution:
$I = 2.0$ A $l = 10$ cm $= 0.10$ m
$B = 0.20$ Wb/m² $= 0.20$ N/A·m

$F = IlB$
$F = 2.0$ A $\times 0.10$ m $\times 0.20$ N/A·m $= 0.040$ N

2. A wire 0.5 m long is put into a uniform magnetic field. The force exerted upon the wire when the current in it is 20 A is 3 N. What is the magnetic induction of the field acting upon the wire?

Solution:
$F = 3$ N $l = 0.5$ m $I = 20$ A

$$F = IlB$$
$$3 \text{ N} = 20 \text{ A} \times 0.5 \text{ m} \times B$$
$$B = \frac{3 \text{ N}}{20 \text{ A} \times 0.5 \text{ m}}$$
$$B = 0.3 \text{ N/A·m} = 0.3 \text{ Wb/m}^2$$

Test Yourself Problems

1. A wire carrying 0.20 A of current is at right angles to a uniform magnetic field for which $B = 2.0 \times 10^{-5}$ Wb/m². What is the force on 0.30 m of this wire?
2. A wire 0.25 m long carrying a current of 4.0 A experiences a force of 2.0×10^{-3} N when placed at right angles to a uniform magnetic field. What is the magnetic induction of the field?
3. A current-bearing wire 0.15 m long is acted on by a force of 3.0×10^{-5} N when it is at right angles to a magnetic field whose magnetic induction is 0.010 Wb/m². What current is flowing in the wire?

26-5 Force Between Two Parallel Currents

We have seen that magnets exert forces upon electric currents and electric currents exert forces upon magnets. We note next that *currents exert forces on other currents. Two long, straight parallel wires carrying current in the same direction attract each other. The same two wires carrying current in opposite directions repel each other.*

The currents in the wires exert these forces upon each other

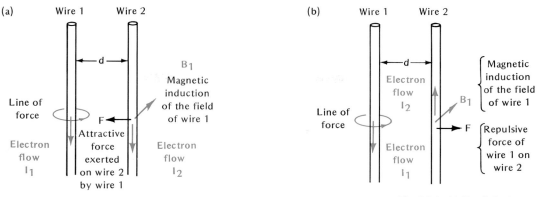

Fig. 26-3. (a) Parallel wires carrying current in the same direction attract each other. **(b)** Parallel wires carrying current in opposite directions repel each other.

through their magnetic fields. In Fig. 26-3 (a), consider the force exerted on wire 2 by the magnetic field of wire 1. The currents in both wires are in the same direction shown by I_1 and I_2 and the magnetic induction of the field of wire 1 is in the direction shown by B_1. Applying the left-hand rule to B_1 and I_2, the force F on wire 2 is seen to be toward wire 1. In a similar way, the force on wire 1 exerted by the magnetic field of wire 2 can be seen to be toward wire 2. Thus, the wires attract each other.

In Fig. 26-3 (b), the currents are shown in opposite directions. Using the left-hand rule, you can see that the magnetic field of wire 1 repels wire 2, and the magnetic field of wire 2 repels wire 1.

The force exerted on wire 2 by wire 1 may be obtained from the relationship $F = I_2 l B_1$ where I_2 is the current in wire 2, l is the length of wire 2 in the field of wire 1, and B_1 is the magnetic induction of wire 1 at wire 2. If I_1 is the current in wire 1 and d is the distance between the wires, the magnetic induction of wire 1 at d is $B_1 = (2 \times 10^{-7} \text{ N/A}^2)\, I_1/d$. Substituting this value of B_1 in $F = I_2 l B_1$, we obtain for the force on length l of wire 2:

$$ \mathbf{F} = \left(2 \times 10^{-7}\, \frac{\mathbf{N}}{\mathbf{A^2}} \right) \frac{\mathbf{I_1 I_2 l}}{\mathbf{d}} $$

In a similar way, it is seen that the force exerted by the field of wire 2 on length l of wire 1 is the same expression. Thus, as expected from the law of action and reaction, the force exerted by wire 1 on wire 2 is equal and opposite to the force exerted by wire 2 on wire 1.

26-6 Definition of the Ampere

The above relationship is used as a means of defining the ampere and has replaced the definition based on electrolysis given in Section 21-5. In the International System (*SI*), a current of one *ampere* is defined as that current which, when flowing in each of two parallel wires one meter apart, causes them to exert a force on each other of 2×10^{-7} newtons for each meter of their length.

Sample Problem

What force per meter do two parallel wires exert upon each other at a distance of 0.1 m apart when one carries 1 A of current and the other carries 2 A of current? The currents are in the same direction.

Solution:
$I_1 = 1$ A $I_2 = 2$ A
$d = 0.1$ m $l = 1$ m

$$F = (2 \times 10^{-7} \text{ N/A}^2) \frac{I_1 I_2 l}{d}$$

$$F = (2 \times 10^{-7} \text{ N/A}^2) \frac{1 \text{ A} \times 2 \text{ A} \times 1 \text{ m}}{0.1 \text{ m}} = 4 \times 10^{-6} \text{ N}$$

It is an attractive force because the currents are in the same direction.

Test Yourself Problems

1. Two parallel electric lines are 1.0 m apart. Each carries 5.0 A of current in the same direction. What force per 100 m do the wires exert on each other?

2. What would the force between the two wires be if (a) the distance between them were halved; (b) the current in one wire was decreased to 2.0 A?

26-7 Turning Effect of Magnetic Forces on a Wire Loop

The force exerted by a magnetic field on a wire carrying electric current is used on a vast scale in the home and in industry to do tasks varying from running a sewing machine to moving an electric train. Two of its most important applications are in electric meters and electric motors. A basic element in each of these is a rectangular loop of wire that is mounted on an axle and put between the poles of a magnet or electromagnet as shown in Fig. 26-4.

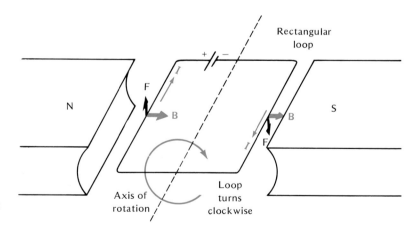

Fig. 26-4. The magnetic field exerts a pair of opposite forces on the wire loop carrying current.

When a current is sent through this loop, the magnetic field exerts a force on those two sides of the loop that are at right angles to the field. Using the left-hand rule, you can see that the force exerted on the left side of the loop is upward and the force exerted on the right side of the loop is downward. This pair of opposite forces turns the loop clockwise until it is in the vertical position.

The rotating part of meters and motors consists of a coil of wire made up of many loops of this kind. The forces exerted by a magnetic field on each of these loops when current is passed through the coil combine to turn the coil.

26-8 Galvanometer

A *galvanometer* is a sensitive meter used *to detect and measure very small electric currents*. It is the basic meter for measuring direct current. By the addition to it of simple parts, it can be used as an ammeter or voltmeter as desired.

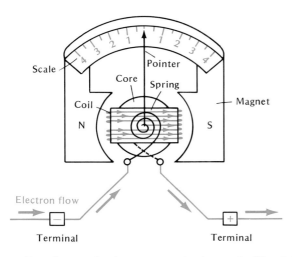

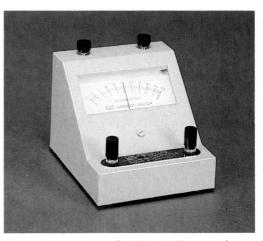

Fig. 26-5. (a) Diagram of a galvanometer. **(b)** A typical lecture-table galvanometer.

One form of galvanometer is shown in Fig. 26-5 (a). It consists of a rectangular coil of wire mounted on an axle supported by jeweled bearings and set between the poles of a permanent horse-shoe magnet. The free turning of the coil is opposed by a pair of spiral springs attached to the opposite ends of its axle. The springs ordinarily hold the coil in position so that the pointer attached to the coil is on the zero mark of the scale. A soft iron cylinder of high permeability is mounted inside the coil to concentrate and strengthen the magnetic field between the poles of the horseshoe magnet. The cylinder is fixed in such a way that the coil is free to turn around it.

The tiny current to be measured is passed into the coil through one of the spiral springs and leaves the coil through the other spiral spring. As soon as current flows through the coil, the magnetic field exerts a pair of opposite forces on each rectangular loop of which it is made. These forces turn the coil against the increasing opposition of the springs until the tension of the springs stops further motion of the coil. The reading of the current is then shown by the pointer's position on the scale. The greater the current in the coil, the stronger are the forces exerted upon it by the magnetic field.

The coil then turns further against the opposition of the springs and moves the pointer over a wider distance on the scale. The scale reading is thus a measure of the current.

When the current stops flowing in the coil, the magnetic field ceases to exert forces on it and the springs restore the coil to its zero position.

26-9 Ammeter

The coil of a galvanometer will heat up and burn out if a large current is passed into it. However, by adding a part called a *shunt* to a galvanometer, it can be changed into an ammeter suitable for measuring larger currents. A shunt is a low-resistance metal strip or wire that is connected in parallel with the coil of the galvanometer. The resistance of the shunt is usually very much smaller than the resistance of the coil of the meter. *An ammeter is thus a galvanometer plus a shunt connected in parallel with its coil,* as shown in Fig. 26-6 (a).

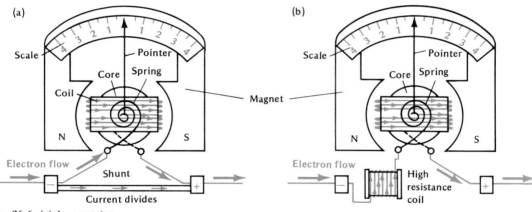

Fig. 26-6. (a) An ammeter showing the parallel connection of its shunt. **(b)** A voltmeter showing the series connection of its high-resistance coil.

As the current to be measured enters the ammeter, it divides into two parts. Most of the current takes the easier path through the low-resistance shunt. The rest of the current passes through the coil. In practice, the resistances of the coil and shunt are so adjusted that a definite part of the total current passes through the coil while the rest passes through the shunt. Thus, if the coil has four times as much resistance as the shunt, only one-fifth of the current entering the ammeter will pass through the coil. The remaining four-fifths will flow through the shunt. The ammeter pointer will, of course, measure only the current that passes through the coil, but since we know that this small current is exactly one-fifth of the total current, the scale is marked off to show the total current that is entering the meter in amperes.

The shunt makes the ammeter a very low-resistance device. It therefore does not appreciably change the resistance of the circuit in which it is inserted. This is desirable because the insertion of the ammeter into a circuit must not change the current it is supposed to measure.

26-10 Voltmeter

A *voltmeter* is made of a galvanometer by *putting a coil of high-resistance wire in series with its movable coil,* as shown in Fig. 26-6 (b). As a result, a voltmeter always has a high resistance. When a voltmeter is connected in parallel with a device whose voltage is to be measured, its high resistance permits only a very small current to pass through it. This tiny current produces a negligible change on the total current flowing in the circuit but it is great enough to cause the coil of the meter to turn and move its pointer across the scale. The voltage can then be read on the scale.

26-11 Principle of Electric Motor

In the electric motor, a current-bearing coil is made to rotate continuously by the forces exerted on it by a magnetic field. To understand how this is done, let us return to the rectangular loop discussed in Section 26-7. There, it was noted that the forces of the magnetic field continue to act upon and turn the loop when current is flowing through it only until the loop is in the vertical position.

The electric motor converts electrical energy into mechanical or kinetic energy.

Suppose that, just as the momentum of the loop carries it through the vertical position, the direction of the current in the coil is reversed. The forces turning the loop will then also be reversed so that the upper part of the loop is pushed downward while the lower part of the loop is pushed upward. These forces will then continue to turn the loop through another half-turn. If, once again, as the loop passes through the vertical position the current in it is reversed, the loop will be turned by the magnetic forces of the field through another half-turn in the same direction. Thus, *by reversing the current in the loop every half-turn, the loop can be kept rotating in the same direction.* This is what is done in the electric motor to keep a coil consisting of many loops rotating continuously.

26-12 Simple Motor

In the type of electric motor shown in Fig. 26-7, the current in a rotating coil is reversed every half-turn by an automatic switching arrangement consisting of a split metal ring called a *commutator* and *two brushes.* The rotating part of the motor is called an *armature* and is a coil of many turns of wire wound on a soft iron core. The armature is mounted on an axle between *two fixed magnetic poles.* The poles may be furnished by permanent magnets or by electromagnets. Each end of the armature coil is attached to one of the halves of the commutator which is mounted on the armature axle and rotates with the armature.

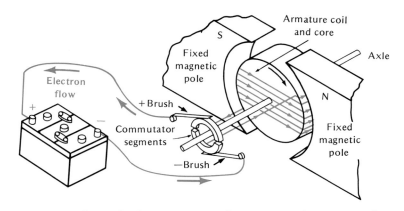

Fig. 26-7. A simple direct current motor.

Current from the battery enters the armature by means of one of a pair of copper or carbon brushes which makes contact with one half of the commutator. Current leaves the armature to return to the battery via the second brush which makes contact with the other half of the commutator. Since the brushes remain fixed while the commutator rotates, each brush is in contact with one half of the commutator during one half-turn and with the opposite half of the commutator during the second half-turn. As a result, the current in the armature reverses its direction every half-turn and provides the condition necessary to keep the armature rotating.

The power of a motor depends upon the size of the magnetic forces acting upon its armature. These forces may be increased by (1) *increasing the strength of the magnetic field* in which the armature moves, (2) *increasing the number of loops or turns in the armature,* and (3) *increasing the current* sent through the armature.

As in the galvanometer, the purpose of the soft iron core is to use its high permeability to concentrate the lines of force of the magnetic field through which the coil moves.

26-13 Magnetic Force on a Single Moving Charge

We have seen that the force exerted by a magnetic field on a wire carrying a current is the resultant of the forces exerted by the magnetic field on each of the moving charges flowing in the wire. When electron beams, proton beams, or other beams of charged particles pass through a magnetic field, the field exerts a force on each of the moving electrical particles in the beam. Since these particles are not confined in a wire, the force exerted upon them by the magnetic field will change their motion and deflect the beam of which they are a part from its original path. The amount of deflection suffered by the beam depends upon the force exerted by the magnetic field on each moving particle in it. Let us see how this force is computed.

Consider the single negatively charged particle of q coulombs in Fig. 26-8 moving at velocity v at right angles to a uniform magnetic field of strength B. Let l be the distance traveled by the charge q

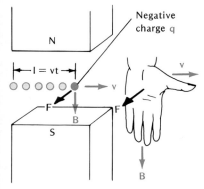

Fig. 26-8. Force exerted by a magnetic field on a moving negative charge, and the left-hand rule for finding its direction.

in t seconds. Then $l = vt$. Since q coulombs pass through l in t seconds, the current I passing through l is q/t coulombs per second. Thus, the moving particle q is equivalent to the current I flowing in length l of a circuit. The force exerted by the magnetic field B on length l of such a circuit is $F = IlB$. Substituting $I = q/t$ and $l = vt$ gives $F = q/t \times vt \times B$ whence:

$$F = qvB$$

Here $\mathbf{F}$ is the force on a charged particle q moving at right angles to a magnetic field B with velocity v. In *SI* units, F is in newtons, q is in coulombs, v is in meters per second and B is in webers per square meter. If the beam consists of electrons or protons, q is equal to the elementary electronic charge and F is the force on each electron or proton in the beam. If the beam consists of ions, q is the total charge on each ion and may be one, two, three, or more electronic charges.

A single moving charge acts like a current on which the magnetic field exerts a force.

26-14 Direction of the Force

The direction of F is found from the left-hand rule in Section 26-2 and *is at right angles to v and B* as shown in Fig. 26-8. In applying this rule to a single moving charge, it is important to remember that the direction of the current is defined as the direction in which electrons move. A moving negative charge is therefore equivalent to a current flowing in the same direction as the charge.

A moving positive charge is equivalent to a current flowing in a direction opposite to that of the charge. It follows that for a positively charged particle q moving at velocity v at right angles to a magnetic field B, the force exerted by the magnetic field is also given by $F = qvB$ but acts in the opposite direction as shown in Fig. 26-9. Note that the direction of this force on a positive charge may be found by using the right hand in the rule instead of the left.

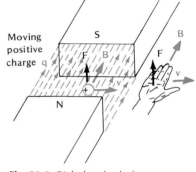

Fig. 26-9. Right-hand rule for finding the direction of the force exerted by a magnetic field on a moving positive charge.

Sample Problems

1. A beam of electrons moving at 3.0×10^7 m/s passes at right angles to a uniform magnetic field in which the magnetic induction is 2.0×10^{-4} Wb/m². What force acts upon each electron in the beam?

Solution:

$q = e = 1.6 \times 10^{-19}$ C $v = 3.0 \times 10^7$ m/s
$B = 2.0 \times 10^{-4}$ Wb/m², or N/A·m
$F = qvB$
$F = (1.6 \times 10^{-19}$ C$) \times (3.0 \times 10^7$ m/s$)$
      $\times (2.0 \times 10^{-4}$ N/A·m$)$

Note that since 1 C = 1 A × 1 s, all the units in the above expression cancel out except N giving:

$$F = 9.6 \times 10^{-16} \text{ N}$$

2. How will the force on a proton moving at the same speed as the electron in problem 1 and in the same direction differ from the force on the electron?

Solution:
It will be the same magnitude since q, v, and B are the same as before. However, F will be in the opposite direction since q is now a positive rather than a negative charge.

Test Yourself Problems

1. An ion of oxygen having 2 elementary negative electric charges is moving at right angles to a uniform magnetic field for which $B = 0.20$ Wb/m^2. If its velocity is 2.0×10^7 m/s, what is the force acting upon the ion?

2. How would the force on the oxygen ion in the above problem be affected by increasing the magnetic induction of the field to 0.30 Wb/m^2?

26-15 Circular Path of a Charged Particle in a Uniform Magnetic Field

The force acting on a charged particle moving at constant velocity at right angles to a uniform magnetic field is always perpendicular to its velocity. Therefore, it cannot change the magnitude of the velocity of the particle. However, since it accelerates the particle at a constant rate always at right angles to its velocity, it causes the particle to move in a circle. The force of the magnetic field on the moving particle is thus a *centripetal force* that compels it to move in a circle at constant speed. If the particle is positively charged, it moves in one direction. If it is negatively charged, it moves in the opposite direction.

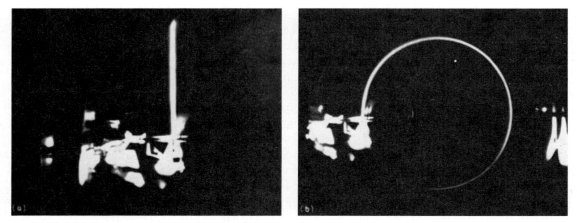

Fig. 26-10. (a) Before the magnetic field is turned on, the electron beam moves straight upward. **(b)** After the magnetic field is turned on, the electron beam is forced to move in a circular path.

In Fig. 26-10 an electron beam is fired upward from an electron gun in a glass tube (not shown) containing hydrogen gas under low pressure. As the electrons pass through the gas, they collide with atoms of gas in their path. The atoms then emit light and reveal the straight-line path of the electron beam (left). At right, a uniform magnetic field has been applied at right angles to the plane of the paper. The field now causes the electron beam to move in a circular path.

26-16 Radius of the Path

To compute the radius of the path of a charged particle q moving at constant speed v at right angles to a field of magnetic induction B, we note that the force exerted upon it by the magnetic field is

$F = qvB$. Since F is also a centripetal force, it is equal to mv^2/r where m is the mass of the charged particle and r is the radius of the circle in which it moves. Setting $mv^2/r = qvB$ gives:

$$mv = qrB \qquad \text{or,} \qquad r = \frac{mv}{qB}$$

Note that the radius of the circle in which the particle moves is proportional to mv, the momentum of the particle. The greater the momentum of the particle, the larger will be the radius of the circle it will describe.

By measuring r for a known charge q in a field of known magnetic induction B, we can obtain mv, the momentum of the particle, from the above relationship. If v can also be measured, we can obtain the mass of the charged particle by dividing mv by v. We shall see that this is essentially the method used in the mass spectrograph to measure the masses of the ions of different atoms.

Sample Problem

An ion of neon having 1 elementary positive charge and moving with a velocity of 2.0×10^4 m/s moves across a magnetic field of $B = 2.0 \times 10^{-2}$ Wb/m². If the mass of the ion is 3.4×10^{-26} kg, what is the radius of its circular path in the magnetic field?

Solution:
$m = 3.4 \times 10^{-26}$ kg $v = 2.0 \times 10^4$ m/s
$q = +1.6 \times 10^{-19}$ C

$B = 2.0 \times 10^{-2}$ Wb/m² or 2.0×10^{-2} N/A·m

$$r = \frac{mv}{qB} = \frac{3.4 \times 10^{-26} \text{ kg} \times 2.0 \times 10^4 \text{ m/s}}{1.6 \times 10^{-19} \text{ C} \times 2.0 \times 10^{-2} \text{ N/A·m}}$$

Note that since 1 C = 1 A × 1 s, all the units in the above expression cancel out except m, giving $r = 0.21$ m

Test Yourself Problems

1. A positively charged ion having 2 elementary charges and a velocity of 5.0×10^7 m/s is moving across a magnetic field for which $B = 4.0$ Wb/m². If the mass of the ion is 6.8×10^{-27} kg, what is the radius of the circular path in which it travels?

2. An electron is moving at 2.0×10^8 m/s in a magnetic field. How strong should the magnetic field be to keep the electron moving in a circle of radius 0.50 m?

26-17 Measuring the Velocity of Charged Particles

One method of measuring the velocity of charged particles in a beam involves passing the beam through both a uniform electric field and a uniform magnetic field. The experimental arrangement is shown in Fig. 26-11.

Enclosed in the left end of an evacuated glass tube is an ion "gun" like that described in Section 22-22. It sends a beam of positively charged ions toward the fluorescent screen on which the beam makes a bright spot. Between the gun and the screen, the beam passes through the uniform electric field E between two oppositely charged plates. At right angles to the charged plates is a uniform magnetic field of strength B supplied by a pair of coils.

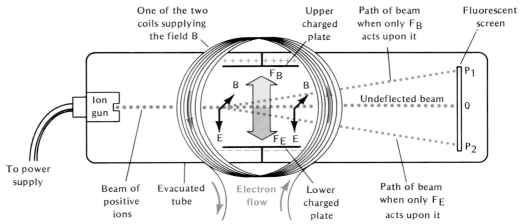

Fig. 26-11. Method of finding the velocity of the charged particles in a beam. B and E are adjusted so that the ion beam is undeflected. Then $v = E/B$.

Since a magnetic field readily passes through glass, these coils are usually mounted outside the glass tube as is shown.

The beam of ions is therefore acted upon by an electric and a magnetic field at right angles to each other. The electric field pushes each of the charges in the beam downward with a force $F_E = Eq$ where q is the positive charge on each ion. The magnetic induction B, at right angles to E, exerts an upward force on each particle in the beam equal to $F_B = qvB$, where v is the velocity of each particle in the beam. The direction of F_B can be checked by the right-hand rule of Fig. 26-9. Pointing the fingers of the open right hand in the direction of B which is perpendicular to the page and away from the reader and pointing the thumb in the direction of motion of the positive charges, you can see that the upturned palm shows the direction of F_B.

If B is reduced to zero and E is allowed to act on the beam alone, the charged particles will be pushed downward by the electric field on their way to the screen and will hit point P_2. If E is reduced to zero while B is allowed to act on the beam alone, the charged particles will be pushed upward by the magnetic field on their way to the screen and will hit point P_1. By adjusting the strength of E or B, it is possible to make the upward force F_B on the particles in the beam equal to the downward force F_E so that they cancel each other. The beam will then move through both fields, undeflected to point O. When this condition exists, $F_E = F_B$ and $Eq = qvB$, whence $v = E/B$.

To measure the velocity of the beam, we need only adjust E and B until these two fields neutralize each other's forces on the beam. The ratio E/B then is equal to v. When E is expressed in newtons per coulomb and B in newtons per ampere-meter, v comes out in meters per second.

This method of obtaining the velocity of charged ions may also be applied to obtain the velocity of the electrons in a beam shot out of an electron gun.

Sample Problem

A beam of positively charged ions passes undeflected through a pair of crossed electric and magnetic fields. When E is 5.0×10^3 N/C and B is 2.5×10^{-2} Wb/m² what is the velocity of the ions?

Solution:

$E = 5.0 \times 10^3$ N/C

$B = 2.5 \times 10^{-2}$ Wb/m² $= 2.5 \times 10^{-2}$ N/A·m

$$v = \frac{E}{B} = \frac{5.0 \times 10^3 \text{ N/C}}{2.5 \times 10^{-2} \text{ N/A·m}} = 2.0 \times 10^5 \text{ m/s}$$

Test Yourself Problems

1. A beam of ions passes undeflected through a pair of crossed electric and magnetic fields. E is 6.0×10^5 N/C and B is 3.0×10^{-3} Wb/m². What is the velocity of the ions?

2. The electrons in a beam are moving at 2.8×10^8 m/s in an electric field E of strength 1.4×10^4 N/C and in an adjustable magnetic field B at right angles to E. For what value of B will the beam pass through the crossed fields undeflected?

26-18 Measuring the Masses of Atoms

The forces exerted on moving charged particles by magnetic and electric fields make it possible to measure directly the masses of individual atoms. The instrument for making such measurements is called a *mass spectograph*. To use this instrument, we must have a beam of ions of the atoms whose masses are to be measured. These ions are usually made by knocking one, two, or three electrons out of each of the atoms being studied. The remainder of the atom is then an ion having a known positive charge usually equal to one or two or three electronic units. The atoms of an element may be ionized by putting the element in gas form and by bombarding the gaseous atoms with electrons from an electron gun. The bombardment knocks out some of the electrons in the gas atoms which then become positive ions. The ions are then accelerated and formed into a beam as explained in Section 22-22.

The mass spectograph measures the mass of the ions in these beams. The mass of each atom from which an ion was formed is equal to the mass of the ion plus the mass of the electrons removed from the atom to form that ion. However, because the mass of an electron is only about one two-thousandth of that of a proton, the difference between the mass of an ion which usually contains many protons and the mass of the neutral atom from which it came is very small.

26-19 Mass Spectrograph

One type of mass spectrograph is shown schematically in Fig. 26-12. A beam of ions having a variety of velocities comes out of the ion gun and passes between a pair of crossed electric and magnetic fields of strength E_1 and B_1. These fields allow all ions whose velocity is equal to E_1/B_1 to pass through undeflected and enter the upper evacuated chamber through a slit S. Ions with velocities greater or smaller than v are deflected and stopped by the sides of the slit. Thus all the ions that enter the upper chamber have the same velocity, $v = E_1/B_1$.

The mass spectrograph can measure the mass of single atoms.

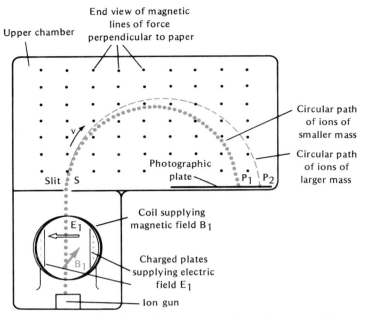

End view of magnetic
lines of force
perpendicular to paper

Upper chamber

Circular path
of ions of
smaller mass

Circular path
of ions of
larger mass

Photographic
plate

Slit S

P_1 P_2

Coil supplying
magnetic field B_1

E_1

Charged plates
supplying electric
field E_1

B_1

Ion gun

Fig. 26-12. Schematic
arrangement of a mass
spectrograph.

In the upper chamber, the ions pass through a second magnetic field at right angles to their path. This field forces them to move in a circular path. On completing half the circle, the ions hit a photographic plate which, when developed, shows a black line at P_1.

The distance SP_1 is the diameter of the circle on which the ions have traveled. The radius of this circle is therefore found by measuring SP_1 and dividing it by two. The speed of the beam is known from $v = E_1/B_1$. The charge q on the ions is also known to be one, two, or some other whole number of electronic units. By substituting these values in the relationship $mv = qrB$, derived in Section 26-16, the mass m of the ion is readily determined.

26-20 Detection of Isotopes

When the mass spectrograph is used to measure the masses of the atoms of different elements, it is found that each element has atoms with the same chemical properties but with masses differing from each other by one or more mass units. For example, when single-charged neon ions are examined in the mass spectrograph of Fig. 26-12, it is found that they make two lines on the photographic plate, one at P_1 corresponding to a mass of 20 atomic mass units and the other at P_2 corresponding to a mass of 22 atomic mass units. This means that the neon consists of at least two types of atoms, one type of mass 20 and another of mass 22 atomic mass units. As was noted in Section 13-20, atoms of the same chemical element that have different masses are called *isotopes* of that element.

The measurement of the masses and the discovery of the isotopes of the atoms of all the elements has been made possible by the mass spectrograph. The instrument may also be used to separate the isotopes of any element. For example, by putting appropriate containers at P_1 and P_2 instead of the photographic plate, the two isotopes of neon may be collected separately. Many pure samples of isotopes are made available in this way.

Sample Problem

In a mass spectrograph, certain sodium ions in the ion beam are made to move in a circular path of radius 0.081 m when the magnetic induction of the deflecting magnetic field is 3.0×10^{-2} Wb/m². If the sodium ions have a single elementary charge of 1.6×10^{-19} C and a speed of 1.0×10^4 m/s, what is the mass of this isotope of sodium in (a) kg; (b) u (1 u $= 1.7 \times 10^{-27}$ kg)

Solution:

(a) $v = 1.0 \times 10^4$ m/s
       $B = 3.0 \times 10^{-2}$ Wb/m² or N/A·m
       $q = 1.6 \times 10^{-19}$ C $r = 0.081$ m

$$mv = qrB \quad \text{whence} \quad m = \frac{qrB}{v}$$

$$m = \frac{(1.6 \times 10^{-19}\ \text{C})\,(0.081\ \text{m})(3.0 \times 10^{-2}\ \text{N/A·m})}{1.0 \times 10^4\ \text{m/s}}$$

$$m = 3.9 \times 10^{-26}\ \text{kg}$$

(b) m (in u) $= \dfrac{3.9 \times 10^{-26}\ \text{kg}}{1.7 \times 10^{-27}\ \text{kg}} = 23$ u

Test Yourself Problem

The following data were obtained for an ion beam analyzed by a mass spectrograph:

$q = 3.2 \times 10^{-19}$ C; $B = 4.0 \times 10^{-2}$ Wb/m²;
$v = 4.0 \times 10^4$ m/s;   and $r = 0.21$ m.
What is the mass of an ion in (a) kg; (b) u?

A magnetic field exerts a force on a current-bearing wire situated in the field. This force acts in a direction perpendicular to that of the magnetic induction B and to that of the current in the wire I. In the special case where I is perpendicular to B, the force F exerted by the magnetic field on a length l of the wire is given by $\mathbf{F = IlB}$. When I is expressed in amperes, l in meters, and B in newtons per ampere-meter, or webers per square meter, F is in newtons.

All conductors carrying currents exert forces upon one another through their magnetic fields. Two parallel conductors in which the currents flow in the same direction attract each other. Two parallel conductors in which the currents flow in opposite directions repel each other. The force in newtons exerted by each of two parallel conductors upon l meters of each other is given by $\mathbf{F = (2 \times 10^{-7}\ N/A^2)\ I_1 I_2 l/d}$, where I_1 is the current in amperes in the first conductor, I_2 is the current in amperes in the second conductor, and d is the distance between them in meters.

The operation of some electric meters and motors is based upon the fact that a rectangular loop carrying a current is acted upon by a pair of opposite forces when placed in a magnetic field. These forces tend to cause the loop to turn until its plane is at right angles to the lines of force. The

CHAPTER REVIEW

Summary

rotating part of many meters and motors is a coil consisting of many rectangular loops. The forces exerted by a magnetic field on each of these loops when current is passed through the coil combine to turn the coil.

The force that a magnetic field exerts upon a wire carrying a current is actually the resultant of the forces exerted by the magnetic field on each of the charged particles moving in the wire. The force exerted by a magnetic field B on a particle having a charge q and moving at right angles to the magnetic field at velocity v is given by $\mathbf{F} = \mathbf{qvB}$. If q is expressed in coulombs, v in meters per second, and B in newtons per ampere-meter, then F comes out in newtons. The direction of F is perpendicular to the directions of both v and B.

The force acting on a charged particle moving at uniform speed at right angles to a uniform magnetic field is a centripetal force. This force causes such a particle to move in a circle whose radius is $\mathbf{r} = \mathbf{mv/qB}$, where m is the mass of the particle, q is its charge, v is its speed, and B is the magnetic induction of the field. This relationship is used in the **mass spectrograph** to determine the masses of atoms.

In this chapter, we have observed the basic fact that a fixed magnetic field exerts a force on a charge moving in the field. We have seen that the direction of the force is at right angles both to the direction of the magnetic field and to the direction in which the charge is moving. In the next chapter, we shall see that a symmetrical relationship is also observed, namely, that a moving magnetic field exerts a force on a stationary charge situated in the field.

Questions

Group 1

1. A horizontal wire carrying a current northward is in a magnetic field whose direction is due east. (a) In what direction is the force exerted by the field upon the wire? (b) Upon what three factors does the magnitude of that force depend? (c) What effect does reversing the current in the wire have on the force?

2. The angle between the direction of the current in a wire and the direction of the magnetic field in which the wire is situated is slowly varied from 0° to 90° and then from 90° to 180°. What changes take place in the magnitude of the force exerted on the wire by the field?

3. Two long parallel wires are carrying currents in the same direction. (a) What kind of force do they exert upon each other? (b) What happens to this force as the wires are moved apart? (c) What happens to this force when the current in one of the wires is reversed? (d) What happens to this force when the current in one of the wires is reduced to zero?

4. Two long vertical wires are carrying current in opposite directions. The current in the first wire is twice as large as that in the second wire. Compare the force exerted by the first wire on one meter of the second with the force exerted by the second wire on one meter of the first.

5. Explain the statement: Two current-bearing wires act upon each other through their magnetic fields.

6. A beam of protons is moving horizontally from left to right. Parallel to it is a fixed horizontal wire. (a) What effect does passing a current through the wire from right to left have on the proton beam? (b) If the proton beam were removed and replaced by an electron beam moving from left to right, what effect would the current in the wire have on the electron beam? (c) Explain the difference in behavior between the proton beam and the electron beam.

7. How is the ampere defined in terms of the force between parallel conductors carrying currents?

8. A rectangular loop of wire is placed in a horizontal magnetic field so that the plane of the loop is parallel to the lines of force. Show by a diagram that when a current flows around the loop, the magnetic field exerts an upward force on one side of the loop and a downward force on the opposite side of the loop.

9. With reference to Fig. 26-5 (a), explain the operation of a galvanometer.

10. (a) How is a galvanometer converted into an ammeter? (b) How is a galvanometer converted into a voltmeter? (c) Why must the resistance of an ammeter be as low as possible? (d) Why must a voltmeter's resistance be as high as possible?

11. With reference to the simple electric motor, explain how the commutator and the brushes make it possible for the armature to continue to rotate.

12. A charged particle is moving to the right in a magnetic field whose direction is upward. Show by a diagram the direction of the force exerted by the magnetic field upon the particle if it is (a) a proton; (b) An electron. (c) If the magnetic field is kept constant, what determines the magnitude of the force exerted upon an electron moving through the field?

13. (a) Why does a charged particle moving at constant speed at right angles to a uniform magnetic field move in a circular path? (b) What effect does increasing the magnetic induction of the field have on the radius of this path for a given particle moving at a fixed speed? (c) What effect does the speed of the charged particle have on the radius of its path in the magnetic field?

14. An electron beam moves through an electric field and a magnetic field at right angles to each other and to the direction of the beam. (a) Under what conditions will the beam pass through the crossed fields undeflected? (b) How can the speed of the electrons in the beam be determined when the conditions mentioned in (a) are established?

15. (a) What are isotopes? (b) How may the isotopes of an element be separated?

Group 2

16. A horizontal beam is formed of the charged particles coming out of a radioactive substance by enclosing it in a cubical lead box having a small hole in the middle of one of its sides. When the beam coming out of the hole is passed through a vertical magnetic field, it is noticed that the beam separates into two parts. One part is turned to the left while the rest of the beam is turned to the right. What does this observation tell about the charges of the particles contained in the beam?

17. An electron beam moving horizontally away from you is deflected toward the right after passing through a certain region of space which contains either a magnetic or an electric field. (a) If the field is magnetic, what is its direction? (b) If the field is electric instead of magnetic, what is its direction?

18. The direction of the electron beam in question 17 is reversed so that it comes toward you. When the beam is passed through the same region of space, it is again deflected to your right. How does this new fact show that the region contains an electric rather than a magnetic field?

19. With reference to the mass spectrograph, explain: (a) How is a beam of ions all of which have very nearly the same velocity obtained? (b) After the ions pass the slit and enter the upper chamber, what causes them to move in circular paths? (c) Strictly speaking, what should be added to the mass of an ion as measured by a spectrograph to give the full mass of the atom from which it came?

20. All the ions entering the upper chamber of the mass spectrograph have the same velocity. (a) If two of the ions have equal masses but different charges, which will move in the larger circular path? (b) If two of the ions have equal charges but different masses, which will move in the larger circular path?

Problems

Group 1

1. A wire 1.0 m long carrying a current of 12 A is at right angles to a magnetic field of 0.20 Wb/m². What is the direction and magnitude of the force on the wire?

2. A current-bearing wire 25 cm long is at right angles to a magnetic field of 0.10 Wb/m² and is acted upon by a force of 0.50 N. What current flows through the wire?

3. A horizontal length of wire 1.0 m long weighs 0.40 N. When it is placed at right angles to a magnetic field and a current of 15 A is passed through it, the magnetic field exerts an upward force on the wire that just supports its weight. What must be the induction of the magnetic field to do this?

4. A current of 10 A is passed through a horizontal wire 40 m long at a place where the downward component of the earth's magnetic field is 5.5×10^{-5} Wb/m². (a) What force will be exerted on the wire by the earth's magnetic field? (b) Will the force be up, down, or sideways?

5. Each of two long parallel conductors carries a current of 5.0 A. The currents are in the same direction. (a) What force per meter do the conductors exert upon each other when the distance between them is 0.020 m? (b) If the current in one of the wires is reversed, what change takes place in the force they exert upon each other?

6. Each of 2 long, parallel conductors carries a current of 10 A. At what distance apart should the wires be placed so that the force exerted by each of them on 0.50 m of the other is 4.0×10^{-5} N?

7. A beam of electrons moving at 2.0×10^{8} m/s is at right angles to a uniform magnetic field of 4.0×10^{-2} Wb/m². (a) What force acts on each electron in the beam? (b) What is the radius of the circular path in which this beam will travel through the magnetic field? (The mass of the electron is 9.1×10^{-31} kg.)

8. The electrons in a beam in a cathode ray tube are moving horizontally at 5.0×10^{7} m/s and pass through a vertical magnetic field of 3.5×10^{-3} Wb/m². (a) What force acts upon each electron in the beam? (b) If this were a beam of protons moving at the same speed, what force would be exerted upon each proton?

9. (a) Compute the radius of the circular path taken by each electron in problem 8 (a). (b) Compute

the radius of the path taken by each proton in problem 8 (b). (The mass of a proton is 1.7×10^{-27} kg.)

Group 2
10. A proton moves at right angles to a magnetic field of 1.7 Wb/m² in a circular path of 0.40 m radius. What is the speed of the proton?

11. What is the speed of a proton that moves in a circle of 0.25 m radius when it enters a magnetic field of 0.20 Wb/m² at right angles to its path?

12. A beam of positively charged ions passes undeflected through a pair of crossed electric and magnetic fields when E is 3.0×10^{3} N/C and B is 0.15 Wb/m². What is the velocity of the ions?

13. A beam of electrons is moving at 1.2×10^{7} m/s in a direction at right angles to crossed electric and magnetic fields. If the magnetic field is 3.0×10^{-4} Wb/m², what must the electric field be so that the electrons pass through both fields undeflected?

14. In a mass spectrograph, certain lithium ions move in a circular path of radius 0.17 m in a magnetic field of 1.2×10^{-2} Wb/m². If these lithium ions are known to have one elementary charge and a speed of 3.2×10^{4} m/s, what is the mass of this isotope of lithium in (a) kg; (b) u? (1 u = 1.7×10^{-27} kg.)

15. Lithium has a second isotope of mass of about 7 u or 1.2×10^{-26} kg whose ions are also present in the beam entering the spectrograph of problem 14. If these ions also have a single elementary charge and have the same speed as the ions of problem 14, (a) what is the radius of their circular paths; (b) what will be the distance by which the spectrograph will separate these two isotopes? (This is $SP_2 - SP_1$ in Fig. 26-12, or the difference between the diameters of their circular paths.)

Applying Physics

1. Make a simple electric meter with a magnetic compass and about 3 m of copper wire, as shown in Fig. 26-13. Using plastic tape, fasten two pencils to the opposite edges of a square of cardboard 8 cm on a side. Fix the compass in the middle of the cardboard square. Leaving about 1 m of wire free at each end, wrap 4 turns

of wire around the compass and the cardboard. Spread the turns to permit a view of the compass needle and fasten the turns to the cardboard with plastic tape so that they will remain fixed. This arrangement of a compass inside of the 4-turn coil of wire is an example of a meter of the type known as a tangent galvanometer.

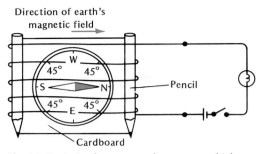

Fig. 26-13. A simple tangent galvanometer which you can make.

To use this meter, first turn it so that when the compass is at rest under the influence of the earth's magnetic field, the wires of the coil are parallel to the compass direction as shown in the figure. Now adjust the angular scale of the compass so that the letter N (which is the zero mark) is in front of the N-pole of the compass. The meter is now ready for use and must not be moved from this position.

Connect a single flashlight cell, a flashlight bulb, and the coil of the meter in series. The compass is now being acted upon by two magnetic fields, that of the earth and that produced inside the coil by the current passing through it. The compass will therefore take a position in the direction of the resultant of these two fields. The angle through which the compass turns is read on the compass scale and is a rough measure of the current flowing through the coil.

This meter can be made sensitive to very small currents by adding many more turns to the coil around the compass. When there are fewer turns on the coil, the meter becomes less sensitive and is therefore suitable for detecting larger currents.

2. The force exerted by a magnetic field on moving electrons can be demonstrated by using a strong bar magnet or electromagnet and the picture tube of your television set. With the set tuned into a station, bring one pole of the magnet close to the screen and move it parallel to a vertical line in the television picture such as the edge of a door. The magnetic field exerts a force on the electrons that are being shot toward the screen. As a result, the vertical line will be distorted as the magnet moves by.

The Aurora: Action! Lights!

The spectacular light displays in the atmosphere toward the north and south magnetic poles were once mysterious **phenomena. Today, from data scientists have gathered through the use of satellites and ground-based observations, we know that the auroral brilliance is an immense electrical discharge similar to that occurring in a neon sign.**

To understand the aurora, picture the earth enclosed by its magnetosphere, a huge doughnut-shaped region created by the earth's magnetic field. Blasting toward the earth is the solar wind, a swiftly moving plasma of ionized gases with its own magnetic field. Protons and electrons in this solar wind speed earthward along the magnetic lines of force with a corkscrew motion. The earth's magnetosphere is a barrier to the solar wind and forces its charged particles to flow around itself. But in the polar regions lines of force of the earth and of the solar wind bunch together. Here many of the charged particles break through and enter the earth's magnetic field. They now spiral back and forth between the earth's magnetic poles very rapidly. In the polar regions protons and electrons ionize and excite the atoms and molecules of the upper atmosphere, causing them to emit auroral radiations of visible light.

The colors of the aurora depend on the atoms emitting them. The dominant greenish-white light comes from low excitation of oxygen atoms. During huge magnetic storms oxygen atoms also undergo high energy excitation and emit a crimson light. Excited nitrogen atoms contribute bands of color varying from blue to violet.

Viewed from space, a dimly glowing belt can be seen wrapped around each of the earth's magnetic poles. These are the auroras. Each aurora hangs like a curtain of light stretching over the polar regions and into the higher latitudes. Violent magnetic storms on the sun, which we see as sunspots, may intensify the solar wind and extend auroral displays as far south as Florida.

Studies of the aurora have given physicists new information about the behavior of plasmas, which sheds light on the nature of outer space and is being applied in our attempts to harness energy from the fusion of atoms.

Electromagnetic Induction 27

27-1 Faraday's Demonstration of Electromagnetic Induction

Oersted discovered that every electric current produces a magnetic field. We now see that the reverse is also true: *A magnetic field can induce a current.* The process whereby this is done is called *electromagnetic induction.* It was discovered by Michael Faraday (1791–1867) in England and independently by Joseph Henry (1797–1878) in America. Such a current is said to be *induced.*

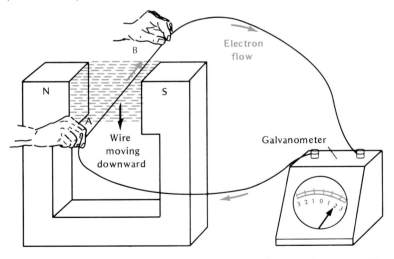

Fig. 27-1. Faraday's experiment showing how to induce a current in a conducting wire.

To study electromagnetic induction, Faraday used a setup like that shown in Fig. 27-1. The ends of a piece of wire are connected to a galvanometer. A straight section *AB* of this wire is held between the poles of a horseshoe magnet. When the wire is moved downward so that *AB* moves perpendicularly across the lines of force of the magnetic field between the poles, a current begins to flow in the wire and makes the galvanometer needle swing to one side. However, the moment the wire stops moving, the flow of current stops and the galvanometer needle returns to zero.

When the motion of the wire is now reversed by moving it upward, current again begins to flow in the circuit, but the galvanometer needle swings in the opposite direction. This shows that reversing the direction of motion in the wire reverses the direction of the current induced in it. Once again, the moment the wire stops moving, the flow of current also stops.

When the wire is moved in the magnetic field in directions differing more and more from that perpendicular to the lines of force, the current induced in it decreases steadily and becomes zero when the wire moves parallel to the lines of force.

These observations demonstrate that *current is induced in a conductor forming part of a closed circuit only when the conductor is moving so that it cuts across the lines of force of a magnetic field.* Furthermore, only that component of the conductor's motion perpendicular to the lines of force is useful in inducing the current. When the conductor moves parallel to the lines of force or stops moving, no current is induced in it.

27-2 Induced EMF

To make a current flow in a circuit, there must be a source of energy or an EMF. In a circuit containing a battery, the battery furnishes the EMF. In the above circuit, the movement of the wire segment *AB* across the lines of force of the magnetic field supplies the energy that moves the electrons through the circuit. It follows that:

> Moving a wire across the lines of force of a magnetic field induces an electromotive force in the wire.

The segment of wire *AB* acts like a chemical battery with terminals at *A* and *B*. The electromotive force induced in it when it moves across the magnetic field sets up a difference of potential between *A* and *B* which is available to provide energy to move electric charges as soon as *A* and *B* are connected by a conducting path. As in the case of a circuit connected to a battery, the size of the induced current depends upon the size of the induced EMF and the resistance of the circuit.

The EMF induced in a wire conductor moving in a magnetic field is proportional to the number of lines of force the wire cuts per second.

Faraday found that the *magnitude of the EMF* induced in a conductor moving at right angles to the lines of force of a magnetic field *depends only upon its velocity and the strength of the field.* The faster the conductor is moved, and the stronger the magnetic field, the greater is the EMF induced in it. *The EMF induced in a wire conductor* moving across a magnetic field therefore *is proportional to the number of lines of force cut by the conductor per second.*

27-3 Source of Induced EMF

To understand how moving a wire across a magnetic field induces the flow of the free electrons in the wire, consider Fig. 27-2. Here, a section l of a conductor is being moved downward through a field of magnetic induction B. If v is the velocity of the conductor, it is also the velocity of every free electron e inside the conductor. The magnetic field therefore exerts a force $F = evB$ on each of the free electrons. Applying the left-hand rule illustrated in Fig. 26-8, it is seen that F is pushing the free electrons in the wire in the direction shown.

The work done by F in moving each electron through the length l of the conductor is Fl. Since $F = evB$, this work is equal to $evBl$. Now, by definition, the work or energy supplied to a circuit per unit of charge is the electromotive force responsible for keeping current flowing in that circuit. The electromotive force induced in l is therefore the work done on each electron, $evBl$, divided by e, the charge on an electron. It follows that the electromotive force induced in a wire of length l that cuts across a magnetic field B at velocity v is:

$$EMF = vBl$$

This relationship is one form of the law known as *Faraday's law of induction*.

In SI units, v is expressed in meters per second, B is in webers per square meter and l is in meters. The EMF then comes out in volts.

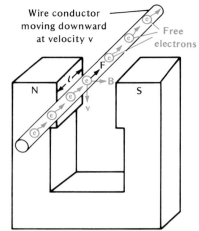

Wire conductor moving downward at velocity v — Free electrons

Fig. 27-2. Free electrons in a conductor moving at right angles to a magnetic field are acted upon by magnetic forces which set up an induced current.

27-4 Factors Affecting the Induced EMF

From the law of induction, it appears that the EMF induced in a wire conductor moving across the lines of force of a magnetic field may be increased in three ways: (1) by increasing the *velocity v* with which the wire is moving, (2) by increasing the *magnetic field B*, and (3) by increasing the *length l* of the wire cutting across the magnetic lines of force.

If the wire segment l is part of a closed circuit, the induced EMF will supply energy and drive a current through it. The size of the induced current depends on the resistance of the circuit just as does the size of a current in the circuit operated by a chemical battery.

If the wire segment is not part of a closed circuit, the induced EMF will pile up a negative charge at one of its ends and an equal positive charge at the other. These ends will then have a difference of potential between them similar to that between the terminals of a battery on open circuit.

Sample Problem

A wire 0.4 m long moves across a uniform magnetic field in which the induction is 2×10^{-2} Wb/m² at the velocity of 5 m/s. (a) What EMF is induced in the wire? (b) If the wire is in a circuit of resistance 0.2 Ω, what current flows through it? (c) What force must be applied to the wire to keep it moving through the magnetic field?

Solution:

(a) $v = 5$ m/s $B = 2 \times 10^2$ Wb/m² $l = 0.4$ m

EMF $= vBl = (5$ m/s$)(2 \times 10^{-2}$ Wb/m²$)(0.4$ m$)$

EMF $= 0.04$ V

(b) EMF $= 0.04$ V $R = 0.2\ \Omega$

$$I = \frac{V}{R} = \frac{0.04 \text{ V}}{0.02\ \Omega} = 0.2 \text{ A}$$

(c) $F = IlB = (0.2$ A$)(0.5$ m$)(2 \times 10^{-2}$ Wb/m²$)$

$F = 2 \times 10^{-3}$ N

Test Yourself Problems

1. A wire 0.40 m long cuts perpendicularly across a magnetic field for which B is 2.0 Wb/m² at a velocity of 8.0 m/s. (a) What EMF is induced in the wire? (b) If the wire is in a circuit that has a resistance of 6.4 Ω, what current flows through it?

(c) What force must be applied to the wire to keep it moving through the magnetic field at this velocity?

2. At what speed must a 0.20-m length of wire cut across a magnetic field for which B is 2.5 Wb/m² to have an EMF of 10 V induced in it?

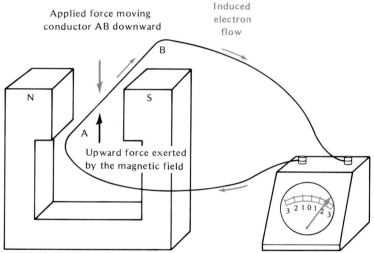

Fig. 27-3. The work done in moving the conductor *AB* through the magnetic field supplies energy to the circuit.

27-5 Where the Energy Comes From

The energy that makes an induced current flow in a conductor comes from the work done by the agent that moves the conductor across the magnetic field. Referring to Fig. 27-3, it can be seen how this work is done. As soon as the wire segment *AB* moves downward across the magnetic lines of force, current is induced in it. The magnetic field then exerts an upward force on this current-bearing wire and resists its downward motion. To keep the wire moving downward, the agent that is moving the wire must exert a force upon it equal and opposite to this upward force. As this downward force moves the wire through the magnetic field, it does mechanical work and feeds the energy into the wire that is responsible for the induced current. The energy change takes place in accordance with the law of conservation of energy. The work done in moving the conductor is equal to the electrical energy used to move the induced current.

27-6 Direction of Induced Current

The direction of the induced current is readily found by the following rule: *If the left hand is held so that the outstretched fingers point in the direction of the lines of force of the magnetic field and the extended thumb points in the direction in which the wire is moved, the palm will face in the direction in which an induced current, consisting of electrons, will flow.*

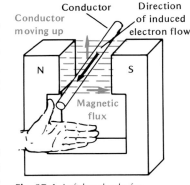

Fig. 27-4. Left-hand rule for finding the direction of the induced electron current.

27-7 Alternating Current Generator

Through the generator, electromagnetic induction provides the abundant quantities of electrical energy available in the modern world. *The generator changes mechanical energy into electrical energy.* It operates on the principle that a wire moving across the lines of force of a magnetic field has an EMF induced in it.

The simple generator shown in Fig. 27-5 consists essentially of a coil of wire called the *armature* that is rotated between the *poles of a magnet.* To help concentrate the magnetic lines of force through the coil, it is wound on a *soft iron core.* As the coil rotates, its individual conductors cut across the lines of force between the poles of the magnet and an EMF is induced in them. The induced EMF's in each turn of the armature add up and set up a difference of potential between the ends of the coil which are connected to two *slip rings* mounted on the axle on which the coil rotates. The slip rings make contact with two carbon or copper *brushes* which connect them with an external circuit (here, it is the lamp) being supplied with electrical energy.

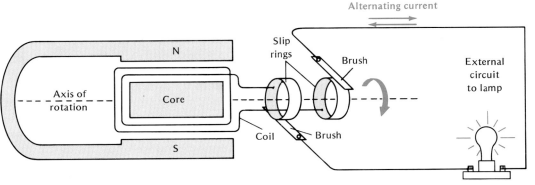

Fig. 27-5. Diagram of an alternating current generator.

This generator supplies an alternating current to the external circuit. An *alternating current* is one that flows in a circuit with a regular rhythm, first in one direction and then in the opposite direction. When the current flows back and forth in the circuit once, it is said to have completed one cycle. The number of cycles it completes per second is called the *frequency* of the alternating current and is measured in *hertz* (Hz).

Lines of force   Armature coil

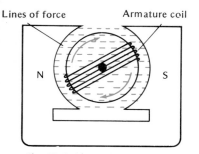

Fig. 27-6. Alternating current is induced in a coil rotating in a magnetic field.

To understand why this generator furnishes an alternating current, consider the left half of the armature coil in Fig. 27-6. During one half-turn of the armature, this part of the coil moves upward through the magnetic lines of force. During the next half turn, this part of the armature coil moves downward. Therefore, as the armature rotates, this part of the coil keeps cutting the magnetic lines of force alternately upward and downward in rapid succession. This is also true of the right half of the armature coil. As a result, the current induced in the armature coil reverses its direction every half-turn of the armature. The frequency of the alternating current of this simple generator is equal to the number of revolutions made by the armature per second.

27-8 Graph of Alternating EMF and Current

It should be noted in Fig. 27-6 that the armature coil cuts lines of force fastest whenever it is passing through its horizontal position. The largest EMF is therefore induced in it at that time. The armature coil momentarily stops cutting lines of force when passing through its vertical position. No EMF is induced in it at that time. Thus, as the armature makes the first half-turn, the EMF induced in it rises from zero to a maximum and then drops back to zero again. On the next half-turn, the induced EMF reverses direction and again goes from zero to a maximum and back to zero again. This results in an alternating EMF that changes in the manner of a sine curve as shown by the graph of Fig. 27-7.

An alternating EMF applied to a resistor drives an alternating current through it. Like the alternating EMF, this current varies in direction and magnitude, as shown in Fig. 27-7. To describe its magnitude, we compare its heating effect with that of a direct current. An alternating current is said to have an effective value of 1 ampere if it generates heat in a given resistor at the same rate as a direct current of 1 ampere. It turns out that the effective value of an alternating current is equal to 0.707 times its maximum value. Similarly, the effective value of an alternating EMF is 0.707 times its maximum value. Alternating current meters measure the effective values of alternating currents and voltages. Thus, a meter that reads 110 volts is measuring an alternating voltage whose maximum value is $110 \div 0.707 = 156$ volts.

Fig. 27-7. Graph of an alternating EMF.

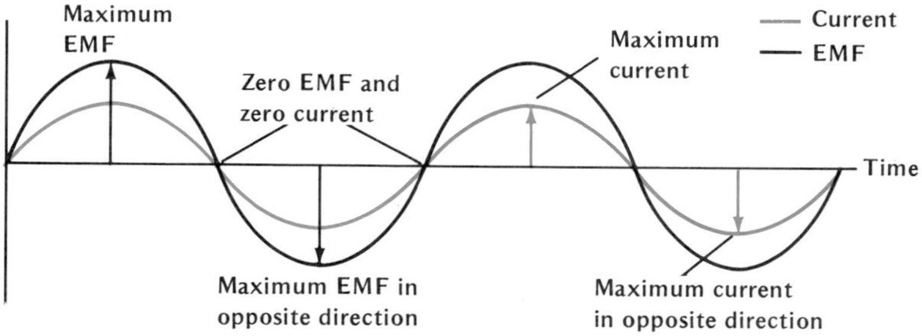

Maximum EMF

Zero EMF and zero current

Maximum current

——— Current
——— EMF

Time

Maximum EMF in opposite direction

Maximum current in opposite direction

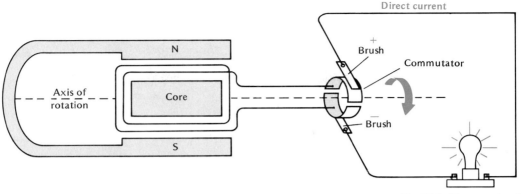

Fig. 27-8. Diagram of a direct current generator.

27-9 Direct Current Generator

To convert an alternating current generator into a direct current generator, it is necessary to remove the slip rings and replace them with a *commutator* like the one used in the motor, as shown in Fig. 27-8. As in the case of the motor, the commutator and brushes are an arrangement for reversing the direction of the current leaving the generator and entering the external circuit every half-turn of the armature. The reversal takes place just as the current in the armature is also about to change direction. As a result, the current in the external circuit always flow in one direction even though the current in the armature continues to alternate.

27-10 Factors Affecting EMF of a Generator

The magnitude of the EMF supplied by a generator depends upon the rate at which the wire conductors making up its armature coil cut across the lines of force of its magnet. *The EMF can be increased by* (1) *increasing the speed of rotation of the armature*, (2) *increasing the number of turns on the armature* so that more segments of wire will cut the lines of force at the same time, and (3) *using stronger magnets* to provide a greater concentration of lines of force. Further concentration of the lines of force is obtained by means of a high permeability soft iron core on which the armature coil is usually wound.

27-11 Generator and Motor Compared

Comparing the motor of Fig. 26-7 and the direct current generator of Fig. 27-8, it is obvious that their construction is the same. They differ only in use. In the *generator*, *mechanical energy* is supplied to rotate the armature and *is converted into electrical energy*. In the *motor*, the process is reversed. *Electrical energy* is supplied to rotate the armature and *is converted into mechanical energy*. Thus, a generator can be used as a motor and a motor can be used as a generator.

27-12 Back EMF

When a motor is running, its armature coil rotates between the poles of a magnet and an EMF is induced in the coil. This induced EMF resists the applied EMF that is driving the current through the armature coil. It is therefore called a *back EMF*. The net EMF that is effective in sending a current through a motor armature is therefore the applied EMF minus the back EMF. Thus, if the back EMF of a 110-volt motor is 77 volts, the net EMF effective in driving current through the motor amounts to only $110 - 77$, or 33 volts.

27-13 Relative Motion and Induction

Let us redo Faraday's experiment (Section 27-1), this time keeping the wire conductor of length *l* fixed but moving the magnetic field by moving the horseshoe magnet, as shown in Fig. 27-9. As the magnet is moved to the left with velocity *v*, its lines of force are cut by the wire and the galvanometer shows that an EMF is induced in the wire. When the magnet stops moving, the induced EMF drops to zero. When the magnet moves to the right at the same speed *v* as before, an EMF is again induced equal but opposite in direction to the previously induced EMF. Thus, it makes no difference whether the magnetic field moves past the stationary wire or the wire moves through a stationary magnetic field. Only the relative velocity between the wire and the magnetic field is necessary to induce an EMF. In both cases, the EMF is equal to *vBl*.

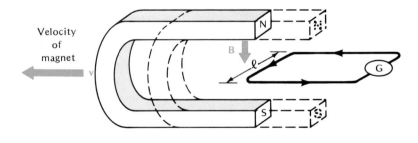

Fig. 27-9. A current is induced in the fixed conducting circuit by moving the magnet.

27-14 Changing Magnetic Flux and Induced EMF

So far, we have expressed the EMF induced in a circuit in terms of the rate at which a moving conductor that is part of the circuit cuts across magnetic lines of force. Another useful way to express the EMF induced in a circuit is in terms of the rate at which the number of lines of force or the magnetic flux passing through an entire circuit changes.

In Fig. 27-10, a rectangular circuit containing the conductor *l* and the galvanometer is moving to the right with constant velocity *v* through a fixed uniform magnetic field of strength *B*. During each second, the conductor *l* moves a distance equal to *v* and cuts across all the lines of force that pass perpendicularly through the shaded rectangle in the figure between its initial and final positions. This rectangle has an area equal to *lv*. Since *B* is the magnetic flux passing through a unit of area at right angles to the lines of force, the flux passing through the area *lv* is *Blv*. This is the part of the total flux that passed through the circuit at the beginning of the given second but is outside the circuit at the end of that second. It therefore represents the change that takes place in the flux passing through the circuit during every second of the circuit's motion. But *Blv* is equal to *vBl* which is the value of the EMF induced in the circuit.

It follows that *the EMF induced in the circuit is equal to the rate of change of magnetic flux through the circuit:*

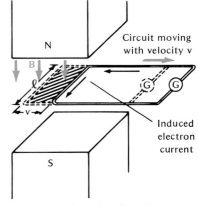

Fig. 27-10. The induced EMF is equal to the rate of change of magnetic flux through the circuit.

Induced EMF = change in magnetic flux per second.

This relationship is also known as *Faraday's law of induction.* It is a more general form than that expressed in Section 27-3.

27-15 Applying the Law of Induction

In using the law of induction, the rate of change of flux is expressed in webers per second. The induced EMF then comes out in volts. Thus, if the magnetic flux through a circuit changes at the rate of 10 webers per second, an EMF of 10 volts is induced in the circuit.

Faraday's law of induction provides the basis for defining the weber in the *SI* metric system. A *weber* is defined as the magnetic flux which, passing through a circuit of one turn, produces in it an EMF of 1 volt as it is reduced to zero at a uniform rate in 1 second.

Note that the law of induction makes it unnecessary to know the shape of the circuit or whether it is the circuit that is moving through the magnetic field or vice versa. Whatever the cause, a change in the magnetic flux passing through any circuit induces an EMF in the circuit.

The circuit considered in Fig. 27-10 consists of a single turn of wire. When the circuit consists of a coil of *N* turns of wire, a given rate of flux change through the coil induces an equal EMF in each turn. The total EMF is therefore *N* times the EMF induced in each turn. See the sample problem 2 below.

Sample Problems

1. The magnetic induction passing through a rectangular loop of wire 0.2 m long by 0.1 m wide increases from 0 to 0.4 Wb/m² in 10^{-3} s. What EMF is induced in the loop?

   Solution:
   The area of the loop is $A = 0.2$ m $\times$ 0.1 m = 0.02 m². The maximum magnetic flux passing

through the loop is the magnetic induction *B* times the area of the loop, or $B \times A = 0.4$ Wb/m² $\times$ 0.02 m² = 0.008 Wb.

The induced EMF is the change in the flux per unit of time. Since the flux changes from zero Wb to 0.008 Wb in 10^{-3} s, the flux change per second is 0.008 Wb $\div$ 10^{-3} s = 8 Wb/s. The EMF induced by a rate of change of magnetic flux of 8 Wb/s is 8 V.

2. The magnetic flux through a coil of 50 turns is increasing at the rate of 0.5 Wb/s. (a) What is the EMF induced in each turn of the coil? (b) What is the total EMF induced in the coil?

Solution:

(a) The EMF induced per turn by a magnetic field changing at the rate of 0.5 Wb/s is 0.5 V.

(b) The EMF induced in the entire coil of 50 turns is 50 turns × 0.5 V per turn or 25 V.

Test Yourself Problems

1. The magnetic flux through a single loop of wire changes from 0.20 Wb to zero in 0.40 s. What EMF is induced in the loop?
2. The magnetic induction through a loop of wire enclosing an area of 0.010 m² drops from 5.0 × 10^{-2} Wb/m² to zero in 2.5 × 10^{-2} s. (a) What was the maximum magnetic flux passing through the loop? (b) What was the rate of change of flux through the loop? (c) What EMF was induced in the loop?
3. The magnetic flux through a coil of wire having 40 turns is increasing at the rate of 0.30 Wb/s. What EMF is induced in the coil?

27-16 EMF Induced by a Moving Magnetic Field

To illustrate how a changing magnetic flux through a circuit induces an EMF in it, consider Fig. 27-11. Here, the ends of a circular coil of wire are connected to a galvanometer. As the north pole of a bar magnet moves into the coil, the galvanometer indicates that a current is being induced in the coil. The induced current stops flowing as soon as the magnet stops moving. As the north pole of the magnet then withdraws from the coil, a second current is induced in it, but this time it is in the opposite direction. Again, the induced current stops flowing as soon as the magnet stops moving.

These observations are readily explained. As the magnet moves forward toward the coil, the number of its lines of force that pass through the coil increases. The magnetic flux passing through the coil is then increasing, and an EMF equal to the rate of increase of flux times the number of turns in the coil is induced in the coil. When the magnet moves out of the coil, its lines of force are being withdrawn and the flux through the coil is decreasing. Again, the change in the flux through the coil induces an EMF in it equal to the rate of decrease of the flux times the number of turns in the coil. Finally, when the magnet is not moving, the change in the flux through the coil is zero so that no EMF is induced in it.

27-17 Lenz's Law

The German scientist Lenz (1804–1865) discovered a law that relates the direction of an induced current to the change that produced it. It states: *An induced current always flows in such a direction as to oppose by its own magnetic field the motion or change that induced it.*

Let us apply Lenz's law to the experiment illustrated in Fig. 27-11. When the north pole of the magnet moves toward the front end of the coil, the current induced in the coil sets up a magnetic field that opposes that motion. To do this, the induced current flows in such a direction as to produce a magnetic north pole in

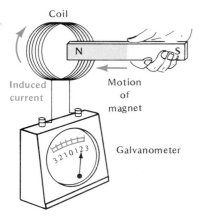

Fig. 27-11. In accordance with Lenz's law, the magnetic field set up by the induced current opposes the motion of the magnet that induces it.

front of the coil that repels the north pole of the approaching magnet. As the diagram shows, the induced current direction must therefore be clockwise.

When the north pole of the magnet is withdrawn from the coil, the induced current reverses direction and sets up a magnetic south pole at the front end of the coil. The south pole of the coil attracts the north pole of the magnet thus opposing its movement away from the coil.

27-18 Lenz's Law and the Conservation of Energy

Lenz's law is a consequence of the law of conservation of energy. If Lenz's law were not true, energy would either be created or destroyed during the process of electromagnetic induction, and the law of conservation of energy would be violated.

To illustrate, suppose that the direction of the current induced in the above coil when the north pole of the magnet approaches it were such as to make the front end of the coil a south pole instead of a north pole. The coil would then pull the north pole of the magnet into itself with increasing speed and without the aid of any outside force. At the same time as the north pole sped into the coil, an increasing current would be induced in the coil. Energy would thus be supplied to make the magnet move faster as well as to make the electrons flow faster in the coil, but no outside source would be supplying this energy. This is contrary to the law of conservation of energy.

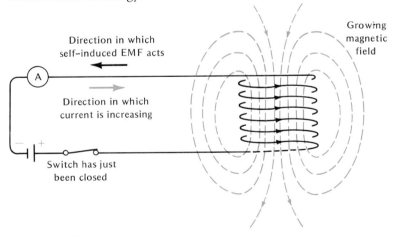

Fig. 27-12. The self-induced EMF opposes the change in the current that produces it.

27-19 Self-Induction

An EMF is induced in any circuit in which the magnetic flux is changing. Since any change in the current flowing in a circuit automatically produces a corresponding change in the magnetic flux surrounding the circuit, any increase or decrease in the current flowing in a circuit induces an EMF in that circuit. This EMF is said to be *self-induced*.

Every change in the current in a circuit induces an EMF in the circuit that opposes the change.

Consider the circuit in Fig. 27-12 in which a coil of wire, a switch, an ammeter, and a battery are connected in series. When the switch is open, there is no magnetic field inside or around the coil. At the moment the switch is closed, current begins to flow through the coil and builds up a magnetic field inside and around it. As the magnetic flux through the coil increases rapidly, it induces an electromotive force in it. By Lenz's law, the direction in which this self-induced EMF will drive a current is such as to oppose the change that induced the EMF.

In this case, the change is the increase in current. The self-induced EMF therefore acts in a direction opposite to the battery EMF and opposes the increase of current in the circuit. The effect of the self-induced EMF will be to delay the growth of the current in the circuit after the switch is closed. When the current finally reaches its maximum value, so will the magnetic field around the coil. Since the magnetic flux in the coil is then no longer changing, the self-induced EMF drops to zero.

If the switch in the circuit is now opened, the magnetic field around the coil begins to collapse, going rapidly from its maximum value to zero. The decrease in the magnetic flux through the coil again induces an EMF in it. As before, the direction in which this self-induced EMF acts is such as to oppose the change that induced it.

The change here is the stoppage of the flow of current. The self-induced EMF therefore acts in such a direction as to tend to keep the current flowing in its original direction. When the coil has very many turns of wire wound on a soft iron core, the self-induced EMF may be great enough to cause a spark to leap across the air gap made in the circuit when the switch is opened.

27-20 Unit of Inductance

Although self-induction is much more noticeable in coils than in straight wires, every circuit or part of a circuit has the ability to set up self-induced EMF's. This ability is called *inductance* and is measured in a unit called the *henry* (symbol, H). A conductor with an inductance of one henry is such that an EMF of one volt is self-induced in it when the current through it changes at the rate of one ampere per second.

27-21 Current Induced by a Changing Current

The changing magnetic field associated with a circuit in which the current is changing induces an EMF not only in its own circuit but also in any other circuits or conductors that may be nearby. In Fig. 27-13 two insulated wire coils are mounted near each other. The first coil, called the *primary coil*, is connected to a battery and a switch. The second coil, called the *secondary coil*, is completely separated from the first and has its ends connected to a galvanometer.

When the switch is closed in the primary circuit, the galvanometer shows that a current is induced in the secondary coil. However, this induced current flows only for a moment and then drops to zero even though a steady current is now flowing in the primary circuit. When the switch is opened in the primary circuit, a momentary surge of current is again induced in the secondary circuit, but this time it flows in the opposite direction.

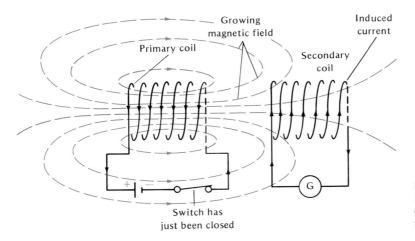

Fig. 27-13. A current is induced in the secondary coil by any change in the magnetic field of the primary coil.

Summarizing these observations, a momentary current is induced in the secondary coil only at the moment the current in the primary circuit is switched on and again at the moment that current is switched off. These observations are readily explained by considering the changes that occur in the magnetic flux passing through the secondary coil.

At the moment the primary current is switched on, the current in the primary circuit begins to grow to its final steady value and so does the magnetic field associated with it. This growing magnetic field spreads out into space and passes through the secondary coil. The increased magnetic flux through that coil then induces an EMF in it. Once the current in the primary circuit reaches its steady value, its magnetic field also no longer changes and the current induced in the secondary coil therefore drops to zero.

When the switch in the primary circuit is opened, both its current and its magnetic field rapidly return to zero. The magnetic flux passing through the secondary coil therefore decreases rapidly and once more causes an EMF to be induced in that coil. Since this EMF is induced by a decreasing instead of an increasing magnetic flux, the induced current flows in a direction opposite to the flow that occurred when the primary current was switched on.

27-22 Principle of Transformers

If we replace the battery and switch in the primary circuit of Fig. 27-13 by an alternating current generator, the primary current will reverse its direction every half cycle. Furthermore, every time the primary current reverses direction, the magnetic field it sets up around the primary coil will also reverse direction. An alternating magnetic field will therefore be continually sweeping in and out of the primary coil. This changing magnetic field also moves in and out of the secondary coil and induces an alternating EMF in each of its turns equal to the rate of change of the magnetic flux through the coil.

The total EMF induced in the secondary coil is therefore the sum of the EMF's induced in each of its turns. If the number of turns and the applied EMF in the primary coil are kept the same, it follows that the EMF induced in the secondary coil can be varied by changing the number of its turns. The more turns put on the secondary coil, the greater will be the EMF induced in it. The fewer turns put on the secondary coil, the smaller will be the EMF induced in it.

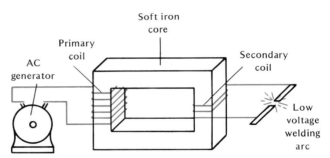

Fig. 27-14. A step-down transformer.

27-23 Step-up and Step-down Transformers

This principle is used in the transformer to change the EMF of a given alternating current to some desired higher or lower value. The simple transformer in Fig. 27-14 consists of a *primary coil* and a *secondary coil* wound on a soft iron ring called a *core*. The given alternating EMF is applied to the primary coil. Depending on the number of turns in the secondary coil, a higher or lower induced EMF is obtained from it. The purpose of the high permeability iron core is to guide the alternating magnetic field set up by the primary current efficiently through the secondary coil.

A transformer that converts the applied EMF to a higher one is called a *step-up transformer*. A transformer that converts the applied EMF to a lower one is called a *step-down transformer*. A step-up transformer is used to obtain the high voltages needed to operate television tubes, neon signs, and X-ray machines. A step-down transformer is used to obtain the low voltages needed in such devices as welding machines and electric doorbells.

To transmit electric power over long distances, step-up transformers are used to raise the voltage but keep the current as low as possible. The lower the current, the lower is the power loss in heating the lines.

27-24 EMF Relationships in a Transformer

The alternating magnetic field that sweeps back and forth through the core of a transformer passes through both the secondary and the primary coils. It therefore induces an equal EMF in each turn of both coils. The total EMF induced in the secondary coil is proportional to the number of turns in the secondary coil, while the total EMF self-induced in the primary coil is proportional to the number of turns in the primary coil. It follows that:

$$\frac{E_s}{E_p} = \frac{N_s}{N_p}$$

where E_s and E_p are the induced EMF's in the secondary and primary coils, and N_s and N_p are the numbers of turns in those coils. In a well-designed transformer where heat and other losses are negligible, the self-induced EMF in the primary coil is equal and opposite to the EMF applied to the primary coil by the generator. Hence, E_p in the above relationship also stands for the EMF applied to the primary coil.

According to this relationship, *the number of times a transformer steps up or steps down the alternating EMF applied to its primary coil is equal to the ratio of the number of turns in its secondary coil to the number of turns in its primary coil.* A transformer with five times as many turns on its secondary coil as on its primary coil supplies an EMF from its secondary coil five times as large as the EMF applied to its primary coil.

Fig. 27-15. Three step-down transformers on local utility poles.

27-25 Current Relationships in a Transformer

In a transformer in which losses caused by heating of the coils and the core are negligible, *the electrical power put into the primary coil is equal to the electrical power obtained from the secondary coil.* This follows from the law of conservation of energy. If I_p is the current flowing in the primary coil, the power put into the primary is $E_p I_p$. If I_s is the current flowing in the secondary coil, the power obtained from the secondary coil is $E_s I_s$. Since $E_p I_p$ is equal to $E_s I_s$, we have:

$$\frac{I_p}{I_s} = \frac{E_s}{E_p}$$

Substituting:

$$\frac{E_s}{E_p} = \frac{N_s}{N_p}$$

gives:

$$\frac{I_p}{I_s} = \frac{N_s}{N_p}$$

It is seen that a transformer that steps up the applied EMF reduces the current proportionately. In a step-up transformer having a turn ratio of 1 to 3, an alternating EMF of 100 volts applied to

Fig. 27-16. A step-up transformer at a power distribution point.

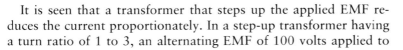

the primary coil will induce an EMF of 3 × 100, or 300 volts in the secondary coil. However, when one ampere is flowing in the primary coil, only one-third as much current, or ⅓ ampere, will flow in the secondary coil.

In the step-down transformer, these relationships are reversed. Here, the secondary EMF is smaller than the primary EMF, but the secondary current is larger than the primary current by the same factor.

Sample Problem

An alternating EMF of 110 V is applied to a step-up transformer having 150 turns on its primary and 3000 turns on its secondary. The secondary current is 0.100 A. (*a*) What is the secondary EMF? (*b*) What is the primary current? (*c*) What is the power input? (*d*) What is the power output? Assume no losses.

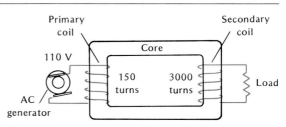

Fig. 27-17. Sample problem. Step-up transformer

Solution:

(*a*) $E_p = 110$ V $N_p = 150$ $N_s = 3000$

$$\frac{E_s}{E_p} = \frac{N_s}{N_p}; E_s = \frac{N_s}{N_p} E_p$$

$$E_s = \frac{3000}{150} \times 110 \text{ V}$$

$$E_s = 20 \times 110 \text{ V} = 2200 \text{ V}$$

(*b*) $I_s = 0.100$ A

$$\frac{I_p}{I_s} = \frac{N_s}{N_p}; I_p = \frac{N_s}{N_p} I_s$$

$$I_p = \frac{3000}{150} \times 0.100 \text{ A}$$

$$I_p = 20 \times 0.100 \text{ A} = 2.00 \text{ A}$$

(*c*) The power input is the power furnished to the primary coil and is equal to $E_p I_p$.

$$E_p I_p = 110 \text{ V} \times 2.00 \text{ A}$$
$$E_p I_p = 220 \text{ W}$$

(*d*) Assuming no losses, the power output is equal to the power input, or 220 W.

Test Yourself Problems

1. An alternating EMF of 220 V is applied to a step-down transformer having 1200 turns on its primary coil and 300 turns on its secondary coil. What is the secondary EMF?
2. (*a*) If the secondary current in the above transformer is 8.00 A, what is the primary current? Assume no losses. (*b*) What is the power input?

3. A transformer is used to step up an EMF of 110 V to 55 000 V. What is the ratio of primary to secondary turns?

27-26 Efficiency of Transformers

In practical transformers, some of the power input is lost in heating the coils and the core. As a result, the useful power output is always smaller than the power input. The ratio of the actual power output to the power input expressed as a percent is the efficiency of the transformer. Well designed transformers often have efficiencies as high as 98 percent.

Faraday showed that if a wire conductor forming part of a circuit is moved so that it cuts across the lines of force of a magnetic field, a current is induced in it. The EMF induced in the conductor is proportional to the rate at which it cuts across the magnetic lines of force; the more lines of force cut per second, the greater the **induced EMF.** It is given by:

$$EMF = v\,B\,l$$

where EMF is the electromotive force in volts, v is the velocity of the wire in meters per second, B is the magnetic induction in webers per square meter, and l is the length of the wire in meters.

It does not matter whether it is the wire conductor or the magnetic field that is moving. As long as the lines of force are cut by the conductor, an EMF is induced in it. Electric **generators** utilize this principle to convert mechanical energy into electrical energy.

Faraday's method of inducing an EMF in a conducting circuit may be described in a second more general way. Known as **Faraday's law of induction,** it states: Every change in the total magnetic flux passing through a circuit induces an EMF in that circuit whose magnitude is equal to the rate at which the total magnetic flux through the circuit changes.

This law explains why a **self-induced EMF** is set up in any circuit in which the current is changing and therefore changing the circuit's magnetic field. The magnitude of the self induced EMF depends upon a property of the circuit called its **inductance.** The unit of inductance is the **henry (H)** which is the inductance of a closed circuit in which an EMF of 1 volt is self-induced when the current in the circuit changes uniformly at the rate of 1 ampere per second.

Faraday's law of induction also explains how a **transformer** steps up or steps down the voltage of an alternating current according to these relationships:

$$\frac{E_s}{E_p} = \frac{N_s}{N_p} \quad \text{and} \quad \frac{I_p}{I_s} = \frac{N_s}{N_p}$$

Here, E_p is the EMF in the primary coil of the transformer, E_s is the EMF in the secondary coil of the transformer, N_p is the number of turns in the primary coil, N_s is the number of turns in the secondary coil, I_p is the primary current and I_s is the secondary current.

Lenz discovered that the direction of flow of the current set up in a circuit by an induced EMF is always such as to set up a magnetic field that opposes the motion or change that induced that EMF. **Lenz's law** is a consequence of the law of conservation of energy.

CHAPTER REVIEW

Summary

Questions

Group 1

1. The ends of a straight length of wire are connected to a galvanometer and the wire is then held between the poles of a horseshoe magnet. (a) Assuming the wire is moved at constant speed, in what direction must it be moved in order to have the maximum EMF induced in it? (b) What effect does changing the speed of the wire have on the magnitude of the induced EMF? (c) In what direction in the magnetic field may the wire be moved without inducing an EMF in it?

2. State the rule that tells how the direction of an induced current is determined from the direction of motion of a conductor and the direction of flux of the magnetic field across which it moves.

3. (a) Where does the force come from that sets the electrons in motion in a conductor moving at right angles to a magnetic field? (b) Draw a diagram showing the relationship between this force, the direction of the conductor, and the direction of the field.

4. Where does the energy come from that does the work of maintaining an induced current in a conductor which is part of a closed circuit moving across a magnetic field?

5. The earth's magnetic field at the equator is horizontal and runs from south to north. (a) If an airplane carrying a vertical conductor flies due west, in what direction will the EMF induced in the conductor be? (b) What change takes place in the size of the induced EMF when the airplane increases its speed? (c) How does the EMF change as the airplane changes its direction?

6. (a) Referring to the alternating current generator, why does the EMF induced in the armature change direction every half-turn? (b) Why does increasing the rate of rotation of the armature increase the maximum EMF induced in it? (c) What is the source of the energy converted by the generator into electrical energy? (d) What is meant by the effective current? (e) What is the relationship between the effective current and the maximum current? (f) What is meant by the effective voltage?

7. (a) How does the direct current generator change the alternating current in its armature to direct current in its external circuit? (b) How may a direct current motor be used as a direct current generator? (c) Why does a back EMF develop in an operating motor?

8. (a) Using a coil of wire, a galvanometer and a bar magnet, describe an experiment illustrating the principle that changing the magnetic flux passing through a circuit induces an EMF in the circuit. (b) What is the relationship between the changing flux and the EMF it induces in the circuit? (c) How does Lenz's law determine the direction of the induced current?

9. Two similar circular coils A and B are mounted vertically. Coil A has its ends connected to form a closed circuit. Coil B does not. The north pole of a bar magnet is now moved at constant speed v into coil A. The north pole of the same magnet is then moved at constant speed v into coil B. Explain why it takes more force to move the magnetic north pole into coil A than it does to move it into coil B.

10. A steady direct current is flowing in a series circuit consisting of a battery, a coil of wire and a switch. (a) What causes an EMF to be induced in the coil when the switch is opened? (b) Why is a second EMF induced in the coil when the switch is closed? (c) In which of these cases is the self-induced EMF in the same direction as the current supplied by the battery? (d) Explain why the EMF self-induced in the coil upon breaking the circuit will be very large if the coil has many turns.

11. Two separate but similar coils of wire are mounted vertically a few inches apart. The first coil has its ends connected to a source of EMF and has a direct current flowing through it. The second coil has its ends connected to a galvanometer. (a) How does the galvanometer show when the current in the first coil is increasing, decreasing, or remaining steady? (b) What must be happening to the magnetic field of the first coil in order for it to induce an EMF in the second coil?

12. (a) Referring to the transformer, what determines how many times the applied voltage is raised or lowered? (b) Assuming no losses, how does the power put into the primary coil compare with that obtained from the secondary? (c) Why is no EMF induced in the secondary coil of a transformer when steady direct current passes through the primary coil?

Group 2

13. A horizontal length of wire conductor is placed in a vertical magnetic field between the poles of a horseshoe magnet. Compare the magnitude and direction of the EMF induced in the wire when both the wire and the magnet are moving horizontally to the right at constant speed v with the magnitude and direction of the induced EMF when the magnet is moving to the right at speed v while the wire is moving to the left at the same speed.

14. A boy is turning the armature of a hand-operated generator that is connected to a switch-controlled lamp. (a) Compare the EMF induced in the armature when the lamp circuit is open with the EMF induced when the lamp circuit is closed. (b) Compare the power output of the generator in the same two cases. (c) Why must the boy work harder to turn the armature when the lamp is on than when it is off?

15. A circular ring of metal is lying horizontally on a table. A bar magnet with its north pole facing downward is held several inches above the center of the ring and then allowed to fall. (a) What effect does the magnetic field of the current induced in the ring have on the rate of fall of the magnet? (b) What change takes place in the induced current as the magnet falls toward the ring?

16. In Fig. 27-18, the U-shaped bare metal wire ABCD is mounted horizontally in a uniform vertical magnetic field whose lines of force pass upward through it. A metal slide rod EF con-

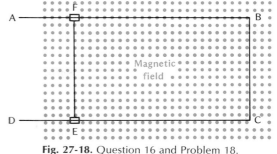

Fig. 27-18. Question 16 and Problem 18.

necting AB and CD is moving to the right at constant velocity. (a) What change does the motion of EF produce in the flux passing through the circuit BCEF? (b) In what direction does the induced current flow in BCEF? (c) As the induced current flows through EF, what is the direction of the force that acts on it? (d) What effect will moving EF faster have on this force? (e) If the direction of EF is reversed, what happens to the direction of the force acting upon it?

Problems

Group 1

1. A wire 1.0 m long is moved at right angles to the earth's magnetic field where the induction is 5.0×10^{-5} Wb/m² at a speed of 4.0 m/s. What EMF is induced in the wire?

2. An EMF of 0.002 V is induced in a wire 0.1 m long when it is moved perpendicularly across a uniform magnetic field at a velocity of 4 m/s. What is the magnetic induction of the field?

3. At what speed must a wire conductor 0.5 m long be moved at right angles to a magnetic field of induction 0.2 Wb/m² to induce an EMF of 1 volt in it?

4. (a) If the wire conductor of problem 3 is in a circuit having a resistance of 0.2 Ω, what is the current induced in it? (b) What force must be exerted upon the wire to keep it moving?

5. The magnetic flux through a coil of 100 turns is changing at the rate of 0.05 Wb/s. (a) What is the EMF induced in each turn? (b) What is the total EMF induced in the coil?

6. The magnetic induction through a loop of wire having an area of 0.02 m² decreases from 0.5 Wb/m² to zero in 0.02 s. What EMF is induced in the loop?

7. At what rate must the magnetic flux change through a coil of 80 turns to induce an EMF of 20 V in it?

8. A transformer having coils of 300 turns and 75 000 turns is used to step up an alternating EMF of 220 V. What is the secondary EMF?

9. A step-down transformer is to be connected to a 120-V alternating current source in order to deliver 6.0 V at the secondary terminals. (a) If the secondary coil has 80 turns of wire, how many turns should the primary coil have? (b) If the secondary current is 30 A, what is the primary current? Assume there are no losses. (c) What is the power output of the transformer?

10. The primary coil of a step-up transformer is connected to a 110-V alternating current line and delivers a 55 000-V, 0.020-A current at the secondary coil. (a) What power is delivered at the secondary coil? (b) What power is supplied to the primary coil, assuming no losses? (c) What is the current flowing in the primary?

11. A step-up transformer having 250 turns on the primary coil and 6000 turns on the secondary coil delivers 24 000 V at its secondary terminals. The secondary current is 0.50 A. Assuming no

losses, find (a) the primary voltage; (b) the primary current; (c) the power input.

12. A transformer operating on a 125-V alternating current line draws 1000 W of power and delivers a current of 25 A in its secondary coil. (a) What is the secondary voltage? (b) What is the transformer's ratio of primary to secondary turns? (c) If there are no losses in the transformer, what is the primary current?

13. A step-down transformer has 12 000 turns on its primary coil. (a) If it is to change a primary alternating EMF of 220 V into 5.5 V, how many turns should be on its secondary coil? (b) If there are no losses in the transformer and the secondary current is 10 A, what is the primary current?

14. The power obtained from a transformer is 2100 W. The primary EMF is 110 V and the primary current is 20 A. (a) What is the efficiency of the transformer? (b) At what rate is electrical energy being expended in the transformer in heat and other losses?

15. A 95% efficient transformer changes a primary EMF of 220 V to 10 V. If the secondary current is 19 A, what is the primary current?

Group 2

16. A circular metal ring held vertically between the poles of a horseshoe magnet has a magnetic flux of 0.04 Wb passing through it. If the ring is then pulled out of the magnetic field in 0.1 s, what is the EMF induced in it?

17. The induction of a magnetic field passing through a square loop 10 cm on a side increases from zero to 0.09 Wb/m² in 0.03 s. What EMF is induced in the square loop?

18. The rectangular circuit of Fig. 27-18 on page 527 is 0.2 m wide and the induction of the magnetic field through it is 0.2 Wb/m². (a) If the wire EF moves to the right at 5 m/s, what EMF is induced in it? (b) What is the rate of change of magnetic flux through the rectangular circuit?

19. A square wire loop 0.10 m on a side is rotated around a horizontal axis in a horizontal magnetic field of strength 5.0×10^{-2} Wb/m². (a) If the plane of the loop turns from horizontal to vertical in 0.20 s, what is the average rate of change of magnetic flux passing through it? (b) What is the average EMF induced in the loop during each half-turn?

Applying Physics

Make a simple galvanometer by mounting a compass at the middle of a vertical coil of 50 turns of No. 26 enameled wire (see Fig. 27-19) and put it in position for making a reading as described in the Applying Physics section at the end of Chapter 26.

Make a second coil of wire consisting of 50 turns and connect its ends to the coil of the galvanometer. Holding this second coil horizontally, move it slowly downward over the north pole of a vertical bar magnet. Then move it slowly upward to its original position. For each movement of the coil, note the size and direction of the swing of the compass needle. Repeat the experiment moving the coil rapidly instead of slowly. Compare the size of the swing of the compass needle with those observed before.

Tape a similar bar magnet to the first so that like poles of the magnet are adjacent to each other. Repeat the above experiment. How does the swing of the compass needle compare with that observed when only one magnet was used? Explain your observation.

Substitute a 100-turn coil of wire for the 50-turn coil. Repeat all of the above experiments. Explain your observations.

Remove the bar magnets. Lay the second coil flat

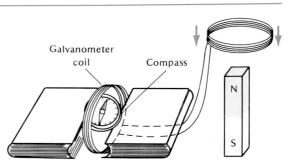

Fig. 27-19. Apparatus for studying the induction of EMF in a coil.

on the table and rest a book on top of it. Now connect another 50-turn coil in series with a battery and a switch. Put this coil on the book directly above the other coil. Close the switch and leave it closed. What happens to the compass needle? Open the switch and leave it open. What happens to the compass needle? Explain your observations.

Slip a sheet of aluminum foil under the upper coil and repeat the experiment. Does the induction process take place through the foil? Substitute sheets of copper and iron in turn for the aluminum foil and repeat the experiment. Explain your observations.

Electromagnetic Waves 28

Aims

1. To learn how electromagnetic waves are generated and transmit energy.
2. To learn how electromagnetic waves are used in radio reception and transmission.
3. To evaluate the successes and failures of Maxwell's theory of the origin, nature, and behavior of electromagnetic waves.

28-1 Transmission of Changes in Electric and Magnetic Fields

We have noted that the electric field around an electric charge extends throughout all space. When the charge moves, the entire field around it must change and rearrange itself in accordance with the new position of the charge. How is the fact that the charge has moved transmitted to all points in the electric field around it? How quickly does it happen?

A similar situation exists for a magnet and the magnetic field around it. Any movement of the magnet causes its magnetic field to change and rearrange itself in accordance with the new position of the magnet. How is this change transmitted through the magnetic field and how fast does it travel?

We are going to see that changes in magnetic and electric fields are transmitted by electromagnetic waves that travel with the speed of light. To understand how such waves are produced, we shall first note that moving magnetic fields generate electric fields and that moving electric fields generate magnetic fields. These two effects acting together are responsible for generating and propagating electromagnetic waves.

28-2 A Moving Magnetic Field Generates an Electric Field

The demonstration of Fig. 27-13 showed that a change in the magnetic field of the primary coil moves through space and induces an EMF in the secondary coil. This change in the magnetic field must move very rapidly since its effect on the secondary coil is observed immediately. We shall now examine how such changes travel through space.

First, let us examine the induction process from the point of view of the individual free electrons inside a conductor situated in a moving magnetic field. In Fig. 28-1, a magnet having a field of strength B between its poles is moving to the right at constant

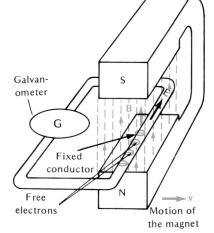

Fig. 28-1. A moving magnetic field generates an electric field in the conductor.

Galvanometer

G

Fixed conductor

Free electrons

Motion of the magnet

velocity v. Between its poles is a fixed conductor in which several free electrons are shown. According to Faraday's law of induction, an EMF is induced in the conductor by the moving magnetic field. Since this EMF makes the free electrons move, it must set up an electric field E inside the conductor that exerts a force on each of the free electrons. The field E acts at right angles to both B and v, as shown in the diagram. Since the direction of an electric field is that in which it pushes a positive charge, the direction of E in the figure is opposite to that in which it drives the electrons.

Thus, from the point of view of a free electron in the path of a moving magnetic field, Faraday's law of induction means that *a moving magnetic field generates an electric field at right angles to itself and to the direction of motion.*

28-3 Electromagnetic Radiation from a Circuit Carrying a Varying Current

Now consider the electric circuit of Fig. 28-2, consisting of the long wire conductor XY connected to an open switch and battery. Here, 1, 2, and 3 represent test electrons free to move in space. The moment the switch is closed, current begins to flow in the circuit and grows from zero to a certain steady value. At the same time, a magnetic field also begins to grow around wire XY, spreading out in all directions in circular lines of force like ripples in a pond.

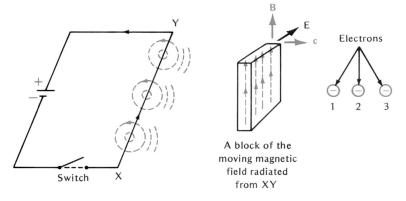

A block of the
moving magnetic
field radiated
from XY

Fig. 28-2. When the switch is closed, a change in the magnetic field is radiated into space.

The figure shows a block representing a small increase B of this growing magnetic field moving to the right at a velocity designated by c. As this magnetic block passes over electrons 1, 2, and 3, it generates an electric field E that exerts a force upon each of them in turn. According to Faraday's induction law, the field E is perpendicular to both B and c as shown in the figure and moves along with B in the same direction and at the same velocity c.

As the current in the circuit reaches its steady value, the magnetic field in the space around XY also reaches its steady condition. All parts of it then stop changing or moving and therefore generate no further electric field.

Suppose that, after the current in XY becomes steady, the switch is opened. Both the current in XY and the magnetic field around it will then begin to decrease to zero. We can think of the steady magnetic field being reduced to zero by an oppositely directed change in the magnetic field that leaves the wire XY, moves outward at velocity c, and grows until it cancels the original steady field completely. As before, a block of this moving field will generate an electric field that moves along with it at velocity c and acts on all charges over which it passes. However, the direction of E will be opposite to that produced earlier when the magnetic field around the wire was increasing instead of decreasing.

28-4 A Moving Electric Field Generates a Magnetic Field

We have already had occasion to note that symmetrical relationships exist between magnetic and electrical effects. For example, the fact that a current produces a magnetic field has its symmetrical counterpart in the fact that a magnetic field can induce a current. It has been seen from Faraday's law of induction that a change in a magnetic field moving out into space at velocity c generates an electric field. Maxwell reasoned from symmetry that there should be a similar law of induction that applies to electric fields. According to this law, called *Maxwell's law of induction, a change in an electric field moving out into space should also travel at velocity c and generate a magnetic field*. Furthermore, this magnetic field should be perpendicular to both c and the moving electric field that generates it.

Maxwell had no direct way of testing his law of induction experimentally because the magnetic fields generated by moving electric fields are usually so small as to make measuring them very difficult. However, by assuming his induction law to be true and combining it with Faraday's law of induction, Maxwell was able to formulate his electromagnetic theory that offers an explanation of all electric and magnetic phenomena.

From this theory Maxwell predicted many electrical and magnetic effects that could be tested experimentally. Most important among these was the prediction that any disturbance of a magnetic or electric field sets up an electromagnetic pulse or wave that travels through empty space at the speed of light. Thus, *the velocity* c *at which changes in a magnetic or electric field travel turns out to be the velocity of light.* The fact that this and many other predictions turned out to be in complete agreement with experiment confirmed the correctness of Maxwell's law of induction and strengthened confidence in his electromagnetic theory.

> Any change in an electric or magnetic field is radiated through space at the velocity of light.

28-5 Electromagnetic Waves or Pulses

The electromagnetic theory enables us to understand the nature of an electromagnetic wave or pulse. Let us return to the small block of magnetic field that moves away from XY in the circuit in Fig. 28-2 when the switch is closed. We have seen that as the block moves on with velocity c, the magnetic field in the block generates an electric field at right angles to itself and to c. Therefore, it is not only the magnetic field in the block that moves through space but a combination of electric and magnetic fields at right angles to each other. If the block is small enough and far enough from the circuit that produced it, the magnetic field B and the electric field E inside of it are uniform and are parallel to the faces of the block as shown. The energy associated with B and E therefore remains constant inside the block as it moves on.

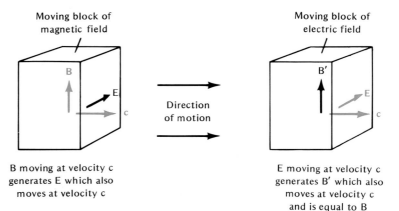

Fig. 28-3. Moving magnetic and electric fields regenerate each other.

Moving block of magnetic field

Moving block of electric field

B moving at velocity c generates E which also moves at velocity c

E moving at velocity c generates B′ which also moves at velocity c and is equal to B

Now, in accordance with Faraday's law of induction, the magnetic field B moving at velocity c generates an electric field E inside the block at right angles to both B and c. This is shown in the left part of Fig. 28-3. In accordance with Maxwell's law of induction, this new electric field E, also moving at velocity c, generates a magnetic field B' inside the block at right angles to E and c. This is shown in the right part of Fig. 28-3. Since no energy enters or leaves the block, it follows that the generated magnetic field B' must not only be in the same direction as the original magnetic field B but also must be equal to B. Then, as the block moves forward, B will generate E which will regenerate B which will then regenerate E, and so forth.

This moving combination of electric and magnetic fields regenerating each other as they move forward together at the speed of light constitutes an *electromagnetic pulse* or *wave*. Maxwell's prediction that such waves exist had to wait until eight years after his death to be confirmed. At that time, Heinrich Hertz discovered a way to produce the electromagnetic waves commonly known as radio waves. Later, their velocity was shown to be that of light, as Maxwell had predicted.

28-6 Transmission of a Change in an Electric Field

In the above discussion, it was explained how a change in a magnetic field is transmitted through space by means of electromagnetic waves or pulses. It is clear, however, that if a change takes place in an electric field such as that caused by a sudden movement of an electric charge, the change in the electric field will generate a magnetic field which, in turn, will generate an electric field, and so on. This change is also transmitted through space by an electromagnetic wave traveling at the speed of light. Thus, *any change in an electric or a magnetic field is transmitted through space by means of electromagnetic waves.*

28-7 Origin of Electromagnetic Waves

Maxwell's theory did more than predict the existence of electromagnetic waves. It offered an explanation as to how all such electromagnetic waves come into being. According to Maxwell's theory, electromagnetic waves are produced whenever an electric charge undergoes an acceleration.

To illustrate, let us return to Fig. 28-2. Here, at the moment of closing the switch, electrons in the circuit begin to flow and increase their velocity from an average value of zero to the value corresponding to the steady current that is soon set up in the circuit. During the time that the current is growing to this steady value, so is the magnetic field around it. As has been noted, the changes in this growing magnetic field are radiated into space as electromagnetic pulses or waves. A moment after the current in the circuit stops increasing and reaches its steady value, no further changes take place in the magnetic field so that the radiation of electromagnetic waves from the circuit also stops. Thus, this circuit radiates electromagnetic pulses or waves only while the electrons flowing in it are being speeded up or accelerated to the average velocity corresponding to the steady value of the current. Once the current is steady and the acceleration of the electrons stops, no further radiation of electromagnetic waves takes place.

If the switch is then opened, the current in the circuit quickly decreases to zero and so does the magnetic field around it. Once again, the change in the magnetic field is radiated into the space around the circuit as electromagnetic waves or pulses. Again, it is the acceleration of the electrons in the circuit that generates these electromagnetic waves. This time, however, the electrons are being slowed down instead of being speeded up. The acceleration is negative rather than positive.

28-8 Periodic Electromagnetic Waves

We have seen how periodic waves can be produced on a water surface by dipping the bead of a wave generator into the surface at a regular frequency. According to Maxwell's theory, periodic electromagnetic waves are generated in analogous fashion. When an electric charge moves in a circle or oscillates back and forth in a circuit at a regular frequency, the charge is accelerated in such a way that it generates *periodic electromagnetic waves* that radiate into space. The frequency of the waves is the same as the frequency of the motion of the circling or oscillating charges that radiate them.

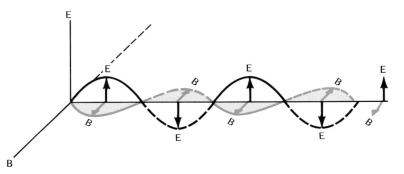

Fig. 28-4. A train of plane polarized electromagnetic waves consisting of an oscillating electric field and an oscillating magnetic field at right angles to each other and advancing at the velocity of light.

Fig. 28-4 is a diagram of a train of plane polarized periodic electromagnetic waves. It consists of an oscillating electric field that moves forward at the velocity of light in the vertical plane and is accompanied by an oscillating magnetic field that moves forward in the horizontal plane, also at the velocity of light. When these electromagnetic waves pass over a vertical conductor, their oscillating electric field causes the free electrons in the conductor to oscillate up and down at the same frequency as that of the waves. Electric oscillations are set up this way in the *antennae* of radio and television sets by the radio waves emitted by broadcasting stations. The sets receive programs by detecting these oscillations.

Since, in an alternating current circuit, electrons oscillate back and forth at a regular frequency, it follows that every alternating current circuit radiates periodic electromagnetic waves having that same frequency. For example, the typical 60-hertz alternating-current circuits available in many homes radiate electromagnetic waves at a frequency of 60 waves per second. However, low frequency electromagnetic waves like these do not travel very far because they have little energy and are easily absorbed by the matter in their paths. Electromagnetic waves normally used in radio and television communication have frequencies varying from a few thousand hertz to millions of hertz. To radiate these radio frequency electromagnetic waves, special oscillating circuits are used to produce alternating currents of these frequencies.

28-9 Role of Capacitor in Oscillating Circuit

In its simplest form, an oscillating circuit consists of an *inductance* such as a coil of wire connected to a device called a *capacitor,* as shown in Fig. 28-5. A capacitor consists of any two conductors separated by an insulator. A simple form of capacitor consists of two parallel metal plates separated by air. A capacitor has the ability to store equal and opposite electric charges.

One way to put charges into a capacitor is to connect its plates to the terminals of a battery or generator. The plate connected to the negative terminal acquires a negative charge while that connected to the positive terminal acquires an equal positive charge. Putting charges into the capacitor in this fashion is called charging the capacitor. Once the opposite charges are in the plates of the capacitor, they attract each other across the insulator and tend to prevent their mutual escape.

Removing the charge from a capacitor is called discharging it. To discharge a capacitor, its plates are connected by a conductor. The excess electrons are then attracted out of the negatively charged plate and pass through the conductor into the positively charged plate until there is no difference of potential between both plates of the capacitor.

The quantity of charge a capacitor can hold depends upon the area of its plates, the distance between them, the nature of the insulator between them, and the EMF or voltage applied to the capacitor plates. The ability of a capacitor to hold electric charges is called its *capacitance* and is measured in the unit called the *farad* (symbol, F). A capacitor rated one farad is one that stores one coulomb of positive and negative charges on its plates when it is charged by an applied EMF of one volt. Capacitors are usually very much smaller than one farad and are measured in *microfarads* (10^{-6} F) or in *picofarads* (10^{-12} F).

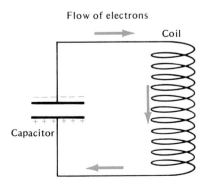

First the discharging capacitor sends electrons through the coil until the self-induced EMF charges the capacitor in the opposite direction

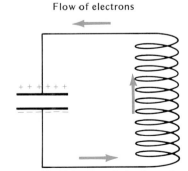

Then the oppositely charged capacitor discharges sending electrons in the opposite direction

Fig. 28-5. In the oscillating circuit, the charged capacitor discharges only to be charged again in the opposite direction by the EMF induced in the coil. The capacitor then discharges again in the opposite direction. Repetition of this cycle keeps the electrons shuttling back and forth in the circuit.

28-10 Principle of Oscillating Circuit

Let the capacitor in the oscillating circuit of Fig. 28-5 be charged by a battery and then connected to the coil. The capacitor will begin to discharge as electrons flow out of the negatively charged plate, through the coil, and into the positively charged plate. This electron flow sets up a magnetic field in and around the coil.

As soon as the capacitor becomes discharged, the electron flow through the coil decreases rapidly and so does the magnetic field around the coil. In accordance with Lenz's law, this change in the magnetic field induces an EMF in the coil that acts in such a direction as to oppose the change in the electron flow that produced it. The self-induced EMF therefore acts in the same direction as the electron flow and prevents further decrease in that flow. Instead, it keeps the electrons flowing in the same direction even after the capacitor is fully discharged so that the capacitor now becomes charged opposite to its previous direction.

The newly charged capacitor once again begins to discharge, sending electrons through the coil from its negatively charged plate to its positively charged plate. As the electrons begin to flow opposite to their previous direction, the process described above is repeated. In this manner, the capacitor is discharged and charged first in one direction and then in the opposite direction and the electrons shuttle back and forth in the circuit at a regular frequency.

28-11 Frequency of Radiated Waves

The frequency at which electrons oscillate in the circuit depends upon the size of the capacitor and the inductance of the coil, and is called the natural or *resonant frequency* of the circuit. This is also the frequency of the radio waves radiated by this circuit. By selecting an appropriate capacitor and inductance, radio waves of any desired frequency can be generated.

The resonant frequency of an oscillating circuit having an inductance of L henries and a capacitance of C farads is given by the relationship:

$$f = \frac{1}{2\pi\sqrt{LC}}$$

The frequency is expressed in hertz.

An oscillating circuit slowly loses its energy by radiating some of it as electromagnetic waves and converting some of it to heat generated by the current as it flows back and forth through the resistance of the circuit. The oscillations in the circuit will therefore gradually die out. To keep the circuit oscillating, it is therefore necessary to feed it electrical energy at the same rate at which it is losing energy. This is done in oscillating circuits used to radiate commercial radio waves.

Sample Problem

What is the resonant frequency of a circuit having an inductance of 8.0 × 10⁻² H and a capacitance of 2.0 × 10⁻¹⁰ F?

Solution:
$L = 8.0 \times 10^{-2}$ H $C = 2.0 \times 10^{-10}$ F

$$f = \frac{1}{2\pi\sqrt{LC}}$$

$$f = \frac{1}{2\pi\sqrt{8.0 \times 10^{-2}\,H \times 2.0 \times 10^{-10}\,F}}$$

$$f = 4.0 \times 10^4\,Hz$$

Test Yourself Problems

1. What is the resonant frequency of a circuit having an inductance of 2.0 × 10⁻³ H and a capacitance of 2.0 × 10⁻¹¹ F?
2. An oscillating circuit has an inductance of 1.0 × 10⁻⁴ H and a capacitance of 1.0 × 10⁻¹² F. What is its resonant frequency?
3. An oscillating circuit has an inductance of 1.0 × 10⁻³ H. What should its capacitance be if it is to resonate at 1.6 × 10⁶ Hz?

28-12 Reception of Radio Waves

Radio waves can be received by means of a length of conducting wire called an antenna. On passing over such a conductor, the radio waves induce an alternating EMF in it having the same frequency as the waves. A more effective receiver for radio waves of a given frequency is a second oscillating circuit having the same natural frequency as the circuit radiating the waves. When the radio waves pass over this receiving circuit, they induce particularly strong oscillations in it. The effect is analogous to the sympathetic vibrations or resonance set up by one vibrating body in a second body having the same natural frequency of vibration. (See Section 15-26.) The matched emitting and receiving oscillating circuits are said to be in *electrical resonance.*

28-13 Electromagnetic Theory of Light

The frequencies of the waves associated with light are far too high to be generated by oscillating circuits. However, the fact that all electromagnetic waves not only travel at the velocity of light but undergo reflection, refraction, interference, diffraction, and polarization like light waves suggests that light itself consists of electromagnetic waves. Thus, Fig. 28-4 may represent the vibration pattern of the electric and magnetic fields in a beam of plane polarized light.

A beam of unpolarized light consists of a combination of countless polarized beams whose electric and magnetic fields take all possible directions perpendicular to the direction in which the light is moving. That light waves actually are made of such crossed oscillating electric and magnetic fields can be demonstrated by passing polarized light through electric and magnetic fields and noting that each of these fields produces changes in it.

Maxwell's theory thus answered the question, *What is it that oscillates when light travels through space?* His reply was crossed

Fig. 28-6. A movable radio telescope designed to receive short-wave radio signals from outer space, which are concentrated at its focus.

electric and magnetic fields. Light now was identified as but one of the members of the large family of electromagnetic waves listed in Section 16-31 and differing from one another principally in wavelength.

28-14 Successes and Limitations of Maxwell's Electromagnetic Theory

Maxwell's theory was brilliantly successful in its prediction of the existence of electromagnetic waves and in its description of their nature and behavior. It correctly identified light as a member of the electromagnetic family and explained its wavelike properties in terms of the oscillating electric and magnetic fields of which it was composed. Its assertion that all electromagnetic waves are produced by accelerated charges also found considerable but not complete support from experiment.

As was noted, the oscillating electrons in a high frequency alternating current actually do radiate radio waves having the same frequency as the current, as predicted by the theory. Certain X rays can also be attributed to accelerated charges. These rays are made in the X-ray vacuum tube where a beam of electrons is allowed to fall upon a metal or other solid target. The sudden slowing up or deceleration of the electrons on striking the target is accompanied by the radiation of the very short electromagnetic waves known as X rays.

However, Maxwell's mechanism of accelerated charges as the sources of electromagnetic waves had very serious limitations. In the next two chapters, we shall see that it did not give a correct picture of the manner in which the energy is distributed in and transmitted by electromagnetic waves. Neither could it correctly explain the emission or absorption of light by the atoms and molecules of matter.

CHAPTER REVIEW

Summary

From Faraday's law of induction, it follows that as a change in a magnetic field moves out into space, it generates an electric field at right angles to itself and to the direction of motion. Maxwell discovered a law of induction for the electric field similar to Faraday's law of induction for the magnetic field. From **Maxwell's law of induction,** it follows that as a change in an electric field moves out into space, it generates a magnetic field at right angles to itself and to the direction of motion.

From these two laws of induction, Maxwell formulated his **electromagnetic theory** explaining electric and magnetic phenomena. One of the consequences of this theory is that any change in a magnetic or an electric field is transmitted through space by means of **electromagnetic waves or pulses** traveling at the speed of light. Electromagnetic waves consist of moving electric and magnetic fields at right angles to each other. They include radio waves, infrared rays, light waves, ultraviolet rays, X rays, and gamma rays. According to Maxwell's electromagnetic theory, all these electromagnetic radiations are produced when electric charges are accelerated or decelerated. In particular, **periodic electromagnetic waves** are produced by electric charges that vibrate or oscillate at a regular frequency.

The electrons in an alternating current circuit oscillate in the circuit at the frequency of the alternating current. The circuit therefore radiates electromagnetic waves of the same frequency. To produce the high frequency waves used in radio and television, special oscillating circuits consisting of an **inductance** L and a **capacitance** C are used. The frequency in hertz generated by such oscillating circuits is given by $\dfrac{1}{2 \pi \sqrt{LC}}$ where L is in **henries** (H) and C is in **farads** (F).

Maxwell's theory successfully explains the velocity, reflection, refraction, interference, diffraction, and polarization of light, and thus identifies light as a member of the electromagnetic wave family. However, the theory's mechanism of accelerated charges as the sources of electromagnetic waves does not adequately explain the nature of the light and other electromagnetic radiations coming from the atoms and molecules of matter. Nor does it explain the manner in which energy is transported and distributed in electromagnetic waves. These limitations led to the new approach to the theory of light discussed in the next two chapters.

Questions

Group 1

1. A horizontal wire conductor is in a vertical magnetic field in which the direction of the flux is downward. Show by a diagram the direction of the electric field that is generated in the conductor when the magnetic field moves (a) to the right; (b) to the left.

2. Why are no electromagnetic pulses or waves generated around a wire in which a steady direct current is flowing?

3. A direct current flowing upward in a vertical wire is increasing. Explain how a small block of the increasing magnetic field around the wire generates an electromagnetic pulse as it moves away from the wire. Show by means of a diagram the directions of B, E, and the velocity of the pulse c.

4. A vertical electric field is moving to the right at the velocity of light. Show by two diagrams how a small block of this field in which E is downward generates a magnetic field B which, in turn, generates E. Show the directions of E, B, and c.

5. Two negative charges are 1 m apart. One of the charges is suddenly moved toward the other. (a) How long will it take before the second charge experiences a change in the force exerted upon it by the electric field of the first charge? (b) Explain how the change thus produced in the electric field of the first charge is transmitted to the second charge.

6. A compass is 0.3 m from a vertical wire carrying a steady current. If the current is suddenly decreased (a) how long will it take before the compass experiences the change that takes place in the magnetic field around the wire; (b) how is this change transmitted through space?

7. (a) According to the electromagnetic theory of Maxwell, how does a change in a magnetic field or an electric field result in an electromagnetic pulse or wave that moves through space at the speed of light? (b) What is the source of electromagnetic radiations? (c) How are periodic electromagnetic waves generated?

8. (a) How is a high-frequency alternating current set up when the switch is closed in an oscillating

circuit such as that pictured in Fig. 28-5? (b) What determines the frequency of this current? (c) Why does this circuit radiate electromagnetic waves? (d) What effect does the passage of electromagnetic waves from a radio station have on the antenna of a radio receiver?

9. (a) What are some of the considerations that support the idea that visible light consists of electromagnetic waves? (b) State one way in which light waves differ from radio waves.

10. In an X-ray tube, high-speed electrons from an electron gun are shot at a target and thus quickly brought to a halt. How does the electromagnetic theory explain why X rays are emitted in this process?

Group 2

11. A flash of lightning is a sudden surge of electric current in which both positive and negative charges in the air are set into motion. Why does a flash of lightning produce the crashing noise known as static in a radio receiver?

12. Sunspots are associated with magnetic storms originating on the sun. These storms produce violent changes in the magnetic field around the earth. (a) What effect do such changes have on the antennae of radios and television sets? (b) How does this influence the reception of the regularly broadcast programs?

13. Two radio stations are broadcasting the same signal by means of radio waves having the same wavelength and amplitude. These two sets of waves interfere with each other as they are superimposed. A radio receiver is located 100 wavelengths from each of the stations. How does the strength of the combined signal that the receiver gets from both stations compare with that which it would receive from one station when the two stations are (a) in phase; (b) in opposite phase?

14. The two radio stations of question 13 are ½ wavelength apart. A receiver is located on the line joining the stations at a distance of 5 wavelengths from the nearer station. Will this receiver get a maximum combined signal when the two stations are in phase or in opposite phase? Explain.

15. Plane polarized radio waves in which E is vertical and B is horizontal as shown in Fig. 28-4 will induce an alternating EMF in a vertical antenna over which it passes but not in a horizontal antenna. Explain.

Applying Physics

1. A simple transmitter that sends out an electromagnetic pulse may be made by connecting one end of a 30-cm length of wire to one electrode of a dry cell. The other end is free. Whenever this free end is used to close or open the circuit, an electromagnetic pulse is sent out from the circuit.

   To detect this pulse, a portable AM radio receiver is turned on and its dial set midway between two stations so that no station is heard. Now bring the transmitter near the radio receiver and make and break the transmitter circuit. Clicks will be heard on the radio loudspeaker. Repeat the experiment several times, each time moving the transmitter further from the radio receiver. What is the greatest distance from the receiver at which the transmitter's signal can be detected?

2. Are there substances opaque to radio waves just as there are substances opaque to ordinary light? Here is a way to test a few substances such as paper, aluminum foil, and wool.

   Turn a small portable AM radio on to some station. While it is playing, wrap it completely in one sheet of aluminum foil. What happens to the sound? Repeat the experiment using a sheet of paper and then a woolen cloth. Are any of these materials opaque to radio waves?

Biological Effects of Low-Level Electromagnetic Fields

Traditionally, power lines have been viewed as a safe, relatively reliable, and inexpensive means to distribute electrical energy. As the demand for electricity increased with our expanding economy and technological development, thousands of more miles of transmission lines were constructed to meet that need. Today the United States has a network of more than 2 million miles of power lines, including 600,000 miles of overhead lines used to deliver electricity to consumers. However, in the past decade, plans to construct new overhead transmission lines have met with opposition. The public is concerned about the possible health hazards of the electromagnetic waves emitted by such lines.

Research conducted during the 1970s and '80s suggests that there may be a link between health problems and long-term exposure to extra-low frequency (ELF) electromagnetic fields such as those coming from electrical transmission lines, power stations, home wiring, and electrical appliances. Statistical studies show an association between chronic ELF exposure and an unusually high incidence of brain tumors, breast cancer, leukemia, and miscarriages and birth defects. Although no direct cause-and-effect relationship has been established, some experimental studies suggest that ELF radiation weakens the body's immune system, our principal defense against cancer. Other researchers have found that ELF radiation causes the genes to produce proteins associated with tumors.

In response to the findings of independent researchers, the utility industry has financed its own research efforts and published public

information on the subject. For 20 years the Electric Power Research Institute (EPRI) has used utility company funds to sponsor research that aims to show there is no significant risk from power lines. Publications of both EPRI and the Edison Institute, an association of electric companies, assert that the public health is not at risk from power lines.

The true extent of the ELF threat has yet to be assessed. Electrical appliances are probably not a cause for concern, since they are generally operated for only short time periods and their ELF fields extend only a few inches from the appliance. The exceptions are electric blankets, water-bed heaters, and computer screens. The health effects of power lines could be reduced by keeping them farther away from businesses, homes, and schools. Overhead lines could be buried, at significant extra expense, or their ELF fields cancelled out by various special techniques.

Nuclear Magnetic Resonance (NMR) Scanner

A new medical tool has revolutionized diagnosis. The NMR scanner not only visualizes chemical changes in the body, but it does not use X rays or substances that might harm tissues. On the other hand, the CT, or "cat" scan, produces cross-sectional X rays and PETT, another scanner, measures emissions from a radioisotope introduced into the body.

NMR works by activating the nuclei of hydrogen atoms (protons) in compounds within the body. Protons spin about their own axes, thus producing a small magnetic field. If an external magnetic field is applied, the spinning protons line up and then begin to precess—rotate around the axis of the applied magnetic field. If a second magnetic field is then applied, some of the protons flip and precess at a different angle. These protons return to their original positions when the second magnetic field is shut off, emitting electromagnetic radiation in the process. This signal is then detected and analyzed.

From the time required for the protons to return to their original positions, the NMR computer determines their density and produces a cross-sectional image.

Suppose you have injured your back and a disc problem is suspected. For diagnosis by an NMR scanner, you will lie down on a narrow table, which will then be passed through the scanner. All you will see is a structure that looks like a large vertical metallic doughnut, with a few leads running to a power supply. There are no moving parts in the scanner, and no flashing lights or sounds will disturb you. From the image produced, the doctor will be able to identify the disc affected.

The NMR scanner can watch the brain pulse in rhythm with the heart's beat, follow the ebb and flow of fluids in tissues, trace tissue changes after treatment, and even follow certain molecular changes.

Doctors find NMR scanners to be useful in diagnosing many different problems, including swelling of brain tissues, aneurysms (ballooning) in blood vessels, and even cancer. Watch for reports of developments in this exciting medical field.

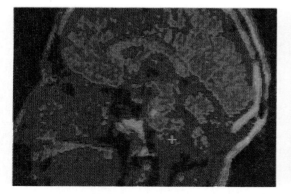

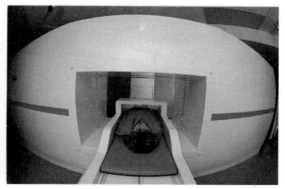

and Careers

Electrical and Electronics Engineers

Does it surprise you to learn that electrical engineering is the largest branch of engineering? In 1988 there were over 400,000 jobs in the United States alone. Electrical engineers design, develop, test, and supervise the manufacture of a host of electrical and electronics devices. Such devices include power generating and transmitting equipment for use by utilities, electric motors, machinery controls, and lighting and wiring in buildings, automobiles, and aircraft. Some electronics devices are TVs, radios, stereos, computers, radar, communications equipment, and medical diagnostic instruments. Engineers who work with electronics equipment are often called electronics engineers.

Some electrical engineers work at a desk in an office. Others work in research laboratories, in industrial plants, or at construction sites to inspect, supervise, or solve problems. Most jobs for electrical and electronics engineers are with firms that manufacture equipment and parts, business machines, professional and scientific equipment, and aircraft devices.

Job opportunities for electrical and electronics engineers are expected to grow much faster than the average for all occupations at least through the year 2000. Most openings, however, will result from the need to replace people who have transferred to other occupations or retired.

A bachelor's degree is usually required for entry-level electrical and electronics engineering jobs. Many colleges and universities offer such programs. Admissions requirements usually include advanced high school mathematics and physical science courses. At some institutions, cooperative plans allow a student to earn the degree by means of combined work and formal study over five or six years. The ability to finance part of the cost of education is an advantage of cooperative plans.

Because of rapid growth in technology, electrical and electronics engineers must continue their education throughout their careers. Those who fail to keep up risk technological obsolescence, which makes them susceptible to layoffs or being passed over for advancement.

Some other occupations in which the work is similar, at least in part, to that of electrical and electronics engineers are physical scientists, mathematicians, engineering and science technicians, and architects.

UNIT 7

Quantum Theory and Nuclear Physics

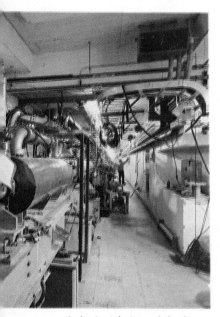

(Left) Aerial view of the linear accelerator at Stanford University. (Above) Interior view of MIT's Bates Linear Accelerator at Middleton, Massachusetts.

Our investigation of the nature of energy and matter now leads us into the atomic and subatomic world. We learn that both energy and matter have a dual nature. Both have wavelike and particle-like properties. To reconcile these seemingly opposing characteristics, the quantum of energy, first suggested by the German physicist Max Planck (1858–1947) is introduced.

The quantum idea is that the energy of electromagnetic waves is not spread continuously over their wave fronts but is bundled into tiny separate packets called photons, or quanta. All radiant energy is emitted, transmitted and absorbed in these units.

The quantum idea soon had remarkable successes in explaining phenomena that Maxwell's electromagnetic theory had failed to explain. In 1900, Planck used it to explain the distribution of energy in the radiation of the so-called black body, a theoretically perfect absorber and emitter of radiation. Five years later, Albert Einstein adapted the quantum idea to propose a new theory of light that successfully accounted for the hitherto unexplained particle-like behaviors of light. In the next twenty years, the Danish physicist, Neils Bohr (1885–1962) and others incorporated the quantum idea into an atomic model first proposed by the British physicist, Lord Rutherford, (1871–1937). Their atomic model succeeded brilliantly in explaining the emission and absorption of light by atoms.

These discoveries about the dual nature of light energy were matched by similar discoveries about the dual nature of matter. In 1924, the French physicist, Louis de Broglie, predicted on theoretical grounds that tiny particles of matter can behave like waves under certain conditions. This prediction was confirmed in the laboratory three years later when it was shown that beams of particles undergo diffraction and other wave phenomena.

Physicists next turned their attention to the structure of the atomic nucleus. All atomic nuclei were found to consist of different combinations of the same two particles—protons and neutrons. Furthermore, the nucleus turned out to be the seat of enormous quantities of energy. This discovery has had profound consequences for every aspect of human society.

But the nucleus was not the end of the search. A new world of subnuclear particles has opened up before us. It is revealing entirely new facets of the nature of matter and energy.

29 Quantum Theory of Light and Matter

Aims

1. To examine the evidence that light has particle-like properties.
2. To study the evidence that matter has wavelike properties.
3. To learn how the quantum theory provides an explanation for both particle-like and wavelike properties of electromagnetic radiation as well as of matter.

29-1 Quantum Theory and the Photon

We found in Fig. 20-5 that when light falls on metal surfaces electrons are ejected from them. This *photoelectric effect* is a fundamental property of light which cannot be explained in terms of the electromagnetic waves of Maxwell. Under the proper conditions, this effect can occur with any substance, but our study of it will be limited to metals. To explain the photoelectric effect, we shall have to invoke the quantum theory. According to this theory, light and all other forms of electromagnetic radiation are emitted and absorbed by matter only in separate bundles of energy called *quanta* or *photons*. The theory thus reveals a new aspect of radiant energy; namely its particle-like characteristics.

29-2 Apparatus to Study the Photoelectric Effect

We may study the photoelectric effect using an apparatus arrangement like that shown in Fig. 29-1. A cathode and an anode made of two metal plates are enclosed in an evacuated quartz container. The plates are connected to a source of variable difference of potential consisting of a battery of voltage V_B connected to a resistor R. The battery sends a current I through R such that $V_B = IR$. The potential difference V_B now divides equally over each unit length of R. Any difference of potential between zero and V_B may be put across the tube by connecting the cathode to the left end of R and the anode, by means of a sliding contact, to any point of R to the right of it.

Thus, if the sliding contact is connected to the point one-quarter of the distance from the left end to the right end of R, the difference of potential across the tube will be $\frac{1}{4} V_B$. If the sliding contact is moved up to the middle of R, the difference of potential across the tube will be $\frac{1}{2} V_B$. When the sliding contact is at the three-quarter

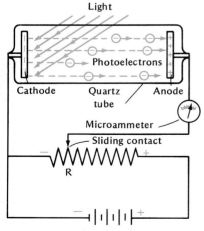

Light

Photoelectrons

Cathode   Quartz   Anode
            tube

Microammeter

Sliding contact

R

V_B = battery voltage

Fig. 29-1. Arrangement for measuring the rate of emission of photoelectrons from the cathode.

582

mark, the difference of potential across the tube will be ¾ V_B, and so on. This arrangement of a battery and a resistor used to obtain varying potential differences is called a potentiometer or a potential divider.

When the cathode is illuminated by an appropriate light source, it emits electrons. These are attracted into the positively charged anode and form a current that is measured by the microammeter, which can measure currents as small as a millionth of an ampere. To make sure that all the electrons emitted by the cathode are collected by the anode, the difference of potential applied to the tube is raised until the current in the microammeter reaches a maximum. The ejected electrons are called *photoelectrons* and the current they form is known as the *photoelectric current*.

29-3 Measuring Emission of Photoelectrons

The above apparatus enables us to measure two things: the *rate* at which electrons are ejected from the cathode and the *maximum kinetic energy* with which they escape. The rate at which the electrons are ejected is proportional to the photoelectric current and is measured by the reading of the microammeter. Knowing the size of this current and the charge of each electron, the number of ejected electrons per second that make up this current is readily computed.

29-4 Measuring Maximum Kinetic Energy of Photoelectrons

It is observed that the photoelectrons ejected by the light incident on the cathode have many different speeds and therefore different kinetic energies. The reason for this is that the electrons situated at the surface of the metal cathode are easier to eject than those situated deeper in the metal. Electrons at the surface are therefore ejected with maximum speed and maximum kinetic energy while those deeper in the metal are ejected with smaller speeds and kinetic energies.

We are interested in measuring only the maximum energy, namely, that associated with the fastest photoelectrons. To do this, we determine how much work must be done to bring these fastest photoelectrons to a halt. The slowing up of the ejected electrons is brought about by reversing the battery connections to the cathode and anode, as shown in Fig. 29-2. The former cathode is now positively charged while the former anode is now negatively charged. Between them is a retarding difference of potential and its accompanying electric field that opposes and slows up the rightward motion of the photoelectrons. The fact that some of the slower photoelectrons have been stopped by the retarding field before they can traverse the tube is indicated by a decrease in the microammeter reading.

A television camera uses the photoelectric effect to convert a light image into electronic signals that can be transmitted over radio waves to home television receivers.

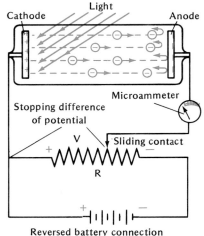

Fig. 29-2. The maximum kinetic energy of the photoelectrons is equal to the work done on them by the stopping difference of potential.

The stopping potential difference measures the maximum kinetic energy of the ejected electrons.

As the retarding difference of potential is now steadily increased, some of the faster photoelectrons are also brought to a halt before crossing the tube and the microammeter reading drops still further. At a certain value V the retarding difference of potential is just great enough to stop the fastest photoelectrons just short of the right plate of the tube. No photoelectrons then traverse the tube and the microammeter reading just drops to zero.

The *work done* by the retarding V in bringing each of the fastest electrons e to a halt is Ve. This work is equal to the kinetic energy with which these electrons were ejected from the metal and therefore *represents the maximum kinetic energy possessed by photoelectrons*. V is called the *stopping potential difference* and is measured with a voltmeter. Since e is known, Ve is readily obtained.

Sample Problem

It takes a stopping difference of potential of 5.0 V to return all of the electrons ejected from a metal by a given source of light. What is the maximum kinetic energy of the photoelectrons (*a*) in electronvolts; (*b*) in joules?

Solution:
(*a*) $V = 5.0$ volts $e = $ charge on an electron

The work done by the stopping difference of potential in bringing the fastest photoelectrons to a halt is:

$$Ve = 5.0 \text{ V} \times e = 5.0 \text{ eV}$$

This is therefore the maximum kinetic energy of those electrons.

(*b*)   Since $1.0 \text{ eV} = 1.6 \times 10^{-19}$ J
                $5.0 \text{ eV} = 8.0 \times 10^{-19}$ J

Test Yourself Problems

1. The stopping difference of potential needed to return all the electrons ejected from a metal by a given source of light is 4.5 V. What is the maximum kinetic energy of the photoelectrons (*a*) in eV; (*b*) in J?
2. What difference of potential is needed to stop electrons having kinetic energy of 4.8×10^{-19} J?

29-5 Electromagnetic Theory Interpretation of Photoelectric Effect

Using this apparatus, we are now prepared to measure how the nature of the light falling upon a metal surface determines the number of electrons ejected from that surface as well as the maximum kinetic energy of those electrons. However, before we observe what actually happens to the ejected electrons, it is useful to see what the electromagnetic theory of light predicts ought to happen and where this theory fails.

As electromagnetic waves from a light source fall on the electrons in a metal, the electromagnetic theory predicts that the rapidly alternating electric fields of these waves (see Fig. 28-4) will exert forces on each electron over which they pass, shaking the electron back and forth. In this manner, an electron may absorb energy from the incident light waves until it has enough kinetic energy to escape from the attractive forces that hold it in the metal.

Now the intensity of the light is a measure of the energy present in its waves. When the intensity of the light incident on a metal surface is increased, more energy will fall upon each part of the surface. The electromagnetic theory therefore makes these two predictions: (1) The greater the intensity of the light incident on a metal surface, the more photoelectrons will it eject from that surface. (2) The greater the intensity of the incident light, the greater will be the maximum speed and kinetic energy of the photoelectrons ejected from the surface.

29-6 Actual Effect of Intensity of Light

Experiment shows that the first prediction is correct. *If light of essentially one frequency is shone upon the cathode and its intensity is increased steadily, the photoelectric current also increases steadily showing that more electrons are being ejected from the cathode.* The second prediction, however, turns out to be false. Increasing the intensity of the light does not affect the maximum energy with which the electrons are ejected. It does not matter whether the light is strong or weak. *As long as it is light of the same frequency, the maximum kinetic energy of the electrons it ejects remains the same.*

This is a remarkable result. It means that each electron can obtain only a certain maximum quantity of energy from incident light of a given frequency and no more, no matter how intense the beam is made. Making the beam more intense simply makes more energy available for ejecting more electrons.

An increase in the intensity of the incident light increases the number of photoelectrons but not their maximum energy.

29-7 Effect of the Frequency of Light

If changing the intensity of the light has no effect on the maximum kinetic energy of the photoelectrons, what characteristic of the light does determine the maximum energy imparted to the photoelectrons? Experiment shows that it is the *frequency*. First of all, it is found that only light whose frequency is equal to or greater than a certain minimum value is able to eject electrons from any particular metal. This minimum is called the *threshold frequency*, f_0, of that metal. Different metals have different threshold frequencies. For sodium and potassium for example, the threshold frequency is relatively low and the light corresponding to it lies in the visible part of the spectrum. For most other metals, the threshold frequency is considerably higher and lies in the invisible ultraviolet part of the spectrum. *Light whose frequency is lower than the threshold frequency of a given metal cannot eject electrons from that metal no matter how intense that light is made.*

For each metal there is a minimum frequency of light called the threshold frequency below which light cannot eject photoelectrons from that metal.

When the frequency of the light falling upon a metal cathode is steadily increased above the threshold frequency of that metal, it is found that the maximum kinetic energy of the electrons it ejects increases in direct proportion, as shown in the graph of Fig. 29-3 (a). Note that for the threshold frequency f_0, the kinetic energy of the ejected electrons is exactly zero.

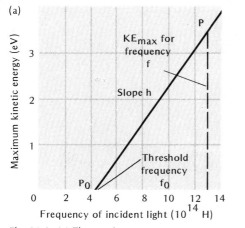

(a)

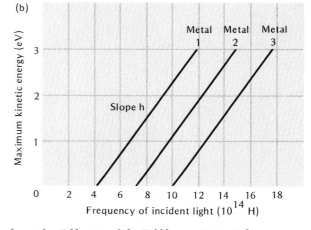

(b)

Fig. 29-3. (a) The maximum kinetic energy of the photoelectrons ejected from sodium, a typical metal, is proportional to the frequency of the incident light. **(b)** The photoelectric effect for three metals. The graphs have the same slope (h) but different threshold frequencies, f_o

29-8 Photoelectric Effect with Different Metals

When the metal used in the photoelectric tube as the cathode is replaced by different metals, a straight line graph similar to that of Fig. 29-3 (a) is obtained for each metal. Since the different metals have different threshold frequencies their graphs begin at different points on the horizontal axis. However, they all have exactly the same slope, as shown by the parallel lines in Fig. 29-3 (b). The value of this slope turns out to be a universal constant known as *Planck's constant* and designated by the letter h. It was discovered by Max Planck (1858–1947), the father of the quantum theory.

29-9 Determination of h

To illustrate how h is measured, consider the straight-line graph corresponding to a given metal in Fig. 29-3 (a). In Section 5-13, we found that the slope of a line can be found from any two of its points by dividing the difference between their ordinates by the difference between their abscissas. Applying this rule to points P and P_0, we find that the ordinate of P is KE_{max}, representing the maximum kinetic energy of an electron ejected by light of frequency f. The ordinate of P_0 is 0. The difference between the ordinates of P and P_0 is $KE_{max} - 0 = KE_{max}$. The abscissa of P is f, and that of P_0 is f_0, and the difference is $f - f_0$. The slope of P_0P is therefore $h = KE_{max}/f - f_0$. Substituting in this equation the measured values of KE_{max}, f, and f_0, we find that:

$$h = 6.62 \times 10^{-34} \text{ J} \cdot \text{s}$$

The relationship $h = KE_{max} / f - f_0$ may be written:

$$KE_{max} = hf - hf_0$$

This is known as the *photoelectric equation*. Every metal has a similar equation in which the slope h is the same for all but f_0 is different for different metals.

29-10 Einstein's Photon Theory of Light

One other observation is of the greatest importance in trying to understand the behavior of light in the photoelectric effect. That is the fact that even the *faintest* light of a frequency higher than the threshold frequency of a given metal is able to eject electrons from that metal *the moment it falls upon it*. There is no lag between the arrival of the light and the release of electrons from the metal surface.

Now it can be calculated that if the energy of a very weak incident light beam were distributed evenly over its successive wave fronts as the wave theory of light says it should be, it would take a considerable time before even a single electron absorbed enough energy from the tiny part of the wave fronts that pass over it to escape from the metal. This fact suggested to Albert Einstein that the energy in a light beam is not distributed evenly over the wave fronts but is packaged in little bundles of energy each of which is called a *quantum* or *photon*.

In this view, *every light beam consists of a stream of separate photons moving along at the speed of light*. Since the maximum energy received by an ejected electron from such a beam depends only on the frequency of the light, Einstein inferred that the *energy of each photon* in the beam must be *proportional to the frequency of the light*. Then, taking a clue from certain work done earlier by Planck, he made the brilliant guess that Planck's constant h is the proportionality factor linking the energy of a photon with its frequency. That is, the energy E in a single photon of frequency f is given by the relationship:

$$E = hf$$

According to this relationship, all photons having the same frequency must have equal quantities of energy.

29-11 Photon Explanation of Threshold Frequency

The idea of photons makes it possible to understand all aspects of the photoelectric effect. Let us begin by explaining the meaning of the threshold frequency f_0 and noting that a photon of this frequency has energy equal to hf_0.

Consider an electron at the very surface of a metal and therefore in the best position to be ejected when light falls upon the surface. Since it is observed experimentally that no light of frequency less

Fig. 29-4. Albert Einstein is remembered for his major contributions to the quantum theory as well as for his theories of relativity and his famous equation relating mass and energy.

than f_0 can eject electrons from the metal, it follows that the minimum energy a photon of incident light must have to release a surface electron is hf_0. Thus, hf_0 represents the minimum energy or work needed to free a surface electron from the metal and is called the *work function* of the metal. Because different metals differ in internal structure, the work needed to free a surface electron from them varies. For this reason, different metals have different threshold frequencies and different work functions.

29-12 Photon Explanation of Effect of Light Intensity

The fact that increasing the intensity of incident light of constant frequency increases the number of electrons ejected from a metal but not their maximum kinetic energies is also now readily explained. The more intense the incident beam, the more photons it contains and the more electrons they eject from the metal on which they fall. However, each electron is acted upon by only one photon. Therefore, *the maximum energy that any electron can receive from a light source of frequency f is equal to hf, the energy of one photon of that light.* It does not depend on the intensity of the light source.

We can now also understand why light of sufficiently high frequency ejects photoelectrons from a metal surface immediately regardless of whether the light is strong or weak. Since every photon of the same frequency has an equal quantity of energy, it makes no difference what light source it comes from. The moment a photon of high enough frequency encounters an electron, it will transfer all its energy to the electron and promptly eject it from the metal.

29-13 Photon Explanation of Effect of Light Frequency

When the frequency f of an incident photon is greater than f_0, an electron at the surface of a metal receives more energy from it than the electron requires to escape from the metal. The electron therefore has kinetic energy left over when it comes out of the metal surface. This is the maximum kinetic energy that was measured in the apparatus of Fig. 29-2 by determining the stopping potential difference. It follows that as the frequency of the incident photon increases, the maximum kinetic energy imparted by it to the ejected electron also increases.

Now the total energy supplied to the ejected electron by an incident photon of frequency f is hf. The part of this energy used in merely freeing the electron from the surface of the metal is hf_0. It follows from the law of conservation of energy that the difference, $hf - hf_0$, is therefore the maximum kinetic energy with which the electron comes out of the metal, or:

$$\mathbf{KE_{max} = hf - hf_0}$$

The maximum energy a light photon of frequency f can impart to an electron is equal to hf.

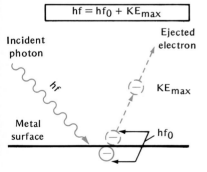

$$hf = hf_0 + KE_{max}$$

Incident photon

hf

Ejected electron

KE_{max}

Metal surface

hf_0

This is exactly in agreement with the photoelectric equation obtained experimentally from the graphs in Fig. 29-3 (b).

Sample Problem

The threshold frequency of the metal sodium is 4.4 $\times 10^{14}$ Hz. (a) What is the work needed to free an electron from the sodium surface? (b) If the frequency of the light falling on the sodium is 1.0×10^{15} Hz, what is the maximum kinetic energy it gives to the photoelectrons?

Solution:

(a) $f_0 = 4.4 \times 10^{14}$ Hz $b = 6.6 \times 10^{-34}$ J·s

work $= b\,f_0$

work $= (6.6 \times 10^{-34}$ J·s$)(4.4 \times 10^{14}$ Hz$)$

work $= 2.9 \times 10^{-19}$ J

(b) $f = 1.0 \times 10^{15}$ Hz

$bf = (6.6 \times 10^{-34}$ J·s$)(1.0 \times 10^{15}$ Hz$)$

$bf = 6.6 \times 10^{-19}$ J

$KE_{max} = bf - bf_0$

$KE_{max} = 6.6 \times 10^{-19}$ J $- 2.9 \times 10^{-19}$ J

$KE_{max} = 3.7 \times 10^{-19}$ J

Test Yourself Problems

1. The threshold frequency of a certain metal is 8.0×10^{14} Hz. What work is needed to free an electron from this metal? (This is the work function of the metal.)

2. Light of frequency 1.6×10^{15} Hz falls on the metal in problem 1. What will be the maximum kinetic energy with which it will eject electrons from the metal?

3. The work function of potassium is 3.5×10^{-19} J. (a) What is its threshold frequency? (b) Will light of frequency 4.5×10^{14} Hz eject electrons from this metal?

29-14 Historical Note

The first to infer that the energy of electromagnetic radiations is packed in quantum or photon units was the German physicist, Max Planck. He advanced the idea in solving successfully the so-called black body radiation problem. A black body is one that absorbs completely all the heat, light, and other radiation that falls upon it. Such a body is not only a perfect absorber but also the best radiator of energy. While there are no perfect black bodies, it is possible to design very close approximations of them. When a black body is hot, it radiates its energy in the form of a continuous spectrum beginning in the infrared region and extending into the shorter wavelengths.

The explanation of the emission of this continuous spectrum of the black body was a critical test for Maxwell's theory of the radiation process. According to this theory, it should have been possible to explain black body radiation by assuming that, like all periodic electromagnetic radiation, it is produced by oscillating electric charges. In this case the oscillating electric charges are simply parts of the molecules making up the walls of the black body. Since these charges oscillate with all possible frequencies and all possible energies, they can be expected to radiate the electromagnetic waves that make up the black body spectrum. However, this

Max Planck, the father of the quantum theory, was the first to postulate that radiant energy is packaged in quanta.

picture of the radiation process proved unable to account fully for the black body spectrum, and it became necessary to find a new theory of the radiation process.

In 1900, Planck came forth with such a theory. Rejecting the Maxwell idea that oscillating electric charges are the radiators of the black body spectrum, he proposed an entirely new kind of radiator to replace them. A major feature of the Planck radiator was that it could radiate and absorb energy in certain units or quanta and no others. It was in this pioneering work of Planck that first Einstein and later Bohr found the clues for their great contributions to the quantum theory.

29-15 Dual Nature of Light

The photoelectric effect reveals that *light energy is granular in nature* and is emitted by a light source of frequency f in separate photons of energy each equal to hf. *Photons* act essentially like *particles of energy*. However, it is also true that in reflection, refraction, interference, diffraction and polarization, light acts like waves and its wave behavior is very successfully explained by the electromagnetic theory. Moreover, what is true for light is also true for the rest of the electromagnetic family of radiations to which light belongs. Clearly, light and the other electromagnetic waves have two sets of behaviors, one that is particle-like and one that is wavelike. It will now be seen how the quantum theory of light reconciles these apparently contradictory behaviors.

29-16 Conditions for Observing Photons

Let us begin by asking under what conditions light shows its photon nature. For this purpose, consider the case of the parallel beam of light and its plane wave fronts in Fig. 29-6. The quantum theory of light pictures each wave front as consisting of separate photons distributed evenly over that wave front and moving with the velocity of light. Ordinarily, each single photon has so little energy, and there are such enormous numbers of them in each wave front, that the granular nature of the light forming the wave front is not noticed. The even distribution of countless photons on each wave front gives the same general effect as though their energy were spread continuously over the wave front. Thus, both the photon picture and the wave picture of the parallel beam predict correctly that the intensity of the beam is the same all over each wave front.

The fact that the energy in a wave front is not spread continuously over it but is in the form of separate photons is noticed only when the light falls upon an object small enough to detect the energy of an individual photon. In Fig. 29-6, the parallel beam is shown as a stream of photons arranged in parallel planes corresponding to the plane wave fronts. It is approaching an electron and the surface of a larger object in succession.

As far as the electron is concerned, the wave fronts of the beam are not continuous but are made up of individual photons, only one of which may hit it. As far as the surface of the large object is concerned, all of the innumerable photons in each wave front fall upon the surface in such rapid succession that the effect of any one photon is utterly lost. Only the average effect produced by all the photons together can be observed. As a result, the intensity of the light falling on the surface of the object is exactly the same as it would be if the energy were distributed continuously over the wave fronts instead of among individual photon packets.

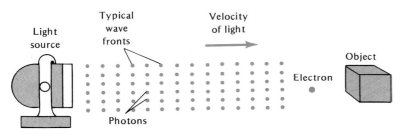

Fig. 29-6. Average distribution of photons on the wave fronts of a beam of light.

Thus, when we are observing the effect of a light beam upon tiny bodies of matter like electrons and atoms, we notice the granular or photon nature of light. When we are dealing with the effect of a light beam on a large object as a whole, and enormous numbers of photons are observed at the same time, the individual photons go unnoticed. Only their combined average effect is observed so that the light beam acts as though it were composed of continuous wave fronts instead of individual photons.

29-17 Effect of Photon Frequency

Another factor that determines whether the behavior of any particular electromagnetic radiation will appear to be wavelike or photon-like is its frequency. Certain electromagnetic radiations, like radio waves, have relatively low frequencies compared to other electromagnetic radiations. Single photons of these low frequency radiations therefore have very low energies which generally escape detection. Only when enormous numbers of these low energy photons are present is the radiation they represent intense enough to be observable. Radio and other low frequency radiations therefore show only the combined effects of innumerable photons and appear to be wavelike in their behavior.

On the other hand, the individual photons of relatively high-frequency electromagnetic radiations like those of X rays and gamma rays have very large quantities of energy and their individual effects are therefore readily observable. Such high-frequency radiations usually reveal their photon nature. The sample problem below compares the energy in one photon of a radio wave with that in a gamma ray.

Sample Problem

A transmitter generates radio waves of a frequency 1.0×10^6 Hz. Gamma rays from a certain radioactive source have a frequency of 1.0×10^{20} Hz. Compare the energies in a photon of each.

Solution:

$$f_{radio} = 1.0 \times 10^6 \text{ Hz}$$
$$f_{gamma} = 1.0 \times 10^{20} \text{ Hz}$$
$$h = 6.6 \times 10^{-34} \text{ J} \cdot \text{s}$$

$E_{radio} = h f_{radio}$
$E_{radio} = (6.6 \times 10^{-34} \text{ J} \cdot \text{s})(1.0 \times 10^6 \text{ Hz})$
$E_{radio} = 6.6 \times 10^{-28}$ J per photon

$E_{gamma} = h f_{gamma}$
$E_{gamma} = (6.6 \times 10^{-34} \text{ J} \cdot \text{s})(1.0 \times 10^{20} \text{ Hz})$
$E_{gamma} = 6.6 \times 10^{-14}$ J per photon

Note that E_{gamma} per photon is 10^{14} times as large as E_{radio}. The energy of the individual gamma ray photon is therefore readily observed. That of the individual radio photon is too small to be detected.

Test Yourself Problem

Compare the energy per photon in yellow light of frequency 5.0×10^{14} Hz with the energy in the radio and gamma photons in the above sample problem.

29-18 Quantum Theory Explanation of Wave Behavior

To explain the wavelike behavior of light in diffraction and interference experiments, the quantum theory assumes that each photon in a beam of light is associated with a guiding wave that determines the path taken by that photon through space. This wave differs from the waves studied in Chapter 15 in that it is not a wave that transports energy. Instead, its task is to accompany the photon as it moves through space and to determine the chance or *probability* that the photon will be at any point of the wave at any given time. It is called a *probability wave*.

The theory assumes that, at any instant, a photon may be *at any point* of its associated probability wave. However, it has a greater chance of being at one of those points of the wave where the amplitude of the wave is greater and a smaller chance of being at those points where the amplitude of the wave is smaller.

The probability waves associated with photons make interference and diffraction patterns exactly like those of the light waves of the wave theory. At any moment, their combined effects at a given point determine how many of the photons associated with them arrive at that point. Therefore, provided that there is a very large number of photons, the photons distribute themselves on the average so as to produce typical interference and diffraction patterns of lighter and darker areas. These turn out to be the same patterns that would be formed if the energy of the light were really distributed continuously over the wave fronts instead of being packed into separate photons.

Photons are guided through space by probability waves.

29-19 Explanation of Single-Slit Diffraction Pattern

To illustrate how probability waves account for a typical wave behavior of light, consider how a diffraction pattern is formed when a light beam passes through a narrow slit. In Fig. 29-7, photons of frequency f pass through a slit and fall on a screen. As each photon passes through the slit, it carries a fixed amount of energy equal to hf with it but *it does not have a fixed path or destination*. Instead, its associated probability wave gives it a chance of falling on any part of the screen.

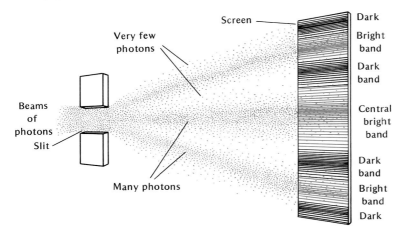

However, its chance of falling on those parts of the screen where there is a bright band in the diffraction pattern is very great. Its chance of arriving at a part of the screen at which a dark band forms is very small. The result is that as very large numbers of photons pass through the slit, they distribute themselves on the screen so that most of them fall where the bright bands are seen and very few fall where the dark bands are. The pattern they make on the screen is therefore exactly the same as the diffraction pattern predicted by the wave theory and is in agreement with observation.

29-20 Momentum of Photons

It has long been known that a light beam *exerts a pressure* on the surface on which it falls. According to the photon theory, this pressure is brought about in a manner similar to that in which the bombardment of a container wall by the molecules of an enclosed gas results in a pressure on that wall. Like each molecule of gas, each photon imparts momentum to the surface. The sum of the impacts of all the photons that strike the surface per unit of time results in the pressure on that surface.

The pressure exerted by a light beam on a surface can be measured and the momentum imparted to the surface can therefore be

Fig. 29-7. Photons guided by their probability waves pass through a narrow slit and distribute themselves so as to produce the single-slit diffraction pattern.

Each photon has a momentum

$$p = h/\lambda$$

and obeys the law of conservation of momentum.

determined. The number of photons striking the surface per unit of time can be calculated from the known intensity of the beam. By dividing the total momentum received by the surface by the number of photons striking it, the momentum of a single photon is obtained. Thus it is found that *each photon of frequency f has a momentum p equal to hf/c* where *h* is Planck's constant and *c* is the velocity of light.

The first of many experimenters to confirm that a single photon actually has the momentum $p = hf/c$, was the American, A. H. Compton. He observed that when a high energy X-ray photon collides with an electron, it sends the electron flying off in one direction while it moves off in another. One effect of the collision, known as Compton's effect, is to cause the photon's frequency to decrease abruptly.

Such a frequency decrease means a loss of momentum and can be computed from the relationship $p = hf/c$. When Compton computed the momentum lost by the photon in the collision, he found it to be equal to the momentum gained by the electron. The photon's momentum was conserved, showing that photons obey the law of conservation of momentum like material bodies.

Writing *p* for the momentum of a photon and noting that $c = f\lambda$ where λ is the wavelength associated with the photon, we have $p = hf/c = hf/f\lambda$ whence:

$$p = \frac{h}{\lambda}$$

In SI units, *h* is expressed in joule-seconds, λ is expressed in meters, and *p* will then be in newton-seconds.

Sample Problem

What is the momentum of a photon of yellow light whose wavelength is 6.0×10^{-7} m?

Solution:
$h = 6.6 \times 10^{-34}$ J · s $\lambda = 6.0 \times 10^{-7}$ m

$$p = \frac{h}{\lambda}$$

$$p = \frac{6.6 \times 10^{-34} \text{ J} \cdot \text{s}}{6.0 \times 10^{-7} \text{ m}} = 1.1 \times 10^{-27} \text{ N} \cdot \text{s}$$

Test Yourself Problems

1. What is the momentum of an X-ray photon having a wavelength of 2.5×10^{-9} m?

2. Compare the momentum in problem 1 with the momentum of a radio photon having a frequency of 7.5×10^5 Hz.

29-21 Uncertainty Principle

A consequence of the momentum of photons is the *uncertainty principle* discovered by Werner Heisenberg. It states that it is not possible to measure exactly both the position and momentum of an atomic particle at the same time because the very act of observing the particle changes its position or its momentum or both.

To illustrate, let us imagine how we would obtain the position and momentum of a moving electron. To observe the electron we must shine light on it. We can then see where it is by the light it reflects to us. However, we cannot use ordinary light because its wavelength is so much larger than the electron that it would simply be diffracted around the electron and not reveal it at all.

To detect the electron, we must use an X ray or gamma ray of very short wavelength. However, the photon of such a ray has a large momentum. On colliding with the electron, it will change both the electron's position and momentum just as it does in the Compton effect. According to Heisenberg, this introduces uncertainties in our measurements of the position, Δx, and of the momentum, Δp, such that $\Delta x \times \Delta p$ is of the order of h, Planck's constant. Since h is very small, this relationship is important only when observing very tiny particles.

The uncertainty principle actually applies to all bodies but its effects on bodies of ordinary size are negligible. The reason for this is that the momentum of the photons we use to observe a body like a moving baseball is far too small to have any effect on its motion.

The act of observing an atomic or subatomic particle changes its position or its momentum or both.

29-22 Matter Waves

The fact that light energy which so often behaves like waves has a particle nature suggested the possibility that ordinary particles of matter might have a wavelike nature. The idea was first advanced by the French physicist Louis de Broglie and was subsequently shown to be correct by an abundance of experimental evidence. De Broglie suggested that *each particle of matter m moving with velocity v has a wavelength λ associated with it which is determined by its momentum according to the relationship p = mv = h/λ.* Here again we find Planck's constant h. It should be noted that this is the same relationship that connects the momentum of a photon and its wavelength. Solved for λ, it gives

$$\lambda = \frac{h}{mv}$$

from which the wavelength of any mass m moving at velocity v can be determined.

A particle having a momentum *mv* has a wavelength

$$\lambda = h/mv.$$

29-23 Evidence for Matter Waves

To demonstrate the existence of matter waves, it must be shown that particles of matter have measurable wavelengths and that beams composed of them have wavelike properties such as diffraction and interference. We have seen how light waves are diffracted on passing a sharp edge or going through a double slit and how the wavelength of the light can be determined from the patterns of light and dark bands on a screen. Very similar diffraction patterns are obtained using electron beams instead of light beams.

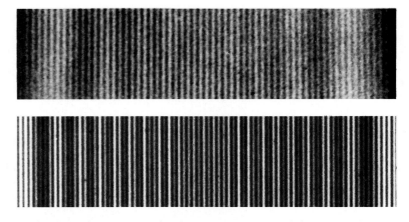

Fig. 29-8. Remarkably similar diffraction patterns made by (**a**) a beam of electrons, and (**b**) a beam of light, in passing through the equivalent of a double slit.

Electrons having equal velocities are emitted from an electron gun and are directed past a sharp edge or through the equivalent of a double slit onto a photographic plate. Here they form diffraction patterns like those made by monochromatic light. See Fig. 29-8. As in the case of light, the wavelength associated with the electron beam is measured from the diffraction pattern. It turns out to be equal to h/mv, thus confirming de Broglie's prediction. When the velocity of the electrons is increased, their momentum increases and their wavelength therefore decreases, again in agreement with de Broglie's theory.

Many other experiments involving the diffraction and interference of electrons, neutrons, gas molecules, protons, and other particles lead to two conclusions. First, all material particles behave like waves under certain circumstances. Second, the wavelength associated with a particle of matter is always equal to Planck's constant divided by its momentum.

29-24 Implications of Matter Waves

We ask now, as we did for light, under what conditions does matter reveal its wavelike nature? Looking back on our experience with light waves, you should remember that when the wavelength of light is very small compared to the dimensions of the object or opening on which it falls, its behavior is raylike. Its photon nature will then be prominently noticed. It will travel in straight lines, form sharp shadows, and show little interference or diffraction. Conversely, when the wavelength of light is nearly equal to or longer than the dimensions of the object or opening on which it falls, the wavelike properties will predominate. The light will then make diffraction and interference patterns.

This is also the case with matter waves. *Bodies that have matter waves of wavelength that is short compared to the dimensions of other bodies or openings which they approach act like particles with respect to those bodies and do not reveal their wave nature.*

Because the wavelengths of electron beams are much shorter than those of light beams, electron microscopes can obtain much higher resolution than optical microscopes.

Bodies that have matter waves of wavelength nearly equal to or longer than the dimensions of other bodies or openings show diffraction and other wave properties prominently.

When enormous numbers of such bodies are in a matter beam, they form interference and diffraction patterns similar to those produced by waves. The waves associated with particles of matter are also probability waves like those accompanying photons. Matter waves carry no energy but simply determine what chance each of the particles they control has to follow each of the many possible paths that particle may take as it moves through space.

Sample Problems

1. What is the wavelength associated with a body of mass 1.0 kg that is moving at a velocity of 10 m/s?

   Solution:

   $$h = 6.6 \times 10^{-34} \text{ J} \cdot \text{s} \qquad m = 1.0 \text{ kg}$$
   $$v = 10 \text{ m/s}$$

   $$\lambda = \frac{h}{mv}$$

   $$\lambda = \frac{6.6 \times 10^{-34} \text{ J} \cdot \text{s}}{1.0 \text{ kg} \times 10 \text{ m/s}} = 6.6 \times 10^{-35} \text{ m}$$

   This wavelength is extremely small and is not observable.

2. What is the wavelength associated with an electron (mass = 9.1×10^{-31} kg) moving at 1.0×10^{6} m/s? (An electron with this velocity has about 3 eV of kinetic energy.)

   Solution:

   $$m = 9.1 \times 10^{-31} \text{ kg} \qquad v = 1.0 \times 10^{6} \text{ m/s}$$

   $$\lambda = \frac{h}{mv}$$

   $$\lambda = \frac{6.6 \times 10^{-34} \text{ J} \cdot \text{s}}{9.1 \times 10^{-31} \text{ kg} \times 1.0 \times 10^{6} \text{ m/s}}$$

   $$\lambda = 7.3 \times 10^{-10} \text{ m}$$

   This wavelength is of the order of X-ray wavelengths and is observable by means such as those used to measure X-ray wavelengths.

Test Yourself Problems

1. A neutron (mass = 1.7×10^{-27} kg) is moving at 2.0×10^{8} m/s. What is its wavelength?
2. (a) What would the momentum of a particle have to be for it to have a wavelength of 2.0×10^{-10} m? (b) If the particle has a mass of 6.8×10^{-27} kg, what velocity does it have?

29-25 Detection and Measurement of Matter Wavelengths

The difficulties of detecting matter waves and measuring their wavelengths can be understood from the two sample problems above. The 6.6×10^{-35}-meter wavelength associated with the 1.0-kilogram mass is extremely small even when compared with the dimensions of the smallest atoms, which are of the order of 10^{-10} meter. There is no body small enough and no slit narrow enough to detect so small a wave. The wave properties of the 1.0 kilogram mass are therefore never observed and it will always behave like a body moving as a whole.

On the other hand, the wavelength associated with the moving electron in the second problem is 7.3×10^{-10} meter. This is of the order of atomic dimensions. Hence, waves of this length are large enough to be diffracted by atoms and by orderly layers of atoms forming crystals of all kinds. As a result, the matter waves associated with electrons moving at this speed and their wavelike properties can be detected and measured.

29-26 Penetrating Ability of High Velocity Particles

A proton traveling at 10^6 meters per second has a matter wavelength about one two-thousandth as great as that of the electron in sample problem 2 because its mass is about two thousand times that of the electron. Its wavelength is therefore about 3.7×10^{-13} meter. Since this is about one thousand times smaller than atomic dimensions, such waves will not be diffracted by atoms but will pass right through them. However, these waves are large compared to the nucleus of the atom, which we will find in the next chapter to have dimensions of the order of 10^{-14} meter. Protons at this velocity will therefore not be able to penetrate the nuclei of atoms. Instead they will be diffracted around the nuclei in their paths.

The wave nature of matter also gives us an idea of how fast a particle must be moving to penetrate and remain in an atomic nucleus. Only a particle whose wavelength is smaller than the size of the nucleus would not be diffracted around the nucleus and would therefore have a chance of entering it. The wavelength of such a particle would have to be less than 10^{-14} meter. (See the following sample problem.)

Sample Problem

Assuming the mass of a neutron to be 1.7×10^{-27} kg and the diameter of an atomic nucleus to be 1.0×10^{-14} m, what is the least velocity a neutron must have to be capable of entering the nucleus?

Solution:
To enter the nucleus, the wavelength λ of the neutron must be 1.0×10^{-14} m or smaller.

$m = 1.7 \times 10^{-27}$ kg $\lambda = 1.0 \times 10^{-14}$ m

$$\lambda = \frac{h}{mv} \quad \text{whence,} \quad v = \frac{h}{m\lambda}$$

$$v = \frac{6.6 \times 10^{-34} \text{ J} \cdot \text{s}}{(1.7 \times 10^{-27} \text{ kg})(1.0 \times 10^{-14} \text{ m})}$$

$$v = 3.9 \times 10^7 \text{ m/s}$$

Test Yourself Problems

1. Which of the following particles will be able to enter an atomic nucleus whose diameter is 4.0×10^{-14} m: (a) A particle with a wavelength of 1.5×10^{-15} m; (b) a particle with a momentum of 1.1×10^{-20} kg · m/s?

2. (a) What must be the minimum momentum of a particle if it is to penetrate an atomic nucleus having a diameter of 6.6×10^{-14} m? (b) If the particle is a neutron, what must be its velocity?

The action of light in ejecting electrons from metals and other bodies of matter is called the **photoelectric effect.** It reveals that the energy associated with light and other electromagnetic waves is not distributed continuously over their wave fronts but occurs in tiny packets or bundles of energy called **photons.** The energy E associated with a photon is proportional to its frequency f and is given by the relationship:

$$E = hf$$

where h is a universal constant known as **Planck's constant.**

The discovery of photons explains the particle-like behaviors of light. Among these is the fact that each photon of frequency f has a momentum p whose value is given by its energy hf divided by the velocity of light c; that is,

$$p = \frac{hf}{c} \quad \text{or} \quad p = \frac{h}{\lambda}$$

where λ is the wavelength of the photon.

The momentum of the photons in a beam of light accounts for the pressure the beam exerts on the surface on which it falls. The transfer of momentum by photons that strike electrons and other tiny bodies changes the motion of those bodies. It explains why, according to the **uncertainty principle,** it is not possible to measure exactly both the position and momentum of a particle at the same time.

The wavelike behaviors of light are explained by the **probability waves** that are associated with photons. These waves do not transport energy but act as guides that determine the probable paths of the photons.

Like all waves, the probability waves associated with photons combine to make interference and diffraction patterns. The photons associated with the waves have the greatest chance of entering those parts of the pattern corresponding to the bright bands and the least chance of entering those parts of the pattern corresponding to the dark bands. As a result, when enormous numbers of photons are involved, the distribution of light energy in the interference and diffraction patterns made by the photons is the same as is correctly predicted by the wave theory of light.

The wave and particle properties of photons of energy are matched by corresponding wave and particle properties of bodies of matter. Every particle of matter is associated with a probability or **matter wave** whose wavelength λ is given by:

$$\lambda = \frac{h}{p}$$

where h is Planck's constant and p is the momentum of the particle. Like the waves associated with photons, the waves associated with particles of matter determine the paths that the particles take as they move through space. Under similar conditions, they produce interference and diffraction patterns similar to those produced by the waves associated with photons.

CHAPTER REVIEW

Summary

Questions

Group 1

1. (a) In the apparatus for observing the photo-electric effect in metals, how is the number of photoelectrons ejected per second measured? (b) What does the stopping potential difference mean? (c) How is it used to measure the maximum kinetic energy of the photoelectrons?

2. Referring to the photoelectric effect in metals: (a) What provides the energy that ejects the photoelectrons? (b) What is meant by the threshold frequency of the metal? (c) In what two parts is the energy given to the photoelectrons divided?

3. The intensity of a source of light of constant frequency is steadily increased. (a) Assuming that the frequency is greater than the threshold frequency of a given metal, what happens to the number of photoelectrons ejected from the metal by that light? (b) What happens to the maximum kinetic energy acquired by the photoelectrons?

4. Assuming that the frequency of the light is less than the threshold frequency, answer questions 3 (a) and 3 (b).

5. A series of light sources of increasing frequency is directed upon a metal surface one at a time. If the frequencies of all the sources except the first two are above the threshold frequency of the metal, how does the maximum energy of the ejected photoelectrons vary as each source is permitted to illuminate the metal in turn?

6. Explain how the value of Planck's constant may be measured from a graph plotting, for a given metal, the maximum kinetic energies of the photoelectrons against the frequency of incident light.

7. (a) According to the electromagnetic theory of light, what effects should increasing the intensity of monochromatic light have on the maximum kinetic energy of the photoelectrons ejected from the metal surface on which it falls? (b) To what extent does this prediction agree with experimental observation?

8. (a) How is the energy associated with a photon of a beam of monochromatic light calculated? (b) What effect does increasing the intensity of the beam have on the energy of each of its photons? Explain.

9. Explain why the maximum kinetic energy with which a photoelectron is ejected from a surface by light of fixed frequency is the same whether the source of light falling on the metal is very weak or very strong.

10. (a) What is meant by the work function of a metal? (b) How is the work function of a metal obtained from its threshold frequency? (c) Why may different metals be expected to have different work functions?

11. (a) Why do photoelectrons emerge from a metal surface with velocities varying from zero to a maximum when light of constant frequency is shone upon it? (b) Which electrons in the metal are ejected with maximum velocity?

12. (a) Why are the individual photons of radio waves never detected? (b) Why are gamma rays usually observed to act like particles of energy rather than waves?

13. (a) According to the quantum theory of light, what is the role of the probability wave accompanying a photon? (b) How do the probability waves associated with the photons in a beam of light cause a diffraction pattern to form when the light passes through a narrow slit?

14. (a) How is the momentum of a photon of given frequency determined? (b) How does the momentum of photons account for the fact that light exerts a pressure on the surfaces on which it falls?

15. (a) How is the wavelength of the matter wave associated with a moving mass determined? (b) Why is the wave nature of most ordinary moving masses not noticed?

Group 2

16. (a) Compare the function of the matter waves associated with an electron beam with the function of the probability waves associated with a beam of photons. (b) In each case, how does the ratio of the wavelength to the size of a slit or obstacle in its path determine whether the behavior of the electron beam or of the photon beam will be particle-like or wavelike?

17. (a) How do the matter waves associated with a beam of electrons cause a diffraction pattern to form when the electrons are passed through the

equivalent of a narrow slit and allowed to fall on a photographic plate? (b) If only a single electron passed through the slit instead of a beam, to what part of the photographic plate would the electron be most likely to go? (c) To what parts would it be least likely to go? Explain.

18. How does the momentum of a tiny particle such as a proton or a neutron determine whether it can penetrate an atomic nucleus whose diameter is about 10^{-14} m?

19. A gamma ray collides with a stationary electron. What change takes place in the gamma ray as a result of the collision? Explain.

20. A spotlight is directed upon the bob of a swinging pendulum in a dark room. Does this introduce uncertainties in the bob's position and momentum? Explain.

Problems

Group 1

1. The stopping difference of potential of a metal for light of a certain frequency is 8.0 V. What is the maximum kinetic energy of the ejected photoelectrons (a) in eV? (b) in J?

2. (a) What is the energy in J of a photon of frequency 3.3×10^{14} Hz? (b) What is the energy in eV?

3. (a) What is the frequency of a photon of wavelength 2.0×10^{-7} m? (b) What is its energy in J?

4. The photoelectric threshold of copper is light whose frequency is 9.4×10^{14} Hz. How much work is needed to free an electron from the surface of the copper (a) in J? (b) in eV.

5. Light of wavelength 3.0×10^{-7} m is shone upon a copper cathode. (a) What is the frequency of this light? (b) What is the total energy in J that this light can transfer to an electron? (c) Using the result of problem 4, what is the maximum kinetic energy this light can give to a photoelectron?

6. The energy of each photon incident upon a metal is 9.0×10^{-19} J. The work function of the metal is 6.7×10^{-19} J. What is the maximum kinetic energy of the electrons ejected from the metal?

7. A light source radiates at the rate of 1.0 J/s. If the average frequency of the light is 5.0×10^{14} Hz, (a) what is the average energy per photon? (b) How many photons does the light source emit per second?

8. The energy needed to remove an electron from aluminum is 4.2 eV. (a) What is this energy in J? (b) What is the threshold frequency of aluminum? (c) Will light of wavelength 3.0×10^{-7} m be able to eject electrons from aluminum?

9. What is the energy of (a) a photon of a radio wave of frequency 3.0×10^6 Hz; (b) a photon of yellow light of frequency 5.0×10^{14} Hz? (c) How many photons of the radio waves does it take to transmit as much energy as one photon of yellow light?

10. What is the momentum of a photon whose wavelength is 3.3×10^{-10} m?

Group 2

11. (a) What is the momentum of a photon in an X-ray beam whose wavelength is 1.5×10^{-10} m? (b) Assuming that all such photons are absorbed by a mass of 0.10 kg that is free to move, what is the total momentum transmitted to the mass by a burst of 10^{15} photons? (c) If the mass was at rest before the burst, what velocity did it acquire after the burst?

12. (a) What is the de Broglie wavelength associated with a bullet of mass 0.050 kg whose velocity is 500 m/s? (b) Will the wave nature of this mass be observable? Explain.

13. What is the momentum of an electron having a de Broglie wavelength of 1.0×10^{-10} m? (b) What is the velocity of this electron?

14. What is the wavelength of the matter wave of a proton ($m = 1.7 \times 10^{-27}$ kg) moving at 1/10 the velocity of light?

15. The diameter of a hydrogen atom containing 1 proton and 1 electron is 1.0×10^{-10} m. (a) What must be the minimum wavelength of the proton so that it can remain confined within the atom? (b) What is its velocity at this wavelength? (This is the maximum velocity it can have without escaping from the atom.)

Applying Physics

The role of probability waves in predicting where photons or electrons will be found after leaving their source is illustrated by the following analogous situation.

Toss a coin and let it fall to the floor. It has an even chance of turning up heads. We may say that its "probability wave" predicts that when the coin is tossed very many times, it will average as many heads as tails. However, on any single toss, the coin may fall either head or tail and no prediction can be made about which it will be.

Toss a coin 30 times and keep count of the number of times it turns up heads. How well does this number agree with the predicted probability of 15 heads? Repeat the experiment by tossing the coin (a) 50 times; (b) 150 times.

Repeat the above experiment tossing two coins simultaneously. Verify that the "probability wave" applying to this situation predicts that, for a very large number of tosses, ½ of them will come up 1 head and 1 tail, ¼ of them will come up 2 heads and ¼ of them will come up 2 tails. Again, note that we can make no prediction for any single toss which may come up 2 heads, 2 tails, or 1 head and 1 tail.

Toss a pair of coins 40 times and count the number of times the result is 2 heads, 2 tails, and 1 head and 1 tail. Compare this distribution with the distribution predicted by the "probability wave." Repeat the experiment by making (a) 80 tosses; (b) 160 tosses.

What is the effect of increasing the number of tosses on the difference between the observed distribution and the predicted distribution?

Discovery of the Atomic Nucleus

30

1. To examine the experimental evidence that revealed the nature of the atomic nucleus.
2. To learn how this evidence led to the Rutherford model of the atom.
3. To evaluate the successes and failures of the Rutherford atomic model.

Aims

30-1 Toward a Picture of the Atom

In this chapter and the next, we shall develop the modern picture of the atom. Two major contributors to our understanding were Sir Ernest Rutherford (1871–1937) in England and Niels Bohr (1895–1962) in Denmark. Rutherford probed the interior of atoms with tiny projectiles called alpha particles. By studying the paths of these particles he discovered that every atom has a massive central nucleus that contains all its positive charges. This left the rest of the atom consisting of its electrons somewhere outside of the nucleus. Bohr discovered the principles governing the arrangement of these electrons from his analysis of the bright line spectrum of hydrogen.

30-2 Rays from Radioactive Substances

Before describing Rutherford's experiments, we should familiarize ourselves with the alpha particles that he used as atomic projectiles.

In 1896, Henri Becquerel, a Frenchman, discovered that the element uranium naturally emits invisible rays which, like light rays, blacken a photographic plate and can therefore be detected by it. Further research showed that several other elements, among which were thorium, actinium, polonium, and radium, emit similar rays. The natural emission of these invisible rays by elements is called *radioactivity*.

When the rays from different radioactive elements are passed through a strong electric field between two oppositely charged parallel plates as shown in Fig. 30-1, they separate into three types called *alpha, beta,* and *gamma* rays. The *alpha rays are positively charged particles* and are deflected toward the negatively charged plate. The *beta rays are high-speed electrons* and are deflected toward the positively charged plate. The gamma rays have no charge and pass between the plates undeflected. On further study, *the*

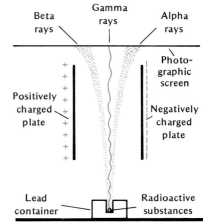

Fig. 30-1. Method of separating alpha, beta, and gamma rays emitted by a radioactive source.

603

gamma rays turn out to be electromagnetic waves similar to X rays but of much shorter wavelength and much more penetrating. Some of them are able to pass readily through thick barriers opaque to ordinary light, such as thick slabs of concrete.

30-3 Nature of Alpha Rays

Alpha rays can be formed into a beam by enclosing the radioactive substance emitting them in a lead box having a hole drilled in one side. The alpha beam issuing from the box can then be studied by passing it through electric and magnetic fields by the methods described in Section 26-19. In this way it is determined that alpha particles have a charge twice as great as that of the proton and a mass about equal to that of a helium atom. Further study reveals that *alpha particles are in fact doubly charged helium ions.* This is verified by sealing a radioactive element emitting alpha particles in a bottle and noting the continuous accumulation of helium gas as more and more alpha particles are trapped in the bottle.

Rutherford used alpha particles, which are doubly charged helium ions, as atomic projectiles.

For his atomic projectiles, Rutherford used a beam of alpha particles emitted by a small mass of polonium enclosed in a lead box. Using the methods described in Section 26-17, he passed the beam through crossed electric and magnetic fields and determined the velocity of the alpha particles to be 1.60×10^7 meters per second. Thus, he had at his disposal a supply of projectiles of known mass, charge, and velocity.

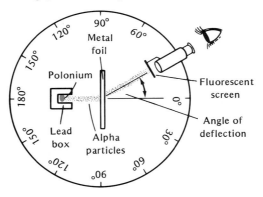

Fig. 30-2. Rutherford's apparatus for studying the deflections of alpha particles by a thin metal foil.

30-4 Rutherford's Experiment

Rutherford's apparatus is shown schematically in Fig. 30-2. A beam of alpha particles emitted by polonium is directed upon a target consisting of a thin metal foil. In the first experiments, gold foil was used as the target because gold can be hammered into thin sheets only about 500 atoms thick.

As the alpha particles move through the foil, some of them collide with parts of atoms in their paths or come close to the electrically charged particles contained in those atoms. In either case, the forces which are exerted upon the alpha particles deflect them from their original direction of motion. While most of the alpha particles succeed in passing through the foil, a very small number do not penetrate it and are turned back.

To see what directions the individual alpha particles take after leaving the foil, the particles are allowed to fall on a translucent fluorescent screen attached to a microscope. Each alpha particle striking the screen produces a tiny flash of light called a *scintillation*. The screen and microscope are mounted on an arm so that they can be moved to any point on a circular path around the metal foil. They can thus be used to detect any alpha particles deflected in that particular direction.

The observer sets the screen at a series of regularly spaced points on the circular path. At each position, the number of alpha particles deflected in that direction during a constant time interval is counted. In this way the observer determines what fraction of the total number of alpha particles deflected by the foil goes in each direction.

Using this apparatus, Rutherford and his coworkers made two important observations, each of which gave a basic insight into the structure of atoms. The first was that the vast majority of the alpha particles directed at the metal foil pass through it with little or no change. The second was that some alpha particles are deflected by the foil through quite large angles, and in some cases do not penetrate the metal foil at all. Instead, they rebound from it like a steel ball hitting a stone wall.

Fig. 30-3. Ernest Rutherford discovered that practically all the mass of an atom is concentrated in its very small positively charged nucleus around which its negative electrons circulate.

30-5 Emptiness of Atoms

Since the metal foil is known to be many hundreds of atoms thick, the first observation means that alpha particles pass through hundreds of atoms in their paths without being seriously obstructed. This suggests that the *atoms of the metal foil are largely empty space* and that the parts of which an atom is made take up only a very small part of its total volume. The relative emptiness of atoms may be compared to the emptiness of the night sky in which the stars, planets and other bodies occupy only a tiny part of the available space.

30-6 Discovery of Atomic Nucleus

The fact that, occasionally, alpha particles rebound from the metal foil and even reverse their direction suggests that these alpha particles have collided head-on with some part of an atom. Since the interior of an atom appears to be mainly empty space through which alpha particles usually pass practically undeflected, such head-on collisions must be very rare. This would explain why so few alpha particles are observed to rebound from the metal foil.

Moreover, once an alpha particle entering the metal foil has made such a rare head-on collision, it is extremely unlikely that it will "hit the bull's eye" again and make a second head-on collision with another atom. These considerations led Rutherford to conclude that the rebounding of an alpha particle from the metal foil is the result of a single major collision with some part of the atom it has entered.

What part of the atom is involved in this major collision? It seems evident that it is not one of the atom's electrons. An alpha particle has about 7500 times the mass of an electron. Its collision with an electron is like the collision of a rapidly thrown baseball with a fly in mid-air. In both cases, the collision has a negligible effect on the speed and direction of the more massive body. Even several successive collisions between an alpha particle and an atom's electrons are not likely to produce a significant change in the original motion of the alpha particle.

On the other hand, a collision with a proton can seriously alter the motion of an alpha particle. Two properties of the proton account for this. The first is its positive charge that repels the oncoming alpha particle with rapidly increasing force as it comes closer. The second is the appreciable mass of the proton, which is about 2000 times that of an electron and one-fourth that of an alpha particle.

The mass is still not great enough to cause an alpha particle colliding with a single proton at rest to reverse its direction. However, if many protons are bundled together, their combined mass and charge can readily account for the observed reversal. Rutherford therefore was led to suspect that *the protons of the atom are packed together in a separate core called the atomic nucleus.* It follows that, since the nucleus contains all parts of the atom except the practically massless electrons, *the nucleus of the atom contains nearly its entire mass.*

30-7 Explanation of Alpha Particle Deflections

From the above considerations, Rutherford could assume that alpha particles entering a metal foil undergo significant deflections only when they pass close to the nucleus of an atom and are relatively unaffected by electrons in or near their paths. Picturing each atom as a central nucleus around which, at relatively large distances, are distributed the atom's electrons, he could then explain the deflections of the alpha particles directed at a metal foil.

In Fig. 30-4, three particles, alpha 1, alpha 2, and alpha 3 are shown approaching the nucleus of an atom. As each positively charged alpha particle nears the positively charged nucleus, the force of repulsion upon it increases in accordance with Coulomb's law. For alpha 1, this repulsive force never becomes large because the path of alpha 1 never brings it very close to the nucleus. Alpha 1 therefore suffers almost no deflection as it passes through

The nucleus occupies only a very small part of the volume of an atom.

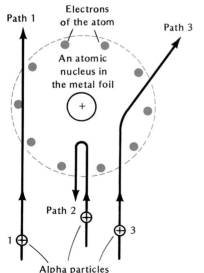

Fig. 30-4. Paths taken by alpha particles on nearing an atomic nucleus.

the atom. Most of the alpha particles entering the foil behave like alpha 1.

Alpha particle 2 approaches the nucleus head-on. The nucleus repels it with increasing force until its motion is stopped and its direction is reversed. Alpha 2 represents those rare alpha particles that rebound from the metal foil and reverse their direction.

Alpha particle 3 approaches on a path that brings it relatively near the nucleus. As it nears the nucleus, the increasing repulsion deflects it markedly from its original path. Assuming that the nuclear force acting upon the approaching alpha particle follows Coulomb's law, and applying Newton's laws of motion, Rutherford computed the path the repulsive force would make an alpha particle take. All such paths turn out to be members of the family of mathematical curves known as hyperbolas. The particular kind of hyperbola on which an incoming alpha particle travels depends upon how closely it approaches the nucleus. A typical path is that taken by alpha 3.

30-8 Factors Determining Angle of Deflection

Assuming this explanation of the deflection of alpha particles to be correct, Rutherford was able to derive a formula predicting that the number of alpha particles deflected by a metal foil through a given angle in a constant time interval should depend upon three factors. It should *increase directly with the thickness of the metal foil.* It should *decrease inversely as the fourth power of the velocity* of the alpha particle. Finally, it should *increase directly as the square of the charge in the nucleus of each atom* of the metal foil.

Using the apparatus described in Section 30-4, Rutherford's co-workers showed that the first two predictions of his formula were in excellent agreement with experiment. The third prediction could not be directly verified at that time because no information about the charge in the nucleus of each atom of the metal foil was available.

30-9 Determination of Nuclear Charge

The remarkable success of Rutherford's formula in predicting the effects of the thickness of the foil and of the velocity of the alpha particles on the deflection pattern gave confidence that the formula was also correct in its prediction about the effect of the charge of the nucleus on deflecting the alpha particles. By substituting in this formula the observed number of alpha particles deflected through a given angle, the measured thickness of the foil, and the known value of the velocity of the alpha particles, the charge in the nucleus of an atom of a given metal foil is computed. By repeating the experiment with thin foils of many different elements, information is obtained from which the nuclear charge in the atoms of these elements is also determined.

The atomic number of an element is the number of protons in its nucleus.

In this way, it is found that *the atomic nucleus of each element has a specific number of protons in it that is always the same for that element.* This number is called the *atomic number* of that element. As one goes from the lightest element to the heavier elements, the atomic number increases in unit steps from 1 to over 100. Hydrogen, the lightest element, has the atomic number 1 because it has 1 proton in its nucleus. Helium, which is the second lightest element, has 2 protons in its nucleus and has the atomic number 2. Lithium, with 3 protons in its nucleus, has the atomic number 3, while uranium, with 92 protons in its nucleus, has the atomic number 92.

Since every complete atom is electrically neutral, it must contain as many electrons as protons. *The atomic number therefore not only tells how many protons an atom has in its nucleus but also how many electrons are distributed around its nucleus.*

To the chemist, the atomic number of an atom has a special meaning because it determines the chemical properties of that atom. Any two atoms that have the same atomic number have the same chemical properties and are given the same name. As we saw in Section 13-20, an element may have atoms of different masses called isotopes. However, the atomic number of the isotopes of a given element must be the same.

30-10 Mass of Nucleus

The mass of an atom is equal to the sum of the masses of its electrons and its nucleus. Since the mass of an electron is extremely small, compared to that of a proton, *practically all the mass of an atom is in its nucleus.*

If the nucleus of an atom contained only protons, its mass would be equal to the sum of the masses of its protons. However, this is not true for any element except the common form of hydrogen. The atomic masses of all other elements are generally about two times as great as the sum of the masses of their protons. This suggests that, in addition to protons, neutral particles are present in the nuclei of atoms and make up the additional mass. These particles were discovered by Chadwick in 1932 and named *neutrons*. A neutron has about the same mass as a proton but has no electric charge.

The mass of the nucleus of each atom of an element is the sum of the masses of the protons and neutrons it contains. This mass, expressed in atomic mass units, is also practically equal to the atomic mass of the element known from chemistry. Since each proton and each neutron contributes one atomic mass unit to the total mass, *the whole number nearest the atomic mass of an element is the number of protons and neutrons in its nucleus.* The number of protons in the nucleus of an atom is known from its atomic number. The number of neutrons in each nucleus can therefore be obtained by subtracting the number of protons from the total number of protons and neutrons it contains.

To illustrate, the whole number nearest the mass of the common isotope of helium is 4. This means that a helium nucleus has a total of 4 protons and neutrons. Since the atomic number of helium is 2, the nucleus of a helium atom contains 2 protons. It therefore has 4 − 2, or 2 neutrons.

One form of uranium atom has a mass very close to 235 atomic mass units. This means that it contains a total of 235 neutrons and protons. Since uranium has the atomic number 92, its atomic nucleus contains 92 protons. It must therefore have 235 − 92, or 143 neutrons.

30-11 Nuclear Composition of Isotopes

According to this picture of atomic nuclei, isotopes are simply atoms whose nuclei have the same number of protons but different numbers of neutrons. They therefore have the same atomic numbers and the same chemical properties, but have different atomic masses.

Hydrogen, for example, has 3 isotopes. The common form of hydrogen has an atomic mass of 1. A second form has an atomic mass of 2 and is called deuterium (doo·TIR·ee·um). A third form has an atomic mass of 3 and is called tritium (TRIT·ee·um). As shown in Fig. 30-4, all of these isotopes are alike in that they have only 1 proton in the nucleus. They differ in that the nucleus of the most common form of hydrogen contains no neutrons, that of deuterium contains 1 neutron, and that of tritium contains 2 neutrons.

Uranium has 3 isotopes of masses 234, 235, and 238, known as U-234, U-235, and U-238. The nuclei of these isotopes all have 92 protons, but that of U-234 has 238 − 92 = 142 neutrons; that of U-235 has 235 − 92 = 143 neutrons; that of U-238 has 238 − 92 = 146 neutrons.

In Section 26-19, we saw how the mass spectrograph separates two isotopes of neon of atomic masses of 20 and 22 atomic mass units. Since neon has the atomic number 10, this means that the nucleus of the first isotope of neon contains 10 protons and 10 neutrons. The nucleus of the second isotope of neon also contains 10 protons but includes 12 neutrons to make up a mass total of 10 + 12, or 22 atomic mass units.

30-12 Size of Nucleus

Another very important bit of information obtained from Rutherford's experiments with alpha particles is an estimate of the size of the nucleus. As an alpha particle approaches a nucleus head-on,

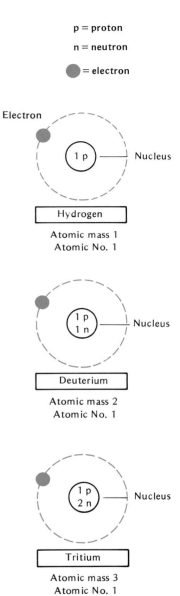

p = proton

n = neutron

= electron

Hydrogen
Atomic mass 1
Atomic No. 1

Deuterium
Atomic mass 2
Atomic No. 1

Tritium
Atomic mass 3
Atomic No. 1

Fig. 30-5. Nuclear composition of the three isotopes of hydrogen.

the electric repulsion exerted upon it by the nuclear charge slows it down to a halt before reversing its direction. At the moment the alpha particle is stopped, its kinetic energy is completely transformed into electric potential energy associated with its position in the electric field of the nucleus.

We saw in Section 22-18 that the potential energy associated with two electrical particles is equal to Kq_1q_2/r where q_1 and q_2 are the charges on the particles, K is Coulomb's constant, and r is the distance between the centers of the particles. Since we know that this potential energy is equal to the known initial kinetic energy of the alpha particle, and since q_1 and q_2 and K are also known, r can be calculated. The distance r is the nearest the alpha particle comes to the nucleus of the atom.

If we assume the alpha particle and the nucleus to be spheres, r cannot be smaller than the sum of the radius of the nucleus and the radius of the alpha particle. This distance turns out to be in the neighborhood of 10^{-14} meter as compared to 10^{-10} meter which is about the diameter of an atom.

According to this calculation, the diameter of the nucleus of the atom is only about one ten-thousandth of the diameter of the atom. Since the nucleus contains practically the entire mass of the atom, this means that the mass of an atom is contained in a very tiny fraction of the total volume of the atom. The rest of the atom is empty except for the electrons that are distributed about the nucleus. The size of each atom appears to be determined by the positions of its outermost electrons.

30-13 Planetary Model of the Atom

From the results of his experiments, Rutherford was able to propose a picture or model of an atom. In this model, each atom consists of a small central positively charged nucleus containing all of its protons as well as practically all of its mass. As was shown later, the total mass of the nucleus is the sum of the masses of its protons and its neutrons. Around the nucleus of the atom are distributed a number of electrons equal to the number of protons it contains. The atom is thus electrically neutral.

Rutherford assumed that each electron of an atom revolves about the nucleus in an orbit just as the planets in the solar system revolve about the sum. The centripetal force needed to keep the electron in orbit is supplied by the electrical attraction between the positively charged nucleus and the negatively charged electron.

The ordinary hydrogen atom, known to contain one proton and one electron, has the simplest structure. In this atom, the single proton is the nucleus and the single electron revolves about it in an orbit that may be circular or elliptical. Models of lithium and neon atoms are shown in Fig. 30-6.

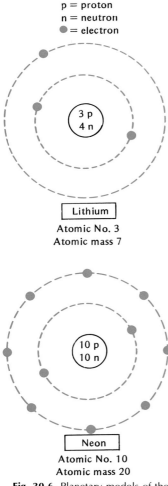

p = proton
n = neutron
● = electron

Lithium
Atomic No. 3
Atomic mass 7

Neon
Atomic No. 10
Atomic mass 20

Fig. 30-6. Planetary models of the atoms of lithium and neon.

Rutherford's planetary model of the atom had some outstanding successes. It accounted for the very large amount of empty space inside atoms and explained accurately how alpha particles are deflected on passing through atoms. It identified the nucleus of the atom as the core containing nearly all of its mass and all of its positive charge. It provided a method of determining the charge of the nucleus as well as its approximate size.

30-14 Failures of Rutherford's Atomic Model

Rutherford's model failed in two critical ways. First, an atom in which electrons orbit like planets around the nucleus cannot remain in existence very long. According to the electromagnetic theory of light, an accelerated electron radiates electromagnetic waves. An electron moving in an orbit is under constant centripetal acceleration and is therefore radiating away its energy continually. This steady loss of energy causes the electron's orbit to change by spiraling closer and closer toward the nucleus. Eventually the electron must crash into the nucleus and the atom is destroyed.

The second failure of the Rutherford model was its inability to explain why the light emitted by the atoms of electrically excited gases forms a bright-line spectrum instead of a continuous spectrum. It was stated in Section 17-28 that each of the bright lines in the spectrum of an excited gas has a definite wavelength and frequency. Furthermore, because each element has its own set of bright lines by which it can be identified, it can be inferred that its light comes from its individual atoms.

Now, if atoms were made according to Rutherford's model, they would produce a continuous spectrum instead of the bright-line spectrum that is actually observed. To illustrate why this is so, consider the simple hydrogen atom. As its single electron revolves around the proton that is its nucleus, it radiates electromagnetic waves in the form of light. According to Maxwell's electromagnetic theory, the frequency of this emitted light is equal to, or is a multiple of, the frequency with which the electron completes its orbit. However, because the electron steadily loses energy by radiation, it gradually spirals in toward the nucleus. As it does so, the number of revolutions it makes per second increases steadily and so does the frequency of the light it emits.

At any given moment, a light source of excited hydrogen gas would have many billions of atoms in all possible stages of their lives. In some atoms, the electrons would be relatively far from the nucleus and in others, relatively near. In all cases, however, the electrons would be spiralling around the nucleus at increasing frequencies. The gas as a whole ought therefore to emit light waves of all possible frequencies and produce a continuous spectrum. This is contrary to the observation that hydrogen has a line spectrum.

CHAPTER REVIEW

Summary

Rutherford's experiments on the deflection of **alpha particles** by matter revealed the existence of a central core in each atom known as its **nucleus.** Rutherford showed that the nucleus contains practically the entire mass of the atom and all of its **protons.** He also was able to infer that the electrons of the atom are situated in the space around the nucleus at relatively large distances from each other and from the nucleus so that much of the atom is empty space.

Rutherford's planetary model pictures the electrons in an atom revolving about the nucleus in orbits like those of the planets around the sun. It explains the emission of light by atoms by noting that such revolving electrons emit light having a frequency equal to or related to the frequency of their motion. However, Rutherford's model of the atom is unsatisfactory because it predicts that every atom will eventually destroy itself and that the atoms of excited gases will emit a continuous spectrum of light. Both of these predictions are contrary to the facts.

These failures of the Rutherford model indicated the need for a new model of the atom that would not only be stable but would also emit light in the form of the observed bright line spectra. We shall now see how the great Danish physicist, Niels Bohr, laid the foundation for such a model.

Questions

Group 1

1. What are (a) alpha rays; (b) beta rays; (c) gamma rays? (d) How is their presence detected?
2. What happens to a beam consisting of alpha, beta, and gamma rays when it is passed through the electric field between two oppositely charged plates?
3. If a beam consisting of alpha, beta, and gamma rays passes through a magnetic field at right angles to the beam, what deflection, if any, does each of the rays undergo?
4. Referring to Rutherford's apparatus of Fig. 30-2, explain: (a) how the beam of alpha particles is obtained; (b) how the alpha particles are detected after they pass through the metal foil.
5. (a) What observations suggested that the atoms of the metal foil are mostly empty space? (b) Why can't the reversal of the path of an alpha particle on nearing or entering an atom be explained as the result of its collision with one of the atom's electrons? (c) How is it explained?
6. (a) According to Rutherford, what parts of the atom are contained in its nucleus and what parts are outside the nucleus? (b) Utilizing this picture of the atom, explain why most alpha particles

pass through the metal foil undeflected. (c) Explain why the rest are deflected through angles of varying size up to and including 180°. (d) What part does Coulomb's law of force between charges play in the deflection of the alpha particles referred to in (c)?
7. (a) According to Rutherford's theory, how does the number of alpha particles deflected through a given angle vary with the charge of the atomic nucleus that causes the deflection? (b) How does this relationship make it possible to find out how much positive charge is present in the nucleus of an atom of the metal of which the foil is made?
8. What does the atomic number of an element tell about (a) the composition of its nucleus; (b) the number of electrons it contains?
9. (a) What does the whole number nearest the atomic mass of an element tell about the composition of its nucleus? (b) How can the number of neutrons in an atom of sodium be determined from the fact that its atomic mass is 23 u and its atomic number is 11?
10. (a) How does Rutherford's experiment make it possible to estimate the size of the nucleus of

the atom? (*b*) What is the approximate ratio of the diameter of an atom to the diameter of its nucleus?

Group 2

11. (*a*) Describe Rutherford's planetary model of the atom. (*b*) What provides the centripetal force needed to keep the electrons in their orbits? (*c*) Why is it possible for most alpha particles to pass through such an atom undeflected? (*d*) Why are the electrons of such an atom easier to remove from it than its protons?

12. (*a*) Why could not an atom as imagined by Rutherford continue to exist indefinitely? (*b*) Why should the light emitted by hydrogen atoms built on this plan always consist of a continuous spectrum? (*c*) What kind of spectrum do discharge tubes containing hydrogen give?

Applying Physics

The principle of determining the nature of the atomic nucleus by firing projectiles at it can be studied by directing marbles at targets of various shapes and noting the directions in which the marbles go after collision with the target. A relatively massive target such as a cylindrical glass tumbler is put on a sheet of paper on which a line is drawn around the tumbler to show its position, as in Fig. 30-7. Marbles are then rolled toward the target one at a time along a series of parallel lines. This can be done by letting the marbles roll down a small inclined plane having a straight groove in it to guide the marbles.

After each "shot," the inclined plane is moved a short distance parallel to itself and another marble is rolled toward the target. The direction of each marble before and after collision is marked on the paper. The pattern formed can then be studied to see what it reveals about the width and shape of the target. Try this procedure with a rectangular block and other objects used as targets.

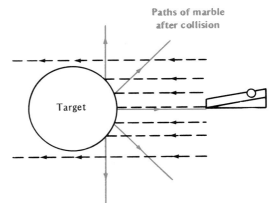

Fig. 30-7. After the marbles rebound from their target, their paths reveal its size and shape.

Nobel Laureates: Marie Curie and Maria Goeppert-Mayer

The Swedish Royal Academy of Science awarded the first Nobel Prize in Physics in 1901. The prize, according to the Nobel Foundation, was to be given to ''the person who has made the most important discovery or invention within the field of physics.'' The first recipient was Wilhelm Conrad Röntgen (1845-1923), discoverer of X rays.

In the eight decades since, only two women have received the Nobel Prize in Physics. The first was Madame Marie Sklodowska Curie (1867-1934). Born in Poland, Marie Curie studied and worked in France because Polish laws forbade higher education for women. She was stimulated by Henri Becquerel's discovery of radioactivity in uranium salts in 1896. She began a systematic search for other radioactive substances and in 1898 discovered that thorium was also radioactive. Teaming with her husband, Pierre, Marie went on to discover the radioactive elements, polonium and radium.

An entire new field of research and knowledge was born. From that point in time studies of radioactive materials dominated the first half of the 20th century. The Nobel Prize in Physics in 1903 was shared by Marie and Pierre Curie and Henri Becquerel for their discoveries. In 1911, Marie Curie received a second Nobel prize, an unprecedented honor for any scientist. This time it was in chemistry. Albert Einstein said of Marie Curie: ''Her strength, her purity of will, her austerity toward herself, her objectivity, her incorruptible judgment—all these were of a kind seldom found joined in a single individual.''

Maria Goeppert-Mayer (1906-1972) received the Nobel prize in physics in 1963 for her discoveries on the nucleus of the atom. Maria Goeppert-Mayer was also born in Poland, but her major research was done in the United States. Her contribution to physics was the concept that the neutrons and protons in atomic nuclei are grouped in accordance with a definite series of numbers: 2, 8, 20, 28, 50, etc. These were called magic numbers. The idea of magic numbers was independently arrived at by Hans D. Jensen at about the same time. Jensen shared the 1963 Nobel prize with Maria Goeppert-Mayer. They explained magic numbers by a model of shells or layers within the atom's nucleus similar to the electron shells outside the nucleus.

They theorized that each nucleon in these shells has a spin that is strongly coupled to its orbital angular momentum. This shell model of the nucleus has had remarkable success in solving many nuclear problems. It has opened new doors to the mysteries of the incredibly small world of the atomic nucleus.

(Far left) Marie Curie holding a conference at the Conservatory of Arts and Techniques of Paris. (Near left) Maria Goeppert-Mayer at the University of California.

Quantum Theory Model of the Atom

31

1. To learn how by applying the quantum idea to the Rutherford model of the atom, Bohr developed a model of the hydrogen atom that fully explained its emission and absorption of light.
2. To evaluate the successes and failures of the Bohr atomic model.
3. To learn how the Bohr model was modified to eliminate its weaknesses.
4. To examine the experimental evidence supporting the current model of the atom.

31-1 Balmer Series in Hydrogen Spectrum

Bohr set out to find a way of correcting the basic defects in the Rutherford picture of the atom. He found there were major clues in the line spectrum of ordinary hydrogen. When the entire line spectrum formed by the visible and invisible light emitted from an electric discharge tube containing hydrogen is studied in the spectroscope, it is found to consist of several distinct *series* of lines. A typical series, shown in Fig. 31-1, is formed by the visible lines in the spectrum and some of the ultraviolet lines. You can see that as the lines of this series go from the red wavelengths to the shorter violet and ultraviolet wavelengths of the spectrum, they are closer and closer together and approach a minimum wavelength known as the limit of the series. The other series, which are found entirely in the invisible infrared and ultraviolet parts of the spectrum, have a similar structure.

The Swiss physicist Jakob Balmer discovered that all the wavelengths of the series in Fig. 31-1 can be represented by the formula

$$\frac{1}{\lambda} = R \left(\frac{1}{2^2} - \frac{1}{n_i^2} \right)$$

where R is a constant having the value 1.097×10^7 wavelengths per meter, n_i is a whole number greater than 2 such as 3, 4, 5, and so on, and λ is the wavelength of a line in the series corresponding to each particular value of n_i. The series of lines represented by this formula is called the *Balmer series*. The wavelength of the first line

The formula for the Balmer series was the key to the Bohr theory of the atom.

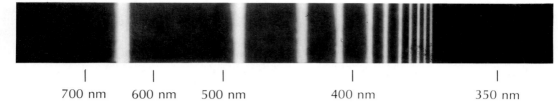

700 nm	600 nm	500 nm	400 nm	350 nm

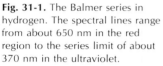

Fig. 31-1. The Balmer series in hydrogen. The spectral lines range from about 650 nm in the red region to the series limit of about 370 nm in the ultraviolet.

of the series is calculated by setting $n_i = 3$; the wavelength of the second line of the series is calculated by setting $n_i = 4$; and so on. The limit of the series is obtained by setting n_i equal to infinity. When the wavelengths of the Balmer series are computed by this formula and compared with the actual values of the wavelengths as measured in a spectroscope, the agreement is excellent.

31-2 General Formula for Bright-Line Series in Hydrogen

Following Balmer's lead, other scientists found that very similar formulas apply to all the other series of lines in the hydrogen spectrum. These take the general form:

$$\frac{1}{\lambda} = R \left(\frac{1}{n_f{}^2} - \frac{1}{n_i{}^2} \right)$$

where, for any particular series, n_f is a constant and may be any whole number beginning with 1. For the Balmer series, for example, $n_f = 2$. Once n_f is selected, the wavelengths of all the lines in the series corresponding to that value of n_f are obtained by giving n_i all the whole number values greater than n_f; that is, $n_f + 1$, $n_f + 2$, and so on.

Sample Problem

Find the wavelength of the first line in the Balmer series of the hydrogen spectrum.

Solution:
For the Balmer series, $n_f = 2$. For the first line in this series, $n_i = 3$. $R = 1.097 \times 10^7$ wavelengths/m.

$$\frac{1}{\lambda} = R \left(\frac{1}{n_f{}^2} - \frac{1}{n_i{}^2} \right)$$

$$\frac{1}{\lambda} = 1.097 \times 10^7 \left(\frac{1}{2^2} - \frac{1}{3^2} \right) \text{ wavelengths/m}$$

$$\frac{1}{\lambda} = 1.097 \times 10^7 \left(\frac{5}{36} \right) \text{ wavelengths/m}$$

$$\lambda = 6.563 \times 10^{-7} \text{ m}$$

This is 656.3×10^{-9} m or 656.3 nm.

Test Yourself Problems

1. Find the wavelength of the line in the Balmer series for which $n_i = 6$.
2. Find the limit of the Balmer series ($n_i = \infty$).

31-3 Bohr's Idea of Stationary Orbits

Hydrogen is not the only element that has its spectral lines arranged in series. It is found that the line spectra of many other elements also form series that obey formulas similar to those that apply to hydrogen. This suggests that there are common features in the structures of different atoms that are responsible for the similarity in the series formed by their spectral lines. Bohr set out to discover those common features. He began by modifying Rutherford's model of the hydrogen atom, in which a single electron orbits around a nucleus composed of a single proton, by making two assumptions or postulates.

Bohr's first postulate is designed to explain how the hydrogen atom remains stable. *It assumes that the electron cannot travel around the nucleus in all possible orbits, but only in certain selected ones and no others.* Moreover, it assumes that, contrary to the predictions of the electromagnetic theory of light, an electron *does not radiate electromagnetic waves while it is in one of these selected orbits.* The electron therefore loses no energy while in any of the selected orbits and can remain in that orbit just as a planet remains in its orbit around the sun.

The greater the energy possessed by the electron, the larger is the selected orbit in which it travels. Normally, the electron is in the orbit closest to the nucleus where it has the lowest energy it can have. The atom is then said to be in its *ground state.* However, the electron can acquire energy in such ways as absorbing a photon of light or being hit by an electron from an electron gun. It may then obtain enough additional energy to permit it to jump into one of the higher energy orbits, as shown in Fig. 31-3. When its electron is in a higher energy orbit, the hydrogen atom is said to be in an *excited state.*

31-4 Bohr's Idea of Light Emission

Bohr's second postulate is designed to explain how the hydrogen atom emits its observed line spectrum. According to this postulate, *an atom emits light or electromagnetic energy only when its electron happens to be in a higher energy orbit and jumps to an orbit of lower energy,* as shown in Fig. 31-3. *The energy is radiated only at the moment the electron jumps, and is equal to the difference between the electron's energy before and after the jump.* Such jumps from a higher energy to a lower energy orbit occur naturally.

This situation is analogous to that of a ball held a few feet above the floor where it has gravitational potential energy. When the ball is released, it naturally drops to the floor, giving up some of its potential energy in the process. If E_i is the initial higher energy associated with the orbit from which the electron starts, and E_f is

Fig. 31-2. Niels Bohr's analysis of atomic line spectra improved upon Rutherford's planetary atomic model but succeeded in explaining the structure of only simple atoms such as hydrogen.

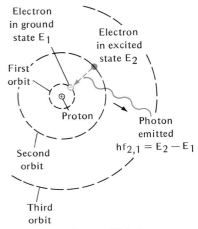

Fig. 31-3. A photon of light is emitted by a Bohr atom when an electron in a higher energy orbit drops back into a lower energy orbit.

the final energy associated with the orbit into which the electron falls, the energy emitted as light or electromagnetic radiation during the jump is $E_i - E_f$. Moreover, if f is the frequency of the emitted radiation, it is known from the photoelectric effect that hf is the energy associated with 1 photon of this radiation. Bohr assumed that exactly 1 photon is emitted during each electron jump so that:

$$E_i - E_f = hf$$

The frequency of the emitted light is therefore given by:

$$f = \frac{E_i - E_f}{h}$$

31-5 How Line Series Are Formed

In Bohr's model of the hydrogen atom, light photons are emitted only when an electron jumps from a higher to a lower energy orbit.

Bohr's picture of the hydrogen atom as shown in Fig. 31-3 explains in a qualitative way how the Balmer and other series of lines are produced. For simplicity, the selected orbits in which the electron can travel are chosen to be circles. The orbit in which the electron has the lowest energy E_1 is labeled $n = 1$. The circles labeled $n = 2, 3, 4, 5$, etc., represent orbits of progressively higher energy, E_2, E_3, E_4, E_5, etc. According to Bohr, no light or electromagnetic energy is radiated by an electron as long as it remains in any of these selected orbits. However, when an electron jumps from an outer orbit of energy E_i to an inner one of energy E_f, the energy it loses is emitted as light of frequency $f = (E_i - E_f)/h$.

For example, if the electron happens to be in the E_3 orbit, it can emit light in two different ways. It can jump to either the E_2 orbit or to the E_1 orbit. In the first case, it emits light of frequency $f_{3.2} = (E_3 - E_2)/h$. In the second case, it emits light of frequency $f_{3.1} = (E_3 - E_1)/h$. It can be seen that the frequency $f_{3.1}$ emitted in the larger jump from E_3 to E_1 is greater than the frequency $f_{3.2}$ emitted in the jump from E_3 to E_2.

Now, when the atoms of hydrogen gas in a discharge tube are made to emit light by applying an appropriate difference of potential to the tube, the individual atoms are excited in different ways. At any instant, the electrons in some atoms will be in the orbit of lowest energy E_1; in other atoms, they will be in the orbit of energy E_2; in still others, the electrons will be in the orbit of energy E_3, and so on.

Let us imagine a series of atoms in the gas such that the electron in each is in a different orbit represented by E_1, E_2, E_3, E_4, E_5, and so on. Suppose that in each atom the electron jumps spontaneously, as shown by the arrow in Fig. 31-4, to the lowest energy orbit E_1. Then, a series of photons of light of increasing frequency is emitted corresponding to the electron jumps from E_2 to E_1, E_3 to E_1, E_4 to E_1, E_5, to E_1, and so on. The electron that is already in the ground state E_1 will remain there and emit no light. It is seen therefore that these particular atoms will radiate a series of specific lines of increasing frequency and decreasing wavelength as each of their electrons drops to the orbit E_1. This series corresponds to one of the series of bright lines actually found in the hydrogen spectrum.

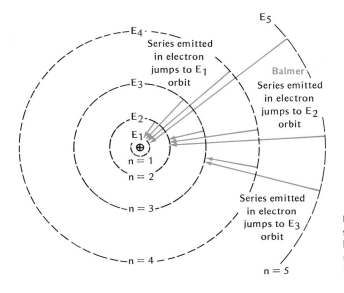

Fig. 31-4. How different line series are produced in the hydrogen spectrum. Each arrow represents one of the bright lines in the spectrum.

In the same way, the lines of a second series, which happens to correspond to the Balmer series, are emitted by the hydrogen atoms in the different excited states whose electrons fall to the E_2 orbit instead of the E_1 orbit. Similar series of lines are emitted by all those excited atoms whose electrons jump from outer orbits to the E_3 orbit, the E_4 orbit, or any other orbit of higher energy than E_1 and E_2.

31-6 Test of Bohr's Model of Hydrogen Atom

So far, Bohr's picture of the atom has explained in a *qualitative* way how the atom remains stable and yet produces the different series of lines observed in the hydrogen spectrum. It now remains to be seen in a *quantitative* way how the stationary orbits in which an electron is allowed to be are determined, and how the energy associated with these orbits may be calculated. Once these energies are obtained, all the frequencies of the radiations that are emitted by hydrogen atoms can be calculated from the relationship $f = (E_i - E_f)/h$. The wavelengths corresponding to these frequencies can then be readily obtained and compared with the wavelengths of the lines actually observed in the hydrogen spectrum. Bohr's idea of the structure of the hydrogen atom can thus be tested.

31-7 Quantum Condition Determining Allowable Orbits

First, Bohr had to decide which of all the possible orbits are the ones in which an electron may be. He found that the spectrum of the hydrogen atom could be successfully explained by his atomic model if he assumed that an electron could only be in those circular orbits for which the momentum of the electron times the radius of its orbit is equal to a whole number times the quantity $h/2 \pi$. This condition may be written $mvr = nh/2 \pi$, where m is the mass of the electron, v is its velocity, r is the radius of its orbit, h is Planck's constant, and n is any whole number equal to or greater than one. The quantity mvr is called the *angular momentum* of the electron.

Although Bohr chose this assumption on the basis of other considerations, the wave nature of matter, developed after Bohr's model of the atom, provides an interpretation of the meaning of this condition. Recall from Section 29-22 that mv, the momentum of the electron, is equal to h/λ, where λ is the electron's associated wavelength. Substituting this value for mv in the above relationship, gives $hr/\lambda = nh/2 \pi$, whence $2 \pi r = n\lambda$. This equation tells us that the circumference $2 \pi r$ of an orbit in which the electron is permitted to be is always such that the *wavelength of the electron occupying that orbit fits into it exactly a whole number of times.* The smallest orbit in which an electron can be is that for which $n = 1$. For this orbit, the circumference is exactly equal to one wavelength associated with the electron. The next larger orbit has a circumference equal to two electron wavelengths; the next, three, and so forth.

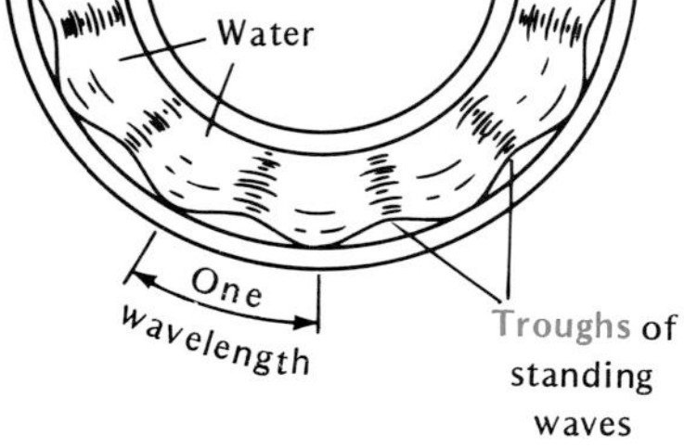

31-8 Water Analogy of Orbit-Wave Relationship

The condition that decides which orbits the electron can travel in can be visualized by imagining each allowed orbit to be a circular tank containing water, as in Fig. 31-5. At one point in the tank, there is a wave generator sending water waves, representing the waves associated with the electrons, around the tank in opposite directions. Only if the waves are of a wavelength such that they fit exactly around the channel a whole number of times will the waves moving in opposite directions from the generator continue to interfere in exactly the same way and thus to make a stable or unchanging interference pattern. Oppositely directed waves that interfere in this stable way are said to make stationary wave patterns. In the figure, exactly twelve full wavelengths fit into the tank to make a stationary wave pattern.

The orbits that Bohr was led to select as the possible ones in which an electron could be are the ones for which the waves associated with the electron make stationary wave patterns. A little more will be said about the manner in which stationary waves are set up in Section 31-21.

Fig. 31-5. Standing water waves are produced in a circular tank by waves passing over each other in opposite directions.

31-9 Radii of Hydrogen Orbits

Using the above condition for selecting the allowable orbits of the electron in the hydrogen atom, Bohr found the following formula for the radii of the different orbits:

$$r = \frac{n^2 h^2}{4 \pi^2 K m e^2}$$

Here n assumes whole number values beginning with 1, e is the unit of elementary charge, K is the constant 9.0×10^9 newton-meters^2/coulombs^2, and m is the mass of the electron. How this formula is obtained is shown in the next section.

The radius of the innermost orbit is calculated by letting $n = 1$, the radius of the next orbit by letting $n = 2$, and so forth. For $n = 1$, the sample problem 1 below shows that the calculated radius of the orbit is 5.3×10^{-11} meter. Thus, the diameter of the innermost electron orbit is about 10^{-10} meters, a value that agrees well with the estimated size of the hydrogen atom obtained from other research.

When n is equal to infinity, r also becomes infinite. Because the electron is infinitely far from the nucleus, it is free and no longer bound to the atom. This represents the condition in which the electron escapes completely from the atom. The atom is then said to be ionized. In the case of ordinary hydrogen, the ionized atom is simply the proton that is its nucleus.

31-10 Deriving the Expression for Orbital Radius

Optional derivation for honor students

Let $+e$ be the charge of the proton forming the nucleus of the hydrogen atom, $-e$ the charge of its electron, m the mass of the electron, and v its orbital velocity. The centripetal force needed to keep the electron in a circular orbit of radius r around the proton is $F = -mv^2/r$, where the minus sign means that F is directed toward the proton. This centripetal force is supplied by the electrical attraction between the proton and the electron, which is $F = K(+e)(-e)/r^2$, where K is Coulomb's constant. It follows that $-mv^2/r = -Ke^2/r^2$, or that $mv^2 = Ke^2/r$.

Now Bohr's condition for selecting the allowable orbits is $mvr = nh/2\pi$, whence $v = nh/2\pi rm$. Substituting this value of v in:

$$mv^2 = \frac{Ke^2}{r}$$

gives,

$$m\left(\frac{n^2 h^2}{4 \pi^2 r^2 m^2}\right) = \frac{Ke^2}{r}.$$

whence,

$$r = \frac{n^2 h^2}{4 \pi^2 K m e^2}$$

Sample Problems

1. Find the radius of the innermost orbit of the hydrogen atom.

   Solution:

   $n = 1$ $h = 6.6 \times 10^{-34} \text{ J} \cdot \text{s}$
   $e = 1.6 \times 10^{-19} \text{ C}$ $m = 9.1 \times 10^{-31} \text{ kg}$
   $K = 9.0 \times 10^{9} \text{ N} \cdot \text{m}^2/\text{C}^2$

   $$r = \frac{n^2 h^2}{4 \pi^2 K m e^2}$$

   $$r = \frac{(1)^2 (6.6 \times 10^{-34} \text{ J} \cdot \text{s})^2}{4 \pi^2 (9.0 \times 10^{9} \text{ N} \cdot \text{m}^2/\text{C}^2) \times (9.1 \times 10^{-31} \text{ kg})(1.6 \times 10^{-19} \text{ C})^2}$$

   $$r = 5.3 \times 10^{-11} \text{ m}$$

   This is 53×10^{-12} m or about 50 pm.

2. What is the radius of the orbit for which $n = 3$?

   Solution:
   Since r is proportional to n^2 in the above formula, the radius of the orbit for which $n = 3$ is 3^2, or 9 times as great as that for $n = 1$. The radius for the orbit for which $n = 3$ is therefore $9 \times 5.3 \times 10^{-11}$ m $= 4.8 \times 10^{-10}$ m

Test Yourself Problem

What is the radius of the hydrogen orbit for which (a) $n = 2$; (b) $n = 10$?

31-11 Energies of Orbits

Having determined the orbits in which an electron may be, Bohr calculated how much energy an electron possesses when it is in each of these possible orbits. In expressing the energy associated with an electron while in any orbit, it has been found convenient to compare it with the energy possessed by the electron when it is in the orbit for which n is equal to infinity. This represents the condition when the electron barely escapes from the atom, and the atom is therefore ionized. The free electron is then arbitrarily said to be at the *zero level of energy*, provided it is at rest.

If the free electron also happens to have kinetic energy, it is said to be at a *positive level* of energy equal to its kinetic energy. If the *electron is not free but in one of its allowed orbits inside the atom, it is said to be at a negative level of energy.* The numerical value of an energy level is the energy an electron in that level must be given to raise it to the zero level and thus free it from the atom. This energy is therefore a measure of how tightly bound an electron in that level is to its atom.

By the method shown in the next section, Bohr found the values of the energy levels E_n associated with each orbit to be:

$$E_n = \frac{-2 \pi^2 K^2 m e^4}{n^2 h^2}$$

The constants K, m, e, and h are the same ones that appear in the formula for the radii of the orbits. If their numerical values are substituted in the expression for E_n, it becomes:

$$E_n = \frac{-2.17 \times 10^{-18}}{n^2} \, J$$

This energy is often expressed in electronvolts as:

$$E_n = \frac{-13.6}{n^2} \, eV$$

The energy associated with any orbit of the hydrogen atom can now be obtained by substituting the value of n associated with that orbit.

The energy associated with the innermost orbit is found by setting $n = 1$, and is $E_1 = -13.6/1^2 = -13.6$ electronvolts. This means that 13.6 electronvolts is the energy that must be given to the electron in the lowest or ground state of the atom to free it from the hydrogen atom and thus ionize it. It is called the *binding energy* or *ionization energy* of the atom.

The energy E_2 associated with the second orbit is found by setting $n = 2$. It is $E_2 = -13.6/2^2 = -3.4$ electronvolts. This means that 3.4 electronvolts is the energy needed to free the electron from the hydrogen atom when the electron is in the second orbit.

The energy E_3 associated with the third orbit is found by setting $n = 3$. It is $E_3 = -13.6/3^2 = -1.5$ electronvolts. Note that E_3 is numerically smaller than E_2 which, in turn, is smaller than E_1. This is explained by the fact that it takes less work to separate an electron that is in an outer orbit from the atom than it takes for an electron in an inner orbit. It follows that the higher n is, the nearer E_n is to zero. When n is set equal to infinity, E_n comes out equal to zero. This is as expected because for n equal to infinity the electron is just free of the atom and therefore is at the zero level of energy.

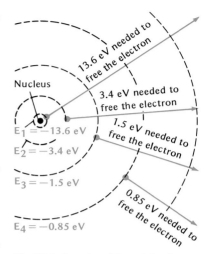

Fig. 31-6. Energies of the orbits of the Bohr hydrogen atom expressed in electronvolts.

31-12 Deriving the Expression for Orbital Energies

Optional derivation for honor students

The energy of an electron in any allowed orbit consists of two parts, the *kinetic energy* of its orbital motion and the *potential energy* resulting from the electric attraction between the nucleus and the electron. The kinetic energy is $\frac{1}{2} mv^2$ where m is the electron's mass and v its orbital velocity. The potential energy of an electron at a distance r from a proton is given by the expression in Section 22-18 as $U = K(-e)(+e)/r = -Ke^2/r$, where e is the elementary unit of charge and K is Coulomb's constant. Note that U is negative, reflecting the fact that the potential energy is taken as zero when the electron is detached or infinitely far from the nucleus. It is therefore a negative number when the electron is in any of the allowed orbits.

The total energy of an electron in an orbit of radius r is its kinetic energy plus its potential energy, or $E_n = \frac{1}{2} mv^2 + (-Ke^2/r)$. Now it was shown in the derivation of r in Section 31-10 that $mv^2 = Ke^2/r$. Substituting this value of mv^2 in E_n gives:

$$E_n = \frac{1}{2}\frac{Ke^2}{r} - \frac{Ke^2}{r} = -\frac{1}{2}\frac{Ke^2}{r}$$

Substituting:

$$r = \frac{n^2h^2}{4\,\pi^2\,Kme^2}$$

we have,

$$E_n = -\frac{Ke^2(4\,\pi^2\,Kme^2)}{2\,n^2h^2}$$

whence,

$$E_n = \frac{-2\,\pi^2\,K^2me^4}{n^2h^2}$$

31-13 Predicting Emitted Frequencies and Wavelengths

Bohr now used the relationship $hf = E_i - E_f$ to predict the frequencies of light that should be emitted by hydrogen atoms. Substituting $E_i = -13.6/n_i^2$ electronvolts and $E_f = -13.6/n_f^2$ electronvolts in this relationship, he obtained $hf = 13.6/n_f^2 - 13.6/n_i^2$ electronvolts, whence $hf = 13.6(1/n_f^2 - 1/n_i^2)$ electronvolts. Now $\lambda f = c$, where λ is the wavelength corresponding to the frequency f, and c is the velocity of light. Substituting $f = c/\lambda$ above gives:

$$\frac{hc}{\lambda} = 13.6\left(\frac{1}{n_f^2} - \frac{1}{n_i^2}\right)\ \text{eV}$$

whence,

$$\frac{1}{\lambda} = \frac{13.6}{hc}\left(\frac{1}{n_f^2} - \frac{1}{n_i^2}\right)\ \text{wavelengths/m}$$

If h is expressed in electronvolt-seconds, this relationship predicts the wavelengths that should be emitted by hydrogen atoms. To be correct, this relationship should agree with that obtained in Section 31-2, which is known to agree with observations; namely,

$$\frac{1}{\lambda} = \mathbf{R}\left(\frac{1}{\mathbf{n}_f^2} - \frac{1}{\mathbf{n}_i^2}\right)$$

For these two expressions to be the same, the constant R must be equal to $13.6/hc$. On computing the value of $13.6/hc$, it is found that this is so. Thus, Bohr's model of the hydrogen atom is remarkably successful in predicting not only the existence of the Balmer and other series in the hydrogen spectrum, but also in giving the exact quantitative relationships that these series obey.

31-14 Explanation of Absorption Spectra

Another important success of the Bohr model of the atom is its ability to explain the absorption of light by atoms. Recall that the atoms of a gas absorb light of the same frequencies that they emit when they are excited. To understand how this happens, consider a hydrogen atom whose electron is in the innermost orbit corresponding to its minimum energy E_1. If this electron is somehow moved into the next orbit of energy E_2, it will soon fall back to the lower energy orbit, E_1. The atom will then emit a photon of frequency $f_{2,1}$, and of energy equal to $hf_{2,1} = E_2 - E_1$.

Now to raise the electron from the E_1 orbit to the E_2 orbit, energy must be given to it equal to $E_2 - E_1$. One way of supplying this energy is to send a photon of light into the atom, as shown in Fig. 31-7. A photon of frequency equal to $f_{2,1}$ has just enough energy to raise the electron from the E_1 to the E_2 orbit. The atom therefore absorbs a photon of this frequency and, in doing so, transfers its electron from the E_1 to the E_2 orbit. In this condition, the atom is excited and its electron soon drops back to the E_1 level, thereby re-emitting the photon it absorbed.

Photons of frequency less than $f_{2,1}$ do not have enough energy to raise the electron from the E_1 to the E_2 orbit. Therefore, they cannot be absorbed by the atom. On the other hand, photons of frequency greater than $f_{2,1}$ have more energy than is needed to raise the electron from the E_1 to the E_2 orbit. However, *unless the energy of a photon is exactly large enough to raise the electron into the third, fourth, fifth, or any other allowable orbit, the atom cannot absorb it. That is to say, the atom absorbs only a whole photon of energy,* not a part of it.

In general, an atom can absorb only those photons of light whose energies are exactly the right quantity to move its electron from any orbit in which it happens to be into any orbit of higher energy. The photons that can be absorbed by an atom are therefore those whose frequencies are the same ones that the atom emits after it is excited.

Thus the Bohr atomic model explains how the atoms of the vapor of an element absorb exactly the same frequencies of light that they emit when excited. The dark lines of absorption spectra observed in the laboratory and the Fraunhofer absorption lines of the sun's spectrum are produced in this manner.

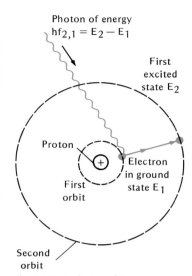

Photon of energy
$hf_{2,1} = E_2 - E_1$

First excited state E_2

Proton

Electron in ground state E_1

First orbit

Second orbit

Fig. 31-7. A photon of frequency $f_{2,1}$ is absorbed by a hydrogen atom and raises its electron from the E_1 to the E_2 level. The energy of the photon is $hf_{2,1} = E_2 - E_1$.

31-15 Successes and Limitations of Bohr's Atomic Model

Bohr's model of the atom was brilliantly successful in explaining precisely the emission and absorption of light by hydrogen atoms and by certain other atoms and ions whose structure approximates that of a hydrogen atom. However, the model was unable to deal with the more complicated atoms in which there are many electrons to be considered instead of just one. The task of finding allowable orbits for each of the many electrons proved formidable.

Furthermore, although the idea of each electron moving in one of its allowable orbits is easily imagined, it does not agree with what is known of the wave nature of the electron. As we found in Sec. 29-24, an electron's motion and position are determined by its associated matter wave. Since, at any instant, an electron has a definite possibility of being at any of many different points of its wave front, it is impossible to assign a specific path or orbit to the electron in an atom. Finally, the idea of electron orbits themselves is actually of no practical value because the physicist has no way of observing or measuring them.

Bohr's atomic model therefore did not provide a completely acceptable picture of the atom, but it was a major stepping stone to the more satisfactory current model of the atom. Just as Bohr incorporated into his model those elements of the Rutherford atomic model that agreed with the experimental facts, so the current model includes those ideas of the Bohr atom that proved so successful.

31-16 Current Atomic Model

In the current atomic model, the nucleus is surrounded by an electron cloud in which individual electrons are at different energy levels.

An atom is now conceived to have a central positively charged nucleus containing all its protons and neutrons. Surrounding the nucleus are the atom's electrons, equal in number to the number of protons in the nucleus. *No specific orbits or paths are assumed for the electrons.* Instead, the electrons are considered to be distributed around the nucleus in a sort of *electron cloud* in which the position of any particular electron is *indefinite.*

While the idea of specific electron orbits is therefore abandoned, the idea of an atom having only certain possible *energy states* or *levels* formerly associated with the electronic orbits is retained. The electron cloud of an atom is believed to be able to exist only in certain specific series of allowable conditions or states. Each of these states is associated with a very definite quantity of energy and is described as an energy level. An atom in its lowest possible energy level is in its *ground state* and its *most stable* condition. An atom in the ground state may absorb energy and get into one of the energy levels higher than the ground state. The atom is then in an *excited state.* If the atom acquires so much energy that one of its electrons is able to detach itself completely, the atom is ionized and is said to be at its *ionization energy level.*

As in the Bohr model, an atom in an excited state E_i can undergo a spontaneous change to a lower energy state E_f. When an atom undergoes such a drop in energy level, the energy difference between its initial state and its final state is emitted as light or radiant energy of frequency f such that $hf = E_i - E_f$. Thus, the familiar

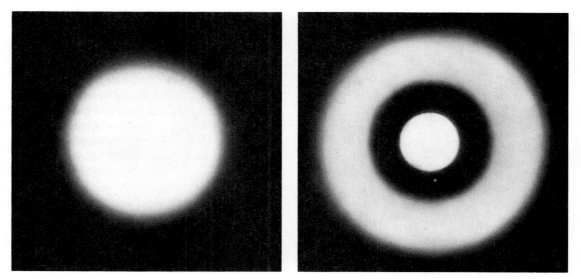

Fig. 31-8. Visualizations of the electron cloud surrounding the nucleus of a hydrogen atom (**a**) in the gound state and (**b**) in an excited state. The single electron has the greatest chance of being where the figure is light and the least chance of being where it is dark.

Bohr postulate that explained the series of spectral lines emitted by hydrogen and hydrogen-like atoms is now applied to all atoms.

31-17 Absorption of Radiant Energy

As in the Bohr model, the absorption of radiant energy by an atom is explained by its passage from a lower energy state E_f to a higher one E_i. Again the relationship between the absorbed frequency f and the two energy levels is $f = 1/h(E_i - E_f)$, so that an atom can absorb the same frequencies that it emits when excited.

The minimum energy required to detach an electron from an atom in its ground state is again called its ionization energy. When a photon having more than the ionization energy enters an atom, it not only ejects an electron from the atom but imparts its surplus energy to the ejected electron as kinetic energy.

31-18 Energy Level Diagram

The energy states or levels of an atom are conveniently shown by means of an *energy level diagram*. Fig. 31-9 shows a simplified diagram for the hydrogen atom. Similar diagrams may be made for the atoms of the other elements. Each allowed energy level of the atom is represented by a horizontal line. The lowest line represents the *ground state* and each line above it represents an *excited state*. The top line represents the *ionization level* at which an electron just escapes from the atom. Above the ionization level is a continuous shaded region representing the fact that after an electron escapes from the atom, it can have any value of kinetic energy. Thus, energy levels above the ionization level are continuous.

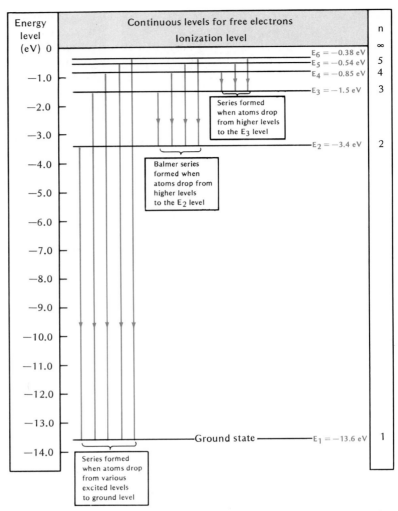

Fig. 31-9. Energy level diagram for hydrogen.

The energy levels are spaced on a vertical scale showing the energy associated with that level. In this diagram, the energy is expressed in electronvolts. The values of *n* for each energy level are also shown. The ionization level of the atom is taken as the zero level of energy. Each of the energy levels below it is represented as a negative number and represents the energy the atom lacks to become ionized when in that state. Thus, the energy level of the ground state is −13.6 electronvolts. This means that when the hydrogen atom is in the ground state, it takes 13.6 electronvolts of energy to pull its electron out of the atom and thus ionize the atom. The energy level of the next higher state is −3.4 electronvolts, meaning that the hydrogen atom in this excited state needs to absorb 3.4 electronvolts of energy to become ionized.

The emission of light or radiant energy by an excited atom is shown by an arrow going from the atom's initial excited state to a state of lower energy. The figure shows how the different series of lines in the hydrogen spectrum are formed. The first series is formed when atoms of hydrogen in different excited states drop to the ground state for which $n = 1$. The second series of lines is formed when atoms of hydrogen in different excited states drop to the energy level just above the ground level for which $n = 2$. This is the Balmer series. The other series are formed by atoms that drop from higher excited states to the ones for which $n = 3$, $n = 4$, $n = 5$, and so on. Each arrow represents one spectral line whose frequency is given by the relationship $hf = E_i - E_f$, where E_i is the energy associated with the excited state and E_f is the energy associated with the final state.

31-19 How the Atom Absorbs Energy

The atom can absorb only such quantities of energy as will just raise it from the energy level in which it happens to be to one of the higher energy levels. It can receive energy either when a photon enters it or when it is struck by a moving particle such as an electron or another atom. In the first case, the atom can absorb only radiant energy of the same frequencies that it emits. In the second case, the atom will absorb only such parts of the energy of a moving particle that strikes it as are just the right size to raise the atom from the energy level it is in to one of the higher energy levels.

If the energy of the photon or of the moving electron is greater than the ionization energy of the atom, the atom will be ionized. The electron that is thus set free will then have kinetic energy equal to the difference between the total energy absorbed and the energy used in ionizing the atom. (See Fig. 31-10.)

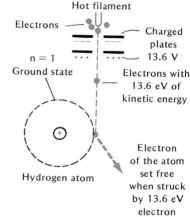

Fig. 31-10. A hydrogen atom is ionized by collision with an electron having a minimum of 13.6 eV of kinetic energy.

Sample Problems

1. Referring to Fig. 31-9, suppose that a hydrogen atom is in the state $E_2 = -3.4$ eV. (a) How much energy would an electron colliding with the atom need to raise the atom to the E_3 state? (b) What will be the energy of the photon emitted when the atom then drops from the E_3 state to the E_1 state? (c) What is the frequency of this photon?

Solution:
(a) To raise the atom from E_2 to E_3 requires energy equal to $E_3 - E_2$.

$$E_3 - E_2 = -1.5 \text{ eV} - (-3.4 \text{ eV}) = 1.9 \text{ eV}$$

(b) When the atom drops from E_3 to E_1, it emits a photon of frequency $f_{3.1}$, whose energy is

$$hf_{3.1} = E_3 - E_1$$
$$hf_{3.1} = -1.5 \text{ eV} - (-13.6 \text{ eV}) = 12.1 \text{ eV}$$

(c) $$f_{3.1} = \frac{E_3 - E_1}{h}$$
$$f_{3.1} = \frac{12.1 \text{ eV}}{h}$$

Since h is expressed as 6.6×10^{-34} J·s, 12.1 eV must be expressed as $12.1 \times (1.6 \times 10^{-19})$ J.

$$f_{3.1} = \frac{12.1 \times (1.6 \times 10^{-19}) \text{ J}}{6.6 \times 10^{-34} \text{ J·s}}$$

$$f_{3.1} = 2.9 \times 10^{15} \text{ Hz}$$

2. A photon of energy 14.0 eV enters a hydrogen atom in the ground state and ionizes it. With what kinetic energy will the electron be ejected from the atom?

Solution:

It takes 13.6 eV to ionize the atom. The photon is supplying 14.0 eV. The difference:

$$14.0 \text{ eV} - 13.6 \text{ eV} = 0.4 \text{ eV}$$

is transmitted to the electron as kinetic energy.

3. A photon of energy 10.8 eV enters the hydrogen atom in the ground state. Will it be absorbed by the atom?

Solution:

No. It takes $E_2 - E_1 = -3.4 \text{ eV} - (-13.6 \text{ eV}) = 10.2 \text{ eV}$ to raise the atom to the E_2 level and $E_3 - E_1 = 12.1 \text{ eV}$ to raise the atom to the E_3 level. The atom can absorb only photons that have exactly the right energy to raise it to a higher level. This photon of 10.8 eV has too much energy to raise the atom to the E_2 level and not enough energy to raise it to the E_3 level.

Test Yourself Problems

1. (a) What is the energy of the photon emitted when the electron in the hydrogen atom drops from the E_5 state to the E_1 state? (b) What is the frequency of this photon? (Refer to Fig. 31-9.)
2. A hydrogen atom is in the E_2 state. (a) What is the least amount of energy it must absorb to become ionized? (b) If this energy is supplied by a photon, what is its frequency?
3. Each of the following photons enters a different hydrogen atom in the ground state. What change, if any, will each photon produce in the energy state of the atom it enters? (a) a 12.1 eV photon; (b) a 9.0 eV photon; (c) a 14.6 eV photon.

31-20 Demonstrating Atomic Energy Levels

The importance of the idea of atomic energy levels is that they can actually be detected and measured in the laboratory. To illustrate the principle by means of which this is done, consider a tube containing hydrogen at low pressure and having an electron gun built into it, as shown in Fig. 31-11.

To begin with, the hydrogen atoms in the tube are all in the ground state. Now let us adjust the electron gun so that the velocity of its electrons is nearly zero at first and then is gradually increased. As the electrons leave the gun and pass through the gas, they collide with some of the hydrogen atoms in their paths. At first, these collisions do not supply enough energy to the atoms to raise them out of their ground state. However, as the electrons speed up, they reach a point at which they strike atoms hard enough to raise their energy level from the ground state to the next higher energy state. These excited atoms then drop spontaneously back to the ground state, emitting light as they do so.

The emission of light by the hydrogen in the tube, therefore is a signal that the bombarding electrons are giving up enough energy to raise the atoms they strike to their first excited state. The light emitted by the excited atoms will consist of the single frequency corresponding to the energy difference between the ground and the first excited state. This light can be observed and identified by examining it with a spectroscope.

Thus, to measure the energy level associated with the first excited state, the speed of the electrons leaving the gun is increased until the hydrogen just begins to emit light consisting of a single frequency. The electrons colliding with the hydrogen atoms are giving

Hydrogen atoms can be raised from their ground state to higher states by bombarding them with electrons. When these excited atoms drop back to the ground state, they emit the spectral lines associated with the transitions.

up just enough energy to them to raise the hydrogen atoms to the first energy level above the ground state. The kinetic energy of these electrons is known from the voltage applied to the gun. This energy is equal to the energy difference between the ground state and the next higher energy level. Thus, the existence of the energy level above the ground state is demonstrated and its level above the ground state is determined.

The higher energy levels may be determined by applying the same principle. The speed of the electrons in the gun is increased gradually until their kinetic energy is sufficient to raise the atoms with which they collide from the ground level to the second higher level. The atoms will then emit light as they drop back to each of the two levels below the second level.

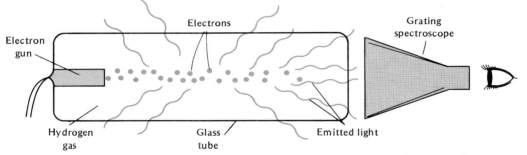

Fig. 31-11. Schematic arrangement for observing the light emitted when electrons bombard hydrogen atoms.

The spectroscope will show that the spectrum now consists of three lines instead of one. One line corresponds to the drop of the electron from the second energy level above the ground state to the first level. The second line corresponds to the drop from this same second energy level to the ground level. The third line corresponds to the drop from the first energy level above the ground state to the ground level. Thus, the kinetic energy of the bombarding electrons that are just capable of stimulating the hydrogen in the tube to emit these three lines is a measure of the energy associated with the second energy level above the ground state. As before, this kinetic energy is determined from the known value of the gun voltage that accelerates the electrons.

31-21 Standing Waves

In the Bohr atom, the energy levels were determined by the sizes of the orbits in which the electron was permitted to travel. In the current atomic model, the energy levels associated with the atom are determined by the wave properties of its electrons. Since atoms are stable and unchanging, it is suggested that the wave pattern associated with their electrons must also be stable and unchanging. Waves that make unchanging patterns are called stationary or *standing waves*.

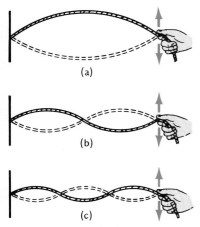

(a)

(b)

(c)

Fig. 31-12. Only those waves whose half wavelengths fit into the length of the rope a whole number of times can make a standing wave pattern.

Standing waves are formed when two sets of waves of equal amplitude and wavelength move through each other in opposite directions. Standing waves can be simply demonstrated with a rope, one of whose ends is tied to a wall while the other is held in the hand and moved up and down as in Fig. 31-12. If the free end of the rope is made to vibrate at the right frequency, the entire rope will move up and down in the pattern shown in (a). If the frequency of vibration is doubled, each half of the rope will move up and down as shown in (b). If the frequency of the vibration is tripled, each third of the rope will move up and down as shown in (c). These patterns are produced by the superposition of and interference between the waves sent down the rope toward the wall by the experimenter and the reflections of those waves after they hit the wall.

Since each loop represents a half wavelength, it is evident that the only waves that can make a standing pattern between the ends of the rope are those whose half wavelength fits into the length of the rope a whole number of times. In (a), 1 half wavelength fits into the length of the rope. In (b), 2 half wavelengths fit into the length of the rope. In (c), 3 half wavelengths fit into the length of the rope. Thus, the distance between the ends of the rope determines the length of the waves that will form standing wave patterns.

This example illustrates the general fact that when the waves of the proper wavelength are reflected back and forth between the walls of a container, they form standing wave patterns. The dimensions of the container decide which wavelengths will form standing waves in it.

31-22 Standing Electron Waves in Atoms

In an analogous way, we can imagine the matter waves associated with the electrons of an atom setting up stationary wave patterns inside the atom. Because the electrons of the atom are kept within a certain volume by the attractive force of the nucleus, the waves associated with the electrons may be considered to be trapped in a "container" surrounding that volume. As the waves are reflected back and forth inside this volume, they form stationary wave patterns just as such patterns are formed by the waves moving back and forth between the ends of a rope. Each of these stationary wave patterns corresponds to an energy state or level in which it is possible for an atom to remain stable.

Thus, *the standing wave patterns of the electron waves in the atom determine which of all possible energy levels are the ones characteristic of that particular atom.* In this connection, we have already seen how the orbits and therefore the energy levels allowable in Bohr's model of the hydrogen atom are the ones for which the matter waves of the electrons in them form standing waves.

To explain the line spectrum of the hydrogen atom, Bohr modified Rutherford's planetary model of the atom by making two assumptions. The first was that an electron in an atom could only be in certain orbits in which it had definite fixed quantities of energy. The second was that while an electron remains in one of its allowed orbits, it does not radiate energy. An atom radiates energy only when its electron jumps from an orbit associated with a higher energy to one associated with lower energy. Each photon of energy then radiated is equal to the difference between the energy levels between which the jump occurred according to the relationship: $hf = E_i - E_f$.

Using these assumptions and a method for determining which were the allowed orbits of the electron in the hydrogen atom, Bohr was able to predict in both qualitative and quantitative fashion how the **Balmer** and other **series** are formed in the spectrum of hydrogen. He was also able to show why a hydrogen atom, and atoms in general, absorb the very same frequencies of light that they emit. He explained that an atom cannot accept photons of any energy but only those that are just capable of raising the electron in the atom from one of its lower energy orbits to one of its higher energy orbits. Therefore, only those photons whose energy is equal to the difference between the energy levels associated with any two of the different orbits can be absorbed by the atom.

In spite of its remarkable successes, Bohr's atomic model was unsatisfactory because the wave nature of electrons makes it impossible to assign them to specific orbits. The current model of the atom therefore eliminates the idea of orbits while retaining the idea of the energy levels associated with them. It pictures each atom as a positively charged nucleus containing all the neutrons and protons of the atom surrounded by a **cloud of electrons.** The electrons have no specific positions in the atom but are distributed in the space around the nucleus so that the atom as a whole may have certain specific energy levels or states, and no others.

An atom at its lowest energy level is said to be in the **ground state.** When an atom receives a large enough quantity of energy either by absorbing light or other radiant energy or by colliding with a fast moving particle such as an electron or another atom, the energy level of the atom may be raised to the first, second, third or higher levels above the ground state. The atom is then said to be in one of its **excited states.** When the atom receives just enough energy to eject one of its electrons, the atom becomes ionized and is said to be at its **ionization energy level.**

Once an atom is in an excited state, it will fall spontaneously into lower energy states and finally into the ground state. In each of these falls, the atom emits a photon of light whose energy is equal to the difference between its initial energy level and its final energy level. In this way, when large numbers of the atoms of an element are in the different excited states, they emit the series of lines that are observed in the spectrum of that element. As in the Bohr model, atoms absorb the same frequencies of light that they emit.

CHAPTER REVIEW

Summary

The energy states in which any atom can be are determined by the wave nature of its electrons. The allowed energy states of the atom are those for which its electron waves form **stationary wave patterns.** Thus, the energy level structure of the atom is a consequence of the wave nature of matter.

Questions

Group 1

1. (a) How were the formulas for the Balmer series and the other series in the hydrogen spectrum first obtained? (b) What is meant by the limit of a series? (c) What does the existence of similar series in the spectra of other atoms suggest about their structure?

2. (a) According to the electromagnetic theory, why should the electron circling in an orbit about the hydrogen nucleus be expected to spiral into the nucleus? (b) How did Bohr's model of the hydrogen atom explain why the electron in any of its allowable orbits does not do this?

3. (a) How does Bohr's model of the hydrogen atom explain the process of emission of light? (b) What is the ground state of the atom? (c) What is an excited state?

4. (a) The electron in the Bohr hydrogen atom goes from the E_5 orbit to the E_2 orbit. Write an expression for the frequency of the light emitted. (b) Compare this frequency with that emitted when an electron jumps from the E_6 to the E_2 orbit.

5. (a) Using a diagram, explain how the Bohr model of the hydrogen atom explains the emission of the Balmer series. (b) How are the other series emitted?

6. (a) What relationship did Bohr use to determine the radius of each of the orbits in which an electron in a hydrogen atom is allowed to be? (b) What is the relationship between the circumference of an allowed orbit and the wavelength associated with the electron that is in that orbit?

7. (a) What is the value assigned to the energy level associated with an electron that is just free of the atom? (b) What is the value of the energy level associated with the electron in the hydrogen atom, when in its ground state, in joules? (c) In electronvolts? (d) Why is this energy expressed as a negative number?

8. (a) What is meant by the binding energy of the hydrogen atom? (b) How is the binding energy related to the ionization energy? (c) How many joules does it take to ionize the hydrogen atom when it is in its ground state? (d) How many electronvolts is this?

9. How does the Bohr model explain why the absorption spectrum of hydrogen contains exactly the same frequencies as its emission spectrum?

10. (a) What were some of the successes of the Bohr model? (b) What were some of its limitations?

11. (a) Why is the concept of specific orbits for electrons in the atom abandoned in the current model of that atom? (b) What takes the place of the energy levels associated with each orbit?

12. (a) What is the process by which the current model of the atom emits light? (b) How does the atom absorb light?

13. How can it be shown experimentally that atoms actually contain definite energy states that determine the nature of the light they emit when excited?

14. (a) How are standing waves produced? (b) In the case of standing waves on a taut string, what is the relationship between the wavelengths that form standing waves and the length of the string?

15. How do the waves associated with the electrons in an atom determine what energy states the atom can have?

Group 2

16. A hydrogen atom is in the E_4 energy state. Make an energy level diagram showing all the possible ways in which this atom can emit light as it drops to lower energy levels. How many different frequencies can atoms in this energy level emit?

17. A photon has energy twice as great as the ionization energy of hydrogen. What happens to the energy of the photon when it enters a hydrogen atom in the ground state?

18. Determine from Fig. 31-9 the smallest quantity of energy that a hydrogen atom in the ground state can accept from a bombarding electron. What happens to the energy of an electron having less than this minimum quantity of energy, on striking a hydrogen atom?

19. When a gas is heated to a high enough temperature, its atoms will get into their first excited state and will emit light as they drop back to the ground state. How does the kinetic theory of gases explain where the atoms acquire the energy that excites them?

20. Referring to Bohr's picture of the atom, explain why excited hydrogen atoms may be expected to be larger than atoms in the ground state.

Problems

Group 1

1. What is the wavelength of the line in the Balmer series for which $n_i = 4$?

2. (a) Given that the radius of the first orbit in the Bohr model of the hydrogen atom is 5.3×10^{-11} m, what is the radius of the orbit for which $n = 5$? (b) How many times larger is the radius of the hydrogen atom when it is in this excited state than when it is in the ground state?

3. What is the energy in electronvolts associated with the energy levels of the hydrogen atom for which (a) $n = 4$; (b) $n = 5$?

4. (a) What is the energy in electronvolts of the photon emitted when the hydrogen atom in problem 3 moves from the state for which $n = 5$ to that for which $n = 4$? (b) What is this energy in joules?

5. What is the frequency of the photon emitted in the transition described in problem 4?

6. Fig. 31-13 shows the energy level diagram of mercury. (a) What energy is needed to ionize a mercury atom in the ground state? (b) How many eV of energy must be supplied to a mercury atom to raise it from its ground state to its first excited state? (c) To its second excited state?

7. (a) Referring to Fig. 31-13, what is the energy of the photon emitted when the mercury atom drops from the energy level -3.7 eV to the level immediately below it? (b) To the ground state?

Group 2

8. (a) An electron is accelerated by a voltage of 8.8 volts. If the electron strikes a mercury atom in its ground state, to which excited state will the atom be raised? (b) If the energy of the electron is raised to 9.0 volts, in which state will the atom be after the collision?

9. A photon having energy of 12.0 eV enters a mercury atom in its ground state. (a) What part of this energy will be used in ionizing the atom? (b) How much kinetic energy in eV will the electron ejected from the mercury atom have?

10. A mercury atom is in the excited state for which the energy level is -5.5 eV. It absorbs a photon that raises it to the next higher energy level. (a) What is the energy of the photon? (b) What is its frequency?

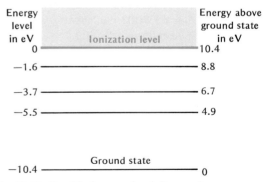

Fig. 31-13. Simplified energy level diagram for mercury.

Applying Physics

Standing waves can be made and studied with a long helical spring. One end of the spring is fixed by attaching it to a hook screwed into the wall. The opposite end of the spring is held in the hand and extended horizontally. As the free end of the spring is moved up and down at different rates, the spring may be made to vibrate as a whole, in halves, in thirds, in quarters, and so forth. These standing waves result from the superposition of the waves sent along the spring by the motion of the free end and the waves reflected from the fixed end. How does the distance between the ends of the spring determine the possible wavelengths of the standing waves that can be set up on it?

Nuclear Waste Disposal

A nuclear power plant produces electricity for homes and businesses. A hospital performs diagnostic tests using radioactive isotopes as "tracers" and provides radiation treatment for cancer patients. A military installation manufactures plutonium triggers for nuclear weapons. A laboratory uses radioactive materials to conduct scientific research. While each facility has a different function, they all share a common problem—the disposal of nuclear, or radioactive, waste—considered by many to be the most pressing environmental issue facing the United States in the 1990s.

Nuclear waste falls into two categories. *Low-level nuclear waste* consists of such items as radioactive medical wastes and the protective clothing worn by those who work with radioactive materials. Federal law requires that by 1993 each state establish, either individually or in collaboration with other states, a low-level nuclear waste dump. *High-level nuclear waste,* produced by weapons factories and power plants, is by far the more dangerous and difficult to dispose of. More than 40 years after the nuclear age began, permanent disposal of high-level nuclear waste is just beginning.

Top priority is being given to the waste produced by military installations from 1944 through the mid-1960s. During this period, millions of gallons of radioactive liquid was stored in leaky underground tanks at risk of explosion. In addition, toxic nuclear sludge was poured into open pits for decades in the mistaken belief that it would become "harmless" through dilution in the ground. In this way, hundreds of square miles of soil, and water, have become contaminated.

For over 30 years scientists searched for a

Low-level nuclear waste will be stored in federally-mandated state or regional dumps, but high-level nuclear waste must be specially processed, sealed up, and either buried or covered with concrete.

way to properly dispose of the highly radioactive liquid. What they came up with is the conversion of the liquid to a solid through a process called *vitrification* or *glassification*. The solidified waste is stored in specially designed steel canisters welded shut to await permanent burial. The solid material cannot leak out and has a very high melting point. It should remain intact for several thousand years, by which time its radioactivity should be down to a safer level.

The volume of radioactively polluted soil in the United States is too great to simply be dug up and buried elsewhere. There are currently two methods under development for dealing with the problem—on-site vitrification and processing of soil in a special apparatus called a *plasma centrifugal reactor*. On-site vitrification is done by passing an enormous electric current through the ground. The hardened material can be left in place and covered over. With the plasma centrifugal reactor, contaminated soil must be removed for treatment. The first reactor, scheduled to be built this decade, will process a ton of soil per hour.

The disposal of nuclear reactor fuel rods from power plants is a major problem yet to be solved. Approximately one third of the rods in a reactor need to be replaced each year because they are "spent," that is, their fuel is used up. However, these spent rods are actually more radioactive than when they were first inserted in the reactor. Spent fuel accounts for over 90 percent of the total radioactivity in the waste in the United States.

Until a final burial site is ready in the year 2010, 22,500 tons of spent fuel is being stored temporarily in water-filled cooling ponds beside reactors. A national dump for spent fuel and vitrified high-level waste is being prepared at Yucca Mountain in Nevada. Once tests are conducted to be sure the area

is safe from the threat of earthquake or volcanic eruption, tunnels will be cut out of volcanic rock a half mile underground. The waste that has accumulated already is expected to fill almost a third of these tunnels.

After a lifetime of more than 30 years, the first commercial nuclear reactor was buried in a trench at a military dumpsite in 1989. In the coming years, many more nuclear plants will have to be dismantled. With no known effective means of burial, the reactors may need to be encased in concrete where they stand. An alternative is to take the plants apart and bury the pieces along with low-level waste.

It will be a monumental task to restore all radioactively contaminated areas and safely dispose of radioactive waste so no further contamination occurs. The clean-up may take several decades.

Vitrification, or the embedding of high-level nuclear waste in glass disks like those shown here, is considered a promising means of storage.

32 Applications of Quantum Theory

Aims

1. To learn how quantum principles explain the structure of atoms and some of their chemical properties.
2. To understand the mechanism of conduction in solids that explains the differences in conductivity between conductors, semiconductors, and insulators.
3. To learn how semiconductor diodes may be used to rectify alternating currents and how semiconductor transistors may be used to amplify currents and voltages.
4. To understand the principle of the laser that enables it to produce concentrated, coherent, single-frequency beams of light.

Energy Levels in Atoms

32-1 Quantum Numbers and Atomic Structure

The quantum theory model of the atom has been successful not only in accounting for the spectra of atoms but also in explaining their structure as well as many of their chemical and physical properties. To do this, it assigns four numbers called *quantum numbers* to each electron of an atom. They are the *principal quantum number, n,* the *angular momentum quantum number, l,* the *magnetic quantum number, m_l,* and the *electron spin quantum number, s.* These quantum numbers place restrictions on where an electron may be in the space around an atomic nucleus. Their meaning may be understood by referring to the Bohr model of the atom. However, while the electrons in the Bohr model move only in a single plane, the electrons in the quantum theory model may move in the three dimensional space surrounding the nucleus.

32-2 Principal Quantum Number, n

The principal quantum number, *n,* corresponds to the number *n* of the Bohr model and is related to the distance of an electron from the nucleus. The greater *n* is, the greater is this distance. As in the Bohr model, *n* can have only whole number values beginning with $n = 1$, which represents the position of the electron closest to the nucleus and the one of lowest energy. Electrons with values of $n = 2, 3, 4$, etc. are progressively further from the nucleus and are

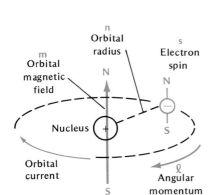

Fig. 32-1. Each quantum number is related to one aspect of the electron's position or motion, or to its associated magnetic fields.

at higher energy levels. Each value of n represents a shell-like space around the nucleus of the atom in which electrons may move. Electrons cannot occupy the space between two shells.

32-3 Angular Momentum Quantum Number, l

This number is related to the motion of an electron around the atom's nucleus. In the Bohr model, an electron moving in a circular orbit has an angular momentum that is defined as the product of its momentum and the radius of its orbit. Although the quantum theory does not assign electrons to specific orbits, it does assume that they have an angular momentum. The quantum number l refers to the electron's angular momentum. It may have only positive whole number values, including 0, and less than the principal quantum number n. That is, if $n = 1$, l can only be 0; if $n = 2$, l can be 1 or 0; while if $n = 3$, l can be 2, 1, or 0.

The quantum number l can be any positive whole number from 0 to $n - 1$.

32-4 Magnetic Quantum Number, m_l

In the Bohr model, the orbital motion of an electron results in a circular current. This produces a magnetic field perpendicular to the plane of the orbit and it is equivalent to a tiny magnet. The quantum theory assumes the existence of these "orbital" magnets and assigns the quantum number m_l to them. This number restricts the directions that orbital magnets may take when an outside magnetic field is imposed upon the atom. The specific directions are determined by the restricted values of m_l which include the whole numbers from $+l$ to $-l$ and 0. Thus, for $l = 0$, m_l can only be 0. For $l = 1$, m_l can be $+1$, 0, and -1. For $l = 2$, m_l can be $+2$, $+1$, 0, -1, and -2.

The quantum number m can be zero or any positive or negative integer from $+l$ to $-l$.

32-5 Spin Quantum Number, s

This number refers to the evidence that every electron spins on its axis. Since the electron is charged, its spin results in a circular current that makes the electron behave like a tiny magnet. The electron's magnet may have only two positions when in a magnetic field. One is in the same direction as the field and the other is opposite to the field. These positions are designated by the two values the spin quantum number can have: $s = +\frac{1}{2}$ and $s = -\frac{1}{2}$.

The quantum number s can be only $+\frac{1}{2}$ or $-\frac{1}{2}$.

32-6 Pauli Exclusion Principle

Given the four quantum numbers for each electron of an atom, how can we assign each electron to its proper place? The German physicist Wolfgang Pauli (VOLF·gahng POW·lee) (1900–1958) discovered the principle for doing this. The *Pauli exclusion principle states that no two electrons in an atom can have the same four quantum numbers.* Let us see how this principle works to determine the place of each electron in an atom of argon.

32-7 Structure of Argon

Since argon has atomic number 18, it has 18 electrons outside of its nucleus. To place them, we begin with the innermost shell for which $n = 1$. In this shell $l = 0$ and $m_l = 0$. However, s can be $+\frac{1}{2}$ or $-\frac{1}{2}$. Thus there are only two places for electrons in this shell, one with the quantum numbers $n = 1$, $l = 0$, $m_l = 0$, and $s = +\frac{1}{2}$, and the other with the quantum numbers $n = 1$, $l = 0$, $m_l = 0$, and $s = -\frac{1}{2}$. The other 16 electrons will have to find places in higher shells.

The next shell is the one for which $n = 2$. In this shell l has two possible values: 0 and 1. Each of these may be considered to form a subshell of $n = 2$. Consider first the subshell for which $l = 0$. Here m_l must also be 0. However since s can be $+\frac{1}{2}$ or $-\frac{1}{2}$, this subshell has room for another two electrons.

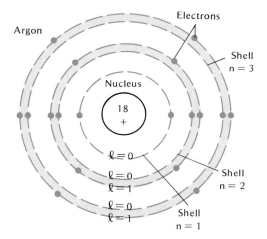

Fig. 32-2. Distribution of electrons among the first three shells of the argon atom.

Now let us go to the $n = 2$ subshell for which $l = 1$. Here m_l can have three values: $+1$, 0, and -1, and for each of these the electron spin can have its usual two values. There are therefore 3×2, or 6, places for electrons with different quantum numbers in this subshell. We now have filled the $n = 2$ shell by putting two electrons in its $l = 0$ subshell and six electrons in its $l = 1$ subshell.

So far we have placed ten of argon's electrons. To accommodate the remaining eight electrons, we go to the $n = 3$ shell. This shell has three subshells: one for which $l = 0$, one for which $l = 1$, and one for which $l = 2$. As we have seen in the previous two shells, a subshell for which $l = 0$ has places for two electrons and one for which $l = 1$ has places for six electrons. Thus we have placed all eighteen of argon's electrons: two in the $n = 1$ shell, eight in the two subshells of $n = 2$, and eight in the first two subshells of $n = 3$.

32-8 Structure of Other Atoms

Following this procedure, we can determine the distribution of the electrons in the atoms of other elements. These are shown in Table 32.1. Note that for the first 18 elements, the subshells are filled in

Table 32.1 Distribution of Electrons in Elements 1 to 36

ELEMENT		NO. OF ELECTRONS	1st SHELL $n = 1$	2nd SHELL $n = 2$		3rd SHELL $n = 3$			4th SHELL $n = 4$	
			$l = 0$	$l = 0$	$l = 1$	$l = 0$	$l = 1$	$l = 2$	$l = 0$	$l = 1$
H	hydrogen	1	1							
He	helium	2	2							
Li	lithium	3	2	1						
Be	beryllium	4	2	2						
B	boron	5	2	2	1					
C	carbon	6	2	2	2					
N	nitrogen	7	2	2	3					
O	oxygen	8	2	2	4					
F	fluorine	9	2	2	5					
Ne	neon	10	2	2	6					
Na	sodium	11	2	2	6	1				
Mg	magnesium	12	2	2	6	2				
Al	aluminum	13	2	2	6	2	1			
Si	silicon	14	2	2	6	2	2			
P	phosphorus	15	2	2	6	2	3			
S	sulfur	16	2	2	6	2	4			
Cl	chlorine	17	2	2	6	2	5			
A	argon	18	2	2	6	2	6			
K	potassium	19	2	2	6	2	6		1	
Ca	calcium	20	2	2	6	2	6		2	
Sc	scandium	21	2	2	6	2	6	1	2	
Ti	titanium	22	2	2	6	2	6	2	2	
V	vanadium	23	2	2	6	2	6	3	2	
Cr	chromium	24	2	2	6	2	6	5	1	
Mn	manganese	25	2	2	6	2	6	5	2	
Fe	iron	26	2	2	6	2	6	6	2	
Co	cobalt	27	2	2	6	2	6	7	2	
Ni	nickel	28	2	2	6	2	6	8	2	
Cu	copper	29	2	2	6	2	6	10	1	
Zn	zinc	30	2	2	6	2	6	10	2	
Ga	gallium	31	2	2	6	2	6	10	2	1
Ge	germanium	32	2	2	6	2	6	10	2	2
As	arsenic	33	2	2	6	2	6	10	2	3
Se	selenium	34	2	2	6	2	6	10	2	4
Br	bromine	35	2	2	6	2	6	10	2	5
Kr	krypton	36	2	2	6	2	6	10	2	6

order one at a time. No electron goes into a higher subshell until the one below it is completely filled.

Thus the single electron of hydrogen goes into the $n = 1$ shell. The two electrons of helium fill this first shell. The three electrons of lithium fill the $n = 1$ shell and take up one place in the $n = 2$, $l = 0$ subshell. The four electrons of beryllium fill the $n = 1$ shell as well as the $n = 2$, $l = 0$ subshell. The next atom, boron, has five electrons. Four of them fill the $n = 1$ shell and the $n = 2$, $l = 0$ subshell. The fifth now begins to fill the next subshell for which $n = 2$ and $l = 1$. This continues up to and including argon.

After that, we find that some electrons go into the $n = 4$ shell before the $n = 3$, $l = 2$ subshell is full. Thus, in potassium, after the first two subshells of the $n = 3$ shell are filled, the remaining electron skips the third $n = 3$ subshell and goes into the first subshell of the $n = 4$ shell. This behavior is explained by the fact that some of the energy levels in the first $n = 4$ shell are lower than some of those in the third $n = 3$ subshell. In general, the electrons occupy those vacant places that have the lower energy levels before entering the higher energy levels.

32-9 Chemical Activity of Atoms

The electronic structure of atoms determines much of their chemical behavior. Table 32.1 shows you that the atoms of helium, neon, argon, and krypton are the only ones that have all of their

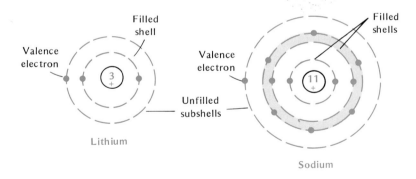

Fig. 32-3. Lithium and sodium atoms, with one valence electron each, have similar chemical properties.

electrons in filled shells and subshells. Chemists have long known that these elements are practically inert; that is, they do not combine with other elements to form compounds. It appears from this fact, that full shells and subshells in atoms are usually very stable arrangements which are not easily disturbed by nearby atoms. This accounts for the chemical inactivity of these elements.

On the other hand, all the other atoms except hydrogen, consist of an inner core of filled subshells with one or more outer partly filled subshells. Since all these atoms form chemical combinations, it appears that the electrons in their unfilled outer shells and subshells are responsible for their chemical activity. These outer electrons are called *valence electrons.*

It is noteworthy that the atoms of lithium, sodium, and potassium, which have a single valence electron outside their core of filled subshells, have similar chemical properties. This is also true of the atoms of fluorine, chlorine and bromine, each of which lacks one electron to fill its outermost subshell. These are examples of the similarity in chemical properties that the chemist finds in families of elements whose atoms have similar arrangements of electrons in their outer subshells.

32-10 Chemical Bonding

The behavior of the outer electrons of atoms helps us to understand how they combine with each other. There are two basic types of bonding between atoms: ionic bonding and covalent bonding. In *ionic bonding,* one atom transfers one or more electrons to another atom. In *covalent bonding,* two atoms share one or more pairs of electrons. Most bonds between atoms are neither wholly ionic nor wholly covalent but somewhere between the two pure types of bonding. However, some may be predominantly ionic and others predominantly covalent. We shall illustrate these types with examples.

32-11 Ionic Bonding

This is illustrated by the combination of sodium and chlorine. The sodium atom shown in Fig. 32-4 (a) has one electron outside of its core of filled shells. The chlorine atom shown in Fig. 32-4 (b) has

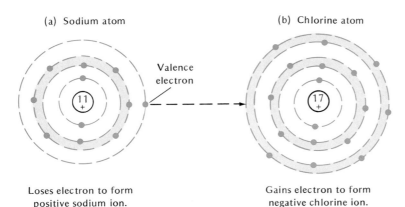

(a) Sodium atom   (b) Chlorine atom

Valence electron

Loses electron to form positive sodium ion.

Gains electron to form negative chlorine ion.

Fig. 32-4. Formation of an ionic bond between sodium and chlorine atoms.

a core of filled subshells but lacks one electron to fill its outer $n = 3$, $l = 1$ subshell. When these two atoms come together, the sodium atom gives up its outer electron to the chlorine atom. This fills the outer subshell of the chlorine atom and leaves the sodium atom with the stable structure of two full shells. The sodium atom, having lost an electron, becomes a positive sodium ion. The chlorine atom having gained an electron becomes a negative chlorine ion. The atoms are now held together by the electric attraction of their oppositely charged ions. Two other examples of ionic bonding are combinations of magnesium and chlorine, and calcium and fluorine.

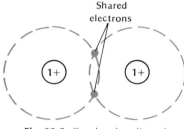

Shared electrons

Fig. 32-5. Covalent bonding of two hydrogen atoms to form a molecule of hydrogen.

32-12 Covalent Bonding

An example of covalent bonding is the hydrogen molecule which consists of two hydrogen atoms. Each hydrogen atom has one proton as its nucleus and one electron circulating around it in the $n = 1$ shell. By sharing their single electrons, the hydrogen atoms have part time use of both electrons in filling their $n = 1$ shells. Also, as the electrons move about both protons, they spend a much greater time in the space between the two protons than elsewhere. While in this region, the electrons attract the protons on either side of them and hold them together to form the hydrogen molecule. Other examples of molecules held together by covalent bonding are the elements oxygen, nitrogen, and chlorine. In each of these a pair of electrons holds the atoms together in a covalent bond to form a diatomic molecule.

Electrical Conductivity in Solids

32-13 Conductors, Insulators, and Semiconductors

The quantum theory provides an explanation for the differences in the ability of different solids to conduct electricity. Solids may be divided into three classes: good *conductors* like copper and silver, poor conductors or *insulators* like mica and rubber, and *semiconductors* like silicon and germanium whose conductivity lies between that of good conductors and insulators.

The carriers of current in solids are generally electrons that are free to move within the solids. In good conductors there is an abundance of free electrons. In insulators, there are very few free electrons. In semiconductors, the number of free electrons is intermediate between the number available in good conductors and in insulators.

32-14 Energy Level Structure in a Solid

According to the quantum theory, the energy level structure of a solid determines how many electrons will be free to carry current within it. A typical solid is crystalline and its atoms are closely bound in regular patterns. Consider a single atom in such a crystal. If it were isolated from the other atoms, it would have a set of narrow energy levels which its electrons are able to occupy. However, when a second atom is brought very near the first, the atoms act upon each other in such a way that each energy level splits into two slightly different levels. When a third atom approaches close to the previous two, each allowed energy level in which an electron moves now splits into three close levels. As more atoms are brought together, this splitting of energy levels continues.

Since the crystal contains enormous numbers of atoms crowded together, each of their allowed energy levels is split into so many closely packed levels that each original level becomes a band of practically continuous levels. Thus, the available energy states in a crystal which electrons can occupy consist of a series of bands separated from each other by gaps of varying size. The energy associated with the bands increases in quantum steps from the lowest band to the higher ones.

If all the energy levels within a band are filled with electrons, those electrons cannot move freely because there are no vacant levels to which they can go. On the other hand, if a band has many empty energy levels, the electrons in it may move freely from their own energy levels to the empty ones. Electrons may not be in the gap between bands. Furthermore, an electron cannot jump from one band to a higher energy band unless it receives the energy needed to make the jump from some source.

32-15 Energy Bands in a Conductor

Generally, all the lower bands in a crystalline solid are full of electrons which are therefore not free to move. Hence it is the condition of the first unfilled upper band that determines how many electrons will be free to carry current.

Fig. 32-7 illustrates the energy situation in a good conductor. Only the two upper bands are shown. The lower one is full of electrons. The upper band is only partly filled. It is called the *conduction band*. Since the energy levels within the conduction band are so close to each other as to be practically continuous, an electron in this band can move into the adjacent higher level on receiving even the smallest amount of energy. The electrons in the conduction band are thus practically free and can be set moving as a current by even the smallest applied voltage. In a good conductor there are a great many of such free electrons in the conduction band.

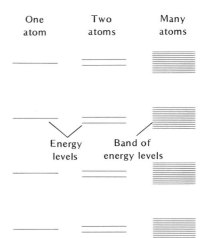

Fig. 32-6. As many atoms come together in a crystal, the splitting of energy levels forms bands of electron energy levels.

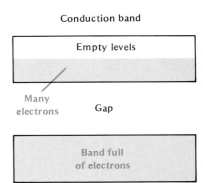
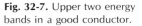

Fig. 32-7. Upper two energy bands in a good conductor.

Conduction band

Some electrons have jumped the gap

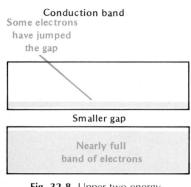

Smaller gap

Nearly full band of electrons

Fig. 32-8. Upper two energy bands in a semiconductor.

Conduction band

Very few electrons

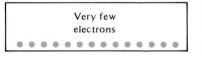

Larger gap

Band full of electrons

Fig. 32-9. Upper two energy bands in an insulator.

Germanium atom core   Valence electrons

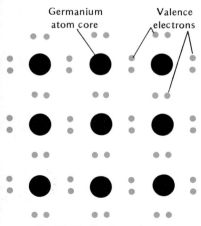

Fig. 32-10. Structure of a pure germanium crystal.

32-16 Energy Bands in a Semiconductor

Fig. 32-8 shows the energy situation in the upper two bands of a semiconductor. Here the lower band is again full but only a small gap separates it from the nearly empty conduction band. A small amount of energy is now needed to enable electrons to jump from the lower band into the conduction band. This is supplied by the internal thermal energy of the semiconductor and by electromagnetic radiation that it absorbs from its surroundings. As a result many electrons make the jump. The more electrons that get into the conduction band, the better will the semiconductor be able to carry current. Thus, heating a semiconductor drives more electrons into its conduction band and improves its conductivity. The semiconductor differs from the good conductor in that it has a lot fewer electrons in the conduction band than does the good conductor.

32-17 Energy Bands in an Insulator

Fig. 32-9 shows the energy situation in the upper two bands of an insulator. Again the lower band is full of electrons and the conduction band is nearly empty. However, a large gap now separates the upper from the lower band. A relatively large quantum of energy is therefore needed to enable an electron to jump from the lower band into the conduction band. Normally, very few electrons are able to acquire the necessary energy and to make the jump into the conduction band. Since the conduction band has far fewer free electrons than in a semiconductor or a conductor, the insulator conducts extremely poorly.

32-18 *n*-Type Semiconductors

The conductivity of a semiconductor can be greatly improved by introducing a small amount of a specific impurity. The process is called *doping*. A common example of doping is the introduction of a small amount of arsenic into a crystal of germanium. The arsenic atoms take the place of germanium atoms in the crystal structure since they are about the same size. Let us see what happens inside the crystal.

First consider the pure germanium crystal. Table 32.1 shows that each germanium atom has four valence electrons in its two outer subshells. To complete its outermost subshell it.needs four more electrons. It gets these in the crystal by sharing its four electrons covalently with four neighboring atoms, as shown in Fig. 32-10. It is this bonding that holds the germanium crystal together and restricts the free movement of the electrons.

Suppose now an atom of arsenic is introduced into the crystal.

Arsenic has five valence electrons in its outermost shell. Four of these form covalent bonds with four neighboring germanium atoms. The fifth electron has no bonding partner and is free to roam about the crystal and move into its conduction band. By introducing large numbers of arsenic atoms into the crystal, we introduce large numbers of free electrons into the conduction band and make the crystal a better conductor.

A crystal that has been doped so that it has acquired free electrons is called an *n-type semiconductor*. The *n* stands for *negative* and refers to the fact that the current carriers in the crystal are negative charges, that is, electrons. Another common type of *n*-semiconductor is made of silicon doped with phosphorus. Here silicon, which has four valence electrons, plays the same role as germanium. Phosphorus, which has five valence electrons, plays the same role as arsenic in furnishing the free electrons.

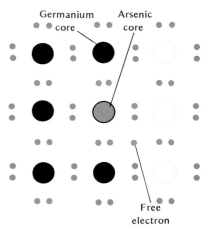

Fig. 32-11. Structure of an *n*-type semiconductor.

32-19 *p*-Type Semiconductors

A semiconductor may be doped in a way that results in producing vacant spaces for electrons called *holes*. For example, atoms of gallium instead of arsenic may be introduced into a germanium crystal. The gallium atom whose atomic number is 31 has about the same size as a germanium atom whose atomic number is 32. It therefore fits well in the crystal's structure.

You can see in Table 32.1 that gallium has only three valence electrons in its outer shell. A gallium atom can therefore make covalent bonds with only three of its germanium neighbors. As a result, there is an empty space or *hole* into which an electron can move. When any electron in the crystal breaks the bond to its atom and moves into a nearby hole, it leaves a hole in the place it left. The hole therefore appears to move in the direction opposite to that of the electron.

The process continues so that as electrons move in one direction to fill holes, the holes appear to move in the opposite direction. In effect, the holes act like positive charges. By putting a large number of gallium atoms into a germanium crystal, many holes are produced. These holes behave like positive current carriers in the crystal.

A crystal doped to produce holes is called a *p-type semiconductor*. The *p* stands for *positive* and refers to the holes that act like positively charged carriers. Another *p*-type semiconductor can be made of silicon doped with aluminum atoms. Here the silicon, with four valence electrons, takes the place of the germanium and the aluminum, with three valence electrons, takes the place of the gallium in supplying the holes.

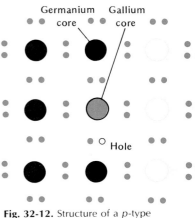

Fig. 32-12. Structure of a *p*-type semiconductor.

32-20 The *p-n* Junction or Diode

When a *p*-type and an *n*-type semiconductor are put together, we have a device called a *p-n junction* or *diode*. A diode allows electric current to pass through it readily in one direction but strongly resists its flow in the opposite direction. This action is called *rectification* and may be *used to change alternating current into direct current*.

To understand how a diode works, consider the *p-n* diode in Fig. 32-13. Assume that the battery is disconnected at first. There is an abundance of free electrons in the *n*-type material and an abundance of holes in the *p*-type material. Due to thermal activity, the electrons and holes drift about in their respective halves of the diode. When some electrons and holes meet at the junction between the *n*-type and *p*-type materials, the electrons cross the junction and move into the holes. Since the atoms in the *n*-material near the

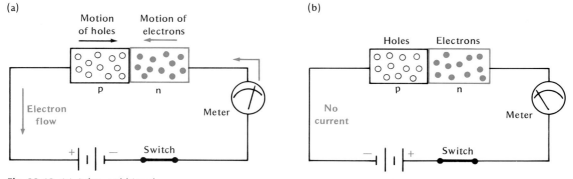

Fig. 32-13. (a) A forward biased *p-n* diode allows current to pass through it. (b) A reverse biased diode does not allow current to pass.

junction have lost these electrons, they become positively charged ions. Similarly, the atoms on the *p*-side of the junction have gained these electrons and become negatively charged ions. Thus a negative charge builds up on one side of the junction and a positive charge builds up on the other side. These charges form an electric barrier that prevents further movement of electrons and holes across the junction.

Let us now connect the positive terminal of the battery to the *p*-side of the diode and the negative terminal of the battery to the *n*-side of the diode. The diode is then said to be *forward biased*. Electrons from the *n*-side now move readily across the junction and a current will flow from the negative terminal through the diode into the positive terminal of the battery. At the same time, the holes in the *p*-side of the diode move in the opposite direction across the junction and add to the current.

However if we now connect the positive terminal of the battery to the *n*-side of the diode and the negative terminal of the battery to the *p*-side, the diode will be *reverse biased*. In this case, the electrons on the *n*-side and the holes on the *p*-side of the junction are attracted away from the junction. As a result, there are no electrons or holes in the vicinity of the junction to carry current across it, and no current flows.

32-21 Diode as a Rectifier

If, instead of a battery, we insert a source of alternating current into the diode circuit, current will flow only during those half cycles when the alternating voltage provides a forward bias on the diode. Electrons will then flow from the n to the p side of the diode while holes move in the opposite direction. Together they will form the pulsating direct current shown graphically in Fig. 32-14. Thus, the diode changes the alternating current input into a pulsating direct current output.

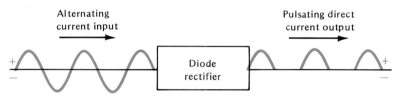

Alternating current input →

Pulsating direct current output →

Diode rectifier

Fig. 32-14. A diode rectifier converts an alternating current input into a pulsating direct current output.

32-22 Transistors

Transistors are semiconductor devices that can amplify currents or voltages applied to them. There are two types of transistors. One consists of a p-type semiconductor sandwiched between two n-type semiconductors. It is known as an *n-p-n transistor*. The other consists of an n-type semiconductor sandwiched between two p-type semiconductors. It is known as a *p-n-p transistor*.

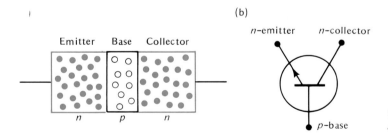

(b)

Emitter   Base   Collector

n p n

n-emitter n-collector

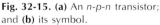

p-base

Fig. 32-15. (a) An *n-p-n* transistor; and **(b)** its symbol.

The middle semiconductor of a transistor is called the *base*. The semiconductor on one side of it is called the *emitter* and the one on the other side is called the *collector*. The emitter furnishes the carriers of current. In the case of the *n-p-n* transistor, the carriers are the free electrons in the n-type emitter. In the case of the *p-n-p* transistor, the carriers are the holes in the p-type emitter. The emitter is generally doped much more heavily than the base or the collector so that it will furnish a large number of electron or

(a)   (b)

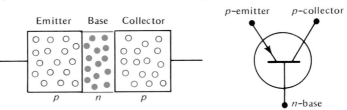

Emitter   Base   Collector

Fig. 32-16. (a) A *p-n-p* transistor; and **(b)** its symbol.

hole carriers of current. The base is also very thin so that the carriers will pass through it easily. The purpose of the collector is to receive the electrons or holes that come out of the emitter.

32-23 Transistor as an Amplifier

Fig. 32-17 shows a circuit in which an *n-p-n* transistor is used to amplify a small changing electrical current known as the signal. Such a signal might be the small current set up in a radio antenna. Note that the emitter-base unit of the transistor is an *n-p* junction that is forward biased. The base-collector unit of the transistor is a *p-n* junction that is reverse biased. When the signal is zero, the voltage applied to the emitter-base unit by battery E_e causes large numbers of electrons to move from the emitter to the base. Some of these complete the circuit and return to the positive terminal of the battery E_e. Most of them however go through the base and are attracted through the emitter into the positive terminal of the battery E_c which provides a much higher voltage than battery E_e. There are thus two circuits, the emitter-base circuit and the emitter-collector circuit.

Now suppose a small increase in the emitter-base voltage is produced by the signal. This causes an increase in the electron current passing from the emitter into the base. A small number of these electrons then leave the base and enter the positive terminal of battery E_e. Most of them, however pass through the base and are attracted through the collector and the load resistor R_L into the positive terminal of the battery E_c. Thus a small increase in the current in the emitter-base circuit produces in the emitter-collector circuit a much larger increase of current that may be as large as 200 times the change in the emitter-base current.

In a similar way, a small decrease in the current in the emitter-base circuit produces a much larger decrease in the current in the emitter-collector circuit. Used in this way, the *n-p-n* transistor is a current amplifier.

A *p-n-p* transistor may be used instead of the *n-p-n* transistor in the above circuit. The only change that is necessary is the reversal of the battery connections so that the emitter-base *p-n* junction is forward biased and the base-collector *n-p* junction is reverse biased.

An important development made possible by the discovery of transistors is the *integrated circuit*. This is a compact circuit

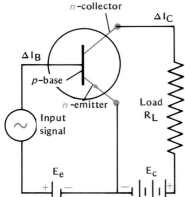

Fig. 32-17. An *n-p-n* transistor used to amplify a signal current. A small change in the base current, ΔI_B, produces a large change in the collector current, ΔI_C.

consisting of many tiny interconnected transistors, diodes, resistors, and other elements engraved on a small silicon chip. Integrated circuits are designed to accomplish specific purposes such as doing arithmetical operations in computers. Transistors and integrated circuits are widely used today in many electronic devices, including radios, televisions, hearing aids, and computers. They make possible the miniaturization of many electronic appliances.

Lasers

32-24 Principle of the Laser

The quantum theory enables us to explain how the laser (see Section 19-25) produces and amplifies coherent light. When an electron in an atom or molecule is in an excited state, it can drop spontaneously to a lower more stable state and emit a photon of electromagnetic radiation. An atom in an excited state may also be "stimulated" to drop into a lower more stable state by having a photon of the proper frequency pass over it. That frequency is the same as the one that the stimulated atom emits. Furthermore, the incident photon and the photon of the atom that it stimulated are in the same direction and phase. When many photons from many atoms are stimulated in this manner, they form a beam of coherent electromagnetic radiation. If the radiation is light, the result is a laser beam of a single frequency. The term, **laser,** summarizes the method by which laser light is produced. It stands for *Light Amplification by Stimulated Emission of Radiation.*

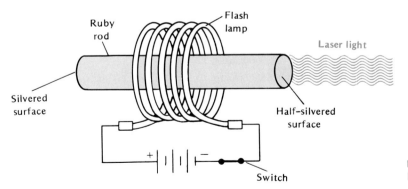

Fig. 32-18. Principle of the ruby laser.

32-25 Ruby Laser

The arrangement in a typical laser is shown in Fig. 32-18. The substance containing the atoms to be excited is a ruby rod half-silvered at one end and fully silvered at the other. Light of an intense source is flashed upon the rod and raises many of the rod's atoms from the ground state to a selected excited state. Some of the excited atoms now spontaneously emit photons as they drop back to the ground state.

Those that are not parallel to the axis of the rod simply leave it through the sides. Those photons that are parallel to the axis stimulate other excited atoms in their paths to return to their ground states and emit photons. These new photons are in the same phase

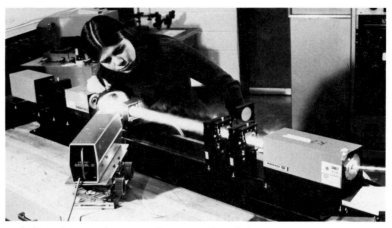

Fig. 32-19. Scientist using a laser beam in the control of photochemical reactions

and direction as the ones that stimulated them. They in turn stimulate still more atoms to emit photons which join the others to produce a coherent beam of increasing intensity traveling parallel to the axis of the rod.

Each time this beam reaches the end of the rod, it is reflected by the silvered or half-silvered surface and turned back into the rod to continue the process of stimulating more excited atoms to emit photons. As the beam is reflected back and forth, its intensity soon builds up enormously until it flashes out of the half-silvered end of the rod as a concentrated beam of **coherent** red **light**.

32-26 Other Kinds of Lasers

The ruby laser emits its light in pulses. This kind of emission is useful in such applications as spot welding and radar range measurements. For other applications, it is desirable to have lasers that emit their light continuously. There are gas-filled lasers that can do this. Other developments of the laser include liquid and semiconductor lasers.

The *semiconductor laser* is particularly useful because it is small and simple in operation. Unlike other lasers, it does not require an external pumping source. Instead, it operates by converting electric current directly into laser light. Since the beam that it emits can be varied by varying the current, the beam can transmit signals and thus can be used for communication purposes. The light weight, small size, and simplicity of the semiconductor laser make it ideal for space communications.

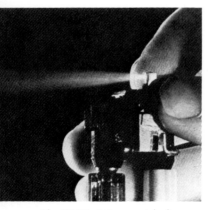

Fig. 32-20. A tiny crystal of gallium arsenide acts as an injection laser in the infrared region. The beam is made visible by infrared film.

Summary

The quantum theory explains the structure of atoms as well as many of their chemical and physical properties. It assigns four **quantum numbers** to each electron in an atom. They are the **principal quantum number n,** the **angular momentum quantum number l,** the **magnetic quantum number m_l,** and the electron **spin quantum number s.** The quantum numbers n, l, and m_l may have only certain whole-number values. The spin quantum number s may have only two values, $+\frac{1}{2}$ or $-\frac{1}{2}$. In addition, all electrons of an atom are subject to **Pauli's exclusion principle** which states that no two electrons in an atom may have the same four quantum numbers. These restrictions on the quantum numbers determine how the electrons of an atom are distributed among its energy shells and subshells.

The innermost energy shell of an atom is filled with electrons first; then the next higher one, and so on. The filled shells of an atom are very stable arrangements, and the electrons in them do not interact with the electrons of other atoms. The electrons in the unfilled outer shell of an atom do react with the electrons of other atoms and take part in chemical reactions. They are called **valence electrons.**

Atoms may combine chemically with each other by ionic bonding, covalent bonding, or a combination of both. In **ionic bonding,** an atom transfers one or more electrons to another atom so that both atoms are left with filled shells or subshells. In **covalent bonding,** two atoms effectively fill their unfilled outer shells or subshells by sharing one or more pairs of valence electrons.

Conductors, insulators, and semiconductors differ in the availability of free electrons. Each of them contains several bands of energy levels filled with electrons. Above these filled bands is an empty or partially filled energy band called the **conduction band.** The number of electrons that are in the conduction band or that can get into it from one of the lower bands determines how well the solid will conduct electricity.

In a **conductor,** the conduction band contains many electrons but is not filled. These electrons are therefore free to move within the band and to carry current readily. In an **insulator,** there are practically no electrons in the conduction band. Furthermore, there is very little possibility of moving electrons from the next lower band into the conduction band. An insulator has therefore very few free electrons and conducts poorly.

In a **semiconductor,** the conduction band contains very few electrons to begin with. However, small amounts of energy can raise many electrons from the band below the conduction band into the conduction band. The introduction of these additional electrons into the conduction band increases the ability of the semiconductor to conduct electricity.

A semiconductor may be made more conducting by introducing an appropriate impurity into it, a process called **doping.** If the semiconductor is doped so that it acquires more free electrons, it is called a negative or **n-type semiconductor.** If the semiconductor is doped so that it has unfilled spaces for electrons called **holes,** it is called a positive or **p-type semiconductor.** Holes behave like unit positive charges.

A **p-n junction** or **diode** is a device made by bonding an *n*-type and a *p*-type semiconductor together. It has the property of allowing electric current to flow through it much more readily in one direction than in the opposite direction. It can therefore be used to change an alternating current applied to it into a direct current. When used in this way the diode acts as a **rectifier.**

A **transistor** is a device made by bonding three doped semiconductors together. Transistors can be used to amplify electric signals. Transistors are major parts of miniature circuits known as **integrated circuits,** which are engraved on small silicon chips and are used to perform specific functions such as computing.

In the **laser,** the principles of quantum theory are used to generate powerful beams of **coherent light.** The process begins by bringing the atoms of a selected substance like ruby to a specific excited state. This is done by having the atoms absorb energy from a powerful light source. Some of the excited atoms then fall back to the ground state and emit photons of a certain frequency. These photons stimulate other excited atoms to drop to the ground state and emit still more photons of the same frequency. This process continues, rapidly increasing the number of photons released. The laser is so designed that the photons produced in this way are in the same direction and phase. They therefore add their effects to produce a concentrated beam of coherent light of a single frequency.

Questions

Group 1

1. (a) What are the four quantum numbers assigned to each electron in an atom? (b) What restrictions are put on the values that each of them may have?

2. How does Pauli's exclusion principle limit the places that an electron can occupy in an atom?

3. The atoms of magnesium and calcium have similar chemical properties. Referring to Table 32.1, explain how this can be accounted for by their electronic structure?

4. Find an element among the first ten listed in Table 32.1 that has chemical properties similar to (a) silicon; (b) sulfur; (c) argon.

5. What is the difference between ionic bonding and covalent bonding?

6. Explain how ionic bonding takes place in a combination of sodium and fluorine.

7. Explain how two fluorine atoms each having 7 electrons in the n = 2 shell form a stable molecule by making a covalent bond.

8. How do the energy levels in a crystal of an element differ from the energy levels of a single atom of the crystal?

9. (a) What are the carriers of current in a solid crystalline substance? (b) What is the conduction band? (c) What determines how well a solid will conduct electricity?

10. How does the size of the gap between the conduction band and the next lower energy band determine whether a solid will be a semiconductor or an insulator?

11. Why does heating a semiconductor improve its conductivity?

12. How does the number of electrons in the conduction band of a semiconductor compare with the number of electrons in the conduction bands of (a) a conductor; (b) an insulator?

13. What are the carriers of current in an *n*-type semiconductor? (b) What effect does the introduction of arsenic atoms into a germanium crystal have on the conductivity of the crystal? Explain.

14. (a) What is meant by a hole? (b) How are holes produced in a germanium crystal? (c) What determines the number of holes produced?

15. (a) What is the main current carrier in a *p*-type semiconductor? (b) In which direction do these

current carriers go when a voltage is applied to a *p*-type semiconductor?

16. (a) What are the carriers of current in an *n-p* diode? (b) What property of the diode enables it to rectify an alternating current?

17. (a) What are the two types of transistors? (b) Why are the emitters in each doped more heavily than either the bases or the collectors?

18. Explain how a transistor may be used to amplify a signal such as a weak alternating current.

19. (a) How are the atoms in a ruby laser brought to a selected excited state? (b) In what two ways can excited atoms return to the ground state?

20. Many of the atoms in a ruby rod laser have been raised to a selected excited state. One of them now drops spontaneously to the ground state and emits a photon in a direction parallel to the axis of the ruby rod. What chain of events does this photon set off that quickly produces a strong beam of coherent light?

Group 2.

21. (a) What is the total number of electrons that can be fitted into the $n = 4$ shell of an atom? (b) How many of them will be in each of the sub-shells?

22. Explain why heating a metal conductor increases its electrical resistance while heating a semiconductor generally reduces its electrical resistance.

23. Explain how a crystal of silicon may be converted into a *p*-type semiconductor.

24. What is one similarity and one difference between a hole and a unit positive charge?

25. Diamond is ordinarily a very poor conductor of electricity. However, when a high-speed charged particle passes through it, diamond becomes momentarily conducting. Explain.

26. A 60-Hz alternating current is passed through an *n-p* diode. (a) How many pulses of direct current are transmitted per second? (b) During what part of each second does current flow? (c) How much time elapses between two pulses?

27. (a) If a *p-n-p* transistor were substituted for an *n-p-n* transistor in Fig. 32-17, what other changes would have to be made in the circuit? (b) What would now be the main current carriers going from the emitter to the collector?

28. A transistor may be used to amplify a small signal current. Where does the energy come from that results in the amplified current?

29. An excited atom in a ruby laser is stimulated to emit a photon when an incident photon of the right frequency passes over it. In what three ways is the emitted photon similar to the one that caused its emission?

30. The ruby laser is said to amplify light. (a) Exactly what light does it amplify? (b) What is the original source of the energy that is accumulated in the laser beam?

Applying Physics

What kind of a conductor of electricity is graphite, the form of carbon used in pencils? Remove a rod of graphite from a pencil. Connect it in series with a 3.0-volt battery and a flashlight bulb, as in Fig. 32-21. Fix one wire connection to the left end of the graphite rod. Make the other wire connection flexible so that it can be moved from the left to the right end of the graphite rod.

Starting with the movable connection at the left end of the rod, move it slowly to the right until it reaches the right end. What change do you observe in the brightness of the bulb? Remove the graphite rod from the circuit and replace it by a copper wire about the same length and diameter. Note the brightness of the bulb. How does the resistance of the graphite rod compare with that of the copper wire?

Replace the graphite rod in the circuit. Adjust the movable connection so that the filament of the bulb glows a dull red. Heat the part of the graphite rod now in the circuit gently by moving a lighted candle under it from end to end. (*Caution:* Be careful not to overheat the graphite since it will burn.) Note the brightness of the bulb.

How does heat affect the resistance of the graphite rod? Does graphite behave like a metallic conductor or like a semiconductor?

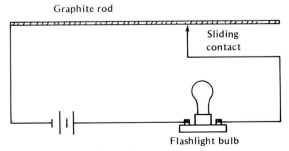

Fig. 32-21. Circuit for student activity.

33 The Atomic Nucleus

Nuclear Structure

33-1 A Promise and a Threat

From a practical point of view, the most important fact about the atomic nucleus is that it can furnish an almost limitless supply of energy for human use. Today the energy of atomic nuclei is being put to work in generating electricity, in providing heat, in powering ships and other means of transportation, in various industrial processes, and in advancing the frontiers of knowledge in every field of scientific research.

However, coupled with this glittering promise are two serious challenges. On the one hand is the problem of safety. Not only must we insure the safety of operation of the nuclear electric power plants themselves but we must also find environmentally safe methods of disposing of their deadly radioactive wastes. On the other hand is the problem of achieving world agreements that will ban the use of the powerful nuclear bombs that have now been developed. They can destroy all that the human race has created and built up through the ages. To realize the promise of nuclear energy, we must successfully meet these two challenges.

33-2 General Picture of the Atom

Let us re-review the picture of the atom that has been developed thus far. We have noted that all atoms are combinations of three basic particles: electrons, protons, and neutrons. Electrons have elementary negative charges, protons have equal positive charges, while neutrons are uncharged. A complete atom has exactly as many protons as it has electrons, hence it is electrically neutral.

All the electrons of an atom circulate outside of its nucleus and are contained in a space about 10^{-10} meter in diameter. All the protons and neutrons of an atom are located in a tiny nucleus that occupies only one-trillionth the volume of the atom. Thus, the atom is mainly empty space, with practically all of its mass located in its nucleus.

33-3 Atomic Number and Chemical Properties

The number of protons in the nucleus of an atom is equal to the number of extra-nuclear electrons contained in that atom and is given by its *atomic number*. Since each proton carries one unit of elementary positive charge, the nucleus of every atom has a total positive charge equal to its atomic number times the elementary unit of charge. It is known from chemistry that atoms having the same atomic number have the same chemical properties. It follows that *the quantity of positive charge in the nucleus of an atom determines its chemical properties*. For this reason, atoms whose nuclei have the same number of protons are given the same name. One way of measuring the nuclear charge of an atom and thus determining its atomic number is by means of Rutherford's alpha particle deflection experiments described in Chapter 30.

33-4 Number of Nucleons and Mass Number

The protons and neutrons contained in an atomic nucleus are called *nucleons*. The number of nucleons in the nucleus of an atom is given by the *mass number* of that atom. Since protons and neutrons each have masses of about one atomic mass unit, and since electrons have relatively negligible masses, *the mass number of an atom is simply the whole number nearest the total mass of the atom expressed in atomic mass units*.

The mass number is obtained by measuring the mass of an atom with the mass spectrograph (Section 26-19) and then taking the whole number nearest to it. When the mass and atomic numbers of a nucleus are known, the number of neutrons present in that nucleus may be obtained by subtracting its atomic number from its mass number. Thus, the nucleus of a sodium atom whose mass number is 23 and whose atomic number is 11 contains $23 - 11$, or 12 neutrons.

33-5 Symbols for Atoms and Nuclides

A *nuclide* is an atomic nucleus with a specific atomic number and a specific mass number. A nuclide is given the same symbol as the atom to which it belongs. Nuclides that have the same atomic number belong to atoms that have the same chemical properties and therefore are designated by the same chemical name. If such nuclides have different mass numbers, they have different numbers

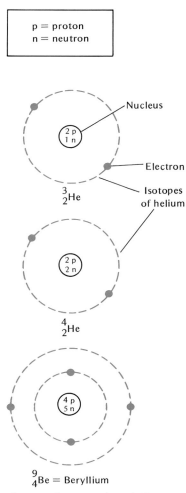

$\begin{array}{l}\text{p} = \text{proton}\\ \text{n} = \text{neutron}\end{array}$

${}^{3}_{2}\text{He}$

${}^{4}_{2}\text{He}$

${}^{9}_{4}\text{Be}$ = Beryllium

Fig. 33-1. Structure of two helium isotopes and beryllium.

of neutrons, but the atoms to which they belong still have the same chemical properties. We have seen that such atoms are called isotopes.

To summarize the facts about a given nuclide, we write *its atomic number as a subscript to the left of its chemical symbol and its mass number as a superscript to the left of its chemical symbol*. The nuclide of the most common form of helium is written ^4_2He. The subscript tells us that the nuclide of this helium isotope contains 2 protons. This also means that there are 2 electrons outside the helium nucleus. The superscript tells us that this nuclide contains a total of 4 nucleons. It follows that the number of neutrons in this nuclide is $4 - 2$, or 2. The nuclide of another isotope of helium is written ^3_2He, meaning that 2 of its 3 nucleons are protons. The remaining one is a neutron.

The nuclide of the most common form of uranium is written $^{238}_{92}\text{U}$. The subscript tells us that the uranium atom has 92 protons in its nucleus and therefore also has 92 electrons circulating outside of its nucleus. The superscript tells us that this particular uranium nuclide contains a total of 238 nucleons. Since 92 of these are protons, the remainder, $238 - 92 = 146$, are neutrons.

33-6 Symbols for the Neutron, Electron, and Proton

Certain symbols deserve special mention. The symbol for the neutron is ^1_0n. This shows that the neutron has zero charge and an atomic mass number of 1. The symbol for the electron is $^0_{-1}\text{e}$. This shows that the electron has a negative unit charge and an atomic mass number of zero. The symbol for the proton is ^1_1H, showing that the proton is simply the nucleus of an ordinary hydrogen atom having both its atomic number and mass number equal to 1.

33-7 Forces Inside the Nucleus

The negatively charged electrons circulating outside of the nucleus of an atom are bound to the atom by the electric attraction of the positively charged nucleus. However, inside the nucleus there are only protons and neutrons. Since the protons repel each other and are very close together, you might expect that their strong repulsions would tear the nucleus apart. The fact that the nuclei of most of the atoms remain intact indicates that there are powerful attractive forces among the nucleons in a nucleus binding these particles together. Since these forces overcome the repulsive forces among the protons, they must be much stronger than the forces between electric charges. It is known that these nuclear forces act only between two nucleons that are close to each other. They exist between two nearby protons, between two nearby neutrons, and between a proton and a nearby neutron.

33-8 Binding Energy of the Nucleus

The source of the energy associated with the nuclear binding forces is the relationship $E = mc^2$, according to which mass and energy are converted into each other.

When we add up the masses of the individual protons and neutrons in the nucleus of any atom, we discover that *the sum of the masses of its parts is greater than the mass of the whole nucleus.* For example, consider the helium nucleus, $_2^4 He$, which is made up of 2 protons and 2 neutrons. The mass of a proton is 1.0073 atomic mass units, and the mass of a neutron is 1.0087 atomic mass units. The sum of 2 protons and 2 neutrons is therefore 4.0320 atomic mass units. Now the mass of the whole helium nucleus as measured with a mass spectrograph is only 4.0016 atomic mass units. This means that, when a helium nucleus is formed by the combination of 2 protons and 2 neutrons, $4.0320 - 4.0016 = 0.0304$ atomic mass unit disappears. According to the relativity theory, this vanished mass is converted into energy in the form of radiation. It is equal to 2.83×10^7 electronvolts!

This lost energy is responsible for the binding forces among the parts of the nucleus and is called the *binding energy* of the nucleus. Before the helium nucleus can be separated again into its original 2 neutrons and 2 protons, the lost energy must be returned to it.

The binding energy of any nucleus is the energy equivalent to the difference between the sum of the individual masses of the protons and neutrons contained in the nucleus and the actual mass of the nucleus itself. It is usually a relatively large amount of energy. The greater the binding energy of an atomic nucleus, the more tightly is it held together and the greater is the energy needed to break that nucleus into its parts.

> The energy that holds a nucleus together is the difference between the sum of the masses of its individual protons and neutrons and the mass of the nucleus.

33-9 Calculating Binding Energy

The difference between the mass of any nucleus and the sum of the masses of the protons and neutrons it contains is called its *mass defect.* If the mass defect, designated MD, is in kilograms, the binding energy to which it is equivalent is obtained from the mass-energy relationship $E = (MD) \times c^2$. When c is in meters per second, E comes out in joules and is then readily converted into electronvolts. It is customary to express nuclear binding energies and the energy equivalents of masses in millions of electronvolts, abbreviated MeV. The energy equivalent to *1 atomic mass unit is 931 million electronvolts.*

To find the mass defect of a nucleus in the manner described above, we must know the mass of that nucleus. However, we do not usually have this information. Instead, we have the mass of the whole neutral atom to which that nucleus belongs. This quantity is the mass of the nucleus plus the masses of the extra-nuclear electrons and is the one measured in the mass spectrograph. We can compute the mass defect using this neutral atomic mass instead of the mass of the nucleus if, in our calculations, we also use the mass of the neutral hydrogen atom instead of the mass of the proton. When we now subtract the masses of the neutral hydrogen atoms from the mass of the neutral atom, the extra masses of the electrons contained in the mass of the neutral atom and in the masses of the neutral hydrogen atoms cancel each other. The difference is therefore equal to the mass defect. The following sample problem illustrates this procedure.

Sample Problem

(a) What is the mass defect of the nucleus of the carbon isotope $^{12}_{6}C$? The mass of the neutral carbon atom is 12.0000 u; the mass of a neutral hydrogen atom is 1.0078 u; and the mass of a neutron is 1.0087 u. (b) What is the binding energy of this nucleus in MeV?

Solution:

(a) The nucleus $^{12}_{6}C$ is made up of 12 nucleons of which 6 are protons and 6 are neutrons. Substituting the mases of neutral hydrogen atoms for the masses of the protons, we have

Mass of 6 $^{1}_{1}H$ atoms $= 6 \times 1.0078 = 6.0468$ u
Mass of 6 neutrons $= 6 \times 1.0087 = \underline{6.0522}$ u

Sum of the above masses $= 12.0990$ u
Mass of the neutral atom of $^{12}_{6}C$ $= \underline{12.0000}$ u

Difference of the above masses =
$\qquad$ Mass Defect $= 0.0990$ u

(b) Binding Energy = Mass Defect in u $\times$ 931 MeV
$\qquad$ Binding Energy = (0.0990 u)(931 MeV)
$\qquad$ Binding Energy = 92.1 MeV

Test Yourself Problems

1. The mass of $^{13}_{6}C$ is 13.0034 u. What is the binding energy of its nucleus in (a) u; (b) MeV?
2. The mass of $^{6}_{3}Li$ is 6.0170 u. What is the binding energy of its nucleus in (a) u; (b) MeV?

33-10 Binding Energy Per Nucleon

The binding energy of a given nucleus divided by the number of nucleons it contains gives its average *binding energy per nucleon*. This quantity measures how tightly each neutron or proton is bound to that nucleus. In Fig. 33-2, the binding energy per nucleon for each atom is plotted against its mass number. Notice that, except for the atoms lighter than carbon, the binding energy per nucleon of all atomic nuclei varies from 7.5 to 8.8 million electronvolts. It rises to its highest value for a mass number of about 60. This means that atoms whose mass numbers are close to 60 are most stable. One of these is the isotope of iron, $^{56}_{26}Fe$, whose binding energy per nucleon is nearly 8.8 million electronvolts.

As the mass number of an atom increases beyond 60, its binding energy per nucleon decreases steadily. This indicates a steady weakening of the attractive forces among the nucleons of these nuclei by the repulsive forces among their protons. The stability of these nuclei therefore decreases as their masses increase. As the mass of a nucleus approaches 240 u, it is likely to be unstable.

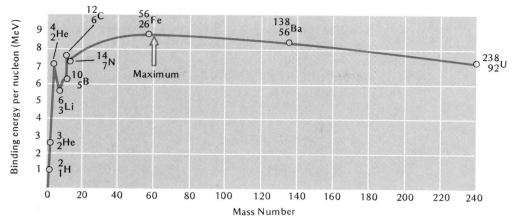

Fig. 33-2. Binding energy per nucleon plotted against the mass number. The binding energy is at a maximum for a mass number of about 60.

33-11 Some Nuclei Are Unstable

The first hint that the nuclei of some atoms do not have sufficient binding energy to hold them together came from the discovery in 1896 by the French scientist Henri Becquerel that the element uranium continuously emits invisible rays. He observed that these rays penetrate objects opaque to ordinary light and that they act upon a photographic plate. Soon afterward, Pierre and Marie Curie and other experimenters showed that several other elements also emit these mysterious rays. This spontaneous emission of rays by elements became known as *radioactivity*. When further study showed that radioactive elements are gradually changing into other chemical elements, it became clear that the rays must be coming from unstable atomic nuclei that are disintegrating spontaneously.

Among the naturally radioactive elements are polonium, thorium, actinium, protactinium, and radium. Of these, radium, discovered by the Curies, is the most active and has played a most important role as a tool of research and in the treatment of cancer.

Recall from Section 30-2 that the nuclei of radioactive atoms emit three different types of rays: alpha rays, beta rays, and gamma rays. Alpha rays consist of positively charged particles, each of which is the nucleus of that isotope of helium represented by the symbol $_2^4$He. Beta rays consist of high velocity electrons and have the same symbol as the electron, $_{-1}^0$e. Gamma rays are electromagnetic waves of short wavelength, similar to X rays but much more penetrating.

33-12 Transmutation of Radioactive Elements

As the nuclei of the atoms of any radioactive element emit either alpha or beta rays, these atoms change into atoms of another element. The process whereby one element changes into another element is called *transmutation*. Uranium, for example, undergoes a series of transmutations during which it changes to radium, then to radon, and after several more changes, to a stable isotope of lead. This behavior is typical of naturally radioactive elements. Each of the nuclei of their atoms is unstable and therefore disintegrates or decays by ejecting parts of itself in the form of alpha or beta particles. The remaining parts of the nucleus then rearrange themselves to form the nucleus of a new element. If this new nucleus is also unstable, it too disintegrates and ejects more parts of itself in the form of rays to become the nucleus of still another atomic element. This process continues until a nucleus is formed that is stable.

33-13 Half-Life

The half-life of a radioactive element is the time it takes for half its atoms to disintegrate.

Different radioactive elements disintegrate at different rates. The time that it takes for half of the atoms in a given mass of a radioactive element to disintegrate is called its *half-life*. The half-life of radium is 1600 years. This means that in 1600 years, half of a given quantity of radium will disintegrate and become radon. In another 1600 years, half of the remaining radium will disintegrate, leaving only one-quarter of the original amount of radium. The half-lives of the radioactive elements differ widely. While that of one isotope of uranium is over 4 billion years, that of radium C' is less than one-thousandth of a second.

33-14 How Transmutation Takes Place

The transmutation of the nucleus of an atom into the nucleus of another atom takes place when it gains or loses protons.

We have seen that the chemical properties of an element depend upon the number of protons in its nucleus. It follows that if the nucleus of any atom loses any of its protons or gains additional protons, it will be transmuted into the nucleus of a new element. This is what happens when a radioactive atom emits an alpha or a beta particle.

(a) *Alpha Emission.* Let us consider the case of an atom like radium that emits an alpha particle. Radium has 88 protons in its nucleus. Now an alpha particle, which is the nucleus of a helium atom, includes 2 protons. Therefore, when a radium nucleus ejects an alpha particle, it loses 2 of its protons and is left with $88 - 2$, or 86 protons. From the table of the Elements in the Appendix, you can see that the element whose nucleus has 86 protons is the gas radon. Thus, in emitting an alpha particle, the radium atom is transmuted into radon.

(b) *Beta Emission.* Let us now see what happens when a radioactive nucleus emits a beta particle, which is simply an electron. Normally, there are no electrons in the nucleus of an atom, only protons and neutrons. However, occasionally a neutron in a nucleus disintegrates to form a proton, an electron, and a *third neutral and almost massless particle called an antineutrino*. Its symbol is v.

The electron is ejected and is readily observed as a beta particle. The antineutrino is also ejected but, because of its lack of charge and its negligible mass, it is not easily observed. The newly formed proton remains in the nucleus and therefore increases the number of protons in the nucleus by one. As a result, the original nucleus is transmuted into the nucleus of the element with the next higher atomic number.

Thus, lead, whose atomic number is 82, has 82 protons in its nucleus. When one of the neutrons in a radioactive isotope of lead disintegrates into a proton, an electron, and an antineutrino, the proton remains in the nucleus while the other two particles are ejected. The number of protons in the nucleus therefore increases by one to 83. From the table of the Elements in the Appendix, you can see that the element with 83 protons in its nucleus is bismuth. Thus, radioactive lead is transmuted into bismuth.

(c) *Gamma Emission.* When a nucleus emits gamma rays, no transmutation occurs because gamma rays have no electric charge and therefore no change takes place in the number of nuclear protons. Gamma rays are the excess energy emitted by a disintegrating nucleus when its parts are rearranging themselves to form a more stable nucleus.

33-15 Equations for Nuclear Reactions

The changes that take place when radioactive nuclei undergo transmutation can be represented by simple equations. For example, the transmutation of radium into radon just discussed can be written:

$$^{226}_{88}\text{Ra} \rightarrow {}^{222}_{86}\text{Rn} + {}^{4}_{2}\text{He}$$
$$\text{radium} \rightarrow \text{radon} + \text{alpha particle}$$

Such equations can be written for all reactions in which the nuclei of atoms undergo changes. In such reactions, two things remain the same on both sides of the equation. They are the arithmetical sum of the electric charges and the total number of neutrons and protons.

The subscripts tell us the number and sign of the electric charges involved in the reaction. Thus, the radium subscript tells us we have 88 protons or positive charges on the left side of the equation. The subscripts on the right side of the equation show that we still have a total of 88 positive charges after the radium has disintegrated, 86 in the radon nucleus and 2 in the alpha particle. In general, *the sum of the subscripts on the left side of a nuclear equation is always equal to the sum of the subscripts on the right side.*

The superscripts tell us the number of nucleons involved in the reaction. Since no nucleons are destroyed in a nuclear reaction, *the sum of the superscripts on the left side of the equation must be equal to the sum of the superscripts on the right side of the equation.* Here, on the left, we begin with radium, whose superscript shows that it has a total of 226 nucleons. The superscripts on the

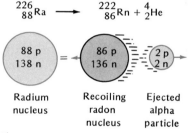

$$^{226}_{88}\text{Ra} \longrightarrow {}^{222}_{86}\text{Rn} + {}^{4}_{2}\text{He}$$

Radium nucleus	Recoiling radon nucleus	Ejected alpha particle
88 p 138 n	86 p 136 n	2 p 2 n

Fig. 33-3. Transmutation of radium into radon by the emission of an alpha particle.

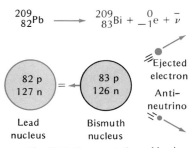

$$_{82}^{209}\text{Pb} \longrightarrow {}_{83}^{209}\text{Bi} + {}_{-1}^{0}\text{e} + \bar{\nu}$$

Ejected
electron

Anti-
neutrino

| Lead | Bismuth |
| nucleus | nucleus |

Fig. 33-4. Transmutation of lead into bismuth by the emission of a beta particle and an antineutrino.

The law of conservation of energy and mass led to the discovery of the antineutrino.

right also show a total of 226 nucleons, 222 in the radon nucleus and 4 in the alpha particle.

To give another illustration, let us write the equation for the transmutation of radioactive lead into bismuth:

$$_{82}^{209}\text{Pb} \rightarrow {}_{83}^{209}\text{Bi} \qquad + {}_{-1}^{0}\text{e} \qquad + \bar{\nu}$$
lead → bismuth + beta particle + antineutrino.

Note that the beta particle has a negative charge represented by -1. The subscript on the left side of the equation is 82 and is equal to the sum of the subscripts on the right; that is, $83 - 1 = 82$. The superscript on the left is 209 and is equal to the sum of the superscripts on the right; that is, $209 + 0 = 209$. The antineutrino with its zero mass and zero charge does not influence the balance of the equation.

33-16 Conservation of Energy and Mass in Nuclear Reactions

The law of conservation of energy and mass applies to all nuclear reactions. The sum of the energy and mass present before a nuclear reaction occurs is equal to the sum of the energy and mass present after the reaction has taken place. As an example, consider the $_{88}^{226}\text{Ra}$ nucleus of radium that emits an alpha particle to become the $_{86}^{222}\text{Rn}$ nucleus of radon. When the radium nucleus ejects the alpha particle, it recoils just as a gun does on firing a bullet. As a result, both the alpha particle and the radon nucleus have kinetic energy.

Let us now use the relationship $E = mc^2$ to convert the masses of all the particles in this nuclear reaction to their equivalent amounts of energy. We find that the energy equivalent of the mass of the radium nucleus is equal to the sum of the energy equivalent of the alpha particle and the energy equivalent of the radon nucleus plus the sum of the kinetic energies of each of them.

33-17 Discovery of the Antineutrino and the Neutrino

In the case of the radioactive nuclei that emit beta rays, the law of conservation of energy and mass first revealed the existence of the *antineutrino*. In all reactions in which an unstable nucleus emits a beta particle, it was found that a very small part of the mass of the unstable nucleus disappears and is apparently not replaced by an equivalent amount of energy or mass. This apparent loss of energy suggested that a hitherto undetected particle is ejected from the unstable nucleus at the same time as the beta particle and carries off the missing energy.

From typical beta-emitting reactions such as $_{82}^{209}\text{Pb} \rightarrow {}_{83}^{209}\text{Bi} + {}_{-1}^{0}\text{e}$, it was inferred that the undetected particle can have no electric charge and must have a negligible mass. These chargeless and practically massless characteristics enabled this particle, now called the antineutrino, to escape detection for a long time. The emission of a beta particle from a nucleus is generally accompanied by the emission of an antineutrino. In this process, one neutron in an unstable nucleus breaks up into a proton, an electron, and an antineutrino according to the equation:

$$ {}_{0}^{1}n \rightarrow {}_{1}^{1}H \quad + \quad {}_{-1}^{0}e \quad + \bar{\nu}$$
$$\text{neutron} \rightarrow \text{proton} + \text{electron} + \text{antineutrino}$$

The antineutrino and the electron are ejected from the nucleus, the latter as a beta particle. The proton is retained in the nucleus, increasing its atomic number by 1. Thus, after emitting a beta particle and an antineutrino, the nuclide ${}_{82}^{209}Pb$ becomes ${}_{83}^{209}Bi$.

A related particle, called the *neutrino* has also been discovered. Its symbol is ν, and it, too, has no electric charge and zero mass. We shall see how it is produced by electron capture in Section 33-18 and in nuclides made artificially radioactive in Section 33-23.

33-18 Electron Capture

Electron capture is another way in which change can occur in the nucleus of an atom. To understand how this change comes about, recall that every atom consists of a central nucleus surrounded by electrons in various energy levels or shells. The shell closest to the nucleus is the *K*-shell. Occasionally, an electron in this shell is so strongly attracted by the positively charged nucleus that it is pulled into the nucleus. This process is known as *electron capture* or *K-capture*.

Once inside the nucleus, the captured electron combines immediately with one of the nuclear protons to form a neutron. At the same time, a neutrino, the massless, uncharged particle given the symbol ν, is ejected. The equation for the reaction is:

$$ {}_{1}^{1}H \quad + \quad {}_{-1}^{0}e \quad \rightarrow \quad {}_{1}^{0}n \quad + \quad \nu$$
$$ \text{captured}$$
$$\text{proton} + \text{electron} \rightarrow \text{neutron} + \text{neutrino}$$

As a result, the nucleus now has one less proton and one more neutron but the same number of nucleons as it had before the electron capture. Its atomic number therefore decreases by 1 while its mass number remains the same. It is now the nuclide of an element with different chemical properties than it had before the electron capture and so is given a different chemical name.

The loss of the captured electron leaves a "hole" in the *K*-shell that is filled by an electron from a higher shell, which jumps in to take its place. This, in turn, leaves a hole in the higher shell that is then filled by an electron leaving a still higher shell. This process continues until all the shells regain their normal numbers of electrons. At the same time, the atom emits a series of spectral lines corresponding to each jump that an electron makes from an outer to an inner shell.

To illustrate electron capture, consider what happens when the beryllium nucleus, $_{4}^{7}Be$, captures one of its K-electrons:

$$_{4}^{7}Be \quad + \quad _{-1}^{0}e \quad \rightarrow \quad _{3}^{7}Li \quad + \quad \nu$$

captured
beryllium + electron → lithium + neutrino

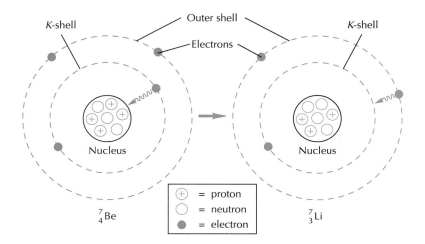

Figure 33-5. $_{4}^{7}Be$ captures a K-electron and becomes $_{3}^{7}Li$ as one nuclear proton becomes a neutron.

Note that the beryllium nuclide with atomic number 4 has been transmuted into lithium with atomic number 3 but that the mass number 7 is the same for both.

33-19 Nuclear Energy Levels

In Chapter 31, it was seen how the line spectrum of the light emitted by atoms led to the discovery that electrons can occupy only certain specific energy levels around the nucleus of an atom and no others. The gamma rays emitted by unstable nuclei also form line spectra in which each line represents radiation of a specific wavelength. These spectra suggest that energy levels also exist inside the nucleus of the atom and that the atom's nucleons may be only in these nuclear energy levels and no others.

Just as the jump of an electron from a higher to a lower energy level outside the nucleus results in the emission of light, so the jump of a proton or neutron from a higher to a lower energy level inside the nucleus results in the emission of gamma rays. A line spectrum of gamma radiation is therefore emitted in a manner similar to the emission of the line spectrum of light that is produced when elec-

There are energy levels within the atomic nucleus similar to those of the electronic cloud surrounding the nucleus.

trons jump from higher to lower energy levels outside the atom's nucleus.

The existence of energy levels inside the nucleus is further confirmed by the fact that the alpha particles emitted by a radioactive element also form "line spectra"; that is, the kinetic energies with which they emerge from a given nucleus have only certain definite values and no others.

33-20 Studying Nuclei by Bombardment

One of the methods of studying the nucleus of an atom is to fire a projectile at it in the hope of destroying its stability. Such a projectile might be a high velocity alpha particle emitted by a radioactive substance. The alpha particle, if successful, enters the target nucleus and makes it unstable. The nucleus then disintegrates as it rearranges its parts into a more stable combination. The fragments and rays resulting from the disintegration are examined in a bubble chamber or other detecting device and provide information concerning the composition of the bombarded nucleus.

Among the other charged particles used for nuclear bombardment are high speed protons and deuterons, which are the nuclei of deuterium, the isotope of hydrogen of mass 2. We shall see how these and other charged particles may be speeded up to very high velocities in the so-called "atom-smashing machines."

The neutron is also an effective projectile for nuclear bombardment. The fact that it is uncharged gives it an important advantage over charged particles as a projectile. When a positively charged particle such as a proton is fired at the nucleus of an atom, its speed is sharply reduced as it nears the nucleus because it is repelled by the positive charge of the nucleus itself. Hence, only a very high speed proton can overcome this repulsion and penetrate the nucleus of an atom. Since a neutron suffers no electrical opposition on nearing the nucleus of an atom, even a slow neutron is capable of penetrating it. In fact, for some purposes, slow neutrons are more effective projectiles than fast ones.

Fig. 33-6. Paths of charged particles through a hydrogen bubble chamber showing the interactions of several subatomic particles.

33-21 Discovery of Artificial Transmutation

Using the method of bombardment, Lord Rutherford was the first to show that elements can be artificially transmuted into other elements. He fired alpha particles at the nuclei of nitrogen atoms and observed the following reaction:

$$^{14}_{7}N + ^{4}_{2}He \rightarrow ^{17}_{8}O + ^{1}_{1}H$$
$$\text{nitrogen} + \begin{array}{c}\text{alpha}\\\text{particle}\end{array} \rightarrow \text{oxygen} + \text{proton}$$

Here, the bombardment of nitrogen increases the number of protons in its nucleus from 7 to 8 and thus transmutes it to oxygen whose atomic number is 8. The gain of one proton by the nitrogen nucleus occurs as follows. When the alpha particle strikes the nitrogen nucleus, it ejects a proton from it but is itself captured. The nucleus thus gains 2 protons from the alpha particle to replace the one ejected, making a net gain of one proton.

This successful transmutation of nitrogen in 1919 paved the way for the transmutation of all the other elements. Today, artificial transmutation is commonplace and is readily accomplished by bombardment and other procedures. It has enabled us to produce many elements and isotopes not found in nature, such as neptunium and plutonium and other elements having higher atomic numbers than uranium.

33-22 Discovery of the Neutron

One outcome of the bombardment of a nucleus by alpha particles was the discovery of the neutron. In 1932, Chadwick bombarded the nuclei of beryllium atoms with alpha particles with this result:

$$^{9}_{4}Be + ^{4}_{2}He \rightarrow ^{12}_{6}C + ^{1}_{0}n$$
$$\text{beryllium} + \begin{array}{c}\text{alpha}\\\text{particle}\end{array} \rightarrow \text{carbon} + \text{neutron}$$

Here the alpha particle is captured as it strikes the beryllium nucleus and ejects a neutron. Since the struck nucleus has gained 2 protons from the alpha particle, this raises its atomic number from 4 to 6, the atomic number of carbon, and converts it into $^{12}_{6}C$. This reaction has been used as a source of neutrons for bombarding nuclei.

33-23 Discovery of Artificial Radioactivity

Another outcome of the bombardment process of nuclei was the discovery in 1934 by Irene Curie and her husband F. Joliot that many bombarded substances continue to emit rays after the bombardment stops. Thus, stable elements are made into artificially radioactive ones. Subsequent research showed that radioactive isotopes of all the elements, from the lowest to the highest atomic numbers, can be made in this way.

The artificial radioactive isotopes resemble the natural ones in that some emit alpha particles and others emit beta particles. They differ from the naturally radioactive isotopes in that some of them emit a positron instead of a beta particle. The *positron*, discovered in 1932 by C. D. Anderson, is a *positive electron*. It has a mass and charge equal to that of an electron, but its charge is positive instead of negative. Its symbol is $_{+1}^{0}e$. In a typical reaction, the aluminum isotope $_{13}^{27}Al$ is bombarded by alpha particles and transmuted to radioactive phosphorus, $_{15}^{30}P$ as follows:

$$_{13}^{27}Al + {}_{2}^{4}He \rightarrow {}_{15}^{30}P + {}_{0}^{1}n$$

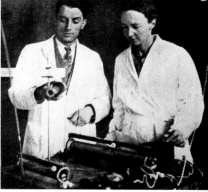

Fig. 33-7. In 1934, Irene Curie and her husband, Frederic Joliot, found a way to make stable elements artificially radioactive.

The radioactive phosphorus then decays into stable silicon, a positron, and a neutrino. The equation for the reaction is:

$$_{15}^{30}P \rightarrow {}_{14}^{30}Si + {}_{+1}^{0}e + \nu$$

There is no positron in the nucleus of any atom. Its ejection from a nucleus is explained by the conversion of one of the nuclear protons into a neutron, a positron, and a neutrino. The reaction is

$$_{1}^{1}H \rightarrow {}_{0}^{1}n + {}_{+1}^{0}e + \nu$$

The neutron remains in the nucleus and the positron and neutrino are ejected. Since this process reduces the nuclear charge by one unit, the disintegrating nucleus is transmuted into an element lower in atomic number by one. It is noteworthy that the emission of a positron from a radioactive nucleus is generally accompanied by the emission of a neutrino.

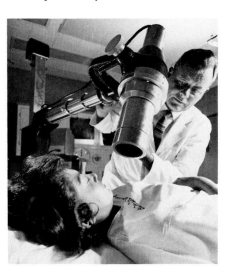

Fig. 33-8. Medical technician using a radiation counter to detect tagged atoms in the thyroid gland in the neck of a patient who has swallowed a radioactive iodine solution.

33-24 Uses of Radioactive Isotopes

Today, the bombardment process and other nuclear processes supply us with radioactive isotopes, also called *radioisotopes,* of nearly every one of the chemical elements. The radioisotopes are finding increasing uses in both research and industry. Doctors use their radiations in the treatment of cancer. Their energy is being harnessed in special long-life atomic batteries and other devices. In many processes, they are used as tracers that make it possible to analyze the process. In biological research, for example, radioactive iodine is fed in harmless quantities to an experimental animal to find out how its body absorbs the element iodine. The progress of the radioactive iodine through the animal's body is then traced by detecting the radiations that the radioisotope emits with a Geiger counter such as is described in Section 34-4.

Fission and Fusion

33-25 Fission

The possibility of tapping the energy stored in atomic nuclei as a source of large scale power was opened up by the discovery of *nuclear fission* by O. Hahn and his coworkers. Nuclear fission is a process whereby *certain nuclei of heavy atoms split up into two nearly equal parts when they are bombarded by neutrons.* At the same time, *each split nucleus ejects one or more neutrons* that often move rapidly enough to cause other nuclei to undergo fission. In 1939, Hahn and his coworkers found that when the nuclei of the isotope of uranium $^{235}_{92}U$, known as U-235, are bombarded by neutrons, they break up into nearly equal pairs of nuclei such as barium and krypton or antimony and niobium.

A typical fission reaction is

$$^{1}_{0}n + {}^{235}_{92}U \rightarrow {}^{141}_{56}Ba + {}^{92}_{36}Kr + 3^{1}_{0}n + Energy$$

Neutron + U-235 → Barium + Krypton + 3 neutrons + Energy

From the standpoint of developing nuclear power, fission is important for two reasons. First, there is a significant loss of mass during every fission reaction so that the sum of the masses of the fragments of the split nucleus is smaller than the original mass of that nucleus. This lost mass is converted into large quantities of energy according to the Einstein relationship $E = mc^2$. The energy is released from the nucleus as powerful gamma rays, heat, and the kinetic energy of the fast-moving nuclear fragments.

Second, each neutron set free by the fission of a U-235 nucleus is moving fast enough to cause another U-235 nucleus to undergo fission. This makes it possible to set up a *chain reaction* in which the fission of one nucleus triggers the fission of other nuclei, which,

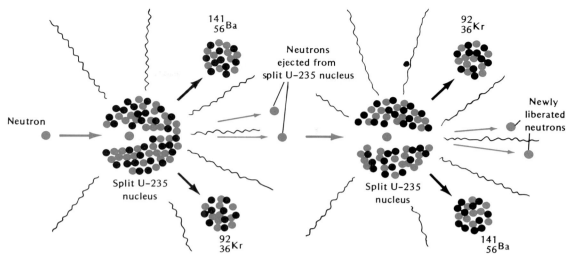

$^{141}_{56}Ba$

Neutrons
ejected from
split U-235 nucleus

$^{92}_{36}Kr$

Neutron

Newly
liberated
neutrons

Split U-235
nucleus

Split U-235
nucleus

$^{92}_{36}Kr$

$^{141}_{56}Ba$

Fig. 33-9. A chain reaction.

in turn, trigger the fission of still other nuclei, and so forth. Such a self-sustaining series of fissions yields enormous amounts of useful energy.

33-26 Conditions for a Chain Reaction

To maintain a self-sustaining chain reaction of fission in a quantity of U-235, it is necessary that, on the average, the neutrons liberated in each fission produce one or more additional fissions. The neutrons will fail to do this if, before they have had a chance to produce fission, too many of them escape from the U-235 mass or are absorbed by nonfissionable elements and impurities present in it. Therefore, to make a chain reaction possible, there must be a reduction in the number of impurities and other neutron-absorbing substances present. There must also be a certain minimum quantity of U-235 present to assure that most freed neutrons will strike a U-235 nucleus before passing through the mass and escaping to the outside. The minimum quantity of U-235 needed to sustain a chain reaction is called the *critical mass*.

In the fission bomb, or A-bomb, a critical mass of U-235 or plutonium is divided into two or more parts. As long as the parts are kept separated, there can be no chain reaction because each part is smaller than the critical mass. When the parts are suddenly brought together, they form a critical mass and a chain reaction begins. The purity of the fissionable material enables the reaction to proceed at an explosively rapid rate and an enormous quantity of energy is liberated almost instantaneously.

In the *nuclear reactor,* the chain reaction can be controlled and made to proceed at a much slower and carefully regulated rate. The energy liberated can therefore be provided at the rate at which it is needed for manufacture of electric power and other uses.

The nuclear reactor is made possible by a self-sustaining and controllable chain reaction.

33-27 Nuclear Reactor

A typical nuclear reactor is the *uranium pile*. This consists of a large mass of carbon blocks in which long cylindrical holes are provided at regular intervals. Chemically pure uranium rods sealed in aluminum cans are inserted in some of these holes while retractable rods of cadmium are inserted in other holes. The uranium rods provide the fuel for the chain reaction. The cadmium rods make it possible to control the rate at which the chain reaction takes place. Cadmium is a good *absorber* of neutrons. Hence, when the cadmium rods are fully inserted in the reactor, they absorb so many neutrons that the chain reaction stops. When the cadmium rods are gradually pulled out of the reactor, they absorb fewer and fewer neutrons, thus allowing the chain reaction to proceed more and more rapidly.

Water is circulated in pipes through the reactor to keep it from becoming dangerously hot. The whole reactor is surrounded by a thick wall of concrete to protect workers from the deadly radiations produced by it.

The reactor serves three major purposes. It maintains a controlled chain reaction in which the fission of U-235 nuclei yields a rich supply of useful energy in the form of heat. It manufactures a new element, plutonium, which is also fissionable and can therefore serve as a good source of nuclear power. Finally, it furnishes an intense source of neutrons which can be used in bombarding nuclei and in producing new artificial radioactive isotopes.

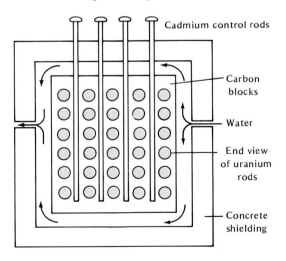

Cadmium control rods

Carbon blocks

Water

End view of uranium rods

Concrete shielding

Fig. 33-10. Diagram of an early type of nuclear reactor.

33-28 How the Reactor Works

To understand how the reactor works, we must note that less than one percent of the ordinary uranium which is put into it consists of the fissionable U-235 atoms which act as the fuel. The remainder is nearly all the denser, more common isotope known as U-238.

The chain reaction begins when stray neutrons, which are abundantly supplied by nuclear disintegration caused by rays coming from outer space, known as cosmic rays (Section 34-14), enter the reactor and cause some U-235 nuclei to undergo fission. Each of the split nuclei emits one or more neutrons which continue to move through the reactor until most of them strike new U-235 nuclei and cause them to undergo fission. More neutrons are then released to produce still more fissions, thus setting up a chain reaction and releasing large quantities of energy.

The continuation of the chain reaction is assured by making the pile greater than the critical size and by the presence of the carbon which is called a *moderator*. The function of the moderator is to slow down the fast neutrons freed in the reactor because slow neutrons are more efficient in producing fission than are fast ones. When fast neutrons pass through the carbon, they collide with many carbon atoms and are slowed to the desired velocities.

33-29 Production of Plutonium

A certain number of neutrons strike U-238 atoms instead of U-235 atoms. These neutrons are absorbed or captured by the $^{238}_{92}U$ nucleus which then becomes the unstable $^{239}_{92}U$ nucleus and disintegrates. Plutonium is now formed in two steps. Each newly formed $^{239}_{92}U$ nucleus first emits a beta particle and becomes the element neptunium. Then, the neptunium nucleus emits a second beta particle and becomes plutonium. The two reactions are:

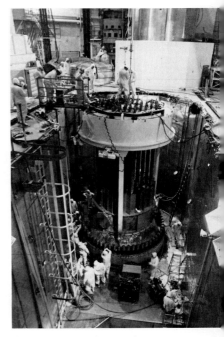

Fig. 33-11. A modern nuclear reactor with core removed for refueling.

$$^{1}_{0}n + {}^{238}_{92}U \rightarrow ({}^{239}_{92}U) \rightarrow {}^{239}_{93}Np + {}^{0}_{-1}e + \bar{\nu}$$

$$\text{neutron} + \text{U-238} \rightarrow \begin{pmatrix} \text{U-239} \\ \text{unstable} \end{pmatrix} \rightarrow \text{neptunium} + \text{beta particle} + \text{antineutrino}$$

$$^{239}_{93}Np \rightarrow {}^{239}_{94}Pu + {}^{0}_{-1}e + \bar{\nu}$$

$$\text{neptunium} \rightarrow \text{plutonium} + \text{beta particle} + \text{antineutrino}$$

The beta particles emitted in these two reactions are accompanied by antineutrinos. However, their zero masses and charges do not influence the balance of the equations.

After the plutonium accumulates in sufficient quantity in the uranium rods, it is removed by chemical means. Plutonium is both radioactive and fissionable. It can be used as the fuel in a reactor or as the explosive in an A-bomb. However, human health and peace considerations sharply limit its usefulness. Its radioactivity makes it a dangerous cancer producing agent when taken into the lungs even in the minute quantities that escape into the air. Furthermore, since it can readily be made into A-bombs, its availability in large quantities increases the danger of its falling into the hands of irresponsible people or governments.

33-30 Nuclear Power Plant

Large scale quantities of useful electric power may be obtained from a nuclear reactor by using its heat to produce steam for a steam turbine that operates an electric generator. Power obtained in this way now provides a substantial proportion of our electric power and drives specially designed ships and submarines. The intense and harmful radiation that accompanies the fission process makes it necessary to take elaborate safety precautions in operating a reactor. These include surrounding the reactor with heavy shielding to absorb most of the radiation and operating the reactor entirely by remote control.

In addition, it is necessary from time to time to remove the large quantities of lighter radioactive atoms produced by fission that accumulate in the uranium rods. The contaminated rods must be removed and the radioactive atoms separated from them chemically. The safe disposition of these highly radioactive and longlived waste products is a problem that is proving very difficult to solve.

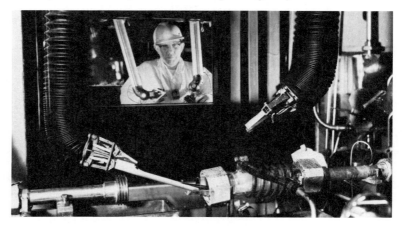

Fig. 33-12. These mechanical hands duplicate the delicate movements of the technician's hands. They are used to handle dangerous radioactive materials on the other side of the thick glass window.

33-31 Solar Energy from Fusion

Although the sun radiates enormous quantities of energy each day, there has been no measurable change in its mass, temperature, or rate of radiation over the centuries. It now seems clear that the sun generates this abundance of energy by a process called *nuclear fusion*. In nuclear fusion, the *nuclei of light elements* such as hydrogen and deuterium *combine to form heavier nuclei*. In a typical series of fusion reactions, four hydrogen nuclei combine to form one helium nucleus. Calculation shows that the mass of the helium nucleus is less than the sum of the four hydrogen nuclei that formed it. Thus, there is a loss of mass in this reaction. As in nuclear fission, the lost mass is converted into energy according to the relationship $E = mc^2$. Thus, each nuclear fusion reaction liberates enormous quantities of energy.

Because nuclear fusion reactions can take place only at very high temperatures, they are called *thermonuclear* reactions. In the sun, the conditions for thermonuclear reactions are ideal. There is a large supply of hydrogen nuclei and the temperature and pressure are extremely high. A series of processes equivalent to the continual fusing of hydrogen into helium nuclei thus takes place, with the release of vast quantities of energy. Since this energy results from the conversion of a negligible percent of the sun's mass, the gradual loss of mass by the sun is not noticeable.

Fig. 33-13. The galaxy N G C 4594. Like our sun, the stars of other galaxies, generate vast quantities of radiant energy by the process of nuclear fusion.

33-32 Fusion and the Hydrogen Bomb

Since the earth has an abundance of hydrogen, nuclear fusion holds out the possibility of providing an almost inexhaustible supply of energy. The chief problem in releasing this energy is to secure the conditions of high temperature and pressure in which nuclear fusion can take place. One method of obtaining these conditions is by exploding a fission or A-bomb. This method is used in the hydrogen or H-bomb to bring about the fusion of hydrogen nuclei into helium. The particular isotopes of hydrogen found suitable for the desired nuclear fusions are deuterium and tritium.

The bomb consists of a casing in which are contained the elements of an A-bomb and the quantities of deuterium and tritium to be fused. The A-bomb is exploded first and provides a temperature of millions of degrees Celsius in which the fusion reactions immediately take place and release enormous quantities of energy. Unlike the fission bomb, there is no limit to the size of the H-bomb and therefore to the quantity of energy that can be released in this manner.

33-33 Controlled Fusion Reactions

The energy obtained from the H-bomb has limited uses because the rate of its production cannot be controlled. It is released in a fraction of a second during which time it must either be put to use or be lost. To take full advantage of the energy liberated in fusion reactions, a way must be found to control the rate at which these reactions proceed. Energy can then be liberated in the quantities needed for particular uses. A possible way to solve this problem is found by the fusion reactions of the nuclei of the hydrogen isotopes, deuterium and tritium. Let us examine them here.

33-34 Deuteron-Deuteron and Deuteron-Triton Fusion

To reproduce the process whereby the sun continuously fuses ordinary hydrogen nuclei into helium is not a practical way to obtain fusion power. The process is too slow in releasing energy when the quantity of hydrogen fuel is small. It works on the sun because the

Fig. 33-14. A thermonuclear explosion liberates tremendous quantities of energy in an uncontrolled fusion reaction.

sun is so massive. Thus, even though only a minute percentage of the sun's hydrogen is undergoing fusion at any instant, an enormous quantity of energy is released.

A more practical fuel for a continuing fusion reaction is a mixture of the nuclei of deuterium and tritium. These nuclei are called *deuterons* and *tritons*. They can be made to combine and release large quantities of energy by means of the following three reactions:

$$^2_1H + {}^2_1H \rightarrow {}^3_2He + {}^1_0n + 3.2 \text{ MeV}$$
$$\text{deuteron} + \text{deuteron} \rightarrow \begin{array}{c}\text{helium-3}\\\text{nucleus}\end{array} + \text{neutron} + \text{energy}$$

$$^2_1H + {}^2_1H \rightarrow {}^3_1H + {}^1_1H + 4.0 \text{ MeV}$$
$$\text{deuteron} + \text{deuteron} \rightarrow \text{triton} + \text{proton} + \text{energy}$$

$$^2_1H + {}^3_1H \rightarrow {}^4_2He + {}^1_0n + 17.6 \text{ MeV}$$
$$\text{deuteron} + \text{triton} \rightarrow \begin{array}{c}\text{helium-4}\\\text{nucleus}\end{array} + \text{neutron} + \text{energy}$$

33-35 Establishing Fusion Conditions

For these reactions to occur, the deuterons and the tritons must be made to collide at velocities high enough to overcome their electrostatic repulsions. Such very high velocities can be attained by heating these particles to temperatures of millions of kelvins. To accomplish this, deuterons and tritons are first produced by ionizing a low density mixture of deuterium and tritium. The gas of ions that results is called a *plasma*.

One way of heating this plasma is to confine it in a container and compress it by means of a very strong magnetic field. The same magnetic field is also used to keep the plasma away from the walls of the container so that it will not lose its heat. The magnetic field thus serves as the actual container of the plasma and is called a *magnetic bottle*. If the plasma can be compressed and therefore heated to a high enough temperature, and if it can be contained long enough, many of its high velocity deuterons and tritons will fuse on colliding and release large quantities of energy. Part of this energy can be returned to heat the next supply of plasma and thus help to sustain a continuing fusion reaction. The rest is available for power.

33-36 Prospects for Fusion Power

So far, efforts to achieve continuous fusion in the laboratory have been unsuccessful. A major difficulty has been that the magnetic bottles have not been able to contain the plasma long enough to permit the high temperature and pressure conditions for the fusion reactions to be established and maintained. Work is continuing in many laboratories to solve this problem.

Fig. 33-15. Toroidal shaped magnetic coils used to confine and compress the high-temperature plasma in an experimental fusion reaction device.

An alternative approach to achieving fusion power has been through the development of powerful laser beams. In this process, a pellet of deuterium and tritium is placed at the center of a sphere and a burst of powerful laser beams is directed at it simultaneously from all directions. The effect is to exert an enormous pressure on the pellet and to cause it to implode violently. This establishes the high temperature needed to enable the contents of the pellet to undergo fusion and release energy.

The effort to achieve fusion power is reaching for a fantastic goal—a practically limitless supply of energy for the world's needs. While fission fuels are in limited supply, fusion fuels are abundantly available in the waters of the world. Furthermore, unlike the fission process, the fusion process does not burden us with large quantities of radioactive wastes.

CHAPTER REVIEW

Summary

The picture of the nucleus of the atom is still incomplete. It is known that the nucleus is composed of protons and neutrons and that these **nucleons** are distributed among nuclear energy levels resembling the atom's electronic energy levels. Jumps of nucleons of a given nucleus from higher to lower energy levels result in the emission of the gamma ray line spectrum characteristic of that nucleus.

The mass of the nucleus of every atom is smaller than the sum of the masses of the nucleons it contains. The difference is called the **mass defect** and is a measure of the **binding energy** that holds each nucleus together. The binding energy of a nucleus is found from the Einstein relationship $E = (MD) \times c^2$. The average **binding energy per nucleon** is found by dividing E by the mass number or number of nucleons. The nuclei of atoms of mass numbers in the neighborhood of 60 have the highest average binding energy per nucleon and are the most stable of the nuclei. As the mass of an atom increases above 60, its binding energy per nucleon decreases so that the nuclei of the atoms with the greatest masses are actually unstable and disintegrate spontaneously. The nuclei of such atoms are naturally **radioactive** and emit alpha, beta, and gamma rays. Nuclei that emit alpha or beta rays undergo a change in the number of protons they contain and are therefore **transmuted** into the nuclei of different chemical atoms. The emission of gamma rays by a nucleus has no effect on the number of its protons and does not result in the transmutation of a nucleus.

Another naturally occurring nuclear transmutation is **electron capture,** in which a nucleus absorbs one of its atom's *K*-electrons. The captured electron combines with one nuclear proton to produce a neutron that remains in the nucleus and a neutrino that is ejected. In this process the original nucleus is transmuted into a new nucleus with an atomic number one less than it had before the capture.

The nuclei of all the varieties of atoms can be made artificially radio-active by bombardment with high velocity charged particles, neutrons, or high energy gamma rays. In these nuclear changes, the following three conservation laws apply: (1) the sums of the energies and masses present before and after the reactions are equal; (2) the sums of the mass numbers before and after the reactions are equal; that is, the total number of nucleons remains constant in all nuclear reactions; and (3) the sums of the electric charges present before and after the reactions are equal.

A study of the particles ejected from natural and artificial radioactive nuclei has led to the discovery of the positron, the neutrino, and the anti-neutrino. The **positron** is a positively charged electron. The **neutrino** and **antineutrino** are neutral, massless particles.

Large quantities of energy for practical purposes may be obtained in nuclear reactions by the processes of fission and fusion. In **fission,** a massive nucleus such as U-235 splits into nearly equal fragments when bombarded by a neutron of appropriate energy. In this process, the energy and mass of the fragments are smaller than the energy and mass of the split nucleus and the neutron that struck it. The mass difference is converted into energy according to the relationship $E = mc^2$.

In **fusion,** lighter nuclei are combined to form a more massive one. Again, the energy and mass of the product nucleus is smaller than the sum of the energies and masses of the nucleons that composed it. The mass difference is again converted into energy.

Questions

Group 1

1. Compare the chemical properties, the masses, and the numbers of extra-nuclear electrons of (a) two atoms having the same atomic number but different mass numbers; (b) two atoms having different atomic numbers but the same mass number.

2. (a) What forces inside a nucleus act in such a way as to tend to tear the nucleus apart? (b) What forces inside a nucleus act in such a way as to hold the nucleus together? (c) Compare the relative size of the forces in (a) and (b).

3. (a) What determines the binding energy of a nucleus? (b) How does the average binding energy per nucleon determine the degree of stability of a nucleus? (c) What shows that nuclei of atoms more massive than uranium may be unstable?

4. (a) What is meant by the transmutation of elements? (b) What nuclear changes result in transmutation? (c) Define radioactive half-life.

5. Explain how transmutation takes place when the nucleus of a naturally radioactive atom ejects (a) an alpha particle; (b) a beta particle. (c) Why is there no transmutation during the emission of gamma rays from an atomic nucleus?

6. Which of the following quantities are equal to each other on both sides of the equation of a nuclear reaction: (a) the number of nucleons; (b) the number of neutrons; (c) the net electric charge; (d) the kinetic energy; (e) the sums of the energy equivalents of the masses and their kinetic energies; (f) the sum of the masses?

7. (a) What changes take place in the nucleus of an atom when it captures a K-electron? (b) Why is electromagnetic radiation emitted in this process?

8. Explain why a slow neutron may be as effective in entering a nucleus at which it is fired as a high velocity proton or alpha particle.

9. (a) How may the artificial transmutation of a stable atom be brought about? (b) How does the positron differ from the electron? (c) What change takes place inside a nucleus that results in the emission of a positron? (d) What other particle is emitted at the same time?

10. (a) Define fission. (b) What is the source of the energy released in a fission reaction? (c) What conditions must be fulfilled to maintain a chain reaction?

11. With reference to a nuclear reactor, explain: (a) the purpose of the cadmium rods; (b) the purpose of the carbon moderator; (c) the manner in which a chain reaction takes place; (d) the manner in which plutonium is produced.

12. (a) How does nuclear fusion differ from fission? (b) What is the source of the energy released during fusion? (c) How are the conditions necessary to bring about the fusion reaction produced in the H-bomb?

13. (a) Why is it not practical to use ordinary hydrogen nuclei as the fuel in attempting to produce controlled nuclear fusion? (b) What nuclei are more suitable for this purpose?

Group 2

14. (a) How does the nature of gamma-ray emission by radioactive atoms suggest the existence of definite energy levels inside the nucleus? (b) How does the emission of alpha particles support the idea of nuclear energy levels?

15. The binding energy of a nucleus may be referred to as the "unbinding energy." Why is this a very good description of this quantity?

16. Explain the statement: the half-life of a radioactive element does not tell us anything about how long it will take before any particular atom in a sample of that element actually disintegrates.

17. When a radium nucleus disintegrates, it is converted into an alpha particle and a radon nucleus. Both of these products have kinetic energy although the original radium nucleus was at rest. What is the source of this kinetic energy?

18. (a) What is a plasma? (b) How is it produced? (c) What two effects can a magnetic field produce in a plasma? (d) Why are they important in producing controlled nuclear fusion?

Problems

In the following problems, use the values: mass of a proton = 1.0073 u; mass of a neutron = 1.0087 u; mass of a neutral hydrogen $^1_1$H atom = 1.0078 u; and 1 u = 931 MeV. Also see the table of the Elements in the Appendix.

Group 1

1. Determine the number of neutrons in the nuclei of the following: (a) $^3_2$He; (b) $^{14}_6$C; (c) $^{22}_{10}$Ne; (d) $^{56}_{26}$Fe; (e) $^{233}_{90}$Th.

2. What is the mass in MeV of (a) a proton; (b) a neutron; (c) a $^4_2$He nucleus whose mass is 4.0016 u?

3. The mass of a $^3_2$He nucleus is 3.0160 u. What is its binding energy in (a) u; (b) MeV?

4. The nucleus of a deuterium atom consists of 1 proton and 1 neutron. (a) If the mass of a neutral deuterium atom is 2.0141 u, find the mass defect of its nucleus. (b) What is the energy in MeV binding the neutron and proton in this nucleus together?

5. The mass of a neutral nitrogen $^{14}_7$N atom is 14.0031 u. What is (a) its mass defect; (b) its binding energy in MeV; (c) its average binding energy per nucleon?

6. The mass of a neutral nitrogen $^{15}_7$N atom is 15.0001 u. (a) What is its average binding energy per nucleon? (b) Does it require more energy to separate a nucleon from a $^{14}_{7}N$ nucleus or from a $^{15}_{7}N$ nucleus? Refer to the answer to problem 5 (c).

7. The mass of carbon $^{13}_{6}C$ is 13.0034 u. Compute its (a) mass defect, and (b) binding energy per nucleon.

8. The mass of nitrogen $^{12}_{7}N$ is 12.0188 u. (a) Compute its binding energy per nucleon. (b) From the result and the answer to problem 7, determine which nucleus is likely to be more stable, $^{12}_{7}N$ or $^{13}_{6}C$.

9. Each of these radioactive nuclei is unstable and is transmuted into another element by emitting a beta particle. Determine the atomic number and the mass number of the new nucleus that is formed in the reaction. Use the table of the Elements in the Appendix to determine the name of the new atomic nucleus formed.
   (a) $^{3}_{1}H \rightarrow ? + _{-1}^{0}e + \bar{\nu}$
   (b) $^{14}_{6}C \rightarrow ? + _{-1}^{0}e + \bar{\nu}$

10. Each of these unstable radioactive nuclei disintegrates by emitting a positron. Determine the name, the atomic number, and the mass number of the new nucleus formed.
   (a) $^{17}_{9}F \rightarrow ? + _{+1}^{0}e + \nu$
   (b) $^{22}_{11}Na \rightarrow ? + _{+1}^{0}e + \nu$

11. Each of these nuclei forms an unstable new nucleus when it absorbs an alpha particle that strikes it. Determine the name, atomic number, and mass number of the new unstable nucleus formed.
   (a) $^{11}_{5}B + _{2}^{4}He \rightarrow (?)$
   (b) $^{27}_{13}Al + _{2}^{4}He \rightarrow (?)$

12. If each of the newly formed unstable nuclei in problem 11 emits a neutron, $_{0}^{1}n$, what is the name, atomic number, and mass number of the nucleus that remains? Write the nuclear equations.

13. In each of the following reactions, a neutron is first absorbed by a nucleus which then disintegrates by emitting a proton. Supply the missing terms.
   (a) $^{14}_{7}N + _{0}^{1}n \rightarrow ? + _{1}^{1}H$
   (b) $^{65}_{29}Cu + _{0}^{1}n \rightarrow ? + _{1}^{1}H$
   (c) $^{24}_{12}Mg + _{0}^{1}n \rightarrow ? + _{1}^{1}H$

14. The following nuclides each capture a K-electron. Complete the reaction.
   (a) $^{41}_{20}Ca + _{-1}^{0}e \rightarrow (?)$
   (b) $^{55}_{26}Fe + _{-1}^{0}e \rightarrow (?)$

15. Write the equation for the transmutation of a nuclear proton into a neutron.

16. What fraction of a sample of a radioactive substance with a half-life of 1 s will have disintegrated at the end of (a) 3 s; (b) 6 s?

Group 2

17. When a deuteron $^{2}_{1}H$ is fired at a nucleus of $^{9}_{4}Be$, the particle is absorbed and a neutron is ejected. Write the equation for the reaction.

18. A proton acquires 10 MeV of kinetic energy in an accelerator. What is the total energy equivalent of this proton?

19. The removal of a neutron from the nucleus of a $^{4}_{2}He$ atom converts that atom into a $^{3}_{2}He$ atom as follows: $^{4}_{2}He \rightarrow ^{3}_{2}He + _{0}^{1}n$. The difference between the energy equivalent of the mass of $^{4}_{2}He$ and the energy equivalent of the mass of $^{3}_{2}He$ and $_{0}^{1}n$ is the energy needed to separate 1 neutron from the $^{4}_{2}He$ nucleus. What is the value of this separating energy in MeV? The mass of a $^{4}_{2}He$ atom = 4.0026 u; the mass of a $^{3}_{2}He$ atom = 3.0160 u.

20. Following the procedure of question 19, find the energy needed to separate a neutron from a deuterium nucleus and thus convert a $^{2}_{1}H$ atom into a $^{1}_{1}H$ atom as follows: $^{2}_{1}H \rightarrow ^{1}_{1}H + _{0}^{1}n$. The mass of a $^{2}_{1}H$ atom = 2.0141 u.

Applying Physics

Make a table showing the percentage of a radioactive element that remains active after a whole number of half-lives from 0 to 10 have elapsed. Thus, at the beginning, when the number of elapsed half-lives is zero, the percentage of active nuclei is 100%. At the end of 1 half-life, the percentage of active nuclei has dropped to 50%. At the end of 2 half-lives, the percentage of active nuclei left is 25%, and so on.

A free neutron outside a nucleus disintegrates with a half-life of 12 minutes. Determine from your table how long it will take for more than 99% of a quantity of free neutrons to disintegrate. Do the same for $^{222}_{90}Th$, with a half-life of 1.9 years, and for $^{238}_{92}U$, with a half-life of 4.5 billion years.

The Superconducting Supercollider

Both physicists who study the world of the atom as well as those who study the beginnings and structure of the universe eagerly await the completion of the Superconducting Supercollider (SSC) in Texas in the late 1990s. With a circumference of about 88 kilometers (53 miles), the SSC will be the world's largest scientific instrument and its most powerful device for accelerating subatomic particles up to speeds close to that of light.

About 10,000 electromagnets will be positioned around the SSC's ring. The electric cables and windings of the magnets will contain 2500 tons of the superconducting alloy niobium-titanium. When maintained at temperatures just a few degrees above absolute zero, the windings will have little resistance and be able to carry enormous currents. These currents, in turn, will generate powerful 6.6-tesla magnetic fields.

The magnetic fields will be used to focus and accelerate two proton beams in the ring. The beams, each less than a millimeter in diameter, will be accelerated in opposite directions to nearly the speed of light. After being accelerated to an energy of 20 trillion electronvolts, the two beams will collide with a combined energy of 40 trillion electronvolts, more than 20 times greater than those achieved by the most powerful particle accelerator currently in operation.

The tremendous collision energy will be transformed into mass in accordance with Einstein's energy-mass equivalence equation, $E = mc^2$. The mass created will be in the form of short-lived particles having a much greater mass but smaller size than any of the subatomic particles currently known. In particular, physicists hope to detect the Higgs boson, whose existence will provide evidence that all of the four known forces of nature (the strong and weak forces that, respectively, bind the atomic nucleus and govern radioactive decay; the electromagnetic force; and gravity) were once unified as a single force that was in existence when the universe began 15 billion years ago.

Although it is certain to add to our scientific knowledge, a degree of controversy has attended the construction of the SSC. Construction costs will be in the billions and annual operating costs are expected to be in the hundreds of millions of dollars. The facility is expected to employ 2500 scientists and technical staff and to attract 500 visiting scientists. Some fear the SSC will drain talent and funds from smaller-scale physics research. It has also been suggested that the United States involve other countries in the project, having them finance one third of its costs.

34

Nuclear Instruments and Particle Physics

Aims

1. To become familiar with the instruments used to detect nuclear and subnuclear particles.
2. To understand the principles underlying the machines used to accelerate high speed projectiles for probing atomic nuclei and subnuclear particles.
3. To classify the subnuclear particles that emerge from nuclear disruptions according to their similarities and differences.
4. To study the four forces that account for all the properties and changes in matter.

Detection Devices

34-1 How Rays and Particles Are Detected

The study of radioactivity and nuclear changes depends upon identifying and detecting the rays and particles that are emitted by atomic nuclei. Gamma rays and charged particles may be detected directly by devices that utilize their three main effects: (1) their ability to affect a photographic plate, (2) their ability to cause certain substances on which they fall to fluoresce, and (3) their ability to ionize gases and other substances. Neutral particles such as neutrons do not produce these effects and cannot therefore be detected directly by them. However, they can be detected indirectly by letting them react with atomic nuclei. The gamma rays and charged particles that result from these reactions are readily detected and serve to identify the neutral particle that released them.

34-2 Photographic Plate

Photographic plates are sensitive to the rays emitted by radioactive substances and were, in fact, the means whereby Becquerel first discovered radioactivity. When charged particles or gamma radiations fall upon a photographic plate, they affect the emulsion so that it will turn black on development where it was exposed to these rays. In one method of detecting charged particles, the particles are permitted to enter a stack of photographic plates piled one on top of the other. When developed, the plates show the actual paths in space taken by the particles.

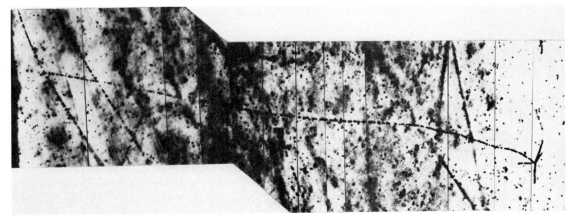

34-3 Scintillation Counter

This is a device that makes use of the fact that when charged particles or gamma rays pass through certain crystalline substances, the substances fluoresce and emit tiny flashes of light called *scintillations*. The counter consists of a block of a suitable fluorescent crystal mounted in an electronic tube called a photomultiplier. Each scintillation ejects some photoelectrons from a metal plate in the photomultiplier. The photomultiplier then uses these electrons to release very large numbers of additional electrons. Thus each scintillation is converted into a strong electrical signal. This signal is then used to operate a counting device which records the actual number of particles that enter the counter.

Fig. 34-1. Pathway of a charged particle entering at left is revealed by a pile of photographic plates. It collided with an atomic nucleus forming a four-pointed star at the extreme right.

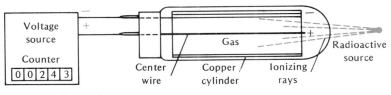

Fig. 34-2. A Geiger counter.

34-4 Geiger Counter

When charged particles or gamma rays pass through substances, they ionize them. Thus, an alpha particle passing through air collides with many atoms in its path and by ejecting electrons from them turns them into positive ions. In this way, positive ions and electrons form along the entire path of the alpha particle. Beta and gamma rays act in a similar manner. Alpha rays are by far the best producers of ions; beta particles are next, while gamma rays produce comparatively few ions.

In the Geiger counter, charged particles and gamma rays are detected by their ability to produce ions. This device consists of a hollow copper cylinder mounted in a thin glass tube and having a fine tungsten wire stretched along its central axis, as shown in Fig. 34-2. The tube contains a gas under low pressure and a voltage of about 1000 volts is applied so that the center wire becomes positively charged and the copper cylinder negatively charged.

When a high speed charged particle or a gamma photon passes through this tube, it frees electrons from the atoms in its path and creates ions. The freed electrons are strongly accelerated by the electric field between the positively charged central wire and the negatively charged cylinder, and acquire a high velocity as they move toward the central wire. These high velocity electrons then ionize the atoms in their paths and thus create more ions and more free electrons. This process is repeated over and over again so that, in a very short time, enormous numbers of electrons are flowing toward the central wire and produce a surge of current. The current is amplified and made to operate a counting device that records each particle entering the tube individually.

The Geiger counter may be adapted to detect neutrons by lining it with a thin layer of uranium. The reaction of the entering neutrons and the nuclei of the uranium atoms produces many charged fission fragments. These are readily detected by the counter and reveal the entry of the neutrons.

34-5 Wilson Cloud Chamber

This device is particularly useful and interesting because it makes the paths of ionizing rays visible. It was invented by the British physicist, C. T. R. Wilson. It makes use of the fact that, under the right conditions of temperature and pressure, water vapor in saturated air will condense upon the ions made by high speed charged particles and reveal their paths in the form of fog trails. The effect is similar to that seen when high-flying aircraft make fog trails in the sky. (See Fig. 5-8 in Section 5-12.

The cloud chamber consists of a cylindrical chamber fitted with a piston at its lower end and a glass window at its upper end. A little water or alcohol is first introduced into the chamber to saturate the air with vapor. The chamber is then ready to detect ionizing rays entering it from the outside or being emitted by a sample of radioactive matter placed inside of it. When a charged particle passes through the chamber, it produces ions along its path. The piston is then suddenly lowered, allowing the air in the chamber to expand and cool, and creating conditions under which the saturated vapor then condenses around the ions. Fog tracks form along the path of the particle and can be photographed or seen through the glass window when properly illuminated from the sides of the chamber.

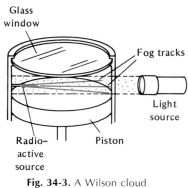

Glass
window

Fog tracks

Light
source

Radio-
active
source

Piston

Fig. 34-3. A Wilson cloud chamber.

34-6 Bubble Chamber

Like the Wilson cloud chamber, the bubble chamber gives a picture of the paths of ionizing rays. It utilizes the fact that bubbles will form in a liquid about to boil around ions present in the liquid.

Fig. 34-4. Installing a 2-meter hydrogen bubble chamber at the Lawrence Radiation Laboratory. Liquid hydrogen fills the bottom of the chamber. Rays to be studied enter the chamber through a window at lower right.

The bubble chamber contains a liquid, such as liquid hydrogen, that is kept at such a temperature and pressure that it is just below its boiling point. When the pressure is suddenly reduced, the boiling point of the liquid is lowered and it begins to boil. Bubbles are then formed on the ions produced along the paths of any ionizing particles that have passed through the chamber. With proper illumination, these tracks can then be seen or photographed.

34-7 Spark Chamber

This is another device that shows the path of ionizing particles. It consists of a set of parallel plates mounted several millimeters apart in a chamber filled with neon gas. Adjacent plates are electrically connected so that, as soon as a particle enters the chamber, a high voltage is applied between them. At the entrance to the chamber is a Geiger counter. A particle entering the spark chamber must first pass through the Geiger counter which detects it and promptly sends an electrical signal that turns on the high voltage. As the particle then proceeds into the spark chamber, it ionizes the neon atoms in its path. The ions thus formed produce a conducting pathway between the plates. The high voltage then drives sparks from plate to plate along this pathway, making it visible. In this way the path of the particle is traced and can be photographed.

Particle Accelerators

34-8 Purpose of Accelerators

We have seen how alpha particles and neutrons have been used to bombard and disrupt atomic nuclei. To expand the possibilities of exploring the structure of nuclei as well as subnuclear particles, projectiles of higher energy and higher speed are needed. Machines called *particle accelerators* have been designed for this purpose.

They are used to accelerate charged particles to speeds close to that of light and to hurl them against the nuclei or other targets being investigated. Typical particle accelerators include the cyclotron, the synchrotron, the linear accelerator, and the Van de Graaff generator which are used mainly to accelerate positively charged particles, and the betatron which is used mainly to accelerate electrons.

Since most of the basic principles involved in particle acceleration are illustrated by the cyclotron, we shall examine this device in some detail.

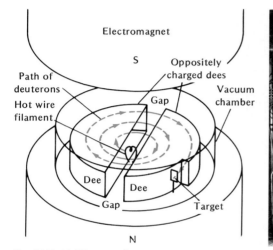

Fig. 34-5. (a) Diagram of a cyclotron. **(b)** A beam of deuterons emerging from a cyclotron becomes visible as it ionizes air and causes it to emit light.

34-9 Cyclotron

The cyclotron was invented in 1931 by the American physicist E. O. Lawrence. In it are two hollow half-cylinders made of metal and called *dees* because of their D-shapes. The dees are enclosed in a *vacuum chamber* situated between the poles of a *powerful electromagnet*. The effect of the electromagnet is to force the particles being accelerated to move in a slowly widening spiral path so that they remain in the cyclotron long enough to be given a series of successive pushes. The dees serve as the opposite plates of a capacitor and are connected to a *source of high frequency alternating voltage that reverses their electric charges every half cycle.*

34-10 Making a Beam of Deuterons

To illustrate how the cyclotron works, suppose it is used to produce high-velocity deuterons. First, a supply of deuterons is obtained by introducing some heavy hydrogen gas, or deuterium, into the middle of the vacuum chamber where it is ionized by an electrically heated wire filament. The positively charged deuterons that are thus formed in the gap between the dees are attracted immediately toward the dee that has a negative charge at that moment and repelled from the other dee that has a positive charge.

Under the influence of these electric forces, the deuterons are accelerated and move rapidly into the hollow negatively charged dee. The electromagnet, however, compels them to move in a circular path so that after making one half-revolution, the deuterons return to the gap between the dees. At this moment, the high frequency alternating voltage reverses the charges on the dees so that the dee toward which the deuterons are moving becomes negatively charged while the dee they are leaving becomes positively charged. Again the deuterons are pushed across the gap and accelerated by the two dees. Once again the magnet keeps the deuterons moving in a circular path. However, this one has a slightly larger radius than before because the deuterons are moving faster.

As this process is repeated many times, the deuterons trace a slowly expanding spiral path. At the same time, the charging and discharging of the dees is so timed that the deuterons are given an additional push each time they cross the gap between the two dees. These pushes add up to give the deuterons a very high velocity. The deuterons are then attracted out of the space enclosed by the dees by a negatively charged deflecting plate and are directed upon the target element whose nucleus is to be bombarded.

34-11 Energy of Deuterons

If the alternating voltage applied to the dees is 100 000 volts, each deuteron will acquire an increase in its speed corresponding to an energy of 100 000 electronvolts every time it crosses the gap between the dees. If a deuteron makes 50 round trips before it leaves the cyclotron, it will cross the gap 100 times. It will therefore acquire a speed corresponding to an energy of $100 \times 100\ 000$, or 10 million electronvolts. Thus, the cyclotron enables us to use the 100 000 volts of our alternating voltage to obtain an effective accelerating voltage 100 times as great.

34-12 Synchrotron

The cyclotron has limited usefulness because it cannot accelerate charged particles beyond speeds corresponding to energies of 15 to 20 million electronvolts. The reason for this is that, as the speed of the particles approaches the speed of light, their masses increase in accordance with the relativity theory. (See Section 7-11.) The increase in mass reduces the acceleration given to the particles as they cross the gap between the dees. As a result, the particles take a longer and longer time during each revolution to reach the gap between the dees. They therefore fall out of step with the accelerating voltage applied to the dees and undergo no further acceleration.

This limitation is overcome in a modification of the cyclotron known as the *synchrotron* or the *bevatron*. The prefix *beva* refers to billion electronvolts. In this machine, as in the cyclotron, the

Fig. 34-6. Construction of the 2-BeV proton synchrotron, or cosmotron, at the Brookhaven National Laboratory.

charged particles are repeatedly accelerated as they circle about in their orbits. However, instead of moving in a spiral path, they are made to move in a circular path of constant radius. This is done by increasing the magnetic field steadily as both speeds and masses of the particles increase. At the same time, the frequency of the alternating voltage is kept in step with the frequency of revolution of the particles. As a result, the particles continue to be accelerated and acquire higher and higher speeds. A modern synchrotron can impart speeds to protons differing from the speed of light by less than 0.0001 percent. The energy corresponding to such speeds is of the order of 10^{12} electronvolts.

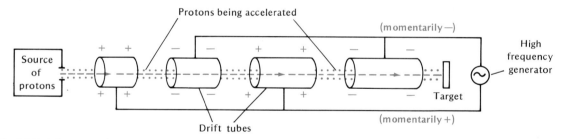

Fig. 34-7. A linear accelerator.

34-13 Linear Accelerator

This machine consists of a long cylindrical vacuum tube having a series of hollow metal *drift tubes* mounted along its central axis, as shown in Fig. 34-7. It is used to give extremely high velocities to positively charged particles such as protons. The protons are usually first brought up to a moderately high velocity by a Van de Graaff generator and then let in at one end of the linear accelerator.

As the protons proceed to pass through one drift tube after another, a high-frequency alternating voltage is applied to the drift tubes so that during one half-cycle, the odd ones are positively charged and the even ones negatively charged, and during the next half-cycle the charges on the drift tubes are reversed. The drift tubes are so spaced and their rate of charging is so timed that the protons always cross the gap from one tube to the next during the half-cycle when the tube they are leaving is positively charged and

Fig. 34-8. Inside the linear accelerator at Berkeley, California.

the one ahead is negatively charged. The tube ahead therefore attracts the protons while the one behind repels them. This gives the protons an accelerating push forward each time they cross a gap between tubes. These pushes add up to give the protons an extremely high velocity when they reach the end of the accelerator.

Here, the protons are allowed to fall upon the target being bombarded. Since the protons speed up as they move through the accelerator, the drift tubes must be made progressively longer to insure that the protons are always in the gaps during the proper half-cycles of the voltage applied to the tubes.

34-14 Cosmic Rays

An invaluable natural ally in the effort to penetrate the atomic nucleus has been the very powerful radiation coming toward the earth from outer space and known as *cosmic rays*. We do not yet know the origin of these rays, but it has been shown that before entering our atmosphere they are composed of nuclei of different atoms, most of which are hydrogen nuclei. The energies they possess are often millions of times greater than can be obtained in particle accelerators.

On entering the earth's atmosphere, the high energy cosmic rays collide with the nuclei of atoms in their paths. Such collisions result in the disintegration of one or both of the colliding nuclei and drive the remaining fragments speedily earthward. The fragments, in turn, collide with the nuclei of other atoms in their paths. This process is repeated until all of the energy in the original rays has been exhausted. One of the results of these cosmic ray collisions is the production of the extremely energetic gamma rays.

Collisions with cosmic rays break open the nuclei of many atoms for our study, and we can observe them by means of such devices as the Wilson cloud chamber, the bubble chamber, the photographic plate, and the Geiger counter.

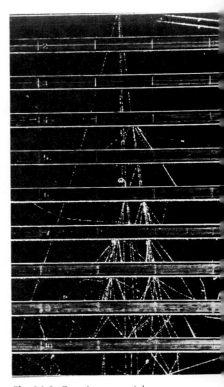

Fig. 34-9. Cosmic ray particles entering cloud chamber at top pass down through several brass plates 1 cm thick. They collide with atomic nuclei in their paths producing a shower of subatomic particles toward the bottom.

Particle Physics

34-15 Subnuclear Particles

The disruption of atomic nuclei by cosmic rays and other nuclear projectiles has revealed the existence of many subnuclear particles other than the proton and the neutron and has expanded the field of particle physics. Some particles have positive charges, some negative charges, and some are neutral. The charges of particles are generally unit charges equal to that of the electron or positron.

Of special interest is the discovery that each particle has an antimatter counterpart, or *antiparticle.* For the electron, the antiparticle is the positron; for the proton it is the antiproton; and for the neutron it is the antineutron. A basic property of the particle and its antiparticle is that they annihilate each other on coming together. Thus, when an electron and a positron meet, they destroy each other and form gamma rays. (See Fig. 34-10.)

Subnuclear particles are divided into four classes: *photons, leptons, mesons,* and *baryons.* (See Table 34.1.) As we have seen, photons are tiny packets of energy having zero mass and zero charge but a definite momentum. Photons are their own antiparticles.

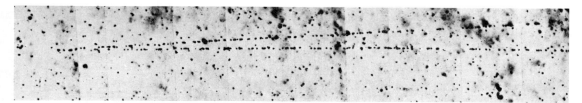

Fig. 34-10. Creation of matter from energy! Paths of an electron and a positron leaving point G where they were created from a gamma ray (not shown). When either of these particles meets its antiparticle again, both will be annihilated and replaced by gamma rays.

34-16 Leptons

These particles include the electron, the muon, and the tau, each of which has a negative charge equal to that of the electron and their equally charged, positive antiparticles. Associated with these leptons are three different neutrinos and their antineutrinos. All of these neutrinos and antineutrinos have no charge and no mass, and are all stable. The electron and its antiparticle, the positron, are also stable, but the muon and the tau and their antiparticles, all of which have masses much greater than that of the electron, are unstable. They decay into ordinary electrons or positrons with the emission of much energy.

Leptons play a particularly important role in radioactive disintegrations that involve the emission of electrons or positrons from the nucleus of an atom. Such emissions are always accompanied by the ejection of a neutrino or an antineutrino. Examples are the decay of $^{209}_{82}$Pb (Section 33-15) and the decay of $^{30}_{15}$P (Section 33-23).

Table 34.1 Particles and Their Antiparticles				
NAME	**SYMBOL**	**ANTIPARTICLE**	**REST MASS** (MeV)	**HALF-LIFE** (s)
photon	γ	γ	0	stable
Leptons				
neutrino (e)	ν_e	$\bar{\nu}_e$	0	stable
neutrino (μ)	ν_μ	$\bar{\nu}_\mu$	0	stable
neutrino (τ)	ν_τ	$\bar{\nu}_\tau$	0	stable
electron	e^-	e^+	0.511	stable
muon	μ^-	μ^+	105.7	1.5×10^{-6}
tau	τ^-	τ^+	1784	4×10^{-13}
Mesons				
pion	π^0	π^0	135.0	7×10^{-17}
pion	π^+	π^-	139.6	1.8×10^{-8}
kaon	κ^+	$\bar{\kappa}^-$	493.8	8×10^{-9}
kaon	κ^0	$\bar{\kappa}^0$	497.8	7×10^{-11}
eta meson	η^0	$\bar{\eta}^0$	548	$< 10^{-16}$
Baryons				
proton	p	$\bar{p}$	938.2	stable
neutron	n	$\bar{n}$	939.5	700
lambda hyperon	Λ^0	$\bar{\Lambda}^0$	1115.4	1.8×10^{-10}
sigma hyperon	Σ^+	$\bar{\Sigma}^+$	1189.2	6×10^{-11}
sigma hyperon	Σ^0	$\bar{\Sigma}^0$	1193.2	$< 10^{-14}$
sigma hyperon	Σ^-	$\bar{\Sigma}^-$	1197.6	1.2×10^{-10}
xi hyperon	Ξ^-	$\bar{\Xi}^+$	1310	1.0×10^{-10}
xi hyperon	Ξ^0	$\bar{\Xi}^0$	1321	9×10^{-11}
omega hyperon	Ω^-	Ω^+	1676	$\sim 10^{-10}$

34-17 Mesons

These are particles with masses between that of the muon and the proton. They include the pions, kaons, eta mesons, and their antiparticles. They are all unstable and decay into lighter particles. The eta mesons and kaons generally decay into pions. The pions decay into muons which then decay into electrons. Pions are of special significance because, as we shall see, they are the carriers of the strong forces that hold the nucleus together. They may be positively or negatively charged, or neutral.

34-18 Baryons

Baryons are particles with masses equal to or greater than the mass of the proton. They include protons, neutrons, and more massive particles called hyperons. The neutron and proton are the major components of the atomic nucleus. Each has an antiparticle that exists only outside the nucleus. While inside the nucleus, they are both stable. Outside the nucleus, only the proton is stable. The neutron decays into a proton, an electron, and an antineutrino. Its half-life is about 700 seconds.

The remainder of the baryons are listed in Table 34.1. They are all unstable, and eventually decay into less massive particles which decay in turn. One of the end products common to these successive decays is the proton. This observation suggests that the hyperons are actually protons in high energy states. It is thought that just as an atom in a high energy or excited state drops back to a lower state with the emission of energy, so does the hyperon drop from its excited state to a lower state with the emission of energy. Ultimately it drops to its ground state which is the proton.

34-19 Four Types of Forces or Interactions

When we studied matter and energy as they are observed outside the nucleus of the atom, we found that particles of matter exert two kinds of forces on each other: electromagnetic and gravitational. The study of subnuclear particles reveals that there are two more types of forces between particles. They are the strong nuclear force and the so-called weak force. These four forces account for all the properties and changes of the matter in the universe. They are called *interactions*.

The electromagnetic interaction may be either attractive or repulsive. The other three interactions are attractive. The *strong nuclear force* is by far the strongest of the interactions. The electromagnetic interaction is next and is about 1/100 as great as the strong reaction. Next comes the *weak interaction* which is only 10^{-13} as great as the strong reaction. Finally comes the gravitational interaction which is 10^{-38} as great as the strong interaction.

34-20 Strong Nuclear Interaction

This refers to the very strong attractive forces that the nucleons in a nucleus exert upon each other. They are forces between two neutrons, two protons, and a proton and a neutron. They are the same for any two nucleons and are independent of their charges. The strong nuclear force has a very short range of the order of 10^{-15} meter. When the distance between two nucleons is greater than this, the strong force drops sharply to zero. Thus, a nucleon in the nucleus attracts only those nucleons that are its close neighbors. However these forces are strong enough to hold the nucleus together against the powerful repulsive forces among its protons.

34-21 Electromagnetic Interaction

This refers to the forces between charged particles. For charges at rest, these are the Coulomb attractions and repulsions that act along the line joining the charges. They follow the inverse square law. For charges in motion, there are magnetic forces in addition to the Coulomb forces. Both the size and the direction of the magnetic forces depend on the relative motion of the charged particles.

The electromagnetic interactions are responsible for binding electrons to their atoms as well as for binding atoms and molecules together to form larger bodies. They determine the chemical properties of substances as well as such physical properties as their state, their elasticity, and their crystalline structure. While they are considerably weaker than the strong nuclear forces, their range is limitless.

34-22 Weak Interaction

The weak interaction is a force that is encountered during the emission of electrons or positrons in the radioactive decay of atomic nuclei as well as in the decay of certain so-called strange particles such as the lambda hyperon. The strange particles that are subject to the weak force take about 10^{13} times as long to decay as they would if they were subject to the strong force. This indicates that the weak force is only 10^{-13} as great as the strong force.

34-23 Gravitational Interaction ·

The weakest of the four interactions among particles is that of gravitation. It is so much smaller than the other three interactions that it is not observable between subnuclear particles. However, its effects are easily observed in the weights of ordinary objects and in its vast influence on the massive heavenly bodies. Like the electromagnetic interaction, its range is limitless.

34-24 Carriers of the Four Interactions

A question that has long challenged physicists is how these forces are transmitted between two particles that are separated from each other. Recall that electric charges are considered to act upon each other through their electric and magnetic fields. Physicists imagine the field to be a cloud of photons that are continually being emitted and absorbed by charges located in the field. When two charges are in their combined fields, they are continually exchanging photons and thus exerting forces on each other. That is why these forces are called exchange forces and the photons are called the carriers of the electromagnetic forces.

A similar exchange force is regarded to be acting in the transmission of the strong nuclear force. Because this is such a great force and has such a short range, the carrier for it could not be a massless particle like the photon. In 1937, the Japanese physicist, Hideki Yukawa, predicted that the carrier of the strong force should be a particle some two hundred times as massive as the electron. Such a particle was soon found. It turned out to be the pion or pi meson. The present view is that pions surround the nucleons in a nucleus. The exchange of pions between neighboring nucleons results in the strong nuclear force between them.

The W and Z particles, carriers for the weak interactions, were finally identified in 1983 and won for their discoverers, Carlo Rubbia and Simon Van der Meer, the 1984 Nobel Prize in Physics. This leaves only the carrier of the gravitational attraction, which is assumed to be the graviton, yet to be discovered.

34-25 Quarks, the Basic Particles

All baryons and mesons are made up of truly basic particles called quarks *and* antiquarks.

As physicists have worked with higher and higher energies, many new baryons much more massive than the proton or the neutron have been detected. To explain this abundance of particles, Murray Gell-Mann and George Zweig were the first to suggest that all baryons and mesons are made of *truly* basic particles called *quarks* and *antiquarks*. These basic particles have the unusual property of having positive or negative electric charges that are either $\frac{1}{3}$ or $\frac{2}{3}$ as large as the charge on the electron. There are six known quarks and their six antiquarks. The quarks are named: up, down, strange, charmed, top, and bottom. Their symbols are u, d, s, c, t, and b. The symbols of their corresponding antiquarks are $\bar{u}$, $\bar{d}$, $\bar{s}$, $\bar{c}$, $\bar{t}$, and $\bar{b}$. Each antiquark has an electric charge equal and opposite to that of its corresponding quark.

All baryons are composed of three quarks. A proton, for example, is composed of two u-quarks and one d-quark to make the combination uud. Each u-quark has a charge of $+\frac{2}{3}$ of the unit electric charge and each d-quark has a charge of $-\frac{1}{3}$ of the unit charge. The total charge on the proton therefore adds up to $(+\frac{2}{3}) + (+\frac{2}{3}) + (-\frac{1}{3}) = +1$ unit charge, the known charge of the proton. The neutron is a combination of one u-quark and two d-quarks, giving a total charge of $(+\frac{2}{3}) + (-\frac{1}{3}) + (-\frac{1}{3}) = 0$, the known zero charge of the electrically neutral neutron. The composition of antibaryons is similar to that of baryons except that they are composed of three antiquarks rather than three quarks.

All mesons are composed of one quark and one antiquark. The pi-plus meson, π^+, for example, is composed of a u-quark and a d̄-antiquark. Its charge is the sum of $+\frac{2}{3}$ (for the u-quark) and $+\frac{1}{3}$ (for the d̄-antiquark), giving a total of 1 unit charge, which is the known charge of this particle.

34-26 Color and the Allowable Quark and Antiquark Combinations

All quarks are arbitrarily assigned one of the three "colors," red, green, or blue, while their antiquarks are assigned the "anticolors," antired, antigreen, or antiblue. These "colors" and "anticolors" have no relationship to light or pigments. They simply identify a property of quarks and antiquarks that limits the ways in which they can combine with other quarks and antiquarks. Thus, the only allowable three-quark combinations that make up all baryons are those in which one is red, one is green, and one is blue. Similarly, the three antiquarks that make up an antibaryon must consist of one antired, one antigreen, and one antiblue. In the quark and antiquark combinations that make up all mesons, the antiquark must be of the anticolor that corresponds to the color of the quark. That is, a red quark can only combine with an antired antiquark.

Fig. 34-11. The ISR (Intersecting Storage Ring) at CERN, the European research facility at which the long-sought W and Z particles were found in 1983.

34-27 Gluons

The carrier of the force between quarks is believed to be a particle called the gluon. Gluons provide an attractive force between quarks that *increases* rapidly as the distance between the quarks becomes greater. This makes it extremely difficult for individual quarks to set themselves free from the baryons and mesons in which they are contained. It is probably the reason no individual quark has yet been isolated. The same is true of antiquarks that are bound in mesons and antibaryons.

CHAPTER REVIEW

Summary

Nuclei are studied by bombarding them with charged particles and examining what remains after a nucleus disintegrates. **Particle accelerators** are used to speed up charged particles and hurl them against nuclear targets. The changes that take place can then be observed and recorded by photographic plates, counters, cloud, bubble, and spark chambers.

The study of nuclear disintegrations and reactions has revealed the existence of four classes of particles: photons, leptons, mesons, and baryons. All of these particles have corresponding **antiparticles.** When a particle meets its antiparticle, both are destroyed and their masses are partly or wholly converted into energy. The **photons** are the energy packets that transmit the energy of electromagnetic waves. The **leptons** are the light particles including the electron, the positron, the neutrinos, the muon, and their antiparticles. The **mesons** are particles with masses between that of the muon and that of the proton. The **baryons** are particles with masses equal to or greater than that of the proton. All of the particles except the proton, electron, neutrino and their antiparticles are unstable and decay into more stable particles. The neutron is stable while in the nucleus, but unstable when free.

There are four forces by means of which particles of matter interact with each other: (1) The **strong nuclear force** acts between protons and neutrons, protons and protons, and neutrons and neutrons; it holds the nucleus together. (2) The **electromagnetic force** acts between charged particles. (3) The **weak force** is involved when radioactive nuclei emit electrons or positrons. (4) The **gravitational force** of attraction acts between any two particles of matter. These forces appear to be transmitted by carriers which are pi mesons for the strong nuclear force and photons for the electromagnetic force. The carriers for the weak force are the W and Z particles, while the carrier for the gravitational interaction is believed to be the hitherto undetected graviton.

All baryons and mesons are believed to be made up of particles called **quarks** and **antiquarks,** which are bound together by carrier particles called **gluons.** All baryons are combinations of three quarks, while all antibaryons are combinations of three antiquarks. All mesons consist of a quark and an antiquark. Each quark and antiquark is assigned a color that limits the ways in which it may combine with other quarks and antiquarks.

Questions

Group 1

1. Explain how photographic plates may be used to detect charged particles and show their paths.

2. Explain how a scintillation counter detects gamma rays and charged particles.

3. (a) How does a Geiger counter increase the number of ions formed by the charged particle that enters it? (b) How may a Geiger counter be adapted to detect neutrons?

4. (a) How does the Wilson cloud chamber reveal the passage of an ionizing particle through it? (b) Why will a Wilson cloud chamber not show the path taken by a neutron passing through it?

5. State one similarity and one difference between the operation of a bubble chamber and a Wilson cloud chamber.

6. With reference to the cyclotron, explain: (a) how it accelerates a charged particle such as a proton to be used as a nuclear projectile; (b) how it keeps the fast-moving proton inside the machine. (c) Why can't a cyclotron be used to accelerate neutrons?

7. (a) What limits the ability of the cyclotron to increase the speed of the particles it is accelerating beyond a certain maximum value? (b) How does the synchrotron overcome this limitation?

8. (a) What is the principle used in the linear accelerator to speed up positively charged particles? (b) Why is no magnet like that in the cyclotron needed in this machine?

9. (a) What are cosmic rays? (b) How are they used in studying the structure of atomic nuclei?

10. (a) Compare the masses of leptons, mesons, and baryons. (b) What is the relationship between a particle and its antiparticle?

11. Compare the strong nuclear force with the electromagnetic force with respect to (a) strength; (b) range; (c) the carrier of the force.

12. Compare the weak force with the gravitational force with respect to (a) strength; (b) range; (c) theoretical carrier of the force.

Group 2

13. A Wilson cloud chamber is placed between the poles of a powerful magnet whose lines of force run vertically upward. An ionizing ray enters the cloud chamber in a horizontal plane. How can the observer tell from the path of the particle ray whether it is a positively or negatively charged particle or a gamma photon?

14. No matter how powerful an accelerator may be, a charged particle in it can never attain the speed of light. Explain.

15. (a) Outside of the nucleus a neutron decays into a proton, an electron, and an antineutrino. Write the equation for the reaction. (a) Assuming that an antineutron decays in a similar manner, write the equation for the reaction.

16. Which of these quark combinations are allowable? (a) uds; (b) $\overline{u}d\overline{s}$; (c) ds; (d) $d\overline{s}$.

Applying Physics

A simple Wilson cloud chamber can be made of a wide-mouthed screw cap glass jar about 8 cm. in diameter and 8 cm. high. A length of old clothesline long enough to fit around the inside of the jar is soaked thoroughly in alcohol and then fastened around the inner surface of the bottom of the jar with a few strips of masking tape. The inside of the screw cap of the jar is then lined with black cloth. To provide a source of radiation, the point of a thumb tack is moistened with plastic cement and then brought into contact with a speck of radioactive luminous paint scraped from a discarded watch dial. (*Caution:*

Do not touch the radioactive material or bring the face near it.)

The thumb tack is set point upward inside the screw cap and the glass jar is then inverted and screwed tightly to the cap. The inverted jar is then put on a cake of dry ice. The alcohol vapor in the jar will begin to fog at first. After about 15 minutes, the mist will clear up and a supersaturated layer of vapor will form just above the lid of the jar. On illuminating the jar from the side by means of a flashlight beam, the tracks of the alpha particles will be seen clearly against the black cloth lining the cap.

Applications

Verifying Computer Programs

Computers are, of course, dependent on the semi-conductor diode and transistor, as well as on many interconnecting wires. What do you do when a computer program with thousands, perhaps millions of lines contains a single error that invalidates it? No one wants to throw out anything as expensive as such a program in terms of hours of development, but the cost of finding the error might be equally high. In the past programs were tested by feeding them data for which the results were known. If a program got the right answers, it was thought to be accurate. Such checks, however, often failed to reveal subtle but important errors that surfaced later, often when the program was in use.

In recent years a new technique to search for hidden errors in programs has evolved. The new approach uses a second program.

The second program is designed to reason about the program being tested. In the last ten years about eight to ten such programs have been developed.

One such program, the Boyer-Moore Theorem Prover, is very good at verifying programs that contain repeated procedures, that is, loops. A simple illustration of a loop is the cooking direction, "Add salt and pepper to taste." This loop involves adding salt and/or pepper, stirring, tasting, and then repeating the process until the desired flavor has been achieved.

A program to verify a program loop or the cooking direction above looks for an "invariant." An invariant is a relationship that remains unchanged throughout a loop. In the cooking direction, the invariant is the well-known flavor produced by adding salt (or pepper). If the direction is in error, say it specifies add sugar instead of salt, an entirely different flavor is produced. The invariant has been violated. As a result, we know there is an error. The Theorem Prover confirms that the mathematical or logical invariants in program loops remain invariant.

All of the new verification programs work in a similar manner. They first compare what the program is designed to do with how it is done. Then they produce a set of verification conditions and prove them. If all work out, the program is error free.

Computer Systems Analysts

Systems analysts plan and develop methods for computerizing scientific, engineering, and business tasks, processes, and applications. A systems analyst begins the process by discussing the data processing problem with specialists or managers in order to break it down into component parts. He or she then uses such techniques as mathematical model building, statistical sampling, and cost accounting to plan the desired system. When a system design has been approved, the systems analyst prepares all specifications and forms and works with computer programmers to develop, refine, and "debug" the needed software (programs).

To accomplish these tasks, systems analysts must be able to think logically, communicate well with other professionals, managers, and technical personnel, and like working with ideas and people. They must have the ability to work with details and handle several tasks simultaneously. In addition, they must have the capability to work independently or as part of a large and complex team.

Systems analysts usually work in offices and keep regular business hours. From time to time, however, evening and weekend work is needed to meet deadlines.

New job opportunities for systems analysts are expected to grow much faster than the average through the year 2000. Demand is expected to rise as technological developments and increased computer capabilities lead to new applications for computers. Scientific research, telecommunications, and factory and office automation are just a few of the areas in which computer applications will expand.

At present there is no accepted standard way to prepare for a job as a systems analyst. However, almost all employers want college graduates, and for some of the more complex tasks, graduate degrees are preferred. For work in scientifically-oriented organizations, a background in the physical sciences, applied mathematics, or engineering is needed. In addition, a strong background in computer science is required. Prior work experience is important. About three-fourths of all people who become systems analysts transfer from occupations such as engineer or scientist, manager, or computer programmer.

Other occupations that require a similar background and use the same thinking and problem solving skills as systems analysts are operations research analysts, mathematicians, engineers, urban planners, and computer programmers.

Mathematics Review

This review is designed to help you brush up on the mathematical operations and information frequently used in solving problems in physics. Additional information may be had by referring to sections and sample problems in the book where each operation is used. These are indicated under each heading where appropriate.

Steps in Solving a Problem

1. READ the problem carefully.
   Problem: A uniformly accelerated car starting from rest travels 100 m in the first 20 s of its motion. Find its acceleration.
2. Identify what is given.
   Given: $d = 100$ m, $t = 20$ s
3. Identify what is to be found.
   Find: a
4. Select the appropriate equation that connects what is given with what is to be found.
   Equation: $d = \frac{1}{2}at^2$
5. Rewrite the equation so that the quantity to be found appears on one side of the equation and all other quantities appear on the opposite side.
   Rewritten equation: $a = \dfrac{2d}{t^2}$
6. Substitute the given values and their units in the equation.
   Substitution: $a = \dfrac{2(100 \text{ m})}{(20 \text{ s})^2}$
7. Solve the equation, giving the answer in the proper units.
   Solution: $a = \dfrac{200 \text{ m}}{400 \text{ s}^2} = 0.500 \dfrac{\text{m}}{\text{s}^2}$
8. Check the final answer to make sure that it is reasonable.
   Check: The unit m/s² is correct for acceleration. The answer, 0.500 m/s², is reasonable, since a car with this acceleration would have an average speed of 5 m/s over the first 20 s of its motion and would therefore cover the given distance of 100 m in that time.

Solving an Equation for an Unknown

The object is to get the unknown quantity on one side of the equation and all other quantities on the other side.

Example 1

Given $\frac{1}{2}mv^2 - \frac{1}{2}mv_0^2 = Fd$, solve for v.

Solution

1. Add $\frac{1}{2}mv_0^2$ to both sides of the equation, giving:
   $\frac{1}{2}mv^2 = Fd + \frac{1}{2}mv_0^2$
2. Multiply both sides of the equation by 2:
   $mv^2 = 2Fd + mv_0^2$
3. Divide both sides of the equation by m:
   $v^2 = \dfrac{2Fd}{m} + v_0^2$
4. Take the square root of both sides of the equation:
   $v = \sqrt{\dfrac{2Fd}{m} + v_0^2}$

Example 2

Given $\dfrac{1}{p} + \dfrac{1}{q} = \dfrac{1}{f}$, find q.

Solution

1. Get rid of the fractions by multiplying both sides of the equation by pqf, which is the common denominator of p, q, and f:
   $\dfrac{pqf}{p} + \dfrac{pqf}{q} = \dfrac{pqf}{f}$, whence $qf + pf = pq$
2. Subtract qf from both sides of the equation:
   $pf = pq - qf = q(p - f)$
3. Divide both sides of the equation by $(p - f)$:
   $\dfrac{pf}{p - f} = q$, or $q = \dfrac{pf}{p - f}$

Significant Figures

Treated fully in Sections 2-20 through 2-22.

Scientific Notation

Treated fully in Sections 2-18 and 2-19.

Arithmetical Operations with Scientific Notation

Treated fully in Sections 2-20 through 2-22.

Changing a Fraction to a Decimal

Example

Change $\dfrac{53}{60}$ to a decimal.

Solution
1. Write 53 as 53.000.
2. Divide it by 60:

$$\begin{array}{r} 0.883 \\ 60\overline{)\ 53.000} \end{array}$$

3. The answer is 0.88 to two significant figures.

Changing a Decimal to a Percent
Example
Change 0.821 to a percent.
Solution
Multiply by 100% by moving the decimal point two places to the right. Drop the zero in front of the number:

$0.821 \times 100\% = 082.1\% = 82.1\%$

Graphing

See Sections 3-7 and 6-2.

Dimensional Analysis

See Section 3-13.
The dimensions of any physical quantity can be expressed as a combination of one or more of the seven basic units—kg, m, s, A, K, mol, and cd. In dimensional analysis, we do the operations called for in a problem on the units only. They combine algebraically to give the combination of units appropriate for the answer.

Example
What force is needed to accelerate a 2-kg mass at 3 m/s²?
Solution
$f = ma$
$f = 2$ kg $\times$ 3 m/s²
By dimensional analysis,
$f = $ kg $\times$ m/s² $= $ kg·m/s²
Thus,
$f = 6$ kg·m/s²

The Right Triangle
General Relationships

See Section 4-17.

Special Right Triangles

1. The 30°-60°-90° triangle
   The side opposite the 30° angle is one-half the hypotenuse. The side opposite the 60° angle is $\sqrt{3}/2$ times the hypotenuse.

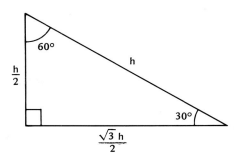

2. The 45°-45°-90° triangle
   Each side is the hypotenuse divided by $\sqrt{2}$.

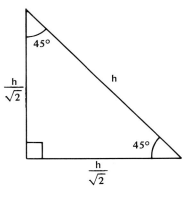

The General Triangle

See Section 4-17.

The Circle

radius $= r$ diameter $= d = 2r$ $\pi = 3.14$

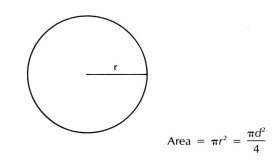

$$\text{Area} = \pi r^2 = \frac{\pi d^2}{4}$$

$$\text{Circumference} = 2\pi r = \pi d$$

702

| The Elements |
| The elements, as found in nature, are generally mixtures of isotopes. Their atomic mass is the average value of the atomic masses of their isotopes based upon carbon-12 = 12 u as a standard. Those in parentheses are the mass numbers of the best known or the most stable isotopes of the elements. |

NAME	SYMBOL	AT. NO.	AT. MASS	NAME	SYMBOL	AT. NO.	AT. MASS
Actinium	Ac	89	(227)	Mercury	Hg	80	200.60
Aluminum	Al	13	26.98	Molybdenum	Mo	42	95.95
Americium	Am	95	(243)	Neodymium	Nd	60	144.26
Antimony	Sb	51	121.75	Neon	Ne	10	20.182
Argon	Ar	18	39.942	Neptunium	Np	93	(237)
Arsenic	As	33	74.91	Nickel	Ni	28	58.71
Astatine	At	85	(210)	Niobium	Nb	41	92.91
Barium	Ba	56	137.35	Nitrogen	N	7	14.007
Berkelium	Bk	97	(249)	Nobelium	No	102	(254)
Beryllium	Be	4	9.013	Osmium	Os	76	190.2
Bismuth	Bi	83	208.99	Oxygen	O	8	15.999
Boron	B	5	10.82	Palladium	Pd	46	106.4
Bromine	Br	35	79.913	Phosphorus	P	15	30.973
Cadmium	Cd	48	112.40	Platinum	Pt	78	195.08
Calcium	Ca	20	40.08	Plutonium	Pu	94	(242)
Californium	Cf	98	(251)	Polonium	Po	84	(210)
Carbon	C	6	12.010	Potassium	K	19	39.098
Cerium	Ce	58	140.12	Praseodymium	Pr	59	140.91
Cesium	Cs	55	132.90	Promethium	Pm	61	(147)
Chlorine	Cl	17	35.455	Protactinium	Pa	91	(231)
Chromium	Cr	24	52.01	Radium	Ra	88	(226)
Cobalt	Co	27	58.94	Radon	Rn	86	(222)
Copper	Cu	29	63.54	Rhenium	Re	75	186.21
Curium	Cm	96	(247)	Rhodium	Rh	45	102.90
Dysprosium	Dy	66	162.50	Rubidium	Rb	37	85.48
Einsteinium	Es	99	(254)	Ruthenium	Ru	44	101.1
Erbium	Er	68	167.26	Samarium	Sm	62	150.34
Europium	Eu	63	152.0	Scandium	Sc	21	44.96
Fermium	Fm	100	(253)	Selenium	Se	34	78.96
Fluorine	F	9	19.00	Silicon	Si	14	28.09
Francium	Fr	87	(223)	Silver	Ag	47	107.875
Gadolinium	Gd	64	157.25	Sodium	Na	11	22.990
Gallium	Ga	31	69.72	Strontium	Sr	38	87.63
Germanium	Ge	32	72.60	Sulfur	S	16	32.064
Gold	Au	79	197.0	Tantalum	Ta	73	180.94
Hafnium	Hf	72	178.49	Technetium	Tc	43	(99)
Helium	He	2	4.003	Tellurium	Te	52	127.60
Holmium	Ho	67	164.93	Terbium	Tb	65	158.92
Hydrogen	H	1	1.0079	Thallium	Tl	81	204.38
Indium	In	49	114.81	Thorium	Th	90	232.04
Iodine	I	53	126.90	Thulium	Tm	69	168.93
Iridium	Ir	77	192.2	Tin	Sn	50	118.69
Iron	Fe	26	55.85	Titanium	Ti	22	47.90
Krypton	Kr	36	83.80	Tungsten	W	74	183.85
Lanthanum	La	57	138.91	Uranium	U	92	238.06
Lawrencium	Lr	103	(257)	Vanadium	V	23	50.95
Lead	Pb	82	207.20	Xenon	Xe	54	131.29
Lithium	Li	3	6.940	Ytterbium	Yb	70	173.03
Lutetium	Lu	71	174.98	Yttrium	Y	39	88.92
Magnesium	Mg	12	24.32	Zinc	Zn	30	65.38
Manganese	Mn	25	54.94	Zirconium	Zr	40	91.22
Mendelevium	Md	101	(256)				

Values of Sines, Cosines, and Tangents

ANGLE	SINE	COSINE	TANGENT	ANGLE	SINE	COSINE	TANGENT
0°	0.000	1.000	0.000	45°	0.707	0.707	1.000
1	0.017	1.000	0.017	46	0.719	0.695	1.036
2	0.035	0.999	0.035	47	0.731	0.682	1.072
3	0.052	0.999	0.052	48	0.743	0.669	1.111
4	0.070	0.998	0.070	49	0.755	0.656	1.150
5	0.087	0.996	0.087	50	0.766	0.643	1.192
6	0.105	0.995	0.105	51	0.777	0.629	1.235
7	0.122	0.993	0.123	52	0.788	0.616	1.280
8	0.139	0.990	0.141	53	0.799	0.602	1.327
9	0.156	0.988	0.158	54	0.809	0.588	1.376
10	0.174	0.985	0.176	55	0.819	0.574	1.428
11	0.191	0.982	0.194	56	0.829	0.559	1.483
12	0.208	0.978	0.213	57	0.839	0.545	1.540
13	0.225	0.974	0.231	58	0.848	0.530	1.600
14	0.242	0.970	0.249	59	0.857	0.515	1.664
15	0.259	0.966	0.268	60	0.866	0.500	1.732
16	0.276	0.961	0.287	61	0.875	0.485	1.804
17	0.292	0.956	0.306	62	0.883	0.469	1.881
18	0.309	0.951	0.325	63	0.891	0.454	1.963
19	0.326	0.946	0.344	64	0.899	0.438	2.050
20	0.342	0.940	0.364	65	0.906	0.423	2.145
21	0.358	0.934	0.384	66	0.914	0.407	2.246
22	0.375	0.927	0.404	67	0.921	0.391	2.356
23	0.391	0.921	0.424	68	0.927	0.375	2.475
24	0.407	0.914	0.445	69	0.934	0.358	2.605
25	0.423	0.906	0.466	70	0.940	0.342	2.747
26	0.438	0.899	0.488	71	0.946	0.326	2.904
27	0.454	0.891	0.510	72	0.951	0.309	3.078
28	0.469	0.883	0.532	73	0.956	0.292	3.271
29	0.485	0.875	0.554	74	0.961	0.276	3.487
30	0.500	0.866	0.577	75	0.966	0.259	3.732
31	0.515	0.857	0.601	76	0.970	0.242	4.011
32	0.530	0.848	0.625	77	0.974	0.225	4.331
33	0.545	0.839	0.649	78	0.978	0.208	4.705
34	0.559	0.829	0.675	79	0.982	0.191	5.145
35	0.574	0.819	0.700	80	0.985	0.174	5.671
36	0.588	0.809	0.727	81	0.988	0.156	6.314
37	0.602	0.799	0.754	82	0.990	0.139	7.115
38	0.616	0.788	0.781	83	0.993	0.122	8.144
39	0.629	0.777	0.810	84	0.995	0.105	9.514
40	0.643	0.766	0.839	85	0.996	0.087	11.43
41	0.656	0.755	0.869	86	0.998	0.070	14.30
42	0.669	0.743	0.900	87	0.999	0.052	19.08
43	0.682	0.731	0.933	88	0.999	0.035	28.64
44	0.695	0.719	0.966	89	1.000	0.017	57.29
45	0.707	0.707	1.000	90	1.000	0.000	

Glossary

A

Absolute temperature. Also called Kelvin temperature; the number of degrees Celsius above absolute zero measured in kelvins (K).

Absolute zero. The lowest possible temperature: $-273.16°$ C or 0 K.

Absorption spectrum. A continuous spectrum, like that of white light, interrupted by dark lines or bands that are produced by the absorption of certain wavelengths by a substance through which the light or other radiation passes.

Acceleration. The rate at which velocity changes with time.

Acceleration of gravity (g). The downward acceleration of a falling body in a vacuum, resulting from the earth's gravitational pull on it.

Adhesion. Attraction between unlike molecules.

Adiabatic system. An insulated system in which heat can neither enter nor leave the system.

Alpha particles ($_2^4$He). The nuclei of helium atoms, consisting of 2 protons and 2 neutrons.

Alternating current (AC). A current that reverses its direction at a regular frequency.

Ammeter. A meter for measuring electric current.

Ampere. The unit for measuring electric current, equal to one coulomb per second.

Amplifier. An electronic device which changes a weak electrical signal into a strong one.

Amplitude. The maximum displacement of a vibrating particle or wave; the height of the crest of a transverse wave.

Angle of incidence. The angle between the incident ray and the normal, or perpendicular, to the surface on which it falls.

Angle of reflection. The angle between the reflected ray and the normal, or perpendicular, to the surface from which it is reflected.

Angle of refraction. The angle between the refracted ray and the normal, or perpendicular, to the surface between the two media at which the refraction takes place.

Angular momentum. For a particle moving in a circular orbit, the product of its momentum and the radius of its orbit.

Anode. The electrode connected to the positive terminal of a battery or other source of EMF.

Antielectron ($_{+1}^0$e). A positive electron or positron.

Antimatter. Matter composed of antiparticles.

Antineutrino. Antiparticle of the neutrino.

Antineutron ($_0^1\bar{n}$). Antiparticle of the neutron.

Antiparticle. The counterpart of a subnuclear particle of matter, whose main property is that it and the particle annihilate each other on coming together, liberating their energy as radiation.

Antiproton ($_1^1$H$^-$). Antiparticle of the proton, having the same mass but a negative charge.

Armature. Usually the rotating coil of an electric generator or motor on an iron core.

Atom. The smallest particle of an element that has all its chemical properties.

Atomic mass. The mass of an atom relative to that of the isotope of carbon $_6^{12}$C, arbitrarily given the value of exactly 12 u.

Atomic mass unit (u). A mass equal to 1/12 that of an atom of $_6^{12}$C, or 1.66×10^{-27} kg.

Atomic number (Z). The number of protons in the nucleus of an atom; also the number of extranuclear electrons.

Avogadro's number. The number of molecules in one mole of any substance, equal to 6.02×10^{23} molecules.

Avogadro's principle. Equal volumes of all gases at the same temperature and pressure contain the same number of molecules.

B

Back EMF. The self-induced electromotive force in the rotating armature of a motor that opposes the EMF applied to the motor.

Balmer series. A series of related lines in the visible part of the spectrum of hydrogen. It is produced by the electrons in excited hydrogen atoms which pass from higher energy levels to the one whose quantum number $n = 2$.

Barometer. A device for measuring atmospheric pressure.

Baryons. Nuclear particles with masses equal to or greater than the mass of the proton.

Beats. A series of alternate reinforcements and cancellations produced by two sets of superimposed sound waves of close but different frequencies heard as a throbbing effect.

Beta rays. Streams of fast-moving electrons ejected from radioactive nuclei.

Betatron. A machine that accelerates electrons.

Binding energy of a nuclear particle. The energy required to free one of the component particles from an atomic nucleus.

Binding energy of a nucleus. The energy required to separate a nucleus into its component neutrons and protons.

Binding energy of a satellite. The energy required to overcome the gravitational attraction of the earth and enable a satellite to escape into space.

Boiling point. The temperature at which a liquid boils freely, typically under normal atmospheric pressure.

Boyle's law. At constant temperature, the volume of a gas is inversely proportional to its pressure.

Brownian movement. The zigzag movement of tiny particles suspended in a gas or liquid resulting from molecular bombardment.

Bubble chamber. A device which makes the paths of ionizing particles visible as trails of tiny bubbles in a superheated liquid.

C

Calorie (cal). The quantity of heat needed to raise the temperature of one gram of water one degree Celsius.

Candela (cd). The *SI* unit for measuring the rate at which a luminous source emits light.

Capacitor. A device consisting of conductors separated by an insulator that stores electric charges.

Capacity. (1) The ability of a body to store electric charges. (2) The volume which a container can hold.

Cathode. The electrode connected to the negative terminal of a battery or other source of EMF.

Cathode rays. Electrons emitted by a cathode.

Celsius scale. Temperature scale having 100 subdivisions between the melting point of ice, fixed at 0°, and the boiling point of water, fixed at 100°.

Centripetal acceleration. The acceleration of a body moving in a circular path that acts radially toward the center of the path.

Centripetal force. The force that must be applied to a body to give it centripetal acceleration and thus to make it move in a circular path. It acts toward the center of the path.

Chain reaction. A self-sustaining reaction which, once started, steadily provides the energy and matter necessary to continue the reaction.

Charles' law. At constant pressure, the volume of a gas is directly proportional to its Kelvin temperature.

Chromatic aberration. The inability of a single lens to refract all the different colors of light to the same focus.

Circuit breaker. A device (essentially an electromagnet) used in place of fuses to protect electric circuits from overloading.

Cloud chamber. A device in which the paths of ionizing particles are revealed as vapor trails.

Coefficient of friction. The proportionality constant for the frictional forces between two surfaces.

Cohesion. The force of attraction between like molecules.

Color. The appearance of visible light to the eye; it depends upon the frequency of the light.

Commutator. A split copper ring whose insulated segments are mounted on the axle of a direct current generator or motor and connected to the ends of the armature coil.

Components of a vector. Two or more vectors whose combined effect equals that of a given vector; also, the parts of a vector that act in directions other than that of the given vector.

Composition of vectors. Process of combining two or more vectors to obtain their resultant.

Compton effect. The increase in wavelength that takes place in a high energy photon such as an X ray or gamma photon on colliding with a subatomic particle such as an electron.

Concave mirror. A curved mirror which causes parallel rays to converge on reflection.

Condensation. (1) Changing of a gas or vapor into a liquid. (2) A compression of a sound wave.

Conduction band. An unfilled band of energy levels of a conductor, semiconductor, or insulator in which electrons are free to move.

Conduction of heat. The transfer and distribution of heat energy from molecule to molecule.

Conductor. A substance such as a metal through which electricity or heat flows readily.

Conservation. The property of indestructibility possessed by certain physical quantities, such as electric charge, momentum, and mass-energy.

Constructive interference. The superposition of two waves approximately in phase so that their amplitudes add up to produce a combined wave of larger amplitude than its components.

Continuous spectrum. A spectrum of a light source or other source of radiation in which all of the wavelengths are present within a wide range; it is produced by incandescent solids, liquids, and highly compressed gases.

Convection. Transfer of heat in a gas or liquid by currents of the heated fluid.

Converging lens. A convex lens which refracts parallel rays passing through it to a focus.

Convex mirror. A curved mirror which causes parallel rays to diverge on reflection.

Cosmic rays. Very highly penetrating radiation from outer space, consisting of charged particles, principally protons and alpha particles.

Coulomb. The unit of electric charge equal to one ampere-second or the charge of 6.25×10^{18} electrons.

Coulomb's law. The force between two point charges varies directly as the product of the charges and inversely as the square of the distance between them.

Covalent bonding. Bonding of atoms resulting from the sharing of pairs of electrons.

Critical angle. The angle of incidence for which a ray passing obliquely from an optically more dense to an optically less dense medium has an angle of refraction of 90° and hence does not pass through the interface.

Critical mass. The minimum quantity of radioactive material in a reactor or in a nuclear bomb that will sustain a chain reaction.

Current. A flow of a liquid or a gas or of particles such as electric charges.

Cycle. One complete vibration of any mechanical oscillation or periodic change.

Cyclotron. A particle accelerator that uses magnetic and electric forces to impart very high velocities to charged subatomic particles.

D

De Broglie waves. The probability waves associated with bodies of matter.

Density. Mass per unit volume of a substance.

Destructive interference. The superposition of two waves approximately in opposite phase so that their combined amplitude is the difference between their amplitudes and smaller than either.

Deuterium ($_1^2H$). The hydrogen isotope whose mass number is 2.

Deuteron ($_1^2H^+$). The nucleus of a deuterium atom consisting of one proton and one neutron.

Diamagnetism. A property of a substance that causes it to be repelled by a magnet.

Diffraction. The spreading of waves around an obstacle and into the region behind it.

Diffraction grating (transmission type). An optically transparent surface on which are ruled thousands of equidistant opaque parallel lines; it uses the diffraction effects of the slits between these lines to separate light passing through it into its spectrum.

Diffuse reflection. The scattering of light rays by the reflection of light from a rough surface.

Diffusion. The continuous random migration of molecular particles of one substance through another resulting from molecular motion.

Diode. A two-element device that conducts electric current more easily in one direction than in the opposite direction.

Direct current (DC). The movement of electrons in one direction around a circuit.

Direct proportion. Relationship between two quantities whose ratio is a constant.

Dispersion of light. The separation of light from a source into its spectrum.

Displacement. A change in a body's position.

Diverging lens. A concave lens which causes parallel rays passing through it to diverge.

Domain. A tiny section of a ferromagnetic substance, such as iron, in which the atoms are lined up with their north poles facing one direction. The entire section acts like a single tiny magnet with its own north and south poles.

Doppler effect. The change in frequency or pitch of sound waves, heard when the source of sound and the observer are moving toward or away from each other. A similar change is observed in the frequency or color of light, when the source of light and the observer are moving toward or away from each other.

E

Echo. A reflected sound wave which is heard separately from the original sound.

Eclipse. The cutting off of the light from a heavenly body so that it is hidden from view.

Effective alternating current. 0.707 times the maximum value of an alternating current.

Effective alternating current voltage. 0.707 times the maximum value of an alternating current voltage.

Efficiency. The percent of the work put into a machine that is converted into useful work output.

Effort. The force applied to a machine.

Elastic collision. A collision in which both the total momentum and the total kinetic energy of the colliding bodies have the same values before and after the collision.

Elastic limit. The largest stress that can be applied to a body without permanently deforming it.

Elastic potential energy. Work that is stored in a spring as potential energy and is related to the spring's elasticity.

Elasticity. The ability of a material to regain its

original size and shape after a deforming force is removed.

Electric current. The rate of flow of electric charges, such as electrons in a metallic conductor, or ions in a liquid or gas, through a circuit.

Electric field. The space around an electrically charged body which exerts an electric force on a charge placed within it.

Electric field intensity at a point. The force exerted by an electric field on a unit charge at that point.

Electrification. The process of giving a body an electric charge.

Electrode. A positively or negatively charged terminal of a device, such as an electrolytic cell, a gas discharge tube, or a vacuum tube.

Electrolysis. The separation of positive and negative ions in an electrolyte or other substance containing free ions by an electric current.

Electrolyte. A solution or a molten substance containing free positive and negative ions by means of which an electric current can pass.

Electrolytic cell. A device in which the positive and negative ions of an electrolyte are separated by passing an electric current between two electrodes immersed in the electrolyte.

Electromagnet. A coil of wire wound around a soft iron core, whose magnetic field is produced by passing an electric current through the coil.

Electromagnetic induction. The process of producing an electromotive force in a conducting circuit by changing the magnetic flux passing through the circuit.

Electromagnetic interactions. Forces of attraction and repulsion between charged particles.

Electromagnetic spectrum. The entire family of electromagnetic radiations ranging from short-wavelength, high-energy gamma rays to long-wavelength, low-energy radio waves.

Electromagnetic waves. Transverse waves moving at the speed of light and consisting of rapidly alternating electric and magnetic fields at right angles to each other and to the direction in which the waves are traveling.

Electromotive force (EMF). The energy given to each unit of charge in a circuit by a source of electrical energy.

Electron ($_{-1}^{0}e$). A particle of negative electricity, having a charge of 1.6×10^{-19} coulomb and a mass of 9.1×10^{-31} kilogram.

Electron capture. A change in an atom's nucleus resulting from a strong attraction between the positively charged nucleus and an electron in the innermost shell. The resultant nuclide has different chemical properties and therefore a new name. Also called *K*-capture.

Electron gun. A device for providing a controllable beam of electrons.

Electron shell. A region around the nucleus of an atom in which electrons having the same principal quantum number may move.

Electronvolt. A unit of energy equal to the work done in moving an electron through a difference of potential of one volt.

Electroscope. A sensitive instrument used to detect and identify small electric charges.

Electrostatic induction. The process of producing an electric charge on a neutral body by bringing it into the electric field of a charged body.

Element. A substance composed of atoms that all have the same atomic number and therefore the same chemical properties.

Emission spectrum. An array of colors made by the dispersion of light from a luminous source.

Energy. The ability to do work.

Energy level. One of a series of regions about the nucleus of an atom in which an electron may move; each region is associated with a specific energy value.

Entropy. A quantitative measure of the disorder of a system.

Equilibrant. The force equal and opposite to a resultant force.

Equilibrium. A condition in which all the forces in a system neutralize each other.

Escape velocity. The velocity a body must have to escape from the earth's gravitational pull.

Evaporation. The changing of a liquid into a gas.

Excited state. Any energy state of an atom above its ground state in which it can emit energy as light or other radiation.

Exclusion principle of Pauli. No two electrons in an atom can have the same four quantum numbers.

F

Farad (F). The *SI* unit of capacity equal to one coulomb per volt.

Ferromagnetism. The ability of iron, nickel, and cobalt to be strongly attracted by magnets.

Fission. The splitting of the nucleus of a heavy atom, such as Uranium 235, into two main parts, accompanied by the release of much energy.

Fixed points. Temperatures such as those of melting ice and boiling water, used as standards in calibrating thermometers.

Fluid. A substance having no definite shape and being able to flow; a liquid or gas.

Fluid friction. The opposing force encountered when an object is moved through a fluid (liquid or gas).

Fluorescence. The process whereby a substance emits radiation (usually as visible light) when struck by charged particles, such as electrons or alpha particles, or when radiation of a higher frequency (usually ultraviolet light) falls on it.

Focal length. The distance between the principal focus of a lens or curved mirror and its center.

Focus. A point to which light rays converge or from which they diverge.

Force. A push or pull that changes the motion of a body unless counteracted by an equal and opposite push or pull, measured in newtons.

Frame of reference. A set of axes used to describe the position and motion of a body.

Fraunhofer lines. The dark lines in the sun's spectrum that result from the absorption of certain wavelengths of sunlight by the vapors and gases in the sun's atmosphere.

Freezing point. The temperature at which a liquid solidifies at normal atmospheric pressure.

Frequency. The number of vibrations or cycles per second, measured in hertz.

Friction. The force that opposes the motion of a body over or through another.

Fuse (electrical). A device made from metal wire or ribbon that has a low melting point and is used to protect electric circuits from overloading.

Fusion (or melting). The changing of a solid into a liquid.

Fusion (nuclear). The process whereby the nuclei of several light atoms combine to form a heavier nucleus with the release of energy.

G

Galvanometer. An instrument used to detect and measure very small electric currents.

Gamma rays (γ). Highly penetrating electromagnetic radiations of very short wavelengths emitted by the nuclei of radioactive atoms.

Gas. The diffuse physical state of a substance in which it has no definite volume or shape.

Geiger counter. An instrument that detects radiations from radioactive substances by their ability to ionize the matter through which they pass.

General gas law. The relationship $PV = RT$, where

R is the universal gas constant, P is the pressure, V is the volume, and T is the Kelvin temperature of one mole of an ideal gas.

Generator. A device that converts mechanical energy into electrical energy.

Geosynchronous orbit. The opposite rotational orbit of a satellite (west to east) from the earth (east to west), resulting in the satellite's staying in the same position relative to the earth.

Gluon. The subatomic particle that carries the force that attracts quarks to each other.

Gram. A small unit of mass, equal to 1/1000 of the standard kilogram.

Gravitational force. The force of attraction that every mass exerts upon every other mass.

Gravitational mass. The mass determined by measuring an object on an equal arm balance.

Graviton. The particle assumed to be the carrier of the gravitational force.

Ground state. The condition of an atom in which its electrons are at the lowest possible energy levels.

H

Half-life. The time it takes for half of the atoms in a sample of a radioactive element to disintegrate.

Heat. The energy which a body at higher temperature transfers to a body at lower temperature at the expense of its internal energy.

Heat engine. A device for converting heat into work. Typical examples include steam, automobile, and jet engines and rockets.

Heat of fusion. The heat needed to melt a unit mass of a solid at its normal melting point.

Heat of vaporization. The heat needed to change a unit mass of a liquid to a gas at its normal boiling point.

Heat pump. A device that operates similarly to a refrigerator by reversing the direction of spontaneous heat flow.

Heat sink. A heat reservoir into which heat flows from a heat source.

Henry (H). The *SI* unit of inductance. A circuit has an inductance of 1 henry when a current through it changing at the rate of 1 ampere per second induces a back EMF of 1 volt in it.

Hole. A vacant space in a semiconductor into which an electron can move thus leaving a hole behind it. The movement of the hole in a direction opposite to that of the electron is equivalent to the movement of a positive charge.

Hooke's law. The strain produced in an elastic body is directly proportional to stress as long as the elastic limit is not exceeded.

Horsepower. A unit of power equal to 746 watts.

Hypothesis. A tentative solution to a problem to be tested by experiment or observation.

I

Ideal gas. An imaginary gas which conforms exactly to the universal gas law.

Ideal machine. A theoretical machine in which there are no frictional or other losses of the work put in so that the output equals the input.

Image. The likeness of an object made by a lens or mirror; it may be real or virtual.

Impulse. The product of the force acting on a body and the time during which it acts; it equals the change of momentum in the body.

Incident ray. The ray which falls upon a surface between two substances.

Inclined plane. A simple machine consisting essentially of a sloping surface.

Index of refraction of a substance. Ratio of the speed of light in air or a vacuum to its speed in the substance.

Induced magnetism. Magnetism produced in a magnetic substance when brought into the field of a magnet.

Induced radioactivity. Radioactivity resulting from the bombardment of a nonradioactive element with high-speed protons, neutrons, etc.

Inductance (L). The property of a circuit or part of a circuit whereby it sets up a back EMF opposing any change in the current flowing through it, measured in henrys.

Inertia. The property of matter by which it resists any change in its state of motion or rest.

Inertial mass. The ratio of the force applied to an object to the acceleration it produces.

Infrared light. Electromagnetic radiations whose wavelengths lie between those of visible light and radio waves; also called *heat rays*.

In opposite phase. The condition in which two waves or vibrating bodies are one-half cycle apart so that when one performs one-half of its vibration, the second performs the opposite half of its vibration, thus reducing their effects.

In phase. The condition in which two waves or vibrating bodies reinforce each other.

Input. The work put into a machine is equal to the effort times the distance it acts.

Insulator. A material that is a poor conductor of electricity or heat.

Integrated circuit. A compact circuit consisting of many tiny interconnected engraved elements (i.e., transistors, diodes, resistors, etc.) on a small silicon chip.

Interference. The mutual reinforcement in some places, and the cancellation in others, of the effects of two sets of superimposed waves.

Internal energy. The kinetic and potential energy associated with the molecules of a body.

Inverse proportion. The relationship between two variables whose product is constant.

Ion. An atom or group of atoms having an unbalanced electric charge.

Ionic bonding. Attraction between atoms due to electron transfer from one to another.

Ionization energy. The energy required to detach one of its electrons from an atom and thus turn the atom into an ion.

Isotopes. Different forms of an element whose atomic nuclei have the same number of protons but different numbers of neutrons.

J

Joule (J). The *SI* unit of energy equal to the work done by a force of one newton acting over a distance of one meter.

K

K-capture. See *Electron capture.*

Kelvin scale. The scale of absolute temperature on which each unit is a kelvin.

Kilogram. The *MKS* and *SI* unit of mass.

Kinetic energy. Energy associated with the motion of a mass and equal to one-half the product of the mass and the square of its velocity.

Kinetic theory of gases. The theory that explains the behavior of an ideal gas as the result of the motions and collisions of its molecules.

Kirchhoff's laws. Two laws that describe current and voltage relationships in circuits consisting of networks of resistances.

L

Laser. A beam of light of a single frequency in which all the waves are in phase; also known as coherent light.

Law of conservation of electric charge. In all exchanges of charges between bodies of matter the net charge remains the same.

Law of conservation of energy and mass. The total

quantity of energy and mass in the universe remains constant.

Law of conservation of momentum. The vector sum of the momenta of all the bodies in a system on which no outside forces are acting remains constant in direction and magnitude.

Law of heat exchange. In any exchange of heat in a closed system, all the heat lost by the warmer bodies is gained by the colder ones.

Law of universal gravitation. Every particle of matter in the universe attracts every other particle with a force that is directly proportional to the product of their masses and inversely proportional to the square of the distance between them.

Law of work. In an ideal machine, the work output is equal to the work input.

Laws of motion. See *Newton's laws of motion*.

Lenz's law. An induced current flows in such a direction as to oppose by its magnetic field the motion by which the current was induced.

Leptons. Subatomic particles of little or no mass including electrons, muons, and neutrinos.

Light. The visible part of the electromagnetic spectrum.

Line of reinforcement. A line consisting of points at each of which two sets of waves from two different sources continue to arrive in phase.

Line spectrum. A spectrum emitted by an incandescent gas under low pressure and consisting of a series of colored lines separated by dark spaces.

Lines of force. Lines drawn to map electric or magnetic fields to show the direction and the intensity of the field from point to point.

Liter. A volume of $1000 \ cm^3$ or $1/1000 \ m^3$.

Longitudinal waves. Waves in which the vibrations are parallel to the direction in which the waves are traveling.

Lumen. The *SI* unit for measuring the intensity of illumination falling on a surface.

Luminous body. A source of light.

M

Machine. A device that can multiply the force applied to it at the expense of distance or can multiply distance at the expense of the applied force.

Magnetic field. The space, usually around a magnet or an electric current, at each point of which a magnetic force is exerted.

Magnetic induction. Strength of the magnetic field expressed in newtons per ampere-meter or webers per square meter.

Magnetization by induction. Magnetizing a magnetic substance by bringing it into a magnetic field.

Mass. The property of matter that resists any change in a body's motion.

Mass defect. The difference between the mass of a nucleus and the sum of the masses of the neutrons and protons it contains. It represents the binding energy holding the nucleus together.

Mass-energy conversion. The changing of mass into energy by the relationship $E = mc^2$, where m is the mass and c is the speed of light.

Mass number. The number of neutrons and protons in the nucleus of an atom; also the whole number nearest the atomic mass of an element.

Mass spectrograph. An electromagnetic instrument used for measuring the masses of atoms.

Matter. Bodies having weight and volume.

Matter waves. De Broglie waves associated with particles, which are responsible for certain wavelike behaviors of those particles.

Mechanical advantage (ideal). The mechanical advantage a machine would have in the absence of friction.

Mechanical advantage (real). The ratio of the resistance overcome by a machine to the effort applied.

Mechanical equivalent of heat. The quantity of mechanical energy equal to 1 unit of heat (1 cal = 4.2 J or 1 J = 0.24 cal).

Melting point. The temperature at which a solid changes to a liquid at normal pressure.

Meson. A particle with a mass between that of the electron and that of the proton. A meson may be positively charged, negatively charged, or neutral.

Meter. The *MKS* and *SI* unit of length.

MKS system. The metric system of measurement in which the fundamental units are the meter, kilogram, and second.

Mole. That quantity of a substance whose mass in grams is numerically equal to the mass of one of its molecules in atomic mass units.

Molecule. The smallest particle of a substance that has all its chemical and physical properties.

Momentum. The product of the mass of a body and its velocity.

Monochromatic light. Light of a single color or narrow range of frequencies.

N

Negative acceleration. Acceleration that acts to slow down a moving body.

Neutrino. A neutral particle having a mass number of zero.

Neutron ($_0^1$n). A neutral particle having about the same mass as a proton and present in all atomic nuclei other than ordinary hydrogen.

Newton (N). The *MKS* and *SI* unit of force; a force that accelerates a mass of 1 kilogram at 1 meter per second per second.

Newton's first law of motion. The law of inertia, which states that a body continues its state of rest or motion at constant velocity, unless acted upon by an unbalanced force.

Newton's second law of motion. The law of acceleration, which states that the acceleration produced by a force acting upon a body is proportional to the force and inversely proportional to the mass.

Newton's third law of motion. The law of action and reaction, which states that to every action there is an equal and opposite reaction.

Nodal line. A line at every point of which the waves from two separate sources continue to arrive in opposite phase.

Normal atmospheric pressure. The pressure exerted by a column of mercury 760 millimeters high, equal to 101.3 kilopascals.

N-type semiconductor. A semiconductor doped so that it acquires free electrons.

Nuclear energy. Energy released during nuclear fission or fusion. *See* fission and fusion.

Nuclear force. The force of attraction between two nucleons. *See* strong nuclear interaction.

Nuclear reactor. A device in which a controlled chain reaction involving nuclear fission generates energy and produces new radioactive elements.

Nucleon. A proton or neutron contained in an atomic nucleus.

Nuclide. An atomic nucleus with a specific atomic number and a specific mass number.

O

Ohm (Ω). The *SI* unit of electrical resistance through which it takes 1 volt to produce a current of 1 ampere.

Ohm's law. At constant temperature, the ratio of the difference of potential across a metallic resistor to the current flowing is constant.

Order of magnitude. The power of ten that is nearest to a given number.

Oscillating circuit. A circuit consisting essentially of an inductance and a capacitor used to generate high-frequency alternating currents.

Output. The work obtained from a machine equal to the resistance times the distance it moves.

P

Parallel connection. Connecting the ends of two or more devices to the same two points of a circuit so as to provide two or more paths for the current to flow between them.

Paramagnetism. The property of a substance that causes it to be weakly attracted by a magnet.

Particle accelerator. An apparatus used to impart very high velocities and energies to charged subatomic particles, such as protons and deuterons, used as projectiles to penetrate atomic nuclei.

Pascal (Pa). The *SI* unit of pressure equal to a force of 1 newton acting on 1 square meter of area.

Pendulum. A mass suspended from a point so that it swings freely by the influence of gravity.

Penumbra. The part of a shadow that is partially illuminated by a light source.

Period. The time required for one vibration, one cycle, or one oscillation; also, the time taken by one complete wave to pass a given point.

Periodic motion. Vibratory, oscillatory, or any other motion which a body repeats at regular time intervals.

Permeability. The ability of a substance to concentrate the lines of force when placed in a magnetic field.

Photoelectric effect. The ejection of electrons from a substance by the incidence of light or higher frequency electromagnetic radiation upon it.

Photoelectron. An electron ejected from a substance when light or other electromagnetic radiation falls upon it.

Photon. The basic quantity in which the energy of electromagnetic radiation of a given frequency is packaged. One photon of radiation of frequency f has a quantity of energy equal to hf, where h is Planck's constant.

Planck's constant. A universal constant h that relates the energy of a photon to the frequency of the radiation from which it originates. $h = 6.63 \times 10^{-34}$ joule-second.

Plasma. An ionized gas containing free electrons

and positive ions that conduct current.

p-n diode. A rectifier consisting of a *p*-type semiconductor and an *n*-type semiconductor bonded together.

Polarization of light. The process of removing from a beam of light the (electromagnetic) vibrations in all directions except one.

Polychromatic light. A mixture of many colors.

Positron. The antiparticle of the electron. It has the same mass as the electron but a positive instead of a negative charge.

Potential difference between two points. The work done in moving a unit charge from one point to another against an opposing electric field.

Potential energy. The stored energy that a body has because of its position with respect to other bodies, its condition, or state of strain.

Power. The rate of doing work.

Pressure. The force per unit of area.

Primary. The coil of a transformer to which the voltage to be stepped up or down is applied.

Principal axis. The line joining the center of curvature of a mirror or of either curved surface of a lens and the center of the mirror or lens.

Principal focus. The point to which incident rays parallel to the principal axis of a lens or mirror converge, or from which they diverge.

Proton ($^1_1H^+$). A positively charged particle having a charge equal to that of the electron and a mass number of 1.

p-type semiconductor. A semiconductor doped so that it acquires many holes whose movement is equivalent to that of positive charges.

Pulse. A single disturbance transferred in wavelike fashion.

Q

Quantum. The unit of energy associated with each given frequency of radiation; a photon.

Quantum numbers. Numbers associated with each of the possible energy levels of an atom. They include the principal quantum number, n, the angular momentum quantum number, l, the magnetic quantum number, m_l, and the spin quantum number, s.

Quantum theory. A theory which assumes that radiant energy is emitted and absorbed by matter in discrete minimum packets equal to Planck's constant times the frequency of the radiation.

Quark. The basic particle out of which mesons and baryons are made.

R

Radar. A method of measuring distance from an observer to an object by means of radio echoes.

Radiant energy. Energy transferred by electromagnetic waves.

Radiation. The transfer of energy by electromagnetic waves.

Radio waves. Electromagnetic radiations having longer wavelengths than infrared waves.

Radioactivity. The spontaneous disintegration of an atomic nucleus accompanied by emission of alpha or beta particles or gamma rays.

Radioisotope. An isotope that is radioactive.

Ray. A straight line of light.

Real image. An inverted image formed by the actual convergence of light rays upon a screen.

Rectifier. A device that changes alternating to pulsating direct current.

Reflection. The rebounding of waves from the surface of a new medium or barrier.

Refraction. The bending of a ray as it passes obliquely from one medium to another.

Regular reflection. Reflection of light from a smooth flat surface so that the parallel rays are not scattered.

Resistance. Opposition of a circuit or part of a circuit to the flow of current, measured in ohms.

Resolving power of a lens. A measure of the ability of a lens to produce clearly distinguishable images of two very close points.

Resolution. The process of dividing a force into two or more components.

Resonance. The setting up of strong vibrations in a body at its natural vibration frequency by a vibrating force or wave having the same frequency; sympathetic vibrations.

Rest mass. The mass that a body has when it is at rest.

Resultant. The single vector whose effect is the same as the combined effects of two or more similar vectors.

Rolling friction. The opposing force encountered when an object is rolled, rather than slid, over another.

S

Scalar. A quantity, such as length and mass, having magnitude but not direction.

Scientific notation. Expression of numbers in terms of powers of ten.

Scintillation. Tiny flash of light when an ionizing

radiation strikes a fluorescent screen.

Second (s). The *SI* unit of time.

Secondary. The coil of a transformer from which the output is obtained.

Self-induced EMF. The EMF induced in a circuit or part of a circuit as a result of a change in the current.

Semiconductor. A solid whose electrical resistance is between that of a good conductor and an insulator.

Series connection. Connecting electrical devices in a line so that they make a single path for the current to pass through them.

Shadow. The space behind an illuminated object from which it excludes light.

Short circuit. An electric circuit in which the resistance is so low as to permit a dangerously large current to flow.

Significant figures. The numbers in the given value of a physical quantity obtained by actual measurement, including the first estimated number.

Simple harmonic motion. A vibratory or periodic motion, as in a pendulum, in which the force acting on the vibrating body is proportional to its displacement from its central equilibrium position and always acts toward that position.

Sliding friction ($f_{f(\text{slide})}$). The frictional resistance of a body once steady sliding has been attained.

Snell's law. A light ray passing from one medium to another is refracted so that the ratio of the sine of the angle of incidence to the sine of the angle of refraction is equal to a constant for all angles of incidence.

Solidification. The change of a liquid to a solid at a specific temperature and a given pressure, usually normal atmospheric pressure.

Sound waves. Longitudinal waves in air and other material media set up by vibrating bodies.

Spark chamber. A device that detects charged subatomic particles by a trail of electric sparks they set off in it.

Specific gravity (of a solid or liquid). The ratio of the mass of the substance to the mass of an equal volume of water.

Specific heat. The heat required to raise the temperature of 1 gram of a substance 1 Celsius degree.

Spectrum of light. The array of colors in the order of wavelengths that results when light from a source is dispersed into its component colors.

Speed. Rate at which the distance traveled by a body increases with time; the magnitude of velocity.

Spherical aberration. The failure of mirrors and lenses with spherical surfaces to bring parallel light rays striking all parts of the mirror or lens to the same focus.

Standard pressure. Normal atmospheric pressure, equal to 760 millimeters of mercury, or 101.3 kilopascals.

Standard temperature. The temperature of melting ice, 0° C, at standard pressure.

Standing waves. Stationary wave patterns formed in a medium when two sets of waves of equal wavelength and amplitude pass through the medium in opposite directions.

Starting friction ($f_{f(\text{start})}$). The maximal frictional resistance of a body, which is reached just as the body begins to move along a surface.

Static electricity. Electric charges at rest.

Static friction (f_f). The frictional resistance that opposes the motion of a stationary object until the outside force on the object exceeds a certain threshold value.

Strain. The deformation produced by a stress.

Stress. A deforming force applied to a body.

Strong nuclear interaction. The force of attraction between two nucleons in which the meson serves as the carrier.

Superconductivity. The complete loss of electrical resistance by certain metals and alloys when cooled to near absolute zero.

Superposition. The process whereby two or more waves combine their effects when passing through the same parts of a medium at the same time.

Sympathetic vibration. The vibration of a body at its natural frequency set up by the vibration of another body with the same natural frequency.

Synchrotron. A particle accelerator in which the frequency of the accelerating voltage is kept in step with the frequency of revolution of the particles.

System. A space singled out for study that has well-defined boundaries.

T

Temperature. The measure in degrees of how hot a body is with respect to other bodies. It determines whether that body will gain or give up heat when put into contact with other bodies.

Terminal velocity. The constant speed reached by a free-falling body that has been dropped from a sufficient altitude for the fluid frictional forces to become equal to the object's weight.

Theory. An imagined conceptual scheme, mechanism, or model that provides a plausible explanation of a series of experimental observations.

Thermal energy. Internal energy.

Thermodynamics. The study of physical processes that involve heat, work, and internal energy. The First Law (law of conservation of energy) states that whenever heat is transformed into work or another form of energy, or *vice versa*, there is no loss of energy. The Second Law states that heat spontaneously travels in only one direction, from a warmer to a cooler body. The Third Law states that it is impossible to bring a body's temperature down to absolute zero.

Thermonuclear reaction. A nuclear reaction that can take place only at very high temperatures involving the fusion of light nuclei into a heavier one with the consequent release of energy.

Thought experiment. An imagined experiment assuming ideal conditions that cannot be achieved experimentally.

Threshold frequency. The minimum frequency of incident light that will eject a photoelectron from a given metal.

Total internal reflection. The total reflection of a beam of light traveling in an optically dense medium when it falls upon the surface of a less dense medium at an angle greater than the critical angle.

Tracer. A radioactive element used to observe the path and distribution of the atoms of that element in a chemical, biological, or physical process in which it takes part.

Transformer. A device for changing the voltage of an alternating current to a desired value.

Transistor. A combination of n-type and p-type semiconductors that can be used to amplify currents and voltages.

Translucent substance. One through which light is transmitted diffusely so that bodies cannot be seen through it.

Transmutation. The conversion of an atomic nucleus of one element into that of another element through a loss or gain in protons.

Transparent substance. One through which light passes without diffusion so that bodies can be seen through it.

Transverse waves. Waves, such as light, in which the vibrations are perpendicular to the direction in which the waves are traveling.

Triple point. The temperature and pressure at which the solid, liquid, and gaseous states of a substance can exist at the same time and remain in equilibrium.

Tritium ($^3_1$H). The isotope of hydrogen having a mass number of 3.

U

Ultraviolet light. The range of invisible radiations in the electromagnetic spectrum between violet light and X rays.

Umbra. The part of the shadow of an object from which all light is excluded.

Uncertainty principle. It is not possible to measure exactly both the position and the momentum of a particle at the same time. The product of the uncertainty in position and the uncertainty in momentum is of the order of h.

Uniform motion. Displacement at constant velocity.

Universal gas constant. The constant R in the general gas law, $PV = RT$.

V

Valence electron. An electron in an outer incomplete shell of an atom that takes part in ionic and covalent bonding.

Van de Graaff generator. A device that builds up a high potential difference by accumulating electric charges on an insulated hollow metal sphere. It is used to accelerate electrically charged particles.

Vaporization. The change of state from a liquid to a gas.

Vector. A quantity having both magnitude and direction, such as force and velocity.

Velocity. A vector whose magnitude is the speed of an object and whose direction is the direction of its motion.

Virtual image. An illusionary upright image that is seen by an observer in a mirror or through a lens but cannot be projected on a screen.

Volt (V). The *SI* unit of potential difference equal to 1 joule per coulomb.

Voltaic cell. A chemical cell consisting of two different metals immersed in an electrolyte which provides a steady EMF. It converts chemical into electrical energy.

Voltmeter. An instrument for measuring potential difference.

Volume. The quantity of space a body occupies.

W

Watt (W). The *SI* unit of power equal to 1 joule per second.

Wavelength. The distance from a point of a wave to the corresponding point of the succeeding wave.

Weak force. Force involved in the weak nuclear interaction; about 10^{-13} times the strong nuclear force.

Weak nuclear interactions. Nuclear reactions in which leptons are emitted.

Weber (Wb). The *SI* unit of magnetic flux.

Weight. The earth's gravitational pull on an object.

Work. The product of a force and the distance it acts in overcoming a resistance.

Work function. The minimum energy needed to eject a photoelectron from a given metal.

W-particle. Carrier of the weak nuclear interaction.

X

X rays. A range of deeply penetrating electromagnetic radiations whose wavelengths lie between those of ultraviolet light and gamma rays.

Answers to Test Yourself Problems

Page	Section	
20	**2-12**	**1.** 0.60 m³; **2.** 904 cm³
27	**2-20**	**(a)** 9×10^5; **(b)** 5.8×10^6; **(c)** 2.1×10^8; **(d)** 7×10^{-3}; **(e)** 3.8×10^{-7}
27	**2-21**	**(a)** 7.2×10^{14}; **(b)** 9.6×10^2; **(c)** 1.6×10^{-8} **(d)** 7.5×10^9
28	**2-22**	**(a)** 4.0×10^3; **(b)** 2.0×10^{-7}; **(c)** 5.0×10^9; **(d)** 2.0×10^2
29	**2-23**	**(a)** 10^2 m; **(b)** 10^{-3} m; **(c)** 10^9 km or 10^{12} m
54	**4-9**	**1.** 310 km/h 15° N of E; **2.** 324 N at 26° to the 150-N force
58	**4-15**	19 N 30° S of W
60	**4-17**	**1.** 310 km/h 15° N of E; **2.** 324 N at 26° to the 150-N force
66	**4-23**	**1. (a)** 47 N; **(b)** 17 N **2.** 7.07 N along each wire
76	**5-9**	**1.** 5.00 m/s²; **2.** 23 m/s
77	**5-10**	64 m
79	**5-11**	**1.** 314 m; **2.** 100 m
100	**6-9**	**1. (a)** 3.00 s; **(b)** 1.50×10^3 m **2. (a)** 10.0 s; **(b)** 20.0 s; **(c)** 8.00×10^3 m
102	**6-11**	6.3 m/s
102	**6-12**	5.00 m/s²
107	**6-18**	3.91×10^8 s
110	**6-22**	1.3 S
111	**6-24**	**1.** 1.40 s **2.** 0.062 m
122	**7-7**	**1.** 15 N; **2. (a)** 0.500 m/s²; **(b)** 750 N
123	**7-8**	**(a)** 490 N; **(b)** 4.90 m/s²
135	**7-21**	**1.** 1.6×10^3 N **2.** 0.57
139	**7-26**	**1.** 76.0 s; **2.** 2.40×10^3 N
140	**7-27**	**(a)** 2.0 kg·m/s; 0 kg·m/s; **(b)** 2.0 kg·m/s; **(c)** 1.0 m/s
148	**8-2**	**1.** 0.480 N; **2. (a)** 0.200 s; **(b)** 1.48×10^3 N
152	**8-9**	2.1×10^{21} N
155	**8-13**	**(a)** 6.22 m/s²; **(b)** 1.55 m/s²
156	**8-15**	**1.** 170 N; **2.** 9.00 m/s²
166	**9-1**	**1.** 750 J; **2.** 147 J
166	**9-2**	752 J
170	**9-7**	**1.** 450 N; **2.** 800 N; **3.** 1.50 m
172	**9-13**	**1. (a)** 640 J; **(b)** 800 J; **(c)** 160 J; **(d)** 80% **2. (a)** 2.500×10^6 J; **(b)** 4.166×10^6 J; **(c)** 1.666×10^6 J
174	**9-14**	**1. (a)** 150 J; **(b)** 100 W; **2.** 312 J; **3.** 2.4×10^3 W or 2.4 kW; **4.** 120 W
180	**10-5**	**1.** 6.7×10^5 J; **2. (a)** 8.0 J; **(b)** 2.0 m/s
181	**10-6**	**(a)** 3.7×10^5 J; **(b)** 7.4×10^3 N
183	**10-9**	**1.** 2.9×10^3 J; **2.** 15.00 m; **3. (a)** 4.9 J; **(b)** 4.9 J
185	**10-11**	**1.** 5.0 N; **2.** 6.3×10^{-2} J
188	**10-13**	20.4 m
188	**10-14**	3.7×10^{10} J
192	**10-20**	0.67 m/s; 2.7 m/s
195	**10-26**	4.0×10^{-11} kg
205	**11-3**	**1.** 1.7×10^2 kJ; **2.** 52 kJ; **3.** 1.5×10^2 kJ
222	**12-6**	**1.** 38.7 kJ; **2.** 22.1°C
225	**12-11**	**1.** 1.67×10^4 kJ; **2. (a)** 419 kJ; **(b)** 3.33×10^3 kJ
237	**12-30**	**1.** 4000 J; **2.** gain of 4500 J
241	**12-35**	**1.** 3.00 kJ/K; **2.** 9.00×10^{-3} J/K
265	**13-22**	**1.** 4.5×10^{23} molecules; **2.** 40 μ

Page	Section	
273	**14-5**	(a) 1.00×10^5 Pa; (b) 6.7×10^4 Pa; (c) 1.2×10^5 Pa
274	**14-7**	1. 0.15 m³; **2.** 40 kPa
276	**14-8**	1. $V_2 = 0.333$ m³; **2.** $T_2 = 819$ K
276	**14-9**	1. 30 kPa; **2.** 250 K
277	**14-10**	8.31×10^4 Pa
285	**14-23**	1. (a) 2.07×10^{-21} J; (b) 2.07×10^{-21} J; **2.** 278 m/s
297	**15-7**	1. 200 Hz; **2.** 5.0×10^{-7} m
313	**15-31**	432 Hz
326	**16-17**	1. 80 lm; **2.** 40 cd
345	**17-7**	(a) 12°; (b) 15°; (c) 18°
348	**17-10**	(a) 38°; (b) 52°; (c) 70°
353	**17-18**	(a) 66°; (b) 24°; (c) 55°
372	**18-17**	1. (a) 15 cm; (b) real, inverted, smaller; (c) 5.0 cm 2. (a) −40 cm; (b) virtual, erect, larger; (c) 50 cm 3. (a) −3 cm; (b) virtual, erect, smaller; (c) 5 cm
380	**18-27**	1. (a) 36 cm; (b) 20 cm; (c) real, larger, inverted 2. (a) −24 cm; (b) 30 cm; (c) virtual, erect, larger 3. (a) −10 cm; (b) 5.0 cm; (c) virtual, erect, smaller
397	**19-10**	1. 5.01×10^{-5} cm; **2.** 0.600 cm
401	**19-16**	1. 4.5×10^{14} Hz; **2.** 4.4×10^{-7} m
404	**19-20**	1. 6.0×10^{-5} cm; **2.** 0.30 cm
452	**22-4**	1. −100 N; attraction; **2.** -9.0×10^{-3} N
454	**22-7**	1. 5.0×10^4 N/C; **2.** 0.020 N
459	**22-16**	1. 2.0×10^3 V; **2.** 3.0×10^{-4} J
459	**22-17**	1. (a) 50 eV; (b) 8.0×10^{-18} J; **2.** (a) 5.0 V; (b) 1.6×10^{-18} J
461	**22-19**	1. 4.3×10^{-18} J; **2.** 9.0×10^{-4} J
463	**22-20**	(a) 2.5×10^{-3} V/m; (b) -4.0×10^{-16} N (opposite to E); (c) 4.0×10^{-16} N (in direction of E)
465	**22-21**	(a) 4.8×10^{-19} C; (b) 3
468	**22-24**	(a) 4.1×10^{-16} J; (b) 2.6×10^3 V
479	**23-10**	1. 180 W; **2.** (a) 2.0 A; (b) 1.2×10^4 J
481	**23-12**	1. 11 Ω; **2.** 6.0 A; **3.** 3 Ω
484	**23-16**	1. 1.8×10^3 J; **2.** 2.0×10^2 s
490	**24-2**	1. (a) 2.00 A; (b) 200 V across 100-Ω resistor; 40.0 V across 20.0-Ω resistor; (c) 80.0 W to 20.0-Ω resistor; 400 W to 100-Ω resistor; (d) 480 W; **2.** (a) 5.0 V across 10-Ω resistor; 7.5 V across 15-Ω resistor; (b) 12.5 V; **3.** 60 Ω
491	**24-3**	(a) 6.00 Ω; (b) 0.250 A; (c) 1.49 V
494	**24-5**	1. (a) 2.00 A, lamp; 10 A, heater; (b) 12 A; (c) 10 Ω; 2. (a) 0.833 A; (b) 144 Ω; (c) 14.4 Ω
495	**24-8**	(a) 50.0 Ω; (b) 125 Ω; (c) 0.800 A; (d) 60V; (e) 40 V; (f) 0.27 A
515	**25-17**	1. 10^{-7} N/A·m or Wb/m²; **2.** 10^{-3} A
526	**26-4**	1. 1.2×10^{-6} N; **2.** 2.0×10^{-3} Wb/m²; **3.** 0.20 A
528	**26-6**	1. 5.0×10^{-4} N; **2.** (a) 1.0×10^{-3} N; (b) 2.0×10^{-4} N
534	**26-14**	1. 1.3×10^{-12} N; **2.** increases to 1.9×10^{-12} N
535	**26-16**	1. 0.27 m; **2.** 2.3×10^{-3} Wb/m²
537	**26-17**	1. 2.0×10^8 m/s; **2.** 5.0×10^{-5} Wb/m²
539	**26-20**	(a) 6.7×10^{-26} kg; (b) 39 μ
548	**27-4**	1. (a) 6.4 V; (b) 1.0 A; (c) 0.80 N; **2.** 20 m/s
554	**27-15**	1. 0.50 V; **2.** (a) 5.0×10^{-4} Wb; (b) 2.0×10^{-2} Wb/s; (c) 2.0×10^{-2} V; **3.** 12 V
560	**27-25**	1. 55.0 V; **2.** (a) 2.00 A; (b) 440 W **3.** 1 to 500
573	**28-11**	1. 8.0×10^5 Hz; **2.** 1.6×10^7 Hz; **3.** 9.9×10^{-12} F
584	**29-4**	1. (a) 4.5 eV; (b) 7.2×10^{-19} J; **2.** 3.0 V
589	**29-13**	1. 5.3×10^{-19} J; **2.** 5.3×10^{-19} J; **3.** (a) 5.3×10^{14} Hz; (b) No
592	**29-17**	3.3×10^{-19} J; E_{radio} is 2.0×10^{-9} smaller; E_{gamma} is 2.0×10^5 larger
594	**29-20**	1. 2.6×10^{-25} N·s; **2.** 1.6×10^{-36} N·s; X ray has 1.6×10^{11} times the momentum of radio photon
597	**29-24**	1. 1.9×10^{-15} m; **2.** (a) 3.3×10^{-24} N·s; (b) 4.9×10^2 m/s
598	**29-26**	1. (a) Yes, $\lambda = 1.5 \times 10^{-14} < 4.0 \times 10^{-14}$ m; (b) No, $\lambda = 6.0 \times 10^{-14} > 4.0 \times 10^{-14}$ m; 2. (a) 1.0×10^{-20} kg·m/s; (b) 5.9×10^6 m/s
616	**31-2**	1. 4.102×10^{-7} m or 410.2 nm; **2.** 3.646×10^{-7} m or 364.6 nm
622	**31-10**	(a) 2.1×10^{-10} m; (b) 5.3×10^{-9} m
630	**31-19**	1. (a) 13.1 eV; (b) 3.2×10^{15} Hz 2. (a) 3.4 eV; (b) 8.2×10^{14} Hz; 3. (a) The electron will go to the $n = 3$ level; (b) The photon will not be absorbed; (c) It will eject the electron with KE of 1.0 eV
660	**33-9**	1. (a) 0.1043 μ; (b) 97.1 MeV 2. (a) 0.0325 μ; (b) 30.3 MeV

Index

Photograph Acknowledgments

Front Cover Martucci Studios   **Back Cover T** Robert Juriet/The Stock Market   **Back Cover M** Pete Saloutos/The Stock Market   **Back Cover B** Harold Sund/The Image Bank   **i** Martucci Studios   **v** Tony Freeman/PhotoEdit   **vi T** Edith G. Haun/Stock Boston   **vi B** Paul Conklin **vii T** Mark Antman/Stock Boston   **vii B** Warren Bolster/Sports Illustrated © Time, Inc.   **viii T** Pat Peticolas   **viii B** Gamma-Liaison   **ix** Dan McCoy/Rainbow   **x** Ellis Herwig/The Picture Cube   **1** Carol Lee/The Picture Cube   **3 T** Paul Conklin   **3 B** Grant Heilman   **6 T** David Scharf/Peter Arnold, Inc.   **6 B** Manfred Kage/Peter Arnold, Inc.   **7** Courtesy of Columbia University, Department of Physics   **8** Courtesy of Philadelphia Museum of Art   **9** Mark Wiklund   **12** Peter Menzel/Stock Boston   **13 L** Ulrike Welsh/PhotoEdit   **13 R** Tony Freeman/PhotoEdit   **15** Courtesy of National Bureau of Standards   **18** Courtesy of Sargeant Welch Scientific Company   **25 L** Courtesy of Hale Observatories   **25 R** Courtesy of Eastman Kodak Company, Kodak Research Labs   **29** Courtesy of National Bureau of Standards   **31** Dr. Harold Edgerton, MIT, Cambridge, MA   **34** Courtesy of National Bureau of Standards   **35** Talbot D. Lovering   **39 T** Richard Megna/Fundamental Photographs   **39 B** Courtesy of Mettler Instrument Corporation   **44** Dave Schaefer   **45** Paul Conklin   **46, 47** NASA   **50** U.S. Air Force photo   **57** Dennis Yeandle/Black Star   **70** U.S. Navy photo   **74** Dr. Harold Edgerton, MIT, Cambridge, MA   **75** J.R. Eyerman/Life Magazine, © 1965 Time Inc.   **81** NASA   **87** BBC Hulton Picture Library   **89** NASA   **90** Jennifer Phillips   **94** Dr. Harold Edgerton, MIT, Cambridge, MA   **97** A. Devaney/E.P. Jones   **101** Harold Waage, Princeton University   **105** Yerkes Observatory photo   **116** Harvard Observatory, courtesy of National Geographic   **120** Dr. Harold Edgerton, MIT, Cambridge, MA   **126** Malcolm Kirk/Peter Arnold, Inc.   **129** NASA   **131** Tony Freeman/PhotoEdit   **137** A. Devaney/E.P. Jones   **140** Professor P.M.S. Blackett and Imperial College of Science & Technology, London, England   **149** H. Armstrong Roberts/E.P. Jones   **152** Hale Observatories   **159** NASA   **164** E.R. Degginger   **166** Ken Duncan   **171 T** UPI/Bettmann   **171 B** Courtesy of Professor Jesse Beams   **173** Mark Antman/Stock Boston   **179** American Red Cross   **182** Dr. Harold Edgerton, MIT, Cambridge, MA   **188, 193** NASA   **194** Y. Arthus Bertrand/Peter Arnold, Inc.   **201** Paul Conklin   **202** Dan McCoy/Rainbow   **203** Peter Menzel/Stock Boston   **206** E.R. Degginger   **213 T** Courtesy of National Bureau of Standards   **213 B** Courtesy of American Iron and Steel Institute   **217 L, R** Dan McCoy/Rainbow   **225** Mark Antman/Stock Boston   **228** Tony Freeman/PhotoEdit   **229** Superstock   **231** U.S. Department of Energy   **238** Ray Pfortner/Peter Arnold, Inc.   **239** E.R. Degginger   **240 T** Superstock   **240 B** Culver Pictures   **243** Kurt Scholz/Superstock   **250** Karen Preuss/The Image Works   **251** AP/Wide World Photos   **253** U.S. Steel photo   **254 T, B** E.R. Degginger   **257** Courtesy of Steelways Magazine   **258** Courtesy of Science Service, Inc.   **259** Courtesy of Bell Telephone Laboratories   **262 T** Dr. Martin J. Buerger   **262 B** BBC Hulton Picture Library   **266** Courtesy of Dr. Erwin Muller, Penn State University   **270** Elizabeth Zuckerman/PhotoEdit   **283** Tony Freeman/PhotoEdit   **290 L, R, 291** Courtesy of General Motors Corporation   **292, 293** E.R. Degginger   **295** Jeff Hornbaker/Sports Illustrated, © Time Inc.   **298 T** Fundamental Photograph, NY   **298 B** PSSC Physics, 2nd edition, 1965, D.C. Heath & Company, Lexington, MA with Education Development Center, Newton, MA   **299, 300** From the film "Reflection and Refraction," Education Development Center, Newton, MA   **301** Courtesy of Education Development Center, Newton, MA   **302 L, R** Reprinted from Light Visible and Invisible by Edward Ruechardt, 1958, by permission of University of Michigan Press   **307** Owen Franken/Stock Boston   **309 L, R** Wide World Photos   **312** Courtesy of Education Development Center, Newton, MA   **317** NASA   **321** Tony Freeman/PhotoEdit   **322** Ellis Herwig/Stock Boston   **327 T** from PSSC Physics, D.C. Heath, 1965   **327 B** Sears and Zemansky, University Physics, 3rd edition, © 1964 Addison Wesley Publishing Company   **329 T, B** Kodansha Ltd   **332 L, R** Courtesy of Polaroid Corporation   **333** E.R. Degginger   **335** David Young-Wolff/PhotoEdit   **340** A. Pasieka/Taurus Photos   **341** E.R. Degginger   **352** The Image Bank   **354 T** The Bettmann Archive   **354 B** Latrobe Barnitz/The Picture Cube   **355** Courtesy of Bausch & Lomb   **357** file photo   **362** E.R. Degginger   **369 T** Dennis Milon   **369 B** David Conklin   **384** Courtesy of Mutual of Omaha's Wild Kingdom/Don Meier Productions   **389** Courtesy of Perkin-Elmer Corporation   **393** Reprinted from Light Visible and Invisible by Edward Ruechardt, 1958, by permission of University of Michigan Press   **394 T, B** Courtesy of Professor Henry A. Hill, University of Arizona, Tuscon   **398** Courtesy of Professor Henry A. Hill, Wesleyan University, Middletown, CT   **399 T, B** Kodansha, Ltd.   **400 T** Applied Research Laboratories   **400 B** BIO-RAD   **401 L, R** Adapted from Fundamentals of Optics, 3rd edition, 1957, McGraw Hill book company, used by permission   **403 T, B, 404** Kodansha, Ltd.   **405 L, M, R** Richards, Sears, Wehr and Zemansky/Modern College Physics, 1952, Addison-Wesley Publishing Company, Inc.   **407** Rainbow   **408** Chuck O'Rear/Westlight   **412** Al Harvey/Masterfile   **413 L** Paul Shambroom/Photo Researchers, Inc.   **413 R** Harriet Casdin-Silver, Equivocal Forks, © 1977, photo by Robin Tooker   **415** Terry McCoy/The Picture Cube   **416, 417, 421** E.R. Degginger   **425** Steven C. Kaufman/Peter Arnold, Inc.   **439** Dave Schaefer   **442** Nimrod/Phototake   **448** Camerique Stock Photography   **456** Courtesy MIT   **475** NASA   **482** Gamma-Liaison   **483** Courtesy Westinghouse Electric Company   **488** Jack Schneider/Energy Research and Development Administration   **495** Talbot D. Lovering   **501** Mike Dobel/Masterfile   **503** K. Kai/The Image Works   **505** Paul Conklin   **506, 507** Dave Schaefer   **509** Courtesy of Sargent-Welch Scientific Company, Skokie, Illinois   **516** E.R. Degginger   **519 L, R** Courtesy of General Motors Research Laboratories   **529** Courtesy of Sargent-Welch Scientific Company, Skokie, Illinois   **534 L, R** PSSC Physics, 2nd edition, 1965, D.C. Heath and Company, Lexington, MA with Education Development Center, Newton, MA   **544** Dr. Helpern, University of Alaska emeritus   **559 T** U.S. Army photo   **559 B** Commonwealth Edison Company, Chicago, Illinois   **573** Camerique Stock Photography   **577** Tony Freeman/PhotoEdit   **578 L, R** Dan McCoy/Rainbow   **579** Tom Campbell/Westlight   **580** Collier/Condit/Stock Boston   **581** Courtesy of Bates Linear Laboratory/MIT, Middleton, MA   **587** Ernst Haas/Magnum Photos   **596 T** Courtesy of Professor G. Mellenstedt, University of Tubingen, Germany   **596 B** Courtesy of Professor Henry A. Hill, Wesleyan University, Middletown, CT   **605** Culver Pictures, Inc.   **614 L** The Bettmann Archive   **614 R** Wide World Photos   **616** From Physics for Engineers and Scientists by Fowler & Meyer, Allyn and Bacon, Inc., 1961   **617** Courtesy of American Institute of Physics   **627 L, R** Bernie Weinstock   **636** AP/Wide World Photos   **637** Department of Energy   **652 T** Courtesy of Brookhaven National Laboratory   **652 B** Courtesy of International Business Machines   **667** Courtesy of Brookhaven National Laboratory   **669 T, B** file photos   **673** Marshall Nunn/Florida Power and Light Company   **674** Courtesy Energy Research and Development Administration   **675 T** Courtesy of Hale Observatories   **675 B** Courtesy of Federal Civil Defense Administration   **677** Courtesy of Energy Research and Development Administration   **681** Superconducting Super Collider Lab, Dallas, TX   **683, 685** Courtesy of Lawrence Berkeley Radiation Lab, University of California   **686** Courtesy of Argonne National Laboratory   **688** Courtesy of Brookhaven National Laboratory   **689 T** Courtesy of Lawrence Berkeley Laboratory, University of California   **689 B** Courtesy of MIT   **690** Courtesy of Lawrence Berkeley Laboratory, University of California   **695** Erich Hartmann/Magnum Photos   **698** Paul Conklin   **699** Druskis/Taurus Photos